AF738816

World Geomorphological Landscapes

Series Editor

Piotr Migoń, Department of Geography, Faculty of Law, University of Wrocław, Wrocław, Poland

More information about this series at http://www.springer.com/series/10852

Olav Slaymaker · Norm Catto
Editors

Landscapes and Landforms of Eastern Canada

Springer

Editors
Olav Slaymaker
Department of Geography
University of British Columbia
Vancouver, BC, Canada

Norm Catto
Department of Geography
Memorial University of Newfoundland
and Labrador
St. John's, NL, Canada

ISSN 2213-2090 ISSN 2213-2104 (electronic)
World Geomorphological Landscapes
ISBN 978-3-030-35135-9 ISBN 978-3-030-35137-3 (eBook)
https://doi.org/10.1007/978-3-030-35137-3

© Springer Nature Switzerland AG 2020
This work is subject to copyright. All rights are reserved by the Publisher, whether the whole or part of the material is concerned, specifically the rights of translation, reprinting, reuse of illustrations, recitation, broadcasting, reproduction on microfilms or in any other physical way, and transmission or information storage and retrieval, electronic adaptation, computer software, or by similar or dissimilar methodology now known or hereafter developed.
The use of general descriptive names, registered names, trademarks, service marks, etc. in this publication does not imply, even in the absence of a specific statement, that such names are exempt from the relevant protective laws and regulations and therefore free for general use.
The publisher, the authors and the editors are safe to assume that the advice and information in this book are believed to be true and accurate at the date of publication. Neither the publisher nor the authors or the editors give a warranty, expressed or implied, with respect to the material contained herein or for any errors or omissions that may have been made. The publisher remains neutral with regard to jurisdictional claims in published maps and institutional affiliations.

This Springer imprint is published by the registered company Springer Nature Switzerland AG
The registered company address is: Gewerbestrasse 11, 6330 Cham, Switzerland

Series Editor Preface

Landforms and landscapes vary enormously across the Earth, from high mountains to endless plains. At a smaller scale, Nature often surprises us creating shapes that look improbable. Many physical landscapes are so immensely beautiful that they received the highest possible recognition—they hold the status of World Heritage properties. Apart from often being immensely scenic, landscapes tell stories which not uncommonly can be traced back in time for tens of million years and include unique events. In addition, many landscapes owe their appearance and harmony not solely to the natural forces. Since centuries, or even millennia, they have been shaped by humans who modified hillslopes, river courses, and coastlines, and erected structures which often blend with the natural landforms to form inseparable entities.

These landscapes are studied by Geomorphology—'the Science of Scenery'—a part of Earth Sciences that focuses on landforms, their assemblages, surface and subsurface processes that moulded them in the past and that change them today. Shapes of landforms and regularities of their spatial distribution, their origin, evolution and ages are the subject of research. Geomorphology is also a science of considerable practical importance since many geomorphic processes occur so suddenly and unexpectedly, and with such a force, that they pose significant hazards to human populations and not uncommonly result in considerable damage or even casualties.

To show the importance of geomorphology in understanding the landscape, and to present the beauty and diversity of the geomorphological sceneries across the world, we have launched a new book series *World Geomorphological Landscapes*. It aims to be a scientific library of monographs that present and explain physical landscapes, focusing on both representative and uniquely spectacular examples. Each book will contain details on geomorphology of a particular country or a geographically coherent region. This volume is the second among the two which show the geomorphology of the second largest country in the world—Canada. This one focuses on the eastern part of this enormous territory, which is often overshadowed by the majestic Rocky Mountains of the west and too hastily simplified to endless plains of the Laurentian Shield. How unjust such a view is, one will realize browsing through this book and reading its individual chapters. Not surprisingly, the theme of Quaternary glaciations features prominently as their impact can be demonstrated almost everywhere. But authors of specific stories included in the book show us much more: fascinating coastal sceneries, shaped in response to land and sea level change, permafrost-dominated tundra landscapes, big lakes, the legacy of long-term landform evolution which occurred at timescales much more protracted than glaciations, karst, devastating mass movements. In addition, various examples of interactions between humans and landforms are shown, in different cultural contexts.

The World Geomorphological Landscapes series is produced under the scientific patronage of the International Association of Geomorphologists—a society that brings together geomorphologists from all around the world. The IAG was established in 1989 and is an independent scientific association affiliated with the International Geographical Union and the International Union of Geological Sciences. Among its main aims are to promote geomorphology and to foster dissemination of geomorphological knowledge. I believe that this lavishly illustrated series, which sticks to the scientific rigour, is the most appropriate means to fulfil these aims and to serve the geoscientific community. To this end, my great thanks go to the Senior Editor, Professor Olav Slaymaker, Past President of the IAG, who enthusiastically responded to the invitation to coordinate both volumes, Western Canada that appeared a few years ago and now Eastern Canada, where he was joined by Professor Norm Catto as Co-Editor and assisted by Dr. Dori Kovanen. I am also grateful to all individual contributors who agreed to add the task of writing chapters to their busy agendas and delivered high-quality final products.

Piotr Migoń

Dedication

We dedicate this volume to the memory of Dr. Ian A. Brookes, former Professor of Geomorphology at York University, Toronto, who passed away in February 2015. Although Ian conducted geomorphological research in many parts of the world, much of his research, beginning with his M.Sc. in 1963, focused on the glacial geomorphology of Newfoundland. Ian's work generated much subsequent work throughout Western Newfoundland, as is evidenced by his co-authorship of Chap. 16 in this volume. His advocacy contributed to the development of research in Gros Morne National Park.

Ian's legacy to geomorphology includes scholarships in his name for undergraduate and graduate geographical field research offered through the Department of Geography, Memorial University of Newfoundland. Ian always considered undergraduate teaching as a core academic activity, blending his passion for geomorphology and his keen sense of social responsibility. His conviction that scientific investigation could never be completely separated from social and political concerns enlivened many of his discussions with colleagues, and finds an echo in the development of the concept of critical physical geography, as discussed in Chap. 26 of this volume.

From 1985 through 1993, Ian Brookes oversaw the publication of a succession of articles detailing 'Canadian Landform Examples' in *The Canadian Geographer*. These concise papers were used, and are still used, by geomorphologists across Canada and beyond to introduce students to the diversity of Canadian landforms and landscapes, several of which are examined in this volume.

—Norm Catto and Olav Slaymaker

Preface for Landscapes and Landforms of Eastern Canada

Olav Slaymaker first set foot on North American soil in Montréal in September, 1961, having set sail on the 'Empress of Canada' from Liverpool, UK 7 days earlier. The previous 2 days had been spent in Canadian waters, firstly skirting 'The Rock' of Newfoundland, then passing through the Gulf of St. Lawrence to the south of the karstic island of Anticosti, Québec, gradually getting squeezed into the estuary of the St. Lawrence River and finally docking at Montréal. Little could he guess that he was sailing past landscapes that would eventually be the subject of a co-edited book on Eastern Canada, almost 60 years later. The contrast between the steep Laurentian Highlands of the Canadian Shield backing the North Shore of the St. Lawrence River and the Gaspé Peninsula forming the gentler South Shore of a younger, folded Appalachian landscape was not lost on him. But the dominant impression was the extravagant welcome to North America provided by the majestic St. Lawrence River itself.

Norm Catto was born in Ottawa, on the Champlain Sea plain directly south of the Grenville front of the Gatineau Hills, in 1956. After spending time appreciating the landscapes of Central, Western and Northern Canada (and elsewhere), arrival in Newfoundland for the first time in 1989 provided a geomorphic and cultural contrast to everything that had gone before. After 30 years studying, publishing on and appreciating the landscapes of the seven provinces and territory discussed in our volume, there is still much to learn, from landscapes both majestic and subtle.

We continue to rely on a traditional definition of landscape as 'the total character of a region of the Earth's surface which includes landforms, vegetation (ecosystems) and fields and buildings (anthropogenically modified land)' (von Humboldt 1845). The landscapes and landforms of Eastern Canada form the subject of our volume, a companion to 'Landscapes and Landforms of Western Canada' (Slaymaker 2017). More than 5 million km^2 of Earth's terrestrial surface, including the provinces of Ontario, Québec, Newfoundland and Labrador, Nova Scotia, New Brunswick, Prince Edward Island, and northeastern Manitoba, and the territory of Nunavut are included. It is impossible to do justice to the varied individual landforms and landform assemblages of such a vast area, but we have endeavoured to provide at least an introduction to all the distinctive geomorphological landscapes of this region. Landscapes at this spatial scale are defined by (a) geological structure, (b) relief and (c) surrogates for geomorphic process. The presence or absence of meteorite craters, highly permeable lithologies, ice sheet cover at the Last Glacial Maximum, permafrost and vegetation are the surrogates for geomorphic process that are used in this volume.

The Canadian Shield contains some of the oldest rocks at the Earth's surface, up to 4 billion years old. However, there are no landscapes that have been preserved unchanged for such a long time. Indeed, only a tiny fraction of the rocks and landscapes that are visible today provides clues to the interpretation of palaeolandscapes older than a few hundred million years. Rocks that underlie the Canadian Shield have undergone numerous mountain building episodes, have been sutured along former plate collision zones and have experienced a series of erosion cycles. Today, such landscapes are topographically subdued and some have suggested that they are some of the most extensive planation surfaces that exist on the planet

(Ambrose 1964). However, they are certainly not featureless as many of the case studies in this volume demonstrate.

This volume is about landscapes that are larger than individual landforms and smaller than continental belts (Slaymaker et al. 2009). Landform assemblages of Eastern Canada that form distinctive landscapes include magnificent arctic fjords and glaciers; spectacular coastlines adjacent to the Arctic and Atlantic oceans; a high incidence of meteorite craters that have been preserved preferentially in the ancient Shield terrain of Québec; youthful, glaciated karst in Ontario, Newfoundland, Québec and Nova Scotia; urban karst in Montreal and Ottawa; wetlands of five subtly distinct varieties with an extent second only to those in Russia; the ubiquitous permafrost terrain (>3 million km^2) in Nunavut, Labrador and northern Québec; and the unique Great Lakes-St. Lawrence basin, that sustains one of North America's largest rivers, is co-managed by Canada and the USA and whose seaway allows modest-sized ships to penetrate into the heart of the continent. Eastern Canada also contains the cradle of the European settler society from which the present-day national governance structure has evolved. Complications that have arisen in relations between First Nations, Inuit, Metis and the more recent 'settlers' have become so urgent in the past few decades that land use, land claims and land governance have become issues that are strongly colouring the landscape, both physical and cultural.

Part I (Chaps. 1–3) is a general overview of Eastern Canadian geomorphological landscapes in relation to geological structure, relief and dominant processes of change.

Part II (Chaps. 4–20) consists of case studies of outstanding representative landscapes from which original, field-based research is reported.

Part III (Chaps. 21–26) provides examples of human impact on, stewardship of, and human perceptions of Anthropocene landscapes in eastern Canada.

This book could not have been completed without the outstanding technical skills of Dori Kovanen, Research Associate in the Department of Geography at UBC. Eric Leinberger, Cartographer in the Department of Geography, UBC should also be acknowledged for emergency assistance. The Book Series Editor, Piotr Migon, provided valuable advice at all stages during the process of book preparation, and Robert Doe and Marielle Klijn of Springer Verlag were consistently supportive. The following expert reviewers gave freely of their time to ensure the scientific quality of the text: Tracy Brennand, Frank Brunton, Evan Edinger, Don Forbes, Derek Ford, Konrad Gajewski, Krister Jansson, Patrick Lajeunesse, Pascale Roy-Leveillee, Dave Liverman, Piotr Migon, Martin Roy, Martin Sharp, and Wendy Sladen. To all these colleagues we extend our warmest thanks.

Vancouver, Canada
St. John's, Canada
2019

Olav Slaymaker
Norm Catto

References

Ambrose JW (1964) Exhumed paleoplains of the Precambrian Shield of North America. Am J Sci 262: 817–857

Humboldt A von (1845) Kosmos: a sketch of a physical description of the universe (translated from the German by Otté EC), vol 1, Harper and Brothers, New York

Slaymaker O, Spencer T, Dadson S (2009) Landscape and landscape scale processes as the unfilled niche in the global environmental change debate: an introduction. In: Slaymaker O, Spencer T, Embleton-Hamann C (eds) Geomorphology and Global Environmental Change. Cambridge University Press, Cambridge, UK, pp 1–36

Contents

Editors and Contributors

About the Editors

Olav Slaymaker is a Member of the Order of Canada, and Professor Emeritus of geography in the Department of Geography at the University of British Columbia, Vancouver, B.C. His 20+ co-authored and edited monographs and 160+ refereed papers reflect three different styles of work. First, he has pursued original field research for 40 years on sediment systems in the Canadian Cordillera and the European Alps, strongly supported and enhanced by his stellar doctoral and postdoctoral students and international colleagues. Second, he has produced papers and books summarizing trends in land use and environmental change in mountain regions worldwide. Third, he has edited monographs with invited international contributions on major environmental and geomorphological themes. He is former President of the Canadian Association of Geographers and former President of the International Association of Geomorphologists.

Lead author of Chaps. 1, 3, 25 and 26.

Norm Catto joined Memorial University in July 1989 and is now the Head and Professor of Geography at that university. His research and teaching interests include coastal landforms, natural hazards, sea level change and impacts in coastal environments; response of river systems to climate and weather events, and flood risk assessment; aeolian geomorphology and response to climate and human factors in boreal and coastal environments; mass movements and slope failures; loess deposition; palaeosol formation; emergency measures policies and effectiveness of human adaptations to climate and weather events; the impacts of climate and weather events on agriculture, transportation, fisheries, and communities; and the relationships among landscapes and people. His research has included projects and investigations in landscapes and environments in 16 countries, and in all Canadian provinces and territories, particularly in Eastern Canada. For 18 years, he served as Editor-in-Chief for *Quaternary International*, a professional international journal dedicated to the study of landscape evolution and climate variation throughout the past 3 million years. He has taught more than 50 different undergraduate and graduate courses focused on numerous aspects of geomorphology, landscape analysis, Quaternary research, natural hazards, ocean sciences, environmental science, climate and climate change, soil science, geoarchaeology, and palaeontology. Study of landscapes forms his central focus for an understanding of the various areas of Eastern Canada, the combination of physical and cultural constituents that contribute to its identities. For most species, and for all human occupants, life in Eastern Canada has been heavily influenced by geomorphology since deglaciation.

Lead author of Chaps. 2, 7, 15, 18, and 21.

Contributors

Michel Beaupré Hatch Engineering, Montréal, Canada; Montréal, Québec, Canada

Antony R. Berger Wolfville, NS, Canada

Pascal Bernatchez Département de Biologie, Chimie et Géographie and Centre for Northern Studies (CEN), Université du Québec à Rimouski, Rimouski, Québec, Canada

Frédéric Bouchette GEOSCIENCES-M, Université Montpellier II et CNRS, Montpellier, France

Ian Brookes (deceased) York University, Toronto, Canada

Etienne Brouard Département de Géographie, Centre d'Études Nordiques, Université Laval, Quebec, Canada

David Burgess Natural Resources Canada, Ottawa, ON, Canada

Mary-Louise Byrne Department of Geography and Environmental Studies, Wilfrid Laurier University, Waterloo, ON, Canada

D. Calvin Campbell Geological Survey of Canada (Atlantic), Natural Resources Canada, Bedford Institute of Oceanography, Dartmouth, NS, Canada

Daniel Caron Société Québécoise de Spéléologie, Montréal, Canada

Norm Catto Department of Geography, Memorial University of Newfoundland and Labrador, St. John's, NL, Canada

Jaclyn M. H. Cockburn Department of Geography, Environment & Geomatics, University of Guelph, Guelph, ON, Canada

Luke Copland Department of Geography, Environment and Geomatics, University of Ottawa, Ottawa, ON, Canada

Pierre A. Cousineau Centre d'Etude sur les Ressources Minérales, Université du Québec à Chicoutimi, Chicoutimi, Canada

Daryl W. Cowell Tobermory, ON, Canada

Fred Dean Department of Geography, Lakehead University, Thunder Bay, ON, Canada

Joseph R. Desloges Departments of Geography and Earth Sciences, University of Toronto, Toronto, ON, Canada

Pierre Dietrich Univ Rennes, CNRS, Géosciences Rennes, UMR 6118, Rennes, France; Department of Geology, University of Johannesburg, Johannesburg, South Africa; Géosciences-Rennes, UMR6118, Université de Rennes 1, Rennes, France

Randy W. Dirszowsky School of the Environment, Laurentian University, Sudbury, ON, Canada

L. A. Dredge Geological Survey of Canada, Ottawa, ON, Canada

Dewey Dunnington Department of Civil and Resource Engineering, Dalhousie University, Halifax, NS, Canada

L. D. Dyke Nipissing University, North Bay, ON, Canada

Jean-François Ghienne Institut de Physique du Globe de Strasbourg, UMR 7516, CNRS, Université de Strasbourg, Strasbourg, France

John Gosse Department of Earth Sciences, Dalhousie University, Halifax, NS, Canada

Bernard Hétu Département de Biologie, Chimie et Géographie & Centre d'Etudes Nordiques (CEN), Université du Québec à Rimouski, Rimouski, QC, Canada

Dori J. Kovanen Department of Geography, University of British Columbia, Vancouver, BC, Canada

Patrick Lajeunesse Département de Géographie & Centre d'Etudes Nordiques (CEN), Université Laval, Quebec City, QC, Canada;
Département de Géographie, Université Laval, Québec, Canada

Scott F. Lamoureux Department of Geography and Planning, Queen's University, Kingston, ON, Canada

Antoine Morissette Département de Géographie & Centre d'Etudes Nordiques (CEN), Université Laval, Quebec City, QC, Canada

David C. Mosher Geological Survey of Canada (Atlantic), Natural Resources Canada, Bedford Institute of Oceanography, Dartmouth, NS, Canada

Monica E. Mulrennan Department of Geography, Planning and Environment, Concordia University, Montréal, QC, Canada

Alexandre Normandeau Geological Survey of Canada (Atlantic), Natural Resources Canada, Bedford Institute of Oceanography, Dartmouth, NS, Canada

Alexis Nutz Centre Européen de Recherche et d'Enseignement des Géosciences de l'Environnement, Aix-Marseille Université, CNRS, IRD, Collège de France, INRA, Aix-en-Provence, France;
Department of Geoscience, Aarhus University, Aarhus, Denmark

Gerald Osborn Department of Geoscience, University of Calgary, Calgary, AB, Canada

Brian Phillips Department of Geography, Lakehead University, Thunder Bay, ON, Canada

Roger T. J. Phillips Department of Geography, University of Toronto, Toronto, ON, Canada

David J. W. Piper Geological Survey of Canada (Atlantic), Natural Resources Canada, Bedford Institute of Oceanography, Dartmouth, NS, Canada

Audrey M. Rémillard Canada Research Chair in Marine Geology, Institut des Sciences de la Mer de Rimouski (ISMER), Université du Québec à Rimouski and GEOTOP, Rimouski, Québec, Canada;
Centre de Recherche sur les Milieux Insulaires et Maritimes (CERMIM), Havre aux Maisons, Québec, Canada

Claude Roquin Institut de Physique du Globe de Strasbourg, UMR7516, CNRS, Strasbourg, France

Ashley C. A. Rudy Northwest Territories Geological Survey, Yellowknife, NT, Canada

Jacques Schroeder Département de Géographie, Université du Québec à Montréal, Montréal, Québec, Canada

Mathieu Schuster Institut de Physique du Globe de Strasbourg, UMR 7516, CNRS, Université de Strasbourg, Strasbourg, France

John Shaw Geological Survey of Canada (Atlantic), Natural Resources Canada, Bedford Institute of Oceanography, Dartmouth, NS, Canada

Olav Slaymaker Department of Geography, University of British Columbia, Vancouver, BC, Canada

Ian Spooner Department of Earth and Environmental Science, Acadia University, Wolfville, NS, Canada

Chris R. Stokes Department of Geography, Durham University, Durham, UK

Guillaume St-Onge Canada Research Chair in Marine Geology, Institut des Sciences de la Mer de Rimouski (ISMER), Université du Québec à Rimouski and GEOTOP, Rimouski, Québec, Canada

Brian J. Todd Geological Survey of Canada (Atlantic), Natural Resources Canada, Bedford Institute of Oceanography, Dartmouth, NS, Canada

Wesley Van Wychen Department of Geography and Environment Management, University of Waterloo, Waterloo, ON, Canada;
Department of Geography, Environment and Geomatics, University of Ottawa, Ottawa, ON, Canada

Kamil Zaniewski Department of Geography, Lakehead University, Thunder Bay, ON, Canada

Part I
Introduction to the Geomorphological Landscapes of Eastern Canada

Eastern Canadian Landscapes as a Function of Structure, Relief and Process

1

Olav Slaymaker, Norm Catto, and Dori J. Kovanen

Abstract

Eastern Canada's landscapes are investigated as a function of structure, relief and process. Eastern Canada can be divided into four roughly concentric megaregions whose boundaries are controlled exclusively by geological structure. These megaregions are (a) the Canadian Shield, which is the craton or core of North America; (b) the surrounding plains arranged in a broken ring around the Shield together with an inlier of similar lithology stratigraphically above the Shield; (c) a broken ring of two mountainous regions that surrounds the plains: Innuitia in the far north and Appalachia in the southeast; (d) two widely separated low-relief continental shelves and coastal plains of Quaternary sediments marginal to the Arctic and Atlantic oceans. These four megaregions are further subdivided into 29 subregions that are sensitive to variations in structure, relief and process. The most spectacular landscapes are arranged around the periphery of Eastern Canada: the magnificent glaciers and snow-capped mountains of the Arctic, the fascinating Atlantic coastline and the unique Great Lakes and St. Lawrence River basin.

Keywords

Landscape • Geological province • Physiographic region • Geomorphological region • Structure • Relief • Process

O. Slaymaker (✉) · D. J. Kovanen
Department of Geography, University of British Columbia, 1984 West Mall, Vancouver, BC V6T 1Z2, Canada
e-mail: olav.slaymaker@ubc.ca

N. Catto
Department of Geography, Memorial University of Newfoundland and Labrador, St. John's, NL A1B 3X9, Canada

1.1 A Word About Landscape

Given that landscapes and landforms are the phenomena of central interest in this volume, a few words about the word 'landscape' are in order. Landscape is a contested word linking the morphology of land with its longer geological past as well as with the man-made features that contribute to the roughness of Earth's boundary (von Humboldt 1845). Landscapes are often envisaged as analogous to palimpsests or medieval parchments on which texts of many different historical time periods are superimposed. Geomorphological landscapes incorporate the aggregate effects of past and present Earth surface processes, and at this point in time the challenge is for us to expose landscapes formed in the geological past; to interpret their relationship with contemporary landscapes and to be aware of the extent to which human activity has transformed Earth's boundary. Human activities have become an integral part of the way that geosystems function. In the context of Eastern Canada we can perhaps be permitted the broadest of generalizations by noting that human activity in its northern regions has left little mark on the morphology of the land, but indigenous people's place names demonstrate that the land is far from empty; by contrast, southeastern Canada has been modified by human activity to such an extent that pristine landscapes are hard to find.

In this chapter and Chap. 2, human impact is excluded from consideration for obvious reasons. But there is a role for humans that affects our understanding of these ancient landscapes, specifically in relation to place names in the Arctic. The term Inuit refers to a group of indigenous peoples of the circumpolar regions of Alaska, Canada and Greenland. The term is also used for the continuum of language varieties spoken by the Inuit people. Three of these languages spoken in Canada and Greenland, are referred to as Inuktitut. The last decade has seen the emergence of a stronger indigenous presence in Canadian place names (Tyman 2008; Inuit Heritage Trust 2015; Canadian

© Springer Nature Switzerland AG 2020
O. Slaymaker and N. Catto (eds.), *Landscapes and Landforms of Eastern Canada*, World Geomorphological Landscapes, https://doi.org/10.1007/978-3-030-35137-3_1

Geographic 2018). This long-overdue acknowledgment of the reality of Canadian indigenous culture complicates place name recognition for English speaking readers, as indicated in a partial list of equivalences between English and Inuktitut names (Table 1.1).

This chapter and Chaps. 2, 3 and 21 deal with landscapes at four different spatial and temporal scales, namely, total geological time scale and very large spaces (chapter 1); Quaternary time scale and moderately large spaces (Chap. 2); Holocene time scale and smaller spaces (Chap. 3); and finally an Anthropocene scale and generally quite small spaces (Chap. 21). The latter chapter is needed to explore the role of human activity in landscape change.

At the largest spatial and longest temporal scales (this chapter), Eastern Canada can be divided into four roughly concentric regions that are controlled exclusively by geological structure. These regions are firstly the central Canadian Shield, which is the craton or core of North America (Stearn 1975). It is by far the oldest and largest of the regions and has a complex history. The shield was formed one to four billion years ago and consists of highly resistant crystalline rocks and planation surfaces. The surrounding stable platforms form a second region, which is characterized by flat-lying and gently sloping Palaeozoic and Mesozoic sedimentary rocks, arranged in a broken ring around the shield as well as an inlier of similar lithology on top of the shield around the margins of Hudson Bay. Thirdly, there is a broken ring of two relatively unstable mountainous regions that surrounds the plains: Innuitia in the north and Appalachia in the southeast. These Phanerozoic orogens consist of highly folded and faulted sedimentary and metamorphic rocks. Finally, there are two widely

Table 1.1 English place names and their Inuktitut equivalents. *Source* Inuit Heritage Trust (2015)

English Canadian names	Inuktitut names (roman type)	English Canadian names	Inuktitut names (roman type)
Arctic Bay	*Ikpiarjuk*	Makkovik	*Marruuvik*
Baffin Island	*Qikiqtaaluk*	Mealy Mountains	*KakKasuak*
Baker Lake	*Qamanituaq*	Melville Peninsula	*Sanirajak*
Bathurst Inlet	*Kingoak*	Meta Incognita Peninsula	*Katannilik*
Bathurst Island	*Qausuittuq*	Mount Asgard	*Sivanitirutinguak*
Belcher Islands	*Sanikiluaq*	Mount Thor	*Qaisualuk*
Boothia Peninsula	*Kingngailap Nunanga*	Nain	*Nunajnguk*
Broughton Island	*Qikiqtarjuaq*	Navy Board Inlet	*Nalluata imanga*
Cambridge Bay	*Iqaluktuuttiaq*	Nouveau Québec	*Nunavik*
Cape Chidley	*Killiniq*	Oliver Sound	*Kangiqluruluk*
Cape Dorset	*Kinngait*	Ottawa Islands	*Arviliit*
Chesterfield Inlet	*Igluigaarjuk*	Pelly Bay	*Kugaaruk*
Clyde River	*Kangiqtugaapik*	Pond Inlet	*Mittimatalik*
Coppermine	*Kugluktuk*	Poste-de-la-Baleine (Whale River)	*Kuujjuaraapik*
Coral Harbour	*Salliq*	Postville	*Qipuqqaq*
Devon Island	*Tatlurutit*	Prince of Wales Island	*Kingailik*
Eclipse Sound	*Tasiuja*	Rankin Inlet	*Kangiqtiniq*
Ellesmere Island	*Umingmak Nuna*	Repulse Bay	*Naujaat*
Gjoa Haven	*Usqsuqtuuq*	Resolute	*Qausuittuq*
Grise Fjord	*Ausuiktuq*	Rigolet	*Kikiak*
Hall Beach	*Sanirajak*	Schwartzenbach Falls	*Qulitasaniakvik*
Hopedale	*Aqvituq*	Somerset Island	*Shugliaq*
Bay Hudson	*Wînipekw*	Southampton Island	*Shugliaq*
Inoucdjouac (Port Harrison)	*Inukjuak*	Spence Bay	*Taloyoak*
Keewatin (part of)	*Kivalliq*	Ungava Bay	*Ungava kangiqluk*
King William Island	*Qikiqtaq*	Victoria Island	*Kitlineq*
Lancaster Sound	*Tallurutiup Timanga*	Wales Island	*Shartoo*

separated low-relief continental shelves and coastal plains of Quaternary sediments marginal to the Atlantic and Arctic oceans. They are the youngest and smallest of the four regions. At Quaternary scale (Chap. 2), the focus of interest is the role of ice sheets and location in relation to those expanding and contracting ice covers. Such features cut across the boundaries created by geological structure. At Holocene scale (Chap. 3), paraglacial processes (non-glacial processes conditioned by glaciation) are emphasized and the gradually increasing importance of human activity is traced. And finally at Anthropocene scale (Chap. 21) environmental change induced by direct and indirect human activity holds centre stage.

1.2 Structure, Relief and Process

1.2.1 The Geological Provinces of Eastern Canada

A geological province is defined as an extensive region characterized by rocks that are relatively homogeneous in type and age in comparison with neighbouring regions. Wheeler et al. (1996) identified 14 geological provinces in Eastern Canada (Fig. 1.1).

The geological assembly of Eastern Canada has resulted from the convergence, collision, and separation of distinctive continental and oceanic fragments at various times over the last four billion years. Canada's geological architecture is dominated by its central foundation—the Precambrian shield—the largest area of Archaean rocks (>2.5 billion years ago) in the world and containing its oldest rocks, dated at 4.0 billion years. The shield consists of several Archaean fragments comprising granitic rocks and gneiss surrounded by greenstone volcanic belts and broader tracts of sedimentary rocks. Orogenic belts between the Archaean fragments, such as those flanking the Superior Province or Slave Province, contain younger, Palaeoproterozoic (2.5–1.6 billion years ago) rocks representing continental, oceanic, and collisional deposits and exotic fragments. Although most of the Precambrian shield was consolidated by the end of Palaeoproterozoic time, its southeast portion (the Grenville Province) was stabilized about one billion years ago. The addition of Mesoproterozoic (1.6–1.0 billion years old) and older gneisses and granitic rocks in the Grenville orogenic belt thus completed the assembly of the Precambrian shield. Two younger deformed belts, mainly of Phanerozoic rocks (<545 million years old) surround the shield. The Appalachian belt, in the southeast, contains large tracts of continental and oceanic fragments that were attached to ancestral North America in early- and mid-Palaeozoic times, about 475 and 375 million years ago, respectively. By contrast, in the Innuitian belt in the Arctic Islands, only a small foreign fragment was attached in the mid-Palaeozoic about 400 million years ago. After these belts were deformed, they were superimposed by extensive less deformed basins dominated by sedimentary rocks in the Appalachian and Innuitian belts. Large parts of the Precambrian shield are covered by a thin veneer of undeformed sedimentary rocks which are mainly of early Palaeozoic age. Locally, where the Palaeozoic rocks are overlain by Cretaceous strata, they form basins, as in Hudson Bay, and fault-bounded troughs in and near Hudson Strait and north of Baffin Island, related to the rifting and opening of Labrador Sea and Baffin Bay. Crustal spreading and ocean opening, east of the passive eastern continental margin, ceased west of Greenland in the early Cenozoic, nearly 40 million years ago (Hodgson 1989), but still continues in the mid-Atlantic Ocean (Wheeler et al. 1996).

1.2.1.1 The Canadian Shield

The Canadian Shield includes seven geological provinces, namely, Bear, Churchill, Grenville, Nain, Slave, Southern and Superior (Fig. 1.1). Each geological province is distinguished by its unique internal structural trend and style of folding. Collectively, they are the Earth's most extensive area of exposed Archaean rock, covering >70% of Eastern Canada. Major rock types of the shield are intrusive and metamorphic (blue and mustard in Fig. 1.2) by contrast with the dominance of sedimentary rocks in the platforms and orogens (purple in Fig. 1.2).

This vast area of stable Precambrian rocks is largely the end product of many orogenies or mountain-building episodes that occurred prior to 500 million years ago. The scenery is uniform for great distances, nevertheless, within this large area there is considerable variety at local and regional scales. The exceptional uniformity of the shield landscape occurs because 80% of the surface of the shield is underlain by granitic gneissic rocks. The remaining 20% consists of folded and deformed metasediments and metavolcanics which produce ridge and valley scenery in southeastern Ontario and Québec, the Nouveau Québec Orogen, the Povungnituk Hills and the Belcher Islands (Sanikiluaq). In the Nouveau Québec Orogen glacial scouring of the landscape eroded remains of an ancient mountain chain that once extended from Ungava Bay almost as far south as the Gulf of St. Lawrence (now folded and faulted) (Fig. 1.3a).

Three major factors account for such variability of landscape as does exist: warping of an ancient land surface; major faulting both defining the edge of the shield and grabens within the shield; and regional joint patterns giving strong linear patterns (Fig. 1.4). In general, the north to northeastern sector of the shield is occupied by the mountains of southern Ellesmere Island, Devon Island, Baffin Island and Labrador. This is some of the most spectacular

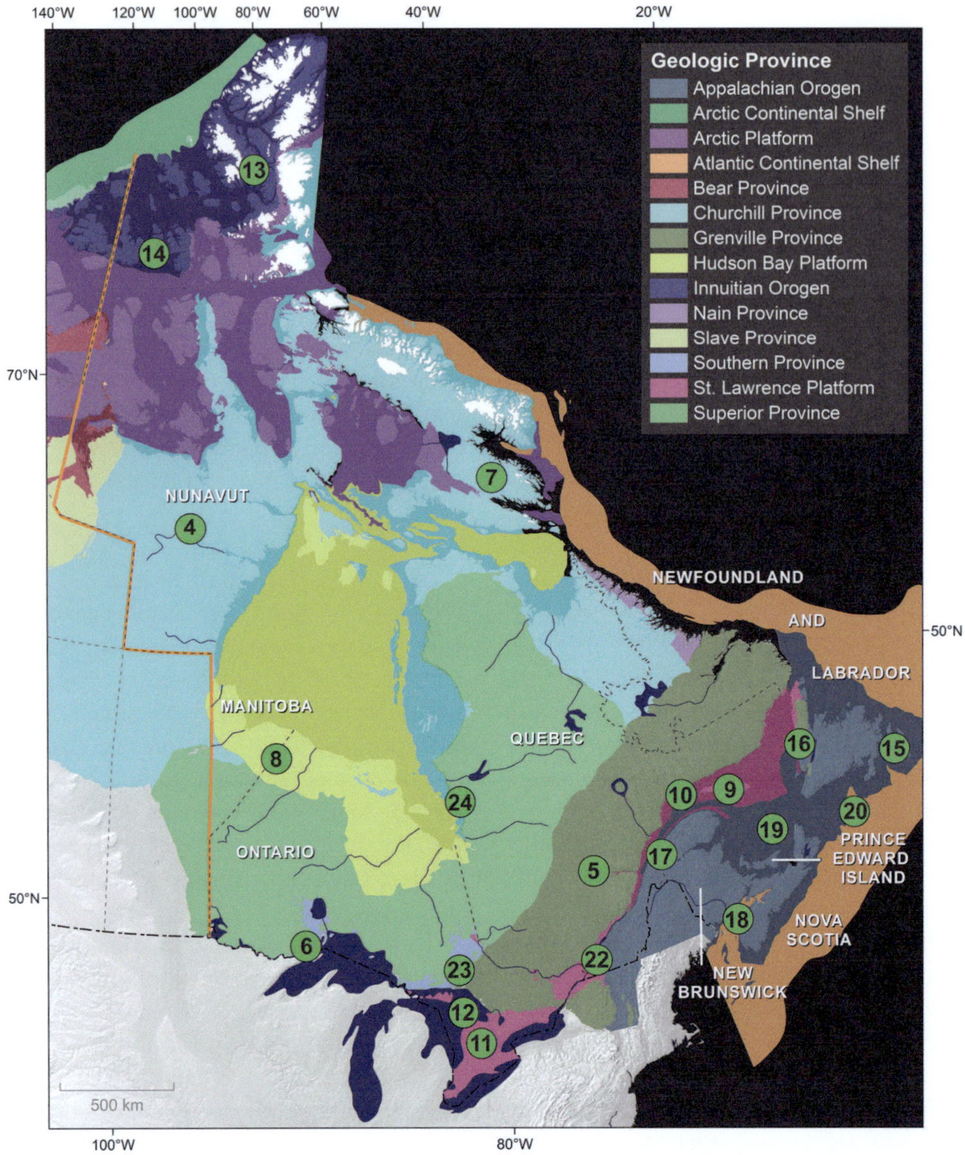

Fig. 1.1 The geological provinces of Eastern Canada (adapted from Wheeler et al. 1996). Numbers correspond to the chapters in which case studies are presented in this book. *4* Mainland Nunavut, Churchill Province; *5* Lake Saint-Jean Lowlands, Grenville Province; *6* Thunder Bay region, Superior Province; *7* Baffin, Labrador, and the Strait of Belle Isle, Churchill, Grenville, Nain provinces and Appalachian Orogen; *8* Hudson Bay Lowlands, Hudson Bay Platform; *9* Anticosti Island landscapes, St. Lawrence Platform; *10* North Shore, St. Lawrence River, St. Lawrence Platform; *11* Great Lakes Lowlands, St. Lawrence Platform; *12* Bruce Peninsula, St. Lawrence Platform; *13* Arctic Archipelago, Innuitian Orogen; *14* Melville, Bathurst and Cornwallis islands, Innuitian Orogen; *15* Avalon Peninsula, Appalachian Orogen; *16* Long Range Mountains, Appalachian Orogen; *17* Témiscouata-Madawaska Valley, Appalachian Orogen; *18* Atlantic coastlines, Appalachian Orogen; *19* Magdalen Islands landscapes, Atlantic Continental Shelf; *20* Submarine geomorphology, Atlantic Continental Shelf; *22* Urban caves in Montréal, St. Lawrence Platform; *23* Sudbury mining landscapes, Southern Province; *24* Wemindji Cree environment, Superior Province. Not shown are Chap. 1, 2, 3, 21, 25 and 26 that deal either with introductory or broad scale issues from more than one location. The orange line is the boundary between western and eastern Canada

scenery in Canada, with glaciers and 2,000 m cliffs rising almost vertically out of the sea. The southern margin of the shield is generally lower but along the north shore of the gulf and estuary of the St. Lawrence the shield rises to 1,000 m directly from the water. For a distance of nearly 1,500 km the edge of the shield is broken by deep valleys and gorges. Between Ottawa and Sault Ste. Marie, the edge of the shield dips gently under the younger sedimentary rocks and the general relief is moderate. Further west along the north shore of Lake Superior the edge of the shield is higher (La Berge 1994). However, in general the west and southwest edges of the shield are relatively low compared with the north and northeast. The scenery of the northern arctic boundary of the shield is rather different. Several fingers of Precambrian

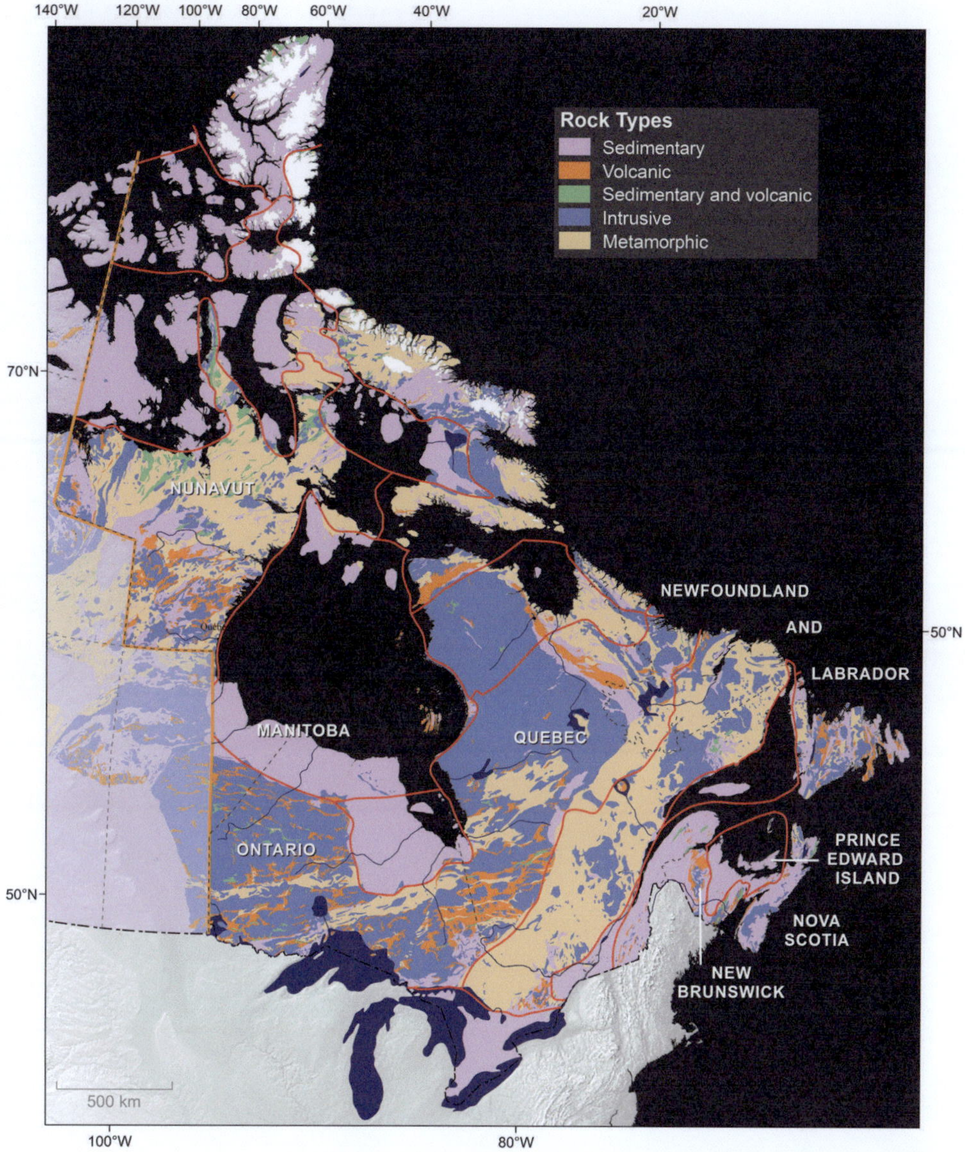

Fig. 1.2 Distribution of major rock types in Eastern Canada (after Wheeler et al. 1996). The red lines represent geomorphological landscape regions and the orange line is the boundary between western and Eastern Canada

rocks penetrate northwards into the arctic islands, making more variable scenery.

1.2.1.2 Platforms

The Arctic, Hudson, and St. Lawrence platforms are underlain by a basement of Precambrian shield rocks, overlain by beds of much younger, unmetamorphosed Palaeozoic and Proterozoic rocks. Sediments were deposited in shallow seas in the intervening areas between the shield and the mountains. Large areas of the interior of the Canadian Shield were depressed below sea level and marine sediments were deposited over parts of Québec, Ontario, Manitoba and NWT between 450 and 350 Ma. The Arctic Platform extends under the islands of the arctic archipelago, between the Innuitian Orogen and the Canadian Shield. The St. Lawrence Platform consists of three geographically disjunct parts separated by two southward extensions of the Canadian Shield. The Hudson Platform surrounds and extends beneath Hudson Bay (Tasiujarjuak).

1.2.1.3 Orogenic Belts

Orogenic belts are areas that have undergone relatively recent (post-Proterozoic) mountain-building episodes accompanied by volcanic eruptions and metamorphism. The rocks in these belts have also undergone folding (Fig. 1.3), faulting and uplift. Eastern Canada has two orogens, namely, Appalachia and Innuitia. Appalachia extends south of the St. Lawrence River, through the Atlantic Provinces. Innuitia extends from Ellesmere Island (Umingmak Nuna) through the Queen Elizabeth Islands in Nunavut and is underlain by highly deformed sedimentary, metamorphic and volcanic rocks.

Fig. 1.3 Surface expression of folded Precambrian and Upper Devonian strata. Precambrian covers 'a' and 'c', but not 'b'. The Arctic Fold Belt in Innuitia, including Cornwallis Island, includes younger orogenies. **a** Complex folded rock layers that represent the Nouveau Québec Orogen, Québec. **b** Part of the Arctic Fold Belt, Innuitia, Cornwallis Island, Nunavut. **c** Folded mafic sill (dark) emplaced in psammite (light) of the Lake Harbour Group, Meta Incognita Peninsula, Baffin Island, Nunavut. *Source* a—Mosa_landsat_qc-2015-lcc-MFFP. b—Government of Canada; Natural Resources Canada (2007). Contains information licensed under the Open Government License—Canada. c—NRCan 2001-017 by M. R. St-Onge/Natural Resources Canada. Inset image here and elsewhere in this chapter is extracted from NASA Visible Earth (Stöckli et al. 2005)

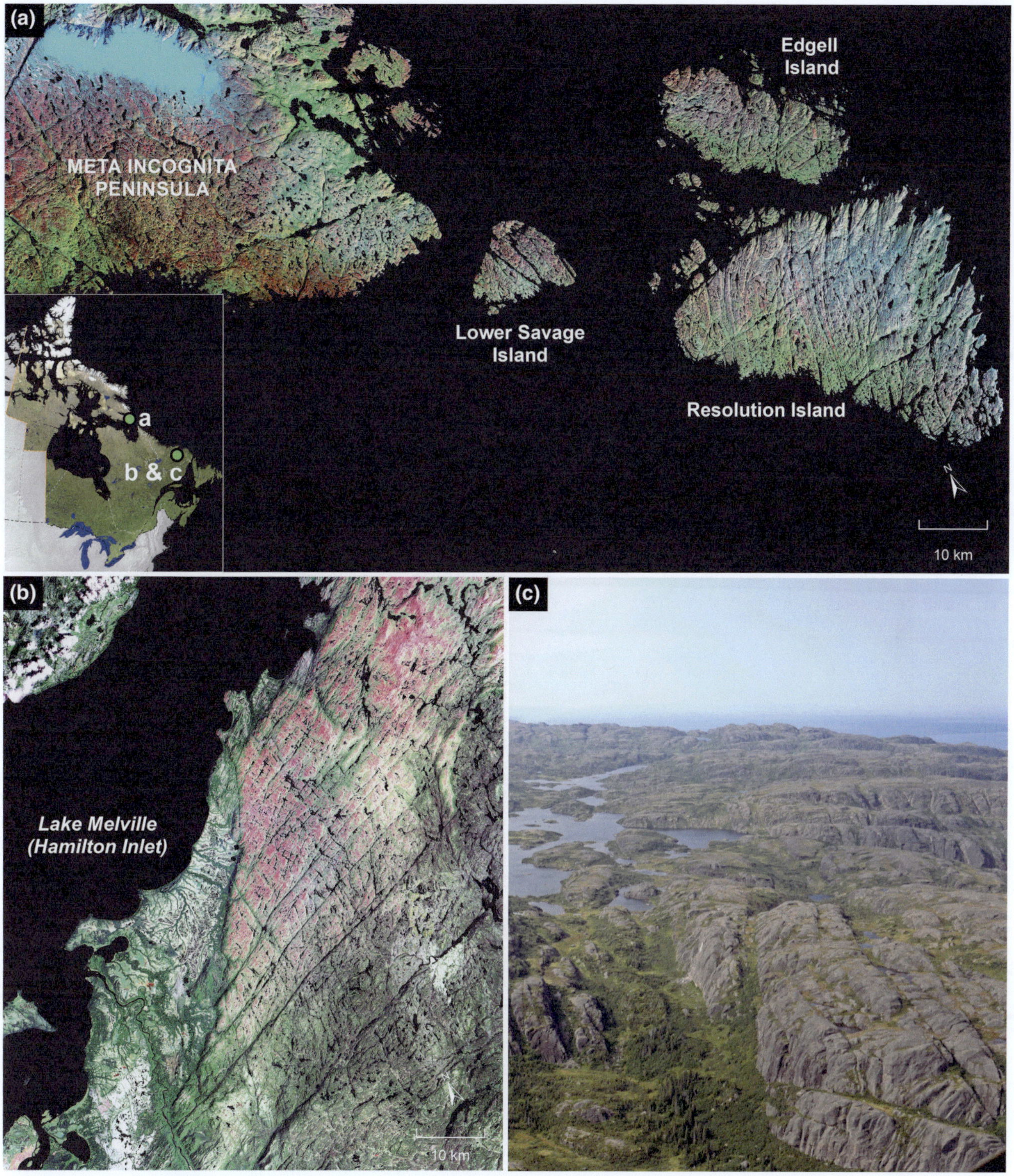

Fig. 1.4 Fractures and lineaments; long, straight, topographic alignments carved by erosion. **a** Meta Incognita (Katannilik) Peninsula, southern tip Baffin Island, Nunavut. **b** Happy Valley-Goose Bay, Labrador. The rugged terrain of the Mealy Mountains is visible in the right third of the image. The Churchill River flows into Lake Melville, lower left. **c** Mealy Mountains, Labrador. Locations are shown in the inset image. *Source* a and b—MDA Federal (2004). c—P. Gierszewsk/Wikimedia Commons, CC BY-SA 4.0

1.2.1.4 Continental Shelves

The Arctic and Atlantic continental shelves are two geological provinces that extend below the Arctic and Atlantic oceans and have experienced multiple cycles of sedimentation and erosion during the glacial and interglacial cycles of the past 2.59 million years.

1.3 Physiography

Physiographic subdivision of Canada is based on a system summarized by Bostock (1970, 2014; see also Graf 1987; Fulton 1989). Eastern Canada has been divided into eight physiographic regions. Whereas the geological provinces are defined by structure and lithology alone, the physiographic regions are defined by structure, lithology and relief and do not extend to submarine areas as shown in Fig. 1.5. These are Canadian Shield, Arctic, Hudson Bay and St. Lawrence lowlands, Innuitia, Appalachia, Arctic Coastal, and Maritime plains.

1.3.1 Canadian Shield

Seventy percent of Eastern Canada is underlain by the Canadian Shield. The present surface of that shield appears to be an exhumed pre-Palaeozoic erosion surface. The shield was eroded and the valley system cut into the bedrock before the Palaeozoic. The Precambrian surface has been little changed since the removal of the Palaeozoic cover, and glaciation has only superficially modified it (Ambrose 1964). The sequence of events envisaged is that first there was the development and partial dissection of an erosion surface on the Archaean rocks in pre-Proterozoic time, then the burial and exhumation of this surface once or perhaps twice during the Proterozoic, followed by the exhuming of these several surfaces from beneath the cover of Palaeozoic strata, and finally the scouring of the surface during the surges of Pleistocene glaciation (see Chap. 2).

During the evolution of the Canadian Shield, many cycles of erosion and sedimentation reduced generations of mountain ranges to low relief such that the direct control of

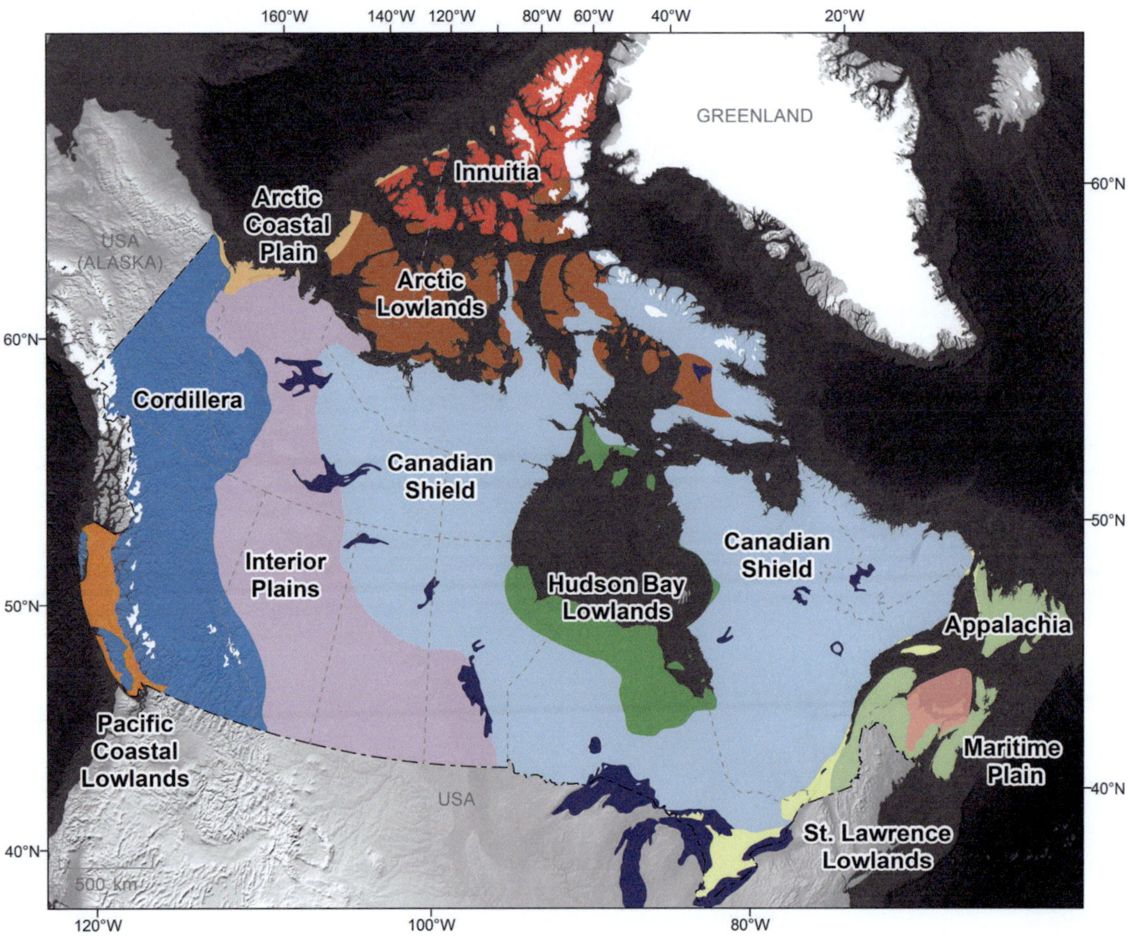

Fig. 1.5 Physiographic regions of Canada (adapted from Bostock 2014)

lithology on these landforms and landscapes is rarely seen. Over the greater part of the shield local relief is less than 100 m. In spite of its overall low relief, a number of different terrains can be identified in the shield. Plains have formed on areas of unmetamorphosed rocks, more or less flat lying, such as the sandstone and conglomerate terrains of the Cobalt Plain, Ontario, and the terrain of flat-lying gabbro sills of Nipigon Plain, northwestern Ontario. Low to moderate relief hills are associated with low grade metamorphic rocks and generally tilted or gently folded sediments or sills such as the Port Arthur, Penokean and Labrador hills (Ontario and Québec). Moderate to high relief uplands are present in the Laurentian Highlands, Québec. High relief mountains are associated with particularly resistant massive rocks, such as the anorthosites of the Mealy Mountains of Labrador and stand high above the surrounding terrain. Highlands formed of broad uplifted areas of deeply incised massive crystalline rocks characterize the Torngat Mountains and glacier-capped Baffin Island, where the greatest relief of up to 1,800 m is found.

1.3.2 Lowlands

Fourteen percent of Eastern Canada comprises lowlands and plateaus: Arctic, Hudson Bay and Great Lakes-St. Lawrence lowlands.

1.3.2.1 Arctic Lowlands

These lowlands are formed on the flat lying or nearly flat Palaeozoic and late-Proterozoic sedimentary rocks, lying between the Canadian Shield and the Innuitian Orogen.

1.3.2.2 Hudson Bay Lowland

The Hudson Bay Lowland is a low, swampy, marshy plain with subdued glacial features and a belt of raised beaches that border the bay. Palaeozoic strata mantle the massive or stratified Precambrian rocks and slope gently northeastward and eastward into Hudson and James bays.

1.3.2.3 Great Lakes-St. Lawrence Lowlands

These lowlands border the Canadian Shield on the southeast, extending from the west end of Lake Huron and the head of Lake Erie northeasterly to the Strait of Belle Isle. Three separate parts can be distinguished: (a) The Gulf of St. Lawrence, that includes Anticosti Island, the Mingan Islands, and several small areas bordering the Gulf of St. Lawrence, the Strait of Belle Isle, and western Newfoundland (see Fig. 9.1). Most of the area is lower than 100 m asl; (b) The St. Lawrence-Ottawa Valley subregion which straddles the St. Lawrence River from Pembroke to Québec City. The land is rarely higher than 150 m asl, except for the Monteregian Hills formed of intrusive igneous rocks; and (c) The Great Lakes Lowland subregion, which has at its centre the Niagara Escarpment cuesta that varies in elevation from 200 m asl at Niagara Falls in the southeast to 500 m asl at Manitoulin Island in the northwest.

1.3.3 Mountains

Sixteen percent of eastern Canada is underlain by the hilly to mountainous areas of Innuitia and Appalachia-Acadia. They are the sites of two great geosynclines, the Franklinian and Appalachian-Acadian geosynclines where orogenesis occurred during the Palaeozoic and Mesozoic eras. The sedimentary rocks are highly folded and faulted with a dominant southwest to northeast strike, locally becoming north-northeasterly.

1.3.3.1 Innuitia

The highest summits of northwest Ellesmere Island and central and western Axel Heiberg Island reach 2,500 m asl and are nearly buried by ice caps with projecting nunataks. Northeastern Ellesmere Island is a second major mountain belt. Eureka Upland extends the length of Innuitia from northeastern Ellesmere Island to the west end of the Parry Islands and is developed on Palaeozoic carbonates, shales and sandstones. Altitudes in this upland are generally less than 1,000 m asl. Low dissected plateaus and gently rolling low uplands have developed on soft late Mesozoic and Palaeogene sandstone and shale.

1.3.3.2 Appalachia

Appalachia consists of three subregions in Newfoundland, Québec and Nova Scotia/New Brunswick. (a) a Newfoundland subregion consists of highlands, up to 850 m asl; uplands between 200 and 300 m asl; and central lowlands extending from sea level to 150 m asl.; (b) a Québec subregion, which includes the Notre Dame Mountains, exceeding 1,200 m asl, the Sutton Mountains that reach 1,000 m asl, and a variety of hills and uplands that reach about 300 m asl; and (c) a Nova Scotia and New Brunswick subregion, consisting of alternating uplands and highlands along the Bay of Fundy to Cape Breton Island to the northeast.

1.3.4 Plains

1.3.4.1 Arctic Coastal Plain

The Arctic Coastal Plain is remarkably flat and uniform. Included are parts of Meighen Island, Ellef Ringnes Island, and Borden Island.

1.3.4.2 Atlantic Maritime Plain

The Atlantic Maritime Plain stretches along the coast of New Brunswick and Nova Scotia from Chaleur Bay to Cape George. It includes Prince Edward Island and the Magdalen Islands.

1.4 Permafrost and Vegetation Cover as Geomorphic Drivers

Much of the 'big picture' landscape regions can be defined in terms of structure and relief. But some account of process is also necessary in order to incorporate a sense of the dynamism of geomorphological landscapes. At the macro-scale we have to look to surrogates for agents of change. We have chosen two such surrogates, namely, presence or absence of vegetation and permafrost, when considering the geomorphological landscape regions of Eastern Canada. Subaerial geomorphic processes, which will be considered more fully in Chap. 3, are greatly influenced by presence/absence and type of vegetation cover. There is also a delicate interplay of vegetation and permafrost which strongly influences the sensitivity and resilience of individual landscape regions to environmental change. Hence the need to define geomorphological landscape regions taking into account ecozones and permafrost zones.

1.4.1 Permafrost

Permafrost is a thermal condition defined as 'ground that remains below 0 °C for two successive summers or longer'. The thermal condition of the ground should be known in order to identify accurately the presence of permafrost. Unfortunately, the thermal condition of the ground is not widely monitored, and surrogate parameters commonly have to be employed. A further problem is that the nature of the materials in the ground (whether massive ground ice, inter-particulate ice, bedrock or snow) has substantial influence on the thermal condition of the ground. This means that there is no simple formula to predict the presence or absence of permafrost. The whole of Nunavut and substantial portions of Labrador and northern Québec are underlain by continuous permafrost. The most serious effects of recent climate change leading to landscape instability are, in general, found in the discontinuous permafrost zones, where the ground is closer to melting and/or thawing. But engineering in the continuous permafrost zone presents its own distinctive challenges.

Four permafrost regions are commonly recognized in Canada (Fig. 1.6). The regions are defined by the proportion of the landscape that is underlain by permafrost.

1.4.1.1 Continuous Permafrost

Continuous permafrost is defined as a region where >90% of the land is underlain by permafrost (Smith et al. 2001). All of Nunavut, Nunavik (Ungava, Québec), the Torngat Mountains (Labrador) and a coastal strip on the western side of Hudson Bay are the areas of eastern Canada that are included.

1.4.1.2 Extensive Discontinuous Permafrost

Extensive discontinuous permafrost is defined as a region where 50–90% of the land is underlain by permafrost. A narrow strip of land south of Ungava, the Torngat Mountains and the Hudson Bay continuous permafrost zone is included.

1.4.1.3 Sporadic Discontinuous Permafrost

Sporadic discontinuous permafrost is defined as a region where 10–50% of the land is underlain by permafrost. Northern Québec, northern Ontario and northeastern Manitoba are included.

1.4.1.4 Isolated Patches of Permafrost

Isolated patches of permafrost are defined as regions where <10% of the land is underlain by permafrost. Such patches are common in central Québec, the Chic-Choc Mountains of Gaspésie, southern Labrador, the Great Northern Peninsula of Newfoundland and a thin strip of northern Ontario (Fig. 1.6).

Geomorphological landscapes with permafrost are radically different in terms of response to changing climate from those that have no permafrost. They display different landforms (discussed in Chap. 3) and experience different processes of change simply because of the presence of permafrost.

1.4.2 Ecozones

A terrestrial ecozone map is presented in Fig. 1.7. In this map, Southern and Northern Arctic ecozones are not differentiated. The importance of ecozones, especially the distinction between arctic and alpine tundra and the forested ecozones, is primarily in terms of the associated processes of landscape change.

1.4.2.1 Arctic Cordillera, Northern Arctic and Southern Arctic Ecozones

The Arctic Cordillera Ecozone consists of glaciers and ice caps, marked by extreme cold, and high winds. The regolith is thin and plants are scarce. Frost-shattered materials in soils are characteristic. At lower elevation, tundra meadows are present, dotted with arctic flowers and prostrate or stunted shrubs.

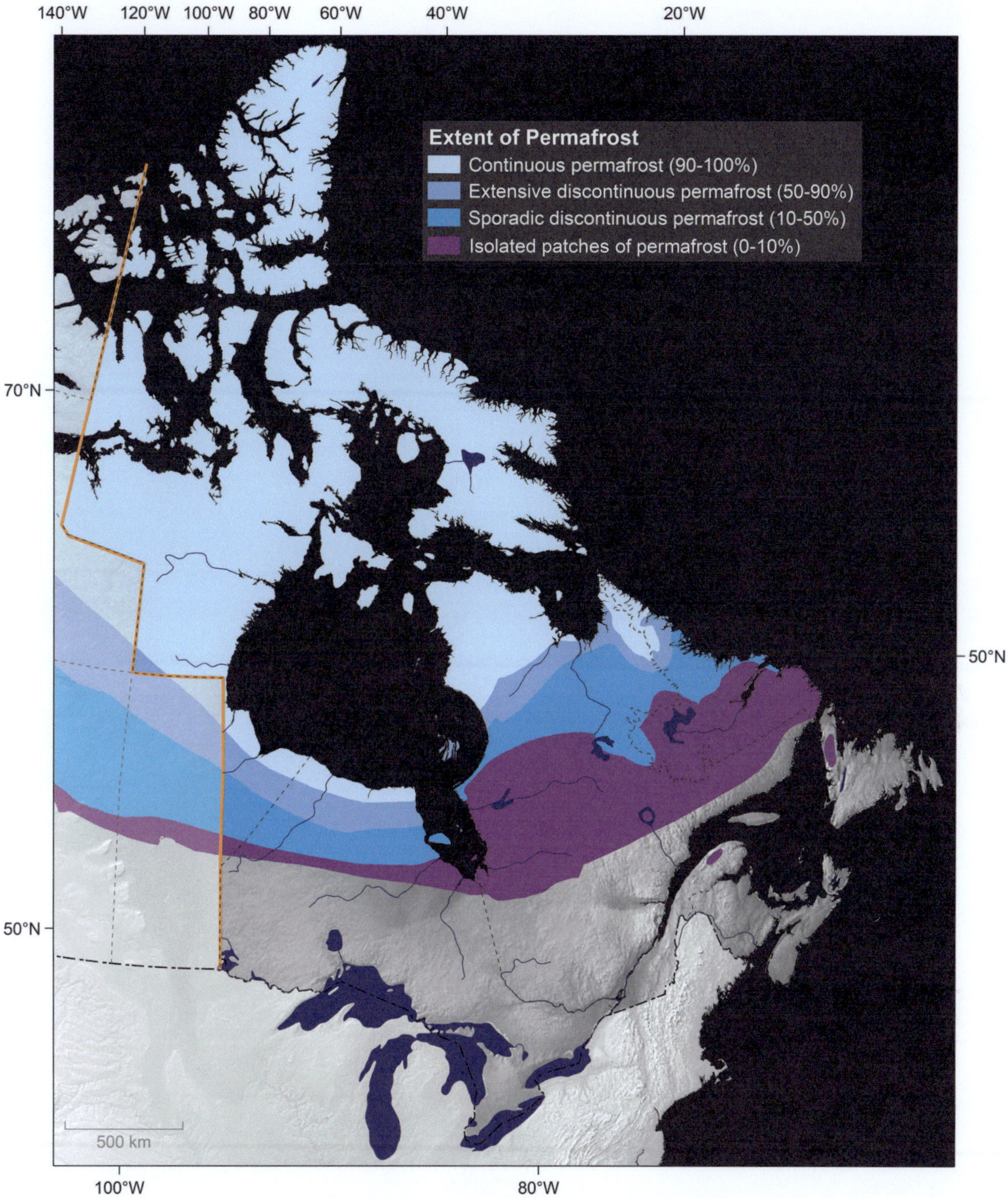

Fig. 1.6 Distribution and extent of permafrost in Eastern Canada (adapted from Smith et al. 2001; Government of Canada 2008). % in legend refers to percentage of land underlain by permafrost

The Northern Arctic Ecozone includes most of the non-mountainous areas of the arctic islands, western Baffin Island, and northern Québec (Nunavik). When not covered in snow, it is typified by barren plains covered by frost-controlled patterned ground and scattered rock outcrops. Moss and lichens are abundant. *Salix arctica* (Dwarf Arctic willow) and prostrate and cushion plants or mats are characteristic. The Southern Arctic Ecozone contains shrublands, wet sedge meadows and other wetlands (Zoltai 1988). Superimposed are the patterned ground, and frost-controlled soils. It is known as the barren lands and demarcates the northern limit of treeline (Bliss et al. 1981).

Baffin Island comprises Arctic Cordillera and Northern Arctic ecozones, all of which lie within the continuous permafrost zone. There are two distinctive vegetation communities that characterize the inland flanks of the mountains:

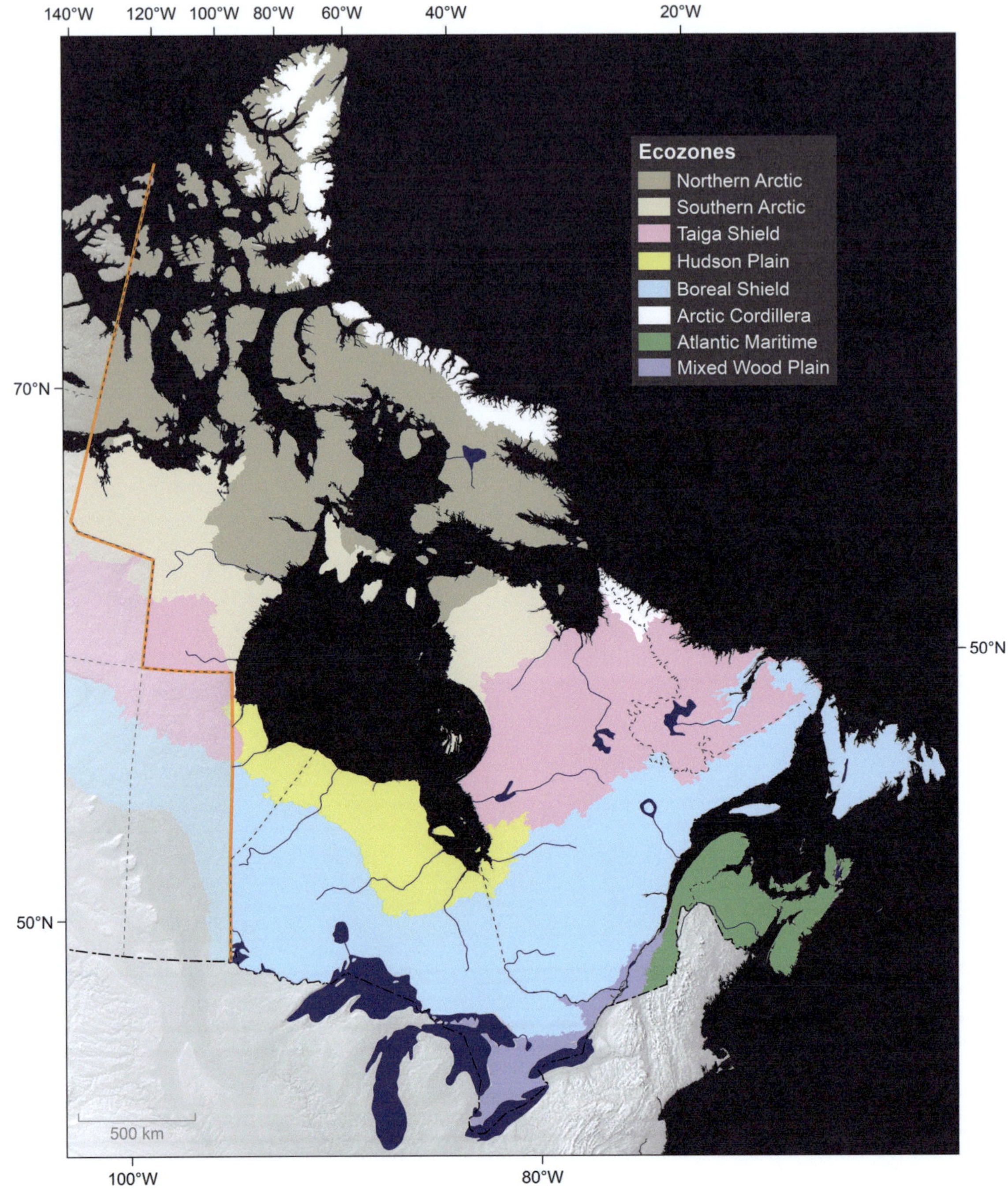

Fig. 1.7 Terrestrial Ecozones of eastern Canada (derived from Environment Canada 1995)

fell-field and tundra. Fell-field assemblages develop in areas where frost heaving has disturbed the surface of the land, producing areas of shattered and thrusted felsenmeer, and areas of cryogenic hummocks developed in Quaternary sediment. The most noticeable plants are *Silene* (campion) and *Papaver nudicale* (arctic poppy), along with scattered lichens and bryophytes. Poorly developed regosols and turbic cryosols are present. Tundra communities are commonly encountered throughout Auyuittuq National Park below approximately 1500 m asl, and in Sirmilik National Park below 1000 m asl. Woody shrubs include *Salix* (willow), *Betula glandulosa* (dwarf birch), and *Betula nana* (arctic birch), all of which commonly do not exceed 60 cm in height but may survive more than 100 years. *Salix* leaves were important supplements to the traditional Inuit diet, as they are rich in Vitamin C. Although *Salix* and *Betula glandulosa* are common in the low- and mid-arctic, in Canada *Betula nana* is confined to Baffin Island. Many of the tundra plants are edible, and some are of great medicinal value, especially sorrel, which prevents scurvy and was known to European visitors as 'arctic salad'.

1.4.2.2 Taiga Shield, Boreal Shield and Hudson Plains Ecozones

The open stunted forests of the Taiga Shield Zone are dominated by *Picea mariana* (black spruce) and *Pinus banksiana* (jack pine), innumerable bogs and wetlands (Halsey et al. 1997), scattered stands of *Betula papyrifera* (paper birch) and *Populus tremuloides* (trembling aspen), lichens and ground-hugging shrubs (Table 1.2). 'Drunken forests' emphasize the role of permafrost gradual subsidence on fringes of bogs as opposed to central heave. The Hudson Plains Ecozone, which lies mostly in Ontario, Manitoba and Québec, is where the boreal forests and the tundra merge and vegetation resembles that of the arctic tundra. Treeless areas extend about 30 km from the coast but stands of trees can penetrate further north in sheltered sites. Vegetation includes willow shrubs and tussocks of sedges, including *Eriophorum* (cottongrass). The Boreal Shield Ecozone largely corresponds with the Canadian Shield between Newfoundland and Alberta. *Picea mariana*, *Picea glauca* and *Abies balsamea* (balsam fir) are the dominant species, and wetlands are also common. In its southern regions, larger wetlands have been converted to cranberry and blueberry farms.

1.4.2.3 Mixed Wood Plains and Atlantic Maritime Ecozones

These southernmost ecozones were largely deciduous forests before the arrival of European settlers. The Mixed Wood Plains Ecozone lies adjacent to the St. Lawrence River and the Great Lakes and has fertile soils that are agriculturally highly productive. Very little of the original forest (*Pinus strobus* (eastern white pine), *Tsuga Canadensis* (eastern hemlock), *Betula alleghaniensis* (yellow birch) and *Pinus resinosa* (red pine) remains today, and farms, woodlots, urban forests and protected areas have taken their place. The Atlantic Maritime Ecozone contains mixed wood Acadian forests, sand dunes, and coastal islands. There are few pockets of old growth forest, as centuries of forestry, agriculture and natural disturbance have produced dominantly secondary and tertiary growth.

1.4.3 The Significance of Arctic Treeline and Alpine Timberline

The distribution of the ecozones of Eastern Canada (Fig. 1.7) provides a sense of the north-south gradation of vegetation. At local scale, however, the presence of the arctic treeline or the alpine timberline can be of greater significance than the north-south gradation of vegetation.

1.4.3.1 Arctic Treeline

The arctic treeline in Eastern Canada can be divided into two parts (Hare and Ritchie 1972). The northern of these is the 'forest tundra zone' consisting of scattered patches of stunted trees set in a matrix of tundra. The forest tundra zone extends across northern Québec from the Hudson Bay coast to the Atlantic Ocean where it is largely cut off by the coastal tundra of the Torngat Mountains. The lichen woodland zone extends south from the arctic tree line to the open boreal forest or 'lichen woodland'. The proportion of trees to lichen mat gradually increases southward towards the forest line where trees cover >50% of the landscape (Larsen 1980). East of Hudson and James bays, the ground lichens often make up the most conspicuous understory vegetation near forest line. The open boreal forest or lichen woodland is a very extensive zone in Québec and central Labrador. At its northern extreme, long stringers of lichen woodland stretch north along the major rivers, while intervening uplands are entirely without trees. Farther south in the main lichen woodland, *Picea mariana*, *Picea glauca*, *Larix laricina* (tamarack), and *Abies balsamea* stand several metres apart in a sea of lichens. A muskeg of stunted *Picea mariana*, *Rhododendron groenlandicum* (Labrador tea) and sphagnum covers the intervening wet sites. The boreal forest zone consists of closed crown conifer forests with a conspicuous deciduous element (Ritchie 1987). The proportions of the dominant conifers *Picea glauca*, *Picea mariana*, *Pinus banksiana*, *Larix laricina* and *Abies balsamea* vary greatly in response to climate, topography, soil, fire and pests.

1.4.3.2 Alpine Timberline

In the highlands of southern Labrador and eastern Québec, there is an alpine timberline at around 1,000 m asl, upslope from the open boreal forest. In southeastern Québec, the Laurentians have some 1,000 m asl peaks that also have treeline communities. In the Long Range Mountains of Newfoundland and the Cape Breton Highlands, *Picea mariana*, *Picea glauca* and *Abies balsamea* form a dense krummholz called 'tuckamore' at about 600 m asl. Although the highest summit in the Cape Breton Highlands reaches barely 500 m asl, several of the ridge tops are covered with caribou lichen. The Notre Dame Mountains extend along the Gaspé Peninsula, with a dozen peaks above 900 m asl.

Geomorphological landscapes above timberline or north of the treeline are radically different in terms of response to changing climate than those that have forest cover. They

Table 1.2 Vegetation[a]

Scientific names	Common names	Typical habitats
Conifers		
Abies balsamea	Balsam Fir; Var	Boreal Forest
Juniperus communis	Common Juniper	Boreal Forest
Larix laricina	Tamarack; Larch	Boreal Forest
Picea glauca	White Spruce; Cat Spruce	Boreal Forest
Picea mariana	Black Spruce	Boreal Forest
Picea rubens	Red Spruce	Acadian Forest
Pinus banksiana	Jack Pine; Princess Pine	Boreal Forest
Pinus resinosa	Red Pine	SE Boreal Forest
Pinus strobus	Eastern White Pine	Eastern Boreal Forest
Thuja occidentalis	Eastern White Cedar Arbor vitae	Eastern Boreal Forest
Tsuga canadensis	Eastern Hemlock	Acadian and Temperate
Deciduous		
Acer pennsylvanicum	Striped Maple; Moosewood	Acadian and Temperate
Acer rubrum	Red Maple	Acadian and Temperate
Acer saccharum	Sugar Maple	Acadian and Temperate
Alnus incana rugosa	Grey Alder; Hoary Alder	Boreal
Alnus viridis crispa	Green Alder; Mountain Alder	Boreal
Betula alleghaniensis	Yellow Birch	Acadian and Temperate
Betula glandulosa	Dwarf Birch	Northern Taiga
Betula nana	Arctic Birch	E Baffin Island
Betula papyrifera	White Birch; Paper Birch	Boreal; Temperate
Carpinus caroliniana	Hornbeam; Bluebeech	Acadian and Temperate
Carya ovata	Shagbark Hickory	Temperate
Fagus grandifolia	Beech	Acadian and Temperate
Fraxinus americana	White Ash	Acadian and Temperate
Fraxinus nigra	Black Ash	Temperate
Juglans cinerea	Butternut	Temperate
Juglans nigra	Black walnut	Temperate
Populus balsamifera	Balsam Poplar	Boreal Forest
Populus tremuloides	Trembling Aspen	Boreal Forest
Prunus pennsylvanica	Pin Cherry	Temperate; S Boreal
Prunus serotina	Black Cherry	Temperate
Prunus virginiana	Choke Cherry	Temperate; S Boreal
Quercus alba	White Oak	Temperate
Quercus rubra	Northern Oak; Red Oak	Acadian and Temperate
Salix bebbiana	Beaked Willow	Boreal Forest
Salix discolor	Pussy Willow	Boreal Forest
Sorbus americana	Dogberry, Rowan, Mountain ash	Boreal Forest
Sorbus decora	Showy Mountain Ash, Dogberry	Boreal Forest
Tilia americana	Basswood; lime	Temperate
Ulmus americana	White Elm; American Elm	Temperate

[a]These taxa are listed in alphabetical order by scientific name. Listing does not imply taxonomic affinity

display different landforms (discussed in Chap. 3) and experience different processes of change simply because of the incomplete ground cover.

1.5 Special Landscapes

Landscapes dominated by the presence of impact craters and thick sequences of carbonate lithology are totally structurally controlled and do not respond to the regional zonation that has been described in Sects. 1.1–1.3. They require separate discussion because of their intrinsic interest, widespread occurrence and distinctive contribution to Eastern Canadian landscapes.

1.5.1 Impact Crater Landscapes in Eastern Canada

Recognition of the importance of impact processes in Eastern Canadian landscapes has changed radically since the 1990s. The recognition that nearly all planetary surfaces are heavily cratered from the impact of meteorites, asteroids and comets has to be balanced with the observation that most of Earth's impact craters have been destroyed by erosional and tectonic processes. Some examples remain preserved because of their young age, their large size, their occurrence in a relatively stable geological region, or their relatively rapid burial by younger protective sediments which, in some cases, erosion has removed. Approximately 160 impact craters have been identified on Earth, of which 18 are located in Eastern Canada (Table 1.3). There is a strong concentration in areas like the Canadian Shield whose relatively uneventful geological history over the last 500 million years has allowed their preservation.

Impact crater types include simple craters to 4 km in diameter with uplifted and overturned rim rocks surrounding a bowl-shaped depression. Larger complex impact structures and basins, generally 3 km or greater in diameter, have a distinct central uplifted peak or ring, a ring-shaped trough, and a slumped rim (Fig. 1.8). The interiors of impact craters typically are partly filled with breccia (Grieve 2001).

The Clearwater Lakes (Eau Claire) consists of two roughly circular lakes separated by a barrier of islands.

Table 1.3 Impact craters

Feature	Location	Dimensions	Age (Ma)
Pingualuit	Nunavik, Québec	3.4 km	1.4 ± 0.1
Bloody Creek	Annapolis Co. NS	400 m maximum	<2.6
Haughton	Devon Island, Nunavut	23 km	23.5 ± 2
Mistastin	Central Labrador, NL	28 km	36.4 ± 4
Wanapitei	Central Ontario	7.5 km	37.2 ± 1.2
Montagnais	Scotia Shelf	45 km	50.5 ± 0.8
Manicouagan	Central Québec	85 km	214 ± 1
Eau Claire (Clearwater) West	Nunavik, Québec	36 km	290 ± 20
Ile Rouleau	Central Québec	4 km	<300
Charlevoix	Central Québec	54 km	342 ± 15
West Hawk	Eastern Manitoba	2.4 km	351 ± 20
Nicholson	Kivalliq, Nunavut	12 km	<400
La Moinerie	Nunavik/Labrador	8 km	400 ± 50
Couture	Nunavik, Québec	8 km	430 ± 25
Brent	Algonquin Park, ON	3.8 km	>453
Slate Islands	Lake Superior, ON	30 km	<450
Eau Claire (Clearwater) East	Nunavik, Québec	26 km	460–470
Presqu'ile	Central Québec	24 km	<500
Holleford	Eastern Ontario	2.35 km	550 ± 110
Sudbury	Central Ontario	130 km	1850 ± 3

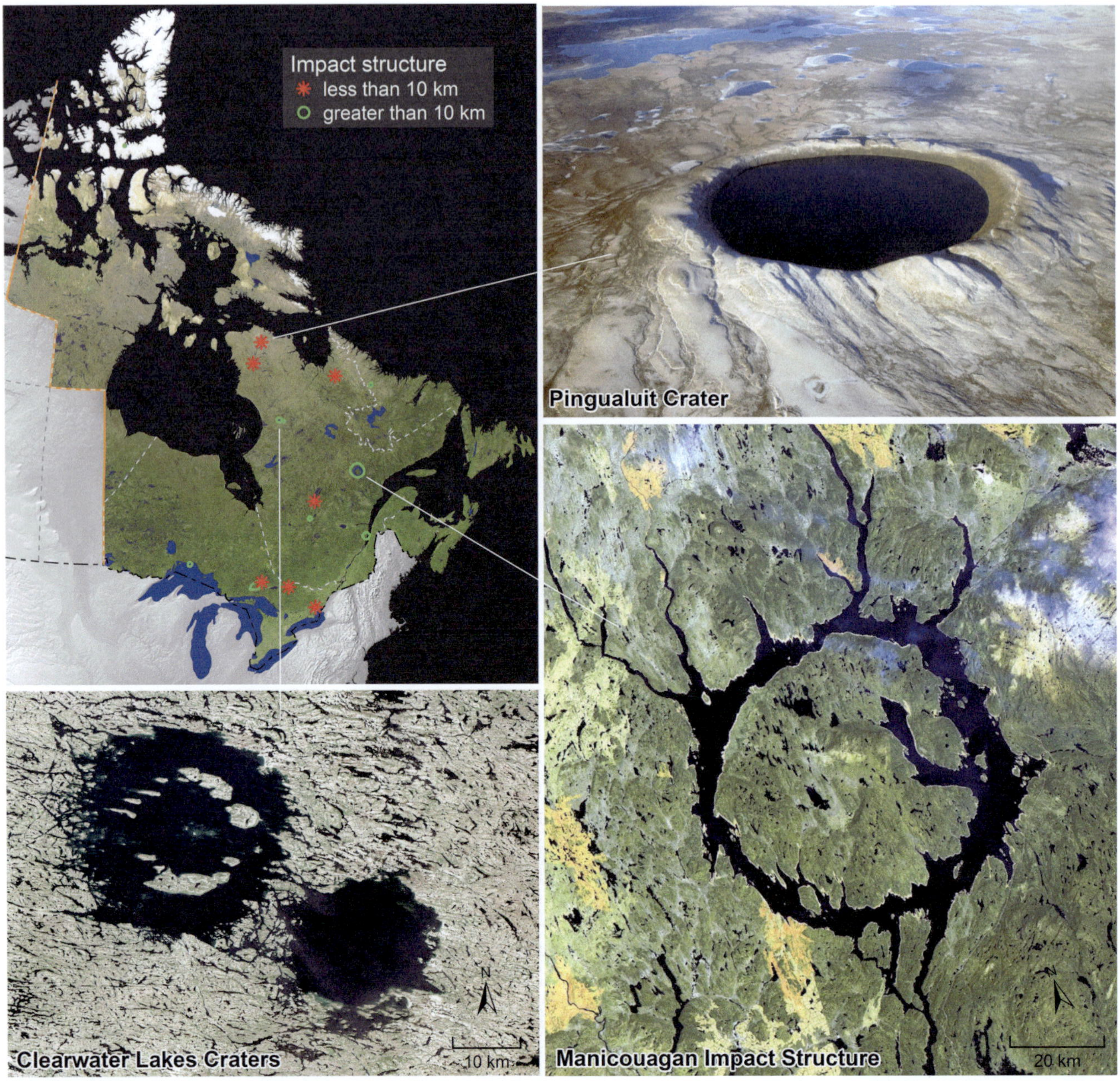

Fig. 1.8 Distribution and selected images of impact structures in Québec. Like many other impact craters, the Clearwater Lakes, Pingualuit crater, and Manicouagan impact structure, formed as a result of meteorites or asteroids striking Earth's surface. Manicouagan Crater is one of the largest and best-preserved craters on the planet. The 100 km wide crater is about 214 million years old. *Source* Earth Impact Database (2017), NASA Earth Observatory images by J. Allen and R. Simmon, NASA/photograph D. Sarrazin, and U.S. Geological Survey

These two impact structures (25 and 24 km in diameter, respectively) are located 200 km north of Poste-de-la-Baleine (Kuujjuaraapik) and 110 km east of Hudson Bay in northern Québec. The Pingualuit (Nouveau Québec) impact structure is a simple crater, 3.4 km in diameter, with a nearly circular 2.8 km diameter lake. It is located about 150 km southeast of Salluit. The Manicouagan impact structure is in central Québec, 180 km northwest of Sept-Iles. Although the reservoir is only 65 km in diameter, the original impact structure diameter is closer to 100 km. It is a common photographic subject for space shuttle astronauts.

1.5.2 Karst Landscapes of Eastern Canada

Karst landscapes are characterized by distinctive hydrology and landforms arising from a combination of high rock

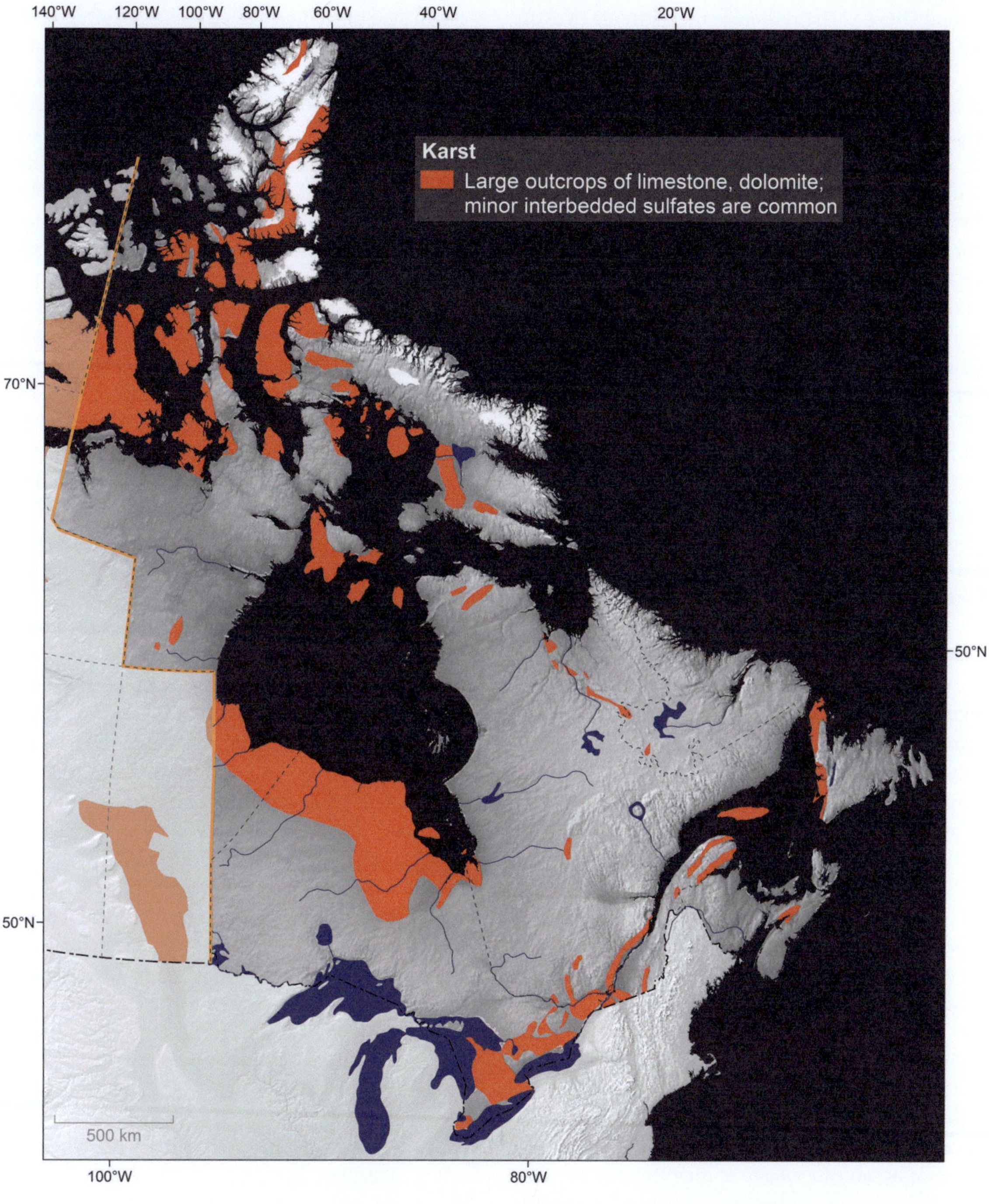

Fig. 1.9 Distribution of karst rocks in eastern Canada (Williams and Fong 2017; also see Ford and Williams 2007)

solubility and well-developed secondary porosity (Ford and Williams 2007) (Fig. 1.9). The occurrence of pure carbonate rocks with high solubility is insufficient to produce karst, because the structure, density and thickness of the rocks are also important. The carbonate rocks in which karst reaches its best expression are thick, dense, pure and massive with low primary porosity, but also with well-developed secondary porosity along fissures such as joints, faults and bedding planes.

The most widespread surface karst landforms are small solution grooves and runnels, collectively called karren. Individual features are rarely longer or deeper than 10 m, but frequently they are densely clustered and dissect larger areas, referred to as limestone pavements. Limestone pavements are particularly well developed on Anticosti Island, Québec, the Bruce Peninsula (Cowell and Ford 1980) (see Chaps. 9 and 12) and Manitoulin Island, Ontario. Small patches may be seen within the city limits of Hamilton, Montréal

(Chap. 22), and Ottawa. The carbonate rock terrains of arctic Canada are commonly covered with frost-shattered rubble fields.

The most recognizable karst landform is the doline. Dolines result from combinations of surface and subsurface dissolution and erosion of carbonate or gypsum rock. The maximum dimension ranges from 10 to 1,000 m. Dolines forming lines or clusters can indicate subsurface bedrock structure, or channels carved by combinations of subsurface and surface erosion. In some karsts their frequency exceeds 500 per km^2. Thousands of dolines exist in Southern Canada, especially in limestone and gypsum terrains in western Newfoundland and Nova Scotia.

Large karst landforms include dry valleys, gorges and poljes, with flat floors and steep walls. Because of repeated glaciation in the past much karst in Eastern Canada has been eroded or infilled (Ford 2006).

1.6 Geomorphological Landscapes of Eastern Canada

In this volume, Eastern Canada has been divided into twentynine geomorphological landscape regions based on (i) geological structure (provinces), (ii) physiography (relief), (iii) permafrost (presence or absence), and (iv) vegetation (% cover and type) (Fig. 1.10). Alternative ways of defining Eastern Canada's geomorphological landscapes are discussed by Bird (1972), Graf (1987) and Dawes and Christie (1991).

1.6.1 Permafrost-Dominated Shield (Region 1)

Region 1 extends either side of Hudson Bay and includes parts of the Bear, Churchill, Slave and Superior geological provinces (Wheeler et al. 1996; Bostock 2014).

1.6.1.1 Keewatin (1a)

The area to the west of Hudson Bay was part of the former NWT district of Keewatin which is now divided between two administrative units of Nunavut known as Kitikmeot (to the north) and Kivalliq (to the south). Proceeding from west to east, the westernmost part of the region includes the Bear-Slave (Tu Nedhe, NWT) Uplands and Bathurst (Kingoak) Hills. The uplands are marked by moderately rolling relief (100–250 m). Granite, gneiss, migmatite, metavolcanics, and metasediments were formed during the Archaean Eon and deformed c. 2.4 billion years ago. Northeast-southwest oriented ridges of folded sedimentary rocks and diabase sills with conspicuous cuestas form the Bathurst Hills along the western side of Bathurst Inlet. Relief locally exceeds 250 m. with rocks ranging in age from 2.2 to 1.4 billion years. Sandstones and quartzites form resistant units, and softer units have been eroded to form embayments. Younger diabase dykes and sills (c. one billion years old) have intruded the older rocks forming resistant ridges and ledges. Immediately to the northwest of Hudson Bay, moderate relief (100–250 m) occurs in the Wager Plateau but lower relief, gently rolling terrain, and plains (50–100 m) characterize areas to the north and northeast (Back River Lowland, Thelon Plain, and Boothia Plateau) (Fig. 1.11) and the Kazan Uplands (see Chap. 4). The oldest rocks are Archaean granites, gneisses, and schists. Granite and ultrabasic plutons have been disrupted by faults, which have been exploited to form Ukkusiksalik Inlet within Ukkusiksalik National Park and other embayments along the western side of Hudson Bay. The northeastern part of the region includes the Melville Peninsula and adjacent Baffin Island. Relief varies from 50 to 200 m.

1.6.1.2 Nunavik (1b)

The eastern part of the permafrost-dominated shield region is located in Nunavik (northern Québec) and adjacent Nunatsiavut (Labrador). Physiographically and geologically, this subregion is split into western and eastern components. The western component (Larch Plateau) consists of planar to gently rolling (50 m relief) surfaces made of granites and gneisses The most prominent geomorphic feature in the Larch Plateau is the 1.4 million-year-old Pingualuit astrobleme (61°17′ N, 73°40′ W; Grieve 2001; Guyard et al. 2011) (Fig. 1.9). The eastern component of the region lies directly to the west of the Torngat Mountains, straddling the Labrador-Québec border. For >1.7 billion years, the original mountains have been weathered and eroded away leaving the deep crustal roots of the mountain system exposed on the surface. Relief varies from 50 to 250 m.

1.6.2 Forested Shield: Low to Moderate Relief (Region 2)

Region 2 extends in an arc from east to west to the south of the Hudson Bay Lowland, from Labrador to Ontario (Wheeler et al. 1996; Bostock 2014). From western Labrador, across Québec to Lake Superior in Ontario, the terrain has low to moderate relief with rocks formed during the Archaean, and deformed c. 2.48 billion years ago. A characteristic pattern involves a series of belts of alternating resistant acidic gneiss and granite, and weaker volcanic rocks and volcaniclastic sedimentary rocks. The weaker

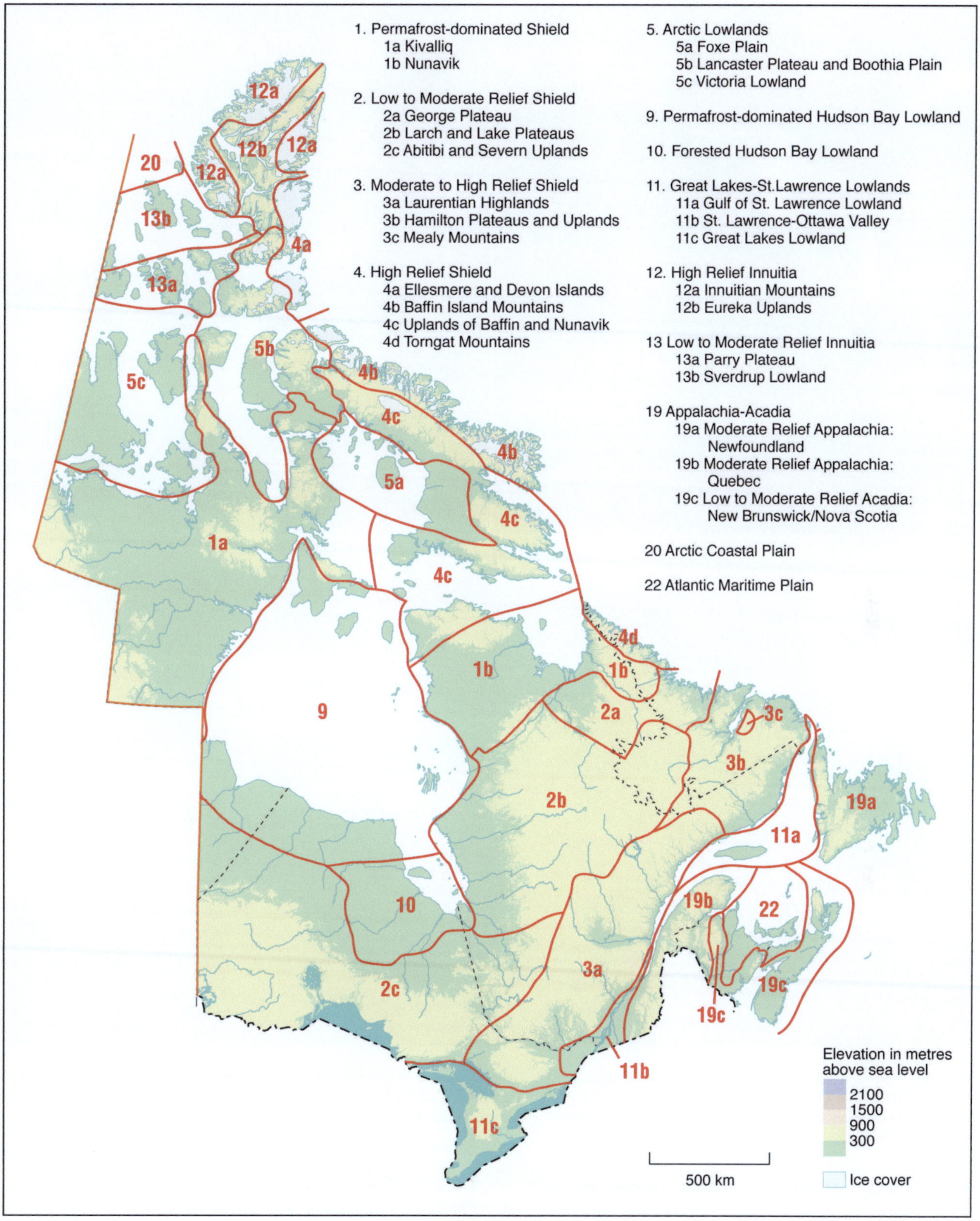

Fig. 1.10 The twentynine geomorphological landscape regions of eastern Canada. *Cartography* by Eric Leinberger (based on Bostock 2014; Environment Canada 1995 and Smith et al. 2001)

Fig. 1.11 Permafrost-dominated Shield. **a** Hoare Point, Beverly Lake, Nunavut. **b** Gibson Peninsula (upper right) on King William Island. **c** Ungava Peninsula, Québec. Locations are shown in the inset image *Source* a—Government of Canada; Natural Resources Canada (2007). Contains information licensed under the Open Government License—Canada. b—MDA Federal (2004). c—AGG_74001—Government of Québec, MFFP (2017), CC BY 4.0

volcanic rocks weather to produce valleys, and thus a characteristic corrugated topography marked by ridges and valleys has developed. Faults have accentuated the pre-existing ridge-and-valley topography. The volcanic rocks commonly weather to a dark green on the surface, and are referred to as 'greenstone belts'.

1.6.2.1 George Plateau (2a)

The northeasternmost area in Labrador, along the coastline near Nain and inland from Hopedale lies physiographically in the George Plateau. Inland, relief in the George Plateau is low (typically 50 m), but deep faulted (i.e. active tectonics) valleys modified by glaciation mark the coast, with local relief exceeding 250 m. Lakes and rivers are aligned predominantly W–E in the northern and NNW–SSE in the southern part of the plateau.

Geologically, this area includes gneisses, granites and anorthosite. There is a NNW–SSE trending belt of deformed metasedimentary, metavolcanic, volcanic, metamorphic and igneous rocks. Thrust and fold belts are responsible for the crenulated topography evident along the Labrador City-Schefferville railway line.

1.6.2.2 Larch and Lake Plateaus (2b)

The Larch and Lake plateaus are marked by gentle to moderately undulating relief (50–300 m), generally higher in the southeastern area (Fig. 1.12). The Mistassini Hills form a NE–SW trending ridge system. Local relief exceeds 800 m, and peak elevations exceed 1000 m asl. Metasedimentary units have been deformed and are intruded with gabbro and diabase sills and dykes. North- to northwest-facing cuestas have been modified by glaciation from the northeast.

1.6.2.3 Abitibi and Severn Uplands (2c)

The westernmost part of Region 2 includes the Abitibi and Severn uplands in Ontario and Manitoba typically having relief of 100–200 m. Development of the alternating greenstone and granite/gneiss belts is most evident in this area, particularly where accentuated by lakeshore or fluvial erosion.

The low-relief Nipigon and Cobalt Plains are developed on horizontally-bedded metasediments with thick gabbro sills, resulting in relatively low relief. These areas consist of sedimentary and metamorphosed sedimentary rocks (1.37 billion years old) that underlie much of the eastern half of Lake Superior, and their erodibility accounts for the location and great depth of the lake. Resistant sills account for much of the relief of the Thunder Bay Hills (Port Arthur Hills of Bostock 1970) and Penokean Hills, locally exceeding 280 m. Chapter 6 and La Berge (1994) provide more information about this region.

1.6.3 Forested Shield: Moderate—High Relief (Region 3)

Region 3 is characterized as moderate to high relief forested shield. It extends diagonally northeastward from Georgian Bay (Lake Huron) across Ontario and Québec to southern Labrador (Wheeler et al. 1996; Bostock 2014).

Fig. 1.12 Low to moderate relief—forested Shield, Québec. Locations are shown in the inset image. *Photographs* a—AAA1570 (51°53′ N, 77°11′ W). b—P1010761 (55°16′ N, 64°50′ W). c—INV_8051 (52°24′ N, 72°6′ W). d—AGG_46782 (51°52′ N, 76°26′ W)—Government of Québec, MFFP (2017), CC BY 4.0

1.6.3.1 Laurentian Highlands (3a)

The Laurentian Highlands (Laurentides) are bounded by distinct escarpments on the southeast, the combined products of tectonic history and differential resistance to erosion. This subdivision of the Canadian Shield is dominated by granites and highly metamorphosed gneisses and schists. Orogenic activity began in this region approximately 1.6 billion years ago and culminated between 1050 and 995 million years ago. Metamorphism and mountain-building resulted but 995 million years of weathering and erosion has ground away the peaks, leaving only the roots of the mountain system exposed. Locally, thicknesses up to 10 km of rock have been removed by erosion, leaving the gneiss and granites which formed the subsurface core of the ranges exposed. Glaciation during the past two million years removed very little material, probably a maximum of about 20 m from any individual location. Following the Precambrian, younger Cambrian and Ordovician sandstones, limestones and shales were deposited to the southeast of the escarpment, in the St. Lawrence and Ottawa Valleys. Faulting caused the St. Lawrence basin to drop, thereby enhancing the prominence of the escarpment.

Areas in the interior of the Laurentian Highlands, such as Algonquin Park (ON) and Parc des Laurentides (QC) typically have lower relief (150–200 m), reflecting repetitive erosion over the past billion years. The Laurentian Highlands are marked by many steep cliffs, numerous lakes with ragged shorelines, and isolated hills (Fig. 1.13). The physiography, and particularly the configuration and alignment of the lakes and rivers, is almost exclusively a product of the bedrock geology. Local relief can exceed 300 m, but typically is less, modified in part by glacilacustrine (e.g. Glacial Lake Algonquin in the Georgian Bay-Lake Huron basin) and glacimarine deposition (Champlain Sea in the Ottawa Valley-St. Lawrence Lowlands; Laflamme Gulf in the Lake St.-Jean basin). The low relief Lake St. Jean area is flanked by the moderate to high relief forested shield and is discussed in detail in Chap. 5.

1.6.3.2 Hamilton Plateau and Upland (3b)

Along the Québec North Shore of the St. Lawrence and in Labrador, the Hamilton Plateau is marked by gently undulating terrain, with relief to 250 m. The physiography of this terrain is controlled by the underlying bedrock structure.

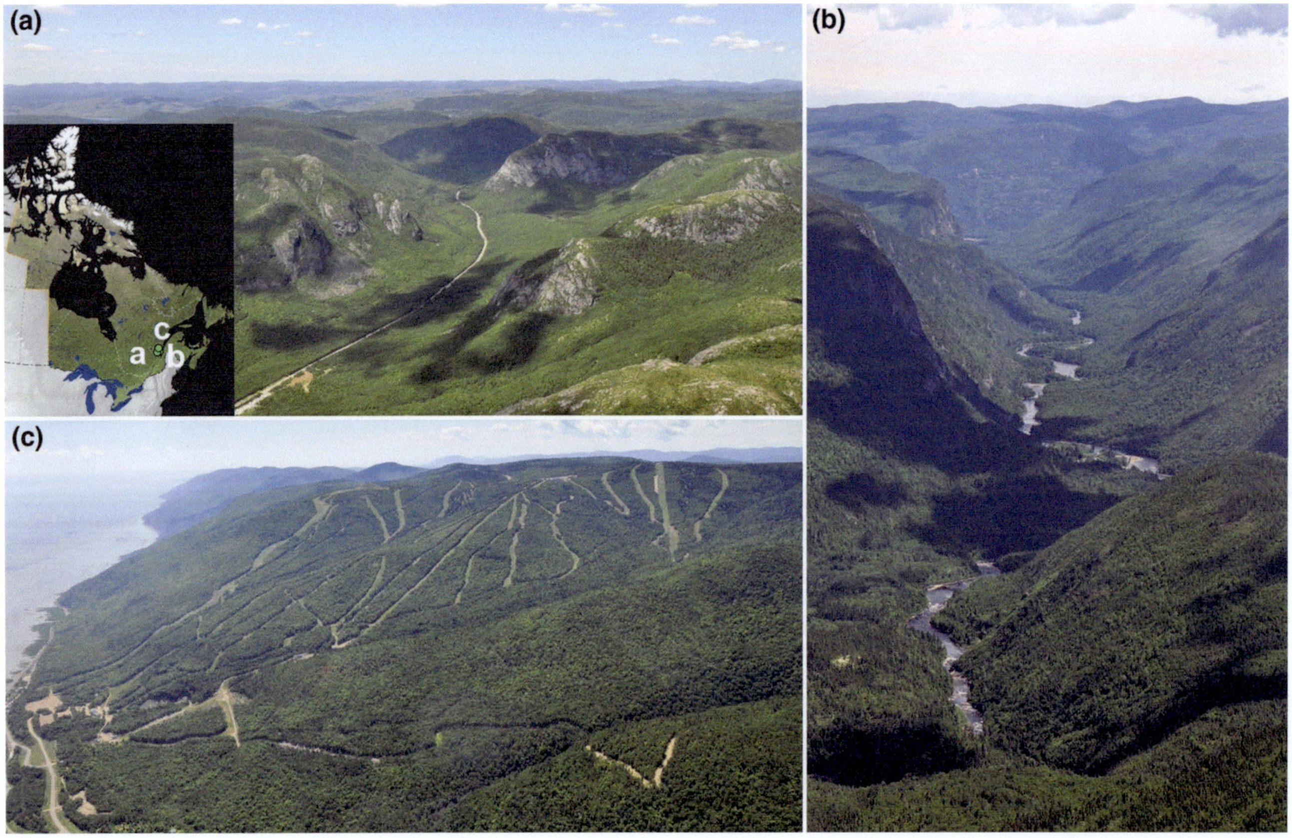

Fig. 1.13 Moderate to high relief forested Shield. Rolling hills and mountains adjoining the St. Lawrence River from the Charlevoix region of Québec, including Grands-Jardins Parc National (**a**, **b**, **c**). Locations are shown in the inset image. *Photographs* a—DDD2938. b—DDD2943. c—DDD2924—Government of Québec, MFFP (2017), CC BY 4.0

The highest parts of the region, above 600 m asl, are referred to as the Hamilton Upland. The numerous joint patterns result in the formation of a somewhat irregular drainage pattern in the Hamilton Plateau region. The underlying bedrock structure is responsible for the location of the Lake Melville Plain. Igneous and metamorphic rocks are blanketed by younger sandstones and conglomerates (c. 700–600 million years old). These sediments infilled the topographic irregularities and depressions on the surface of the down-dropped graben to the northwest of the Mealy Mountains. The Lake Melville Plain area was inundated by marine waters following deglaciation, to maximum elevations varying locally from 125 to 140 m asl. The northeasternmost corner of the Laurentian physiographic region is called the Mecatina Plateau and includes the north flank of the Benedict Mountains. Rocks are dominantly granite, granodiorite and high-grade metamorphic units. The Mecatina Plateau borders the Mealy Mountains and Hamilton Plateau to the southeast, extending along the Labrador Sea coastline north of Battle Harbour, with limited relief, generally less than 250 m.

1.6.3.3 Mealy Mountains (3c)

The Mealy Mountains form a distinct physiographic division in the Laurentian region. Maximum elevations of the summits reach 1190 m asl. The mountains are composed of resistant granitic and gneissic rocks and extremely resistant anorthosites. Irregular knobs of bedrock mark the terrain to the west of the Kenamu River, with local relief to 300 m. Streams are short and irregularly aligned, with gradients locally in excess of 100 m/km, and numerous small waterfalls and rapids. The eastern part of the Mealy Mountains has the highest elevations, with local areas of permafrost (generally above 900 m asl). More than 20 cirques are present in the headwaters of the Etagulet and English Rivers, with floor elevations at 600 m asl or greater. Cirques facing northwards have linear or undulatory slopes, whereas tarns and riegels are common in cirques developed on the southern and eastern flanks of the Mealy Mountains. No active glaciers are present. The southern limit of the Mealy Mountains is transitional to the gently rolling and undulatory terrain of the Hamilton Upland.

1.6.4 Shield: High Relief (Region 4)

The high relief eastern shield region consists of mountain belts of Labrador and Nunavut, flanked by rolling upland terrain to the west in Nunavut and hilly terrain in northernmost Nunavik, Québec (Wheeler et al. 1996; Bostock 2014). From north to south, the region includes southeastern Ellesmere Island, eastern Devon Island, Bylot Island, Sirmilik National Park through Auyuittuq National Park on Baffin's Cumberland Peninsula and ending in the Torngat Mountains of Labrador (Fig. 1.14).

1.6.4.1 Ellesmere and Devon Islands (4a)

Lancaster Sound (Fig. 1.15) separates Baffin and Bylot Islands from Devon Island to the north and is the eastern entrance to the Northwest Passage. North of Lancaster Sound, mountainous terrain extends northward along the eastern edge of Devon Island and southeastern Ellesmere Island. Much of this area is occupied by active glaciers, including the Manson and Sydkap glaciers and icefields on Ellesmere, and the Devon Ice Cap (with maximum ice thickness 880 m). Archaean granites and gneisses dominate.

1.6.4.2 Baffin Island (Qikiqtaaluk) Mountains (4b)

Eastern Baffin Island is part of the Churchill geological province of the Canadian Shield. The granites and gneisses of the Churchill Province were originally formed during the Archaean Eon, c. 3000 Ma, and were then extensively deformed and further metamorphosed during the Hudsonian orogeny at the conclusion of the Early Proterozoic, c. 800 Ma. This area is one of the most intensively deformed and modified zones of the Canadian Shield. With the exception of a relatively narrow coastal plain at Clyde River (Kanngiqtugaapik), the northeast flank of Baffin and Bylot Islands is marked by mountainous glacierized terrain with deeply incised fjords (Ives 2016).

- **Sirmilik National Park**

Sirmilik National Park ('the place of the glaciers'; 22,252 km^2) in northern Qikiqtaaluk, encompasses part of the Borden Peninsula, the Oliver Sound area southeast of Bylot Island and Bylot Island itself (Fig. 1.15). The coastlines alternate between steep granite cliffs and coarse shingle beaches developed on flat-lying or planed surfaces. Glaciers cover more than 50% of Bylot Island. Active accumulation of glacial ice occurs above c. 700 m asl, and several glaciers extend to tidewater. Glaciers on Bylot Island have been in recession for the past 150 years.

In Sirmilik, relief reaches 1800 m, and the terrain in the interior of Bylot Island, the Byam Martin Range, is rugged. Embayments and fjords follow faults and structural weaknesses in the bedrock, and have been further modified by glacial activity. Sirmilik was glaciated multiple times during the Quaternary. The pre-MIS 5 Baffin Glaciation was marked by a major eastward-flowing ice stream occupying Lancaster Sound. The coastal areas were glaciated during the MIS 2 although the montane glaciers appear to have advanced only minimally.

Fig. 1.14 High relief Shield. **a** Miriam, Torngat Mountains, Labrador. *Photograph* N. Catto. **b** Torngat Mountains. *Source* A. Willard/flickr, CC BY-SA 2.0. **c** Midnight view of The Inugsuin Pinnacles', Nuksuklorolu Mountain, Baffin Island, Nunavut. Photograph reproduced with the permission of J. D. Ives (also see Ives 2016)

1.6.4.3 Uplands of Baffin and Nunavik (4c)

To the south and west of the mountainous terrain, upland areas form a crescent through Baffin Island. Geologically, these areas are underlain by Archaean granites and gneisses as well as Palaeozoic carbonate sequences. The Baffin, Hall and Frobisher uplands all have a relief typically around 150–300 m. Directly across the Hudson Strait the northern flank of the Sugluk Plateau is located in northernmost Nunavik and has been modified by glacial erosion, resulting in steep cliffs locally in excess of 600 m height. But the interior of the Sugluk plateau consists of planar to gently rolling (50 m relief) surfaces. The Povungnituk Hills form part of a broad arc of deformed metasedimentary, metavolcanic, volcanic, metamorphic and igneous rocks that extends northwards through Nunavik, west-southwestward to Cape Smith along Hudson Bay, through Povungnituk and the Inoucdjouac Hills, and southward through Hudson Bay.

Baffin Island was glaciated several times during the Quaternary. During the extensive pre-MIS 5 Baffin Glaciation, ice reached the shoreline along all fjords. Glacial retreat followed and. associated marine deposits occur up to 90 m asl. MIS 2 glaciation in the Auyuittuq-Pangnirtung region involved montane glaciers and ice streams in the fjords, extending to the edge of the continental shelf. In addition, continental glaciation originated from the Foxe Basin to the west. Foxe Basin ice overran almost all of Baffin Island, forming drumlins and bringing Palaeozoic limestone erratics to the granitic and gneissic terrains. The maximum late glacial (MIS 2) extent was reached c. 13 ka BP. The Foxe Basin ice centre collapsed following deglaciation of Hudson Strait, and the basin was ice-free by c. 7 ka BP.

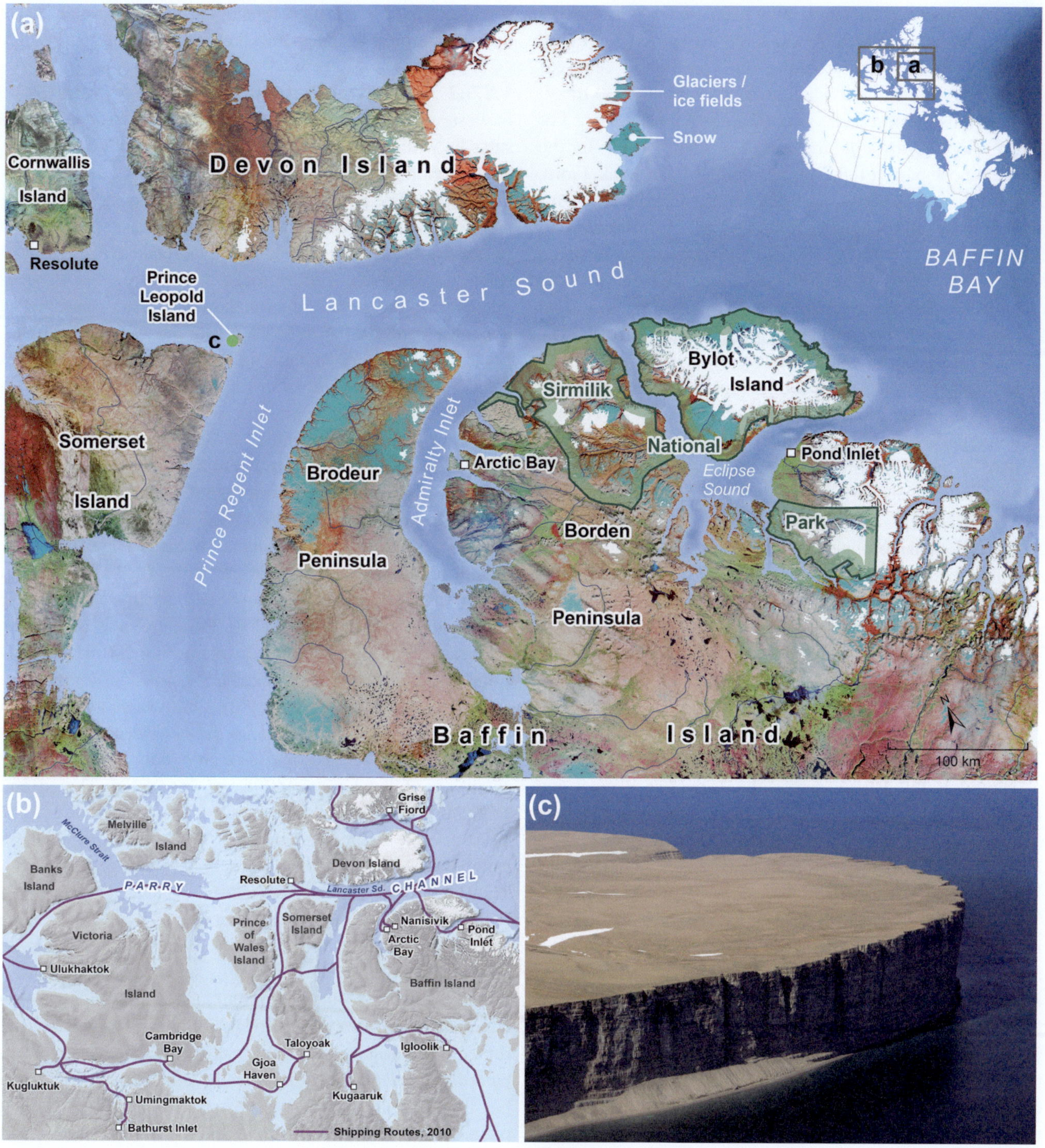

Fig. 1.15 Lancaster Sound, the eastern entrance to Parry Channel and the Northwest Passage. **a** Satellite image of structurally controlled trough located between Devon Island and Baffin Island, Nunavut. **b** Shipping routes based on vessels that reported in 2010 to the Canadian Coast Guard Marine Communications and Traffic Services (NORDREG). **c** Tip of Prince Leopold Island, Nunavut. Location is shown in (**a**). Sirmilik is an Inuktitut word meaning "the place of the glaciers." *Source* a—MDA Federal (2004). b—"Arctic Voyage Planning Guide," available at http://geoportal.gc.ca/eng/Gallery/MapProfile/5, and Natural Resources Canada (2012). Contains information licensed under the Open Government License—Canada. c—T. Kal/Wikimedia Commons, CC BY-SA 3.0

1.6.4.4 The Torngats (4d)

The Torngat Mountains of northern Labrador are the southernmost of the mountain belts (see Chap. 7 for more information). The Torngats form a series of rugged NW–SE ranges, cut by deep E–W fjords. The northern part is more rugged than the southern one. Fjords extend up to 15 km inland from the open coast, and main valleys dissect the ranges, extending across the drainage divide and Labrador/Québec boundary. Relief along the coastline is approximately 1200 m, and individual peaks rise above 1500 m asl. The summits of most peaks in the Torngats are flat-topped or gently sloping towards the coastline (Fig. 1.14). In the northernmost parts of the range, north of Nachvak Fjord, four distinctive erosional surfaces can be recognized: Torngat (highest and oldest), Komaktorvik, Koroksoak and Saglek (lowest and youngest). The Torngat surface shows no signs of glaciation, but each of the three younger surfaces contains striated areas, erratic boulders derived from distant localities to the west in Ungava, and glacial landforms. The Torngat surface nunatak probably has been subaerially exposed since the end of the Cretaceous. Currently, about 70 active glaciers are present in the Torngats, most in Torngat National Park. Most are very small (less than 1 km in length) and appear to be retreating at present. Rock glaciers are also present on the flanks of some of the higher summits. Most of the Torngat rocks are gneisses (3.8–3.6 billion years old). Additional metamorphism c. 2.8 billion years ago further modified many of these rocks and they are cut by numerous faults, generally trending NNW–SSE. Most of the fjords cut across the structural grain of the fault systems.

1.6.5 Arctic Lowlands (5)

The Arctic Lowlands extend westward from the Foxe Plain of central Baffin Island, across the Lancaster Plateau and Boothia Plain, to the Victoria Island (Kitlineq) Lowland (Wheeler et al. 1996; Bostock 2014).

1.6.5.1 Foxe Plain (5a)

The Foxe Plain is exposed along the margins of the Foxe Basin, including the northeast Melville Peninsula (Sanirajak), and the Great Plain of the Koukdjuak on Baffin Island. The Foxe Plain is underlain by Ordovician and Silurian carbonates, shales and sandstones. The Great Plain of the Koukdjuak is marked by abundant wetlands (Zoltai 1988). Shallow thermokarst lakes develop from local melting of ice-rich silty or sandy sediments, induced by combinations of thermal and mechanical erosion. The usual pattern, particularly developed in the Great Plain of the Koukdjuak, involves a myriad of circular to oval lakes of multiple generations, including incipient and drained features along with actively developing basins (Bird 1967). Relief is very limited, generally less than 20 m. Inland from Cape Dominion, the 15 m contour line is as much as 80 km from the modern Foxe Basin shore (Fig. 1.16). Raised beaches indicate former sea-level positions. Sanirajak (formerly Hall Beach), is one of the oldest known permanently inhabited communities north of the Arctic Circle. The community and surrounding terrain are spread across a series of gently sloping raised marine beaches with intervening wetlands and thermokarst features.

1.6.5.2 Lancaster Plateau and Boothia Plain (5b)

In westernmost Baffin Island, the Borden and Brodeur peninsulas lie within the Lancaster Plateau of the Arctic Lowlands geomorphological landscape region. Relief is generally less than 200 m. The Borden Peninsula is composed of Ordovician and Silurian carbonates, shales, and sandstones. Locally, the red sandstones have been modified and eroded to form isolated spires, or hoodoos. Similar rocks crop out on the Brodeur Peninsula to the west, and throughout Lancaster Plateau. The youngest rocks in the area are Cretaceous and Palaeocene-Eocene sedimentary units, which are preserved as isolated patches that escaped erosion during the Neogene prior to Quaternary glaciation. The most conspicuous feature is the deep fjord of Admiralty Inlet, between the Brodeur and the Borden peninsula to the east. Admiralty Inlet locally is more than 200 m deep, and served as a major outlet valley for Foxe Basin glacial ice during MIS 2. The northeastern part of Somerset Island also lies within the Lancaster Plateau. Despite this terminology, relief here is generally greater (to 400 m) than in the areas of the Canadian Shield. The most spectacular feature is the unbroken cliff along Prince Regent Inlet at the northeast tip of the island, locally exceeding 200 m in height (Fig. 1.15). Most of the Brodeur Peninsula and northeastern Somerset Island are underlain by Ordovician and Silurian carbonates, shales and sandstones. Lower Palaeozoic carbonate and clastic successions also crop out in the northern extension of the Lancaster Plateau on western Devon and southwestern Ellesmere islands. Deep canyons, with cliffs to 1090 m high, cut the Lancaster Plateau. The alignment of these canyons does not conform to the modern topographic slope. They represent drainage channels excavated during the Palaeogene, part of the Barrow Plain which sloped gently to the northeast. The canyons are remnants of an integrated drainage system, involving streams flowing eastward towards Lancaster Sound. Wind gaps and other fluvial features indicate that development occurred over several millions of years, uninterrupted by glaciation or tectonic uplift. The southeasternmost part of Somerset Island lies within the Boothia Plain. This area has limited relief (c. 50 m), and consists of an east-sloping plain. Ordovician and Silurian dolomites, limestones, and shales underlie the Gulf of Boothia and the Boothia Plain.

1.6.5.3 Victoria Lowland (5c)

Prince of Wales, King William (Qikiqtaq) (Fig. 1.15), and most of Victoria (Kitlineq) islands lie within the Victoria Lowland (Bird 1967). The area consists of several gently sloping plains. Relief is low, generally less than 200 m. In the most gently sloping regions, such as the Wollaston Peninsula, relief is less than 50 m. The Victoria Lowland is underlain by undeformed limestones and shales of Cambrian

Fig. 1.16 Arctic Lowland landscapes. **a** Gravelly raised beaches, Gjoa Haven, Nunavut. **b** Gjoa Haven of King William Island. **c** Grassy areas with tundra pond form habitat for snow geese and other wildlife in coastal lowlands adjacent to Foxe Basin, Nunavut. *Source* a—N. Catto. b—MDA Federal (2004). c—NRCan 2007-197 by L. A. Dredge/Natural Resources Canada

to Devonian age, with the youngest rocks located in the northeast on Prince of Wales Island. The northwest part of Victoria Island consists of rolling to rugged hilly terrain, referred to as the Shaler Mountains. The highest point within the Shaler Mountains is 655 m asl. Proterozoic metasediments with intruded gabbro sills forming cuestas crop out. These rocks were formerly overlain by Ordovician and Silurian sediments, but erosion during the late Palaeozoic and Mesozoic stripped off the younger sediments, leaving the Precambrian rock exposed.

1.6.6 Permafrost-Dominated Hudson Bay Lowland (Region 9)

The Hudson Bay Lowland consists of gently sloping plains surrounding Hudson and James bays, developed on sedimentary bedrock (Smith et al. 2001; Bostock 2014). Discontinuous permafrost distinguishes this region from Region 10 to the south. The southern and western margins are covered by glacimarine sediments of the Tyrrell Sea. The southern and southwestern shores of Hudson Bay are included in the Hudson Bay Lowland. The plain slopes almost imperceptibly towards the bay, and is covered with wetlands (primarily fens) (Zoltai 1988) and sluggish meandering and wandering rivers. Total relief throughout the Hudson Bay Lowland is approximately 200 m. In coastal areas, such as Wapusk National Park, Manitoba (Fig. 1.17), relief is considerably less, and there are no elongate ridges or isolated hills to disturb the gently northeastward slope of the plain. The most conspicuous topographic features are the numerous beach ridges, aligned parallel to the modern shore, indicating the successive withdrawal of the marine waters of the Tyrrell Sea following glaci-isostatic recovery of the land.

Fig. 1.17 Permafrost-dominated Hudson Bay Lowland landscapes. **a, b, c** Thermokarst ponds and patterned ground in Wapusk National Park. *Source* a and b—Parks Canada. c—Government of Canada; Natural Resources Canada (2007). Contains information licensed under the Open Government License—Canada

Rebound continues today at a rate of approximately 30 cm/100 y, and thus the Hudson Bay Lowland is the most rapidly emerging region of Canada.

The major rivers, including the Nelson, Severn and Winisk, are generally braided and wandering streams, marked by high volumes of spring discharge and greatly reduced summer flow. The streams are dominantly sand-transporting, although finer sediments are provided by reworking the marine deposits of the Tyrrell Sea. Originally, the Nelson River mouth was to be the site of Canada's Hudson Bay port, but the plan was changed when it was realized that the area is marked by shifting river and tidal bars, and is surrounded by permafrost terrain that poses great difficulties for construction. Consequently, the position of the railroad was shifted (as construction was ongoing) to Churchill, at the northern margin of the Hudson Bay Lowland.

The bedrock geology of the region involves two major geological elements. The western shore of Hudson Bay (and the bay itself) is underlain at depth by granites, gneisses, and other metamorphic and igneous rocks of Proterozoic age. In most regions, these rocks are buried beneath hundreds of metres of younger sedimentary rocks. In northernmost Ontario southwest of Cape Henrietta Maria, the major physiographic feature is the Sutton Hills inlier, an arcuate band of topography rising to 290 m asl. The Sutton Hills are a topographic high from Precambrian time which was initially buried by Silurian rock and subsequently exhumed after the Cretaceous (but prior to glaciation). The prominence of the Sutton Hills on the topography is an indication of the general featurelessness of the landscape. The younger rocks are limestones and shales of Early and Middle Silurian and Devonian age. These Palaeozoic sedimentary rocks cover the floor of Hudson and James bays. In a few places, limestones, dolomites and shales crop out. The youngest rocks in the region, of Cretaceous age, crop out near the Ontario/Manitoba border. The Silurian rocks continue beneath the waters of Hudson Bay. In the centre of the bay, Devonian and Cretaceous sedimentary rocks overlie the Silurian deposits. The presence of this relatively young, soft rock in the centre of Hudson Bay indicates that glacial erosion was minimal in the area. On the eastern side of the bay, intensely folded Proterozoic metasediments are exposed on the Sanikiluaq (Belcher) and Arviliit (Ottawa) islands. Topographically and geologically, these archipelagos are related to the Povungnituk Hills of the Canadian Shield. Southampton Plain formed on flat Palaeozoic strata and is generally less than 100 m asl. The Belcher Islands are formed from resistant strata of elaborately folded Proterozoic sedimentary and volcanic rocks. Richmond Hills are mainly a group of east-facing cuestas of Proterozoic sedimentary and volcanic rocks (see Chap. 8 for more information).

1.6.7 Forested Hudson Bay Lowland (Region 10)

Sporadic and discontinuous permafrost with medium to high ice content that characterizes Region 9 to the north decreases to isolated patches surrounding James Bay. The terrain adjacent to the western and southern shores of James Bay, the forested Hudson Bay Lowland, is similar to that of the permafrost-dominated region to the north, with a plain which slopes gently toward the bay (Smith et al. 2001; Bostock 2014). Relief in the Hudson Bay Lowland is minimal, typically less than 50 m (Fig. 1.18). Most of the Québec section of the region is part of the Eastmain Lowland. The region consists largely of flat, poorly drained plains with subdued fluvial and marine features. Throughout the area, there are gravelly, well-drained belts of raised beaches, resulting from postglacial, isostatic rebound. To the south, the region is dominated by fine-textured lacustrine and marine deposits. Wetlands cover between 50% of the area in the south to over 75% of the area in the north and around James Bay (Zoltai 1988). They are composed largely of northern ribbed fens, northern plateau bogs and palsa bogs (see Chap. 3). Sporadic, discontinuous permafrost with medium to high ice content that characterizes Region 9 to the north decreases to isolated patches surrounding James Bay. Mineral soil profiles exhibit uneven and often discontinuous or distorted soil horizon development as a result of past and present permafrost action.

The major rivers, Harricana, Moose, Albany and Attawapiskat, are dominantly characterized by a wandering habit. Anabranching reaches notably the lower Attawapiskat and lower Moose develop where the banks are stabilized by a combination of resistant Tyrrell Sea marine clay and local, discontinuous permafrost. Large woody debris is transported downstream by freshet and river ice. Islands are composed of cohesive glacimarine clay and glacial till, and are erosional forms, dissected by avulsion. The low surrounding relief allows avulsion channels to depart from the existing river valley gradually expanding the area directly influenced by the Attawapiskat River. In contrast, the reach of the Moose River at Moosonee carries sand and lesser amounts of gravel as bedload. The vegetated islands develop predominantly from accumulation of sediment around sand flats, and are subject to dissection by chute channels. Avulsion channels develop within the existing river valley.

1.6.8 Great Lakes—St. Lawrence Lowlands (Region 11)

The Great Lakes—St. Lawrence Lowlands are marked by low relief. The three parts—Gulf of St. Lawrence, St. Lawrence-Ottawa Valley, and Great Lakes Lowlands—are separated by extensions of the Laurentian region of the Canadian Shield (Chapman and Putnam 1984; Wheeler et al. 1996; Bostock 2014).

1.6.8.1 The Gulf of St. Lawrence (11a)

The Gulf of St. Lawrence includes the Mingan Archipelago, Anticosti Island and the discontinuous fringe of the west coast of Newfoundland from the Port-au-Port Peninsula north (Douglas 1970; Bostock 2014).

- **Mingan Archipelago**

Mingan is an isolated outlier of the St. Lawrence Lowlands, nestled adjacent to the Laurentian Highlands of the Canadian Shield. The Mingan Archipelago initially developed as the Ordovician rocks were subject to dissection forming a plain cut by valleys parallel to the bedding of the rocks and

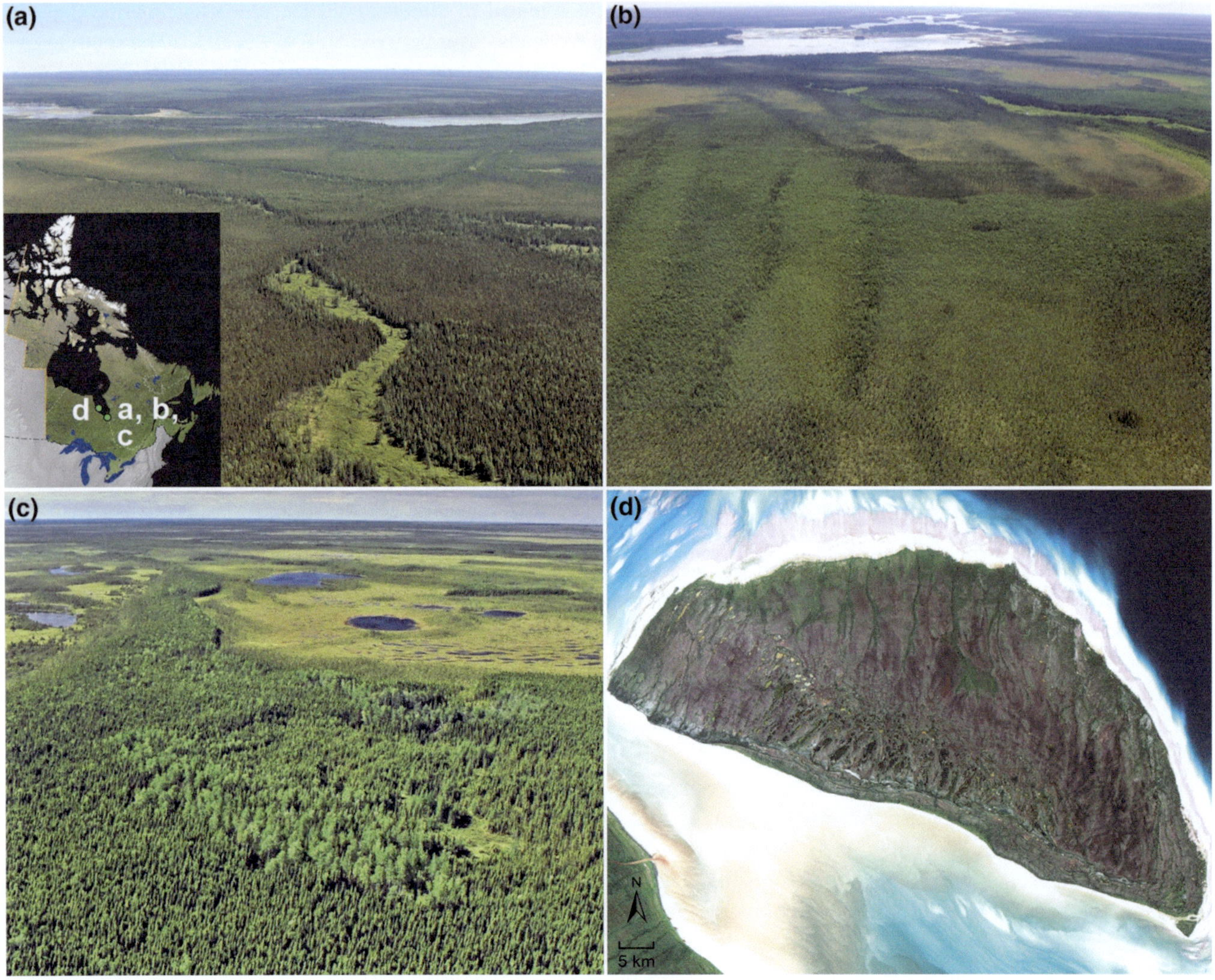

Fig. 1.18 Forested Hudson Bay Lowland landscapes. **a**, **b**, **c** *Photographs* a—_AAA1905. b—DPF_7410. c—2196—Government of Québec, MFFP (2017), CC BY 4.0. d—Akimiski Island in James Bay, Ontario. *Source* NASA image created by J. Allen, using Landsat data provided by the University of Maryland's Global Land Facility

following faults and joints. When sea level rose following deglaciation, the dissected plain was flooded, creating the archipelago. Relief is generally low, approximately 100 m, and the remnants of the plain slope to the southeast. The northern coast of Anticosti Island represents another facet of the pre-existing planar topography. At Mingan, the erosion of the limestone has created a complex network of pillars, steep bluffs, flowerpot structures (with narrow bases, widening upwards), sea caves, and sea arches. Many of these features have attractive descriptive local names, such as the Tete-du-chat (Cat's Head) cliff; the 'Full Dome', developed where a gentle anticlinal fold of limestone is washed by the sea; 'Fallen Dome', marked by erosion of the anticline's crown; Cap-St-Charles and False Pass Pillars; and 'The Chalice', 'The Hen' and 'The Pottery' flowerpots (Fig. 1.19). The Ordovician limestones of the Mingan Archipelago are the northernmost and oldest units of the succession evident in the Gaspé Peninsula. The limestones have been folded into a series of very gentle anticlines and synclines, but major faults are not in evidence within the limestone succession. The youngest Ordovician rocks, assigned to the Gamachian stage, are unique to Mingan and adjacent Anticosti Island (Roberge and Ford 1983; see Chap. 9).

- **Anticosti Island**

Anticosti Island is located in the Gulf of St Lawrence, between the Jacques-Cartier and Honguedo straits. The overall geomorphology of the island, which has a 550 km coastline, is dominated by calcareous limestones and dolomites that have produced extensive Holocene karst. See Chap. 9 for more detail.

Fig. 1.19 Gulf of St. Lawrence Lowland. **a** Anticosti Island in the Gulf of St. Lawrence, Québec. **b** Vauréal Canyon and Waterfall, Anticosti National Park, Québec. At the Vauréal Waterfall, the Vauréal River drops 76 m and then flows through a deep canyon carved in limestone. **c** Mingan Archipelago, Québec features a chain of about 40 islands that formed as the continent rose after the last glaciation. Over the course of the last few thousand years limestone rocks eroded to form the largest group of monoliths in Canada. *Source* a—MDA Federal (2004). b—Courtesy of Jean-Marie Dubois, University of Sherbrooke (World of Images No. 38944). c—M. Villeneuve/Wikimedia Commons, CC BY 2.0

- **Newfoundland Coastal Plain**

The Newfoundland Coastal Plain consists of discontinuous segments underlain by flat-lying to moderately deformed Cambrian to lower Devonian carbonates and shales, extending inland to approximately 100 m asl. From Bonne Bay northeastward to Burnt Cape, all of the coastal plains were inundated post-glacially. The elevation of marine features increases from south to north, from 135 m above modern sea level at St. Barbe, 140 m at Watts Point, and approximately 150 m asl at Burnt Cape (Grant 1989). The plain generally widens to the northeast, to a maximum width of c. 20 km. Pistolet Bay Provincial Park conserves a representative part of the coastline. To the west of a thrust fault, the terrain is underlain by an Ordovician mélange consisting of limestone and shale with inclusions of volcanic rocks and peridotite. The Burnt Cape and Watts Point barrens are developed on limestone eroded by karst action. To the north of Gros Morne, the Arches Provincial Park preserves two sea arches eroded from Ordovician dolomitic limestone. Initially, the area was submerged following deglaciation, with marine waters extending up to 110 m asl. Sea level then fell under the influence of glaci-isostatic rebound reaching its minimum value in the early Holocene. At present, sea level is rising slowly, c. 1 mm per year. At The Arches, the positions of the openings are controlled by the pattern of fractures and joints in the dolomitic limestone.

1.6.8.2 St. Lawrence—Ottawa Valley (11b)

The St. Lawrence-Ottawa section of the Great Lakes-St. Lawrence Lowlands lies south of the Laurentian Highlands, between Pembroke and Québec City. It is gently northward-sloping plain, with little relief (<100 m) (Douglas 1970; Bostock 2014). The cities of Ottawa, Montréal and Québec are located in this region.

Most of the relief is related to the St. Lawrence and Ottawa rivers, or to a few escarpments which cross the region from southeast to northwest. Tributary streams which enter the Ottawa River east of Ottawa from the south (e.g. South Nation River) commonly have their headwaters within a few km of the St. Lawrence. The position of the drainage divide, closer to the St. Lawrence than to the Ottawa, and the flatness of the lowland terrain are both consequences of the glacial history of the region and the development of the Champlain Sea following deglaciation. The southern margin of the Ottawa Valley is marked by the Frontenac Axis at Brockville, and by the Canadian Shield at communities such as Smiths Falls and Calabogie. The southeastern margin of the St. Lawrence Valley is marked by the eastern Québec Uplands. The Ottawa and St. Lawrence valleys are underlain by Cambrian and Ordovician sedimentary rocks, which dip gently towards the south. The oldest of these, the Cambrian Nepean sandstone forms the steep cliffs of Parliament Hill in Ottawa, and has been used extensively as a building stone on Parliament Hill and elsewhere. Ordovician limestones, dolomites and shales underlie glacimarine deposits throughout much of the region. All of the Ordovician rocks are calcareous, and the calcium carbonate derived from these units is responsible for the luvisolic soil development characteristic of the St. Lawrence Lowlands. SE–NW trending normal faults mark the northern and western area, from Montréal to Pembroke, locally forming steep escarpments 10–25 m high. Northeast of Québec City, the 5 February 1663 La Malbaie (Charlevoix) earthquake, with an estimated Richter magnitude of 7.2 ± 0.2 triggered numerous slides and other styles of slope failure along the St. Lawrence and Saguenay valleys. The Charlevoix seismic zone is the most active in Eastern Canada (Adams and Basham 1991).

Mont-Royal, within the city of Montréal, is one of the summits called the Monteregian Hills (Fig. 1.20a, b, c). Ten isolated summits varying in height from 218 m asl (Mont Saint-Bruno) and 233 m asl (Mont-Royal) to 1105 m asl (Mont Mégantic) form the E–W aligned Monteregian chain. With radiometric ages between 118 and 140 million years, the hills are Cretaceous intrusions, dominantly gabbro with associated syenite and flanks of metamorphosed hornfels. The Monteregian Hills formed as the North American tectonic plate slid northwestward across the New England Hotspot during the Cretaceous (see Chaps. 10 and 22 for more information on this region).

1.6.8.3 Great Lakes Lowland (11c)

The Great Lakes Lowland encompasses southern Ontario and lies south of the Canadian Shield. It is the most densely populated region in Canada, including the city of Toronto, and has undergone the greatest human modification of any region in Eastern Canada. Relief is generally low, less than 150 m (Douglas 1970; Chapman and Putnam 1984; Bostock 2014). The lowland is marked by several low escarpments developed parallel to lakes Ontario, Erie and Huron representing former shorelines of higher glacilacustrine stages.

- **Niagara Escarpment**

The Niagara Escarpment cuts diagonally across the lowland from SE–NW, separating the lower plain from Lake Ontario to southern Georgian Bay, underlain by Ordovician carbonates and shales; and a broad, gently undulating upper plain, sloping almost imperceptibly to the southwest, extending south to the shore of Lake Erie and west to Lake Huron, underlain by Silurian and Devonian sedimentary rocks (see Chap. 11). The Niagara Escarpment averages 100 m in relief and varies from a vertical cliff to a series of steep slopes within a 5 km-wide belt. The escarpment, traceable northwest and west to the Bruce Peninsula, Manitoulin Island, and Green Bay (Lake Michigan), is the product of differential erosion and modification of bedrock units. Friable latest Ordovician and earliest Silurian sandstone and shale are overlain by the extremely resistant Silurian Lockport dolomite, which forms the escarpment. The upper plain is developed on younger Silurian carbonates and shales. Karst topographic features in Bruce Peninsula National Park and Manitoulin Island include doline lakes. Lake cliffs of limestone, up to 90 m high, are eroded at the base by the cold waters of Lake Huron and Georgian Bay, forming dissolved notches and caves at water level. The narrow bases of Flowerpot Island and its neighbours are also the result of dissolution of limestone by lake water (see Chap. 12).

- **Niagara Falls**

Erosion by the Niagara River has resulted in a steady southward recession of Niagara Falls, as the river has cut a deep 8 km-long gorge extending from the escarpment margin at Queenston Heights to the current position of the falls (Gilbert 1907; Tinkler 1986, 1997). The total difference in elevation between the upper and lower plain here is

Fig. 1.20 St. Lawrence-Ottawa Valley. **a** The isolated, steep-sided circular features are plutons of the Monteregian Hills, rising about 250 m above the Saint Lawrence plain, Québec. **b** Monteregian Hills: Mont St. Hilaire, Mont Rougemont, and Mont Yamaska. The Monteregian Hills provide contrast with the surrounding tan to brown agricultural fields of the fertile lowlands, Québec. **c** Oblique photograph of Mont Rougemont, Québec. **d** Long Sault Parkway, Ontario connects 11 islands that were created from high points of land left after the flooding of the St. Lawrence River during the construction of the Seaway in the 1950s. *Source* a—Landsat 7 ETM+, 014028_0100_010608_L7. Natural Resources Canada, Geomatic Canada, Centre for Topographic Information. b—Photograph ISS014-E-19807 courtesy of the Earth Science and Remote Sensing Unit, NASA Johnson Space Center. Available at http://eol.jsc.nasa.gov. c—P. Bona/Wikimedia Commons. d—© 2018 St. Lawrence Parks Commission

approximately 100 m, of which the falls themselves account for 55–57 m, the remaining 45 m being represented by the descent through the Whirlpool Rapids of the Gorge (Fig. 1.21a, b). The Niagara River diverges upstream of the waterfalls around Goat Island with the main body of water travelling over the Horseshoe or Canadian falls to the west of the island, and a smaller volume of water travelling to the east of Goat Island and descending over the American Falls. The total discharge of the Niagara River entering Lake Ontario is approximately 8000 m^3/s. The volume of water passing over the falls is considerably less than this figure, as most is used for hydroelectric power generation. Niagara Falls is developed along the Niagara River (see Chap. 2).

Niagara Falls was covered by ice several times during the Quaternary. Evidence of the older glacial and interglacial periods is preserved at several locations along the shores of Lakes Erie and Ontario, such as the Scarborough Bluffs in metropolitan Toronto. To the north of the falls, older gorges cut by stream systems developed prior to the last glaciation, such as St. David's Gorge, have been buried by younger glacial sediments.

Thus, the modern Niagara Falls is less than 4,500 years old. Water has continued to course over the falls almost uninterrupted and undiverted, except by the requirements of hydroelectric power plants, for the past 4,500 years. On 29 March 1848, spring break-up in the Niagara River and strong winds combined to completely dam the river with broken ice floes. As a result, the supply of water to the falls downstream was cut off, and the falls were dry for approximately 30 h. The present rapid rates of erosion of the lips of the falls, so severe as to necessitate the temporary diversion of water away from the American Falls in order to cap the escarpment with concrete reflect only the activity of the past 4,500 years.

Niagara River water is used for hydroelectric power generation. The amount of water diverted to each country's turbines is regulated by treaty, so that the 'scenery' is not impaired during daylight hours. Nighttime diversions

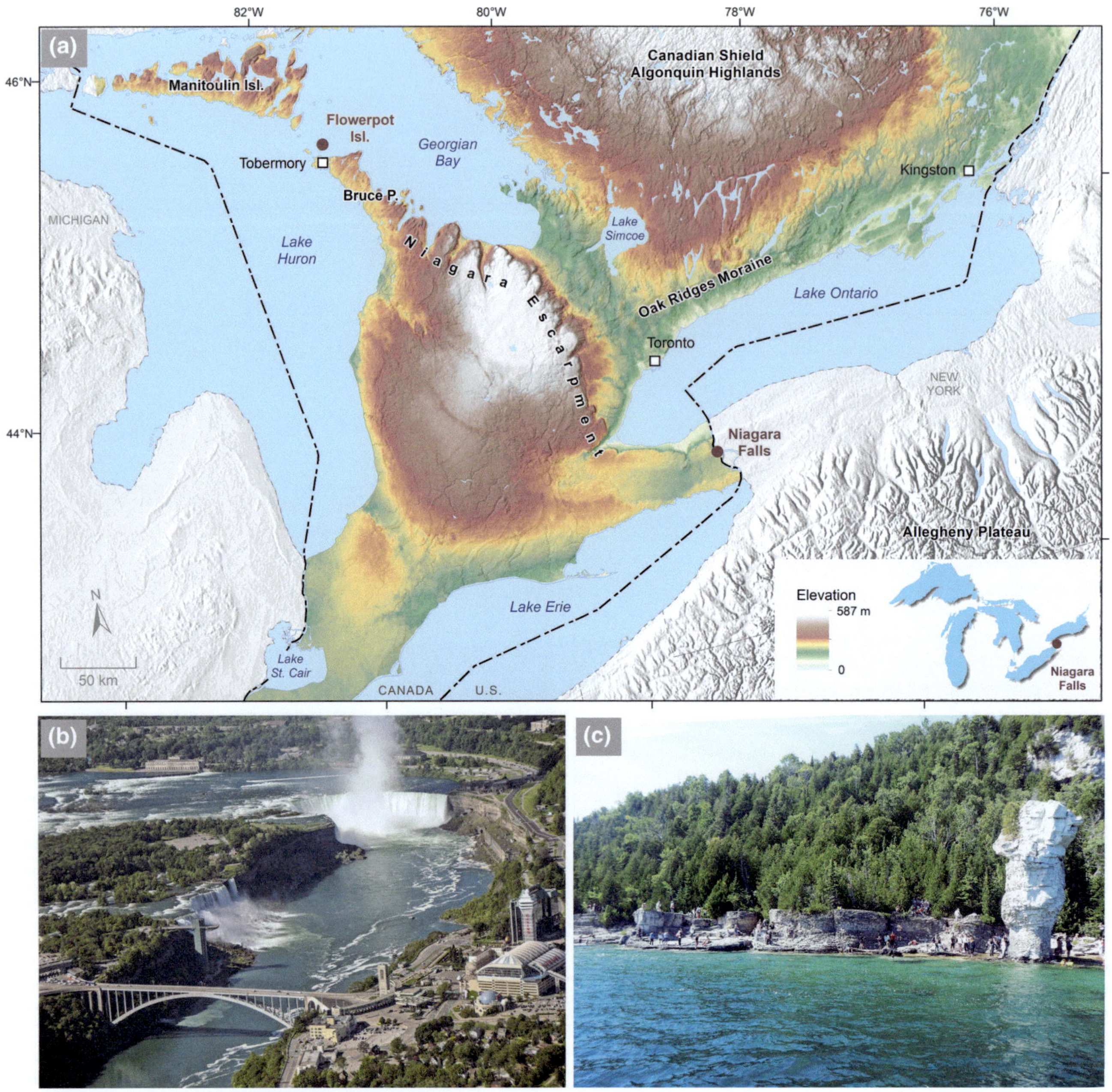

Fig. 1.21 Great Lakes Lowland. **a** The Niagara Escarpment extending(about 725 km long) from Niagara River to the northern tip of Bruce Peninsula, and continuing through the Manitoulin Island, displays vertical cliffs that often outline the edge of the dolostone of the Lockport and Amabel Formation while the slopes below are carved in red shale. The escarpment stands out boldly in the Niagara Peninsula and along the shore of the Georgian Bay. **b** The Niagara River drains Lake Erie into Lake Ontario. Niagara Falls consists of three waterfalls on the border between the Canada and United States and is situated in the Canadian province of Ontario and the American state of New York. The three individual waterfalls are the Horseshoe, American and Bridal Veil falls. Niagara Falls is a 12,000-year-old relic of the last Ice Age. Since then, the falls have moved more than 9.5 km upstream from their original location near Lake Ontario. **c** A massive pillar formed in dolomite by wave erosion on Flowerpot Island, part of Fathom Five National Marine Park on the Bruce Peninsula, Ontario. Locations of (b) and (c) are shown in upper image. *Source* a—Bedrock topography—Gao et al. (2006); @ Queens Printer for Ontario. b—Planet Labs Inc. 2015, CC BY-SA 4.0. c—L. Gpoubran/flickr, CC BY 2.0

greatly exceed the daytime totals. Although the idea of preservation in a national park came to North America too late to totally preserve the falls, the co-existence of power generation and conservation of scenery here represented a positive step.

1.6.9 Innuitia: High Relief (Region 12)

This Innuitian region includes the most northerly parts of Canada and is diverse topographically, including mountains, plateaus, rolling hills, and flat plains. Geologically, it is classified within the Innuitian Orogen (Bird 1967; Wheeler et al. 1996; Bostock 2014). High relief northern Innuitia includes Axel Heiberg and most of Ellesmere Island. Three physiographic regions are readily discernible: two mountain belts, the Grantland Mountains and Princess Margaret Range to the northwest and west, and the Victoria and Albert Range of eastern Ellesmere Island flank the Eureka Upland through the centre of Ellesmere Island and eastern Axel Heiberg.

1.6.9.1 Innuitian Mountains (12a)

The United States Range and the British Empire Range of the Grantland Mountains trend NE–SW across the northwestern coastline of Ellesmere Island. The highest mountain in the British Empire Range is Barbeau Peak rising to 2616 m asl. Most of the area is covered by alpine glaciers and ice fields of the Barbeau Glacier Complex, with ice thicknesses in excess of 900 m. The Ward Hunt Ice Shelf fringes the coast, and locally is 80 m thick. (Lemmen et al. 1988) (see Chap. 13 for details).

The Grantland Mountains are composed of a succession of complexly deformed Proterozoic to Devonian sedimentary rocks, both clastic sandstones and shales and carbonate limestones and dolomites. Volcanic and volcaniclastic units are also present. Devonian and Cretaceous granitic intrusions occur in the southwestern British Empire Range. The rocks of the Grantland Mountains were deformed during the Ellesmere Orogeny, in the Devonian. The northwestern coast of Ellesmere Island is cut by deep fjords oriented northwest-southeast. South of the British Empire Range, fjords such as Greely and Tanquary Fjord in the west and Archer Fjord in the east were developed along SW–NE trending fault valleys. These valleys extend across the width of northern Ellesmere Island. The Princess Margaret Range on Axel Heiberg Island is composed of Carboniferous, Permian and Mesozoic clastic strata deformed by thrust faulting aligned N–S. Coal seams are also present. Local relief exceeds 1200 m. Extensive icefields cover the axial core of the range.

Northernmost Ellesmere Island is separated from Greenland by Robeson Channel with a maximum width of 30 km. During the maximum Quaternary glaciation, ice from Greenland crossed the channel and impinged on northeastern Ellesmere Island. The Victoria and Albert Range, largely buried beneath the Agassiz Ice Cap, isolates the Hazen Plateau and interior Ellesmere from the shoreline along Kennedy Channel (maximum width 50 km) and Kane Basin. Relief in this area approximates 1800 m (Fig. 1.22). Ridge-and-valley topography is aligned NE–SW developed in folded Cambrian to Silurian sedimentary successions.

1.6.9.2 Eureka Uplands (12b)

Between the ranges, the Eureka Upland is a belt of weakly deformed to undeformed sedimentary rocks, generally increasing in age from Mesozoic in the southwest to early Palaeozoic in the northeast. Graben structures allow the preservation of friable rock units in SW–NE trending depressions, notably in Quttinirpaag National Park in the Hazen Lake valley, southwest of Alert. Relief typically is 150–250 m, although cliffs along Archer Fjord are more than 700 m high.

Hazen Lake is developed along the low-lying Hazen Plateau, an extension of the fault valley occupied by Tanquary Fjord. Although the fjords and valleys have been accentuated by glaciation, their original position and basic morphology are controlled by the orientation of the pre-Quaternary fault systems. Sheltered in the fault valley of the Hazen Plateau, Hazen Lake is a thermal oasis, with temperatures as much as 10 °C warmer than on adjacent highlands.

1.6.10 Innuitia: Low to Moderate Relief (Region 13)

Region 13 can be divided into two physiographic subregions: Parry Plateau in the south and Sverdrup Lowland to the north (Bird 1967; Douglas 1970; Dawes and Christie 1991; Bostock 2014).

1.6.10.1 Parry Plateau (13a)

This plateau, with moderate relief, extends across central Melville Island, Bathurst Island, and Cornwallis Island (see Chap. 14) and the Grinnell Peninsula of Devon Island (Douglas 1970). The terrain of Parry Plateau is strongly ridged on Bathurst Island. The ridges are broad, flat-topped and straight-sided, averaging less than 250 m asl. On Melville Island there are a few hills at 800 m asl with four small ice caps and cliff-walled fjord-like bays. On Bathurst Island, the maximum relief is 412 m. Freeze-thaw cycles are not common, and thicknesses of permafrost locally exceed 500 m. The terrain consists of a series of sharply defined ridges and valleys, locally rolling to undulating. In the eastern part of the island, the ridges and valleys are aligned N–S, reflecting the folded terrain of the Cornwallis fold belt. In the western, central and northern parts of the island, the ridges and valleys are more sharply defined than in the east and are aligned WSW–ENE. The Polar Bear Pass wildlife sanctuary extends along one of these fold-created valleys, includes two elongated lakes and numerous small and shallow thermokarst ponds, and is flanked by rolling ridges

Fig. 1.22 High relief Innuitia. **a** Satellite image of the Chapman Glacier (July 29, 2000) located on Ellesmere Island, Nunavut. Formed by the merger of several smaller glaciers, rocky debris on top of the glacier marks the edge of each individual glacier. North is to the right. **b** Tilted Mesozoic sedimentary rocks, Axel Heiberg Island. **c** Bear Strait between Ulvingen Island and Axel Heiberg Island. Wolf Fjord can be seen in the background. *Source* a—NASA/METI/AIST/Japan Space Systems, and U.S./Japan ASTER Science Team.. b—NRCan 2009-5221 by E.C. Turner/Natural Resources Canada. c—@ Natural Resources Canada

to 240 m asl. In the area where the fold belts intersect, in the eastern third of Bathurst Island, the terrain is marked by irregular rolling hills and hummocks. Faulting and folding have shaped the coastline into a series of deep, linear embayments, aligned both parallel and at right angles to the fold systems.

The tectonic history of Bathurst Island is the dominant factor evident in the physiography. The Early Devonian Cornwallis fold belt, which affected the easternmost parts of the island, buckled pre-existing Ordovician, Silurian, and Devonian limestones, dolomites, shales and gypsum beds. This buckling gave rise to synclines and anticlines that were subsequently exploited by surface processes into a series of tightly spaced N–S ridges and valleys. During the folding, thrust and normal faults were also generated accentuating the ridge-and-valley pattern. This folding event was associated with the uplift of the Boothia Arch region to the south creating a spine of disturbed terrain from Bathurst Island through Somerset Island to the Boothia Peninsula. The youngest rocks in the area (Mesozoic sandstones and shales) are found in the northwesternmost points of Bathurst Island and Cameron Island. They are draped over the folded rocks of the Parry Islands fold belt and form the southernmost edges of the geological Sverdrup Basin.

1.6.10.2 Sverdrup Lowland (13b)

Sverdrup Lowland with low relief encompasses numerous islands west of Axel Heiberg and Ellesmere islands, and north of Melville Island (Douglas 1970). These islands are marked by limited relief. In the western Queen Elizabeth

Islands, Sverdrup Lowland is a region of low relief, less than 150 m asl. The islands of the Sverdrup Lowland show little similarity with the rugged terrain that marks Ellesmere and Axel Heiberg, or the linear ridges and valleys of the Parry Plateau. Geologically and physiographically, they form distinctive terrain within the Innuitian region.

Dissected domes and ring structures developed on gypsum anhydrite diapirs and igneous intrusions are distinctive landforms. Relief here is somewhat greater, on the order of 100 m, but no major topographic features are present. Local topography is created by salt domes which have breached the surface on Ellef Ringnes producing annular drainage patterns and areas of dissected terrain where halite, gypsum and anhydrite are exposed. Ellef Ringnes Island is typical of those in the Sverdrup Basin (Fig. 1.23). The topography is controlled by the structural geology, particularly by the disturbances resulting from the salt domes which form the cores of the three peninsulas. The rocks of the Sverdrup Basin form a series of saucer-like units overlying Palaeozoic sediments. Successive strata of Triassic, Jurassic, Cretaceous and Palaeogene sedimentary rocks, predominantly sandstones, overlie the Palaeozoic units. The Palaeozoic units are

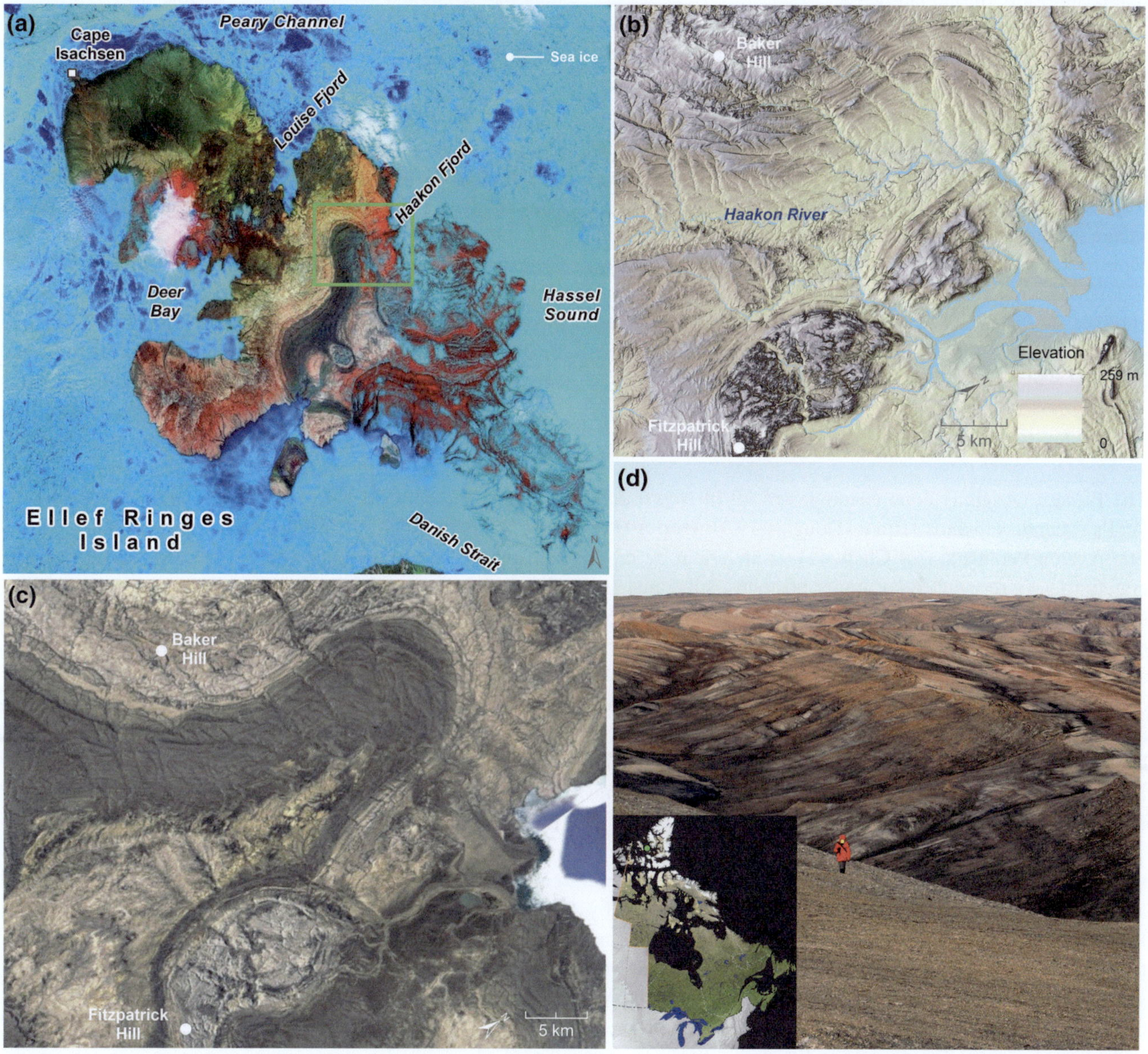

Fig. 1.23 Low to moderate relief Innuitia on Ellef Ringnes Island, Nunavut. **a** Ellef Ringnes Island from space, showing the structural effects of seven salt domes. **b**, **c**, **d** Cretaceous strata on western Ellef Ringnes Island that have been gently folded as a result of movement of salt in the nearby salt domes. *Source* a—NASA. b—digital elevation model; Government of Canada, Natural Resources Canada, Arcticdem_34_30_MNEHR_MNS, http://open.canada.ca/en/open-government-licence-canada. c—Google Earth: Landsat/Copernicus (78°42′ N, 101°40′ W). d—NRCan 2014-234 by S. Grasby/Natural Resources Canada, http://open.canada.ca/en/open-government-licence-canada

only exposed where salt domes and diapirs of Ordovician rock have broken the surface.

1.6.11 Appalachia-Acadia (Region 19)

Appalachia-Acadia is a diverse region, encompassing the Maritime Provinces, Newfoundland and eastern Québec (Wheeler et al. 1996; Bostock 2014). Topography ranges from rugged ridges to flat plains. Apppalachia-Acadia was affected by a succession of orogenic events from the Late Proterozoic to the Late Palaeozoic, which have left substantial imprints on the landscape.

1.6.11.1 Newfoundland (19a)

Physiographic units of the island of Newfoundland include the Atlantic Uplands in the south and east; the Central Lowlands in the northeast, extending NE–SW from Fogo to Buchans; and the Long Range Mountains in the northwest (Fig. 1.24; Bostock 2014). Geologically, the core of the Long Range Mountains on the Great Northern Peninsula (Gros Morne National Park) (Brookes 1993) and the Annieopsquotch Mountains further south are formed of granites and gneisses. The island of Newfoundland is divided geologically into zones: Humber; Dunnage/Gander; and Avalon zones (Colman-Sadd et al. 1990) (see Chaps. 15 and 16).

- **The Eastern (Avalon) Zone**

The Eastern (Avalon) Zone encompasses all of Newfoundland's Atlantic Uplands terrain (Douglas 1970; Bird 1972). The Avalon Peninsula (see Chap. 15) is underlain by moderately to very resistant Ediacaran sandstones, siltstones and shale, most of which have been metamorphosed. Small areas of volcanic rocks, primarily basaltic and volcaniclastic rocks are also present. Physiographically, the Avalon Peninsula contains several areas of rolling uplands, interspersed with small plateaus. The peninsula is connected to the main part of Newfoundland by the Isthmus of Avalon. The peninsula and isthmus encompass approximately 10,000 km^2, and no location is more than 30 km from the coast. The terrain of Terra Nova National Park is marked by fault-created valleys separated by low, rolling uplands up to 229 m asl, veneered and blanketed with glacial sediment. This region is marked by linear ridges, interspersed with isolated areas of undulating terrain and small plateaus, with numerous small islands and skerries. The embayments and valleys of the park are aligned northeast–southwest, reflecting the underlying rock structure and jointing pattern. The deepest embayments are developed following bedrock faults. Flat-topped summits represent remnants of a preglacial landscape which has survived modification by Quaternary ice.

- **The Central (Dunnage and Gander) Zone**

The Central (Dunnage and Gander) Zone includes the Newfoundland Central Plateau (Douglas 1970). This region is marked by linear ridges, interspersed with isolated areas of undulating terrain and small plateaus. Relief in the area is generally less than 200 m. Bedrock includes complicated assemblages of plutonic and ultramafic units, metamorphic rocks and clastic sediments of Cambrian to Silurian age. The Northeast Shelf Coastal Lowland bordering Hamilton Sound and Notre Dame Bay is marked by lower relief. Undulating plains and plateaus slope generally northeasterly to northerly into the Atlantic Ocean. Bedrock in this area consists of marine clastic sedimentary rocks, ophiolites, island-arc basalts and plutons.

- **The Western (Humber) Zone**

The Western (Humber) Zone is well displayed at Gros Morne National Park (Fig. 1.24 a, b; Douglas 1970; Brookes 1993). The park contains three major geological divisions: the Proterozoic Long Ranges; the flanking belt of thrust-faulted sedimentary rocks; and the oceanic crustal ophiolite sequences of the Tablelands. To the west are NNE–SSW trending thrust belts of shale, limestone, dolomite, sandstone and breccia which are exposed throughout the centre of Gros Morne National Park, particularly along the coast. The Tablelands to the southwest are composed of ophiolite complexes consisting of peridotite, dunite, serpentinite and pillow basalts. In southwestern Newfoundland, the Anguille Mountains to the northwest and the Annieopsquotch Mountains to the southeast flank a fault-bounded lowland trough. The sedimentary rocks, including gypsum, are easily erodible by river and wave activity, and locally form karst topography.

1.6.11.2 Québec (19b)

The segment of Appalachia in Québec includes the core Notre Dame Mountains, extending southwestward from Gaspé to Beauce; the Sutton Mountains of Estrie, to the border with Vermont; a pair of flanking uplands, Chaleur and Eastern Québec; and the Mégantic Hills, along the Québec-USA border (Fig. 1.25). The Gaspé Peninsula, and especially the Notre Dame Mountains, are the easternmost extension of the Appalachians of Québec (Bird 1972; Bostock 2014).

- **Notre Dame Mountains**

Gaspé displays the most rugged terrain present in the Appalachian-Acadian physiographic region (Douglas 1970). Relief throughout the area approximates 600 m, with the summit of Mont Jacques-Cartier standing at 1268 m asl. The

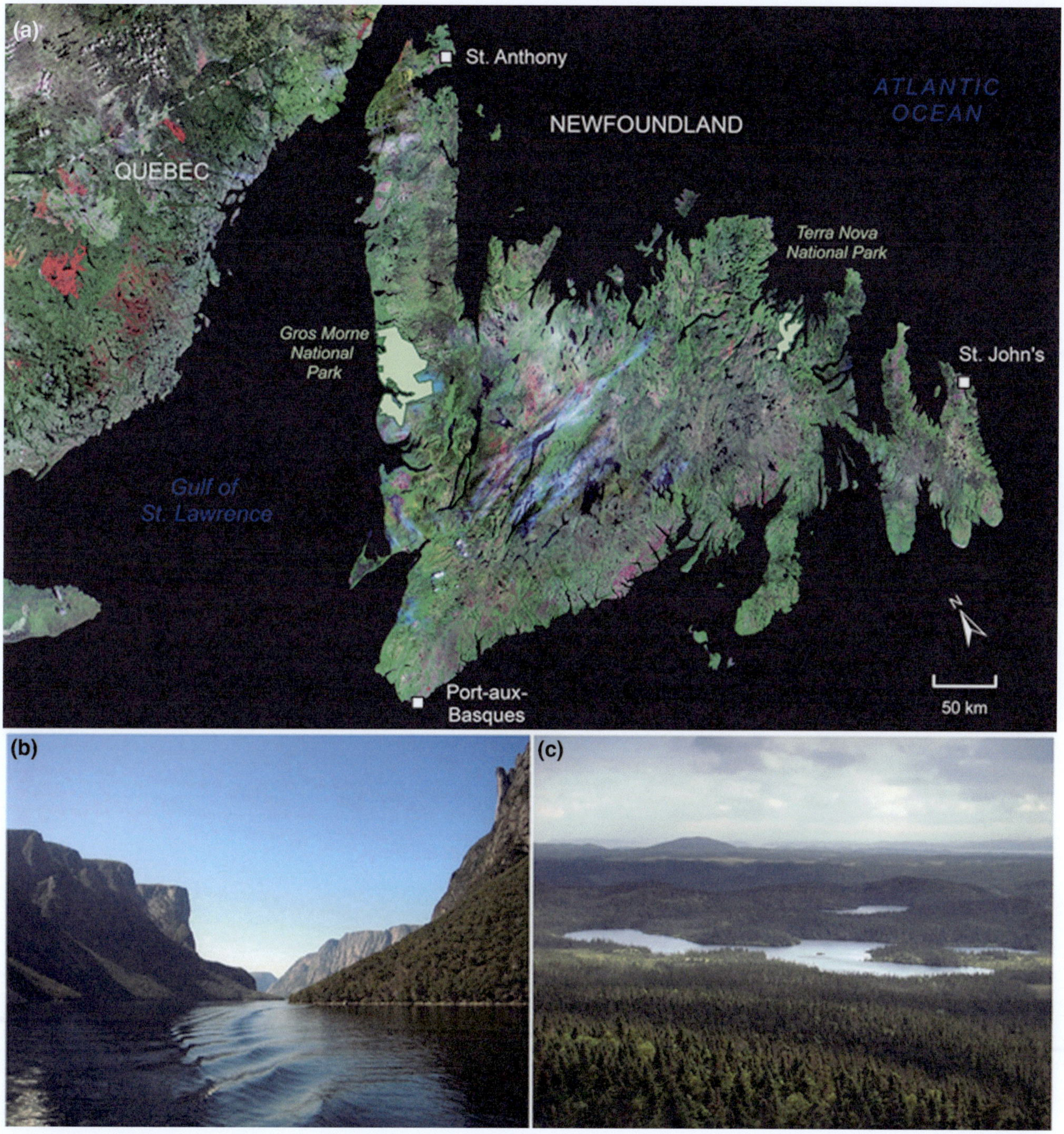

Fig. 1.24 Moderate relief Appalachia, Newfoundland. **a** Newfoundland. **b** Western Brook Pond, Gros Morne National Park, Newfoundland. **c** View from Ochre Hill, Terra Nova National Park, Central Newfoundland. *Source* a—MDA Federal (2004). b—N. Catto. c—Tango7172/Wikimedia Commons, CC BY-SA 4.0

peninsula is composed of several ranges of hills aligned in gently curving arcs, separated by valleys developed along areas of less resistant rock. The Notre Dame Mountains form the spine of the peninsula extending westward from the Baie-de-Gaspé to Causapscal. The highest range on the peninsula is the Chic-Choc Mountains, which form the northern shore of Baie-de-Gaspé (including Forillon Park) and extend west to Mont Jacques-Cartier (Fig. 1.25c). Although these ranges are termed 'mountains', they are actually rugged, rolling hills, with sharp, straight ridges separated by deep valleys. Viewed from above, the peninsula resembles a series of waves pushed to the northwest,

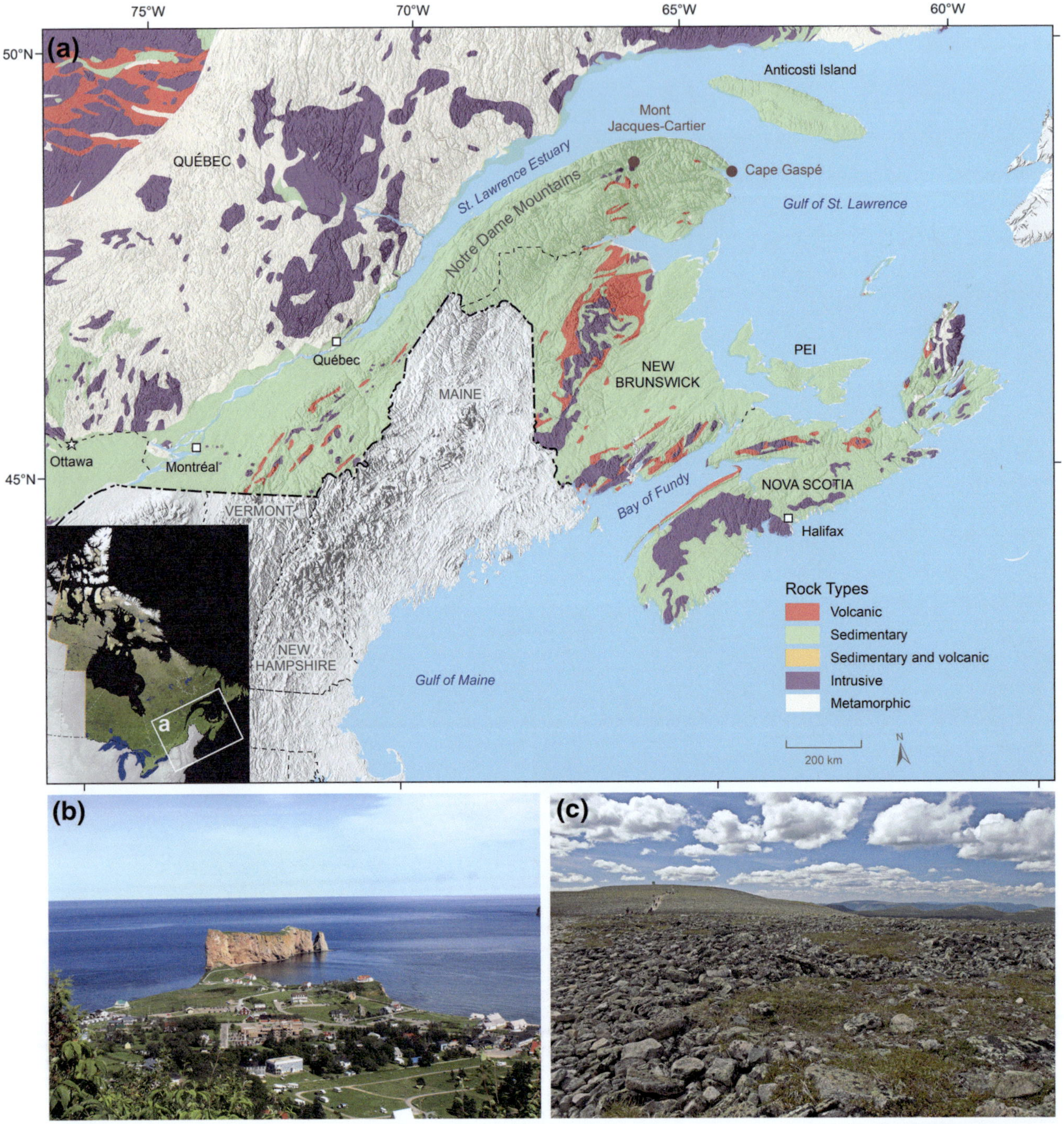

Fig. 1.25 Moderate relief Appalachia: Québec. **a** Overview map of Appalachia-Acadia (zones 19b and 19c). **b** Percé Rock, Cape Gaspé, Québec. **c** Mont Jacques-Cartier (1268 m), Gaspésie, Québec. Locations of (b) and (c) are shown in upper image. *Source* b—Pixabay, CC0. c—Seokhee Kim/flickr, CC BY 2.0

against the rampart of the Canadian Shield north of the St. Lawrence. The major rivers flow either parallel to the rock structure, or at 90° to it along faults and joints. Rivers are braided to wandering, with steep gradients and seasonal variations in flow are extreme. Coastal embayments are partially enclosed by baymouth bars (barachoix) along the eastern side of the peninsula and along the Baie-des-Chaleurs. The northern coast has few embayments along the narrow coastal plain flanking the Chic-Choc Mountains. As in other areas of Appalachian Canada, the physiography and configuration of the coastline are controlled in large measure by the type of bedrock exposed. The ridges of Gaspé represent the results of extensive thrust faulting of sedimentary bedrock. The rocks in Gaspé are limestones, shales and sandstones of Ordovician, Silurian and Devonian age. These units were initially deposited as flat-lying sediments and subsequently, the rocks were subjected to thrusting from the south-southeast during the Late Palaeozoic (Bourque et al. 2000). The strata were pushed towards the north-northwest, where the Canadian Shield north of the St. Lawrence River formed an undeformable barrier. As the strata were thrust to the north, they were stacked atop each other as the leading edges of each unit buckled against the Canadian Shield. Weaker units, such as shales, failed first and were subject to additional folding. More resistant units, especially limestones, moved as blocks over the deformable shales. In many areas, repeated thrusting pushed older, more resistant limestones over younger, weaker shales, thus creating a series of ridges oriented at right angles to the direction of thrusting, with gentle slopes indicating the source of the deformation. The differential resistance of the limestones and shales is evident along the coastline, where coves represent areas of shale outcrops and high cliffs mark limestone exposures. The picturesque sites of Percé Rock, Ile Bonaventure and the cliffs of Forillon are all associated with limestone exposures.

- **The Sutton Mountains**

The Sutton Mountains represent the northern extension of the Green Mountains of Vermont (Douglas 1970). The core is an exposed anticlinorium with deformed metasedimentary and volcanic units flanked by folded and thrusted Cambrian and Ordovician sedimentary successions and mafic intrusions. Maximum relief exceeds 1000 m.

- **Hills and Uplands**

The Mégantic Hills represent the northeastern extension of the sedimentary rocks subsequently deformed into the White Mountains of New Hampshire and Maine. Anticlinally folded Silurian and Devonian sedimentary rocks were intruded by Devonian granites. Mont Mégantic (1105 m asl) is a Cretaceous gabbro-syenite intrusion in the Monteregian Hills chain. The Chaleur Uplands extend diagonally across southern Gaspé and northwestern New Brunswick. Maximum relief is c. 300 m. Folded Ordovician to Devonian carbonate-shale successions are aligned in NE–SW trending ridges, with lesser normal and shallow thrust faults. The Eastern Québec Uplands flank the Notre Dame Mountains to the northwest. Thrust-faulted Cambrian to Ordovician carbonate and clastic units with scattered Proterozoic inliers and isolated Early Palaeozoic mafic intrusions form terrain with relief gradually decreasing from c. 200 m in the southeast to merge with the St. Lawrence Lowlands to the northwest (see Chap. 17).

1.6.11.3 New Brunswick and Nova Scotia (19c)

The low-moderate relief region of Acadia (New Brunswick and Nova Scotia) is divided into the New Brunswick and Nova Scotia Highlands; Annapolis-Avon River valley; and Atlantic Uplands of Nova Scotia (Bird 1972; Bostock 2014).

- **The New Brunswick and Nova Scotia Highlands**

These highlands extend from Bathurst to Moncton in New Brunswick and from Digby to Cape Breton Island in Nova Scotia. The highlands are marked by rolling to rugged hills, with steep cliffs and short valleys. Carbonate and fine clastic successions have been intruded by mafic rocks, monzonite and granite plutons. The Bay of Fundy marks the southern border of the New Brunswick Highlands physiographic region and the southeastern boundary of the Nova Scotia Highlands (Fig. 1.26).

Along the Bay of Fundy in New Brunswick the rocks are dominantly volcanics, sandstones, shales and metasediments, moderately to weakly resistant to erosion. The coastal plain is underlain by red-bed clastic sedimentary rocks, predominantly sandstones and conglomerates. These sandstones and conglomerates form crumbling cliffs along the Fundy shore, notably at Hopewell Rocks where conglomerates are exposed. Minor beds of gypsum and limestone have been partially dissolved by acidic groundwaters to form surface erosional depressions and caves. In Nova Scotia, the Cape Breton Highlands include irregular hills isolated flat-topped to rolling plateaus, up to 532 m asl, and steep escarpments. The plateau is extensively dissected along its margins, and short, steep streams plunge abruptly from the plateau's northern, eastern, and western margins to sea level along cataracts. Waterfalls, notably the 40 m high Ben Beulach, are common along these streams. Three principal bedrock units comprise the Cape Breton Highlands. The oldest rocks, exposed in the northwest are metasediments and igneous. In the southeasternmost corner of the highlands, granites and metamorphic rocks are exposed. Many of these older rocks were intruded by granitic plutons. Both the granite plutons, which have survived environmental modification to form irregularly undulating, low hills on the plateau, and the metamorphic rocks are relatively resistant to erosion. Subsequently, sandstones and shales were deposited over the metamorphosed sediments and granites, remaining as a fringe covering the older rocks at elevations less than 200 m asl around the margins. A fault-controlled escarpment trending SW–NE separates the older igneous rocks to the northwest from the younger rocks to the southeast. The fault, formed in the Ordovician, is not currently active, but at some time during the past 125,000 years, reactivation displaced an MIS 5 marine terrace by 15 m.

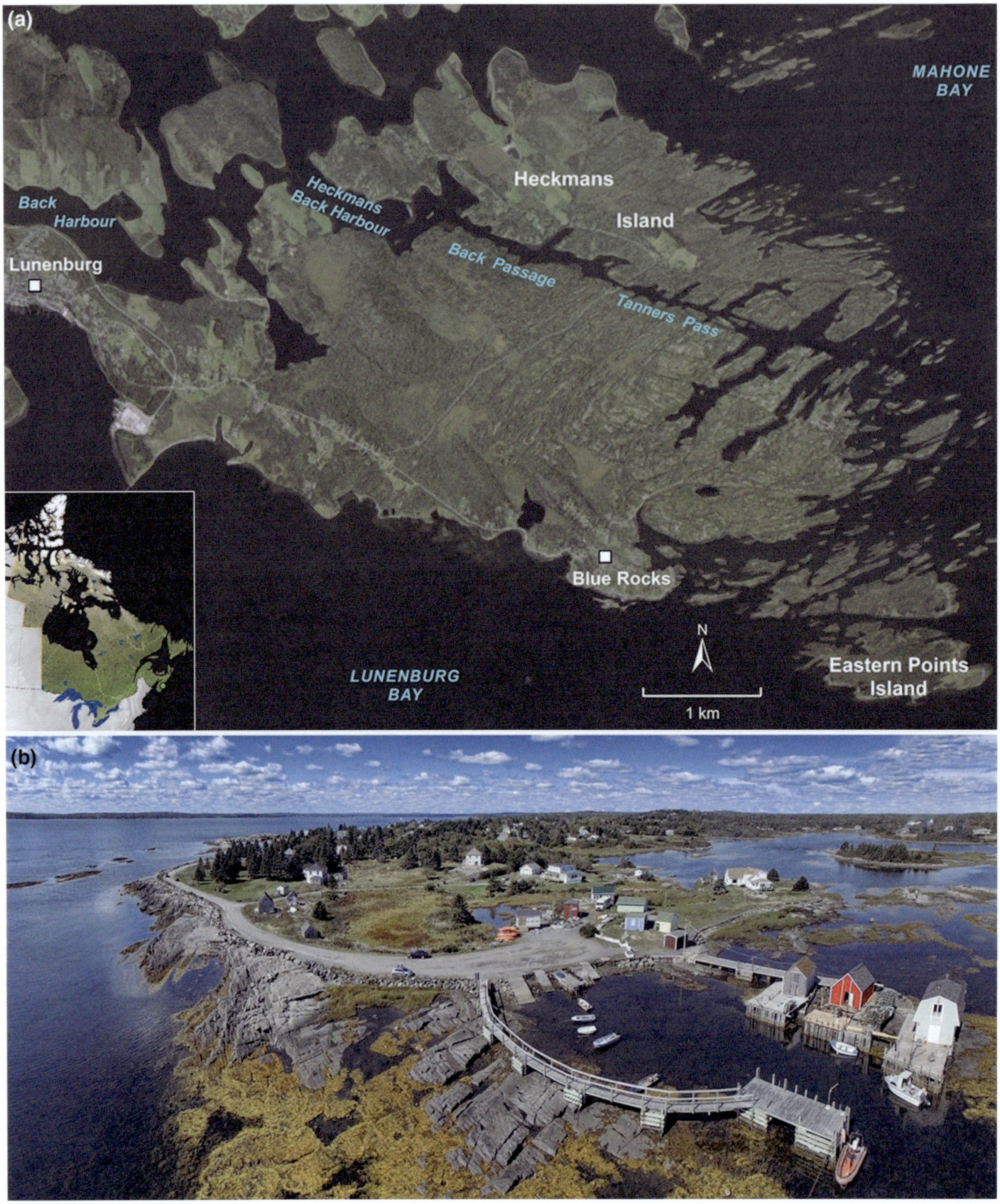

Fig. 1.26 Low to moderate relief Acadia at Tanners Pass, Nova Scotia. **a** *Source* Internal Services Department, Province of Nova Scotia, Orthophotomaps, 1:10,000. Contains information obtained under license from the Province of Nova Scotia which is provided without warranty or liability for errors or omissions. **b** Blue Rocks, Nova Scotia/Pixabay, CC0

- **The Annapolis—Avon River Valley**

The central axis of the Annapolis Valley is a low-lying plain, ranging from 0 to 50 m asl, occupied by the Annapolis River in the southwest and by the Gaspereau and other smaller rivers in the northeast. Flooding of the plains periodically occurred prior to regulation of the rivers. The terrain surrounding the Avon River is also a low-relief plain with its long axis aligned NW–SE. Red Triassic sandstones, conglomerates, and shales, soft and easily eroded, underlie the Annapolis Valley. Mississippian sandstone, conglomerate, shale, gypsum and limestone are exposed along the Avon River Valley.

- **Atlantic Uplands of Nova Scotia**

The southern margins of the Annapolis Valley and the Avon River lowland are bordered by South Mountain, one component of the Atlantic Uplands of Nova Scotia. It is a complex of rolling hills underlain by granite, with a maximum relief of 200 m. Glacial activity centred over South Mountain has resulted in the deposition of tills and glacifluvial deposits, locally impounding lakes. Numerous drumlins, to 50 m in length and 15 m in height, are scattered across the landscape. To the south, the Atlantic Uplands of Nova Scotia are underlain by granites (e.g. at Peggy's Cove) in the north, intruding older metasediments. The older slates and quartzites extend south to the southern shore of Nova Scotia. These rocks are easily eroded, and form gently sloping plains and low, rolling hills.

1.6.12 Arctic Coastal Plain (Region 20)

The northernmost fringe of the Arctic Coastal Plain flanks the Sverdrup Lowland to the northeast, and the tips of Borden, Ellef Ringnes, and Meighen islands (Bird 1967; Douglas 1970). This region is underlain by the Beaufort Formation of unconsolidated Tertiary or early Pleistocene sands and gravels. On Meighen Island, this formation has been uplifted 200 m and eroded to form a hilly terrain that carries an ice cap at around 250 m asl. On Ellef Ringnes and Borden islands the coastal plain is low, remarkably flat and uniform with a network of consequent streams draining seaward. The fringing areas along the margins of the Beaufort Sea lie within the Arctic Coastal Plain and are marked by gently westward-sloping surfaces with shallow ephemeral braided streams and no glacial features. The Arctic Coastal Plain extends beneath sea level to merge with the Arctic Continental Shelf. Offshore from Meighen, Ellef Ringnes and Borden the sea floor is shallow and slopes both seaward and towards the channels which separate the islands. The exposed part of the Arctic Coastal Plain is part of an emerged surface of deposition that antedates the deep channels between the islands.

1.6.13 Atlantic Maritime Plain (Region 22)

The Maritime Plain includes the coastal areas of New Brunswick and Nova Scotia, Prince Edward Island, and the Magdalen Islands (Douglas 1970; Bird 1972) (see Chap. 19). Relief is very low, with all the terrain lying below 150 m asl. Coastlines are dominated by sandy beaches, barrier islands and gently sloping shelves. Most areas are marked by dissipative conditions where wave energy is expended offshore, low-energy beaches and littoral drift transport parallel to the shoreline. Prince Edward Island is characterized by planar to gently rolling topography, with many flat plains. Relief is extremely limited, with the highest point on the island, only 142 m asl. The northern coastal region is generally lower than the southeastern part of the island with a maximum relief of approximately 60 m. Much of the topography is the result of post-glacial downcutting by the numerous small streams (many now diminished in volume or diverted due to agricultural activity) that developed on the terrain following deglaciation. The absence of major river valleys and other significant topographic features inland results in the gently undulating landscape characteristic of much of Prince Edward Island. The shorelines are marked by low cliffs of glacial diamicton and Pennsylvanian-Permian rock, separated by expanses of marshland and tidal flats. A similar landscape is evident along most of the Gulf of St. Lawrence coast of New Brunswick, within the New Brunswick Lowlands, and in Nova Scotia. This lowland forms a gently eastward-sloping plain across the eastern and central parts of New Brunswick, with very limited relief. The coastline is marked by long, shallow embayments which are influenced by tidal activity. The semi-diurnal tidal ranges along the Gulf of St. Lawrence coast of New Brunswick varies from 1.5 to 3.5 m. Salt marshes are developed in most tidally influenced lagoons and estuaries.

In comparison to other areas of Atlantic Canada, the bedrock geology of the Maritime Plain is straightforward. Sandstones, siltstones, and conglomerates of later Carboniferous age extend westward from the Gulf of St. Lawrence shoreline to underlie approximately half of New Brunswick. These rocks represent red-bed clastic sedimentation, associated with alluvial fans, rivers, and lakes formed along the flanks of the Appalachian Mountains. The sedimentary rocks are soft and easily eroded, not exposed at the surface throughout most of eastern New Brunswick and have relatively little influence on the natural landscape of the region. Similarly, all of Prince Edward Island is underlain by sedimentary rocks, dominantly sandstones. Ferric oxides have coloured the rocks bright red giving the cliffs and sediments of Prince Edward Island an unmistakable glowing, warm hue (Fig. 1.27).

Fig. 1.27 Atlantic Maritime Plain. **a** Overview. **b** Aerial view of Magdalen Islands, Québec. **c** Prince Edward Island National Park (PEI NP). **d** Queens Point, Prince Edward Island. **e** Cavendish Beach, Prince Edward Island. Also see Fig. 1.19 for location. *Source* a—MDA Federal (2004). b—D. Searls/flickr, CC BY 2.0. c—D. Bezaire/flickr, CC BY-SA 2.0. d—H. Harvey/flickr, CC BY_SA 2.0. e—M. Rathwell/flickr, CC BY 2.0

The age of the rocks decreases successively from south to north, and thus the youngest rocks on the island outcrop along the north coast. The sandstones and shales exposed along the coast are subject to erosion, and form relatively weak, crumbling cliffs that consistently supply sediment to the shore. Ongoing coastal erosion of bedrock locally exceeds 80 cm per year. Similar Permian clastic units underlie the Magdalen Islands (see Chap. 19).

1.7 Conclusion

Whereas Western Canada is dominated by the mountains of the Cordillera at its centre, Eastern Canada is centred on the low relief, rocky landscapes of the shield. But this impression of monotony is gained simply because the most spectacular landscapes in Eastern Canada are arranged around the periphery and not at its geographical centre: the Atlantic coastline, the glaciers and snow-capped mountains of the Arctic and the unique Great Lakes and St. Lawrence River basin is magnificent landscape by any measure. Eastern Canada's geomorphological landscapes cover over five million square kilometres of Earth's surface that vary from vast unbroken and homogeneous spaces to dramatic mountain scapes that contain the greatest concentration of glacier ice outside Antarctica and Greenland. Although human migration into these landscapes occurred very recently in geological time, the presence of Inuit peoples, Indigenous First Nations and francophones adds cultural interest to these marvellous landscapes as reflected in their place names.

References

Adams J, Basham PW (1991) The seismicity and seismotectonics of eastern Canada. In: Slemmons DB, Engdahl ER, Zoback MD, Blackwell DD (eds) Neotectonics of North America. Geol Soc Am, vol Decade Map 1, pp 261–276

Ambrose JW (1964) Exhumed paleoplains of the Precambrian Shield of North America. Am J Sci 262:817–857

Bird JB (1967) Physiography of Arctic Canada. Johns Hopkins Press, Baltimore

Bird JB (1972) The natural landscapes of Canada: a study in regional earth science. Wiley Publishers of Canada, Toronto

Bliss LC, Heal OW, Moore JJ (eds) (1981) Tundra ecosystems: a comparative analysis. Cambridge University Press, Cambridge, UK

Bostock HS (1970) Physiographic subdivisions of Canada. In: Douglas RJW (ed) Geology and economic minerals of Canada. Economic geology report no. 1, Geol Surv Can, Department of Energy, Mines and Resources, Ottawa, pp 10–30

Bostock HS (2014) Physiographic regions of Canada. Geol Surv Can Map 1254A (1st and 2nd editions), scale 1:5 000 000

Brookes IA (1993) Table Mountain, Gros Morne National Park, Newfoundland: Canadian landform examples 26. Can Geogr 37:60–73

Bourque PA, Malo M, Kirkwood D (2000) Palaeogeography and tectono-sedimentary history at the margin of Laurentia during Silurian to earlies Devonian time: the Gaspé belt, Québec. Bull Geol Soc Am 112:4–20

Canadian Geographic (2018) Indigenous peoples' Atlas of Canada. Royal Canadian Geographical Society, Ottawa

Chapman LI, Putnam DF (1984) The physiography of Southern Ontario, 3rd edn. Ont Geol Surv Spec, vol 2

Colman-Sadd SP, Hayes JP, Knight I (1990) Geology of the Island of Newfoundland. Newfoundland and Labrador, Department of Mines and Energy, Geological Branch, Map 90-01. Open File GS# NFLD/2192

Cowell DW, Ford DC (1980) Hydrochemistry of a dolomite karst: the Bruce Peninsula, Ontario. Can J Earth Sci 17:520–526

Dawes PR, Christie RL (1991) Geomorphic regions. In: Trettin HP (ed) Geology of the Innuitian Orogen and Arctic Platform of Canada and Greenland. Geology of Canada, no 3. Geological Survey of Canada, Ottawa, pp 29–56

Douglas RJW (1970) Geology and economic minerals of Canada. Economic geology report no. 1, Geological Survey of Canada, Department of energy, mines and resources, Ottawa

Earth Impact Database (2017) University of New Brunswick, planetary and space science centre. http://www.unb.ca/passc/ImpactDatabase/faq.html. Accessed 10 May 2017

Environment Canada (1995) Ecozones of Canada (version 1) http://sis.agr.gc.ca/cansis/nsdb/ecostrat/gis_data.html. Accessed 21 Jan 2015

Ford DC (2006) Karst geomorphology, caves and cave deposits: a review of North American contributions during the past half century. Geol Soc Am SP 404

Ford DC, Williams PW (2007) Karst hydrogeology and geomorphology. Wiley & Sons, Chichester

Fulton RJ (1989) Quaternary geology of the Canadian Shield. In: Fulton RJ (ed) Quaternary geology of Canada and Greenland. Geology of Canada Series, no. 1. Geological Survey of Canada, Ottawa, pp 177–178

Gao C, Shirota J, Kelly RI, van Haaften S (2006) Bedrock toporagraph and overburden thickness mapping, southern Ontario. Ontario Geological Survey, Miscellaneous Release—Data 207. https://www.mndm.gov.on.ca/en/mines-and-minerals/applications/ogsearth

Gilbert GK (1907) The rate of recession of Niagara Falls. US Geol Surv Bull 306

Government of Canada; Natural Resources Canada (2007) Canada centre for mapping and earth observation, CanImage—Landsat 7 Orthoimages of Canada, 1:50 000 Tiles: 058f13, 058f14, 058f15, 058g02, 058g03, 058g04, 058g05, 058g06, 058g07, 066c10, 054f11. ftp.geogratis.gc.ca/pub/nrcan_rncan/image/

Government of Canada (2008) Canadian permafrost thickness, Geological Survey of Canada. Natural Resources Canada; Earth Sciences Sector; Canada Centre for Mapping and Earth Observation. http://geogratis.gc.ca/api/en/nrcan-rncan/ess-sst/092b663d-198b-5c8d-9665-fa3f5970a14f.html. Accessed 30 Jan 2015

Government of Québec, Ministry of Forests, Wildlife and Parks (MFFP) (2017) Bank of photos of Northern Québec. www.donneesquebec.ca/recherche/fr/dataset/photographie-oblique-du-nord-quebecois

Graf WL (1987) Regional geomorphology of North America. In: Graf WL (ed) Geomorphic systems of North America, Centennial Special, vol 2. Geol Soc Am. Boulder, Colorado, pp 1–4

Grant DR (1989) Quaternary geology of the Atlantic Appalachian region of Canada. In: Fulton RJ (ed) Quaternary geology of Canada and Greenland. Geology of Canada Series, no. 1. Geological Survey of Canada, Ottawa, pp 393–440

Grieve RAF (2001) Impact cratering on Earth. In: Brooks GR (ed) A synthesis of geological hazards in Canada. Geol Surv Can Bull, vol 548, pp 207–22

Guyard H, St. Onge G, Pienitz R, Francus P (2011) New insights into Late Pleistocene glacial and postglacial history of northernmost Ungava (Canada) from Pinualuit Crater Lake sediments. Quat Sci Rev 30:3892–3907

Halsey L, Vitt D, Zoltai S (1997) Climatic and physiographic controls on wetland type and distribution in Manitoba, Canada. Wetlands 17:243–262

Hare FK, Ritchie J (1972) The boreal bioclimates. Geog Rev 62:333–365

Hodgson DA (1989) Quaternary geology of the Queen Elizabeth Islands. In: Fulton RJ (ed) Quaternary geology of Canada and Greenland. Geology of Canada Series, no. 1. Geological Survey of Canada, Ottawa, pp 441–478

von Humboldt A (1845) Kosmos: a sketch of a physical description of the universe (vol 1) (translated from the German by Otté EC) Harper and Brothers, New York

Inuit Heritage Trust (2015) Where we live and travel: named places and selected routes. Vancouver Foundation, Vancouver, BC

Ives JD (2016) Baffin Island: field research and high arctic adventure, 1961–1967. University of Calgary Press

La Berge GL (1994) Geology of the Lake Superior Region. Geoscience Press, Phoenix, Arizona

Larsen JA (1980) The Boreal ecosystem. Academic Press, New York

Lemmen DS, Evans DJA, England J (1988) Ice shelves of northern Ellesmere Island. N.W.T. Canadian landform examples, 10. The Can Geogr 32:363–367

MDA Federal (2004) Landsat GeoCover ETM+2000 Edition Mosaics Tiles N-14-65, N-14-70, N-16-65, N-16-70, N-16-75, N-18-65, N-18-70, N-18-75, N-19-45, N-19-50, N-20-40, N-20-45, N-20-50, N-20-60, N-21-45, N-21-50, N-22-45 ETM-EarthSat-MrSID, 1.0, USGS, Sioux Falls, South Dakota, 2000. Accessed 2016, at the Global Land Cover Facility

Natural Resources Canada (2012) Atlas of Canada, Northern geodatabase. https://open.canada.ca/data/en/dataset/702ebdea-39ff-50e4-ab5f-de1150d16b7a

Ritchie JC (1987) Postglacial vegetation of Canada. Cambridge University Press, Cambridge, UK

Roberge JS, Ford DC (1983) Karst of the Salmon River, Anticosti Island, Québec. J Hydro 61:159–162

Smith SL, Burgess MM, Heginbottom JA (2001) Permafrost in Canada: a challenge to northern development. In: Brooks GR (ed) A

synthesis of geological hazards in Canada. Geol Surv Can Bull, vol 548, pp 241–264
Stearn CW (1975) Canada. In: Fairbridge RW (ed) Encyclopedia of world regional geology. Dowden, Hutchinson and Ross, Stroudsburg, Pa, pp 139–144
Stöckli R, Vermote E, Saleous N, Simmon R, Herring D (2005) The blue marble next generation—a true color earth dataset including seasonal dynamics from MODIS. NASA, Earth Observatory
Tinkler KJ (1986) Niagara falls: Canadian landform examples 2. Can Geogr 30:367–371
Tinkler KJ (1997) Rockbed wear at a flow convergence zone in Fifteen Mile Creek, Niagara Peninsula, Ontario. J Geol 105:263–274
Tyman J (2008) Inuit: people of the Arctic, Bill Hillman's edutech research project, at Brandon University, Canada. Copyright the Pitt Rivers Museum at the University of Oxford. www.johntyman.com/arctic/
Wheeler JO, Hoffman PF, Card KD, Davidson A, Sanford BV, Okulitch AV, Roest WR (compilers) (1996) Geological map of Canada. Geol Surv Can, Map D1860A
Williams PW, Fong YT (2017) World map of carbonate rock outcrops (version 3.0). http://web.env.auckland.ac.nz/research/karst.shtml. Accessed 9 May 2017
Zoltai SC (ed) (1988) Wetlands of Canada. National Wetlands Working Group, Canada Committee on Ecological Land Classification, Environment Canada, Ecological Land Classification Series 24

Olav Slaymaker is a Member of the Order of Canada, and Professor emeritus of geography in the Department of Geography at the University of British Columbia, Vancouver, B.C. His 20+ co-authored and edited monographs and 160+ refereed papers reflect three different styles of work. First, he has pursued original field research for 40 years on sediment systems in the Canadian Cordillera and the European Alps, strongly supported and enhanced by his stellar doctoral and postdoctoral students and international colleagues. Second, he has produced papers and books summarizing trends in land use and environmental change in mountain regions worldwide. Third, he has edited monographs with invited international contributions on major environmental and geomorphological themes. He is former President of the Canadian Association of Geographers and former President of the International Association of Geomorphologists.

Norm Catto joined Memorial University in July 1989 and is now the Head and Professor of Geography at that university. His research and teaching interests include coastal landforms, natural hazards, sea-level change and impacts in coastal environments; response of river systems to climate and weather events, and flood risk assessment; aeolian geomorphology and response to climate and human factors in boreal and costal environments; mass movements and slope failures; loess deposition; palaeosol formation; emergency measures policies and effectiveness human adaptations to climate and weather events; the impacts of climate and weather events on agriculture, transportation, fisheries and communities; and the relationships among landscapes and people. His research has included projects and investigations in landscapes and environments in 16 countries, and in all Canadian provinces and territories, particularly in Eastern Canada. For 18 years, he served as Editor-in-Chief for *Quaternary International,* a professional international journal dedicated to the study of landscape evolution and climate variation throughout the past three million years. He has taught more than 50 different undergraduate and graduate courses focused on numerous aspects of geomorphology, landscape analysis, Quaternary Research, natural hazards, ocean sciences, environmental science, climate and climate change, soil science, geoarchaeology, and palaeontology. Study of landscapes forms his central focus for understanding the various areas of Eastern Canada, the combination of physical and cultural constituents that contribute to its identities. For most species, and for all human occupants, life in Eastern Canada has been heavily influenced by geomorphology since deglaciation.

Dori J. Kovanen is a professional engineering geologist and Research Associate in the Department of Geography at the University of British Columbia, Vancouver, BC. With an MSc from Western Washington University in Bellingham and a Ph.D. in Quaternary geology from the University of British Columbia she is a specialist in Quaternary geomorphology. She has written extensively on Late Glacial and Holocene history of the Pacific Northwest as well as contemporary environmental changes in the region and was a significant contributor to 'Landscapes and Landforms of Western Canada' (Slaymaker 2017).

Quaternary Glacial, Glacimarine and Glacilacustrine History

2

Norm Catto, Dori J. Kovanen, and Olav Slaymaker

Abstract
Eastern Canada was almost completely glaciated from four major and several lesser ice centres on multiple occasions during the Quaternary. Most of the geomorphology relates to Marine Isotope Stage 2. Areas of resistant bedrock are dominated by glacial erosional features, such as roches mountonées. Glacial depositional landforms are prominent in southern Ontario, Québec, and the Maritime provinces. In Arctic Canada, glaciation had lesser effects on geomorphology, with the exception of the mountainous areas of Baffin and Ellesmere islands. Glacial retreat exposed the isostatically depressed coastlines to marine incursions. Champlain Sea glacimarine sediments cover the St. Lawrence and Ottawa Valleys. In the Great Lakes Lowlands and southern Canadian Shield, blockage of northward and eastward drainage led to the formation of successions of proglacial lakes and other paraglacial landforms during the transition from MIS 2 to the Holocene.

Keywords
MIS 2 • Laurentide ice sheet complex • Innuitian ice sheet complex • Glacimarine • Champlain sea • Glacilacustrine • Great lakes • Lake Barlow-Ojibway

2.1 Introduction to Glacial History

The Quaternary record of eastern Canada is dominated by glaciation. With the exception of nunataks in alpine regions, and limited areas in the Arctic Lowlands and Innuitia, all parts of eastern Canada appear to have undergone multiple glaciations during the Quaternary, particularly during Marine Isotope Stage (MIS) 2. Erosional and depositional landforms are generally used to recognize past glaciations although some areas of cold-based glaciation have been proposed. This chapter outlines the glacial, glacimarine and glacilacustrine events and landscapes of eastern Canada. Paraglacial, periglacial and Holocene landscapes are discussed in Chap. 3 (Table 2.1).

2.2 Pre-MIS 2 Events

The sedimentary record of pre-MIS 2 Quaternary events in eastern Canada is somewhat limited, and the geomorphic record more so. The earliest continental glaciation in Atlantic Canada (Bridgewater deposits, Nova Scotia) is tentatively attributed to either MIS 16 or MIS 22, as there is no numerical dating. Although the available evidence suggests that intensive glaciation occurred at several times in the Quaternary prior to MIS 2, glacial deposits older than MIS 5 have been identified only in isolated locations, including southern Ontario (Karrow 1989), southern Québec (Occhietti 1989), Qikiqtaaluk (Baffin Island; Andrews 1989), New Brunswick (Rampton et al. 1984), Nova Scotia (Grant 1989) and Newfoundland. The deposits are exposed only in sections, without associated surface landforms (Fig. 2.1).

Scattered distributions of glacial sediments and computer-generated ice sheet modelling suggest that extensive glacial ice cover existed throughout Nunavut, Nunavik (northern Québec) and in the Torngat Mountains during MIS 5 (Clark 1988; England et al. 2006; Stokes et al. 2012, 2015; Catto, Chap. 7, this volume; van Wychen et al., this volume). Expansion of continental glaciation from centres in Labrador–Québec and Kivalliq (mainland Nunavut) commenced during MIS 4, and c. 65 ka all of eastern Canada (aside from nunataks) appears to have been ice covered. Along the southern margin, retreat during

N. Catto (✉)
Department of Geography, Memorial University of Newfoundland and Labrador, St. John's, NL A1B 3X9, Canada
e-mail: ncatto@mun.ca

D. J. Kovanen · O. Slaymaker
Department of Geography, University of British Columbia, 1984 West Mall, Vancouver, BC V6T 1Z2, Canada

© Springer Nature Switzerland AG 2020

O. Slaymaker and N. Catto (eds.), *Landscapes and Landforms of Eastern Canada*,
World Geomorphological Landscapes, https://doi.org/10.1007/978-3-030-35137-3_2

Table 2.1 Marine isotope stages

Marine isotope stage	Time (ka)	Events in eastern Canada
1	0–11.7	Holocene, see Chap. 3
2	11.7–29	Widespread glaciation; majority of glacial geomorphology formed; glacimarine and glacilacustrine events
3	29–57	Glacial retreat from southern Ontario and Québec, Maritime provinces, Newfoundland
4	57–71	Glacial advances covering most of eastern Canada
5	71–123 (MIS 5.5, 109–123)	Local preservation of interglacial (MIS 5.5) deposits; scattered MIS 5.1 deposits preserved in southern Ontario and Québec; readvances of Laurentide ISC begin c. 100 ka
6	123–191	Widespread glaciation; local preservation of deposits

Fig. 2.1 MIS 3 deposits preserved in karst developed in Carboniferous gypsum, Cape Anguille area, NL. MIS 2 glacial deposits cap the succession. *Photograph* N. Catto

MIS 3 exposed the western and central Great Lakes–St. Lawrence Lowlands, and much of the Maritime provinces. Newfoundland and Labrador remained glaciated at this time. Subsequent readvances began during the latter part of MIS 3, culminating in the Last Glacial Maximum (LGM) of MIS 2, c. 25 ka (e.g. Gosse et al. 2004; Grant 1989; Karrow 1989; Occhietti 1989; Pico et al. 2018; Vincent 1989).

2.3 MIS 2 Glaciation

MIS 2 glaciation resulted in the complete ice cover of eastern Canada, aside from mountain nunataks. The Laurentide Ice Sheet Complex (LISC) expanded to cover all of Ontario, Québec, Kivalliq, and southern Nunavut, and interacted with alpine glaciers in northern Qikiqtaaluk and the Torngat Mountains. In Atlantic Canada, the southern boundary of the LISC fluctuated due to shifting flow directions and coalescence with local ice caps developed in New Brunswick, Nova Scotia, the Gulf of St. Lawrence, and Newfoundland. The Innuitian Ice Sheet Complex (IISC) developed in northern Nunavut, interacting with Laurentide ice along its southern margin, and with alpine glaciation in Tatlurutit (Devon Island) and Umingmak Nuna (Ellesmere Island; see van Wychen et al., this volume).

2.3.1 Laurentide Ice Sheet Complex (LISC)

The Laurentides (Laurentian Mountains) form the southeastern margin of the Grenville Structural Province across Ontario and Québec. Initially, glacial flow to the Great Lakes–St. Lawrence Lowlands to the south was recognized by the presence of granitic and gneissic clasts derived from the Precambrian rocks, and the term 'Laurentide' applied to the event. Subsequently, Laurentide has been applied to all continental glaciation in eastern Canada which originates from any area of the Canadian Shield, and dates to MIS 2, with final deglaciation during the initial phases of MIS 1 (Dyke 2004; Dyke et al. 2002, 2003; Dyke and Prest 1987; Tarasov et al. 2012; Figs. 2.2 and 2.3). Laurentide origin is commonly attributed to glacial deposits that contain some clasts derived from the Canadian Shield, particularly granitic, mafic, ultramafic, or high-grade metamorphic clasts of distal origin.

The limits of Laurentide glaciation correspond to the most distal occurrence of MIS 2 glacigenic sediments containing Canadian Shield clasts, marked either by transitions into non-glacial Quaternary deposits or coalescence with other ice masses. In eastern Canada, the southeastern boundary of Laurentide glaciation is marked by coalescence with local ice masses in Atlantic Canada and Appalachian Québec (e.g. Stea et al. 2011; see Morissette et al., this volume). Laurentide glacial deposits extend across the entire southern boundary of Ontario, including the Great Lakes basins, and westward into Prairie Canada and mainland Northwest Territories (e.g. Dredge and Cowan 1989; England et al. 2009; Mahaney et al. 2014; Stokes, this volume; Veillette et al. 2017; Zaniewski et al., this volume). The northern limit is marked by areas covered by the IISC and with local alpine glaciation in Nunavut. To the east, Laurentide glaciation extended to the MIS 2 marine shoreline, except in the highest parts of the Torngat Mountains.

The LISC has been divided into several 'centres', 'domes', 'sectors' and ice sheets (the terminology varies greatly amongst authors), which coalesced to produce the Ice Sheet Complex. Several time-transgressive changes in configuration have been noted, designated as different sectors by different authors. In broad strokes, the chief areas of accumulation and expansion were (1) an area centred east of Hudson Bay in western Labrador and adjacent Québec; (2) an area centred west and north of Hudson Bay in Kivalliq and western Nunavut; and (3) an area in Foxe Basin and southern Qikiqtaaluk.

2.3.1.1 Labrador—Western Québec

This accumulation centre is commonly designated the Labrador or Labradorean. The eastern part of the source area is at the current location of the Smallwood Reservoir. The source area of the Laurentide Ice Sheet Complex in westernmost Labrador and adjacent Nouveau-Québec was further confirmed through the work of Hughes (1964; also see Klassen et al. 1992; Klassen and Thompson 1993; Occhietti et al. 2004, 2011). Ice extended eastward into the Labrador Sea (Clark and Fitzhugh 1990), northward to block Hudson Strait and coalescing with ice over Qikiqtaaluk and Foxe Basin (Andrews and MacLean 2003; DeAngelis and Kleman 2007; Stokes 2018; Winsborrow et al. 2004, 2010), westward across Québec and Hudson Bay (Figs. 2.4 and 2.5) to coalesce with ice in eastern Kivalliq and northern Manitoba (Roy et al. 2009; Dredge and Dyke, this volume) and southward to Ontario (Figs. 2.6 and 2.7), Maritime Canada, the Northern Peninsula of Newfoundland and the eastern and midwest USA (Beaupre et al., this volume; Desloges et al., this volume; Normandeau et al., this volume; Nutz et al., this volume; Zaniewski et al., this volume).

The core region of the accumulation centre is well defined by the orientations of striated and sculpted bedrock, and by the distribution of distinctive erratics, including high-grade metamorphic gneiss, iron-bearing units from the Labrador City–Wabush area, and ultrabasic rocks such as anorthosite. Several phases of activity, involving shifting ice divides and variations in coalescence zones with adjacent ice masses, are indicated by erratic distributions in sequential deposits, and by cross-cutting landforms.

Glacial features characteristic of western Labrador include drumlins, flutings and eskers. Eskers serve as passage routes for animals through the taiga, and are commonly used as sources of gravel and sand aggregate for road construction. Although esker systems can extend for hundreds of kilometres, all but the shortest are composed of segments, rather than a continuous ridge. Individual segments of eskers

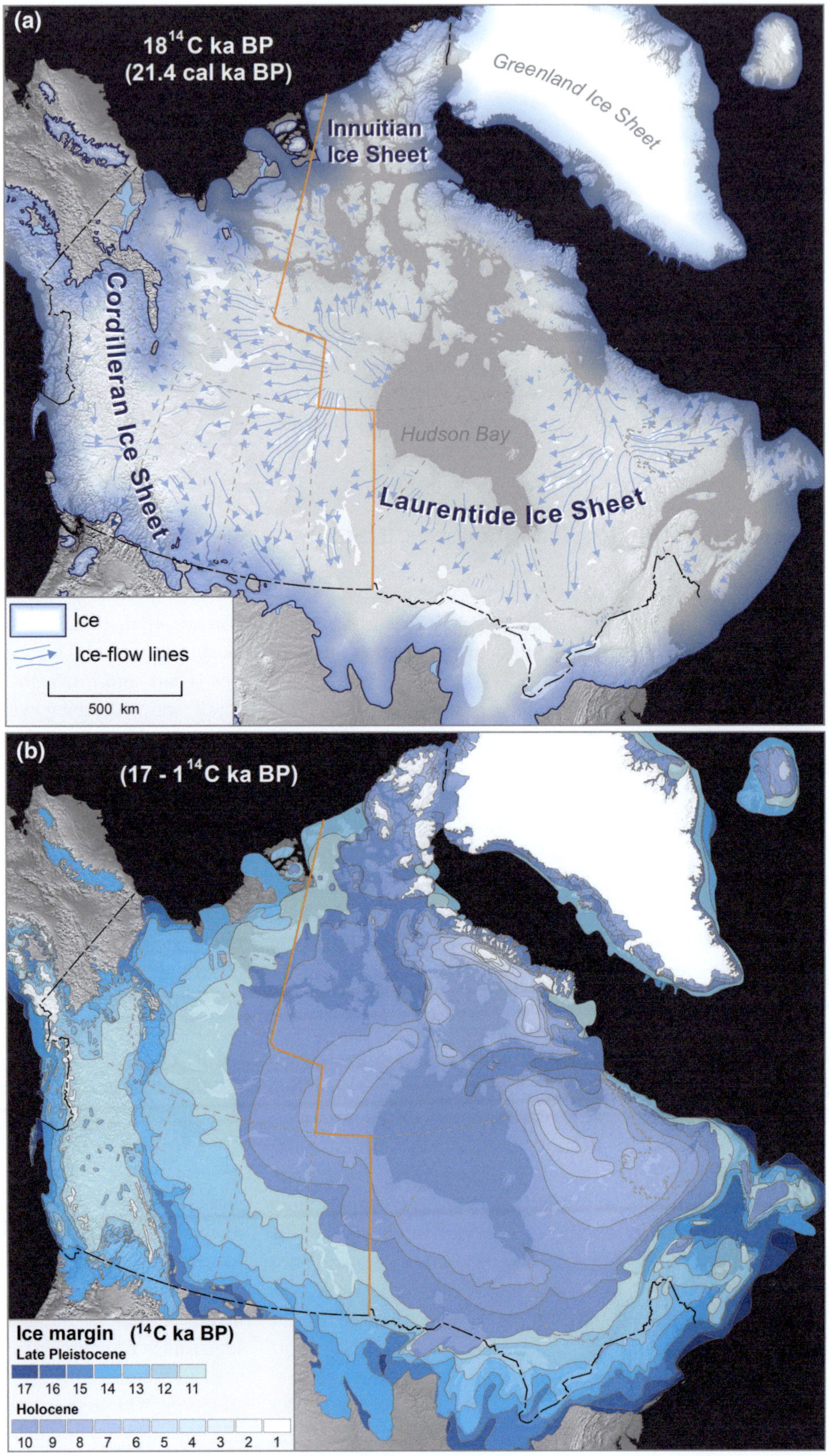

Fig. 2.2 Ice masses in Canada. **a** Distribution of Cordilleran, montane, plateau and continental ice coverage and generalized ice flow direction (modified from Shaw et al. 2000) of MIS 2 (Late Wisconsinan) Glaciation. **b** Extent of ice coverage at various stages of deglaciation. Ice margins from Dyke et al. (2003)

Fig. 2.3 Granitic and gneissic erratics derived from the Canadian Shield indicate Laurentide glaciation. *Photograph* N. Catto

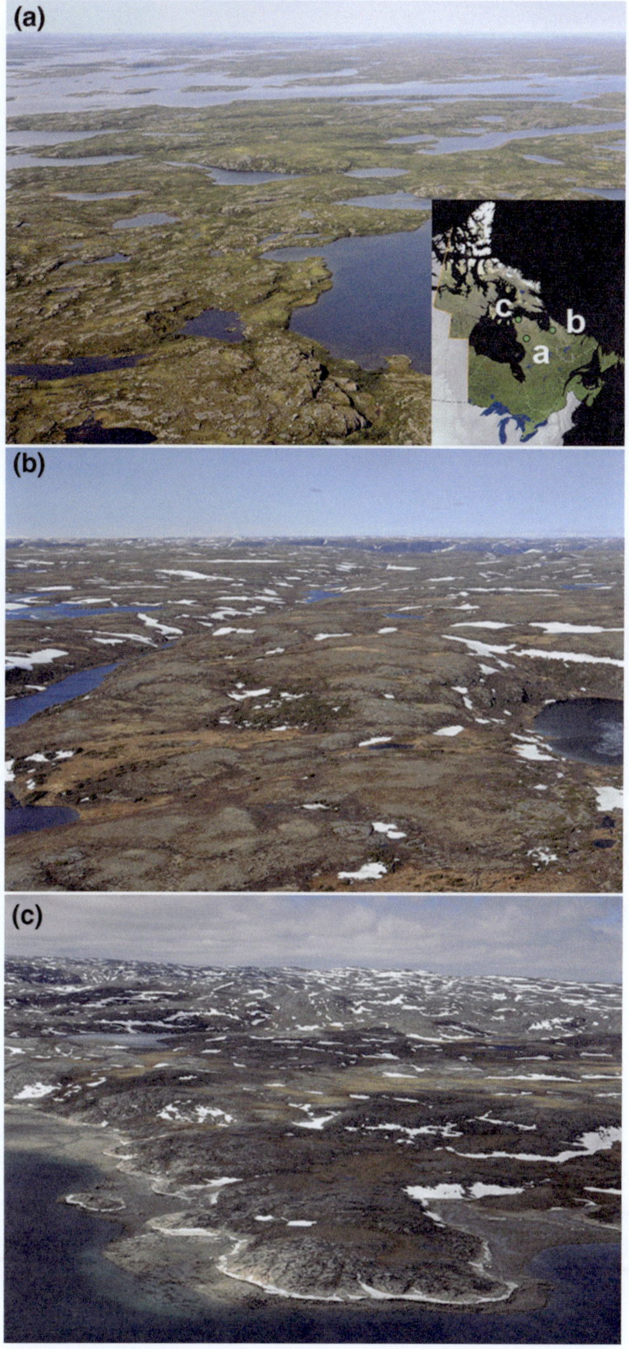

Fig. 2.4 Shield bedrock landscape. **a** Low relief, **b** Low to moderate relief and **c** Moderate to high relief. *Photographs* DER8594, ADD_8922, _ADD0820—Government of Québec, MFFP (2017), CC BY 4.0

may exceed 20 m in height and 20 km in length although most segments are much shorter. They are most commonly single ridges but may be aligned in a braided or reticulate pattern. The eskers of Labrador are attributed to subglacial formation.

Western Labrador remained a source area for glacial ice throughout the Quaternary. It was the last area of continental North America to be deglaciated, approximately 6800 BP (Dyke and Prest 1987; Vincent 1989; Dyke et al. 2002).

2.3.1.2 Kivalliq and Western Nunavut

A succession of glacial accumulation areas developed west and north of Hudson Bay in Kivalliq and western Nunavut. Terms used to describe this accumulation area include 'Keewatin' (the former political name for part of Kivalliq) and 'M'Clintock' (after M'Clintock Channel, east of Victoria Island). Ice expanded to the south and west to cover mainland Northwest Territories, Victoria and Banks islands, and most of the western prairies. To the east and north, zones of coalescence developed with the other Laurentide centres and with the IISC. The exact configuration of the accumulation centre is somewhat controversial, with evidence of shifting divides and positions.

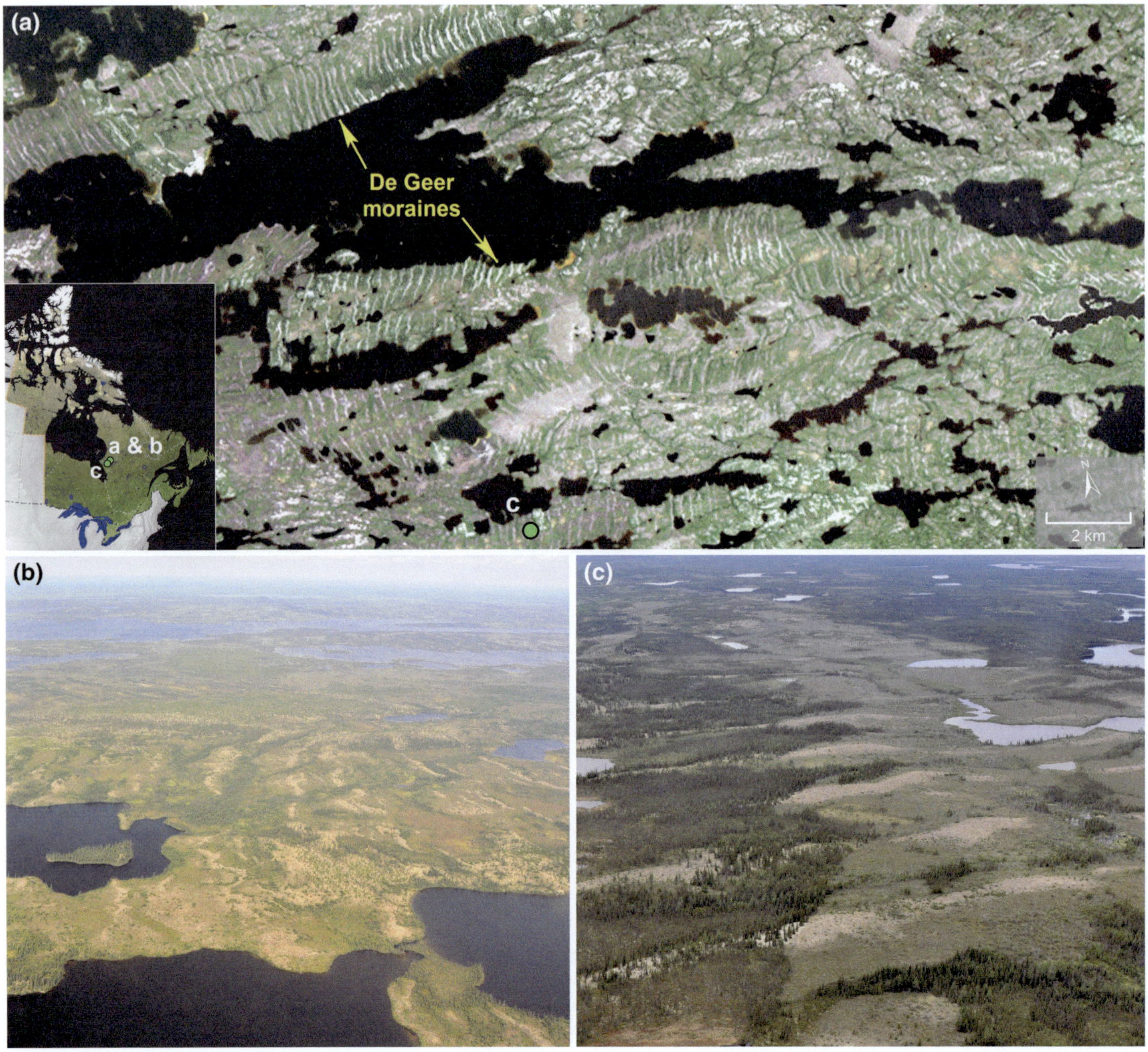

Fig. 2.5 De Geer moraines. **a** Landsat image showing classic De Geer moraines on the lowlands east of Hudson Bay, Québec. **b** and **c** Oblique photographs of fairly regularly spaced ridges of De Geer moraines in the area north of La Grande River, east of Sakami moraine. *Source* a—Government of Canada; Natural Resources Canada (2007). b—DER7018, c—DDDD2408, Government of Québec, MFFP (2017), CC BY 4.0

Flow directions in the marginal areas of this sector are generally well defined (see Stokes, this volume; Figs. 2.8 and 2.9). On Victoria Island, ice from the M'Clintock Dome moved across Victoria Island, diverging into two streams. The northern stream flowed westward through M'Clure Strait, impinging on northern Victoria, northern Banks Island, and southern Melville Island. The southeastern ice stream flowed across southern Victoria Island towards the east-southeast, into the Amundsen Gulf. Drumlins, flutings and accumulations of till mark the areas influenced by the ice streams. Retreat of the ice began approximately 14 ka. Deglaciation of the eastern shore, adjacent to the M'Clintock Dome source area, occurred approximately 9 ka. The isostatic depression produced marine limits of 160–200 m asl along the eastern shores, marked by raised gravel beaches, eroded terraces and fossils of bowhead whales and other marine species (Dyke et al. 1992).

Somerset Island was glaciated by Laurentide ice flowing eastward from the M'Clintock centre, combined with radially flowing local glaciers on the Lancaster Plateau. During MIS 2, the Laurentide ice was prevented from spreading to the northeast by the escarpment separating the Boothia Arch

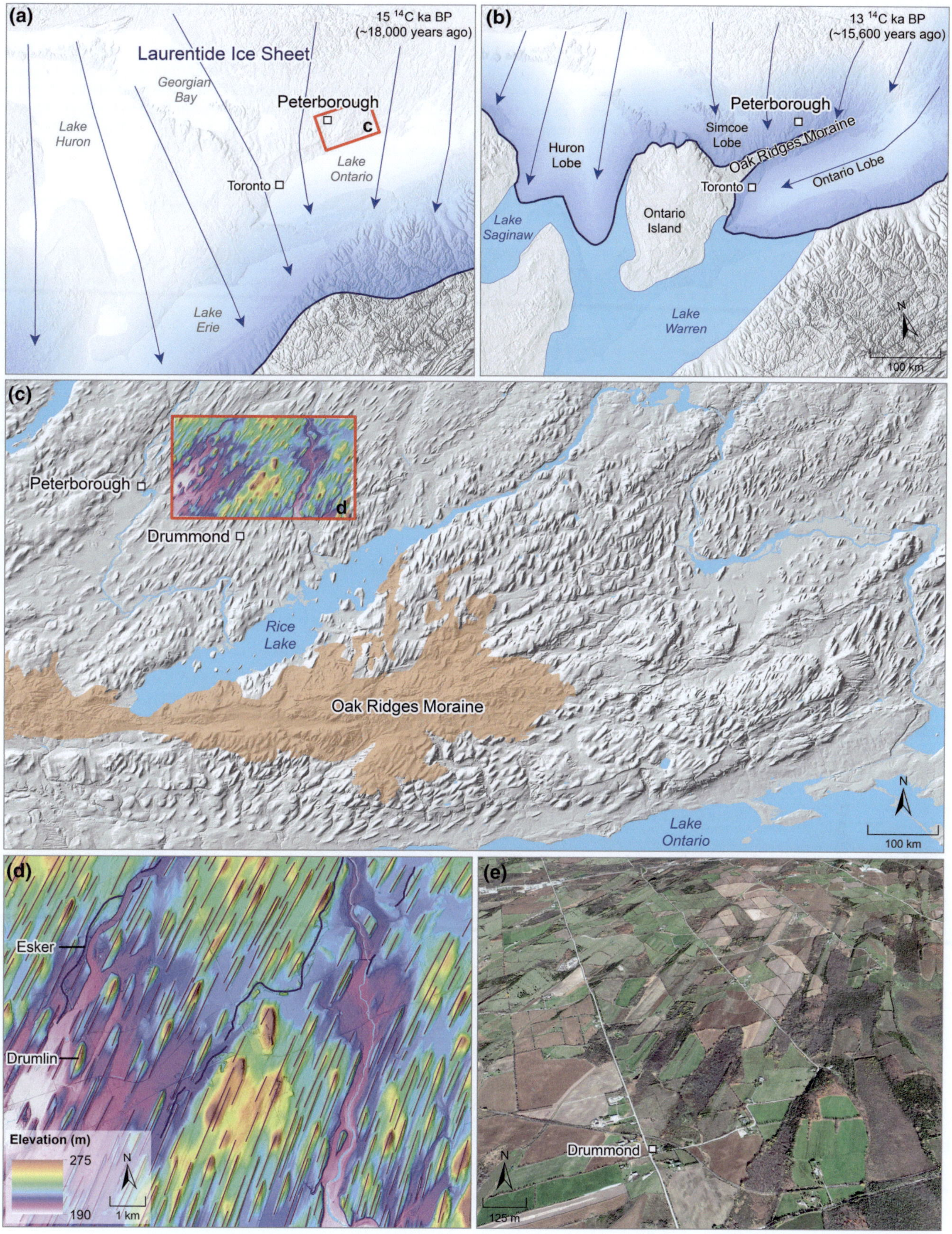
(a)
15 ¹⁴C ka BP
(~18,000 years ago)
Laurentide Ice Sheet
Georgian Bay
Lake Huron
Peterborough
c
Lake Ontario
Toronto
Lake Erie
(b)
13 ¹⁴C ka BP
(~15,600 years ago)
Peterborough
Simcoe Lobe
Huron Lobe
Oak Ridges Moraine
Ontario Lobe
Toronto
Lake Saginaw
Ontario Island
Lake Warren
N
100 km
(c)
Peterborough
d
Drummond
Rice Lake
Oak Ridges Moraine
Lake Ontario
N
100 km
(d)
Esker
Drumlin
Elevation (m)
275
190
N
1 km
(e)
Drummond
N
125 m

Fig. 2.6 The Peterborough Drumlin Field is an example of a Laurentide glaciated landscape in southern Ontario. The field's approximately 3000 drumlins cover the region from Lake Ontario to the north of Peterborough. **a** and **b** Landforms created by glacial advance, retreat, stagnation and final melting of the Simcoe, Ontario and Huron Lobes of the Laurentide Ice Sheet (adapted from Maclachland and Eyles 2013). **c** Drumlin orientations in the northern and eastern parts of the drumlin field record southward flow of the Simcoe lobe. The southern part of the field records northwestward movement of the Ontario lobe from the Lake Ontario basin. *Source* shaded relief image from MRD128-REV; © Queen's Printer for Ontario 2010. **d** and **e** Drumlins can range from approximately 45 m to greater than 5 km in length, a few metres to 500 m wide, and from a few metres to greater than 40 m relief. Also see Sookhan et al. (2018) and Marich (2016) for detailed discussions. *Source* d—Lidar from Ontario Geological Survey (2010). e—Google Earth: Landsat/ Copernicus, DigitalGlobe 2019. Centreed on about 44°16′12″ N, 78°12′36″ W

Fig. 2.7 **a** Landsat image showing ribbed terrain (Rogen moraine) northeast of Lake Mistassini. *Source* Government of Canada; Natural Resources Canada (2007). **b** and **c** Oblique photographs of ribbed moraine ridges. *Photographs* b—DIF_1764, c—DER0231, Government of Québec, MFFP (2017), CC BY 4.0

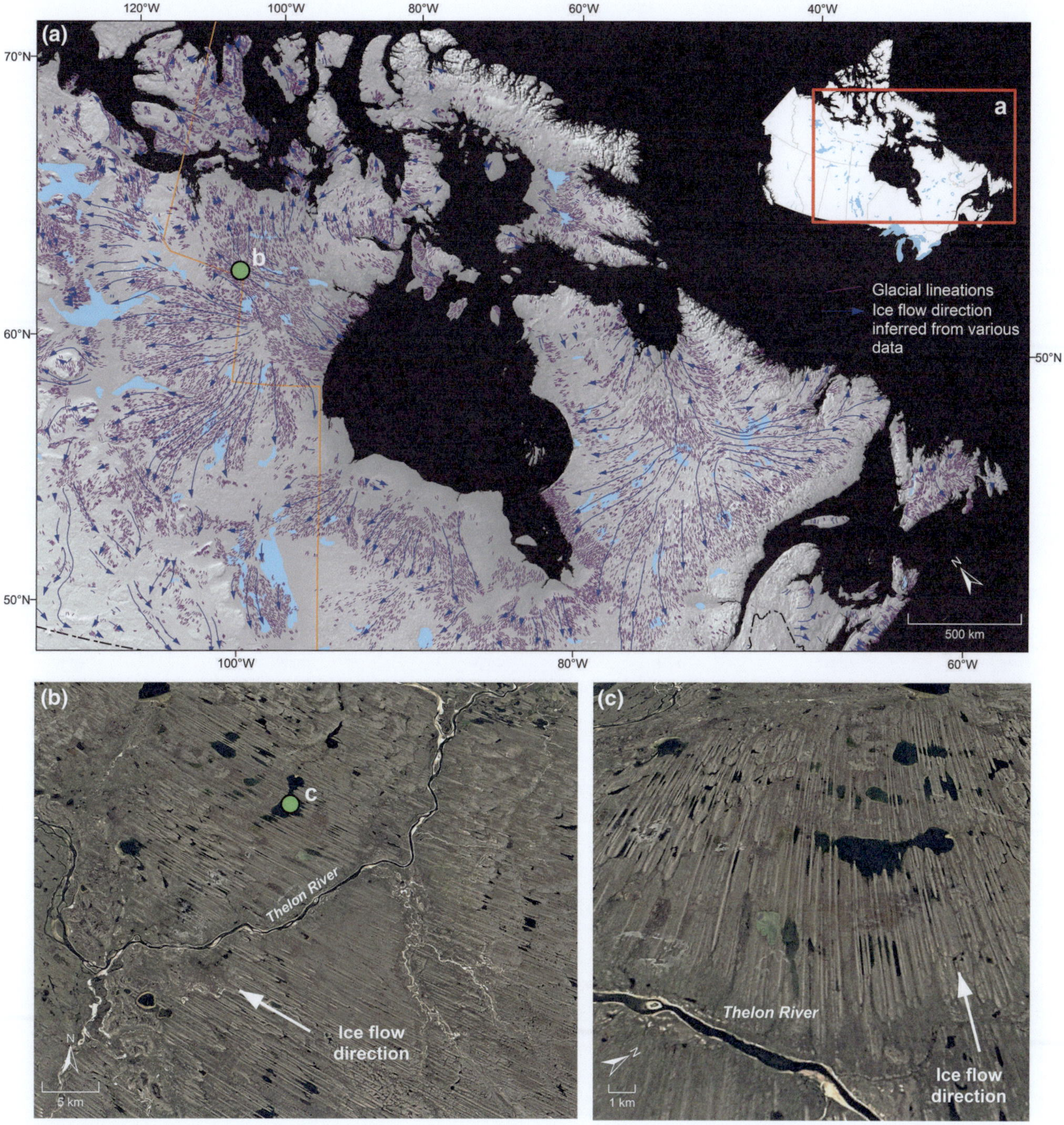

Fig. 2.8 Streamlined bedrock-dominated features are common in glaciated shield regions. **a** Distributions and orientation of streamlined features. *Source* Glacial lineations and ice flow direction from Shaw et al. (2010). **b** Landsat image showing examples of mega-scale (>5 km long) glacial lineations with long axis aligned with ice flow direction; part of the Dubawnt Lake Ice Stream, northwest of Dubawnt Lake in the vicinity of Thelon River, Kivalliq. *Source* Google Earth: Landsat/Copernicus; centred on about 64°15′ N, 102°06′ W. **c** Close-up, oblique view of features looking northwest across the field of mega-scale glacial lineations. *Source* Google Earth: Landsat/Copernicus; centred on about 64°19′ N, 102°09′ W. Also see Stokes et al. (2015)

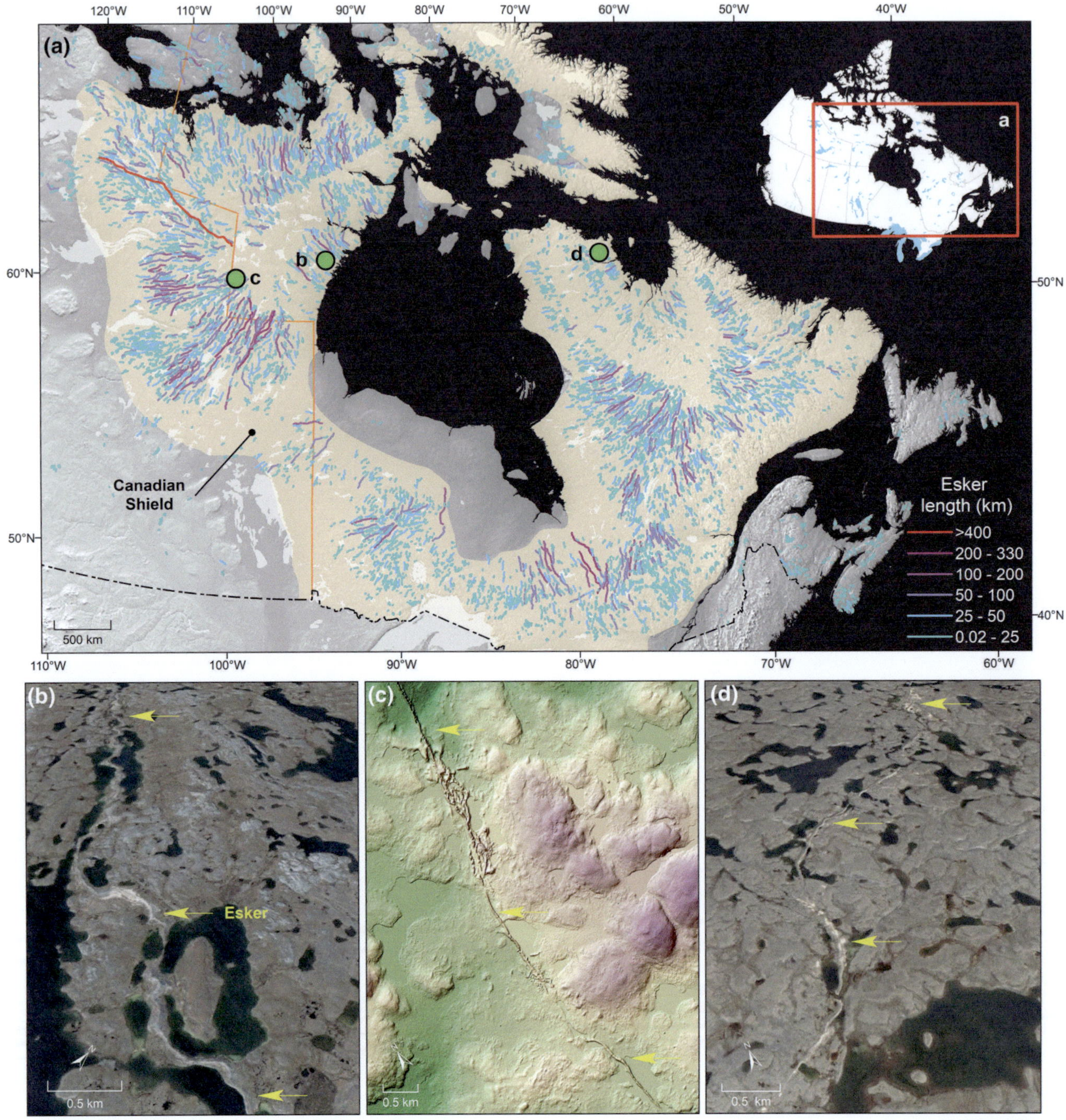

Fig. 2.9 **a** Map showing esker segments associated with deglaciation of the LIS concentrated mainly on the Precambrian Canadian Shield area. The red coloured interpolated esker (i.e. tracing gaps between individual esker segments), the Thelon esker, may be the longest esker in the world (about 750 km; Storrar et al. 2014, 2013). **b** and **c** Typical eskers associated with the Keewatin dome of the LIS. **d** Esker associated with the Labrador dome of the LIS. *Source* a—esker segments modified from Shaw et al. (2010), after Prest et al. (1968). b—Google Earth: Landsat/ Copernicus, DigitalGlobe 2019. c—colour hillshade dsm_5m_polarstereo_23_14_2_2; Government of Canada; Open Government License. d—Google Earth: Landsat/ Copernicus, DigitalGlobe 2019

from the Lancaster Plateau. As a result, a northwest-southeast trending nunatak on the eastern flank of the Boothia Arch, extending from Aston Bay to Creswell Bay, remained unglaciated during MIS 2. An earlier glaciation appears to have covered the entire island (Dyke 1983, 1984).

Marine waters inundated Somerset Island to 120 m asl following deglaciation. Local ice fields persisted in the highest parts of the Lancaster Plateau, and expanded during the Neoglacial c. 1500–1850 A.D. At present, no actively flowing glaciers exist on Somerset Island although remnant bodies of stagnant glacial ice up to 30 m thick are present on the Lancaster Plateau. Today, the landscape is dominated by periglacial features, including tors, cryoturbation hummocks, bedrock felsenmeer, gelifluction stone stripes, and tundra polygons.

2.3.1.3 Foxe Basin

A third Laurentide accumulation centre developed in Foxe Basin (an accumulation area in Foxe Basin and southern Baffin Island (the ‘Foxe’ or ‘Foxe-Baffin’) flowed eastward and northward across Qikiqtaaluk, westward to coalesce with the Kivalliq sector, and southward to meet the Labrador sector ice in Hudson Strait (see Dyke 2004; De Angelis and Kleman 2007; Margreth et al. 2016). Palaeozoic carbonate and shale erratics transported by Foxe Basin ice are conspicuous across the areas of exposed Canadian Shield granites and gneisses. Carbonate detritus from the glacial outflow through Hudson Strait has contributed to the distinctive Heinrich and Dansgaard–Oeschger event layers preserved in North Atlantic sedimentary successions (Andrews and MacLean 2003; Andrews and Voelker 2018). Isolated patches of continental glacial ice, the Barnes and Penny ice caps, persist today on Qikiqtaaluk.

Southampton Island was glaciated by Laurentide ice from the Foxe centre to the northeast during MIS 2 (see Hefter et al. 2017). Flow varied from southeastward (east) to southwestward (centre and southwest coast) to westward (west coast). Coats Island was also glaciated by Foxe Dome ice. Mansel Island was successively glaciated by ice from the Foxe Basin to the northwest and by northwest-moving Labrador ice. All of the islands were affected by the last major phase of Laurentide glaciation, the Cockburn event. A powerful flow southeastward towards Hudson Strait modified the terrain of Southampton Island extensively. As seawater flooded Hudson Strait, the Cockburn event came to an end, and the sea flooded in rapidly. The destabilized ice masses in Foxe Basin and Hudson Bay broke up, and Southampton was deglaciated by 8 ka (Ross et al. 2011, 2012).

2.3.2 Innuitian Ice Sheet Complex (IISC)

In northern Nunavut, many areas outside the alpine-glaciated high mountains (see van Wychen et al., this volume) show relatively little evidence of glaciation. However, the presence of erratic and scattered small glacial landforms indicates that most areas, if not all, were covered by glacial ice at some time during MIS 2 (Lamoureux and Rudy, this volume; Fig. 2.10). Cold-based conditions were extensive in areas such as Ellef Ringnes Island and the Sverdrup Basin. The IISC expanded westward across northern Nunavut, reaching the continental shelf of the Beaufort Sea (see Maclean et al. 2010; Lakeman and England 2012; Nixon et al. 2014). To the south, Innuitian ice coalesced and interacted with Laurentide glaciation (England et al. 2006, 2009).

Bathurst Island, in common with most high arctic areas, also shows little evidence of glaciation apart from scattered erratic boulders across the surface. During MIS 2, the island was covered by locally generated cold-based ice caps, which coalesced and appear to have expanded to blanket the entire island sometime before 20 ka. Glacial deposition influences the landscape along the southeastern and northeasternmost shores. In other localities, bare rocks protrude above the thin soil. Deglaciation occurred c. 9 ka and was followed by a marine incursion into the isostatically depressed coastal areas to approximately 90 m above present sea level. Gravel beaches and erosional marine benches mark the elevated positions.

2.3.3 Atlantic Canada

During MIS 2, local ice caps grew and coalesced throughout Atlantic Canada (Prest 1973; Catto 1998a, b, this volume; Hétu et al., this volume; Morissette et al., this volume; Shaw et al. 2002, 2006; Stea and Mott 2005; Stea et al. 2011; Figs. 2.11 and 2.12). Expansion of the local ice centres blocked intrusion of the Laurentide ice from the north. Laurentide ice coalesced with local ice caps to form ice streams through the Gulf of St. Lawrence, which was glaciated due to marine drawdown to −120 m asl. Offshore, the Grand Banks, Scotian Shelf, and Georges Bank were exposed by lower sea level (Miller 1999; see Normandeau et al., this volume). Periglacial features, environmentally modified clasts, and poorly developed regosols and cryosols indicate subaerial exposure. Locally ice streams extended along the major embayments to the continental shelf. Glacial flow patterns and deglaciation were complex, but after

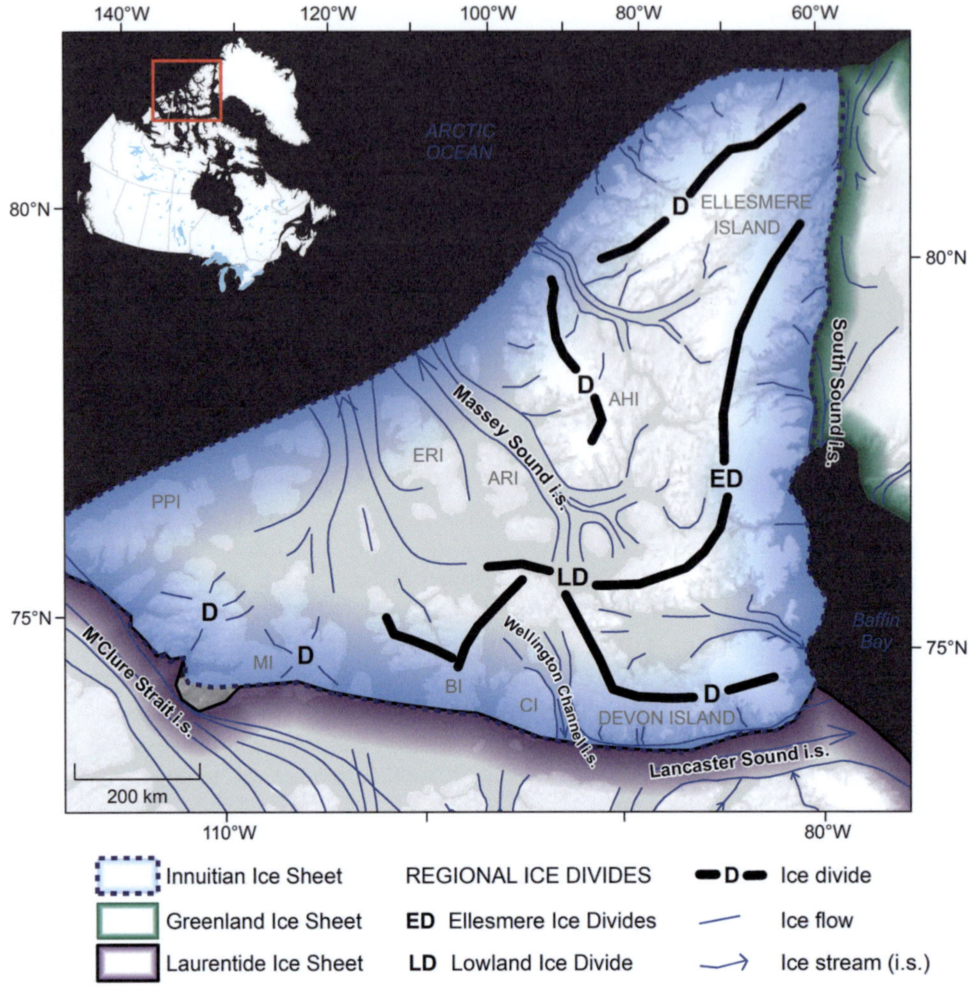

Fig. 2.10 Reconstruction of the Innuitian Ice Sheet over parts of the Canadian High Arctic (Queen Elizabeth Islands) during the LGM, showing ice divides and ice streams (modified from England et al. 2006, 2009; Stokes et al. 2015). AHI—Axel Heiberg Island, ARI—Amund Ringnes Island, BI—Bathurst Island, CI—Cornwallis Island, ERI—Ellef Ringnes Island, MI—Melville Island, PPI—Prince Patrick Island

Younger Dryas readvances throughout the region, rising sea level resulted in rapid destabilization of the ice centres. Deglaciation of the Maritime provinces was complete by 12 ka, and in Newfoundland by 9 ka.

Cape Breton Island displays typical landscapes formed by local glaciation (Grant 1994). Striations and erratics indicate that the island was affected by several glacial events involving ice originating from the west and southwest early in the Quaternary, and all areas were covered by ice at some time. The upland areas of Cape Breton Highlands National Park (above 350 m asl) were largely covered by cold-based ice during MIS 2. Some exposures of weathered bedrock that were formerly considered to represent exposed nunataks have been shown to have developed more recently, and scattered glacial erratics have been found in others. Ice from the Bras d'Or Lakes area was confined to the region south of Mabou Harbour and Wreck Cove, producing a drumlin-dominated coastline along the southeastern coast between Louisbourg and St. Peters.

In southwest mainland Nova Scotia, the current location of Kejimkujik National Park was the source area for ice caps which expanded northward to cover all of the Annapolis Valley, westward to Yarmouth, and southward to the Southern Shore during MIS 2.

The most distinctive glacial features on the Kejimkujik and Southern Shore landscape are its drumlins. From the Halifax Regional Municipality southwest to Bridgewater, almost 2500 drumlins of various shapes are present, including several which have been subject to differing degrees of coastal and/or anthropogenic erosion and modification (Jones 1996). Dimensions of intact drumlins distant from the coastline vary in the range 200–600 m length, 70–400 m width and 8–25 m height. In plan view, the majority are oval or egg shaped although spindle drumlins occur.

Halifax, established as a naval base in 1749, was laid out on the flanks of the steeply sloping Citadel Hill drumlin. Halifax's grid pattern contrasts sharply with the agglomerated patterns characteristic of towns without an initial

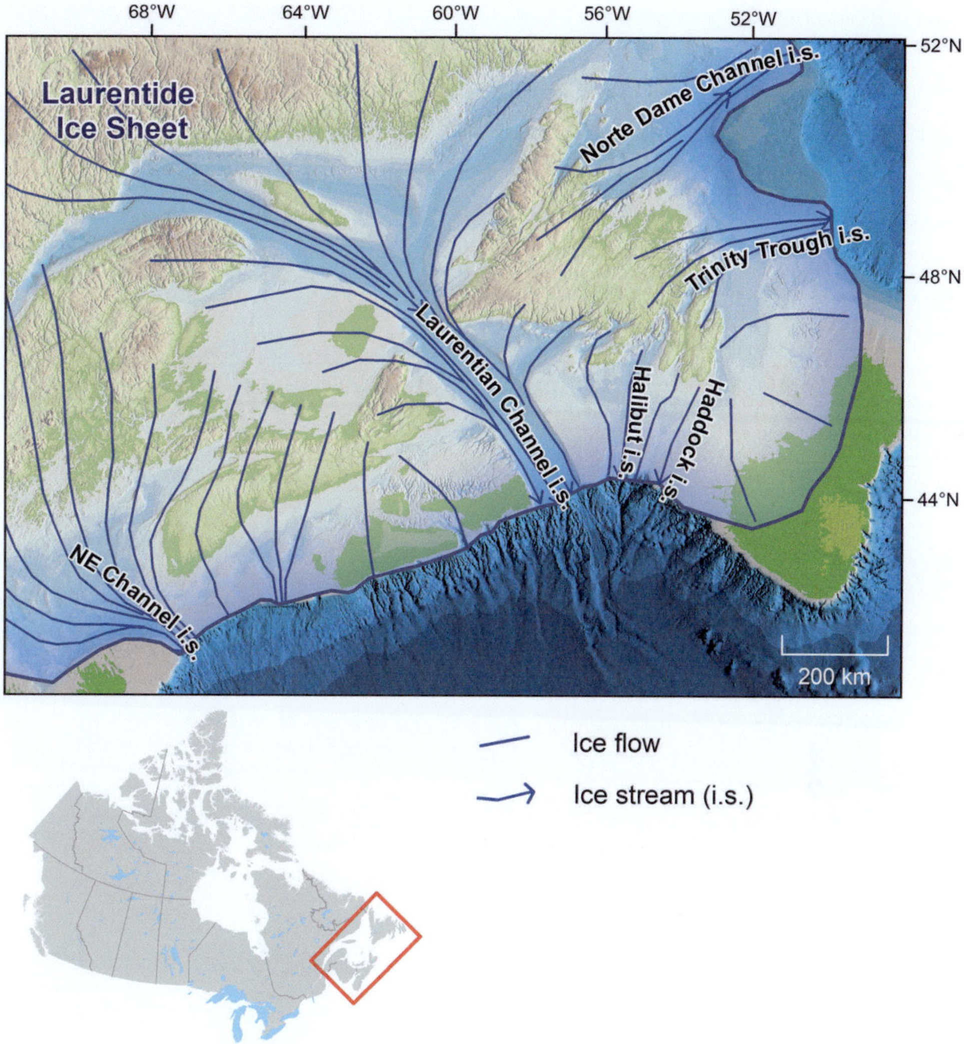

Fig. 2.11 Maximum MIS 2 glacial limits in Appalachian–Acadian region. *Data* background image from Shaw et al. (2002), maximum ice margin from Dyke et al. (2003), and ice flows and streams from Shaw et al. (2006), Stea et al. (2011) and sources cited therein

military presence, such as Boston, also built in a similar topographical setting. Lunenburg, established in 1753 and designated as a World Heritage Site, also displays a grid pattern, with long streets parallel to the drumlin axis and short, extremely steep cross streets.

The 600 drumlins of the Lunenburg field were the subject of research by Jones (1996). The direction of ice flow, as indicated by clast lithology and other adjacent glacial features, was towards the southeast open Atlantic Coast. Jones (1996) noted that 30% of the drumlins had steeper lee slopes than stoss slopes, which would indicate that slope angle is not a reliable guide to flow direction, as has been documented elsewhere (e.g. Spagnolo et al. 2010). Differences between adjacent drumlins in the same field are common, both in relative slope angles and overall shape. In the Lunenburg field and elsewhere in Nova Scotia, many drumlins are essentially symmetrical in form. Many Nova Scotia drumlins show stratified interior successions, with two or more diamicton layers of distinct texture, mineralogy and clast lithology, indicating alternating phases of erosion and deposition and differences in flow direction.

After retreat from the coastlines c. 12 ka, ice persisted for approximately 1,000 more years in Kejimkujik, producing an irregular hummocky topography of glacial diamicton. Remnant glaciers may have persisted through the Younger Dryas, but did not readvance, and final ablation occurred prior to 10 ka.

2.3.4 Alpine Glaciation

Alpine glaciers developed in the Chic-Choc Mountains (QC) and Long Range Mountains (NL; Spooner et al., this volume) of Appalachia–Acadia; in the Mealy (Gray 1969) and Torngat Mountains of Labrador (Way et al. 2014; Catto, Chap. 7, this volume); throughout the High Relief Eastern

Fig. 2.12 Glacial features, Atlantic Canada. **a** Glacially eroded Devonian granite further modified by wave action, Peggy's Cove NS. **b** Glacial till lithology and mineralogy commonly reflects the underlying bedrock. MIS 2 till developed over Permian sandstone, Malpeque Bay PEI. **c** Glacifluvial esker and kame deposits are common. Carmanville, NL. *Photographs* N. Catto

Shield areas of Qikiqtaaluk (Baffin Island; van Wychen et al., this volume; Figs. 2.13 and 2.14) and Tatlurutit (Devon Island); and in Umingmak Nuna (Ellesmere Island; van Wychen et al., this volume). Glaciers disappeared from all areas south of the Torngat Mountains during the early Holocene.

The Chic-Choc Mountains, extending along the spine of the Gaspé Peninsula, include Gaspésie Provincial Park west of Murdochville, surrounding Mont Jacques-Cartier (1268 m asl) and Forillon National Park along the north shore of Baie de Gaspé (Bédard and David 1991). In Forillon, two cirques represent source areas for alpine

Fig. 2.13 Fjord landscapes. **a**. Landsat image of fjords, Baffin Island. *Source* MDA Federal (2004). **b** A researcher standing on the soaring cliffs above the Buchan Gulf in northeastern Baffin Island, Nunavut. *Source:* NRCan 4524 by J. Boles/ Natural Resources Canada, http://open.canada.ca/en/open-government-licence-canada. **c** Inugsuin Fjord, Baffin Island. *Photograph* reproduced with permission of J. D. Ives (also see Ives 2016)

glaciers, and many others are evident throughout Gaspé. Local alpine glaciers covered the Gaspé Peninsula throughout MIS 2, receding from the coastal areas about 13 ka. The last glaciers ablated from the Chic-Choc Mountains prior to 10 ka, after a Younger Dryas readvance. Along the margins of the glaciers, belts of permafrost terrain developed, and traces of the ice wedge casts and frost heave structures are evident today.

2.4 Glacimarine Geomorphology

Changes in sea level throughout the 2.588 million years of the Quaternary, coupled with glaciations, have resulted in the formation of glacimarine deposits and landforms at elevations from more than −120 m to more than 300 m asl in the coastal areas of eastern Canada. Typical coastal

Fig. 2.14 Cirque landscapes. **a** and **b** Cirques, Bruce Mountains, near Dexterity Fiord, on northern Baffin Island, Nunavut. *Source* Government of Canada; Natural Resources Canada (2007). *Photograph* NRCan 4528 by D. Utting/Natural Resources Canada http://open.canada.ca/en/open-government-licence-canada

glacimarine assemblages involve transitions from glacial landforms and sediments inland, to glacimarine terraces, deltas and fan deltas along the constantly shifting coast, to fine-grained laminated or structureless silts and clays, locally with scattered dropstones, offshore. With the exception of isolated exposures of MIS 5 (or older) marine sediments along the shorelines of Qikiqtaaluk and Atlantic Canada (e.g. Langlade sediments, Newfoundland and St-Pierre-et-Miquelon), glacimarine deposits date from the most recent deglaciation.

2.4.1 Atlantic Canada

Following MIS 2 deglaciation, glaci-isostatic depression resulted in a succession of marine incursions in Atlantic Canada (Fig. 2.15). A series of marine incursions occurred along the open Atlantic coast of Newfoundland and Labrador (Liverman 1994), including beaches and deposits at Terra Nova National Park and the Eastport Peninsula (35 m asl), Carmanville (~ 60 m asl), Brown's Arm (~ 65 m asl), Springdale (75 m asl), southern White Bay (75–80 m asl; McCuaig 2003), Roddickton (~ 100 m) and Port Hope Simpson (150 m asl; McCuaig 2002). Along the Gulf of St. Lawrence coast of Newfoundland, the maximum level of marine inundation declines to the southwest, from 150 m asl at Burnt Cape, along the Strait of Belle Isle (Grant 1989, 1992), to 140 m asl at Watts Point, 135 m asl at St. Barbe (Grant 1989), 110–120 m asl between Port au Choix and The Arches (Bell et al. 2005), 100 m asl at Cow Head, Gros Morne National Park (Grant 1989), 75 m asl at Bonne Bay (Grant 1989; Fig. 2.16), 50 m asl at Deer Lake and Corner Brook (Batterson 1998) and 27 m asl in St. Georges Bay (Bell et al. 2001, 2003). At Cape Ray, an eroded rock platform indicates that the maximum postglacial sea level was less than 10 m asl. Marine incursions also occurred from the Bay of Fundy north along the Saint John River Valley (NB; DeGeer Sea) to 60 m asl at Fredericton (Broster and Dickinson 2015); and from the southern Gulf of St. Lawrence across Kouchibouguac National Park (NB) to 45 m asl (Rampton et al. 1984), rising to 60 m asl further north.

Prince Edward Island (PEI; maximum relief 142 m) underwent a complex sequence of sea level changes following MIS 2 glaciation (Prest 1973; also see Hétu et al., this volume). Eastern and central PEI were covered by MIS 2 ice which advanced northwards from Nova Scotia, crossing the then dry Northumberland Strait. Western PEI was influenced by glacial ice which formed to the north of the present coastline, in the area which is now submerged by the Gulf of St. Lawrence. Glacial ice of the Escuminac Ice Centre expanded southeast from the Gulf of St. Lawrence, covering western PEI (Catto 1998b). The ice sheets fluctuated in extent several times, finally departing from the north coast approximately 12.5 ka (see Shaw et al. 2002, 2006).

Following deglaciation, the area adjacent to the Escuminac Ice Centre was isostatically depressed. Sea level rose throughout this region, flooding the terrain below approximately 15–20 m asl. The rise in sea level was sufficient to flood the land between Malpeque and Bedeque Bays, covering Summerside, and also flooded the low terrain between Egmont and Cascumpec Bays. Western PEI consisted of two islands at this time, one extending from North Cape to O'Leary and a second surrounding Mount Pleasant and Wellington, in addition to small areas of land surrounding Kensington and Kinkora. The period of raised sea level was short, with the sea having dropped to its present position approximately 10 ka. Sea levels did not rise above the present values east of New London Bay. Prince Edward Island National Park and the coastline to the east were not submerged by glacimarine waters.

Throughout the early and mid-Holocene, sea level fell steadily along the entire Gulf of St. Lawrence coast of PEI (Shaw et al. 2002). Drowned stumps of deciduous trees mark former forested areas developed at this time (Prest 1973). Sea level began to rise at the beginning of the late Holocene. Currently, sea levels are rising between 3 mm and 4 mm/y at Charlottetown (Shaw et al. 1998; McCulloch et al. 2002; Vasseur and Catto 2008), and transgression has marked the PEI north coast throughout the late Holocene.

Along the eastern shore of New Brunswick, the Gaspé Peninsula, and the St. Lawrence Estuary, glacimarine deposits are attributed to the Goldthwait Sea (Dietrich et al., this volume; Catto, this volume). Elevations range from 60 m asl along the Baie des Chaleurs, 25 m asl at Forillon National Park, more than 100 m asl at Cap-Chat, and 135 m asl at Mingan National Park, to more than 170 m asl along the Strait of Belle Isle. Goldthwait glacimarine silt and clay are subject to periodic slope failure.

2.4.2 Champlain Sea

Continued glacial retreat allowed marine waters to penetrate along the Saguenay and St. Lawrence Valleys. These marine incursions are designed as the Golfe Laflamme and Champlain Sea, respectively (Fig. 2.17). The Golfe Laflamme extended northwestward along the Saguenay River Valley to 50 km north of the present Lac-St-Jean (Dietrich et al., this volume; Nutz et al., this volume).

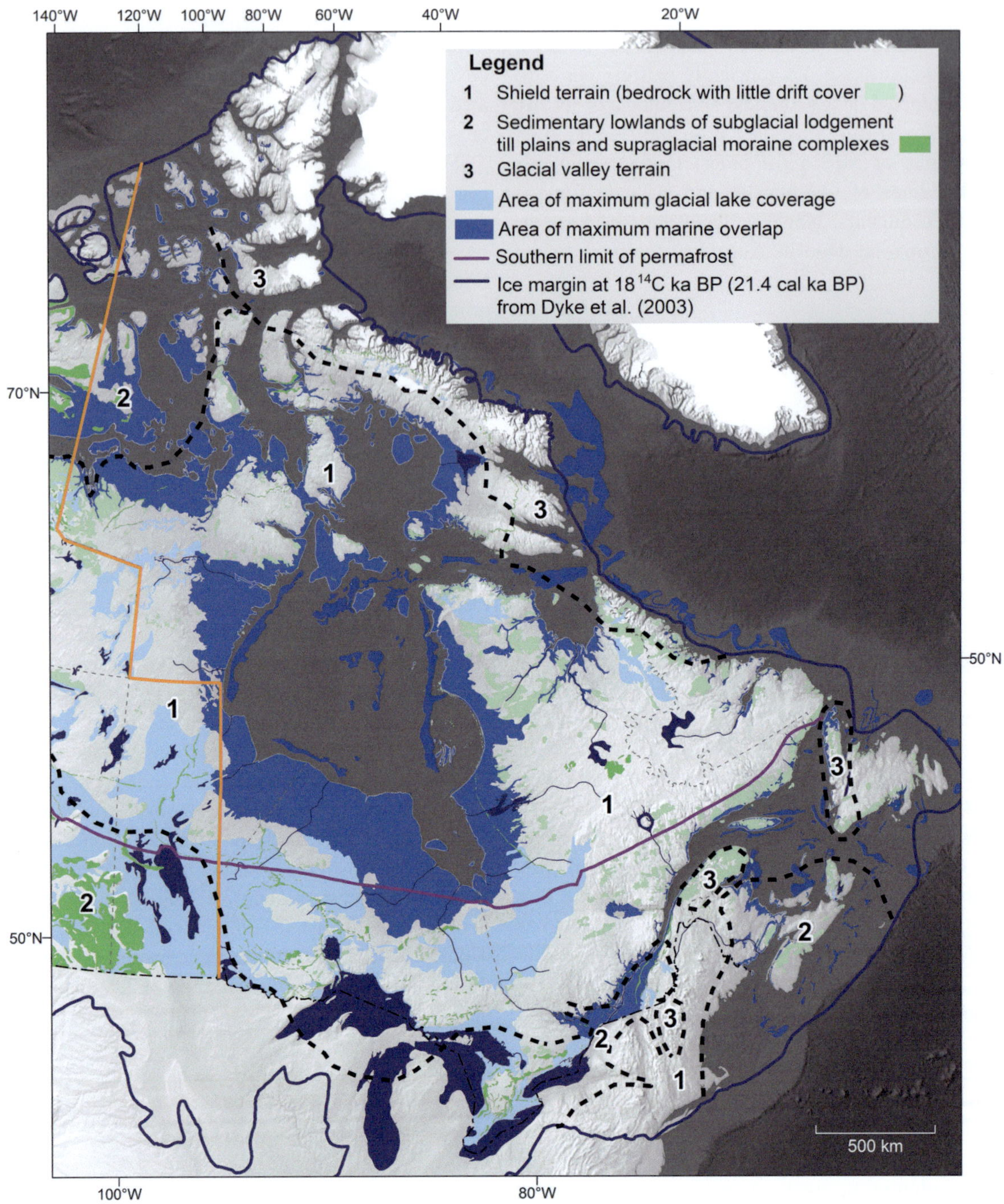

Fig. 2.15 Glacimarine, glacilacustrine and glacial land systems in eastern Canada. Three glacial land systems were defined by Eyles (1983): (1) Shield terrain, dominated by subglacial processes; (2) Sedimentary lowlands, dominated by supraglacial processes and landforms which were draped over the former glacier bed during ice retreat; and (3) glacial valley terrain, which is characteristic of mountain areas (Benn and Evans 2010). *Source* sediment distribution after Prest et al. (1968) and Fulton (1995). Also see Eyles et al. (1983)

Approximately 12 ka, deglaciation at Québec City allowed marine waters to flood the central and western St. Lawrence Lowlands. The resulting Champlain Sea extended along the St. Lawrence Lowlands southwestward to the present day Lake Champlain and Brockville, and northwestward along the Ottawa Valley to Chalk River (Catto et al. 1981; Occhietti 1989; Occhietti et al. 2001). In Gatineau National Park, north of Ottawa, marine deposits and landforms are present to 213 m asl.

A short, sharp readvance of glacial ice c. 11 ka constructed the large frontal St-Narcisse moraine north of the St. Lawrence and Ottawa Valleys from Saint-Siméon near the Saguenay

Fig. 2.16 Glacimarine features, western Newfoundland. **a** and **b** Trout River. Glacimarine terrace and sea stack indicate relative sea level of 70 m asl, c. 12 ka. **c** Distorted glacimarine sediment, Bonne Bay NL, was deposited c. 12 ka. *Photographs* N. Catto

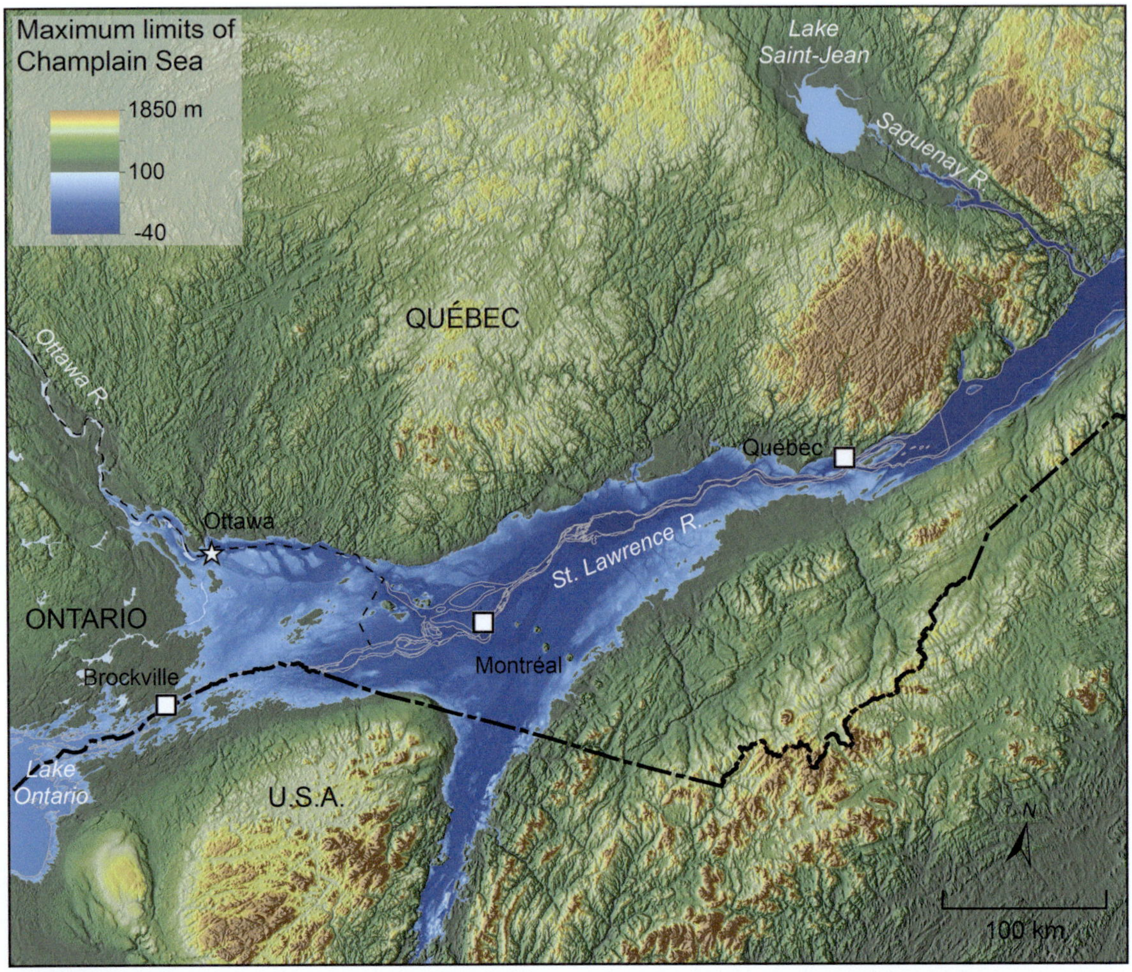

Fig. 2.17 The Champlain Sea was a body of saline to brackish water that occupied the depressed land of the St. Lawrence Lowland between Québec City and Brockville and extended up the Ottawa River Valley from 12 to 10 ka. The modern Ottawa River evolved as the ancestral river and its tributaries adjusted to the retreat of the Champlain Sea

River mouth west to Thurso, east of Ottawa (LaSalle and Shilts 1993; Occhietti 2007). A correlative feature is present near Pembroke, near the western limit of the Champlain Sea. The short duration of this event and its widespread areal extent marks it as a significant episode in the deglacial history of Canada, correlated to the Younger Dryas event.

As the land underwent isostatic recovery, the Champlain Sea was forced to contract from northwest and southwest to northeast. Simultaneously, areas in the southwest (first deglaciated) rebounded, forcing the sea to retreat towards Québec and causing the future tributary streams of the Ottawa River to flow north, down the newly developing slope and away from the St. Lawrence River. By approximately 10 ka, the former Champlain Sea was reduced to a small body of water (Lampsilis Lake, 75–80 m asl) which was isolated from the marine Gulf of St. Lawrence. Continued rebound forced the lake to dissipate completely, and by 9 ka, the ancestral Ottawa and St. Lawrence Rivers flowed through the lowland.

2.4.3 Tyrrell Sea

Glacimarine deposits flank Hudson and James Bays, throughout the Hudson Bay Lowlands and in the Kivalliq region of Nunavut, along the western shore of the permafrost-dominated shield (Dredge and Dyke, this volume). In Wapusk National Park, Polar Bear Provincial Park, and other locations in the Hudson Bay Lowlands, older glacimarine deposits have been identified and assigned to the Bell Sea (MIS 5.5) and the Prest Sea (MIS 5.3).

Following deglaciation and opening of Hudson Strait, the glaci-isostatically depressed Hudson Bay Lowland and adjacent Kivalliq were rapidly flooded by the Tyrrell Sea.

Along southwestern Hudson Bay, the Tyrrell Sea inundated all terrain below approximately 180 m asl. Tyrrell Sea marine clays formed a thick blanket over all of the pre-existing Quaternary deposits and Silurian bedrock, covering everything in the Hudson Bay Lowlands except the elevated (290 m asl) Precambrian Sutton Inlier. South of James Bay and in Kivalliq, maximum Tyrrell Sea levels exceeded 200 m. Withdrawal of the marine waters began shortly after the Tyrrell Sea reached its maximum extent, producing flights of raised beaches. Along the eastern shore line, DeGeer moraines built into the Tyrrell Sea show the progress of the systematic retreat of the Laurentide glacier into Québec.

Isostatic recovery is ongoing, and Hudson Bay can be considered as a remnant of the Tyrrell Sea. As Hudson Bay has a mean depth of 93 m, continued isostatic recovery will cause the bay to shrink significantly through the next 5,000 years, countering the effects of global sea level rise.

2.5 Glacilacustrine Geomorphology

Glacilacustrine landforms dominate much of the landscape of the western Great Lakes Lowlands and the Forested Shield of northern and central Ontario. Smaller glacial lakes formed in Appalachia–Acadia, the Maritime Plain, and the shield areas of Québec and Labrador.

Most glacial lakes in eastern Canada resulted from glacial blockage, interruption or shifting of drainage routes, either by ice impoundment, glaci-isotatic deformation or both in combination. Laurentide ice advancing from western Labrador moved against the topographic gradient as it crossed Québec and Ontario, blocking pre-existing drainage to the Atlantic Ocean. Southward glacial movement from Kivalliq similarly blocked pre-existing drainage routes into the Hudson Bay (Teller 1995; Dredge and Dyke, this volume). Local impoundments resulted as the ice moved upslope into the Appalachian Mountains and other topographic features.

During glacial retreat, ice continued to block drainage into Hudson Bay and the Atlantic Ocean. Glaci-isostatic depression augmented the flow of water towards the retreating ice fronts, causing deep and extensive impoundments in the Great Lakes region and northeastern Ontario. As deglaciation progressed, glaci-isostatic uplift resulted in changes in the configurations of glacilacustrine basins. As areas distant from the ice front were uplifted, the zone of isostatic depression was displaced towards the ice. Glacial retreat exposes new isostatically depressed terrain. A glacial lake thus gradually changes position in synchronization with the retreating ice front. Isolines, contour lines connecting points of equal amounts of isostatic uplift, indicate the gradual progress of glaci-isostatic adjustment. In the lifetime of a large glacial lake, successive outlets may be active, as ice retreat exposes new routes and isostatic recovery closes southern ones. In Manitoba, at various times between 13 and 8 ka, Glacial Lake Agassiz drained southward, into the Mississippi River; southeastward into the Lake Superior basin at Duluth, Minnesota; eastward into Lake of the Woods through Whiteshell Provincial Park; eastward into the Lake Nipigon basin, and thence via the Ottawa River into the Gulf of St. Lawrence; and northwestward via the Clearwater River of Saskatchewan into the Athabasca at Fort McMurray, Alberta, and thence via the Mackenzie Valley to the Beaufort Sea (Teller 1995; Teller et al. 2005, 2018; Leydet et al. 2018). Eventually, continued northward retreat allowed Lake Agassiz to drain directly into Hudson Bay.

2.5.1 Great Lakes

The glacilacustrine history of the Great Lakes is complex, marked by multiple configurations and outlets. Locally, chronological difficulties have led to disputes in interpretation and correlation amongst the basins. A general synopsis, considering each lake basin in turn, is presented as follows: readers seeking more detail are referred to Karrow and Calkin (1985), Karrow (1989) and Desloges et al. (this volume). Figure 2.18 indicates the progressive development of glacilacustrine bodies.

2.5.1.1 Lake Erie Basin

Early in MIS 2, the Lake Erie lobe of the Laurentide Ice Sheet Complex advanced to the southwest c. 27 ka, covering the entire Lake Erie basin and extending southwest into Ohio. The ice retreated in stages, beginning approximately 19 ka. Point Pélée National Park was initially deglaciated c. 16.5 ka, but a subsequent readvance buried the area to the east and built the Pélée moraine system, approximately 16 ka. The ice finally left the Pélée moraine approximately 15.5 ka.

As the ice stood at the Pélée moraine, Lake Erie could not drain eastward via Niagara Falls. Glacial Lake Maumee ponded against the ice front, flooded the Toledo (OH) area, and drained southwestward through several channels into the Ohio River and eventually into the Mississippi. As the ice gradually retreated northwards, progressively lower lakes (Whittlesey, Warren, Wayne, and Arkona) were formed, each one being dammed at its eastern end by the retreating ice. The later stages of this sequence extended north into the southern part of the Lake Huron basin, which had also been deglaciated. These lakes drained to the west via the Grand River of Michigan into Lake Michigan, and thence south through Chicago into the Mississippi system.

When Niagara Falls became ice free, about 13.5 ka, the rapid retreat of the ice in the Lake Ontario basin and in northern Lake Huron opened new eastern outlets for the

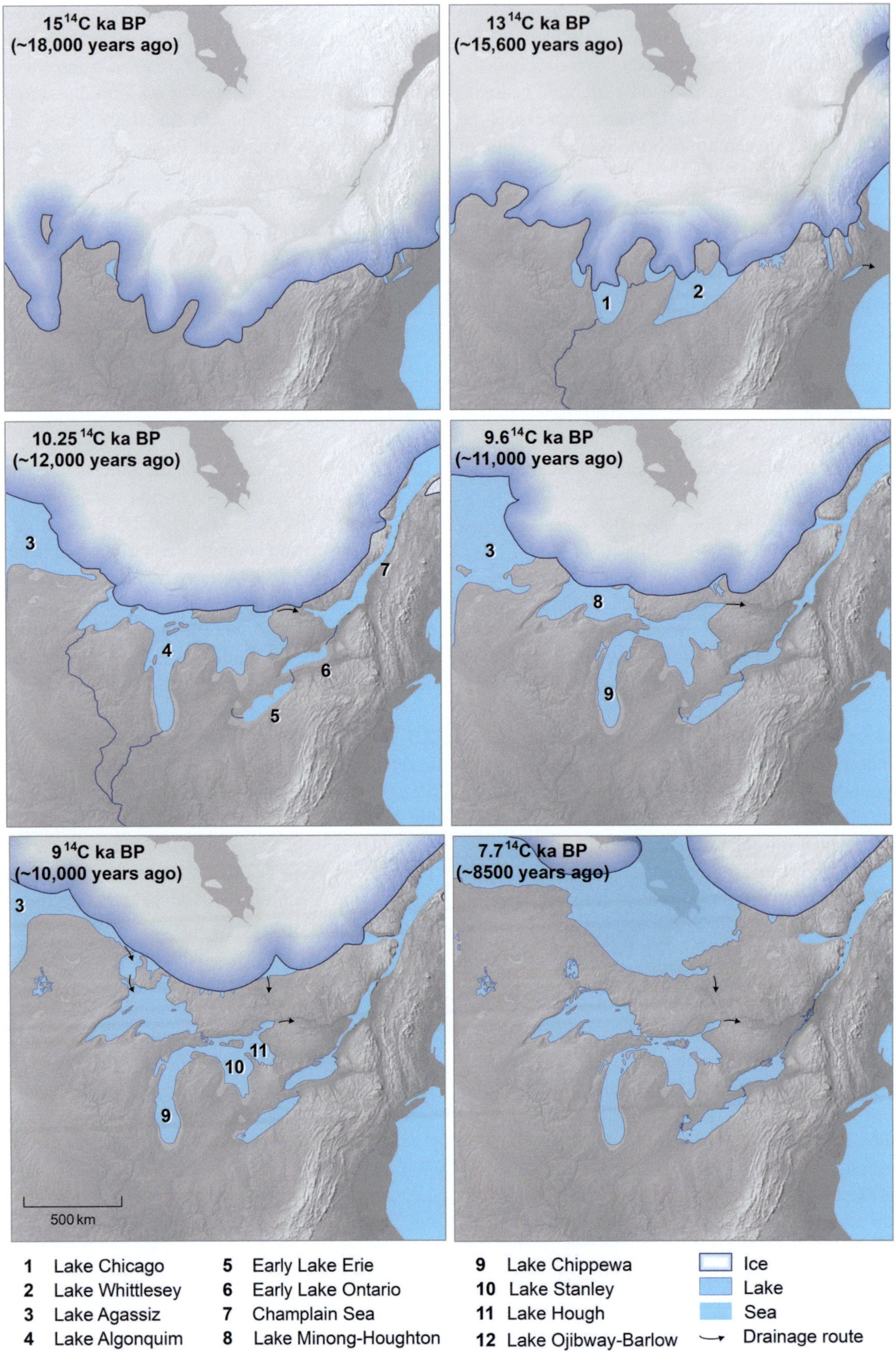

Fig. 2.18 Stages of development of the Great Lakes. Ice margins from Dyke et al. (2003). See text for discussion

upper Great Lakes. Waters drained directly east to Lake Ontario and the Ottawa Valley, thus completely bypassing Lake Erie and causing it to be reduced to several small bodies of water. The Pélée moraine isolated the shallow basin to the west from the main part of the Lake Erie basin to the east. The lake remained at levels much lower than those of the present until approximately 8 ka, when isostatic rebound started to affect the North Bay Outlet and large amounts of water began once again to flow south from Lake Huron via Sarnia into Lake Erie. By approximately 4.5 ka, the modern drainage pattern was completely established, and Lake Erie resembled its present configuration. Point Pélée developed as the lake level rose 11 m over the past 4300 years, causing the sediment to gradually migrate to the northwest. The spit attained its present configuration less than 700 years ago.

2.5.1.2 Lake Huron Basin

The Lake Huron basin was glaciated by Laurentide ice from western Labrador. Glacial ice remained in the lake basin throughout most of the last glaciation (MIS 2).

Retreat began approximately 15 ka with the withdrawal of ice from the southern Lake Huron basin. This part of the basin was occupied by the northern arms of Lakes Whittlesey, Wayne, Warren, and Arkona, which extended north from the Lake Erie basin and drained to the west via the Grand River of Michigan into Lake Michigan. Several glacial advances and retreats occurred throughout the period between 15 and 13 ka, with the ice front standing alternately in the southern part of the Huron basin and as far north of the Straits of Mackinac, joining Lakes Huron and Michigan.

After a readvance, the glaciers again retreated through the Huron Basin, and Glacial Lake Algonquin developed south of the ice front. Lake Algonquin drained alternately to the west (via Lake Michigan) and to the east, via the outlet at Kirkfield, through the Trent River system and into Lake Ontario. The main period of drainage through the Kirkfield outlet into the Lake Ontario basin occurred between 12 and 11.7 ka. As the ice retreated northwards, lower outlets were opened, initially through Algonquin Park and subsequently through North Bay, into the Ottawa Valley. Throughout these events, Lake Algonquin remained present in the Huron basin. High-level shorelines are present along the margins of the Lake Huron basin.

By 11 ka, the North Bay Outlet carried the entire discharge of the Lake Huron and Michigan basins, and the deglaciated part of Lake Superior. Water levels dropped in the Huron and Michigan basins, and the Georgian Bay basin became isolated from the main body of Lake Huron. These low-level lakes, termed 'Lake Hough' (Georgian Bay), 'Lake Stanley' (Huron) and 'Lake Chippewa' (Michigan), remained in the basins as the outlet at North Bay gradually rose. By approximately 8.4 ka, the water level in the Lake Huron basin stood at 176 m asl (at the present elevation of the lake), and Georgian Bay and Lake Huron were joined together again.

Water levels continued to rise as the North Bay Outlet rebounded. By 8 ka, the water level in the Huron, Michigan and Superior Basins stood at the same level, approximately 186 m asl. A single large lake, the 'Nipissing Great Lakes', developed in the basins. Shorelines developed during the Nipissing Great Lakes phase are present at approximately 10 m above the present level of the lake in Georgian Bay Islands and Bruce Peninsula National Park. As the North Bay Outlet continued to rise isostatically, new outlets for the Nipissing Great Lakes developed at Chicago and Sarnia, and all three outlets were functioning simultaneously about 6 ka.

Eventually, the combination of isostatic recovery at North Bay and enhanced downcutting at Sarnia caused the gradual abandonment of the outlets at North Bay (c. 5.5 ka) and Chicago (c. 4.5 ka). The Lake Huron basin drained southward through Sarnia, and became isolated from the Lake Superior basin at Sault Ste. Marie about 3.5 ka.

2.5.1.3 Lake Ontario Basin

Glacilacustrine sediments border much of the Lake Ontario basin. After deglaciation, approximately 13 ka, the ice front remained in the Thousand Islands area, blocking the eastern outlet of Lake Ontario. Waters in the basin were ponded against the ice front to the east, forming Lake Iroquois in the basin, ±140 m above modern Lake Ontario. Lake Iroquois sediment underlies most of the part of Toronto south of Eglinton Avenue. As Lake Iroquois was prevented from draining to the northeast along the St. Lawrence Valley by the retreating glacial ice, it drained to the east through an outlet at Rome (NY). From Rome, the waters followed the Mohawk River valley, and then flowed south along the Hudson River.

As the ice front retreated northeastwards, lower lake levels formed in the Lake Ontario basin as new outlets opened to the northeast. The water surface dropped in stages from the Lake Iroquois level to the Lake Frontenac positions (±90–110 m above modern Lake Ontario level), the Lake Belleville position (±30 m above modern lake level) and finally to levels ±45 m below that of the modern Lake Ontario, approximately 30 m asl. Simultaneously, the waters of Glacial Lake Algonquin, covering the area of the upper Great Lakes, were diverted via newly opened spillway outlets in Algonquin Park and North Bay into the Ottawa River, and consequently the Lake Ontario basin was bypassed. The dry areas of the basin were subjected to frost action, and ice-wedge polygons similar to those currently forming in permafrost regions of Canada developed around the margins of the reduced lake. Although Lake Ontario had retreated to a small area in the deepest part of the basin, it continued to

Fig. 2.19 Niagara Falls. **a** Horseshoe Falls lies along the Canada–USA border. *Photograph* Flickr, CC0 1.0. **b** The resistant Silurian Lockport dolomite caps the Niagara Escarpment, underlain successively by the Silurian Rochester shale and Ordovician Grimsby sandstone. *Photograph* N. Catto

drain northeast into the St. Lawrence Valley. Subsequently, levels began to rise in the Lake Ontario basin. By 10 ka, the configuration of Lake Ontario began to resemble what we see today.

2.5.1.4 Niagara Falls

The most recent glaciation to affect Niagara Falls (Fig. 2.19) advanced from the north and northeast, from the Lake Ontario basin, and covered the area approximately 28 ka. Retreat of the ice began about 17 ka, and by 14 ka the glacial front stood midway between Fort Erie and Niagara-on-the-Lake. A small lake was ponded against the retreating ice, but was unable to drain north because the falls and Lake Ontario basin were still ice covered. When the ice did retreat to the escarpment edge c. 13.7 ka, a deep lake quickly developed between the escarpment and the ice, and the resulting production of icebergs by calving caused the glacier to retreat rapidly through the entire Lake Ontario basin. Lake Iroquois developed at this time, eventually extending northeast to the Bay of Quinte.

Progressive glacial retreat opened new, isostatically depressed outlets for Lake Huron at Kirkfield, Algonquin Park, and North Bay. As a result, Lake Huron drained eastward into the Ottawa Valley, rather than southward into Lake Erie. Niagara Falls was thus deprived of most of its water: the modern discharge at Sarnia is approximately 90% of the discharge at Niagara Falls. The falls shrank to a relative trickle, and the rate of erosion at the lip, controlled by the volume of water passing over the Lockport Dolomite, was extremely small. During the period between 13 and 8 ka, when almost all of the drainage from Huron, Superior and Michigan bypassed Niagara Falls, the falls and gorge remained at essentially the same position.

During the mid-Holocene, the Upper Great Lakes outlet at North Bay began to isostatically rise rapidly. By 6 ka, Niagara Falls was receiving approximately 40% of its modern water discharge, with the remainder draining eastward from Lake Huron via North Bay and southward from Lake Michigan via Chicago into the Mississippi River. The closure of the competing outlets at 5.5 ka (North Bay) and 4.5 ka (Chicago) eventually forced all of Lake Huron's discharge to travel south past Sarnia, into Lake Erie and to Niagara Falls.

2.5.1.5 Lake Superior Basin

As the most northerly of the Great Lakes, Lake Superior was the most recently deglaciated. Glacial ice covered the Canadian part of the lake basin until about 12 ka, when glacial retreat from the Sault Ste. Marie area occurred. A small lake was ponded against the ice in the southwestern part of the basin, in the area north of Duluth (MN). Lake Duluth drained southwestwards into the Mississippi River Basin.

After the Younger Dryas, most of Lake Superior remained covered by ice. Lake Duluth remained impounded by ice to the north of Duluth, and was fed by water from Lake Agassiz (which occupied southern Manitoba and the Lake-of-the-Woods area to the northwest). In the eastern part of the basin, Lake Minong was dammed by ice to the northwest and drained eastwards into Lake Algonquin, which occupied the Lake Huron basin (Zaniewski et al. this volume).

Deglaciation of Pukaskwa National Park did not begin until c. 11 ka, when the ice left the northern shore of Lake Superior. The lake basin was occupied by a later phase of Lake Minong, which received water from Lake Agassiz to the west for a short time and drained into the Lake Huron basin to the east. The Superior basin continued to drain east into the Huron basin and through the North Bay Outlet into the Ottawa River.

During the Holocene, water levels gradually dropped from the high Lake Minong levels, as they simultaneously rose in the Huron and Michigan basins to the south. By 8 ka, as the last glacial ice left northernmost Ontario, the three lakes had reached a common level, creating the 'Nipissing Great Lakes' draining through the North Bay Outlet. As the North Bay Outlet isostatically rebounded, the waters of the Nipissing Great Lakes began to drain simultaneously through Chicago and Sarnia. By 4.5 ka, all the drainage of the upper Great Lakes flowed southwards through Sarnia, and over Niagara Falls.

Lake Superior did not become separated from Lake Huron until 3,500 years ago, when the Sault Ste. Marie area had rebounded sufficiently to isolate the northern lake. Water levels in the Superior Basin continued to fall, as rebound along the northern shore continued. Lake Superior did not reach its present shoreline until c. 500 ce.

2.5.2 Lake Barlow-Ojibway

The Clay Belt of northeastern Ontario reflects the region's glacilacustrine history. By 9 ka, ice had retreated northwards along the Ottawa Valley and Lake Temiscaming, allowing a glacial lake to be impounded in the valley. Ice stood in Hudson and James Bays, and across northern Ontario and Québec, blocking drainage and impounding lakes along the margins of the retreating glacier. By 8.5 ka, the lakes had coalesced into two bodies of water. Glacial Lake Barlow developed to the west, and Glacial Lake Ojibway to the east. As the ice continued to retreat, the lakes amalgamated into one body, Lake Barlow-Ojibway. Through long usage by Quaternarists, the double name has stuck. Lake Barlow-Ojibway extended westward to the headwaters of the Albany River.

Initially, the amalgamated Lake Barlow-Ojibway continued to drain south into the Ottawa River system through Lake Temiscaming, as ice blocked the natural drainage northwards to James Bay. The Cochrane glacial readvance event, involving southward-flowing ice between 8.2 ka and 8.0 ka, firmly impounded the lake. Seasonal discharge of meltwater plumes from the ice front allowed the deposition of yearly rhythmites of alternating layers of silt (spring runoff) and clay (summer settling). Some of the varves of Glacial Lake Barlow-Ojibway were disrupted by earthquake disturbances, which can be traced across the lake basin.

After Hudson Strait was deglaciated, calving rapidly exposed Hudson Bay. Lake Barlow-Ojibway rapidly drained northwards, as the Tyrrell Sea flooded southward. The Tyrrell Sea inundated all terrain below 200 m asl, and Barlow-Ojibway clays locally are overlain by Tyrrell Sea clays.

2.6 Conclusion

The geomorphology of eastern Canada is dominated by the landforms from four major and several lesser ice centres, primarily from Marine Isotope Stage 2. The shield regions of resistant bedrock, and the margins of Newfoundland, are dominated by glacial erosional features. Glacial depositional landforms dominate terrain above glacimarine or glacilacustrine limits in southern Ontario, Québec, and the Maritime provinces. In Arctic Canada, alpine glacial features dominate the mountainous areas of Baffin and Ellesmere islands. Arctic Lowlands show little geomorphic input from glaciation.

Glacial retreat exposed the isostatically depressed coastlines to marine incursions. Champlain Sea glacimarine sediments cover the St. Lawrence and Ottawa Valleys, locally to 213 m asl. The western shore of Hudson Bay and mainland Nunavut were inundated by the Tyrell Sea. Other marine incursions occurred in Atlantic Canada. In most regions, isostatic recovery caused relative sea level to drop to 0 m asl by c. 10 ka.

In the Great Lakes Lowlands and southern Canadian Shield, blockage of northward and eastward drainage led to the formation of successions of proglacial lakes and other paraglacial landforms during the transition from MIS 2 to the Holocene. The complicated glacilacustrine history in the Great Lakes Lowlands, involving shifting ice front positions and multiple outlets, resulted in deposition of thick blankets of clay and silt across the terrain flanking Lakes Ontario, Erie and Huron.

The glacial, glacimarine and glacilacustrine terrains throughout southern Ontario, Québec and Maritime Canada are marked by relatively few geomorphic hazards, with the exception of slope failures. In the sparsely populated northern and arctic regions, Quaternary glacial geomorphology has remained largely undisturbed throughout the later Holocene.

References

Andrews JT (1989) Quaternary geology of the Northeastern Canadian Shield. In: Fulton RJ (ed) Quaternary geology of Canada and Greenland. Geology of Canada Series, no. 1. Geological Survey of Canada, Ottawa, pp 276–301

Andrews JT, MacLean B (2003) Hudson Strait ice streams: a review of stratigraphy, chronology and links with North Atlantic Heinrich events. Boreas 32:4–17

Andrews JT, Voelker AHL (2018) "Heinrich events" (and sediments): a history of terminology and recommendations for future usage. Quat Sci Rev 187:31–40

Batterson MJ (1998) Quaternary history, palaeo-geography, and sedimentology of the Humber River Basin and adjacent areas. PhD thesis, Memorial University of Newfoundland

Bédard P, David PP (1991) La météorisation sur les hauts plateaux de la Gaspésie, Québec: Quelques aspects. Géogr Phys Quat 45:195–211

Bell T, Batterson MJ, Liverman DGE, Shaw J (2003) A new Late-Glacial Sea-level record for St. George's Bay, Newfoundland. Can J Earth Sci 40:1053–1070

Bell T, Smith IR, Renouf MAP (2005) Postglacial sea-level history and coastline change at port au Choix, Great Northern Peninsula, Newfoundland. Nfld Labrador Stud 20:9–31

Benn DI, Evans DJA (2010) Glaciers and glaciation. Arnold, London

Broster BE, Dickinson PE (2015) Late Wisconsinan and Holocene development of the Grand Lake Meadows area and southern reaches of the Saint John River Valley, New Brunswick, Canada. Atl Geol 51:206–220

Catto NR (1998a) The pattern of glaciation on the Avalon Peninsula of Newfoundland. Géogr Phys Quat 52:23–45

Catto NR (1998b) Comparative analysis of striations and diamicton fabrics as ice flow direction indicators, Malpeque Bay region, Prince Edward Island. Boreas 27:259–274

Catto NR, Patterson RJ, Gorman WA (1981) Late Quaternary marine sediments at Chalk River, Ontario. Can J Earth Sci 18:1261–1267

Clark PU (1988) Glacial geology of the Torngat Mountains, Labrador. Can J Earth Sci 25:1184–1198

Clark PU, Fitzhugh WW (1990) Late deglaciation of the Central Labrador coast and its implications for the Age of Glacial Lakes Naskaupi and McLean and for Prehistory. Quat Res 34:296–305

De Angelis H, Kleman J (2007) Palaeo-ice streams in the Foxe/Baffin sector of the Laurentide Ice Sheet. Quat Sci Rev 26:1313–1331

Dredge LA, Cowan WR (1989) Quaternary geology of the south-western Canadian Shield. In: Fulton RJ (ed) Quaternary geology of Canada and Greenland. Geology of Canada Series, no. 1. Geological Survey of Canada, Ottawa, pp 214–235

Dyke AS (1983) Quaternary Geology of Somerset Island, District of Franklin, NWT. Geological Survey of Canada Memoir 404

Dyke AS (1984) Quaternary geology of Boothia Peninsula and northern District of Keewatin, central Canadian Arctic. Geological Survey of Canada, Memoir 407

Dyke AS (2004) An outline of North American deglaciation with emphasis on central and northern Canada. In: Ehlers J, Gibbard PL (eds) Quaternary Glaciations—extent and chronology, Part II: North America. Elsevier, Amsterdam, pp 371–424

Dyke AS, Andrews JT, Clark PU, England JH, Miller GH, Shaw J, Veillette JJ (2002) The Laurentide and Innuitian ice sheets during the Last Glacial Maximum. Quat Sci Rev 21:9–31

Dyke AS, Moore A, Robertson L (2003) Deglaciation of North America, Open File 1574. Geological Survey of Canada

Dyke AS, Morris TF, Green DEC, England J (1992) Quaternary geology of Prince of Wales Island, Arctic Canada. Geological Survey of Canada Memoir 433

Dyke AS, Prest VK (1987) Late Wisconsinan history of the Laurentide ice-sheet. Géogr Phys Quatern 41:237–263

England J, Atkinson N, Bednarski J, Dyke AS, Hodgson DA, O'Cofaigh C (2006) The Innuitian Ice Sheet: configuration, dynamics and chronology. Quat Sci Rev 25:689–703

England JH, Furze MFA, Doupe JP (2009) Revision of the NW Laurentide Ice Sheet: implications for paleoclimate, the northwest extremity of Beringia, and Arctic Ocean sedimentation. Quat Sci Rev 28:1573–1596

Eyles N (1983) Glacial geology: a landsystems approach. In: Eyles N (ed) Glacial geology. Pergamon, Oxford, pp 1–18

Eyles N, Dearman WR, Douglas TD (1983) The distribution of glacial landsystems in Britain and North America. In: Eyles N (ed) Glacial Geology. Pergamon, Oxford, pp 213–228

Fulton RJ (1995) Surficial materials of Canada. Geological Survey of Canada Map 1880A, Scale 1:5 000 000

Gosse JC, Bell T, Gray JT, Klein J, Yang G, Finkel R (2004) Interpreting a landscape record of glaciation with cosmogenic isotopes in weathering zone type localities of Newfoundland, Canada. In: Knight PG (ed) Glaciology and earth's changing environment. Blackwell Publishing Ltd., Oxford

Government of Canada; Natural Resources Canada (2007) Canada centre for mapping and earth observation, CanImage—Landsat 7 Orthoimages of Canada, 1:50 000 Tiles: 033i07, 033i10, 033i11, 033k02, 033k07. ftp.geogratis.gc.ca/pub/nrcan_rncan/image/

Government of Québec, Ministry of forests, wildlife and Parks (MFFP) (2017) Bank of photos of Northern Québec. www.donneesquebec.ca/recherche/fr/dataset/photographie-oblique-du-nord-quebecois

Grant DR (1989) Quaternary geology of the Atlantic Appalachian region of Canada. In: Fulton RJ (ed) Quaternary Geology of Canada and Greenland. Geology of Canada Series, no. 1. Geological Survey of Canada, Ottawa, pp 393–440

Grant DR (1994) Quaternary geology, Cape Breton Island. Nova Scotia, Geol Surv Can Bull, p 482

Gray JT (1969) Glacial history of the eastern Mealy Mountains, Southern Labrador. Arctic 22:106–111

Hefter J, Naafs BDA, Zhang S (2017) Tracing the source of ancient reworked organic matter delivered to the North Atlantic Ocean during Heinrich Events. Geochim Cosmochim Ac 205–211–225

Hughes OL (1964) Surficial geology, Nichicun—Kaniapiscau Map-area. Geol Surv Can Bull 106

Ives JD (2016) Baffin Island: field research and high Arctic adventure, 1961–1967. University of Calgary Press

Jones NK (1996) The internal character and formation of the battery point drumlin, Lunenburg County, Nova Scotia. Can Geogr 40:273–280

Karrow PF (1989) Quaternary geology of the Great Lakes sub-region. In: Fulton RJ (ed) Quaternary geology of Canada and Greenland. Geological Survey of Canada Series, no. 1. Geological Survey of Canada, Ottawa, pp 326–350

Karrow PF, Calkin PE (1985) Quaternary evolution of the Great Lakes. Geological Association of Canada, Special Paper 30

Klassen RA, Paradis S, Bolduc AM, Thomas RD (1992) Glacial landforms and deposits, Labrador (Newfoundland) and eastern Québec. Geological Survey of Canada Map 1814A

Klassen RA, Thompson FJ (1993) Glacial history, drift composition, and mineral exploration, central Labrador. Geol Surv Can Bull 435

Lakeman TR, England JH (2012) Paleoglaciological insights from the age and morphology of the Jesse moraine belt, western Canadian Arctic. Quat Sci Rev 47:82–100

LaSalle P, Shilts WW (1993) Younger Dryas-age readvance of Laurentide ice into the Champlain Sea. Boreas 22:25–37

Leydet DJ, Carlson AE, Teller JT, Breckenridge A, Barth AM, Ullman DJ, Sinclair G, Milne GA, Cuzzone JK, Caffee MW (2018) Opening of Glacial Lake Agassiz's eastern outlets by the start of the Younger Dryas cold period. Geology 46:155–158

Liverman DGE (1994) Relative sea-level history and isostatic rebound in Newfoundland, Canada. Boreas 23:217–230

Maclachland JC, Eyles CH (2013) Quantitative geomorphological analysis of drumlins in the Peterborough Drumlin Field, Ontario, Canada. Geogr Anna A 95:125–144

MacLean B, Blasco S, Bennett R, England J, Rainey W, Hughes-Clarke J, Beaudoin J (2010) Ice keel seabed features in marine channels of the central Canadian Arctic Archipelago: evidence for former ice streams and iceberg scouring. Quat Sci Rev 29:2280–2301

Mahaney WC, Hancock RGV, Milan A, Pulleyblank C, Costa PJM, Milner MW (2014) Reconstruction of Wisconsinan-age ice dynamics and compositions of southern Ontario glacial diamictons, glaciofluvial/lacustrine, and deltaic sediment. Geomorphology 206:421–439

Margreth A, Gosse JC, Dyke AS (2016) Quantification of subaerial and episodic subglacial erosion rates on high latitude upland plateaus: Cumberland Peninsula, Baffin Island, Arctic Canada. Quat Sci Rev 133:108–129

Marich AS (2016) Quaternary geology of the Lindsay and Peterborough areas, southern Ontario. Ont Geol Surv Open File Report 6321

McCuaig SJ (2002) Quaternary Geology of the Alexis River area, and the Blanc-Sablon to Mary's Harbour road corridor, southern Labrador. Current Research, Newfoundland Department of Mines and Energy Geological Survey, Report 02-1, 1–20

McCuaig SJ (2003) Glacial history and quaternary geology of the White Bay region. Geological Survey of Newfoundland and Labrador, Report 03-1, 279–292

McCulloch MM, Forbes DL, Shaw RW, CCAF A041 Scientific Team (2002) Coastal impacts of climate change and sea-level rise on Prince Edward Island. Geological Survey of Canada Open File 4261

MDA Federal (2004) Landsat GeoCover TM+ 2000 Edition Mosaics Tiles N-18-65, N-65-70, N-20-70 ETM-EarthSat-MrSID, 1.0, USGS, Sioux Falls, South Dakota, 2000. Data available at the Global Land Cover Facility, www.landcover.org

Miller AAL (1999) The Quaternary sediments and seismostratigraphy of the Grand Banks of Newfoundland and The Northeast Newfoundland Shelf: foraminiferal refinements and constraints. PhD dissertation, George Washington University

Nixon FC, England JH, Lajeunesse P, Hanson MA (2014) Deciphering patterns of postglacial sea level at the junction of the Laurentide and Innuitian Ice Sheets, western Canadian High Arctic. Quat Sci Rev 91:165–183

Occhietti S (1989) Quaternary Geology of St. Lawrence Valley and Adjacent Appalachian subregion. In: Fulton RJ (ed) Quaternary Geology of Canada and Greenland. Geology of Canada Series, no. 1. Geological Survey of Canada, Ottawa, pp 350–379

Occhietti S (2007) The Saint-Narcisse morainic complex and early Younger Dryas events on the southeastern margin of the Laurentide ice sheet. Géogr Phys Quat 61:89–117

Occhietti S, Govare E, Klassen R, Parent M, Vincent J-S (2004) Late Wisconsinan Early Holocene deglaciation of Québec-Labrador. In: Ehlers J, Gibbard PL (eds) Quaternary glaciations—extent and chronology, Part II: North America. Elsevier, Amsterdam, pp 243–273

Occhietti S, Parent M, Lajeunesse P, Robert F, Govare E (2011) Late Pleistocene-early Holocene decay of the Laurentide ice sheet in Québec-Labrador. In: Ehlers J, Gibbard PL, Hughes PD (eds) Quaternary glaciations—extent and chronology, vol 15. Elsevier, Amsterdam, pp 601–630

Occhietti S, Parent M, Shilts WW, Dionne J-C, Govare E, Harmand D (2001) Late Wisconsinan glacial dynamics, deglaciation, and marine invasion in southern Quebec. Geol Soc Am Special Paper 351:243–270

Ontario Geological Survey (2010) Surficial geology of southern Ontario. Miscellaneous Release—Data 128-REV. ISBN 978-1-4435-2483-4 [DVD] ISBN 978-1-4435-2482-7 [zip file]—Revised

Pico T, Birch L, Weisenberg J, Mitrovica JX (2018) Refining the Laurentide Ice Sheet at Marine Isotope Stage 3: A data-based approach combining glacial isostatic simulations with a dynamic ice model. Quat Sci Rev 195:173–179

Prest VK (1973) Surficial deposits, Prince Edward Island. Geological Survey of Canada Map 1366 A
Rampton N, Gauthier RC, Thibault J, Seaman AA (1984) Quaternary geology of New Brunswick. Geological Survey of Canada Memoir 416
Ross M, Lajeunesse P, Kosar KGA (2011) The subglacial record of northern Hudson Bay: insights into the Hudson Strait Ice Stream catchment. Boreas 40:73–91
Ross M, Utting DJ, Lajeunesse P, Kosar KGA (2012) Early Holocene deglaciation of northern Hudson Bay and Foxe Channel constrained by new radiocarbon ages and marine reservoir correction. Quat Res 78:82–94
Roy M, Hemming SR, Parent M (2009) Sediment sources of northern Quebec and Labrador glacial deposits and the northeastern sector of the Laurentide Ice Sheet during ice-rafting events of the last glacial cycle. Quat Sci Rev 27–28:3236–3245
Shaw J, Gareau P, Courtney RC (2002) Palaeogeography of Atlantic Canada 13–0 ka. Quat Sci Rev 21:1861–1878
Shaw J, Piper DJW, Fader GBJ, King EL, Todd BJ, Bell T, Batterson MJ, Liverman DJE (2006) A conceptual model of the deglaciation of Atlantic Canada. Quat Sci Rev 25:2059–2081
Shaw J, Sharpe DR, Harris J, Lemkow D, Pehleman D (2010) Digital landform patterns for glaciated regions of Canada—a predictive model of flowlines based on topographic and LANDSAT 7 data. Geological Survey of Canada Open File 6190
Shaw J, Faragini DM, Kvill DR, Rains RB (2000) The Athabasca fluting field, Alberta, Canada: implications for the formation of large scale fluting (erosional lineations). Quat Sci Rev 19:959–980
Shaw J, Taylor RB, Forbes DL, Solomon S, Ruz M-H (1998) Sensitivity of the coasts of Canada to Sea-level Rise. Geol Surv Can Bull 505
Sookhan S, Eyles N, Arbelaez-Moreno L (2018) Converging ice streams: a new paradigm for reconstructions of the Laurentide Ice Sheet in southern Ontario and deposition of the Oak Ridges Moraine. Can J Earth Sci 55:373–396
Spagnolo M, Clark CD, Hughes ALC, Dunlop P (2010) The topography of drumlins; assessing their long profile shape. Earth Surf Processes 36:790–804
Stea R, Seaman AA, Pronk, T, Parkhill MA, Allard S, Utting D (2011) The Appalachian glacier complex in Maritime Canada. In: Ehlers J, Gibbard PL, Hughes PD (eds) Quaternary glaciations—extent and chronology, vol 15, pp 631–659. Elsevier, Amsterdam
Stea R, Mott RJ (2005) Younger Dryas glacial advance in the southern Gulf of St. Lawrence: analogue for ice inception? Boreas 34:345–362
Stokes CR (2018) Geomorphology under ice streams: moving from form to process. Earth Surf Process 43:85–123
Stokes CR, Tarasov L, Blomdin R, Cronin TM, Fisher TG, Gyllencreutz R, Hättestrand C, Heyman J, Hindmarsh RCA, Hughes ALC, Jakobsson M, Kirchner N, Livingstone SJ, Margold M, Murton JB, Noormets R, Peltier WR, Peteet DM, Piper DJW, Preusser F, Renssen R, Roberts DH, Roche DM, Saint-Ange F, Stroeven AP, Teller JT (2015) On the reconstruction of palaeo-ice sheets: recent advances and future challenges. Quat Sci Rev 125:15–49
Stokes CR, Tarasov L, Dyke AS (2012) Dynamics of the North American Ice Sheet Complex during its inception and build-up to the Last Glacial Maximum. Quat Sci Rev 50:86–104
Storrar RD, Stokes CR, Evans DJA (2014) Morphometry and pattern of a large sample (>20,000) of Canadian eskers and implications for subglacial drainage beneath ice sheets. Quat Sci Rev 105:1–15
Tarasov L, Dyke AS, Neal RM, Peltier WR (2012) A data-calibrated distribution of deglacial chronologies for the North American ice complex from glaciological modelling. Earth Planet Sci Lett 315–316:30–40
Teller JT (1995) History and drainage of large ice-dammed lakes along the Laurentide Ice Sheet. Quat Int 28:83–92
Teller JT, Boyd M, Yang ZR, Kor PSG, Fard AM (2005) Alternative routing of Lake Agassiz overflow during the Younger Dryas: new dates, paleotopography, and a re-evaluation. Quat Sci Rev 24:1890–1905
Teller JT, McGinn RA, Rajaparac HM, Shuklac AD, Singhvi AK (2018) Optically stimulated luminescence ages from the Lake Agassiz basin in Manitoba. Quat Res 89:478–493
Vasseur L, Catto N (2008) Atlantic Canada. In: Lemmen DS, Warren FJ, Lacroix J, Bush E (eds) From impacts to adaptation: Canada in a changing climate 2007. Government of Canada, Ottawa, pp 119–170
Veillette JJ, Roy M, Paulen RC, Ménard M, St-Jacques G (2017) Uncovering the hidden part of a large ice stream of the Laurentide Ice Sheet, northern Ontario, Canada. Quat Sci Rev 155:136–158
Vincent JS (1989) Quaternary geology of the southeastern Canadian Shield. In: Fulton RJ (ed) Quaternary Geology of Canada and Greenland. Geology of Canada Series, no. 1. Geological Survey of Canada, Ottawa, pp 249–275
Way RG, Bell T, Barrand NE (2014) An inventory and topographic analysis of glaciers in the Torngat Mountains, northern Labrador, Canada. J Glaciol 60:945–956
Winsborrow M, Clark CD, Stokes CR (2004) Ice streams of the Laurentide Ice Sheet. Géogr Phys Quat 58:269–280
Winsborrow M, Clark CD, Stokes CR (2010) What controls the location of ice streams? Earth-Sci Rev 103:45–59

Norm Catto joined Memorial University in July 1989 and is now the Head and Professor of Geography at that university. His research and teaching interests include the following: coastal landforms, natural hazards, sea level change and impacts in coastal environments; response of river systems to climate and weather events, and flood risk assessment; aeolian geomorphology and response to climate and human factors in boreal and coastal environments; mass movements and slope failures; loess deposition; palaeosol formation; emergency measures, policies and effectiveness human adaptations to climate and weather events; the impacts of climate and weather events on agriculture, transportation, fisheries and communities; and the relationships among landscapes and people. His research has included projects and investigations in landscapes and environments in 16 countries, and in all Canadian provinces and territories, particularly in eastern Canada. For 18 years, he served as Editor-in-Chief for *Quaternary International,* a professional international journal dedicated to the study of landscape evolution and climate variation throughout the past 3 million years. He has taught more than 50 different undergraduate and graduate courses focused on numerous aspects of geomorphology, landscape analysis, Quaternary research, natural hazards, ocean sciences, environmental science, climate and climate change, soil science, geoarchaeology and palaeontology. Study of landscapes forms his central focus for the understanding of the various areas of eastern Canada, the combination of physical and cultural constituents that contribute to its identities. For most species, and for all human occupants, life in eastern Canada has been heavily influenced by geomorphology since deglaciation.

Dori J. Kovanen is a professional engineering geologist and Research Associate in the Department of Geography at the University of British Columbia, Vancouver, BC. With an M.Sc. from Western Washington University in Bellingham and a Ph.D. in Quaternary geology from the University of British Columbia, she is a specialist in Quaternary geomorphology. She has written extensively on late glacial and Holocene history of the Pacific Northwest as well as contemporary environmental changes in the region and was a significant contributor to 'Landscapes and Landforms of Western Canada' (Slaymaker 2017).

Olav Slaymaker is a Member of the Order of Canada, and professor emeritus of geography in the Department of Geography at the University of British Columbia, Vancouver, B.C. His 20+ co-authored and edited monographs and 160+ refereed papers reflect three different styles of work. First, he has pursued original field research for 40 years on sediment systems in the Canadian Cordillera and the European Alps, strongly supported and enhanced by his stellar doctoral and postdoctoral students and international colleagues. Second, he has produced papers and books summarizing trends in land use and environmental change in mountain regions worldwide. Third, he has edited monographs with invited international contributions on major environmental and geomorphological themes. He is Former President of the Canadian Association of Geographers and Former President of the International Association of Geomorphologists.

3 Holocene Geomorphic Processes in Eastern Canada

Olav Slaymaker, Norm Catto, and Dori J. Kovanen

Abstract

The paraglacial, periglacial and non-glacial geomorphic processes that characterize eastern Canada since the retreat of the Laurentide and Innuitian Ice Sheets and the resultant landscapes and landforms are reviewed. Ice sheet recession, glaci-isostatic rebound, relative sea level change and glacimarine and glacilacustrine sedimentation are the most distinctive paraglacial processes in the early Holocene Epoch. Hydroclimate, vegetation and wetland evolution, permafrost processes, fluvial erosion and sedimentation, sediment transport, wind action, seismicity, and mass movements are the most distinctive contemporary processes that all show regional variations in type and distribution. Hazards associated with these contemporary processes and the sensitivity of eastern Canada's coasts to sea level change are also identified.

Keywords

Paraglacial • Periglacial • Relative sea level • Permafrost • Fluvial erosion • Sedimentation • Mass movement • Hazards • Sensitivity

3.1 Introduction

The formal starting point for the Holocene Epoch is 10,000 radiocarbon years BP or approximately 11,450 calibrated calendar years BP, but it is well-understood that the actual transition date varies from place to place depending on the presence or absence of ice. No region of the world illustrates more clearly the provisional nature of that starting point than eastern Canada where the earliest dates of deglaciation of the Laurentide Ice Sheet (LIS) range from c. 20–10 ka (Dyke 2004) and the Innuitian Ice Sheet (IIS) began to recede after 11 ka (England et al. 2006). The complexities of two huge ice sheet recessions, and the relative sea level changes, massive lake and river fluctuations, and associated large-scale sediment flux that were conditioned by those glacial recessions influenced every part of eastern Canada. Eastern Canada today still includes the third-largest extent of glacier ice in the world (after Antarctica and Greenland) and all its currently ice-free landscapes have been marked by continental glaciation in the geologically recent past. All of eastern Canada continues to respond to the unloading of the MIS 2 ice sheets and in this sense is a paraglacial environment in its entirety.

The overall postglacial climate sequence has been determined from a variety of records, including ice cores, lake sediments and the thickness of saprolites (Bouchard and Pavich 1989; Gajewski and Atkinson 2003). It can be summarized as follows: (a) a long-term cooling over the past 3–4 ka, with maximum temperatures shortly after deglaciation; (b) millennial-scale climate variability superimposed on these long-term trends; and (c) warming since the 1850s. In the Canadian Arctic, there was a period of very warm temperatures immediately or soon after deglaciation (Kaufman et al. 2004). It seems that during this time, temperatures were warmer than in the first part of the twentieth century although it is not clear if the current rapid warming of the past decades has exceeded the early Holocene temperatures. Maximum temperatures did not occur at the same time in all areas of the Arctic and Subarctic, and indeed in northern Québec occurred after 6 ka. These warm conditions in the early Holocene meant that several plant taxa were growing north of their present-day limit and peat deposits were emplaced in many areas, even as far north as the Fosheim Peninsula on Ellesmere Island. The position of the treeline seems to be associated with the mean position of the Arctic

O. Slaymaker (✉) · D. J. Kovanen
Department of Geography, University of British Columbia, 1984 West Mall, Vancouver, BC V6T 1Z2, Canada
e-mail: olav.slaymaker@ubc.ca

N. Catto
Department of Geography, Memorial University of Newfoundland and Labrador, St. John's, NL A1B 3X9, Canada

© Springer Nature Switzerland AG 2020
O. Slaymaker and N. Catto (eds.), *Landscapes and Landforms of Eastern Canada*, World Geomorphological Landscapes, https://doi.org/10.1007/978-3-030-35137-3_3

front in summer and changes in treeline position are a reflection of climate variations (Ladd and Gajewski 2010). In northwestern Québec the treeline is a broad band of several degrees of latitude, the so-called forest-tundra (Payette et al. 1989). There is no evidence of the treeline being north of the present position in the past 6 ka since deglaciation. Rather, spruce quickly migrated to its present position, perhaps with trees being denser in the region of the present-day forest-tundra before 4 ka. In the past several 1000 years, there has been a progressive deforestation of the forest-tundra (Ritchie 1993). This occurs gradually, as the trees still continue to grow although frequently in krummholz form, a condition too severe to permit reproduction. These shrub-like trees persist until fire kills them and the landscape slowly reverts to tundra (Gajewski et al. 1993).

3.2 Implications of Ice Sheet Recession

The process of disintegration of the Laurentide and Innuitian Ice Sheets (LIS and IIS) is described in Chap. 2 and provides the specific context for paraglaciation in eastern Canada (Briner et al. 2004; Dyke and Hooper 2001). The final disintegration of the LIS led to the incursion of the Tyrrell Sea over the present location of Hudson Bay (Barber et al. 1999) and around 11.5 ka marine waters also flooded the St. Lawrence Valley. The resulting Champlain Sea reached elevations of more than 200 m above the current sea level near Montreal and Ottawa (Fig. 3.1). The sea also reached southwest to Brockville and northwest to Chalk River in Ontario at its maximum extent. By approximately 9 ka, the St. Lawrence River and the ancestral Ottawa River were able to flow northeastward towards the gulf (for further details see Chap. 2).

3.2.1 Paraglacial Landscapes

Landscapes emerging from beneath ice sheet cover have been called paraglacial and have been defined as landscapes characterized by 'non-glacial processes of sediment redistribution conditioned by glaciation' (Fig. 3.2; Church and Ryder 1972). The most distinctive feature of this transition period from Pleistocene to Holocene Epoch is the presence of large masses of glacial melt water and huge volumes of glacigenic sediments in the newly subaerially exposed landscape. These glacigenic sediments, that were eroded, transported and deposited by ice at the base or at the margin of the Laurentian and Innuitian Ice Sheets, are subsequently referred to as 'primary'. These primary glacigenic sediments were subsequently redistributed by non-glacial processes that are called 'secondary' paraglacial processes. Secondary paraglacial processes include fluvial, lacustrine, mass wasting, marine and aeolian erosion, transportation and deposition, and, more controversially, relative sea level change, the net vertical difference between eustatic and isostatic changes (Kovanen and Slaymaker 2015). These secondary paraglacial processes commenced changing the primary glacigenic sediments and the adjacent land surfaces following the retreat of the ice sheets. The exceptional volume of sediment available and the whole range of subaerial processes that became active after the ice sheet recessions have been used to justify the use of the term 'paraglaciation' (Ballantyne 2002). A detailed example of paraglaciation in eastern Canada is available from southern Ontario (Campo and Desloges 1994).

3.2.2 Glaci-Isostasy and Eustasy

Glaci-isostatic rebound is a non-glacial geodynamic process conditioned by glaciation in formerly ice covered landscapes (Walcott 1970). This rebound infused new potential energy into the landscape. Glaci-isostatic rebound has resulted in the formation of marine beaches and terraces at elevations which are now substantially above present mean sea level. Along the coast of Hudson Bay, raised beaches are found at elevations in excess of 250 m above present sea level and indeed so rapid is the current rebound that the local Cree and Inuit inhabitants have incorporated an awareness of the rising land into their folklore (Bell 1896); raised marine features occur at elevations of more than 150 m asl in the St. Lawrence and Ottawa Valleys and more than 100 m asl in northern Baffin Island.

Eustatic or ocean volume change is also a non-glacial process conditioned by glaciation in landscapes that were formerly ice covered. The largest factor responsible for relative sea level change is the interaction between the changing volume of the oceans and glaci-isostatic activity (Church and White 2011). Because absolute sea level changes in the past are difficult to ascertain, relative sea level change is used as a more readily observable effect. The relative sea level change is the net effect of land rebound and sea level change. Future changes in relative sea level along Canadian coasts are determined by combinations of ongoing glaci-isostatic adjustment potentially augmented by increases in volume of the ocean due to glacial melting (IPCC 2013). Tide gauge data show that crustal movements in response to deglaciation are continuing across Canada (Shaw et al. 1998). Relative sea level rise (net land submergence) is occurring along the coast of the Atlantic Provinces and eastern Baffin Island, Nunavut. This trend contrasts with the well known relative sea level lowering (net land emergence) that continues in the area of Hudson Bay as glaci-isostatic readjustment around the core of the former Laurentide Ice Sheet. Some of the most remarkable raised shorelines can be

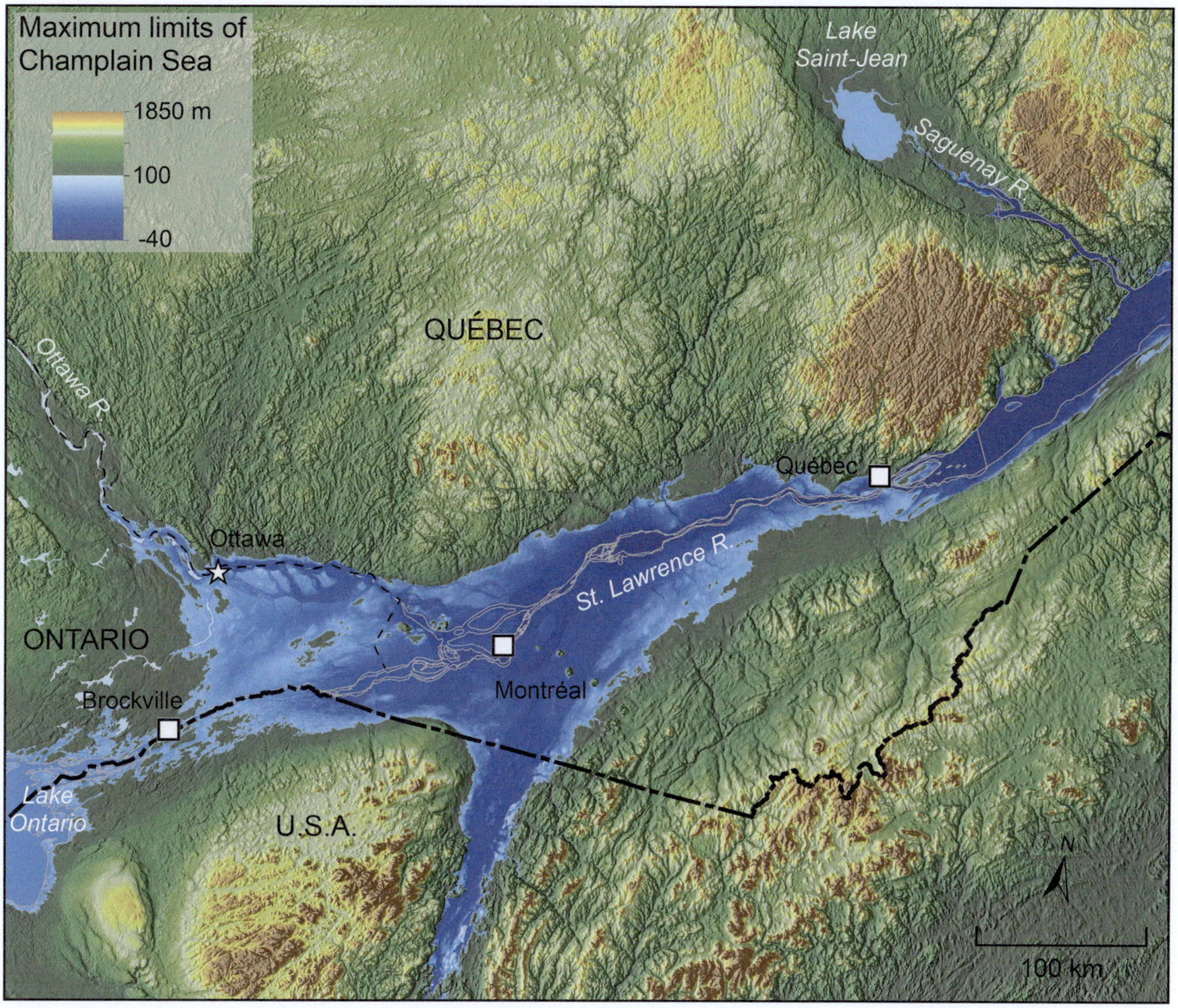

Fig. 3.1 The Champlain Sea was a body of saline to brackish water that occupied the depressed land of the St. Lawrence Lowland between Québec City and Brockville and extended up the Ottawa River valley and the Saguenay Fjord from 13 to 10 ka. The modern Ottawa River evolved as the ancestral river and its tributaries adjusted to the retreat of the Champlain Sea (Richard and Occhietti 2005; Carrivick and Tweed 2013)

found along the shores of Hudson Bay, in Nunavut, and on the Québec North Shore (Fig. 3.3a–d).

Atlantic Canada has a complex, regionally and temporally varying relative sea level history (Shaw et al. 2006). 'Given that postglacial emergence began at times ranging from 14 ka in the Bay of Fundy to 9 ka in the Gulf of St. Lawrence, and reflects gross differences in former ice thickness, a series of local relative sea level histories expresses the regional variation' (Grant 1989, p. 430). Québec and Nunavut show net emergence with a suite of raised shorelines and the Atlantic shelf shows net submergence with all postglacial shores submerged. Shorelines are therefore found at +150 m asl on the St. Lawrence North Shore and at −115 m asl on the outer Atlantic shelf. Marine limit ranges from 150 m asl at the Strait of Belle Isle to 0 m asl through Nova Scotia, Prince Edward Island (PEI) and southern Newfoundland. Each region, from the north shore of St. Lawrence and northern Newfoundland through Gaspésie and Anticosti Island to the Bay of Fundy and Sable Island on the outer Atlantic shelf, has its own local relative sea level history. About 14 ka, glacial ice remained in central Newfoundland and the Notre Dame Mountains in Québec. The Grand Banks, Sable Island Bank, and Georges Bank were prominent islands offshore (Fig. 3.4a). By 10 ka, the offshore islands had contracted and much of the areas around Prince Edward Island, the Magdalen Islands and Cape Breton were exposed because of the variable isostatic effects (Fig. 3.4b). By 7 ka, rising relative sea level had inundated the continental shelf and the

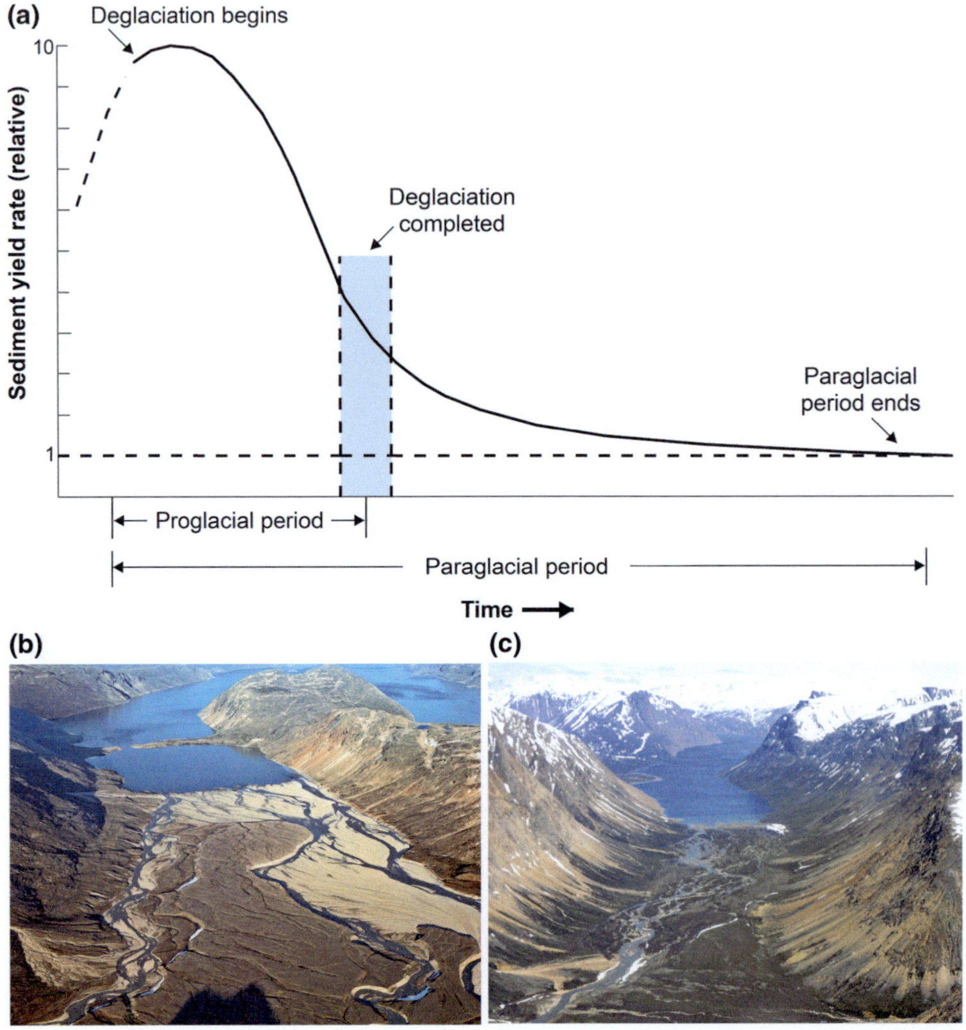

Fig. 3.2 **a** Original paraglacial model according to Church and Ryder (1972) showing the overlapping proglacial and paraglacial periods. **b** Ekalugad sandur, a multilevel outwash plain tributary to Ekalugad Fjord, Baffin Island, that provided crucial data for the paraglacial model (Church 1972) (Photo: July 1965, reproduced with permission of J. D. Ives; also see Ives 2016). **c** Palmer River at the head of the Tallek Arm of the Nachvak Fjord, Torngat Mountains National Park, Labrador (58°52′23″ N, 63°51′41″ W). *Photograph* ADD_9755—of Québec, MFFP (2017), CC BY 4.0

Northumberland Strait, separating Prince Edward Island from the mainland (Fig. 3.4c). The modern situation shows the Laurentian Channel as >400 m deep and extending from the shelf edge northward into the Gulf of St. Lawrence (Fig. 3.4d) (Catto 2015).

3.3 Secondary Paraglacial Sediments and Processes

3.3.1 Marine, Lacustrine and Fluvial Sediments, and Processes

Depending on the newly exposed morphology of the terrain, marine, lacustrine and fluvial processes of erosion, transport and deposition replaced all glacial processes as the ice sheets receded. Glacimarine, glacilacustrine and glacifluvial sediments became the dominant surficial materials as ice sheets, ice caps and glaciers receded and relative sea level changes moved local base levels up and down. The paraglacial landscape was characterized by the most intensive reorganization of its morphology since the onset of glaciation (Forbes and Syvitski 1994). Not only deltas, raised shorelines, alluvial fans, sandurs and deep alluvial valley fills became common but their presence as unconsolidated materials created opportunities for renewed acceleration of local erosional forms. Furthermore, non-glacial processes conditioned by glaciation included the accelerated sedimentation associated with marine and lacustrine flooding of the land. The repeated incursions of the sea during the early Holocene Epoch followed by isostatic uplift of the land and the massive transient lake expansions led to the extensive deposition of marine and lacustrine clays and silts. Over 50% of the area once occupied by the LIS was covered by water at one time or another during MIS 2, either by marine transgression or by proglacial lakes formed around the margins of the LIS. Meltwater was confined between the ice margin and any higher land to the south as the retreating ice prevented drainage to Hudson Bay. These large proglacial lakes had a complex history of changing levels and sizes, as outlets opened and closed in

Fig. 3.3 Raised beaches in eastern Canada. The glaci-isostatic rebound of land formerly covered by the Laurentide Ice Sheet is marked by successive shore line beaches and ridges. **a** Southwest Hudson Bay. **b** Southwest Hudson Bay. **c** Coastal Lowlands on eastern Melville Peninsula, Nunavut. **d** St. Lawrence North Shore, Québec. *Source* a—Doc Searls/flickr CC BY-SA 2.0. b—Government of Canada; Natural Resources Canada (2007). Contains information licensed under the Open Government License—Canada. c—NRCan 2002-535 by Lynda Dredge Onge/Natural Resources Canada. d—R0011778 - Government of Québec, MFFP (2017), CC BY 4.0

response to glaci-isostatic uplift and fluctuations in the positions of the ice margins (Teller 2003). According to the position of the ice terminus, the overflow from these lakes flowed either into the St. Lawrence, Mackenzie, Mississippi or Hudson Bay drainage systems (see Chap. 2 for details). There were three specific implications of the presence of these large volumes of glacimarine and glacilacustrine sediments: (a) high rates of sediment distribution and redistribution; (b) glacimarine sediments that contain much salt have subsequently become sites of sensitive clay flows; and (c) glacilacustrine sediments that are also prone to failure.

Studies of contemporary patterns of erosion in the glaciated and largely undisturbed Canadian landscape reveal spatial patterns that do not conform to the conventional concepts of drainage basin sediment delivery. In particular, stream and valley-side sediment sources dominate over erosion of the land surface and specific sediment yield is often positively related to the drainage area. Much of this is a legacy of glaciation that left large quantities of sediment in the landscape and disrupted regional drainage patterns. This observation has been elaborated for the whole of Canada (Church et al. 1999). Only in the agricultural areas of the Great Lakes basin and the St. Lawrence valley (southern Ontario and Québec) do more conventional patterns of specific sediment yield–area relationships occur (Ashmore 1993; Stone and Saunderson 1996).

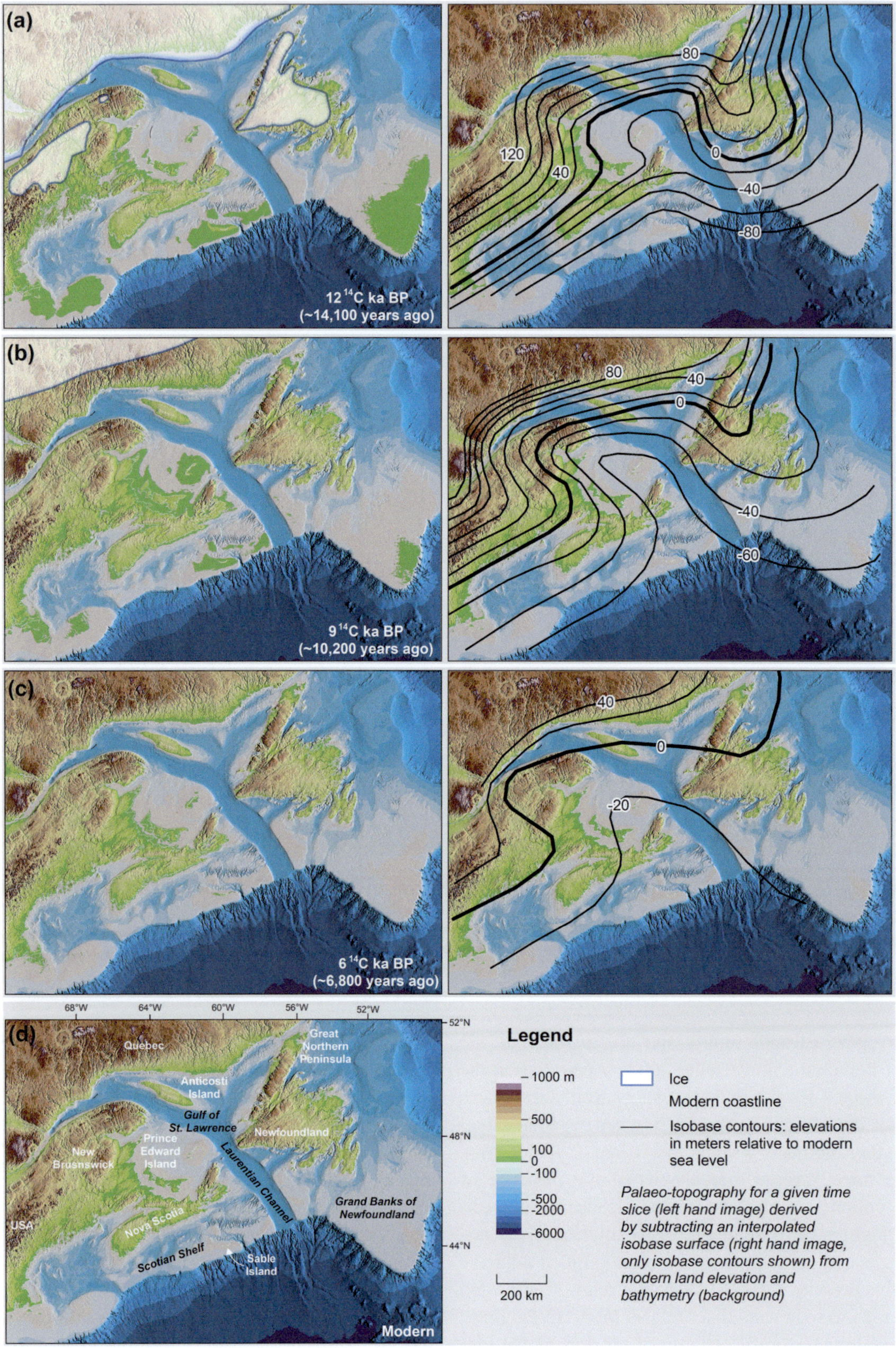

Fig. 3.4 Palaeo-geographic reconstructions of Atlantic Canada. The sequence of maps shows the regional pattern of relative sea level change following retreat of the MIS 2 Laurentide Ice Sheet. They delineate the approximate shoreline and subaerial features (left-hand image), and are a topographic and bathymetric model of the region for a given time. **a** About 12 ^{14}C ka, with lower relative sea level due to glaciation. **b** By 9 ^{14}C ka, variable isostatic effects inundated islands on the continental shelf but led to emergence of land around Prince Edward Island and the Magdalen Islands. **c** By 6 ^{14}C ka, rising relative sea level separated Price Edward Island from the mainland. **d** Modern. Note that relative sea level rise due to glaci-isostasy that continues today in much of Atlantic Canada; the line of zero subsidence extends along the northern Gulf of St. Lawrence. *Source* palaeo-geographic reconstructions and isobase contours adapted from Shaw and Courtney (2002) and contains information licensed under the Open Government Licence—Canada. Also see Shaw et al. (2002) for a detailed discussion and Dyke et al. (2003) for updated ice margin configurations

3.3.2 Littoral Processes: The Case of Paraglacial Coasts

Coastal evolution conditioned by glaciation (Fig. 3.5a, b) occupies 15% of the discussion on paraglaciation in Ballantyne (2002). The implications of changing relative sea level changes for secondary sediment redistribution led to an interesting distinction between two types of paraglacial coasts in terms of sediment production.

3.3.2.1 Coasts Defined by Direct Release of Primary Glacigenic Sediment into the Nearshore Sediment System

The shores of active outwash and paraglacial barriers are the best examples of this kind of coastline and a conceptual model of cyclic formation, destruction and reformation of paraglacial barrier systems at St. George's Bay in Nova Scotia has been prepared by Boyd et al. (1987) (Fig. 3.5c). Other barrier coasts, such as St. George's Bay in Newfoundland, Baie des Chaleurs in New Brunswick/Québec and Chezzetcook Inlet, Nova Scotia are also classic examples of longshore drift redistribution of abundant glacigenic sediments. Two other examples are found, for example, (a) in drumlin coasts such as Story Head and Mahone Bay in eastern Nova Scotia, where the glacial landforms and their glacigenic sediments condition the coasts and (b) in primary glacigenic boulder lag deposit coasts, such as those described from the Gulf of St. Lawrence (Ballantyne 2002).

3.3.2.2 Coasts with Direct Influx of Secondary Fluvial and Mass Wasted Paraglacial Sediment

Such coasts include fjord settings, fluvial deltas and fan deltas. For example, fjord sedimentation on Baffin Island, shallow fjords like Saguenay Fjord, a north shore tributary of the St. Lawrence Estuary or Lake Melville, a large coastal basin in Labrador demonstrate the importance of sandurs and fjordal valley sides in producing secondary paraglacial sediment yield. Another example of enhanced influx is provided by fluvial deltas and fan deltas typical of the north shore of the St. Lawrence Estuary east of Sept-Iles (Syvitski 1993). Glaciated shelf sedimentation has produced distinctive landscapes, well described in Chap. 10 (this volume).

3.3.2.3 Composite Paraglacial Coastlines Have also Been Identified

St. George's Bay, Nova Scotia (Fig. 3.5) provides one example in which four distinct stages of paraglacial evolution are recognized as follows: (a) Primary glacigenic sediments are (b) reworked by coastal processes to form a prograding barrier system. (c) As sediment supply diminishes, barriers retreat and lose sediment to washover, dune and tidal inlet sinks and (d) barrier regrowth occurs from new tertiary sediment sources (Boyd et al. 1987). These secondary, tertiary and Quaternary paraglacial processes are implied in the conceptual model of Fig. 3.5b.

3.3.3 Mass Wasting

One of the most remarkable ways in which secondary paraglacial processes of sediment redistribution by mass wasting have affected eastern Canada's landforms and landscapes is through the extensive distribution of glacimarine and glacilacustrine sediments (Evans 2000). The Champlain Sea deposited thicknesses of silt and clay locally exceeding 70 m. The marine and brackish water sediments left behind by the Champlain Sea are termed 'Leda clay'. Leda clay is a highly unstable thixotropic material, maintaining its strength only when undisturbed or dry. Its open-structured fabric contains significant void space between the particles. When disturbed by a sudden shock (earthquake or blasting) or when subjected to heavy rainfall, the clay fails and liquefies, causing rapid mudflows. Flow scars showing the locations of past slope failures are present along many streams, and any riverbank exposure of the sediment is prone to failure. The development of rapid retrogressive earthflows has been found to be associated with the presence of zones of sensitive clay at the base of the initial slopes. Groundwater leaching can also reduce the original salt content and enhance the sensitivity of the sediments. A number of significant landslides has occurred in the glacimarine sediments of the Great Lakes-St. Lawrence Lowlands.

On 20 June 1993 a large landslide occurred in Leda clay near the former town site of Lemieux, Ontario (Fig. 3.6). This rapid earthflow in Leda clay involved 2.8×10^6 m^3 of sand, silt and clay and retrogressed 680 m from the edge of the river. The debris flowed into the South Nation River channel, inundating 3.3 km of the valley bottom and impounding the South Nation River. Rapid slides and earthflows may also occur in glacilacustrine sediments, as in the cases of the 1990 Nipigon River landslide, westernmost Ontario, and the 1946 Beattie mine disaster at Duparquet, westernmost Québec in the Hudson Bay Lowland (Evans 2000).

Earthquake activity has been identified as a trigger for rotational sliding of sensitive glacimarine Champlain Sea clays in several instances in eastern Canada (Locat et al. 2014). Large slope failures ('fingerprint slides') resulting from earthquakes are preserved in the landscapes of eastern Ontario and southern Québec, indicating that liquefaction has been occurring repeatedly over the past 9,000 years.

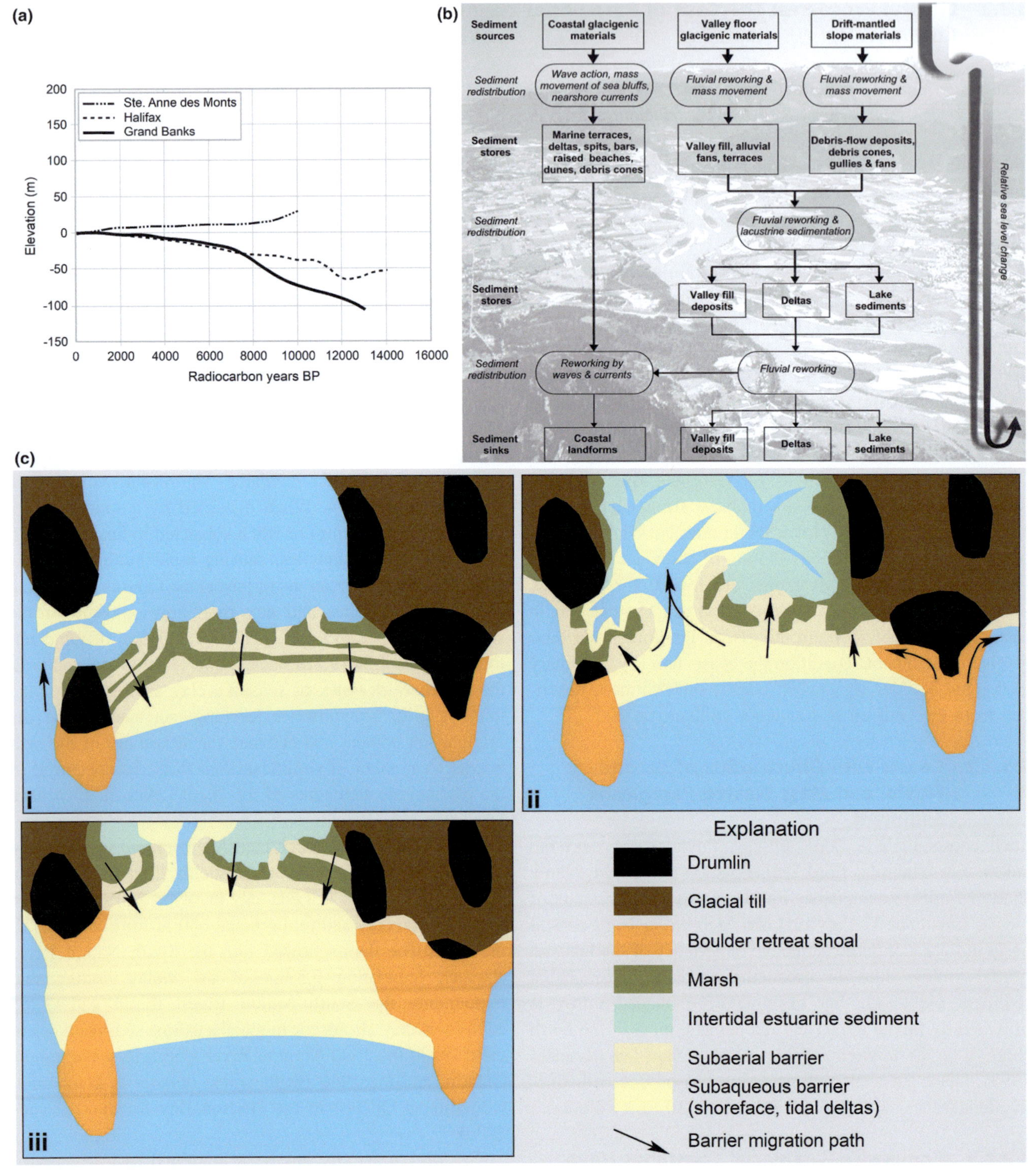

Fig. 3.5 Paraglacial coasts (Forbes and Syvitski 1994) in relation to the Ballantyne (2002) model. **a** Selected inferred Holocene relative sea level curves for Québec and the outer shelf (modified from Shaw et al. 2002). **b** Conceptual flow diagram of paraglacial processes of glacigenic sediment reworking (from Kovanen and Slaymaker (2015), adapted from Ballantyne (2002). **c** Cyclic formation, destruction and reformation of paraglacial barrier systems proposed by Boyd et al. (1987), St. Georges's Bay in Nova Scotia. (i) Marine reworking of glacigenic deposits into a prograding barrier system. (ii) As sediment supply diminishes, barriers retreat, losing sediment to washover, dune and tidal inlet sinks. (iii) Barrier regrowth from new sediment sources

Fig. 3.6 Lemieux landslide of 20 June 1993. This sensitive clay flow occurred near the former town site of Lemieux, Ontario. The failure involved 2.5–3.5 million m^3 of sand, silt and clay which flowed into the South Nation River Valley, inundating 3.3 km of the valley bottom and impounding the South Nation River. *Photograph* NRCan 1993-296 by S. G. Evans/Natural Resources Canada http://open.canada.ca/en/open-government-licence-canada

Cut banks are prone to failure following heavy rainstorms. In 1908, a rotational slide in Champlain Sea clay at Notre-Dame-de-la-Salette (QC) blocked the Lièvre River, flooding the town and drowning 33 people. At Saint-Jean-Vianney (QC), following heavy rain, a rotational slide in Golfe Laflamme clay on 4 May 1971 destroyed a newly constructed community overlooking the Saguenay River along a cut bank, killing 31 people. Following the tragedy, the remaining homes in Saint-Jean-Vianney were abandoned and destroyed, and the former community was designated a memorial site.

A proactive response to this geomorphic problem occurred in the community of Lemieux, 45 km southeast of Ottawa. The geomorphology of Lemieux showed evidence of repetitive slope failure, and a 1970 rotational slide had blocked the adjacent South Nation River. After the Geological Survey of Canada recommended the abandonment of Lemieux, buildings were purchased by the Government of Ontario, the residents were relocated and compensated and the town was abandoned in 1991. On 20 June 1993, the former Lemieux was completely destroyed when a slope failure deposited 2.8 million m^3 of Champlain Sea clay into the river.

3.3.4 Aeolian

Aeolian processes in eastern Canada have been underestimated in the literature, but it seems clear that in comparison with the other secondary paraglacial processes, wind action is of modest importance. Deflation by wind action occurs close to and just outside the northern boreal limit in the zones of sporadic and discontinuous permafrost (Seppala 2004). Extensive present-day wind action has been reported in Québec from Poste-de-la-Baleine, on the east coast of Hudson Bay, where niveo-aeolian deposits are annual events (Filion et al. 1991). In Nunavut, Gilbert (1983) found in his studies of fjords of Baffin Island that in September aeolian deflation on the sandurs was essentially unrestricted. In the Schefferville area of Labrador, during two successive winters, approximately 18 cm of surface peat was eroded from a palsa surface by wind action (Cummings and Pollard 1990).

3.3.5 General

Another set of non-glacial processes conditioned by glaciation was the accelerated sedimentation associated with primary glacigenic sediments exposed for the first time to subaerial conditions and reworked as alluvial sediments in alluvial fans.

Fluvial, lacustrine, marine and mass wasting processes were all accelerated because of the impetus provided by the potential energy of vertical land and sea movements, affecting tidal and wave energy and the kinetic energy of moving water to transport sediment, of gravitational energy to move glacigenic sediments off the steeplands by mass wasting and coastal processes both above and below present sea level.

3.4 Periglacial Landscapes

3.4.1 The Role of Permafrost

There is no ambiguity about the importance of permafrost as a land-forming agent, a process accelerator and as a phenomenon of supreme importance in the northern half of Canada. Geomorphological processes associated with permafrost are complex.

One of the critical features of permafrost that is difficult to capture in a figure like Fig. 1.6 is the variety of forms of ground ice. Structure-forming ice includes pore ice, icy coatings on soil particles, ice veins, ice lenses and intrusive ice. The amount of pore ice in a soil depends on soil porosity which in fine-grained silts may be as much as 40% of the total soil volume. Most ground ice occurs in the form of ice lenses, which range from hair line thickness to more than 10 m (Fig. 3.7). Because permafrost of some kind covers over 60% of eastern Canada, it exercises considerable control on the properties and performance of ground materials, and significantly affects potential human development of the land; it is important to recognize that the extent of permafrost is only one half of the problem. The structure-forming ground ice content is the other half. Ground ice distribution is highly complex, dependent on climate, ground temperature and the nature of the ground materials (Smith et al. 2001).

3.4.2 Permafrost and Periglacial Processes

French (2007) defines periglacial geomorphology as the sub-discipline of geomorphology concerned with cold non-glacial landforms and landscapes. Two criteria are regarded as diagnostic of periglacial environments. First, there is ground freezing and thawing. Second, there is the presence of perennially frozen ground, or permafrost (Fig. 3.7a). The term permafrost refers to the thermal condition of the ground that remains at 0 °C or sub-zero temperatures for more than two summers. Although ice is commonly present in permafrost, bare rock surfaces, and sediments overlain by salt water may be classified as permafrost in the absence of ice. The upper part of the ground that thaws each summer and refreezes each winter constitutes the active layer, which is not part of the permafrost, by definition. The temperature of the atmosphere close to the ground is the main factor determining the occurrence of permafrost. The relation between that temperature and the ground thermal regime depends on site-specific conditions such as vegetation cover, snow cover, surficial materials, the presence or absence of an organic layer, soil moisture content and drainage. Mean annual ground surface temperature is generally 2–4 °C higher than mean annual air temperature and is more variable at a local scale.

3.4.3 Frost Action Processes

Ice segregation and frost heave are two of the most significant aspects of frost action (French 1988). Frost action describes the effects of the freezing of moisture in porous media, particularly soils. Soil freezing causes volumetric expansion while thawing of frozen soil is often accompanied by weakening and consolidation; frost action phenomena are related to the strains that result from these processes. When water freezes, it expands by approximately 9% of its original volume. However, when some soils freeze, the volume may increase by an amount considerably greater than that which can be explained by the volume increase of water on freezing. Frozen soil consists of water frozen in pores but ice is also found in discontinuous layers, referred to as ice lenses. These ice lenses appear to be more or less segregated from the soil grains.

Ice segregation results from the peculiar behavior of water as it freezes within porous materials. The complex nature of soil freezing arises from the fact that frozen soil can contain appreciable amounts of unfrozen water at temperatures down to several degrees below 0 °C. The amount depends on temperature and soil type; in particular the higher the specific surface area, the more unfrozen water there is at any temperature. The amount of frost heave due to ice segregation depends on a variety of factors such as soil water conditions, the rate of heat loss from the soil and soil characteristics. Coarse sands and gravels generally exhibit little or negligible heave on freezing. Silty sands can produce spectacular heave while clays generally develop less segregated ice.

The sorting of particles and stones is one of the most striking features of frost-affected soils (Gray and Seppala

Fig. 3.7 **a** Exposed permafrost in Katherine River Valley, Torngat Mountains, Labrador. **b** and **c** Active layer, retrogressive thaw slump and scar of permafrost disturbances on the Fosheim Peninsula, Ellesmere Island, Nunavut. *Source* a—NRCan A92S0044 by M. J. Van Kranendonk Natural Resources Canada http://open.canada.ca/en/open-government-licence-canada. b and c—A. Cassidy/flickr CC BY 2.0

1991). The process can give rise to both macroscopic (patterned ground) and microscopic features (distinctive microfabrics). Short-term freeze–thaw cycles that occur daily or seasonally create stresses by freezing and thawing of moisture in the pores and voids of soils and sediments and in the pores and cracks in rocks (Fraser 1958). Frost shattering and disintegration of bedrock and heaving, churning and downslope movement of the active layer result. Frost heaving and cryoturbation result from stresses exerted by the growth of ice crystals. Cryoturbation results from differential rates of freezing and heaving and differential rates of thawing and settling induced by soil properties. Frost-related soil structures include contorted soil horizons, earth hummocks and some kinds of patterned ground (Fig. 3.8).

Frost heaving of bedrock in permafrost areas is also important in producing a variety of landforms. Accumulation of weathering products in cracks provides sites for heaving and the growth of ice wedges is also influential in

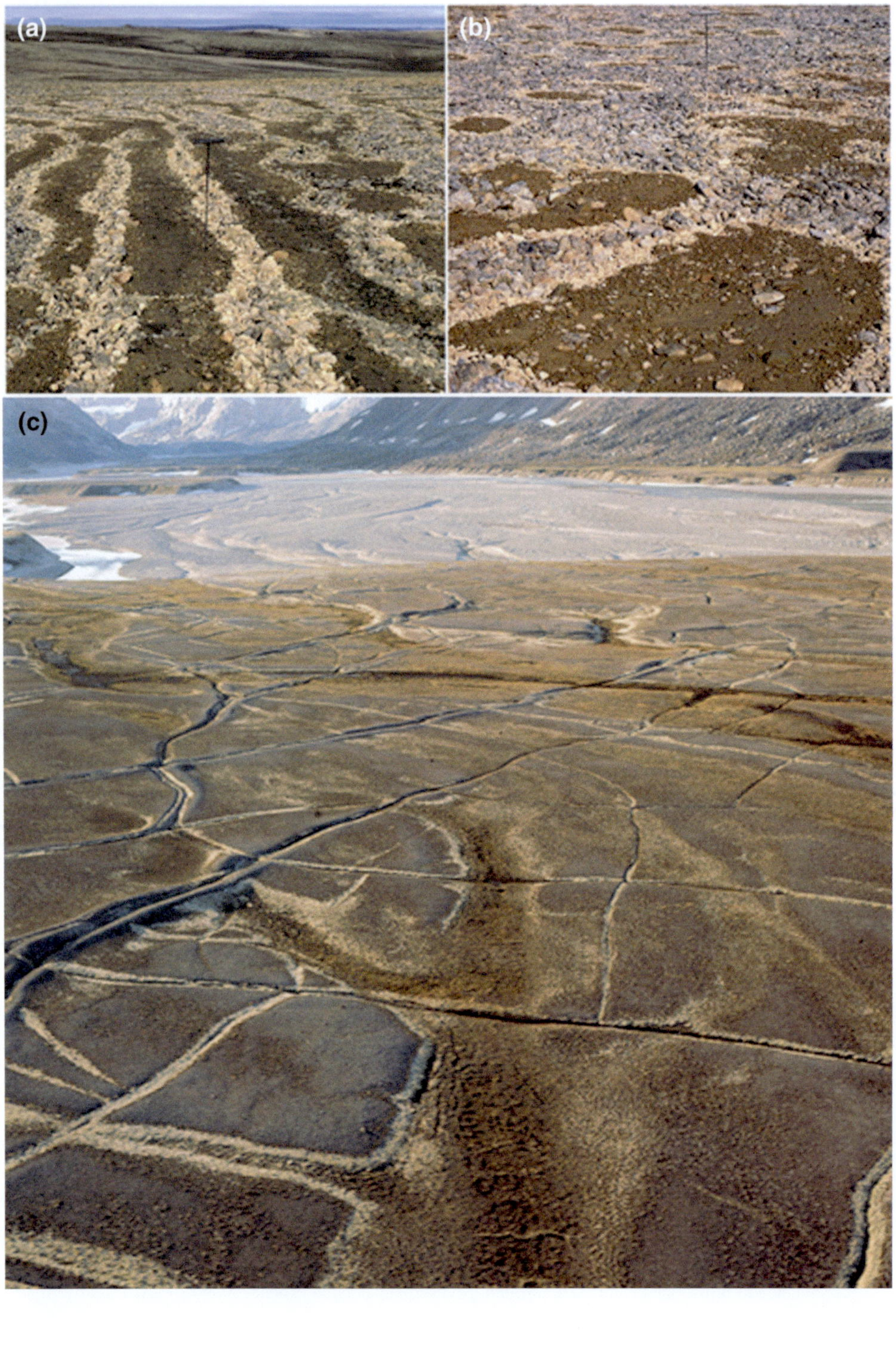

Fig. 3.8 Patterned ground. **a** Striping and **b** polygonal ground in coarse sediments, Devonian sandstone near Hecla Bay, Melville Island, Nunavut. *Source* NRCan 2009-130 and 2009-132 by Lynda Dredge/Natural Resources Canada http://open.canada.ca/en/open. **c** Ice-wedge polygonal patterns in fine sediments, Ekalugad sandur, Baffin Island. Note two levels of the sandur are visible. The older, vegetated surface is split by the ice-wedge polygons. Photograph reproduced with permission of J. D. Ives (also see Ives 2016)

bedrock heave in Arctic Canada (Michaud et al. 2011). The importance of ground ice should not be underestimated. Aggradation of ground ice bodies gives rise to a variety of ice-cored landforms and thaw degradation of ice-rich permafrost is associated with the development of thermokarst terrain (Burn 1992). Solifluction is the downslope movement of soil during freezing and thawing and also during wetting of the soil from melting ice. Solifluction is nearly ubiquitous in permafrost areas, generating distinctive lobes or terraces.

3.4.4 Thermokarst

Retrogressive thaw slumps are the most active geomorphic features of permafrost terrain (Lamoureux and Lafreniere 2009). They consist of a layer of ice-poor overburden, an ice-rich face and a low-angle mudflow downslope (Burn and Lewkowicz 1990). Fluvial incision and an abundance of ground ice favour the development of retrogressive thaw slumps. Once initiated, retrogressive thaw slumps may continue to develop for many years at rates of up to 16 m yr^{-1}, reaching lengths of 650 m. Recent warming and increasing rainfall have given rise to large, long-lived instances of these thermokarst features (Kokelj et al. 2015).

In the context of a warming climate, the most serious implications of permafrost degradation are felt in the sporadic discontinuous permafrost zone and in isolated patches of permafrost where the permafrost itself is relatively warm and thin. Permafrost thawing and thermokarst are serious environmental hazards for settlements, roads, railway construction and maintenance.

Thermokarst (thaw) lakes are water bodies that occupy a closed depression formed by settlement of the ground following melting of ground ice (Fig. 3.9). Such lakes are common in ice-rich sediments throughout the permafrost zone (e.g. Seguin and Allard 1984). They often possess an irregular circumference and bottom profile and, if enlarging in icy sediments, have slumping shorelines. Some have been initiated by artificial terrain disturbance but most result from natural disturbance which leads to soil warming, such as forest fires and hydroclimate change. Thermokarst lakes may be useful as surface expression of the response of permafrost to climate warming but the precise locations are due to site-specific factors (Dredge and Nixon 1992).

3.4.5 Alpine Landscapes: Glaciers and Permafrost

The general factors that influence permafrost distribution at the local scale include elevation, continentality, depth and duration of snow cover, vegetation cover and soils so that local variability is extreme. Because of the absence of monitoring stations, permafrost distribution in alpine regions of eastern Canada can only be estimated at a more local level (Fig. 3.10) through surrogate parameters such as climatic regions previously estimated from the nearest low-lying monitoring station (e.g. Hare and Thomas 1979), climatic parameters, such as precipitation, snowfall, temperature extremes and range (Fisheries and Environment Canada 1978), and presence or absence of landforms such as rock glaciers. An excellent example of the use of surrogate variables to predict permafrost conditions in alpine areas south of the southernmost limit of continuous permafrost is provided by Ryder (1998). She used mean annual precipitation, mean annual snowfall, coldest mean January temperature and annual temperature range at the nearest climate station as surrogate information to map permafrost in Atlantic Canada. She defined five regions which she characterized in the following ways (Fig. 3.10).

3.4.5.1 Torngat Mountains

The Torngat Mountains are in the continuous permafrost zone with significant ice cover. Their high relief imposes moderate to heavy snowfall and moderate to heavy precipitation totals in addition to cold temperatures.

3.4.5.2 North Labrador Plateau

The North Labrador Plateau straddles the continuous/discontinuous permafrost boundary. Temperatures are comparable with those of the Torngats but precipitation of all forms is less.

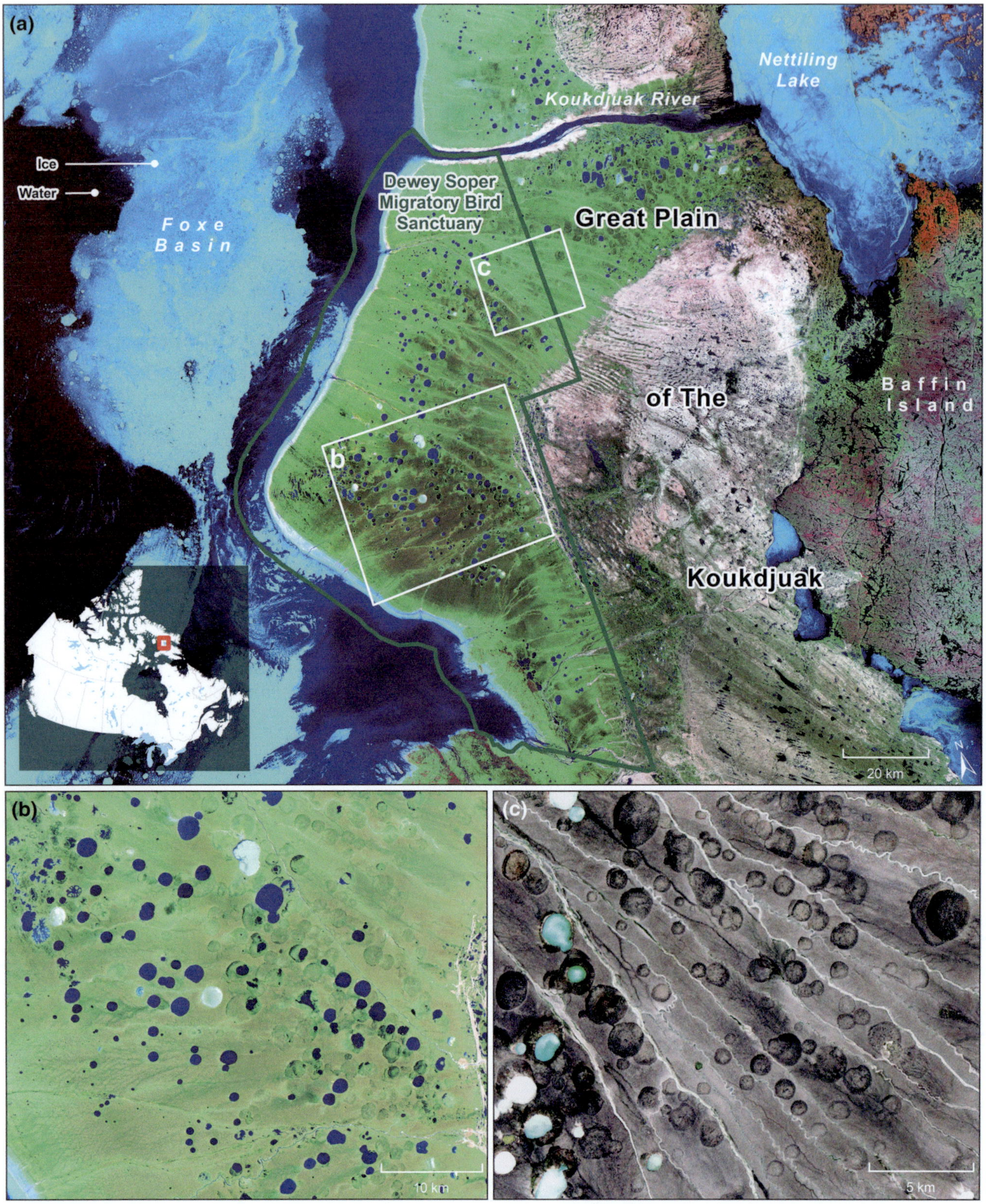

Fig. 3.9 Thermokarst lakes on the Great Plain of the Koukdjuak, Baffin Island. **a** The Great Plain of the Koukdjuak is located on the western side of Baffin Island along the southeastern coast of Foxe Basin (Qikiqtaaluk, Nunavut). It is an intertidal zone on a broad coastal plain, dotted with thermokarst lakes (which result from water released by thawing permafrost) and extensive marshes crossed by numerous small streams that drain the plain. Raised beach ridges 25–80 km from the coast define the boundary between the marshy coastal plain and the drier limestone plateau to the east. False color image captured by the Landsat 7 Enhanced Thematic Mapper (ETM +); spectral bands 7, 4, and 2. **b** Shallow (less than 1 m) thermokarst lakes. Compound lakes are identifiable by their uneven margins. **c** Natural-color image of thermokarst lakes and numerous small streams that flow across the plain from old inland beach ridges. Dark brown spots are probably locations of former thermokarst lakes. *Source* a—MDA Federal (2004). b—MDA Federal (2004). c—Government of Canada; Natural Resources Canada (2007)

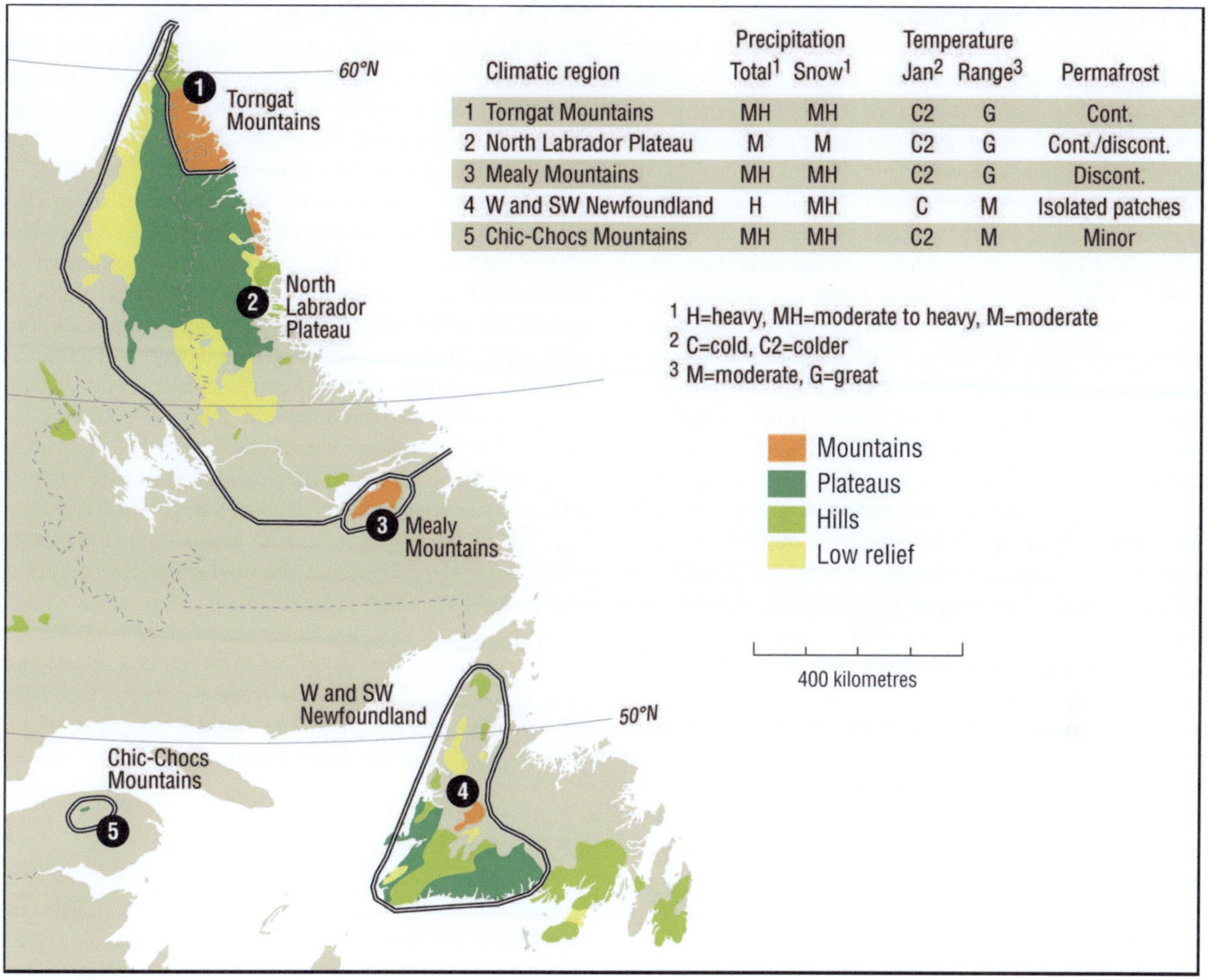

Fig. 3.10 Permafrost conditions predicted by surrogate variables in five Alpine areas of Atlantic Canada (modified after Ryder 1998). Cartography by Eric Leinberger

3.4.5.3 Mealy Mountains

Although the precipitation and temperature conditions of the Mealy Mountains are similar to those in the Torngats their more southerly location produces a discontinuous permafrost zone.

3.4.5.4 Western and Southwestern Newfoundland

Isolated patches of permafrost result from the warmer temperatures and lower elevations.

3.4.5.5 Chic-Choc Mountains

The immediate vicinity of Mt. Jacques Cartier experiences minor isolated patches of permafrost (Payette and Boudreau 1984).

3.5 Contemporary Hydroclimate and Wetlands

3.5.1 Hydroclimate

We turn next to an overview of eastern Canada's regional climates that are impacting landscapes and landforms during the later Holocene. Differences in climate reflect the influence of latitude, elevation, distance from the sea and the extent to which areas can be invaded from the outside by warm and cold air masses (Phillips 1990; Trenhaile 2007). Regional variations are considerable, as illustrated by Newfoundland (Banfield and Jacobs 1998) and Canada in general (Chiotti 1998).

3.5.1.1 Continental Climates

Continental climates in eastern Canada experience frequent penetration into the region by very cold arctic air in winter and warm, often humid tropical air in summer (Fig. 3.11a). The continental climate is modified by frequent mid-latitude cyclones. Maritime air masses from the Atlantic reduce the annual range of temperature. Eastern Canada is also affected by the Great Lakes which moderate temperatures in summer and produce local belts of higher precipitation in the form of snow in winter. Despite Atlantic Canada's maritime location, the preponderance of eastward flowing storms produces a continental type of climate. Nevertheless, the coasts of Newfoundland and Nova Scotia are cooled in summer by the cold Labrador Current. Eastern Canada's continental climate is in the polar front zone: the battle ground of polar and tropical air masses. Ample precipitation throughout the year is increased in summer by the invading maritime tropical air mass. Cold winters are dominated by continental arctic or continental polar air masses from subarctic sources. Most of northern Ontario experiences a continental climate, with warm to hot summers and very cold winters. Mean January temperatures are commonly about −15 °C or lower. The temperature can fall to below −40 °C with the influx of arctic air from the north, but it may be only just below freezing when milder Pacific air enters the region. Summers are warm in southern Ontario and the influx of warmer air from the southwest provides milder winter conditions than those further west. These westerly air masses are responsible for much greater ranges of temperature from day to day and from season to season than could occur in a true maritime climatic region.

3.5.1.2 The Boreal Climate

The boreal climate zone is the climate of the boreal forest, a forest of conifers and some hardy broad-leaved trees distributed around and south of the arctic tundra. Most of the boreal forest climate is classed as humid, with precipitation of 50–100 cm. Arctic air streams also dominate boreal climates in winter and spring but westerly air streams are more common in the summer and fall. The highest precipitation in eastern Canada occurs east of a line running from the Torngat Mountains of Labrador to Winnipeg (Fig. 3.11b). Annual total precipitation with a heavy snowfall component is more than 1000 mm in the Laurentides and in the hillier areas of the Atlantic provinces. Main sources of moisture are the Atlantic Ocean and the Gulf of Mexico. Boreal regions are cold, with mean annual temperatures near freezing. Cooling stops in the spring with the invasion of warmer air although up to 80% of the solar radiation may be lost by reflection from the snow cover. Temperatures quickly increase once the snow finally melts but cooling soon predominates again in the fall, as the length of the nights increases.

3.5.1.3 The Arctic Climate

The arctic climate produces areas of open treeless tundra and permanently frozen ground beyond the treeline. The severity of arctic climates results from the dominance of cold air streams from the pack ice of the Arctic Ocean and the Greenland ice cap. Arctic climates extend further south in eastern than in western Canada because greater snowfall in the east reduces the warming effect of spring sunshine. This climate includes the arctic coastal fringes, dominated by continental Polar, maritime Polar, and continental Arctic air masses and with frequent cyclonic storms. Winters are long and severe. There is a very short mild season. The tundra climate ranges from a humid subtype bordering the Atlantic Ocean to subhumid and semi-arid subtypes bordering the Arctic Ocean. Annual precipitation may be less than 125 mm in the High Arctic islands and it is usually less than 400 mm throughout the arctic and boreal regions (Fig. 3.11b). Nevertheless because evaporation is low in the north, low precipitation does not produce aridity, except in what is known as true polar desert of a few of the Queen Elizabeth Islands and the Lake Hazen depression on the Ellesmere Island. The winter ice does not completely disperse in Hudson Bay until late August or early September and can persist throughout the summer in some of the channels leading to the Arctic Ocean.

3.5.2 Wetlands of Eastern Canada

Wetlands of eastern Canada can be viewed as surrogates for bioclimates in regions where meteorological data are scarce. Wetland is defined as land having the water table at, near or above the soil surface and includes peatland. The major wetlands occur in an area extending from central Labrador to the south of Hudson Bay (Tarnocai et al. 2000).

3.5.2.1 Classifications of Wetlands

Classifications of wetlands are based on the height of the water table. Bogs are defined as poorly drained, peat-filled depressions dominated by sphagnum mosses, evergreen shrubs and conifers like black spruce and larch. They are particularly common in poorly drained boreal forest. Fens are peatlands with a high water table, usually at or above the surface. They are dominated by sedges, grasses, reeds and brown mosses with some shrub cover. Marshes are treeless wetlands with grasses, sedges and reeds. The better known freshwater marshes in Canada include the St. Clair, Erie and Ontario lakes and the shores of the upper St. Lawrence. Saltwater marshes are found mainly in the Maritime Provinces and Québec along the coastlines of Hudson and James Bay. Swamps are characterized by the presence of trees growing on silty to organic soils. They usually occur along river floodplains and in poorly drained basins. Shallow

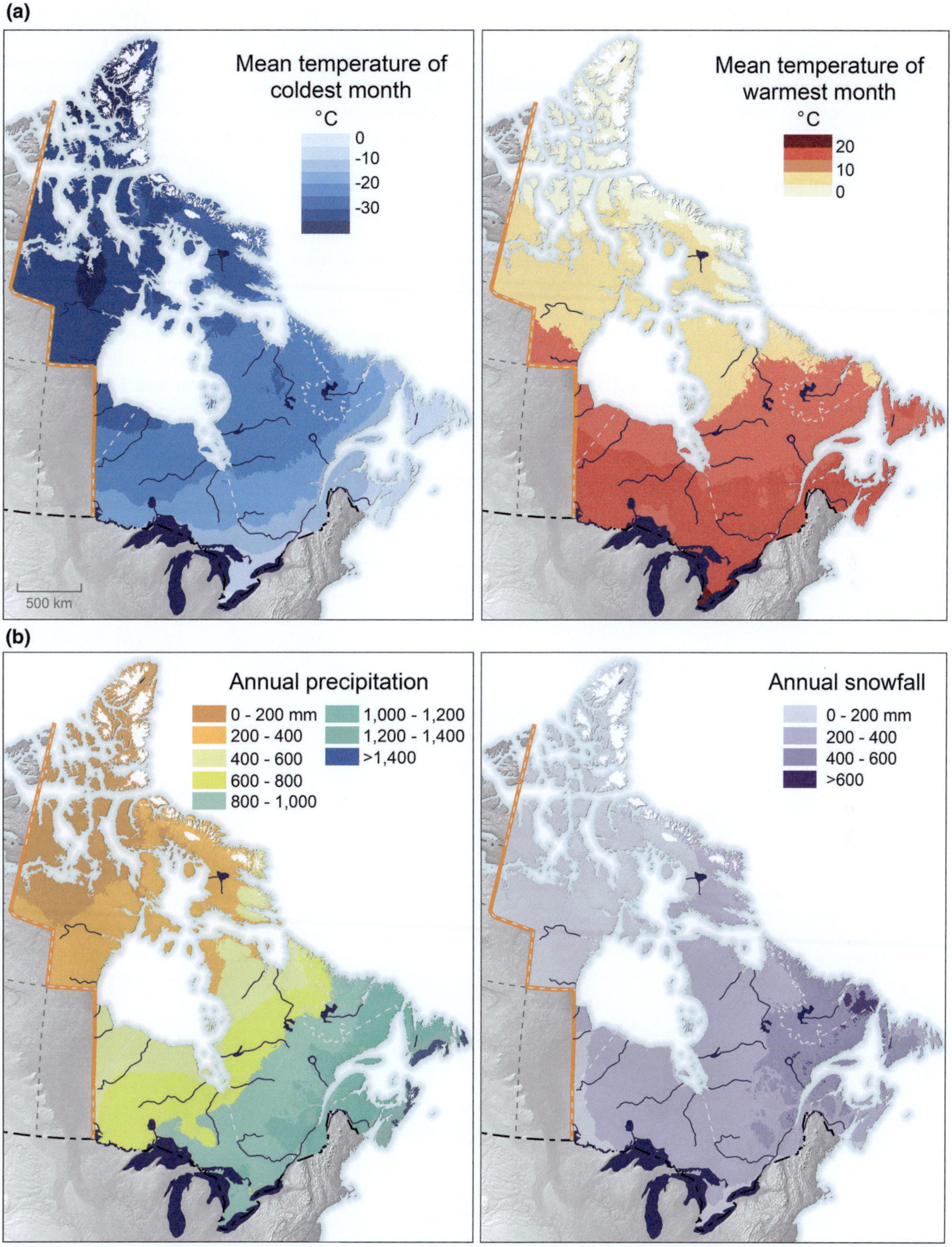

Fig. 3.11 **a** Mean temperature of the coldest and warmest months and **b** annual mean total precipitation and average maximum snowfall for eastern Canada from 1961 to 1990. Climate data have been generated with the Climate NA v5.10 software package available at http://tinyurl.com/ClimateNA, based on methodology described by Wang et al. (2016)

open water, generally less than 2 m deep, is also included in wetlands.

Four broad wetland regions are recognized in eastern Canada and are distinguished by the amount and type of wetland that is most common as follows (National Wetlands Working Group 1988).

3.5.2.2 Eastern Temperate

Eastern temperate wetlands include less-severe winters and slightly warmer summers with modest precipitation and moderate snowfall. Basin and stream swamps are common. Typical depths of peat in swamps are about 2 m and about 3 m in bogs.

3.5.2.3 Atlantic

Atlantic wetlands are found in coastal locations from Labrador to Newfoundland. Winter temperatures are less extreme than the arctic and subarctic wetlands and precipitation and snowfall are much higher (Fig. 3.11b). The southern and southeast coast of Newfoundland receive the highest precipitation (Yoxall 1981). High water tables are associated with extensive wetlands. Atlantic subarctic wetlands include Atlantic plateau bogs with palsa bogs in Labrador but they generate less than 2 m of peat.

3.5.2.4 Boreal

Cold winters; moderate precipitation and high snowfall are characteristic of boreal wetlands. They include peat plateau bogs, palsa bogs and string bogs (Fig. 3.12). Ribbed fens, tree bogs and fens and raised bogs in humid east, and peat margin swamps. They have 2–5 m depth of peat in plateau bogs on average. The boreal wetlands of the Hudson Bay Lowlands have up to 2 m of peat in their fens (Dredge and Dyke 2019) and 2–4 m of peat in their plateau bogs (see

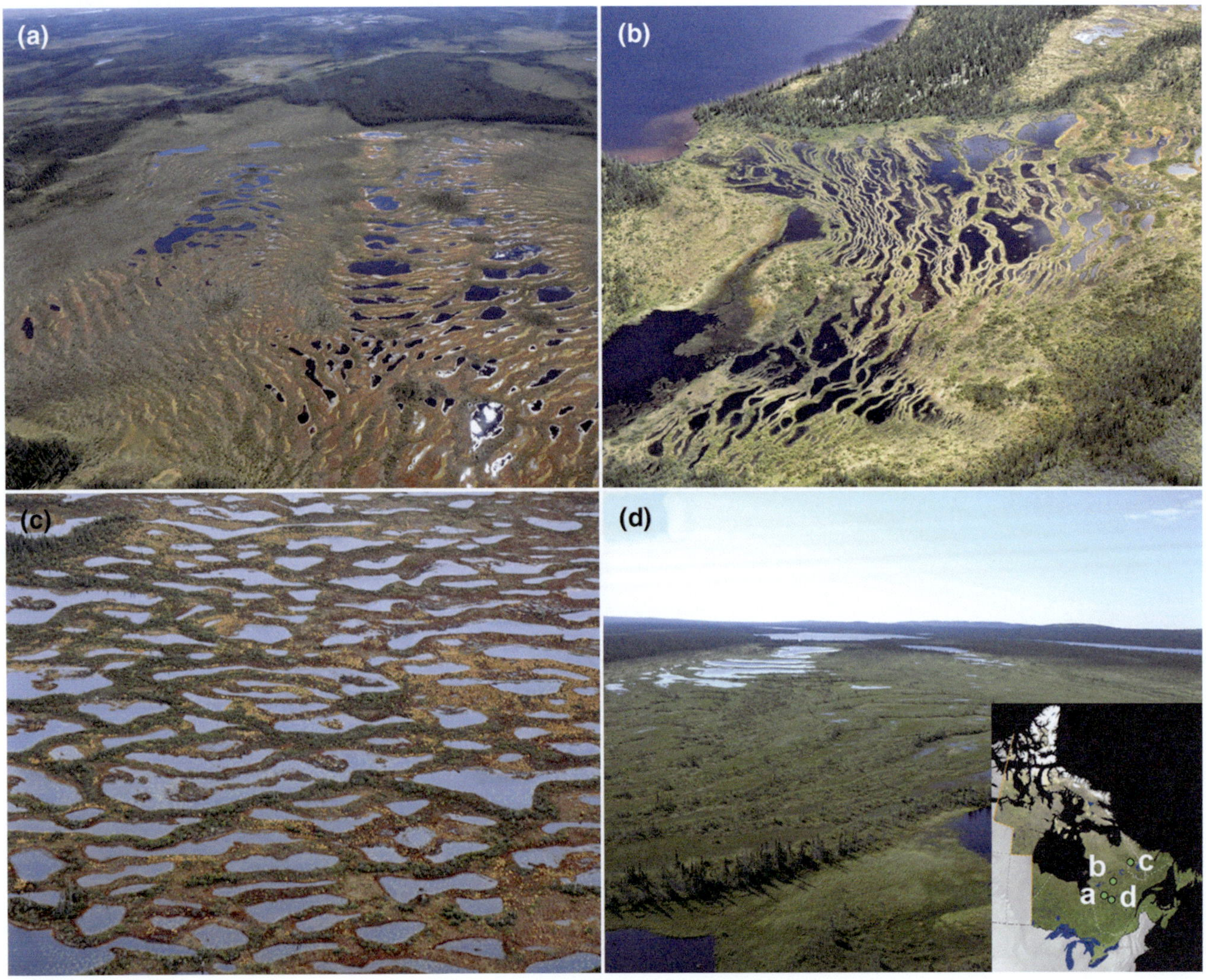

Fig. 3.12 String bogs in Québec, **a**, **b**, **c**, **d**; peat ridges (strings) and hollows (flarks). *Photographs* DPF_7246, _3020703, INV_7470, P1010683 - Government of Quebec, MFFP (2017), CC BY 4.0

Chap. 8). Atlantic boreal wetlands have 5−10 m of peat in domed bogs and 8−10 m of peat in plateau bogs.

3.5.2.5 Arctic and Subarctic

Arctic wetlands include everything north of the arctic treeline and are characterized by extremely low winter temperatures (c. −30 °C in January) and low precipitation (<400 mm including moderate snowfall (up to 30 cms). High Arctic wetlands have relatively few wetland features. Mid-Arctic wetlands consist largely of fens and low-centre polygonal bogs. Low Arctic wetlands are found between the arctic islands and the arctic treeline and have lowland polygonal fens, both high- and low-centre polygonal bogs (Fig. 3.13). Subarctic wetlands are characterized by equally cold winter temperatures but significantly warmer summer temperatures and higher precipitation and high snowfall. Subarctic wetlands have polygonal peat plateau bogs and small palsa bogs. They typically have 2–3 m of peat (Seppala et al. 1991).

3.5.3 Rivers

3.5.3.1 Rivers

Here, we have chosen mean annual discharge (Fig. 3.14) to rank the major rivers of eastern Canada and to comment on the limitations of the data. Numerous streams are tributary to the Arctic Ocean but they are not monitored and/or too small to register at the scale of this map. The mean annual discharge of eastern Canada's largest river, the St. Lawrence River at Québec City, is 12,100 cumecs (Table 3.1).

Although the St. Lawrence River is ranked first in both mean annual discharge and basin area, Table 3.1 identifies a number of seemingly anomalous rankings, notably the Nelson and Thelon rivers. Both rivers drain very dry regions (the Prairies and the Arctic Lowlands) and the Nelson is dammed by Manitoba Hydro for power generation. Discharge intensity is also commonly used to rank rivers as it is the hydrologic parameter that is most closely related to mean annual precipitation.

Fig. 3.13 Palsa fields consist of palsas (permanently frozen mounds, generally < 100 m in diameter) and peat plateaus (permanently frozen peat mounds, generally >100 m in diameter). **a** Aerial and **b** profile views of palsa field near Whapmagoostui, Québec. Photographs courtesy of Jean-Marie Dubois, University of Sherbrooke (World of Images No. 22745 and No. 22761). **c** Emerging palsa in a fen near Churchill, Manitoba. *Source* NRCan 2001-155 by Lynda Dredge/Natural Resources Canada http://open.canada.ca/en/open-government-licence-canada

Fig. 3.14 Major rivers of eastern Canada with river channels depicted proportionate to mean annual discharge (Fish and Environment Canada 1978). Cartography by Eric Leinberger

3.5.3.2 Large River Channels

Large river channels can be effectively characterized in terms of (a) plan-form pattern and (b) number and type of channel islands. Plan-form pattern ranges from straight, sinuous, wandering, regular and tortuous meanders. The hierarchy of channel islands varies from none, through irregular, regular and anastomosing. Increased flows or delivery of coarse sediment increase activity and lateral instability producing a braided pattern, whereas decreased discharge tends to reduce the river into a single, more stable channel (Church 1992).Variability in river morphology within regions may be as great as the variation between regions and specific river morphologies may occur in all regions (Kellerhals and Church 1989, Fig. 3.15). The most general reason for this is that the glaciation of Canada has left extensive glacial, proglacial and paraglacial sediments covering the bedrock and these surficial materials are the dominant source of sediment for eastern Canada's rivers. This fact obscures the distinctive underlying bedrock and produces both local variability and some interregional homogeneity in source materials for rivers.

3.5.3.3 Fluvial Regions in the Context of Hydroclimate Change

In the course of assessing the potential impact of hydroclimate change on Canadian rivers, Ashmore and Church (2001) defined fluvial regions of eastern Canada. These are regions within which the existing streamflow characteristics,

Table 3.1 Ranking of eastern Canadian rivers

River	Mean discharge (m^3 s^{-1})[a]	Basin area (km^2)	Mean discharge intensity (m^3 s^{-1} km^{-2})
St. Lawrence, Ont/Qu/USA	12,100	1,200,000	0.010
Nelson, AB/SK/MB/USA	2,370	1,072,000	0.002
Koksoak–Caniapiscau, Nunavik	2,200	133,400	0.016
Ottawa, Québec/Ontario	1,950	146,300	0.014
Saguenay, Québec	1,750	88,000	0.020
La Grande, Québec	1,690	97,600	0.016
Churchill, Labrador	1,620	79,800	0.020
Albany, Ontario	1,420	135,200	0.011
Thelon, Nunavut	1,400	240,000	0.006
Moose, Ontario	1,370	108,500	0.013
Nottaway, Québec	1,100	66,000	0.017
St. John, NB/USA	990	55,000	0.018
Manicouagan, Québec	815	46,000	0.018
Attawapiskat, Ontario	630	50,200	0.012
Natashquan, Labrador/Québec	410	16,100	0.025
Nastapoca, Québec	175	13,400	0.013
Ekalugad Fjord, Nunavut	10[b]	380	0.027

[a]Milliman and Farnsworth (2011)
[b]Church and Ryder (1972)

river morphology and river dynamics are relatively distinct and the effects of hydroclimate change and consequent fluvial responses are likely to be distinct (Table 3.2).

These fluvial regions describe the characteristics of small to moderate size rivers contained within a single region (Fig. 3.16). However, the drainage basins of very large river systems often cross one or more of the regional boundaries, the Great Lakes-St. Lawrence basin (Frenette 1990) and the Nelson River basin (Newbury 1990) being the most obvious examples in eastern Canada.

3.5.3.4 The Great Lakes-St. Lawrence Lowlands

The Great Lakes-St. Lawrence Lowlands region is divided into three subregions based on differences in surficial materials and hydroclimate: western (Great Lakes) covering southern Ontario south of the Canadian Shield, central (St. Lawrence valley) and eastern (Gulf of St. Lawrence). The western section, covering southern Ontario south of the Canadian Shield, is a region of low relief with extensive till, glacifluvial and lacustrine deposits. Originally forested, the region is now largely agricultural with rapidly expanding urban areas. The central section is flat, apart from occasional isolated volcanic hills and underlain by glacimarine clay, glacifluvial sand and alluvium. Disturbance by agriculture and urban development is extensive. The eastern section is distinguished by its gently sloping to flat carbonate lithology.

Fluvially dominated deltas are the product of fluvial sediment influx and deposition and, as can be seen from Fig. 3.17, sediments generated by agricultural and urban development in Canada and the United States directly impact the St. Clair River delta.

Flood events in eastern Canada are clustered in the south. 111 of the 168 flood disasters occurring in Canada between 1900 and June 1997 were recorded in eastern Canada. Ontario, New Brunswick and Québec were the worst hit (1900−1997). The Saguenay region of Québec experienced widespread flooding in July, 1996 (Brooks and Lawrence 1998). Estimated costs were 10 fatalities and 1.5 billion dollars, the costliest flood in Canadian history at that time. The relatively high level of development on flood-prone lands is a major factor (Brooks 2001). The number of rainfall-generated flood events and their magnitude are greatest in southwest Ontario, decreasing northward and eastward (e.g. Gardner 1977). But the duration and relative importance of snowmelt floods increase northward and eastward (Irvine and Drake 1987). In southern Québec the greatest flood damage costs are associated with ice jams (Watt 1989). Extreme flood events of the past 2 years (2017–2019) have reinforced these patterns.

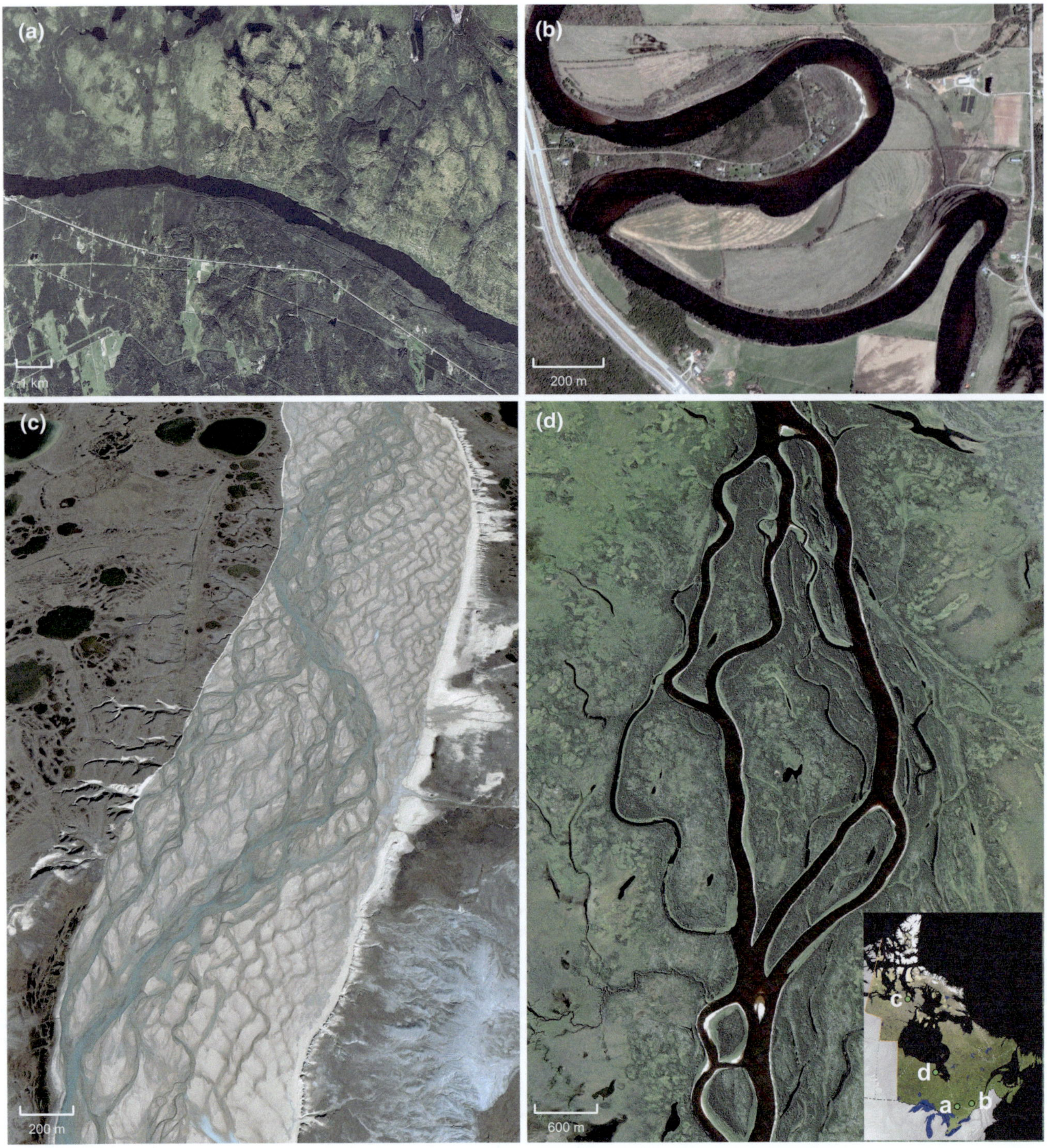

Fig. 3.15 Various river channel patterns in eastern Canada. **a** Straight—overall, the Ottawa River is a constrained, straight river whose main river channel lacks the sinuosity generally observed in unconstrained rivers. **b** Meandering, sand-bed river (River Rouge, near La Conception, Québec). Point bars and submerged sand bars are visible in the channel. Active erosion and migration is evident from the scars of former channel positions visible in the floodplain. **c** Braided gravel-bed rivers (e.g. Atorquait River, Nunavut) are inherently unstable. **d** Anabranching (anastomosing) rivers (e.g. Ekwan River, Ontario) have more than one permanent channels that flow around stable islands. *Source* a—Google Earth: image © 2018 CNES/Airbus. b—Google Earth: image © 2018 DigitalGlobe. c—Google Earth: image/NASA, © 2018 DigitalGlobe. d—Google Earth: image Landsat/Copernicus, © 2018 DigitalGlobe

Table 3.2 Regional subdivisions in eastern Canada and their equivalents in five independent sources

Geomorphological regions[1]	Permafrost zones[2]	Climate regions[3]	Wetlands regions[4]	Fluvial regions[5]
1 Permafrost-dominated Shield	Continuous	Arctic	Arctic	Northern
2–3 Forested Shield	Discontinuous/None	Boreal	Boreal	SSSB[a]
4 Shield	Continuous	Arctic	Arctic	Northern
5 Arctic Lowlands	Continuous	Arctic	Arctic	Northern
9–10 Hudson Bay Lowland	Discontinuous/None	Boreal	Boreal	SSBS[a]
11 Great Lakes-St. Lawrence Lowlands				
11 a Gulf of St. Lawrence	None	Continental	Temperate	GLSLL[b]
11 b, c St. Lawrence/Great Lakes	None	Continental	Temperate	GLSLL[b]
12–13 Innuitia	Continuous	Arctic	Arctic	Northern
19 Appalachia-Acadia				
19 a Newfoundland	Sporadic/None	Continental	Atlantic	Appalachia
19 b, c Québec/NB/NS	Sporadic/None	Continental	Atlantic	Appalachia
20 Arctic Coastal Plain	Continuous	Arctic	Arctic	Northern
22 Atlantic Maritime Plain	None	Continental	Atlantic	GLSLL[b]

1 This book, Sect. 1.6 our Fig. 1.10
2 Smith et al. (2001); Sect. 1.4.1, our Fig. 1.6
3 Trenhaile (2007); Sect. 3.5.1, our Fig. 3.11
4 National Wetlands Working Group (1988); Sect. 3.5.2
5 Ashmore and Church (2001); Sect. 3.5.3, our Fig. 3.16
Abbreviations
[a]SSBS—Shield/Southern Boreal Shield
[b]GLSLL—Great Lakes-St. Lawrence Lowlands

3.5.3.5 Appalachia

Appalachia covers Atlantic Provinces and eastern Québec south of the St. Lawrence River. It is generally a region of low relief plains and uplands up to about 800 m asl, except in the Notre Dame Mountains of Gaspésie where elevations reach 1200 m asl. Thin glacial deposits cover many areas but the bedrock is also extensively exposed, especially in the uplands. River valleys usually contain Quaternary sands and gravels. Forest cover is extensive except in the agricultural lowlands along the Gulf of St. Lawrence and the Bay of Fundy. The region is subdivided on the basis of relief and hydroclimate with the Atlantic coast having substantially higher annual precipitation, larger annual runoff and a larger proportion of precipitation falling as rain. Rainfall generated floods are more common than elsewhere in eastern Canada. New Brunswick and the valley of the St. John River are exceptionally flood prone being the second-most flood-prone province in Canada (Brooks 2001).

3.5.3.6 The Southern Shield and Southern Boreal Shield

The Southern Shield and Southern Boreal Shield regions are distinguished from the Northern Boreal Shield by the absence of permafrost. They are generally of low relief except for the Laurentian Highlands and coastal uplands of Labrador. The terrain is the bedrock with a thin, sometimes discontinuous cover of glacigenic sediments, numerous lake basins and bedrock valleys rimmed by glacilacustrine and fluvial silt, sand and gravel deposits (see King and Martini 1984). Some areas are characterized by glacial lake deposits overlying the bedrock.

3.5.3.7 Northern Regions

Northern regions include Nunavut and NWT, significant areas of northern Québec and some portions of northern Ontario. These regions are strongly influenced by the occurrence of permafrost and their hydrology is dominated by snowmelt/permafrost thaw processes (Woo 2012). The effect of frozen-bank sediments on channel morphology and erosion rates is uncertain (Church 1992). The central issue is the relative effectiveness of hydraulic versus thermal processes of erosion (Fortier et al. 2007). While frozen sediments may resist erosion, erosion rates may be increased by thermal erosion during the thaw season. In the presence of permafrost, erosion accomplished by a flood of only a few hours or days in duration may be determined

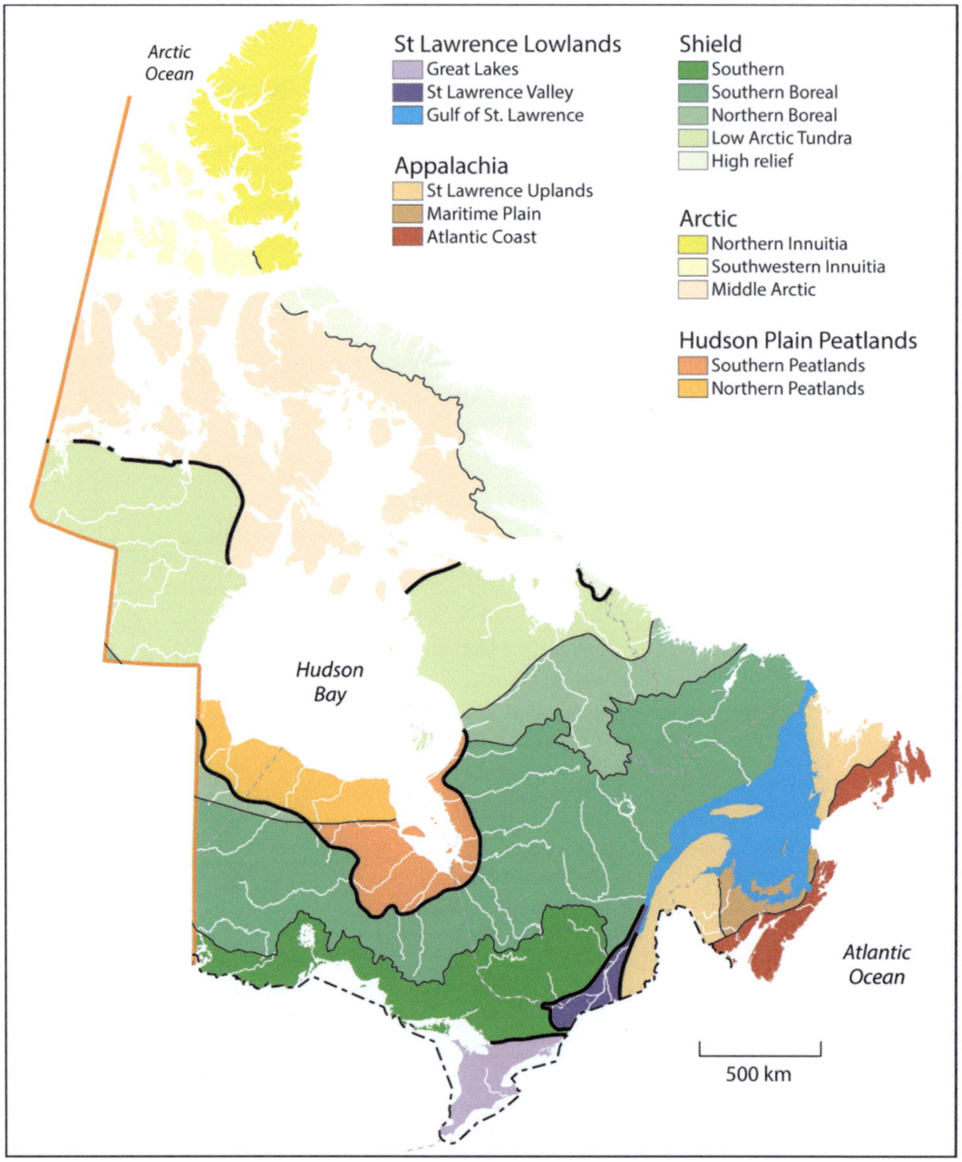

Fig. 3.16 Fluvial regions of eastern Canada (boundaries modified from Ashmore and Church 2001). *Cartography* by Eric Leinberger

by thaw rate rather than mechanical erosion. In such circumstances, the presence of permafrost is a limitation on the erosional effectiveness of large floods (Woo and McCann 1994). On vegetated banks, the thawing of the permafrost may undermine the bank such that an organic mat may drape over the bank face and protect it from hydraulic erosion. Long-term rates of bank erosion appear to be no different in permafrost regions than in temperate regions.

Icing is the result of northern rivers freezing to their bed in winter and expelling water trapped under the bed. Large snow dams may block river channels and dam failure releases exceptionally large discharge. Drift-ice processes are common along Canada's coastlines, giving rise to ice push features such as ice-push ridges, ice-push boulders and scoured furrows (Dionne 1994).

3.6 Hydroclimate Change and Coastal Responses

The Geological Survey of Canada (Shaw et al. 1998) has concluded that there is only one major coastal region in eastern Canada with high sensitivity to physical changes due to sea level rise, namely Maritime Canada, which means parts of the coasts of Nova Scotia, New Brunswick and Prince Edward Island. There are also small areas of high sensitivity in Québec and Newfoundland. Here sensitivity means the degree to which a coastline may experience changes due to sea level rise. Most of eastern Canada and especially Nunavut is continuing to experience isostatic uplift and therefore is relatively insensitive to sea level rise (i.e. the rate of uplift exceeds the rate of sea level rise). For

Fig. 3.17 Great Lakes Lowland. **a** Lake St. Clair (area 1,114 km^2, elevation 175 m, average depth 3.7 m) is fed by the St. Clair River, which flows southwards from Lake Huron to form an extensive river delta where it enters Lake Saint Clair. The lake drains into Lake Erie to the south via the Detroit River. The international boundary between Canada and the United States runs through the lakes, such that urban and agricultural pollutant sources, has to be internationally managed. **b** Lake St. Clair is part of the St. Lawrence Seaway, a significant transportation route stretching from Lake Superior through the Great Lakes to the Gulf of St. Lawrence (see Sect. 3.8.1). Captured in the satellite image are algal blooms visible as swirls of green. **c** The bird's foot delta is the largest delta in the Great Lakes system. *Source* a—MDA Federal (2004). b—NASA Earth Observatory. c—from ArcGIS.com (ESRI)

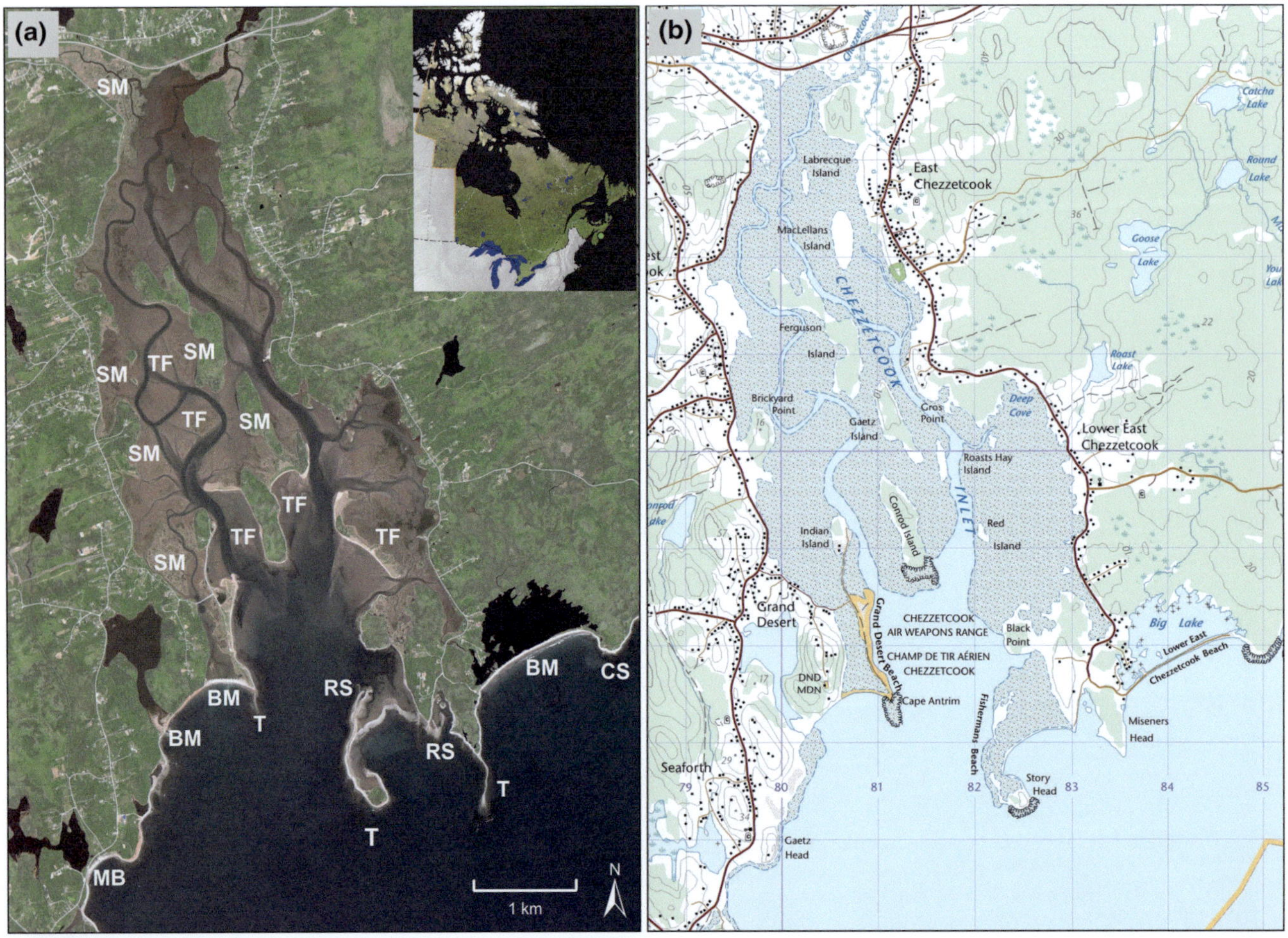

Fig. 3.18 Wave-dominated coast, Nova Scotia. **a** Adjacent segments of a deeply indented, complex coastline may show either drift-aligned or swash-aligned features, as here, Chezzetcook Inlet. The meso-tidal coastline also shows tidal geomorphology. BM—baymouth bar; CS—cuspate spit; MB—mid-bay bar; RS—recurved spit; SM—saltmarsh; T—tombolo; TF—tidal flat. *Source* Data from Google Earth. **b** Topographic map of the Chezzetcook Inlet. *Source* Government of Canada; Natural Resources Canada (2011). © Department of Natural Resources Canada. All rights reserved

those regions that are said to be highly sensitive, the following local distinctions apply.

3.6.1 Barrier Island Coastlines

The development and geomorphology of a barrier island coastline are conditioned by the balance between a number of transportational and erosional processes. Maintenance of a barrier island requires influx of sediment through littoral drift. The necessary sand feed may come from off shore sources such as submerged longshore bars, terrestrial sources such as rivers and deltas or combinations of these. Interruption to the supply of sand may be caused by rapid changes of sea level, reduction of sediment influx, both of which are controlled by hydroclimate change, or by human interference with the feeding mechanism, such as docks or breakwaters. Barrier island development is limited to micro- or meso-tidal situations. Meso-tidal barrier islands do occur along the Gulf of St. Lawrence in New Brunswick and Prince Edward Island. At Kouchibouguac National Park, New Brunswick, the barrier island system developed as sea level rose slowly over the past 2,500 years; similar changes have been observed in the barred littoral drift coastline of Bouctouche Spit, south of the national park (Ollerhead and Davidson-Arnott 1995).

3.6.2 Wave-Dominated Coasts

Many coastlines develop along deeply embayed shorelines common in Atlantic Canada and along the shorelines of the upper Great Lakes. Such coasts are less vulnerable to hydroclimate change except in so far as wave energy is influenced by increasing incidence of storms. Curved or scallop-shaped cusps are diagnostic features associated

Fig. 3.19 Tide-dominated coasts. **a** The Bay of Fundy and the Hopewell Rocks. Tides in the Bay of Fundy are among the most extreme in the world. Chignecto Bay and its branching bays, at the northeastern end of the Bay of Fundy, both have a huge tidal range that varies between 10 and 14 m. **b** The Bay of Fundy is home to interesting geomorphic features, the Hopewell Rocks formation, that produces so-called flower pot shaped stacks. This formation is shaped by the tides, the strength of the rocks (arkosic sandstone and coarse, poorly sorted conglomerates) and the presence of joints in the rocks. **c** Big Cove, Hopewell Rocks. **d** Daniel's Flats, an extensive mud flat at the northwestern side of Chignecto Bay, at low tide. **e** Daniel's Flats at high tide. *Source* a—MDA Federal (2004). b—A. Schaffer/flickr CC BY 2.0. c—M. Sprague/flickr CC BY-SA 2.0. d—A. Schaffer/flickr CC BY 2.0. e—M. Sprague/flickr CC BY 2.0

with wave-dominated coasts. On a high-energy shore-normal coast, baymouth bars and spits develop. On a shore-parallel drift-aligned coast, cells develop from wave refraction and mid-bay and bay head bars may develop. Cuspate spits and tombolos are indicators of high wave energy and laterally confined embayments (Fig. 3.18).

3.6.3 Tide-Dominated Coasts

Tidal range is the most important driver of change and is classified as micro-tidal (0−2 m), meso-tidal (2–4 m) and macro-tidal (>4 m). Micro-tidal conditions dominate along the coastlines of Newfoundland and Labrador, Northumberland Strait, Hudson Bay and Baffin Bay. Meso-tidal conditions dominate along the Atlantic coast of Nova Scotia, the southern Gulf of St. Lawrence coastline of New Brunswick and Prince Edward Island. Macro-tidal areas in eastern Canada are the Bay of Fundy, Cumberland Sound and Ungava Bay. Tide-dominated coastlines with abundant sediment influx and gently sloping or eroded bedrock are commonly flanked by broad tidal flats (Fig. 3.19). Salt marsh development is influenced by sediment influx, the rate of sea level change, ice scouring and human activity.

Sea ice influences coastlines throughout Arctic Canada and in Atlantic Canada in the Gulf of St. Lawrence, north-eastern Newfoundland, and Labrador (Forbes 2011). Sea ice can be protective or act as an erosional agent. Effects are greatest in macro-tidal coasts such as Ungava Bay and Cumberland Sound in Nunavut or Pistolet Bay in Newfoundland (Forbes and Taylor 1994).

3.6.4 Sensitivity to Hydroclimate Change

Changes in sea level are driven by a combination of local, regional, hemispheric and global factors (Church and White 2011). When considering geomorphology and coastal hazards, it is the relative sea level (RSL) that is important rather than the absolute sea level change. The largest region of high sensitivity comprises much of the coasts of Nova Scotia, Prince Edward Island and New Brunswick, but many localized areas of the St. Lawrence north shore in Québec and islands in the Gulf of St. Lawrence show equally high sensitivity. Shaw et al. (2001) have selected three areas of eastern Canada that are especially sensitive to sea level rise for detailed study: Placentia, Newfoundland; Eastern Shore, Nova Scotia; and Head of the Bay of Fundy (Fig. 3.20). Their findings are repeated below, with acknowledgment.

3.6.4.1 Placentia, Newfoundland

Placentia (population 4,500) was founded by French settlers in 1715 but the site was occupied by Basque fishermen prior to that date. It is located on the Avalon Peninsula near St. John's, Newfoundland. Placentia has a history of flooding which has become more of a problem in recent decades as the settlement has expanded into lower areas of the beach-ridge plain. Several times since the 1980s, houses have had to be evacuated and power outages have occurred because of flooding. Flooding is due to high water levels in Placentia Arm that occur when high spring tides and storm surges coincide and waves overtop the beach. A high gravel berm was constructed in 1992 and a sea wall was built in the northern part of the settlement. Maintenance costs and the risks of storm damage will likely increase sharply if sea level rises more rapidly due to global warming. Present rate of sea level rise along the Avalon Peninsula is approximately 30 cm per century.

3.6.4.2 Eastern Shore, Nova Scotia

During the past 10,000 years, sea level has risen 45 m on Nova Scotia's eastern shore and is still rising at 36 cm/century at Halifax. This is mostly due to long-term crustal subsidence. In the Story Head area, sea level rise has created many hazards over the past 50 years. Roads, fishing camps and farmland have been consumed by the ocean. Story Head beach has migrated rapidly landward, especially rapidly between 1966 and 1974 at >10 m/annum. During storms, water levels can be raised up to 1 m above normal levels, causing local flooding. During Hurricane Hortense in September 1996, barrier beaches were affected at elevations of 4.5 m above sea level and the Story Head beach was pushed 38 m landward.

3.6.4.3 Head of the Bay of Fundy

The Bay of Fundy region has one of the largest tidal ranges in the world, up to 16 m on exceptional spring tides. Sea level has been rising at a rate of c. 33 mm/annum. The coastal areas consist of mudflats and salt marshes. A high salt marsh, dominated by *Spartina patens,* is found at elevations >1 m above mean high water. A low salt marsh (*Spartina alerniflora*) occurs at elevations below 1 m above mean high water. These salt marshes have kept pace with sea level rise for the past few 1000 years, but during several phases of slower than average rise of sea level fresh water, systems have spread across these marsh surfaces. Acadians first dyked and drained parts of the salt marshes in the late 1630s. Today about 85% of the former marsh lies behind dykes constructed at various times. The cost of dyke maintenance will obviously increase. Major storms will overtop

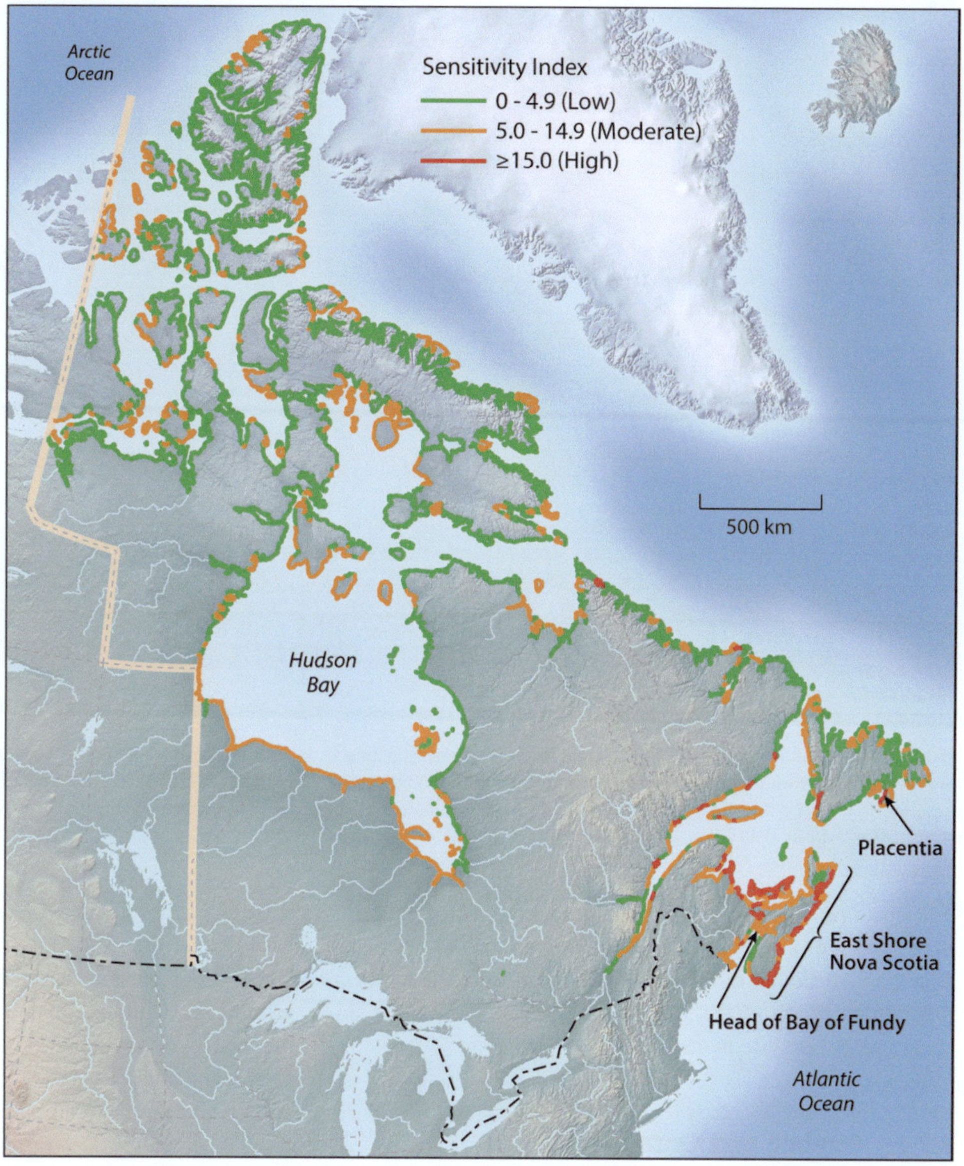

Fig. 3.20 Sensitivity of eastern Canada's coastline. The sensitivity index is a measure of the hazard presented by rising sea level: (a) Low in green, (b) Moderate in yellow and (c) High in purple. Note the locations of Placentia, Newfoundland, Eastern Shore, Nova Scotia and the Head of the Bay of Fundy (Shaw et al. 2001. *Cartography* by Eric Leinberger

the dykes and inundate the lowlands behind them, damaging agricultural land and flooding roads and railways (Shaw et al. 1998).

3.7 Regional Hazards

3.7.1 Arctic

A combination of warmer temperatures, shorter winter seasons and sea level rise are expected to have a serious impact on natural and human systems in Canada's Arctic (Maxwell 1997). Reductions in sea ice thickness and a rise in sea level could threaten the natural habitat that supports unique plants and animals indigenous to the North. Canada's tundra could also shrink by as much of a third of its current size, becoming confined to the arctic archipelago (MacDonald et al. 1993). Wetlands in the southern NWT are projected to be severely impacted by falling water levels and increasing temperatures (ACIA 2005). Traditional lifestyles and the wage economy are expected to be extensively impacted. Severing the historical bond between indigenous people and the land is the long-term threat that is epitomized by the generation gap between Inuit youth and elders. Hence the excitement that has occurred with the publication of 'The Indigenous People' Atlas of Canada' (Canadian Geographic 2018) in which the rich cultural diversity of eastern Canada is explored.

Increased accessibility amongst the arctic islands and the Northwest Passage as a result of less severe ice conditions will prompt increasing surveillance, search and rescue related to maintaining Canadian sovereignty (Assel et al. 1995).

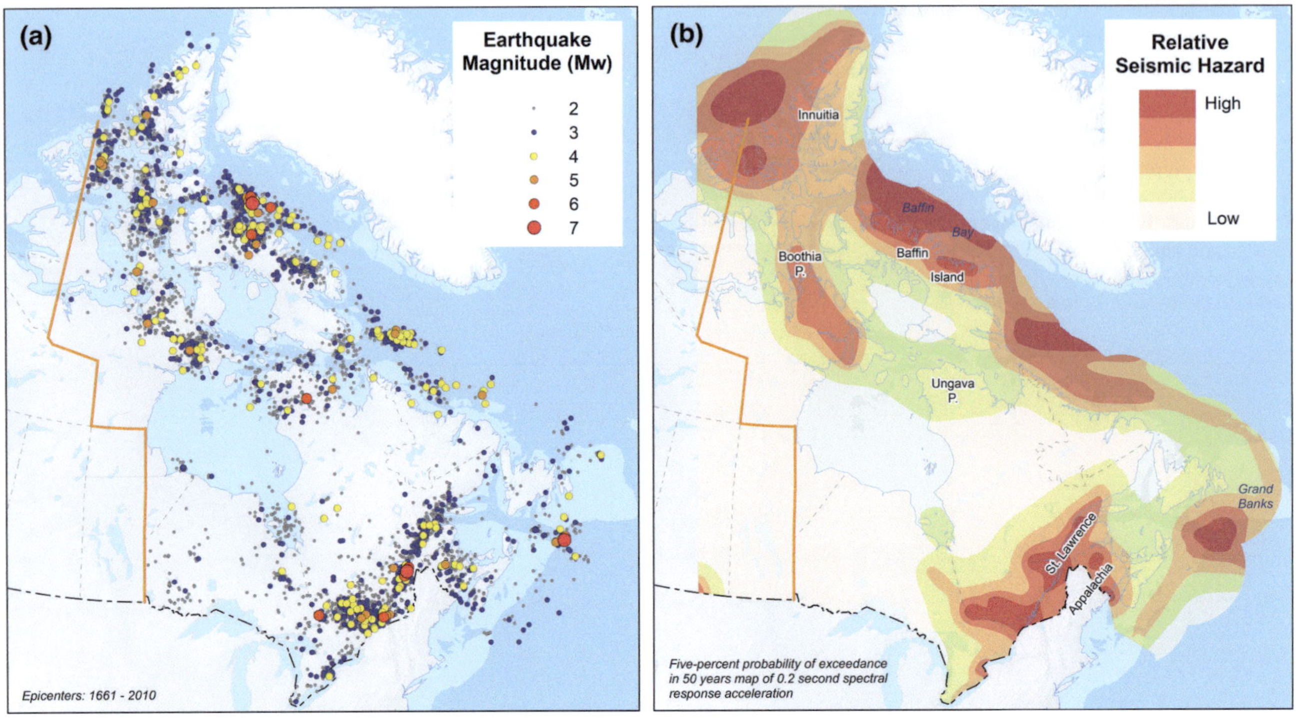

Fig. 3.21 Seismicity and simplified seismic hazards of eastern Canada. **a** Earthquakes occur in most areas of eastern Canada, certain areas have a higher probability of experiencing damaging ground motion caused by earthquakes. **b** The map shows relative seismic hazard zones across eastern Canada. It is based on the National Building Code of Canada and shows the relative seismic hazard for single-family dwellings (1–2 story structures). *Source* data from Halchuk et al. (2015a, 2015b)

The costs of offshore oil and gas operations from hydroclimate change alone are difficult to calculate; when the changing social attitudes towards human-induced environmental deterioration are also factored in, predictions at this stage are impossible to make.

3.7.2 Québec and Ontario

Impacts of hydroclimate change in Québec and Ontario are also expected to be significant (Bergeron et al. 2001). Many of the impacts in these provinces are tied to the concentrations of industrial activity, population and urban centres throughout the region, such as the impacts on human health, building and construction and water resources. In the Great Lakes-St. Lawrence basin, water resources could be severely compromised by intensifying human consumption (Hofmann et al. 1998). Declines in water levels could adversely impact the generation of hydropower in southern Ontario. Impacts in northern Québec and Ontario tend to be more positive, for example, possible increases in power generation potential in Labrador and northern Québec and increases in crop yields and land areas suitable for cultivation (Brklacich et al. 1994).

There is a distinction made between seismicity and seismic hazards, where seismicity is the number and magnitude of earthquakes recorded over a specified time interval and seismic hazard expresses the probability of exceedance of an earthquake of specified magnitude. In this sense, the seismic hazard is greatest in the St. Lawrence Lowlands and offshore around the Grand Banks. There is also moderate seismicity in the St. Lawrence River valley from Montreal to Sept-Iles. There are three clusters of greater than average seismicity, namely western Québec, Charlevoix and the lower St. Lawrence region near Baie-Comeau (Fig. 3.21).

3.7.3 Atlantic Canada

In Atlantic Canada, sea level rise is seen to be the most significant potential impact of hydroclimate change (Abraham et al. 1997). Coastal erosion and associated aeolian activity indicate the severity of this hazard (Fig. 3.22).

Flood hazards, storm damage and overtopped dykes in the Bay of Fundy can prove costly. The data suggest that the number of flood disasters has increased through the twentieth century, with about 70% occurring after 1959 (Beltaos et al. 1996). This trend likely reflects several factors. First, the magnitude of flood events in the second half of the twentieth century has generally increased, at least along some rivers. Second, development has increased on flood-prone lands as Canada's population has grown. Third, flood reporting has improved over the past few decades.

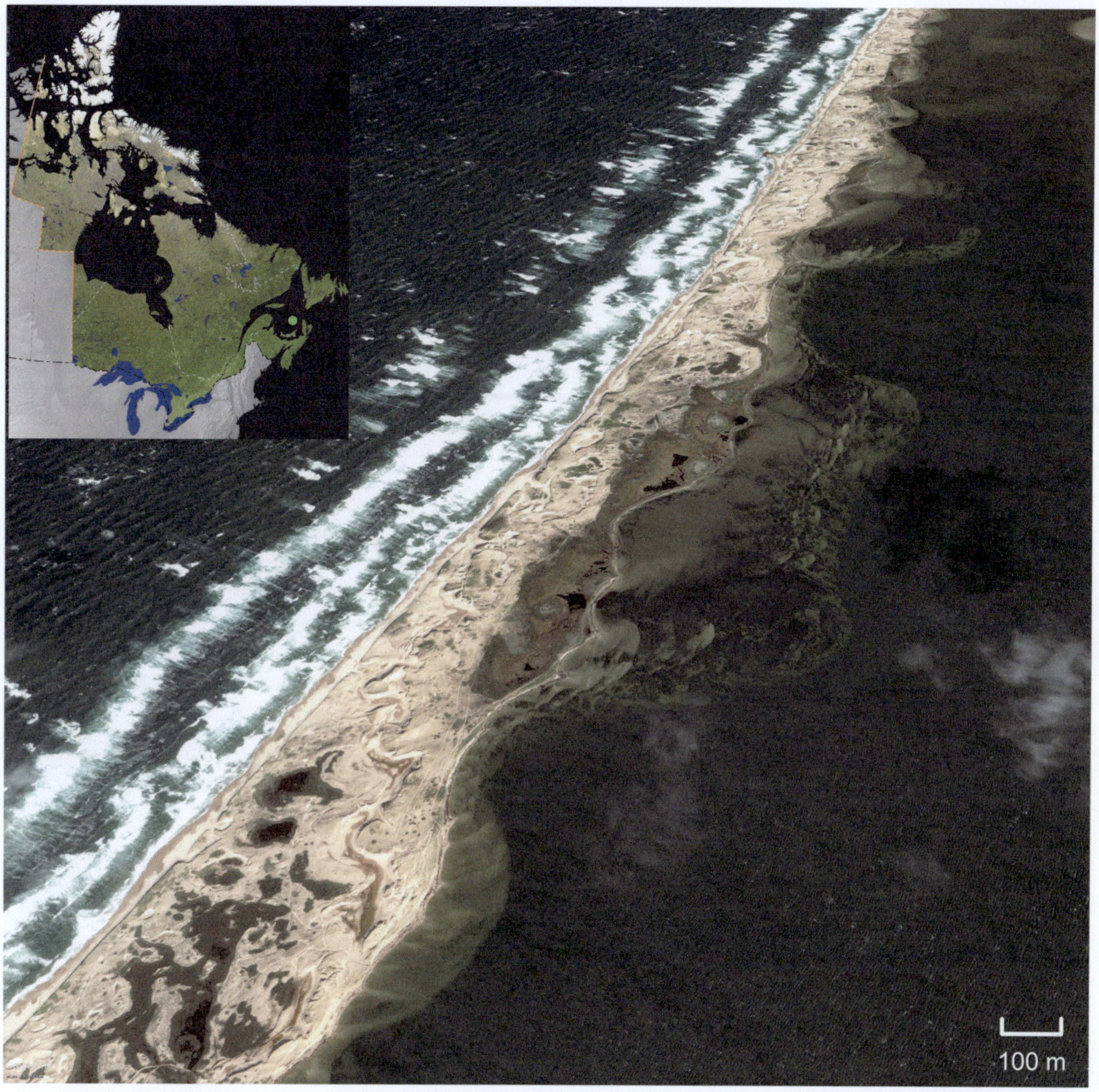

Fig. 3.22 Transverse dunes paralleling the shoreline, Magdalen Islands, Québec. The dunes are subject to coastal erosion, and blowouts are developed on the surface. *Source* Data from Google Earth. Also see Fig. 1.27b

Landslides in sensitive clays at the surface are a hazard (see Sect. 3.3.3, this chapter) but also, in numerous places, tills superimposed on sensitive clays at depth are subject to mass movement of various kinds. At least three recurrences of damaging landslides (pre-2004; 15 April 2007 and 2013) have occurred in the community of Daniel's Harbour on the Great Northern Peninsula of Newfoundland (Fig. 3.23). The slides have been characterized as debris flows: a combination of translational slip failure and rotational slip failure and occurred as a result of the saturation and subsequent loss of strength of a clay-rich layer underlying till (Catto 2015).

3.8 The Great Lakes-St. Lawrence River Basin

The Great Lakes-St. Lawrence River basin is so large and complex that it is rarely thought of as a single connected system. The basin is unusual in a number of ways. All parts of the basin are connected by water but each of the Great Lakes traps sediments, solutes, nutrients and pollutants from the adjacent land surfaces. There are four major elements in the system: (a) the Great Lakes, (b) the riverine corridor, (c) the estuary and (d) the gulf. From the farthest headwater

Fig. 3.23 Debris flows Daniel's Harbour, Newfoundland. **a** A major landslide took place in Daniel's Harbour, Newfoundland over the period 15–20 April 2007. The slide was characterized as a combination of translational-slip failure and rotational-slip failure and occurred as a result of the saturation and subsequent loss of strength of a clay-rich layer underlying till. **b** The houses are constructed on a marine terrace with a surface elevation of about 24 m asl. *Photographs* Government of Newfoundland and Labrador, all rights reserved; available at https://www.nr.gov.nl.ca/nr/mines/outreach/disasters/landslides/april15_2007.html. Also see Spooner et al. (2012)

Table 3.3 The four regions of the Great Lakes-St. Lawrence Basin. After Frenette (1990)

Region	Cumulative Basin Area (km^2)	Cumulative Channel Length (km)	Mean Annual Discharge (m^3/s)
Great Lakes	722,000	3,050	7,000
Riverine	1,120,000	3,550	12,000
Estuary	n.a.	4,150	16,800
Gulf	n.a.	4,750	n.a.

in the Mesabi Range, Minnesota, to the Gulf of St. Lawrence, the St. Lawrence system is shared between eight states of the United States and two Canadian provinces, Ontario and Québec (Table 3.3).

The basin achieved its present form only about five millennia ago after forming its channel in the glacimarine clays of the Champlain Sea. Therefore from a geomorphological perspective, the St. Lawrence is a young river, even though the valley through which it runs is very old. The St. Lawrence drainage system is delimited by the Laurentian Highlands to the north and the Appalachian Mountains to the south. Consequently, a mature landscape parallels the river all along its course to the sea. That mature landscape originated at the beginning of the Ordovician Period, about 500 million years ago, when newly formed sediments derived from erosion of the Appalachians came into contact with the igneous rocks of the Canadian Shield to the north. This line of contact is known as the Logan Fault and is the origin of the very old depression in which the St. Lawrence River sits. The Great Lakes portion of the St. Lawrence basin is largely a creation of the Quaternary Period with repeated advance and retreat of the Laurentide Ice Sheet during the past two and a half million years. Several of the Great Lakes, however, are themselves structurally controlled troughs of much greater age.

Jacques Cartier, the European explorer of much of the Gulf of St. Lawrence and of the St. Lawrence River in the 1530s became disillusioned by the absence of gold, the absence of a sea route, the rigours of a Québec winter and the vigorous attacks by the indigenous Algonquian speaking people from the Ottawa valley, and the Iroquoian speaking people from Lake Huron. In 1608, Samuel de Champlain established a fur trading outpost at Québec City, an easily fortified site at the head of navigation for large ships. It was not until 1634 that Trois Rivières was founded followed by Montréal in 1642. After the Iroquois and Algonquins were displaced into the forest, eastern Canada, and specifically the St. Lawrence valley, became the cradle of the Canadian settler society that has impacted this landscape continuously over the past four centuries (Frenette 1990).

3.8.1 The Great Lakes and St. Lawrence Seaway

Since 1959, the year of its official opening, the St. Lawrence Seaway has been, in comparison with the Mississippi or with Hudson Bay, incontestably the shortest route between Europe and the centre of North America. Thanks to the St. Lawrence and the Great Lakes, the heart of North America is accessible from ports the world over (Fig. 3.24). However, only 10% of current oceangoing ships, which have been built much larger than in the 1950s, can traverse the entire seaway. Much more important today are the environmental issues that have accompanied seaway development. The introduction of invasive species of aquatic animals into the Great Lakes basin has had seriously damaging effects, and agricultural and industrial pollutants from both sides of the border have influenced the lake and water quality negatively. The seaway remains important for Canadian and American international trade and is, for example, the main route for Ontario grain exports to overseas markets (Macfarlane 2014).

3.8.2 The Riverine Section

The riverine corridor from the eastern end of Lake Ontario to Ile d'Orléans north of Québec City is 500 km long. The majority of the shore is in the province of Québec. From the Great Lakes to Lac Saint-Pierre, the St. Lawrence River makes its way through the so-called Thousand Islands, shared by Ontario and New York State. After Lac St. Francois, the St. Lawrence River is joined by its most important tributary, the 1,271 km-long Ottawa River (146,300 km^2 basin area draining the Laurentian Highlands of the Canadian Shield). The Richelieu River, draining Lake Champlain, joins the St. Lawrence upstream from Lac Saint-Pierre. This is a part of St. Lawrence River that was referred to as the 'zone of jamming' between Montréal and Québec. Ice breakers were needed to permit winter navigation as far as Montréal.

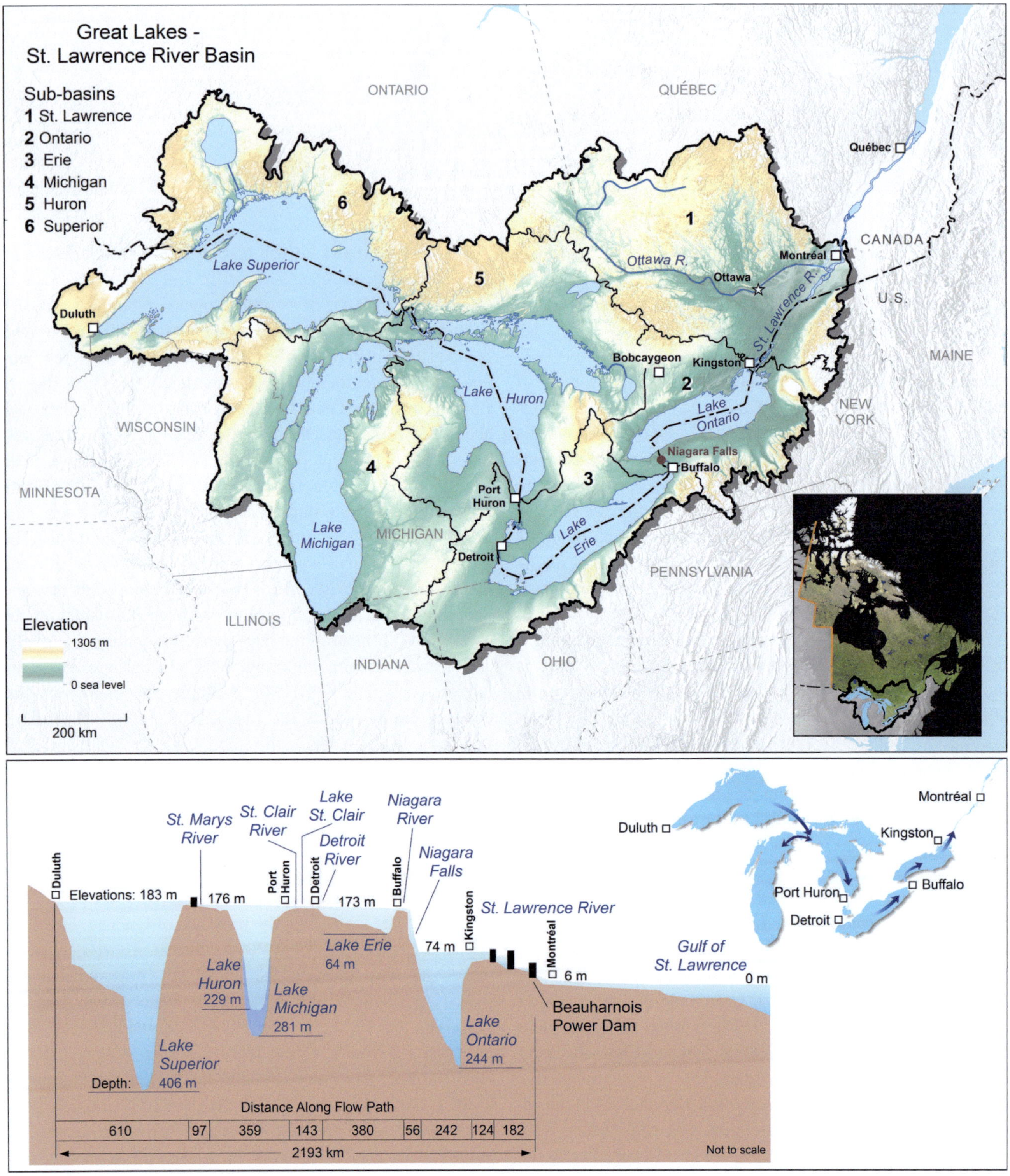

Fig. 3.24 Great Lakes-St. Lawrence River basin: upper image shows the whole basin above Montréal; lower image is a long profile of the Great Lakes-St. Lawrence Seaway

3.8.3 The Estuary

From Ile d'Orléans to a north–south line from Sept-Iles to Matane, the estuary is a further 600 km long, bounded exclusively by the province of Québec to north and south. At the Ile d'Orléans, just below Québec City spectacular changes occur with width changes from 1 to 19 km. At Québec City where the historic site of the Plains of Abraham sits atop imposing cliffs, the river is just 1 km wide. The Rivière Saint-Maurice joins the St. Lawrence from the north shore Laurentian Highlands, a region known as 'La Mauricie'. The lower part of the estuary is characterized by the mixing of fresh and salt water. The eastern limit of the estuary is the Saguenay Fjord where the highlands of the Laurentians reach 1,070 m asl. This part of the St. Lawrence is known as the drift-ice zone which, without the presence of ice breakers was a continuing hazard for shipping prior to the 1960s.

3.8.4 The Gulf

The Gulf of St. Lawrence is not strictly a part of the St. Lawrence River basin, though it is defined as the outlet of the Great Lakes via the St. Lawrence River into the Atlantic Ocean. Downstream from the Saguenay, the river attains more than 500 m depth and widens to more than 90 km between Sept-Iles and the Gaspé coast. At the limit of the gulf, east of Anticosti Island, the continental shelf is reached and it becomes impossible to define the limits of the river (Frenette 1990).

3.8.5 Environmental Hazards

The Great Lakes are suffering ecological strain, an environmental hazard the seriousness of which can be judged from the fact that roughly one-quarter of Canada's population and roughly one-tenth of USA's population drink from the waters of the Great Lakes basin. The environmental hazards are wide ranging, such as the damming of tributaries, the reduced ice cover, the high phosphorus levels in the Great Lakes and the presence of PCBs in Great Lakes sediments even though PCBs have not been manufactured nor imported to North America since around 1980 (Marvin et al. 2004). Water, sediment and nutrient flows are impeded, water temperatures have increased because of reduced ice cover over the past 40 years, agricultural and municipal runoff has led to eutrophication and hypoxia, and the complex food web of the Great Lakes has been affected by the presence of PCBs in Great Lakes sediments.

3.9 Conclusion

Hazards associated with eastern Canada's contemporary geomorphic processes are exacerbated by human activity, especially around the shorelines of the Atlantic Ocean and the Great Lakes. It has been argued elsewhere (Slaymaker and French 2012) that human activity, in the form of land use, land cover change and economic activity, will become the dominant driver of environmental change as the twenty-first century progresses. The effects of contemporary human activity on the landscape are important and more is said about this in Chap. 21.

The changing social order is even less predictable than the future of the biophysical landscape. Population demographics, resource management and the forces of globalization have a direct bearing on environmental change. While the biophysical landscapes of eastern Canada cannot be understood without reference to their coldness and 'nordicity' (French and Slaymaker 2012); neither can they be understood without reference to hydroclimate change (Fig. 3.25) and the interaction of people, the ecology and their environment.

Eastern Canada will continue to be a relatively cold place over the next century in spite of the recent warming trend. The cryosphere, though probably depleted, will continue to be an important part of eastern Canadian reality. The landscape will continue its process of rebounding from the latest glaciation. The newest and most pervasive changes will likely be associated with the process of globalization associated with a shrinking and interconnected northern hemispheric world. We believe that the spatial and temporal variability of the Canadian landscape over the next century will be driven primarily by human activity. Adaptive management of the environment and continuing warming temperatures, as illustrated by the Great Lakes-St. Lawrence basin, will be important components of this changing scenario.

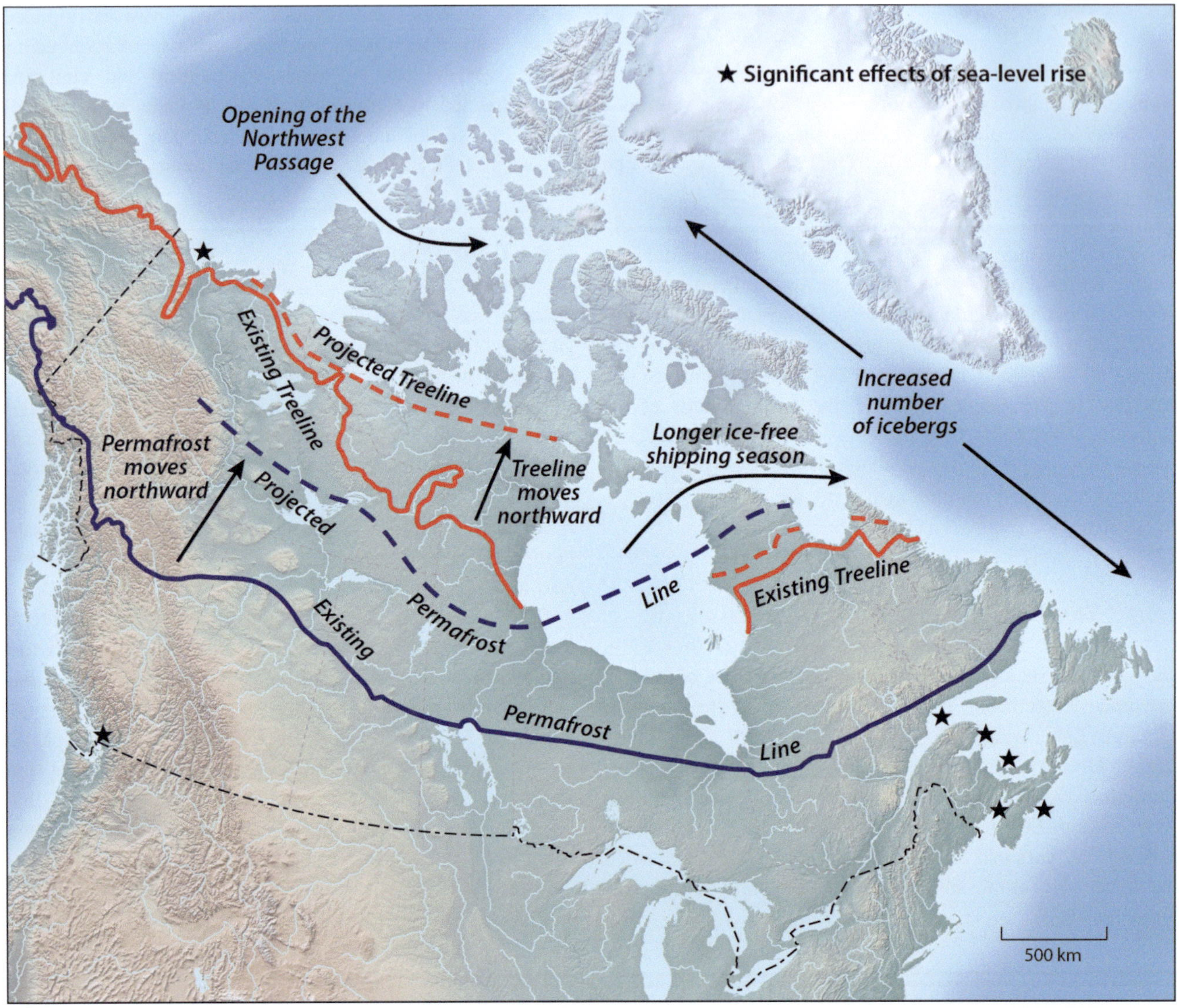

Fig. 3.25 Projected environmental changes in Canada under continuing global warming, time scale unspecified (after Slaymaker and French 1993) Cartography by Eric Leinberger

References

Abraham I, Canavan T, Shaw R (eds) (1997) Climate change and climate variability in Atlantic Canada. The Canada country study on climate change, v. 6. Environment Canada, Bedford Nova Scotia

Ashmore P (1993) Contemporary erosion of the Canadian landscape. Prog Phys Geog 17:190–204

Ashmore P, Church M (2001) Impact of climate change on rivers and river processes in Canada. Geol Surv Can Bull 555

Assel RA, Robertson DM, Hoff MH, Selgeby JH (1995) Climatic change implications from long term (1823–1994) ice records for the Laurentian Great Lakes. Ann Glac 21:383–386

Ballantyne CK (2002) Paraglacial geomorphology. Quat Sci Rev 21:1935–2017

Banfield CE, Jacobs JD (1998) Regional patterns of temperature and precipitation for Newfoundland and Labrador during the past century. Can Geogr 42:354–364

Barber DC, Dyke A, Hillaire-Marcel C, Jennings AE, Andrews JT, Kerwin M, Bilodeau G, Mcneely R, Southon JR, Morehead MD, Gagnon J-M (1999) Forcing of the cold event of 8,200 years ago by catastrophic drainage of Laurentide lakes. Nature 400:344–348

Bell R (1896) Proofs of the rising of the land around Hudson Bay. Am J Sci 151:219–228

Beltaos S, Burrell B, Ismail S (1996) 1991 ice jamming along the St. John River: a case study. Can J Civil Eng 23:381–394

Bergeron Y, Gauthier S, Kafka V, Lefort P, Lesieur D (2001) Natural fire frequency for the eastern Canadian boreal forest: consequences for sustainable forestry. Can J For Res 31:384–391

Bouchard M, Pavich MJ (1989) Characteristics and significance of pre-Wisconsinan saprolites in the northern Appalachians. Zeits Geomorph Supp 72:125–137

Boyd R, Bowen AJ, Hall RK (1987) An evolutionary model for transgressive sedimentation on the eastern shore of Nova Scotia. In: FitsGerald DM, Rosen PS (eds) Glaciated coasts. Academic Press, San Diego, pp 87–114

Briner JP, Michelutti N, Francis DR, Miller GH, Axford Y, Wooller MJ, Wolfe AP (2004) A multi-proxy lacustrine record of Holocene climate change on northeastern Baffin Island, Arctic Canada. Quat Res 65:431–442

Brklacich M, Stewart R, Kirkwood V, Muma R (1994) Effects f global climate change on wheat yields in the Canadian prairie. In: Rosenzweig C, Iglesias A (eds) Implications of climate change for international agriculture: crop modelling study. US Environmental Protection Agency, Washington, DC, pp 1–23

Brooks GR (ed) (2001) A synthesis of geological hazards in Canada. Geol Surv Can Bull 548

Brooks GR, Lawrence DE (1998) Geomorphic effects and impacts of severe flooding: photographic examples from the Saguenay area, Québec. Geological Survey of Canada Misc Report 62

Burn CR (1992) Thermokarst lakes: Canadian landform examples 24. Can Geogr 36:81–85

Burn CR, Lewkowicz AG (1990) Retrogressive thaw slumps: Canadian landform examples 17. Can Geogr 34:273–276

Campo SH, Desloges JR (1994) Sediment yield conditioned by glaciation in a rural agricultural basin of southern Ontario. Phys Geog 15:495–515

Canadian Geographic (2018) Indigenous peoples Atlas of Canada. Royal Canadian Geographical Society, Ottawa

Carrivick JL, Tweed FS (2013) Proglacial lakes: character, behaviour, and geological importance. Quat Sci Rev 78:34–52

Catto N (2015) Geomorphology: landscapes we live in. Oxford University Press, Don Mills, Ontario

Chiotti Q (1998) An assessment of the regional impacts and opportunities from climate change in Canada. Can Geogr 42:380–393

Church JA, White NJ (2011) Sea-level rise from the late 19th to the early 21st century. Surv Geophys 32:585–602

Church M (1972) Baffin Island sandurs: a study of arctic fluvial processes. Geol Surv Can Bull 216

Church M (1992) Channel morphology and typology. In: Carlow P, Petts GE (eds) The rivers handbook. Blackwell Scientific, Oxford, UK, pp 126–143

Church M, Ryder JM (1972) Paraglacial sedimentation: a consideration of fluvial processes conditioned by glaciation. Geol Soc Am Bull 83:3059–3072

Church M, Ham D, Hassan M, Slaymaker O (1999) Fluvial clastic sediment yield in Canada: scaled analysis. Can J Earth Sci 6:1267–1280

Cummings CE, Pollard WH (1990) Cryogenetic characterization of peat and mineral-cored palsas in Schefferville area, Québec. Collection Nordica 54:95–102

Dionne J-C (1994) Les cordons de blocs frangeants (boulder barricades). Mise au point avec exemples du Québec. Rev Géomorphologie Dyn 43:49–70

Dredge LA, Dyke LD (2019) Wetland landscapes and landforms in discontinuous permafrost of the Hudson Bay Lowlands. In: Catto N, Slaymaker O (eds) Landscapes and landforms of Eastern Canada. Cambridge University Press, Cambridge (this volume)

Dredge LA, Nixon FM (1992) Glacial and environmental geology of Northeastern Manitoba. Geological Survey of Canada, Memoir 432

Dyke AS (2004) An outline of North American deglaciation with emphasis on central and northern Canada. In: Ehlers J, Gibbard PL (eds) Quaternary glaciations—extent and chronology, Part II: North America, vol 2b. Elsevier Science and Technology Books, Amsterdam, pp 373–424

Dyke AS, Hooper MJG (2001) Deglaciation of northwest Baffin Island, Nunavut. Geological Survey of Canada Map 1999A. Scale 1:500 000

Dyke AS, Moore A, Robertson L (2003) Deglaciation of Canada. Geological Survey of Canada Open File 1574. Thirty-two digital maps at 1:7,000,000 scale with accompanying digital chronological database and one poster (two sheets) with full map series; 1 CD-ROM

England J, Atkinson N, Bednarski J, Dyke AS, Hodgson DA, O Cofaigh C (2006) The Innuitian Ice Sheet: configuration, dynamics and chronology. Quaternary Sci Rev 25:689–703

Evans SG (2000) The record of disastrous landslides and geotechnical failures in Canada (1840–1999). Workshop on geotechnique and natural hazardous, 53rd geotechnical conference, Montreal 15–18 October

Filion L, St. Laurent D, Desponts M, Payette S (1991) The Late Holocene record of aeolian and fire activity in northern Québec. Holocene 1:201–208

Fisheries and Environment Canada (1978) Hydrological Atlas of Canada. NRC of Canada, Ottawa

Forbes DL (ed) (2011) State of the Arctic Coast 2010: Scientific review and outlook. International Arctic Science Committee, Land-oceans interactions in the coastal zone project, International Permafrost Association, Arctic monitoring and assessment programme working group of the Arctic council, Geesthacht, Germany

Forbes DL, Syvitski JPM (1994) Paraglacial coasts. In: Carter RWG, Woodroffe CD (eds) Coastal evolution: Late Quaternary Shoreline Morpho-dynamics. Cambridge University Press, Cambridge, pp 373–424

Forbes DL, Taylor RB (1994) Ice in the shore zone and the geomorphology of cold coasts. Progr Phys Geog 18:59–89

Fortier D, Allard M, Shur Y (2007) Observation of rapid drainage system development by thermal erosion of ice wedges on Bylot Island, Canadian Arctic Archipelago. Permafrost Periglac 18:229–243

Fraser JK (1958) Freeze-thaw frequencies and mechanical weathering in Canada. Arctic 11:40–53

French HM (1988) The active layer. In: Clark MJ (ed) Advances in Periglacial Geomorphology. Wiley, Chichester, UK, pp 151–177

French HM (2007) The Periglacial environment, 3rd edn. Wiley, Chichester, UK

French HM, Slaymaker O (eds) (2012) Changing Cold Environments: a Canadian Perspective. Wiley, Chichester, UK

Frenette M (1990) The St. Lawrence. In: Wolman MG, Riggs HC (eds) Surface water hydrology: Geology of North America, v. O-1. Geological Society of America, pp 295–304

Gajewski K, Payette S, Ritchie JC (1993) Holocene vegetation history at the boreal forest-shrub tundra transition in northern Québec. J Ecol 81:433–443

Gajewski K, Atkinson DA (2003) Climatic change in northern Canada. Environ Rev 11:69–102

Gardner JS (1977) Some geomorphic effects of a catastrophic flood on the Grand River, Ontario. Can J Earth Sci 14:2294–2300

Gilbert R (1983) Sedimentary processes of Canadian arctic fjords. Sedim Geol 36:147–175

Government of Canada; Natural Resources Canada (2007) Canada Centre for Mapping and Earth Observation. CanImage - Landsat 7 Orthoimages of Canada, 1:50 000 Tiles: 036i07, 043m14. ftp.geogratis.gc.ca/pub/nrcan_rncan/image/

Government of Canada; Natural Resources Canada (2011). Canada Centre for Topographic Information. Cantopo digital map, 011d11, West Chezzetcook—1:50 000. ftp.geogratis.gc.ca/pub/nrcan_rncan/raster/cantopo/50k/

Government of Canada; Natural Resources Canada (2018) The State of Canada's Forests: Annual Report. Ottawa

Government of Québec, Ministry of Forests, Wildlife and Parks (MFFP) (2017) Bank of photos of Northern Québec. www.donneesquebec.ca/recherche/fr/dataset/photographie-oblique-du-nord-quebecois

Grant DR (1989) Quaternary geology of the Atlantic Appalachian region of Canada. In: Fulton RJ (ed) Quaternary Geology of Canada and Greenland. Geological Survey of Canada. Geology of Canada Series no. 1, pp 391–440

Gray JT, Seppala M (1991) Deeply dissected tundra polygons on a glacio-fluvial outwash plain, Northern Ungava Peninsula, Québec. Geogr Phys Quat 45:111–117

Halchuk S, Allen TI, Rogers GC, Adams J (2015a) Seismic hazard earthquake epicentre file (SHEEF2010) used in the fifth generation seismic hazard maps of Canada. Geological Survey of Canada, Open File 7724

Halchuk SC, Adams JE, Allen TI (2015b) Fifth generation seismic hazard model for Canada: grid values of mean hazard to be used with the 2015 National Building Code of Canada. Geological Survey of Canada, Open File 7893

Hare FK, Thomas MK (1979) Climate Canada, 2nd edn. Wiley, Toronto

Hofmann NL, Mortsch L, Donner S, Duncan K, Kreutzwiser R, Kulshreshtha S, Piggott A, Schellenberg S, Schertzer B, Slivitzky M (1998) Climate Change and Variability: Impacts on Canadian Water. In: Koshida G, Avis W (eds) The Canada Country Study: Climate Impacts and Adaption, vol VII. National Sectoral Volume. Environment Canada, Ontario, pp 1–120

IPCC (2013) Climate Change 2013: The Physical Science Basis. Contribution of Working Group 1 to the Fifth Assessment Report of the Intergovernmental Panel on Climate Change. In: Stocker TF, Qin D, Plattner G-K, Tignor M, Allen SK, Boschung J, Nauels A, Xia Y, Bex V, Midgley PM (eds) Cambridge University Press, Cambridge, UK and New York, NY, USA

Irvine KN, Drake JJ (1987) Spatial analysis of snow- and rain-generated high flows in southern Ontario. Can Geogr 31:140–149

Ives JD (2016) Baffin Island: field research and high Arctic Adventure, 1961–1967. University of Calgary Press, Calgary

Kaufman DS, Ager TA, Anderson NJ, Anderson PM, Andrews JT, Bartlein PJ, Brubaker LB, Coats LL, Cwynar LC, Duvall ML, Dyke AS, Edwards ME, Eisner WR, Gajewski K, Geirsdóttir A, Hu FS, Jennings AE, Kaplan MR, Kerwin MW, Lozhkin AV, MacDonald MM, Miller GH, Mock CJ, Oswald WW, Otto-Bliesner BL, Porinchu DF, Rühland K, Smol JP, Steig EJ, Wolfe BB (2004) Holocene thermal maximum in the western Arctic (0° to 180° W). Quat Sci Rev 23:529–560

Kellerhals R, Church M (1989) The morphology of large rivers: characterization and management. In: Dodge DP (ed) Proceedings of the international large river symposium. Canadian Fisheries and Aquatic Science, Special Publication 106, pp 31–48

King WA, Martini IP (1984) Morphology and recent sediments of the lower anastomosing reaches of the Attawapiskat River, James Bay, Ontario. Sedim Geol 37:295–320

Kokelj SV, Tunnicliffe JF, Lacelle D, Lantz TC, Chin KS, Fraser R (2015) Increased precipitation drives megaslump development and destabilization of ice-rich permafrost terrain, northwestern Canada. Global Climate Change 129:56–68

Kovanen DJ, Slaymaker O (2015) The paraglacial geomorphology of the Fraser Lowland, southwest British Columbia and northwest Washington. Geomorphology 232:78–93

Ladd M, Gajewski K (2010) The North American summer arctic front during 1948 to 2007. Int J Climatol 30:874–883

Lamoureux SF, Lafreniere MJ (2009) Fluvial impact of extensive active layer detachments, Cape Bounty, Melville Island. Arctic Alpine Res 41:59–68

Locat J, St-Gelais D (2014) Nature of sensitive clays from Quebec. In: L'Heureux JS, Leroueil S, Demers D, Locat J (eds) Landslides in sensitive clays. Advances in natural and technological hazard research, vol 36. Springer, Dordrecht

MacDonald GM, Edwards TWD, Moser KA, Pienitz R, Smol JP (1993) Rapid response of treeline vegetation and lakes to past climate warming. Nature 361:243–246

Macfarlane D (2014) Negotiating a River: Canada, the US, and the Creation of the St. Lawrence Seaway. UBC Press, Vancouver

Marvin D, Painter C, Williams V, Richardson V, Rossmann R, Van Hoof P (2004) Spatial and temporal trends in surface water and sediment contamination in the Laurentian Great Lakes. Environ Pollut 129:131–144

Maxwell B (1997) Recent climate patterns in the Arctic. In: Oechel WC, Callaghan T, Gilmanov T, Holten JI, Maxwell B, Molay U, Sveinbjornsson B (eds) Global change and Arctic terrestrial ecosystems. Springer, Berlin and Heidelberg, pp 21–46

MDA Federal (2004) Landsat GeoCover ETM+ 2000 Edition Mosaics Tiles N-17-40, N-18-65, N-19-40, N-19-45, N-20-40, N-20-45 ETM-EarthSat-MrSID, 1.0, USGS, Sioux Falls, South Dakota, 2000. Accessed 2016, at the Global Land Cover Facility

Michaud Y, Dionne J-C, Dyke L (2011) Frost bursting: a violent expression of frost action in rock. Can J Earth Sci 26:2075–2080

Milliman JD, Farnsworth KL (2011) River Discharge to the Coastal Ocean: A Global Synthesis. Cambridge University Press

National Wetlands Working Group (1988) Wetlands of Canada. Ecological Land Classification Series # 24 Environment Canada, Ottawa and Polyscience Publications, Montreal

Newbury R (1990) The Nelson River. In: Wolman MG, Riggs HC (eds) Surface Water Hydrology: The Geology of North America, v. O-1. Geological Society of America, Boulder, Colorado, pp 287–292

Ollerhead J, Davidson-Arnott RGD (1995) The evolution of Buctouche Spit, New Brunswick, Canada. Mar Geol 124:215–236

Payette S, Boudreau F (1984) Evolution postglaciaire des hauts sommets alpins et subalpins de la Gaspésie. Can J Earth Sci 21:319–335

Payette S, Filion L, Delwalde A, Bégin C (1989) Reconstruction of tree-line vegetation response to long-term climate change. Nature 341:429–432

Phillips D (1990) The Climates of Canada. Minister of Supply and Services Canada, Ottawa, p 176

Richard PJH, Occhietti S (2005) Radiocarbon chronology for ice retreat and inception of Champlain Sea in the St. Lawrence Lowlands, Canada. Quat Res 63:353–358

Ritchie JC (1993) Northern vegetation. In: French HM, Slaymaker O (eds) Canada's Cold Environments. McGill-Queens University Press, Montreal & Kingston, pp 93–116

Ryder JM (1998) Geomorphological Processes in the Alpine Areas of Canada. Geol Surv Can Bull 524

Seguin MK, Allard M (1984) Le pergelisol et les processus thermokarstiques de la region de la riviere Nastapoca, Nouveau Quebec. Geogr Phys Quat 37:11–25

Seppala M (2004) Wind as a Geomorphic agent in cold climates. Cambridge University Press, Cambridge

Seppala M, Gray J, Richard J (1991) Development of low centred ice wedge polygons in the northernmost Ungava Peninsula, Quebec. Boreas 20:259–282

Shaw J, Courtney RC (2002) Postglacial coastlines of Atlantic Canada: digital images. Geological Survey of Canada Open File Report 4302

Shaw J, Gareau P, Courtney RC (2002) Paleogeography of Atlantic Canada 13-0 kyr. Quaternary Sci Rev 21:1861–1878

Shaw J, Taylor RB, Forbes DL, Solomon S (2001) Sea level rise in Canada: a hazard. In: Brooks GR (ed) A Synthesis of Geological Hazards. Geol Surv Can Bull 548

Shaw J, Taylor RB, Forbes DL, Ruz MH, Solomon S (1998) Sensitivity of the coasts of Canada to sea-level rise. Geol Surv Can Bull 505
Shaw J, Piper DJW, Fader GBJ, King EL, Todd BJ, Bell T, Batterson MJ, Liverman DGE (2006) A conceptual model of the deglaciation of Atlantic Canada. Quat Sci Rev 25:2059–2081
Slaymaker O, French HM (1993) Cold environments and global change. In: French HM, Slaymaker O (eds) Canada's cold environments. McGraw-Hill Queens University Press, pp 313–334
Slaymaker O, French HM (2012) The changing Canadian cryosphere, globalization and global environmental change. In: French HM, Slaymaker O (eds) Changing cold environments: A Canadian perspective. Wiley, Chichester, UK, pp 301–312
Smith SL, Burgess MM, Heginbottom JA (2001) Permafrost in Canada, a challenge to northern development. In: Brooks GR (ed) A synthesis of geological hazards in Canada. Geol Surv Can Bull 548, pp 241−264
Spooner I, Batterson M, Catto N, Liverman D, Broster BE, Kearns K, Isenor F, McAskill GW (2012) Slope failure hazard in Canada's Atlantic Provinces: a review. Atlantic Geol 49:1–14
Stone M, Saunderson HC (1996) Regional patterns of sediment yield in the Laurentian Great Lakes basin. Intern Assoc Hydrol Sci Pub 236:125–131
Syvitski JPM (1993) Glaciomarine environments in Canada: an overview. Can J Earth Sci 30:354–371
Tarnocai C, Kettles IM, Lacelle B (2000) Peatlands of Canada. Geological Survey of Canada Open File 3834
Teller JT (2003) Controls, history, outbursts, and impact of large late-Quaternary proglacial lakes in North America. In: Gillespie AR, Porter SC, Atwater BF (eds) The quaternary period in the United States. Elsevier, p 45–61
Trenhaile AS (2007) The Geomorphology of Canada: an Introduction, 2nd edn. Oxford University Press, Toronto
Walcott RI (1970) Isostatic response to loading of the crust in Canada. Can J Earth Sci 7:716–727
Wang T, Hamann A, Spittlehouse DL, Carroll C (2016) Locally downscaled and spatially customizable climate data for historical and future periods for North America. PLoS ONE 11:e0156720
Watt WE (1989) Hydrology of floods in Canada: a guide to planning and design. National Research Council of Canada, Associate Committee on Hydrology, Ottawa
Woo M, McCann SB (1994) Climatic variability, climatic change, runoff and suspended sediment regimes in northern Canada. Phys Geogr 15:201–226
Woo M-K (2012) Permafrost Hydrology. Springer Verlag, Berlin and Heidelberg
Yoxall WH (1981) The surface waters and associated landforms of the island of Newfoundland. In: Macpherson AG, Macpherson JB (eds) The natural environment of Newfoundland past and present. Memorial University of Newfoundland, St. John's, Newfoundland, pp 154–184

Olav Slaymaker is a Member of the Order of Canada, and professor emeritus of geography in the Department of Geography at the University of British Columbia, Vancouver, B.C. His 20+ co-authored and edited monographs and 160+ refereed papers reflect three different styles of work. First, he has pursued original field research for 40 years on sediment systems in the Canadian Cordillera and the European Alps, strongly supported and enhanced by his stellar doctoral and postdoctoral students and international colleagues. Second, he has produced papers and books summarizing trends in land use and environmental change in mountain regions worldwide. Third, he has edited monographs with invited international contributions on major environmental and geomorphological themes. He is former President of the Canadian Association of Geographers and former President of the International Association of Geomorphologists.

Norm Catto joined Memorial University in July 1989 and is now the Head and Professor of Geography at that university. His research and teaching interests include coastal landforms, natural hazards, sea level change and impacts in coastal environments; response of river systems to climate and weather events and flood risk assessment; aeolian geomorphology and response to climate and human factors in boreal and coastal environments; mass movements and slope failures; loess deposition; palaeosol formation; emergency measures policies and effectiveness human adaptations to climate and weather events; the impacts of climate and weather events on agriculture, transportation, fisheries and communities; and the relationships among landscapes and people. His research has included projects and investigations in landscapes and environments in 16 countries, and in all Canadian provinces and territories, particularly in eastern Canada. For 18 years, he served as Editor-in-Chief for *Quaternary International,* a professional international journal dedicated to the study of landscape evolution and climate variation throughout the past 3 million years. He has taught more than 50 different undergraduate and graduate courses focused on numerous aspects of geomorphology, landscape analysis, Quaternary research, natural hazards, ocean sciences, environmental science, climate and climate change, soil science, geoarchaeology and palaeontology. Study of landscapes forms his central focus for the understanding of the various areas of eastern Canada, the combination of physical and cultural constituents that contribute to its identities. For most species, and for all human occupants, life in eastern Canada has been heavily influenced by geomorphology since deglaciation.

Dori J. Kovanen is a professional engineering geologist and Research Associate in the Department of Geography at the University of British Columbia, Vancouver, BC. With an M.Sc. from Western Washington University in Bellingham and a Ph.D. in Quaternary geology from the University of British Columbia, she is a specialist in Quaternary geomorphology. She has written extensively on late glacial and Holocene history of the Pacific Northwest as well as contemporary environmental changes in the region and was a significant contributor to 'Landscapes and Landforms of Western Canada' (Slaymaker 2017).

Part II
Case Studies

4 Glacial Landform Assemblages of Mainland Nunavut, West of Hudson Bay and Their Palaeoglaciological Significance

Chris R. Stokes

Abstract

This case study focuses on the glacial landscapes of the low-relief tundra of the Canadian Shield of mainland Nunavut, west of Hudson Bay. Having been studied for well over a century, the glacial geomorphology of the region is spectacular and is widely recognised as being of international significance to palaeo-glaciology. At the global Last Glacial Maximum, the region lay close to the centre of the former Laurentide Ice Sheet and was submerged beneath a large (>3 km thick) ice dome, and with a major ice divide orientated roughly north-east to south-west (the Keewatin Ice Divide). This was an important influence on the distribution of glacial landforms and the aim of this chapter is to provide a concise overview of the key landform assemblages and their zonation around the Keewatin Ice Divide with a particular focus on hummocky terrain, ribbed moraines, drumlins, mega-scale glacial lineations, meltwater features, end moraines, and raised beaches and shorelines related to proglacial lakes and marine transgression. It is clear that whilst the distribution of landforms is a complex interplay between the overriding ice sheet properties (thickness, velocity, basal thermal regime) and the underlying geology, a detailed analysis of these landscapes has led to some important advances in our broader understanding of ice sheet dynamics and glacial geomorphology, particularly in the area of drift dispersal, the migration of ice divides, ice streaming and the evolution of subglacial meltwater drainage systems.

Keywords

Glacial geomorphology • Keewatin • Drumlins • Ribbed moraine • Mega-scale glacial lineations • Laurentide Ice Sheet

C. R. Stokes (✉)
Department of Geography, Durham University, Durham, DH1 3LE, UK
e-mail: c.r.stokes@durham.ac.uk

4.1 Introduction

This case study focusses on the glacial landscapes of mainland Nunavut, west of Hudson Bay, which is characterised by the low to moderate relief terrain of the north-western Canadian Shield (see Fig. 4.1). The shield itself is geologically complex, with rocks that were deposited, formed, and deformed over a period that encompasses over three quarters of the known geological history of the Earth (Aylsworth and Shilts 1989a). Most of the region is characterised by massive granitoid gneiss, but this is interrupted in places by plutonic bodies, sedimentary rock and volcanic strata (Aylsworth and Shilts 1989a; Dyke and Dredge 1989; Dyke et al. 1989). Of particular significance to the glacial geomorphological record are the Dubawnt Group of Upper Proterozoic unmetamorphosed sediments that outcrop as the Thelon sedimentary basin near the centre of the study area (Fig. 4.1). Unlike the physically hard rocks that characterise most of the Canadian Shield and that are generally resistant to glacial erosion and yielded very little debris, the Upper Proterozoic sedimentary and volcanogenic rocks are much less consolidated and more susceptible to glacial erosion. As such, these outcrops are known to have exerted a major influence on the nature and distribution of glacial landforms and sediments (Shilts et al. 1987; Aylsworth and Shilts 1989a).

The underlying geology is also an important control on the relief of the study region, which is typically flat to gently undulating topography with low broad hills scoured by glacial erosion of the hard crystalline bedrock, and with negligible relief over the areas of unmetamorphosed sedimentary bedrock, such as the Thelon sedimentary basin (Shilts et al. 1987). Glacial erosion of the low-relief landscape (and/or the melting of ice blocks in areas of thicker drift) has led to myriad lakes (Shilts et al. 1987) and, across most of the study area, >20% of the land surface is covered by water (Dyke et al. 1989). The drainage of the major river systems is generally east towards Hudson Bay (e.g. the

© Springer Nature Switzerland AG 2020
O. Slaymaker and N. Catto (eds.), *Landscapes and Landforms of Eastern Canada*,
World Geomorphological Landscapes, https://doi.org/10.1007/978-3-030-35137-3_4

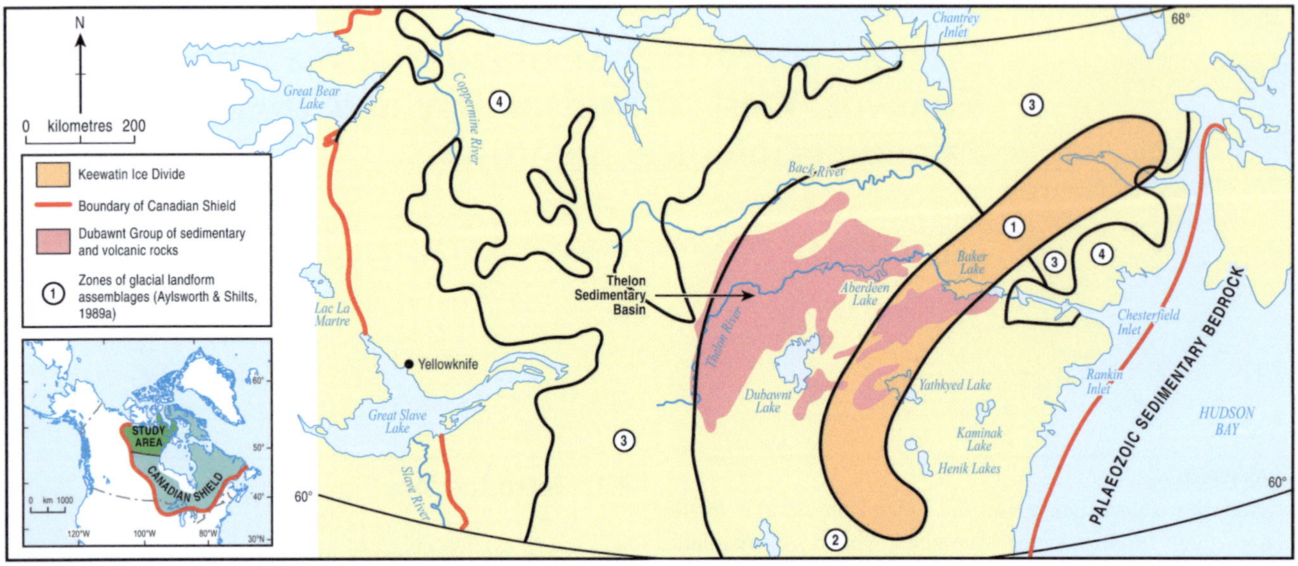

Fig. 4.1 Location map of the study area (inset) showing the four zones of glacial landform assemblages defined by Aylsworth and Shilts (1989a, b). Note the location of the last inferred position of the Keewatin Ice Divide (Zone 1) and the location of the Thelon Sedimentary Basin predominantly within Zone 2

Thelon River) or north towards the Arctic coast (e.g. the Back River) (Fig. 4.1). Most of the study region is classified as tundra with continuous permafrost, such that water-saturated organic deposits (bogs, peats) typically developing in the active layer of low-relief terrain. Patterned ground (e.g. stone circles) is also found throughout the region, although pingos are much rarer on the shield, e.g. compared to the fine-grained sediments to the north-west in the Mackenzie River corridor.

The surficial geology is dominated by glacial geomorphology and the glacial landscapes of the northwestern Canadian Shield are some of the most spectacular and important in the world, and have a long history of research (Tyrell 1897; Bird 1953; Dean 1953; Fyles 1955; Taylor 1956; Lee 1959; Blake 1963; Craig 1964, 1965; Falconer et al. 1965b). The region lay close to the centre of the former North American Laurentide Ice Sheet (LIS), which has traditionally been referred to as the 'Keewatin Sector' (e.g. Tyrell 1897; Dyke and Prest 1987; Dyke et al. 1989). Of particular note is that the distribution of glacial landforms is thought to be intimately associated with the Keewatin Ice Divide (KID) (Shilts et al. 1987; Aylsworth and Shilts 1989a; Tarasov and Peltier 2004), which represented the highest dispersal centre/dome of the ice sheet and which was recognisable as a distinct feature throughout much of deglaciation, even though it may have migrated by as much as 500 km (Lee et al. 1957; McMartin and Henderson 2004). During deglaciation, this region also was subjected to rapid ice flow in the form of ice streaming, which created some of the most spectacular landforms created by the LIS and has resulted in one of its best-studied palaeo-ice streams: the Dubawnt Lake palaeo-ice stream (Stokes and Clark 2003a, b). Superimposed on these landscapes are isolated moraine systems and numerous landforms and deposits related to subglacial meltwater activity (e.g. eskers) and the development of proglacial lakes (e.g. relict shorelines). In addition to its palaeo-glaciological significance, the Keewatin Sector has also been the focus of extensive and ongoing drift prospecting programmes, e.g. by the Geological Survey of Canada since 1970 (Cunningham and Shilts 1977; Shilts 1977; Shilts et al. 1979; Shilts 1980; McMartin and Henderson 2004).

The aim of this chapter is to provide a concise overview of the key glacial landform assemblages that occur on the northwestern Canadian Shield, with a particular focus on hummocky terrain, ribbed moraines, drumlins, mega-scale glacial lineations, meltwater features, end moraines, and glacial lake shorelines. Section 4.2 describes the characteristics of these features and their spatial distribution in relation to the KID, leaning heavily on the instrumental work of William (Bill) Shilts and co-workers who were the first to recognise an obvious zonation of glacial landforms in the region (Aylsworth and Shilts 1985, 1989a, b, c; Shilts et al. 1987). Section 4.3 then briefly highlights how the landform assemblages and patterns described in Sect. 4.2 are of international interest in terms of our understanding of the palaeo-glaciology of the LIS, and with particular reference to the KID (Sect. 4.3.1), meltwater drainage systems (Sect. 4.3.2) and fast-flowing ice streams (Sect. 4.3.3).

4.2 Glacial Landform Assemblages

In a series of publications in the 1980s, Shilts and co-workers noted that a striking feature of glacial landforms on the northwestern Canadian Shield was the apparent arrangement of landforms in broadly concentric zones around the location of the last inferred position of the Keewatin Ice Divide (KID) (e.g. Aylsworth and Shilts 1985, 1989a, b, c; Shilts et al. 1987) (Fig. 4.1). Ice divides are analogous to hydrological (drainage) divides in fluvial systems and they represent broad ridges or domes in the ice sheet surface topography (see e.g. Dyke and Prest 1987). The KID has long been recognised as one of the major dispersal centres in the LIS (Lee et al. 1957; Lee 1959), representing a northeast–southwest orientated feature several hundred kilometres long and lying approximately 200 km west of the coast of western Hudson Bay (Fig. 4.1; Dyke and Prest 1987). It has been inferred as one of the highest parts of the LIS (Prest 1969; Dyke and Prest 1987; Boulton and Clark 1990; Tarasov and Peltier 2004) and recent relative sea level data and glacial isostatic modelling indicate that it was likely around 3.4 km thick during the Last Glacial Maximum (Simon et al. 2014).

Aylsworth and Shilts (1989a) identified four main zones of glacial landform assemblages associated with the KID, with Zone 1 representing the innermost zone that includes the footprint of the ice divide itself (Fig. 4.1). It is generally characterised by featureless till plains (i.e. lacking prominent glacial landforms) or low hummocky terrain and with a general absence of eskers. Zone 2 encircles most of Zone 1 (Fig. 4.1) and is characterised by the presence of well-developed linear trains or ribbed moraine, that radiate out from the location of the former ice divide, but which are interspersed with fields of glacial lineations, again showing radial flow away from the ice divide. Beyond Zone 2, Zone 3 is characterised by a fairly continuous cover of till and numerous drumlin fields that contain some highly elongate glacial lineations, but where till thins to a discontinuous veneer in the outer portion of the zone. The outermost Zone 4 comprises extensive areas of massive crystalline bedrock that is typically devoid of any till cover.

Aylsworth and Shilts (1989a) noted that the arrangement of these zones around the location of the inferred ice divide suggests that they are likely to be broadly related to the dynamics of the ice sheet and the style of deglaciation, but they also noted the importance of the underlying geology in modulating the influence of the ice sheet dynamics, e.g. with the major sedimentary basins typically being associated with the best developed and most spectacular glacial landforms. The following subsections describe the glacial landforms assemblages within each of these zones (Fig. 4.1), followed by a discussion of major moraine ridges and glacial lake shorelines that typically span these zones.

4.2.1 Zone 1

The innermost Zone 1 includes the footprint of the KID (Fig. 4.1) and is characterised by areas of low-relief hummocky terrain and featureless plains of generally thin till cover. The hummocky terrain (see e.g. in Fig. 4.2) is most obvious towards the southern end of the inferred ice divide where it is comprised of small (<100 m across), low (<5 m high), rounded hummocks of till (Aylsworth and Shilts 1989a). Aylsworth and Shilts (1989a) noted that the hummocks appear to be underlain by till similar in composition to material formed in nearby drumlins (e.g. in Zone 2 and sporadically in Zone 1), but do not have the dense surface cover of bounders that typically cover ribbed moraines in Zone 2. Aylsworth and Shilts (1989a) also noted that although the hummocky terrain is best developed in Zone 1, it is by no means restricted to that area and that isolated patches occur elsewhere in the study area, e.g. west of Dubawnt Lake in Zone 2.

A characteristic feature of Zone 1 is the general absence of glacial landforms. Ribbed (sometimes known as 'Rogen') moraines are absent and streamlined features, such as drumlins, are rare. Where drumlins occur they indicate radial ice flow away from the last inferred position of the divide (Aylsworth and Shilts 1989a). Likewise, isolated crag-and-tails and roche mountonnées have been reported in this region and generally indicate radial ice flow away from the position of the ice divide (e.g. Bird 1953; Lee 1959). It has also been noted that Zone 1 is characterised by a marked absence of eskers, which are otherwise very common across the northwestern Canadian Shield (Aylsworth and Shilts 1989a; Storrar et al. 2013, 2014a, b). Storrar et al. (2014a), for example, analysed >20,000 mapped eskers across the bed of the LIS, but noted that they were absent from regions of inferred ice divides in both Keewatin and Labrador. They identified an esker-free area approximately 100–150 km wide and covering an area around 93,000 km^2 in Keewatin, noting its association with the last inferred position of the KID. Indeed, Zone 1 is almost completely devoid of any glacifluvial deposits, except for minor accumulations of outwash in some valleys (Shilts et al. 1987).

4.2.2 Zone 2

Aylsworth and Shilts (1989a) demarcated Zone 2 as a 200–250 km wide, U-shaped belt that wraps around the southern

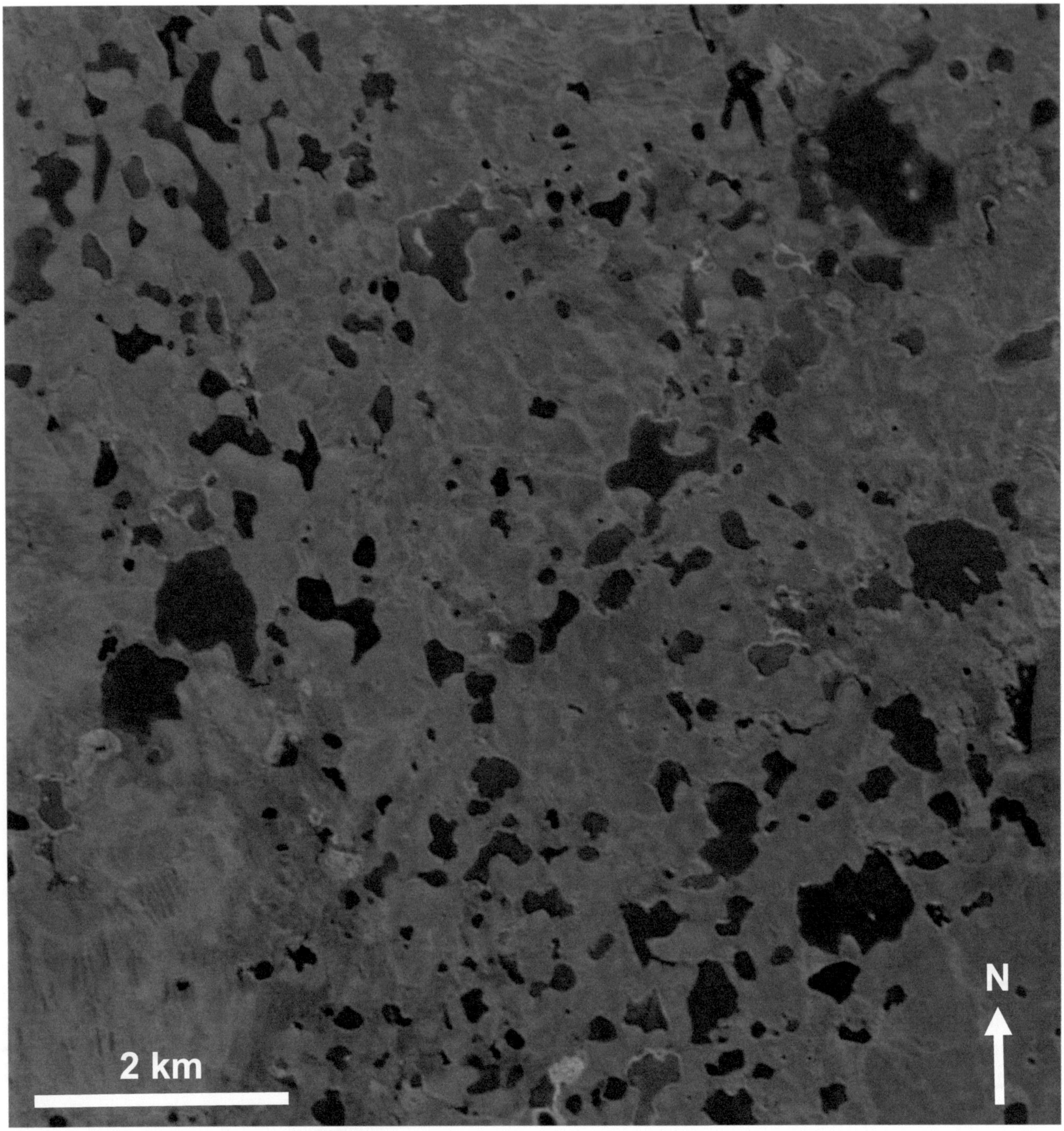

Fig. 4.2 Landsat image (data from Google Earth: Landsat/Copernicus) showing low-relief hummocky terrain in Zone 1 at the southern end of the of the last inferred location of the Keewatin Ice Divide, 20 km north-west of Yathkyed Lake (located on Fig. 4.1). Image is centred on 62°49′46″ N, 99°05′04″ W

end of the last inferred position of the KID (Fig. 4.1). The zone is characterised by extensive linear trains of ribbed moraine which radiate from the location of the inferred ice divide and which are interpreted to extend along former ice flow-lines. The areas of ribbed moraines (e.g. in Fig. 4.3) are typically comprised a nested series of short (<2 km long), low (<10 m high) sinuous ridges of irregular height that trend perpendicular to the inferred ice flow direction (Aylsworth and Shilts 1989a). Aylsworth and Shilts (1989a) noted that they appear to be asymmetric in cross section, with a steeper down-ice side, but this has not been systematically observed or quantified. They also noted that these 'classic' ribbed moraines can also be associated with less well-defined ridges and/or with irregularly shaped hummocks, albeit with a similar composition and relief.

The development of ribbed moraines in Zone 2 is abrupt, with virtually no gradational transition from the featureless terrain in Zone 1. Where the ribbed moraines are extensive, their distribution does not appear to be linked to topography, occurring on both lower and higher ground. That said, Aylsworth and Shilts (1989a) noted that isolated patches tend to occur in lower points in the landscape, which was also noted by Bouchard (1989) who suggested that they are formed by glacitectonic shearing and stacking of subglacial sediments under a compressive flow regime in topographic depressions.

In terms of their composition, Aylsworth and Shilts (1989a) reported that the ribbed moraines are commonly formed in till but are often associated with a dense concentration of boulders, which may mantle the till or be intermixed. In particular, they noted that some large fields of ribbed moraines, for example, west of Rankin Inlet (Fig. 4.1), extend down-ice from outcrops of post-glacially shattered granitic bedrock. They suggested that large quantities of felsenmeer (blockfields) are likely to have existed in these regions prior to the last glaciation and which have subsequently been incorporated into the existing trains of ribbed moraine. They also reported prominent trains of ribbed moraines that are composed of disaggregated debris derived from the friable sandstones and conglomerates of the Thelon sedimentary basin. Aylsworth and Shilts (1989a) also noted that there is little erratic material in these ridges and they are almost exclusively composed of orthoquartzite clasts, quartz sand and kaolin, with so few nutrients that vegetation is scarce. This lack of vegetation, especially on the well-drained ridge crests, enables the ribbed moraines and other streamlined features, such as drumlins, to be easily identified on satellite imagery and aerial photographs throughout the region.

Aylsworth and Shilts (1989a) drew attention to the fact that fields of ribbed moraines are closely associated with linear fields of drumlins and other highly elongate streamlined landforms in Zone 2 that also radiate from the inferred position of the ice divide. In most cases, there is a sharp lateral contact between ribbed moraines and alternating parallel trains of drumlins and other streamlined features (Fig. 4.4), which is common elsewhere on the Canadian Shield (Shilts et al. 1987; Dunlop and Clark 2006). It has also been observed that there are places where the ribbed moraines have been heavily streamlined with drumlins superimposed on ridge crests (e.g. Fig. 4.3), and where streamlined features have themselves been broken up into ribbed moraines (Fig. 4.5). It is important to note that whilst ribbed moraines are more common in Zone 2, the drumlins and streamlined features extend into Zone 3 and, although much less common, Zone 4.

More recently, Greenwood and Kleman (2010) reported a much larger scale of landform in Keewatin than had hitherto been recognised. They mapped a series of large till 'belts' or 'ridges', apparent only on satellite imagery and lying beneath the drumlins, flutes and ribbed moraine (Fig. 4.6). Planform and crestline mapping from remotely sensed imagery allowed them to map a population of >2500 individual landforms, whose dimensions are on average $\sim 10^{\circ}$ km long and $\sim 1.5^{\circ}$ km wide and which form extensive and coherent patterns throughout the Keewatin region, i.e. not just restricted to Zone 2. Some of these features were interpreted as over-printed mega-scale glacial lineations or moraines that predate the LGM (see also Kleman et al. 2002, 2010), but a group, comprising a significant number of the Keewatin population, did not fit any existing category of glacial landforms (Fig. 4.6). They referred to these as 'mega-scale transverse bedforms' and postulated that their close spatial integration with ribbed moraines suggests a similar mode of genesis.

Common throughout Zone 2 is an extensive network of hundreds of eskers that also radiate from the edge of the location of the ice divide (Bird 1953; Shilts 1984; Aylsworth and Shilts 1989a; Storrar et al. 2013, 2014a). Towards the east of the study area and below the marine limit occupied by the much larger ancestral Hudson Bay (known as the Tyrell Sea), eskers have been reworked and appear more subdued (e.g. Shilts et al. 1987; Aylsworth and Shilts 1989a). Elsewhere, most of the northwestern Canadian Shield is covered by an integrated pattern of eskers radiating outward from the area of the KID (Prest et al. 1968; Storrar et al. 2013, 2014a) (Fig. 4.7). Aylsworth and Shilts (1989a) noted that typical esker systems begin as a series of hummocks or short, flat-topped segments which pass down-ice into a relatively continuous and larger esker ridges or (occasionally) beaded eskers. They also noted that discontinuous eskers are common where material has either been removed or, in numerous cases, flooded by lakes (see example in Fig. 4.3a).

More recent work by Storrar et al. (2014a) found that gaps typically represent around 35% of the total length of esker systems that are obviously traceable despite the gaps. The length of the largest unbroken esker ridges can be traced for up to 75 km (Fig. 4.7), but the largest systems can be traced for several hundred kilometres (Storrar et al. 2014a). Indeed, Dyke and Dredge (1989) noted that the 'Thelon Esker', at >800 km long, including gaps, is probably the longest esker in the world (see Fig. 4.7). It is unlikely that some of the very longest eskers were formed rapidly (e.g. in tunnels of several tens of kilometres in length). Rather, it has been argued that it is more likely that eskers were deposited in short meltwater channels a few kilometres in length and

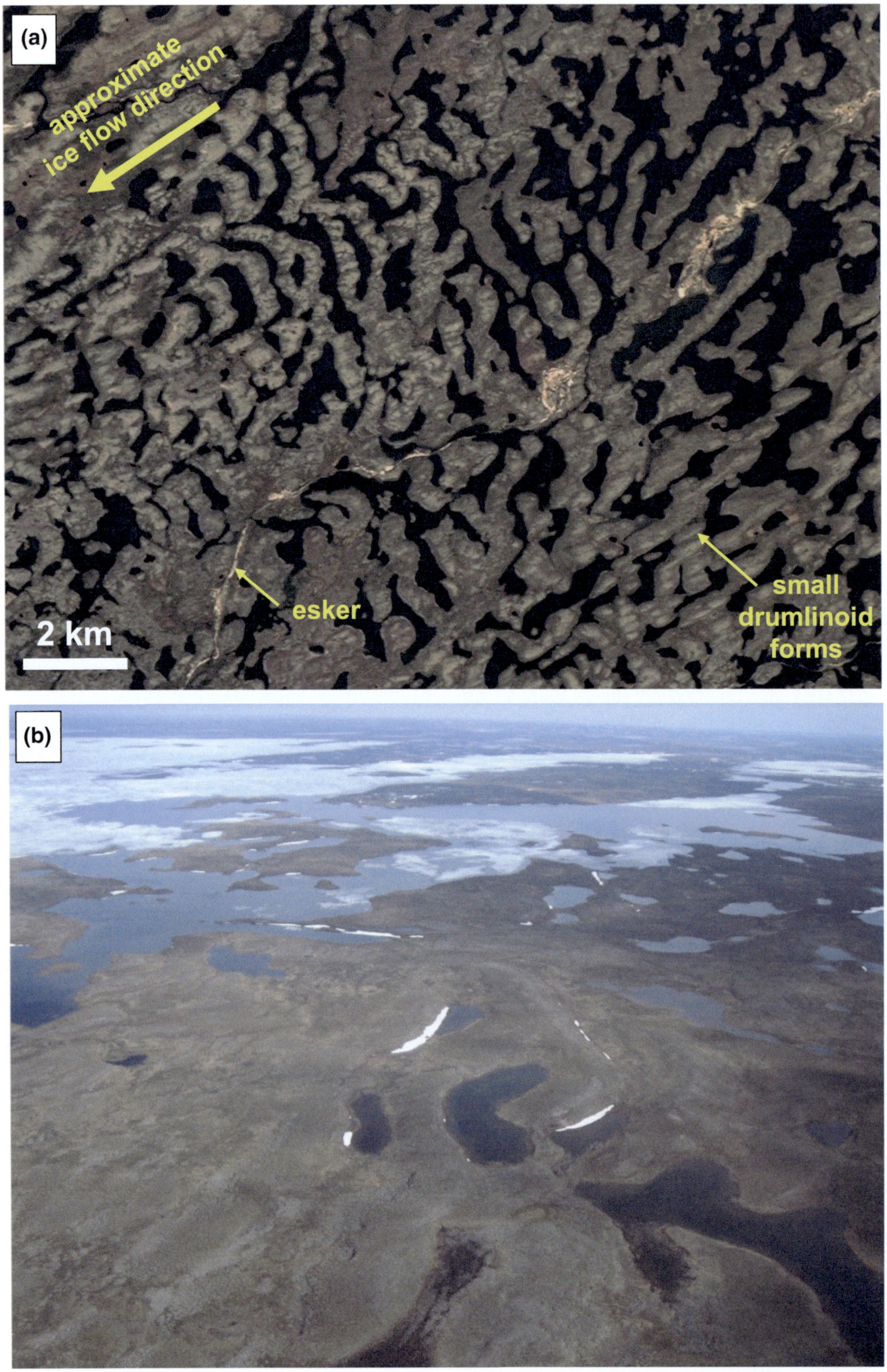

Fig. 4.3 **a** Landsat image (data from Google Earth: Landsat/Copernicus) showing classic ribbed moraines in Zone 2, around 170 km south-west of Dubawnt Lake (located on Fig. 4.1). Image is centred on 61°22′30″ N, 103°54′26″ W. Note the small drumlinoid forms superimposed on the ribbed moraines towards the south-east of the image. Also note the large esker system trending north-east to south-west broadly aligned with the approximate direction of ice flow. **b** An oblique aerial photograph of classic ribbed moraine from Zone 2. *Photograph* C. Stokes

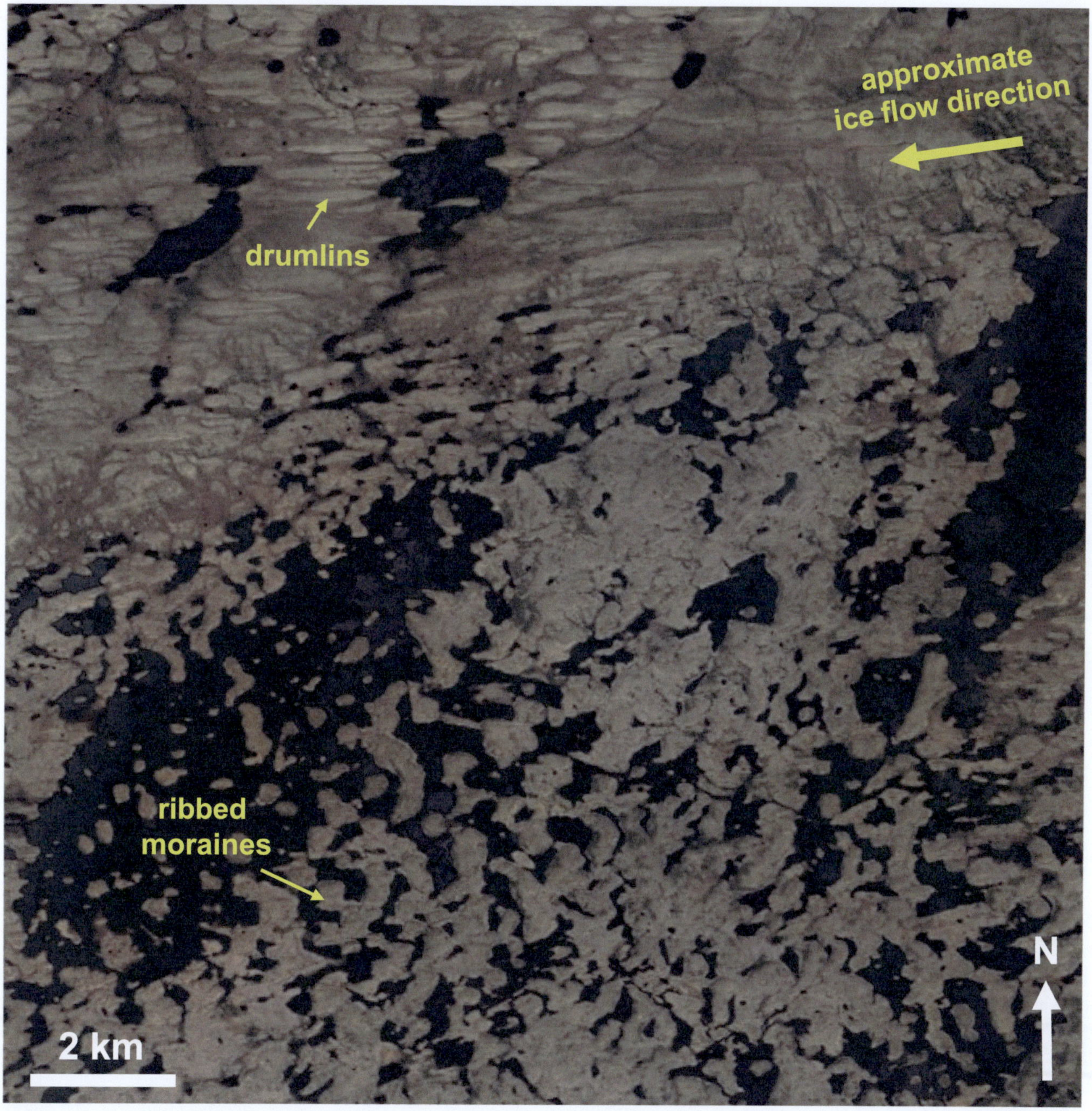

Fig. 4.4 Landsat image (data from Google Earth: Landsat/Copernicus) showing classic ribbed moraines with a sharp lateral contact to a drumlin field in Zone 2, around 70 km south-west of Dubawnt Lake (located on Fig. 4.1). Image is centred on 62°07′40″ N, 102°17′06″ W

largely fed by surface meltwater in the ablation zone of the retreating ice sheet (Aylsworth and Shilts 1989a; Storrar et al. 2014a; Livingstone et al. 2015). The fact that many eskers can be traced over much longer distances likely represents the headward propagation of surface meltwater streams (Livingstone et al. 2015).

Aylsworth and Shilts (1989a) reported that most eskers are sharp-ridged and up to 40 m high, with conical knobs projecting well above the average elevation of the crests in some locations. They also noted that eskers may be interrupted by bulges where a single ridge splits into multiple ridges before coalescing down-ice, which may be related to blockages in subglacial drainage and the formation of bypass tunnels in/under the ice. Aylsworth and Shilts (1989a) noted that above the marine limit, and beyond areas inundated by proglacial lakes (see Sect. 4.2.6), eskers are commonly flanked by glacifluvial outwash terraces that are scarred by channels and disrupted by kettle holes. These have also been

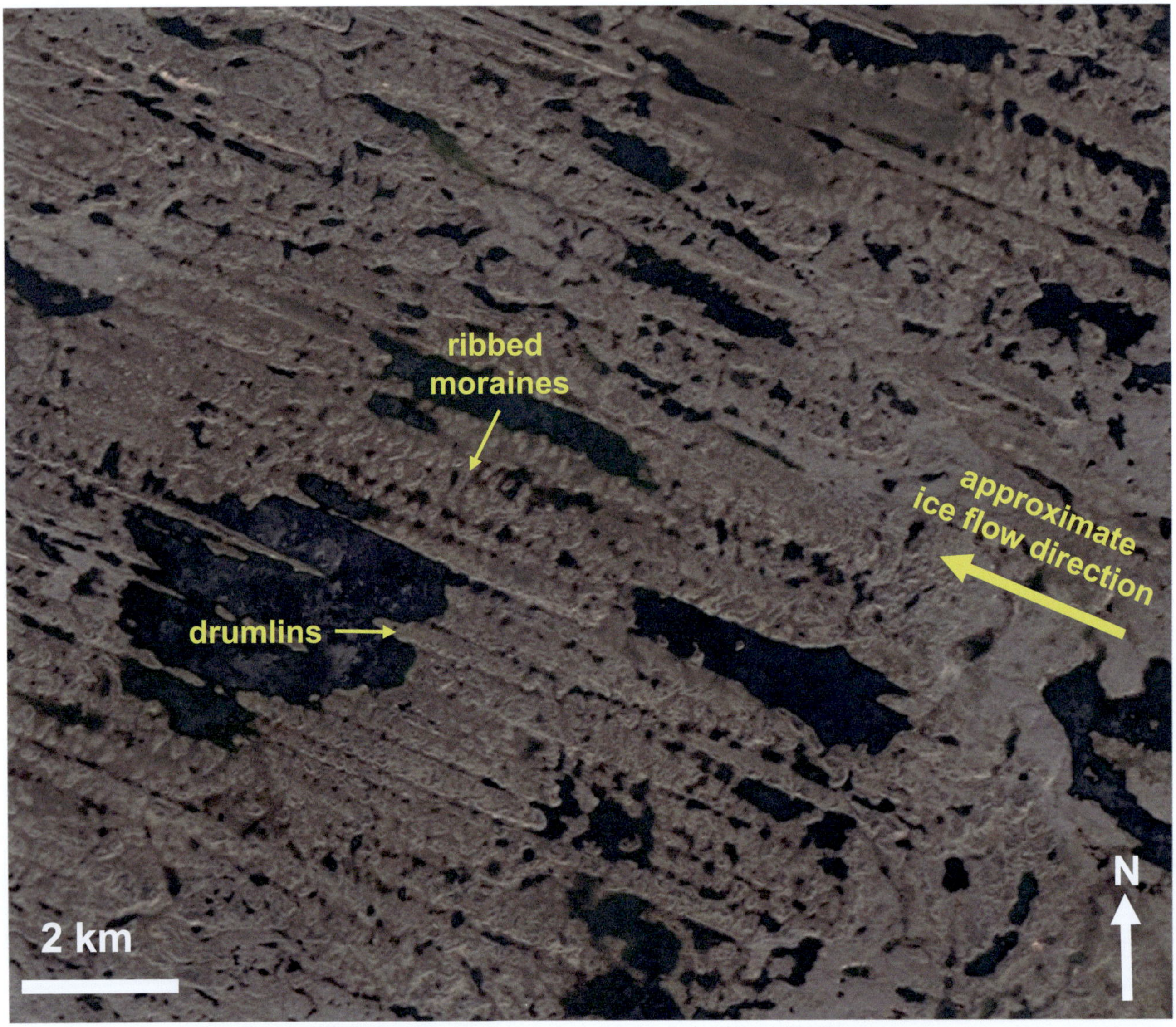

Fig. 4.5 Landsat image (data from Google Earth: Landsat/Copernicus) showing examples of where small ribbed moraines are superimposed on (and have partially 'broken up') glacial lineations in Zone 2, around 70 km north of Dubawnt Lake (located on Fig. 4.1). Image is centred on 63°57′58″ N, 101°33′46″ W. These elongate lineations form a part of the onset zone of the Dubawnt Lake palaeo-ice stream (see Sects. 4.2.3 and 4.3.3)

referred to as 'glacifluvial corridors' (St-Onge 1984) or 'glacifluvial corridor hummocks (GCHs)' (Utting et al. 2009), which are wide (several km) tracts of glacifluvial sand and gravel, often located in erosional corridors (Fig. 4.8) (Utting et al. 2009; Storrar and Livingstone 2017). They have been shown to occur extensively throughout the study area and, like eskers, can occur in quasi-dendritic systems that can extend for several tens of kilometres in length and regularly spaced 5–10 km apart (Utting et al. 2009). Detailed analysis of the hummocks associated with GCHs using both ground penetrating radar and pit excavations has revealed a single lithofacies consisting of coarsely stratified, matrix-supported gravely sand to a depth of several metres (Utting et al. 2009). Utting et al. (2009) suggested that this composition is similar to the 'sliding bed facies' commonly observed in esker sediments and other hyperconcentrated flow deposits, both of which are attributed to high meltwater discharges.

4.2.3 Zone 3

Zone 3 is characterised by a continuation of the dendritic networks of eskers, but here they are more commonly superimposed on extensive fields of drumlins and other streamlined forms, particularly in the inner part of the zone

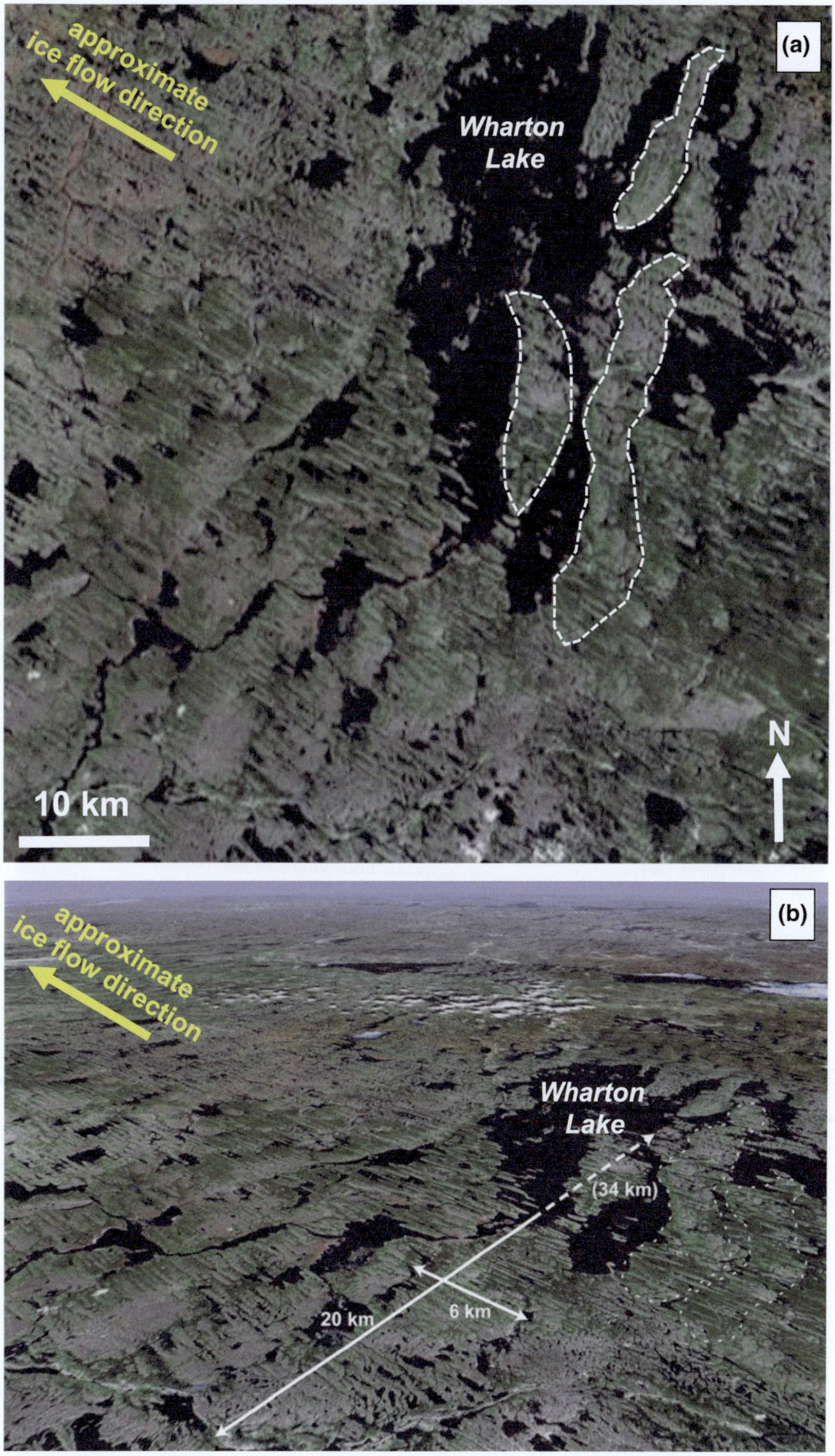

Fig. 4.6 Landsat images (data from Google Earth: Landsat/Copernicus) showing examples (white dashed polygons) of large 'till belts' or 'mega-scale transverse ridges' north-west of Dubawnt Lake. These features were first identified by Greenwood and Kleman (2010) who suggested that they form a coherent pattern in the landscape beneath the much smaller drumlins and ribbed moraines (examples shown in Figs. 4.3, 4.4 and 4.5). **a** Shows a Landsat image centred on 63°56′ 02″ N, 100°00′20″ W and **b** is an oblique view of the same area. Figures modified from Greenwood and Kleman (2010)

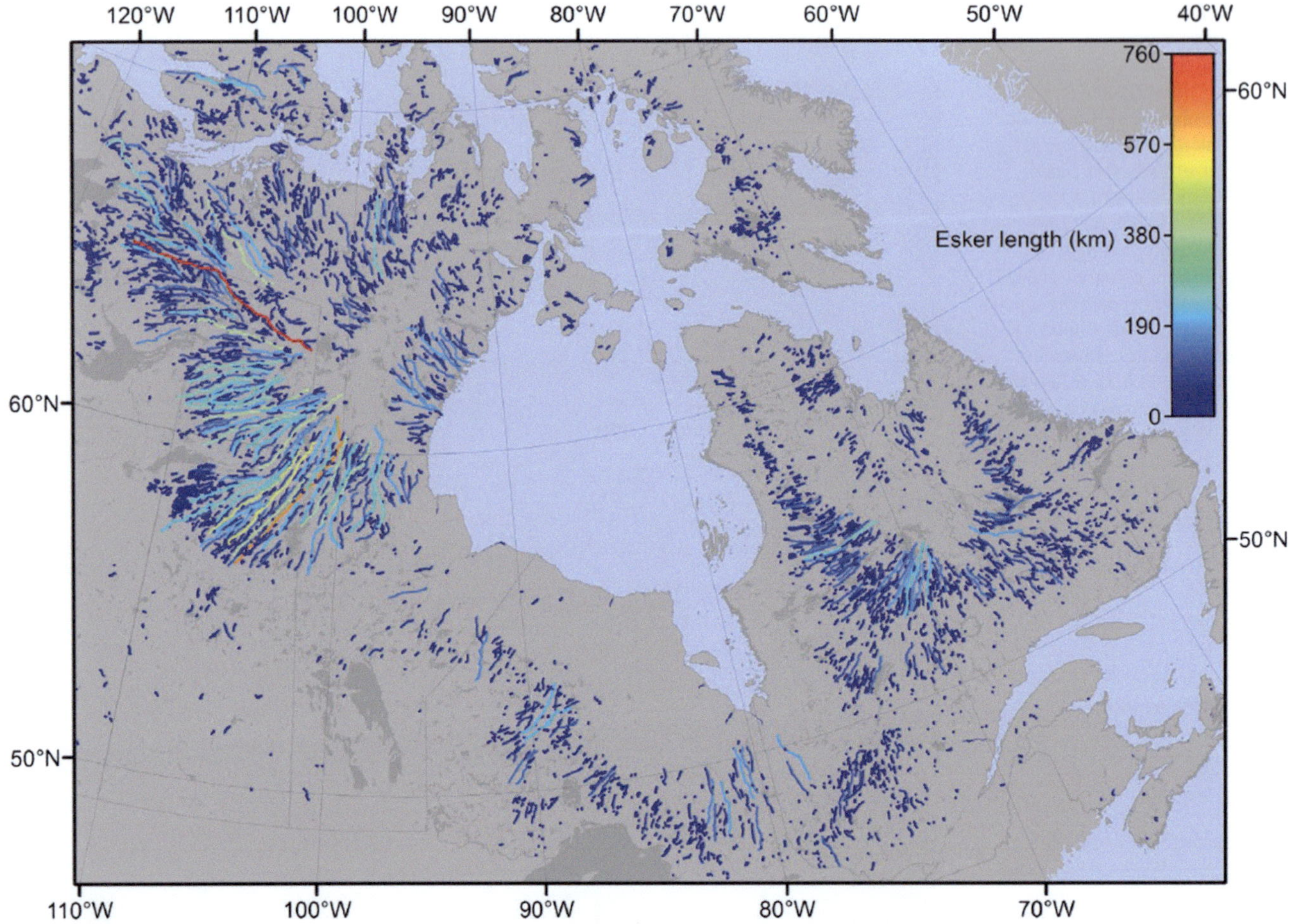

Fig. 4.7 Map of interpolated esker systems in Canada (i.e. tracing gaps conservatively between individual esker segments: from Storrar et al. 2014a). The northwestern Canadian Shield contains a high concentration of long esker systems and the large esker coloured red is probably the longest esker in the world (the 'Thelon esker': see Dyke and Dredge 1989) Also note the absence of eskers from under the inferred position of the Keewatin Ice Divide (Zone 1 on Fig. 4.1). Image taken from Storrar et al. (2014a)

where they are contiguous with similar features in Zone 2 (Shilts et al. 1987; Aylsworth and Shilts 1989a). Zone 3 is also characterised by a general decrease in till thickness from a fairly continuous mantle in the innermost portions to a discontinuous veneer towards the outer limits of the zone (Aylsworth and Shilts 1989a). Ribbed moraines also exist in Zone 3, but generally, occur as widely scattered fields of narrow linear trains in between the streamlined terrain (Shilts et al. 1987).

The streamlined glacial lineations in Zone 3 have attracted much attention in the literature, largely because of their extreme length (Tyrell 1897; Bird 1953; Dean 1953; Craig 1964; Ayslworth and Shilts 1989a; Stokes and Clark 2002; 2003a; Stokes et al. 2013). Although drumlins of classic form exist, it has been recognised for a long time that numerous features in Zone 3 are much more elongate (Bird 1953; Dean 1953; Craig 1964), with lengths approaching 20 km, but with widths <250 m wide (Aylsworth and Shilts 1989a). Some of the most spectacular features occur north of Dubawnt Lake (Fig. 4.9), where elongation (length:width) ratios reach a maximum of 149:1 (Stokes et al. 2013). Indeed, Stokes et al. (2013) found that $\sim$25% of lineations in this region have elongation ratios >10:1. Bedforms with these extreme lengths and elongation ratios are typically referred to as 'mega-scale glacial lineations' (MSGLs: Clark 1993). In this location, MSGLs exist in an area of highly convergent flow, and Aylsworth and Shilts (1989a) noted that individual lineations can appear curved. It is also worth noting that much smaller (drumlin-like) features often occur interspersed with the much larger mega-scale glacial lineations (Stokes et al. 2013). Analysis of sediment exposures cut into the features by fluvial erosion (see Fig. 4.10) revealed a sandy till formed by a combination of non-pervasive subglacial sediment deformation and lodgement and which appears to have been reworked from pre-existing outwash sediments (Ó Cofaigh et al. 2013).

The most spectacular MSGLs on the Canadian Shield form a 'flow-set' that is developed on and down-ice from the

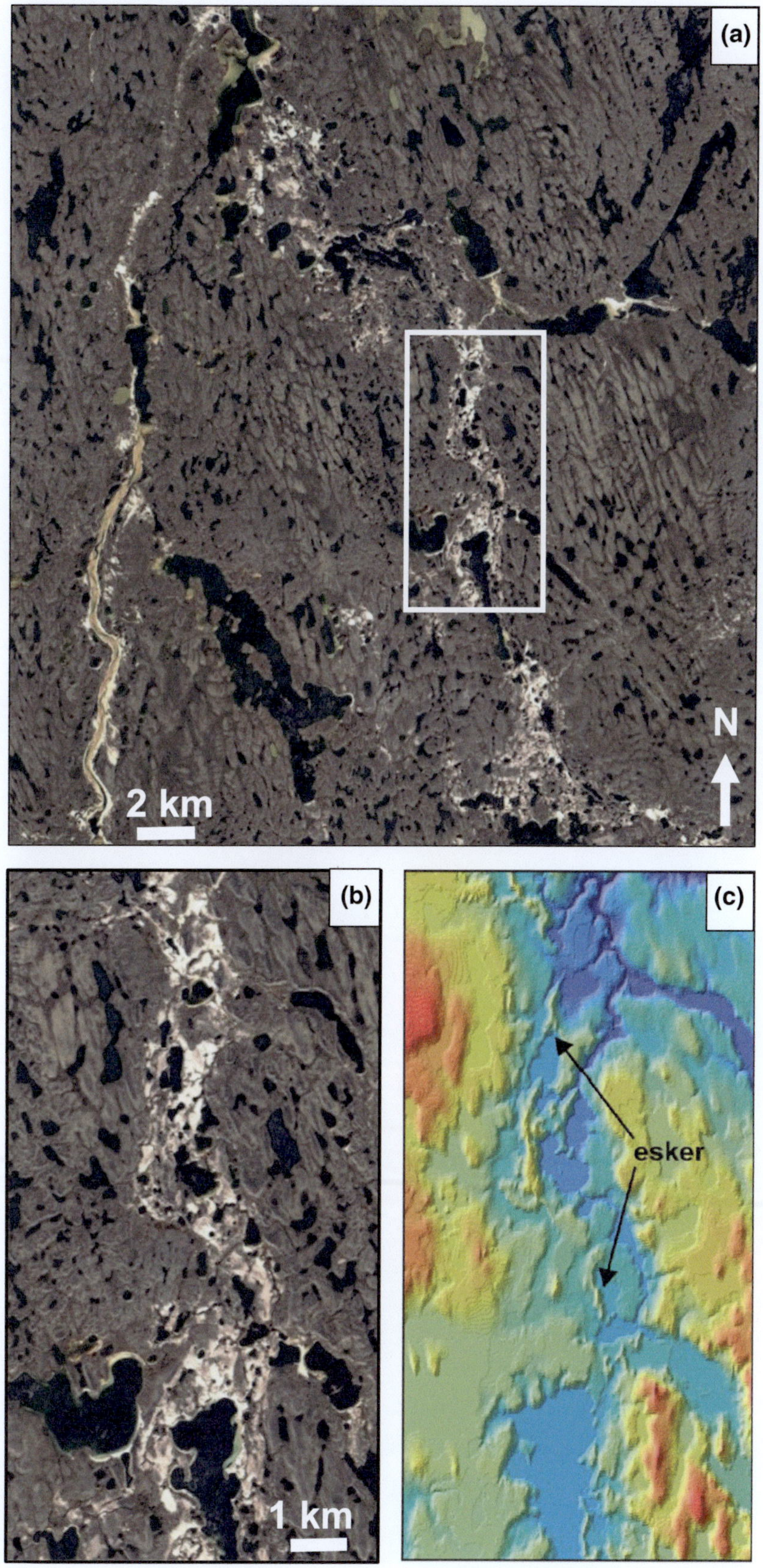

Fig. 4.8 Corridors of glacifluvial material ('glacifluvial corridor hummocks' (GCHs): Utting et al. (2009) characterised by a broad channels of sands and gravel and typically associated with esker ridges throughout Zones 2 and 3 (modified from Storrar and Livingstone 2017). Images in **a** and **b** from Google Earth (Landsat/Copernicus), with image (**a**) centred on 65°36′47″ N, 100° 30′37″ W, just south of the Garry Lake/Back River. Image (**c**) is a digital elevation model taken from Storrar and Livingstone (2017)

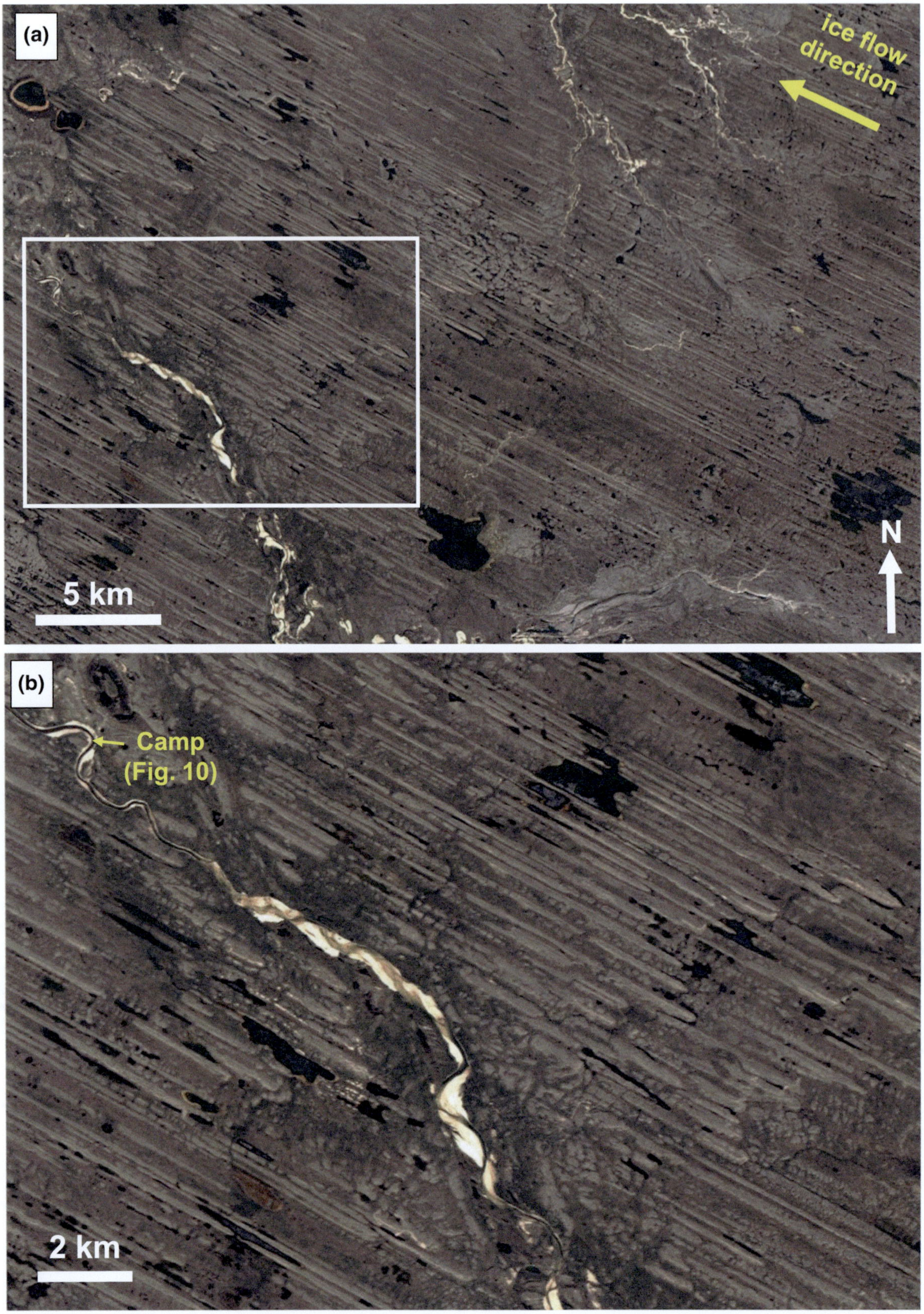

Fig. 4.9 Landsat images (data from Google Earth: Landsat/Copernicus) showing examples of mega-scale glacial lineations that exist in Zone 2 and Zone 3, north and northwest of Dubawnt Lake. **a** Illustrates the a large view of the landscape (image centred on 64°01′20″ N, 10202′51″ W) with **b** showing a close-up of features in the vicinity of the Finnie River (see also Fig. 4.10). Note that some of the features are >20 in length and are the largest glacial lineations identified above sea level anywhere on Earth

Fig. 4.10 **a** Photograph of a field camp on the Finnie River (see Ó Cofaigh et al. 2013) looking southwest across the field of mega-scale glacial lineations that form part of the Dubawnt Lake Ice Stream (location of photograph shown in Fig. 4.9. *Photograph* C. Stokes. **b** Shows an oblique aerial photograph of similar features nearby. *Photograph* C. Stokes

Thelon sedimentary basin (Figs. 4.9 and 4.10) where intense glacial erosion of poorly consolidated, unmetamorphosed clastic sedimentary rocks occurred. Aylsworth and Shilts (1989a) argued that the presence of highly elongate lineations may simply reflect an underlying geological control, in addition to any glacidynamic control related to increasing ice flow velocities away from the KID. In contrast, more recent work has recognised the importance of rapid ice flow, in the form of ice streaming, in helping explain the presence of MSGLS, with the flow-set being referred to as the Dubawnt 'surge fan' (Kleman and Borgström 1996) or the Dubawnt Lake Ice Stream (Stokes and Clark 2003a, b), see discussion in Sect. 4.3.3.

The eskers in Zone 3 are particularly well-developed (Fig. 4.7) and, as noted above, are the longest to be observed anywhere on the bed of the LIS (Storrar et al. 2014a). Shilts et al. (1987) noted that eskers are commonly associated (in space if not in time) with the scattered fields of ribbed moraine in Zone 3 (see also Stokes and Clark 2003a). They also noted that the spacing of eskers in Zone 3 was closer together (around 8 km) than in Zone 2 (13 km) and that although the number of eskers is greater in Zone 3 than Zone 2, there size (height and width) tends to decrease in Zone 3. Storrar et al. (2014a) also noted the remarkably consistent lateral spacing of eskers, which they measured to be around 12 km apart. Aylsworth and Shilts (1989a) observed that eskers in Zone 3 often cross-cut other landforms that were created by actively flowing ice, and that esker tributaries often join esker trunks at steep angles. They argued that this was evidence that the ice sheet had locally stagnated at the time of esker deposition.

4.2.4 Zone 4

The outermost Zone 4 is dominated by extensive areas of massive crystalline bedrock outcrop that is nearly everywhere devoid of till cover (Shilts et al. 1987; Aylsworth and Shilts 1989a) (Fig. 4.11). As such, streamlined features are rare and even eskers become gradually less well-developed and then absent towards the westward limit of Zone 4. Aylsworth and Shilts (1989a) attributed the lack of till cover and the general absence of glacial landforms to the fact that the underlying crystalline bedrock was generally resistant to glacial erosion and sediment production. As such, they argued that although the ice sheet dynamics may have been similar in this zone as in Zone 3, the depositional record is absent because of the scarcity of subglacial debris.

Aylsworth and Shilts (1989a, b, c) also noted that the lack of eskers in the westernmost part of the study area undoubtedly reflects a sediment poor part of the ice sheet related to the resistant bedrock and the lack of far-travelled material from Zone 3. Where eskers do occur (e.g. Fig. 4.11), it has been noted that they tend to be isolated, less-regularly spaced, and with few or no tributaries (Storrar et al. 2014a).

4.2.5 Major Moraine Ridges

Major end moraines, distinct from accumulations or aprons of from glacifluvial outwash, are not common on the northwestern Canadian Shield, although some occur in the northern and northwestern parts of the area (Aylsworth and Shilts 1989a). The largest end moraine system is the MacAlpine Moraine (Blake 1963, 1966; Falconer et al. 1965a). It comprises a complex of end moraine ridges associated with aprons of coalescing glacifluvial fans and deltas and clusters of small eskers (e.g. Fig. 4.12). Blake (1963) noted that in some places it is a sharp ridge that is only a few tens of metres wide, whereas at other points it may be several kilometres wide. It is, however, a conspicuous feature on the otherwise low-relief terrain, with Blake (1963) noting it reaches up to 70 m in height. A similar landform assemblage forms part of the Chantrey Moraine system to the north of the region and beyond onto Baffin Island (Falconer et al. 1965a; Dyke 1984). Together, these moraines are probably the world's longest system of end moraines (Falconer et al. 1965a, b; Dyke and Dredge 1989), which Falconer et al. (1965a, b) interpreted as the last climatically induced stillstand or readvance of the LIS prior to its demise. No other substantial end moraines lie up-ice of these systems in the study region (Dyke and Dredge 1989).

Aylsworth and Shilts (1989a) also drew attention to the Twin Jugs Moraine (Blake 1963), which is a prominent feature of the landscape in the western part of the study area. This is characterised by an extensive area of hummocky terrain with a few large eskers intersecting it. Further west, moraine systems are not obvious and, where present, may resemble eskers and areas of glacifluvial outwash, except for the fact that they are orientated parallel to the assumed ice margin positions (Ayslworth and Shilts 1989a). Aylsworth and Shilts (1989a) argued that the scarcity of major end moraines in the region is probably a function of both the passive nature of ice margin retreat and the paucity of debris available for moraine construction.

4.2.6 Features Related to Glacial Lakes and Marine Transgression

Wave-cut cliffs, beach ridges, spits, bars and other related features are common throughout the study area (e.g. Figs. 4.13 and 4.14), often in association with, but elevated above, present-day lake shorelines such as Dubawnt Lake (Tyrell 1897; Bird 1953; Lee 1959; Craig 1964; Craig and

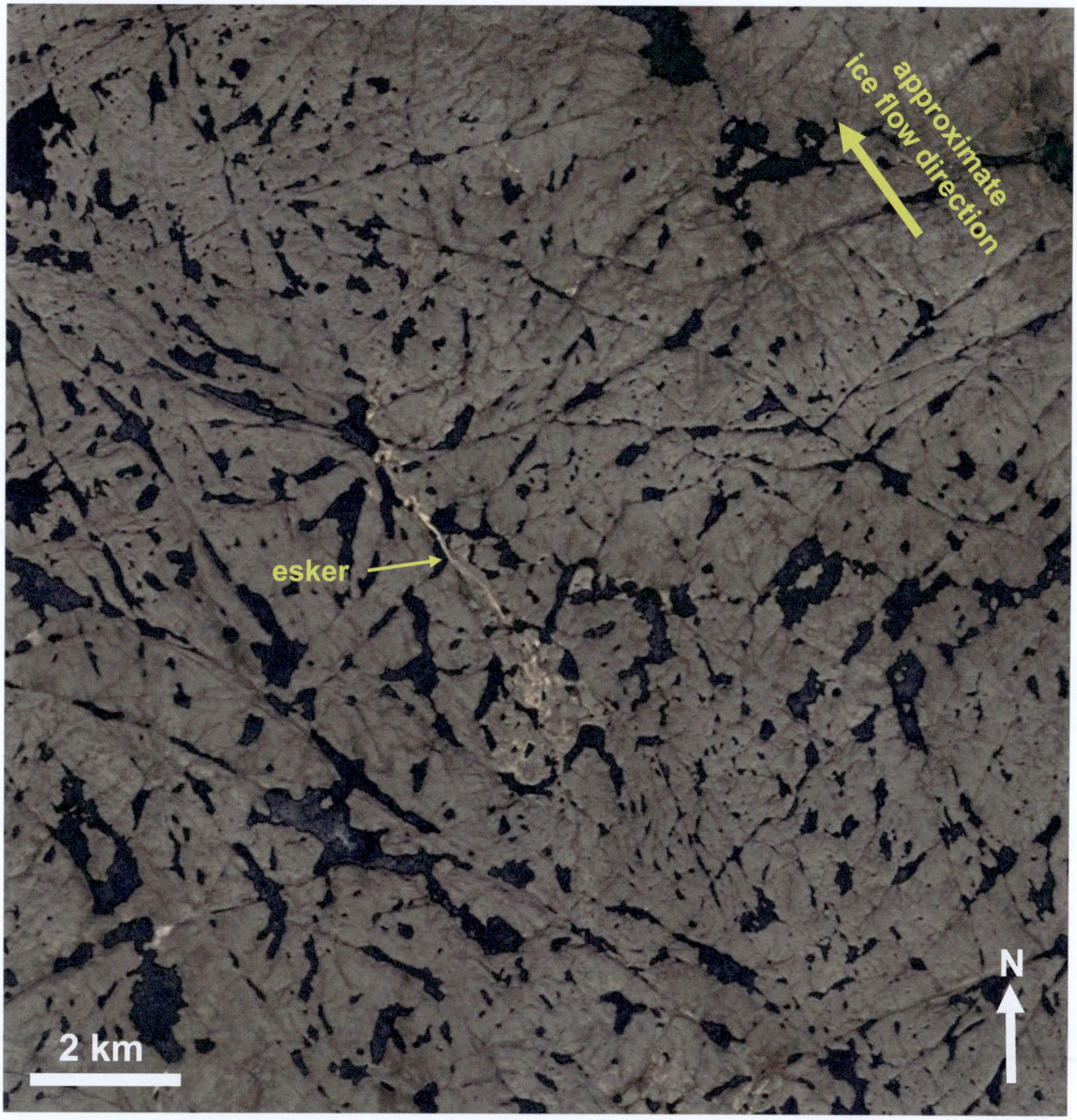

Fig. 4.11 Landsat image (data from Google Earth: Landsat/Copernicus) showing a typical area within Zone 4 of crystalline bedrock with very little till cover in the far north-west of the study area. Image is centred on 66°46′40″ N, 111°35′15″ W. Note the absence of any glacial lineations, but that a short esker likely reflects the ice flow direction towards the north-west during deglaciation

Fyles 1960; Stokes and Clark 2004). For a long time, it was recognised that proglacial lakes formed as the ice front retreated from the north-west to towards Hudson Bay and blocked the eastward drainage of meltwater (e.g. Tyrell 1897; Bird 1953; Craig and Fyles 1960). J. B. Tyrell was the first to record the elevations of some of these lakes, which he suggested were up to 225 m above sea level near Dubawnt Lake (Tyrell 1897). In the Thelon River Valley, Craig and Fyles (1960) noted that the ice sheet would have dammed the natural eastward drainage and that numerous beaches indicate a progressive eastward lowering of an ice-dammed lake as successively lower outlets were uncovered (see also

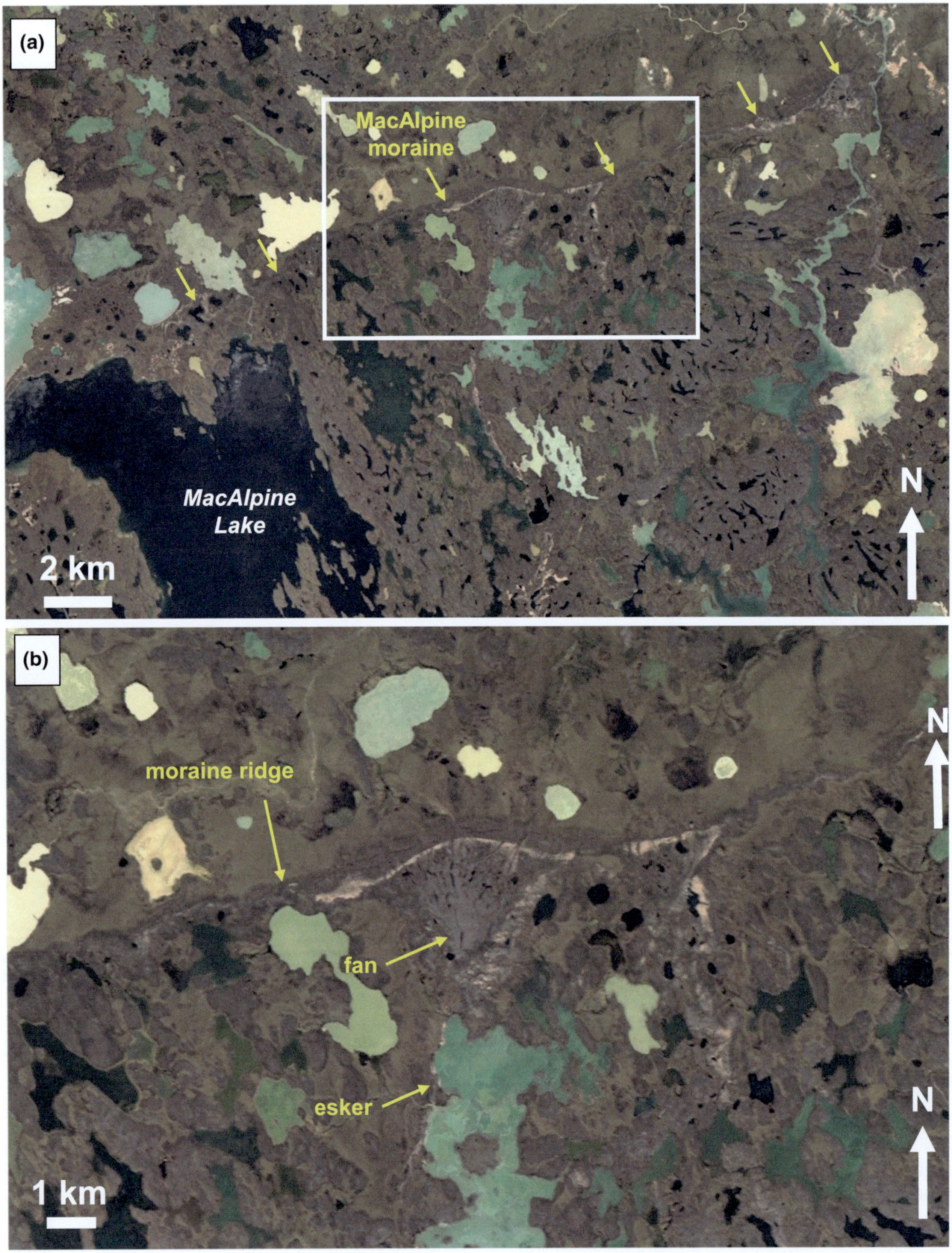

Fig. 4.12 Landsat images (data from Google Earth: Landsat/Copernicus) showing the MacAlpine moraine system immediately north-west of MacAlpine Lake in Zone 3. **a** Illustrates the moraine ridge running approximately south-west to north-east (image centred on 66°43′52″ N, 102°36′18″ W) and **b** is a close-up of an esker and glacifluvial fan that terminates at the moraine ridge

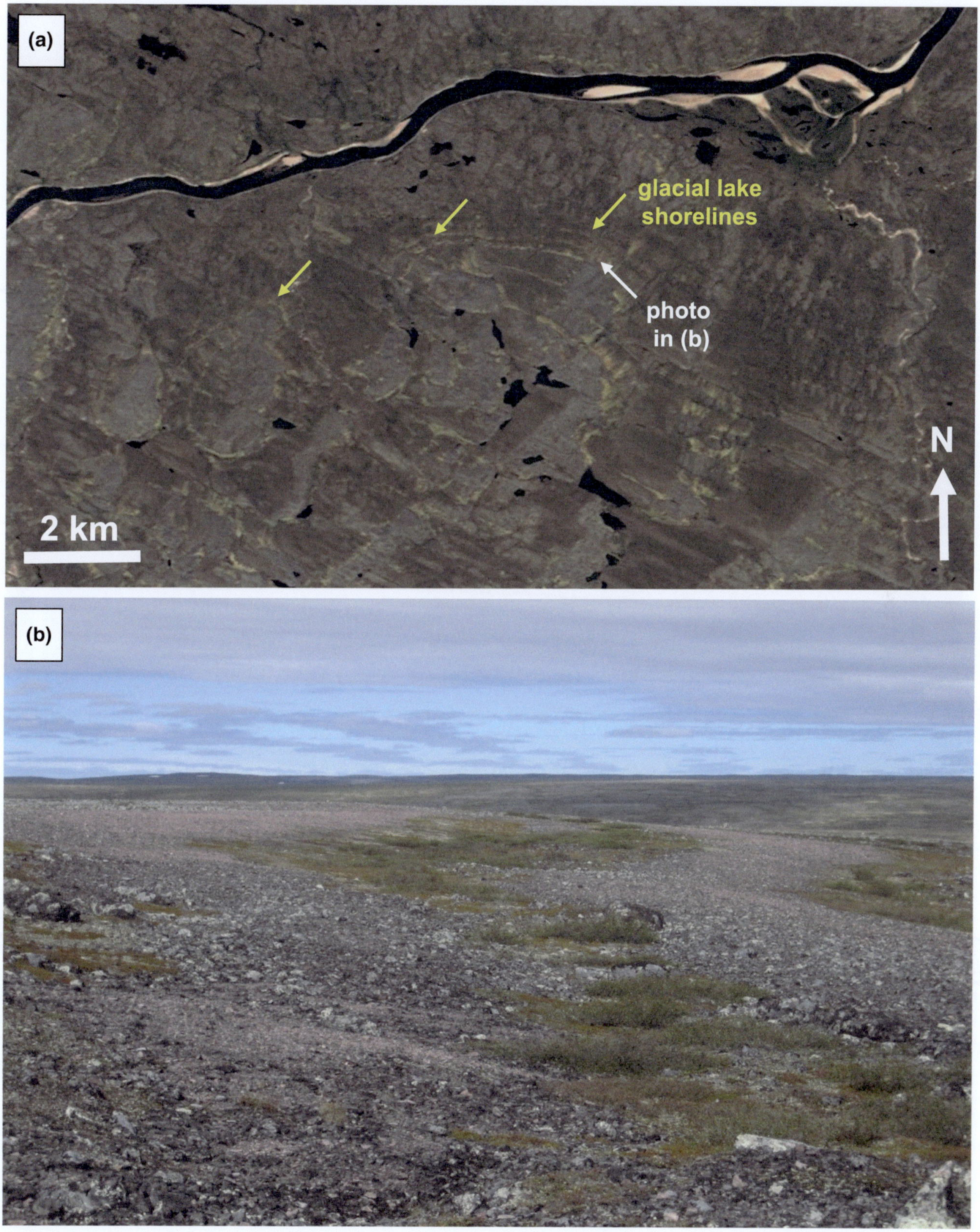

Fig. 4.13 **a** Numerous glacial lake shorelines in Zone 2, just south of the Thelon River (image is centred on 64°29′54″ N, 100°59′05″ W). Note glacial lineations in the bottom right indicating ice flow to the north-west. **b** Photograph of the beaches/shorelines looking north-west (approximate location shown on (a). *Photograph* C. Stokes)

Fig. 4.14 Landsat image (data from Google Earth: Landsat/Copernicus) showing relict shorelines along the west coast of Hudson Bay, around 50 km north of Arviat (image is centred on 61°35′39″ N, 93°55′17″ W)

Bird 1953) (Fig. 4.13). Sediment stratigraphy also supports the existence of proglacial lakes, with several workers noting lacustrine sediments separate by organic layers and which may indicate ice advance through former proglacial lake basin (e.g. Taylor 1956).

The largest lake in the study area has been termed Glacial Lake Thelon (Craig 1964; Stokes and Clark 2004). Evidence from shorelines and spillways indicates that this lake was dammed by an ice lobe that flowed north-westward through the Thelon Sedimentary basin and which contains some of the most spectacular MSGLs described in Zone 3. Craig suggested that the lake may have initially drained to the south but that, in its later stages, abandoned channels indicate spillways to the north were occupied and drained into the Back River basin (Fig. 4.1). More recent work by Stokes and Clark (2004) used a reconstruction of ice margin positions (from Dyke and Prest 1987) and a digital elevation model to predict the evolution of lakes during ice margin retreat. This prediction of where lakes *should* exist supported previous work (e.g. Craig 1964) and the location of lakes matched well with the distribution of lacustrine sediments mapped on the Glacial Map of Canada (Prest et al. 1968). Their reconstruction also suggests that Glacial Lake Thelon was a large (7,500 km^2) and deep ($\sim$ 130 m) lake that would have exerted an important control on ice dynamics through its influence on calving and the drawdown of ice from the KID (see Sect. 4.3.3).

In addition to glacial lake shorelines, the isostatically depressed land surface was subjected to marine incursion during and immediately following deglaciation. The ancestral and much larger Hudson Bay was known as the Tyrell Sea and numerous strandlines and raised shorelines exist in the study area, with particularly remarkable flights of shingly beach ridges encircling present-day Hudson Bay (Dyke and Dredge 1989) (Fig. 4.14).

4.3 Implications for Laurentide Ice Sheet Dynamics

This section briefly highlights how the landform assemblages and patterns described in Sect. 4.2 have increased our understanding of the palaeo-glaciology of the LIS during the last glacial cycle, with particular reference to the Keewatin

Ice Divide (KID) (Sect. 4.3.1), meltwater drainage systems (Sect. 4.3.2), and fast-flowing ice streams (Sect. 4.3.3). It is important to note that many of these advances hold implications that extend well beyond the glacial history of the study area.

4.3.1 The Keewatin Ice Divide (KID)

The zonation of landforms described in Sect. 4.2 has been particularly important in terms of constraining the location and migration of the KID, which was the largest and highest feature in the ice surface topography of the LIS (Dyke and Prest 1987; Tarasov and Peltier 2004). The general absence of features in Zone 1, for example, is one of the main lines of evidence used to infer the presence of the ice divide, where ice thicknesses would have been large, but where ice velocities would have been negligible. Such conditions would have precluded subglacial erosion and transport and, where low-relief hummocky terrain exists in Zone 1 (Fig. 4.2), it has been suggested that it was formed by debris released from basal ice that was almost totally stagnant (Aylsworth and Shilts 1989a).

The general absence of eskers in Zone 1 is also conspicuous, but consistent with the location of an ice divide. For example, Storrar et al. (2014a) argued that any meltwater produced in that location would need to accumulate for a certain distance before it was able to form channels. They noted that the absence of esker ridges at the final location of the ice divide is rather abrupt, and the total width of the esker-free area is relatively constant at 100–150 km. They argued that this might relate to an underlying control on how far from ice divides eskers can form, perhaps related to the hydraulic potential and meltwater supply, or the availability of a sufficient quantity of sediment to build eskers. They also noted that the ice was likely cold-based under the ice divide until the final stages of deglaciation (Kleman and Glasser 2007), thus routing any surface-generated meltwater over the ice surface, rather than reaching the bed to form a channelized system, which may also explain their absence.

Aylsworth and Shilts (1989a) also noted that the presence of ribbed moraines in Zone 2, encircling the divide, may provide clues as to their mode of formation. They argued that the morphology of the ribbed moraines resembled inclined plates of sediment thrust one on top of the other and that this might reflect a compressive flow regime (see also Bouchard 1989) whereby englacially thrusted debris subsequently melted out during ice sheet stagnation. As such, they argued that the distribution of ribbed moraine in Zone 2 reflected the melting of a largely stagnant ice sheet. They also noted that esker tributaries in Zone 2 commonly join trunk eskers at right angles, implying that ice flow as sluggish or non-existent (see also Aylsworth and Shilts 1985). In contrast, Hättestrand and Kleman (1999) have argued that ribbed moraines are formed when a previously frozen till sheet is fractured apart during the transition to a warm-based, extensional, ice flow regime. Thus, Kleman and Hättestrand (1999) suggested that the widespread occurrence of ribbed moraines around the KID (and other core areas of the Laurentide and Fennoscandian ice sheets) implies that large areas of these ice sheets were cold-based at the LGM and, therefore, more likely to be high-domed and stable. In support of these arguments, it has also been recognised that there are numerous swarms of landforms on the Canadian Shield that must predate the LGM and that were protected by cold-based ice (Kleman et al. 2002, 2010). Kleman et al. (2010), for example, have used these relict flow-patterns to piece together the growth and evolution of ice dispersal centres prior to the LGM and they argued that they were dynamically independent for most of the Wisconsinan glaciation.

Although the zonation of landforms around the former location of the KID has long been recognised, more recent work has shown that the ice divide must have migrated to this final position (e.g. McMartin and Henderson 2004). The early work used ice flow indicators, particularly a dispersal train of red Dubawnt Supergroup rock types from the inferred position of the ice divide southeast towards Hudson Bay, to argue that it was a relatively static, long-lived feature (Shilts et al. 1979; Shilts 1980). Several authors have earlier recognised that the ice divide may have migrated towards its final position (e.g. Cunningham and Shilts 1977), but Aylsworth and Shilts (1989a) argued that 'it probably migrated no more than 100 km' (p. 3). More recently, however, McMartin and Henderson (2004) used the orientation of glacial landforms and small-scale glacial erosion features (e.g. striae) from over 800 sites to reconstruct the sequence of ice flows in the region. They argued that the profusion of multifaceted bedrock outcrops and intersecting striations, superimposed on streamlined landforms and stacked till units, was clear evidence of the migration of the main ice divide by as much as 500 km. Thus, whilst Aylsworth and Shilts' (1989a) Zone 1 (Fig. 4.1) is a reasonable approximation of the last position of the central axis of the KID, it is unlikely to reflect its position throughout much of glaciation and likely represents its final position having migrated from a position to the north-west (Cunningham and Shilts 1977; McMartin and Henderson 2004).

4.3.2 Subglacial Meltwater Drainage

As noted above, individual esker systems in the study area can be very long (several hundred kilometres when small gaps are taken into account: Fig. 4.7) and are developed into an integrated pattern with up to fourth-order tributaries that resembles a Hortonian drainage system (Shilts et al. 1987; Storrar et al. 2014a). It is unlikely that this whole system was formed subglacially across such a large area and, rather, several authors have inferred that it developed incrementally during deglaciation as the ice margin retreated (Shilts et al. 1987; Storrar et al. 2014a; Livingstone et al. 2015). They have argued that it was likely fed predominantly by surface meltwater reaching the bed in the ablation zone and that tunnels migrated headward (up-ice) and likely controlled by surface drainage patterns (Shilts et al. 1987; Livingstone et al. 2015).

The pattern and distribution of eskers throughout the study area have led to some important insights regarding the evolution of subglacial meltwater drainage systems during ice sheet deglaciation. Closer inspection of the esker patterns, supported by an extensive quantitative analysis (Storrar et al. 2014b), has also shown that there is a clear increase in esker density during deglaciation (e.g. from Zone 4, through to Zone 3 and into Zone 2: Fig. 4.1), but that the number of eskers with tributaries decreases. Storrar et al. (2014b) suggested that these patterns might be explained by increased surface meltwater production during deglaciation which led to a more efficient drainage system beneath the ice sheet as the ice margin retreated. An important implication, if correct, is that meltwater drainage systems of ice sheets appear to evolve in response to changes in meltwater input over hundreds of years, reminiscent of seasonal changes observed in much smaller valley glaciers (Hubbard and Nienow 1997).

4.3.3 Ice Streaming

The exceptionally long and spectacular glacial lineations north and north-west of Dubawnt Lake (see Sect. 4.2.3; Figs. 4.9 and 4.10) have attracted the attention of glacial geologists for over 100 years. Tyrell (1897) was probably the first to describe them and several studies in the 1950s and 1960s drew attention to their exceptional length and parallel conformity, particularly north and west of Dubawnt Lake (e.g. Dean 1953; Bird 1953; Craig 1964). Their highly elongate appearance was also noted by Aylsworth and Shilts (1989a), who were the first to speculate about the possible role of rapid ice flow (usually referred to as 'ice streaming' in an ice sheet setting) in explaining their great length. They also noted that, together, the highly elongate lineations form a distinctive 'bottle-neck' flow pattern that is clearly identifiable on the Glacial Map of Canada (Prest et al. 1968), i.e. with a broad zone of flow convergence narrowing into a main 'trunk' and then diverging to form a lobate terminus. Aylsworth and Shilts (1989a) also noted that parts of this flow pattern coincide with the relatively 'softer' rock of the Thelon sedimentary basin and they suggested that this geological influence might be more important than any change in the ice dynamics. That is, the highly elongate lineations might simply be a reflection of a softer and more erodible/deformable material beneath the ice, rather than a reflection of more rapid ice flow (ice streaming) in this region.

The distinctive bottle-neck flow was also mapped in Boulton and Clark's (1990) impressive synthesis and they attributed its formation to a late glacial event but did not speculate on the possible role of ice streaming. The interpretation that it represented a late-glacial event was supported by Kleman and Borgström (1996), who went one step further in suggesting that the flow-set was an exemplar of a 'surge fan'. Surge fans are thought to form during the decay stages of an ice sheet, often in relation to proglacial lake basins, and lineations are thought to form nearly synchronously over the whole fan area, rather than time-transgressively during steady ice margin retreat (Kleman and Borgström 1996). They referred to it as the 'Dubawnt fan', after the nearby Dubawnt Lake (see Fig. 4.1).

Building on this body of work, Stokes and Clark (2003a) used remote sensing techniques to undertake a detailed analysis of the landform assemblages comprising the Dubawnt fan/flow-set. They argued that the flow-set exhibited many of the criteria that would be expected to have formed beneath a spatially restricted zone of rapidly-flowing ice, i.e. a palaeo-ice stream (Stokes and Clark 1999, 2001). They named it the Dubawnt Lake Ice Stream and reconstructed its dimensions at ~ 450 km in length and ~140 km wide in the narrowest part of the bottle-neck. They also argued that it probably operated for only a few hundred years during deglaciation, and just prior to 8.2 ^{14}C ka BP, which is the age obtained for the MacAlpine moraines where the flow-set terminates (see Fig. 4.12 and Sect. 4.2.5).

Stokes and Clark (2003a, b) noted that although parts of the ice stream are underlain by the softer sedimentary rocks, including the extensive sandstones of the Thelon sedimentary basin, it is unlikely that the landform assemblage is simply a function of the underlying geology. In part, this is because there are gradual and predictable transitions in bedform elongation that match the expected velocity profile through the bottle-neck flow-set, rather than abrupt changes being observed at underlying geological boundaries (Stokes an Clark 2002). The shape of the flow-set itself is also clear evidence for spatially restricted rapid ice flow, which is the hallmark of ice streaming. Thus, they argued that there was clear evidence for ice streaming taking place during deglaciation and which likely played an important role in driving the south-eastward migration and eventual demise of the Keewatin Ice Divide (cf. McMartin and Henderson 2004).

At that time, the identification of a major palaeo-ice stream on the Canadian Shield challenged the paradigm that ice streams were largely restricted to soft-bedded areas of the ice sheet (see Clark 1992, 1994). Given the low relief, the location of the ice stream begged the question as to what may have triggered the fast flow. Stokes and Clark (2003a, b) had speculated that it might have been related to the development of proglacial lakes along this part of the ice sheet margin (see Sect. 4.2.6) and a more detailed analysis revealed that this was indeed the only part of the western margin of the Keewatin Sector where large and deep lakes formed. The ice stream is, therefore, thought to have been triggered by the development of Glacial Lake Thelon which induced high calving rates and the drawdown of ice from further up-ice (Stokes and Clark 2004).

As noted above, parts of the ice stream bed (e.g. in Zones 2 and 3) are characterised by ribbed moraines that are superimposed on the elongate bedforms produced by the ice stream (Fig. 4.5) (e.g. Aylsworth and Shilts 1989a; Stokes et al. 2006). This presents somewhat of a conundrum because ribbed moraines are not usually associated with rapid ice flow and yet they cover around 7% of the ice stream bed, with individual fields of ribbed moraine ranging in extent from <1 km^2 up to $\sim$2,500 km^2 (Stokes et al. 2006). Their patchy appearance on the ice stream bed is intriguing and their presence implies a marked change in ice dynamics following ice stream activity. Using observations of their morphometry, internal structure and sedimentological characteristics, Stokes et al. (2008) suggested that they formed in areas of the ice stream bed that stiffened as a result of dewatering of the till, e.g. through changes in subglacial meltwater drainage pathways or as a result of basal freeze-on (Christoffersen and Tulaczyk 2003). As such, they argued that patches of ribbed moraines represent sticky spots that developed during ice stream shut-down.

In summary, the identification of the Dubawnt Lake Ice Stream has proved important to a number of areas of research, with implications that extend beyond the glacial history of the Keewatin Sector of the LIS. Firstly, it is one of the largest and best-preserved palaeo-ice stream imprints that has so far been discovered on land and it has proved a useful template to identify other ice streams, including on the Canadian Shield (see review by Margold et al. 2015), but also further afield. Secondly, its large size and unusual location on the shield has enabled a broader understanding of the triggers and controls on ice streaming in ice sheets (cf. Winsborrow et al. 2010), demonstrating that ice streaming can occur over relatively hard bedrock terrain, likely triggered by proglacial lakes and facilitated via rapid basal sliding. Thirdly, the large size and reasonably well-constrained age bracket of the ice stream offers a useful analogue to understand the duration and propagation of ice sheet instabilities (Kleman and Applegate 2014) and the role of ice streams during ice sheet deglaciation (Stokes et al. 2016). It also offers a useful test for numerical ice sheet models that generate ice streams and can which can be tested against their 'known' location (e.g. Stokes and Tarasov 2010).

4.4 Conclusions

The glacial landform assemblages of mainland Nunavut, west of Hudson Bay, are some of the most spectacular and best-studied anywhere on Earth. At the Last Glacial Maximum, this region lay close to the centre of the former Laurentide Ice Sheet and was covered by a large ice dome (the Keewatin Dome) that was >3 km thick (Simon et al. 2014). This high point in the ice surface was associated with a major ice divide (the Keewatin Ice Divide) that was orientated north-east to south-west and which gradually migrated toward the south-east as deglaciation progressed (Lee et al. 1967; McMartin and Henderson 2004). It has, for a long time, been recognised that the Keewatin Ice Divide exerted an important control on the distribution of glacial landforms on the northwestern Canadian Shield (Ayslworth and Shilts 1985, 1989a, b, c; Shilts et al. 1987). These have been identified as four broadly concentric zones that include and encircle the location of the last inferred position of the ice divide (Fig. 4.1). Zone 1 is the innermost zone that includes the footprint of the ice divide itself. It is generally characterised by featureless till plains that lacking any obvious glacial landforms and/or low hummocky terrain and with a general absence of eskers. Zone 2 encircles most of Zone 1 and is characterised by the presence of well-developed linear trains or ribbed moraine that radiate out from the location of the former ice divide, but which are interspersed with fields of glacial lineations. Major esker systems dominate the landscape in Zone 2, where they reflect an extensive dendritic network in a pattern that also radiates away from the location of the ice divide. Individual esker ridges are some of the longest in the world with many tens of kilometres in length, and with some traceable for hundreds of kilometres with gaps. These esker systems continue into Zone 3, which is characterised by a fairly continuous cover of till and numerous drumlin fields that contain some highly elongate mega-scale glacial lineations. Towards the outer limits of Zone 3, till cover thins to a discontinuous veneer. Ribbed moraines also occur in Zone 3 but they are much less frequent than in Zone 2, and typically occur in isolated patches as linear trains interspersed with or superimposed on drumlins and mega-scale glacial lineations. The outermost Zone 4 comprises extensive areas of massive crystalline bedrock that are typically devoid of any till cover and with far fewer obvious glacial landforms and only sporadic eskers. Major moraine ridges are not common within the study area, but the largest and most obvious moraine systems (the MacAlpine and Chantrey Inlet moraines) can be traced for several hundred kilometres and likely represent a pause or readvance of the ice sheet during overall retreat (Falconer et al. 1965a, b). Throughout all four zones, relict shorelines, beaches, strandlines and spillways occur that demarcate the growth and drainage of numerous proglacial lakes, in addition to marine transgression of the ancestral Hudson Bay (Tyrell Sea).

Aylsworth and Shilts (1989a) noted that the distribution of glacial landforms on the northwestern Canadian Shield reflects "a complex and yet imperfectly understood interplay between dynamic conditions within the ice sheet and the geology of the bed across which it flows" (p. 19). It is clear that the location of the Keewatin Ice Divide was an important control on the broad-scale zonation of landforms, including both their formation and preservation (e.g. in relation to cold-based ice); but it is also clear that the underlying geology influenced the amount of sediment that could be eroded and remobilised at the ice sheet bed, with some of the most spectacular features forming over the relatively less resistant sedimentary outcrops (Aylsworth and Shilts 1989a). More recent work has reinterpreted some of these landscapes within a new paradigm that recognises the importance of ice streams in palaeo-ice sheet dynamics (Stokes and Clark 2001). Specifically, the Dubawnt Lake palaeo-ice stream (Stokes and Clark 2003a, b) is one of the largest and best-studied ice streams in the LIS and its rapid ice flow is now known to have created a distinctive bottle-neck flow-set through the centre of the study area. Its activity is able to explain the formation of the spectacular mega-scale glacial lineations north of Dubawnt Lake (Stokes et al. 2013) and the final southwestward migration of the ice divide (McMartin and Henderson 2004). The impressive network of eskers has also been subjected to detailed quantitative analysis (Storrar et al. 2014a, b), which has revealed that increased surface meltwater production during deglaciation led to a more efficient drainage system beneath the ice sheet.

Acknowledgements I would like to acknowledge several colleagues who I have had the pleasure to work with on the glacial history of the northwestern Canadian Shield (and some of whom have shared in some memorable fieldwork in the region), namely, Chris Clark, Olav Lian, Slawek Tulaczyk, Colm Ó Cofaigh, Matteo Spagnolo, Rob Storrar and Martin Margold. These collaborations were largely funded through grants from the NERC (NER/M/S/2003/00050 and NE/J00782X/1), and a Philip Leverhulme Prize awarded to CRS.

References

Aylsworth JM, Shilts WW (1985) Glacial features of the west central Canadian Shield. Curr Res Part B Geol Surv Can Paper 85-1B, 375–381

Aylsworth JM, Shilts WW (1989a) Glacial features around the Keewatin Ice Divide: Districts of Mackenzie and Keewatin. Geol Surv Can Paper 88–24, p 21

Aylsworth JM, Shilts WW (1989b) Glacial features around the Keewatin Ice Divide: Districts of Mackenzie and Keewatin. Geol Surv Can Map 24-1987, scale 1:1,000,000

Aylsworth JM, Shilts WW (1989c) Bedforms of the Keewatin Ice Sheet, Canada. Sediment Geol 62:407–428, 25:215–230

Bird JB (1953) The glaciation of central Keewatin, Northwest Territories, Canada. Am J Sci 251:215–230

Blake W Jr (1963) Notes on glacial geology, northeastern District of Mackenzie. Geol Surv Can, Paper 63-28

Blake W Jr (1966) End moraines and deglaciation chronology in northern Canada, with special reference to southern Baffin Island. Geol Surv Can Paper 66–26, p 31

Bouchard MA (1989) Subglacial landforms and deposits in central and northern Quebec, Canada, with emphasis on Rogen moraines. Sediment Geol 62:293–308

Boulton GS, Clark CD (1990) A highly mobile Laurentide ice sheet revealed by satellite images of glacial lineations. Nature 346:813–817

Christoffersen P, Tulaczyk S (2003) Signature of palaeo-ice stream stagnation: till consolidation induced by basal freeze-on. Boreas 32:114–129

Clark CD (1993) Mega-scale glacial lineations and cross-cutting ice-flow landforms. Earth Surf Process 18:1–29

Clark PU (1992) Surface form of the southern Laurentide Ice Sheet and its implications for ice sheet dynamics. Geol Soc Am Bull 104:595–605

Clark PU (1994) Unstable behavior of the Laurentide ice sheet over deforming sediment and its implications for climate change. Quat Res 41:19–25

Craig BG (1964) Surficial geology of east-central District of Mackenzie. Geol Surv Can Bull 99

Craig BG (1965) Glacial Lake McConnell, and the surficial geology of parts of Slave River and Redstone River map-areas, District of Mackenzie. Geol Surv Can Bull 122:33

Craig BG, Fyles JG (1960) Pleistocene geology of Arctic Canada. Geol Surv Can, no 60-10, p 21

Cunningham CM, Shilts WW (1977) Surficial geology of the Baker Lake area, District of Keewatin. Geol Surv Can, Paper 77-1B, pp 3111–3314

Dean WG (1953) The drumlinoid land forms of the Barren Grounds. Can Geogr 1:19–30

Dunlop P, Clark CD (2006) The morphological characteristics of ribbed moraine. Quat Sci Rev 25:1668–1691

Dyke AS (1984) Quaternary geology of Boothia Peninsula and northern District of Keewatin, Central Canadian Arctic. Geol Surv Can Memoir 407:26

Dyke AS, Prest K (1987) Late Wisconsinan and Holocene history of the LaurentideIce Sheet. Géog Phys Quat 41(2):237–263

Dyke AS, Dredge LA (1989) Quaternary geology of the northwestern Canadian Shield. In: Fulton RJ (ed) Quaternary Geology of Canada and Greenland. Geology of Canada Series, no. 1. Geological Survey of Canada, Ottawa, pp 189–214

Dyke AS, Vincent J-S, Andrews JT, Dredge LA, Cowan WR (1989) The Laurentide Ice Sheet and an introduction to the Quaternary geology of the Canadian Shield. In: Fulton RJ (ed) Quaternary Geology of Canada and Greenland. Geology of Canada Series, no 1. Geological Survey of Canada, Ottawa, pp 178–189

Falconer G, Ives JD, Loken OH, Andrews JT (1965a) Major end moraines in eastermn and central Arctic Canada. Geogr Bull 7:137–153

Falconer G, Andrews JT, Ives JD (1965b) Late-Wisconsin end moraines in Northern Canada. Science 147:608–610

Fyles JG (1955) Pleistocene features. In: Wright GM (ed) Geological notes on central District of Keewatin, Northwest Territories. Geological Survey of Canada Paper, 55-17

Greenwood SL, Kleman J (2010) Glacial landforms of extreme size in the Keewatin sector of the Laurentide Ice Sheet. Quat Sci Rev 29:1894–1910

Hättestrand C, Kleman J (1999) Ribbed moraine formation. Quat Sci Rev 18:43–61

Hubbard B, Nienow P (1997) Alpine subglacial hydrology. Quat Sci Rev 16:939–955

Kleman J, Borgström I (1996) Reconstruction of palaeo-ice sheets: the use of geomorphological data. Earth Surf Process 21:893–909

Kleman J, Glasser NF (2007) The subglacial thermal organisation (STO) of ice sheets. Quat Sci Rev 26:585–597

Kleman J, Hättestrand C (1999) Frozen-bed Fennoscandian and Laurentide ice sheets during the Last Glacial Maximum. Nature 402:63–66

Kleman J, Applegate PJ (2014) Durations and propagation patterns of ice sheet instability events. Quat Sci Rev 92:32–39

Kleman J, Fastook J, Stroeven AP (2002) Geologically and geomorphologically constrained numerical model of Laurentide Ice Sheet inception and build-up. Quat Int 95–96:87–98

Kleman J, Jansson K, De Angelis H, Stroeven AP, Hättestrand C, Alm G, Glasser N (2010) North American ice sheet build-up during the last glacial cycle, 115-21 kyr. Quat Sci Rev 29:2036–2051

Lee H (1959) Surficial geology of southern District if Keewatin and the Keewatin Ice Divide, northwest Territories. Geol Surv Can Bull 51

Lee HA, Craig BG, Fyles JG (1957) Keewatin Ice Divide (abstract). Geol Soc Am Bull 51, 42 pp

Livingstone SJ, Storrar RD, Hillier JK, Stokes CR, Clark CD, Tarasov L (2015) An ice sheet scale comparison of eskers with modelled subglacial drainage routes. Geomorphology 246:104–112

Margold M, Stokes CR, Clark CD (2015) Ice streams in the Laurentide Ice Sheet: identification, characteristics and comparison to modern ice sheets. Earth Sci Rev 143:117–146

McMartin I, Henderson PJ (2004) Evidence from Keewatin (central Nunavut) for paleo-ice divide migration. Géog Phys Quat 58 (2/3):163–186

Ó Cofaigh C, Stokes CR, Lian OB, Clark CD, Tulaczyk SM (2013) Formation of mega-scale glacial lineations on the Dubawnt Lake Ice Stream bed: 2. Sedimentology and stratigraphy. Quat Sci Rev 77:190–209

Prest VK (1969) Retreat of Wisconsin and Recent Ice in North America. Geological Survey of Canada Map, 1257A, scale 1:5,000,000

Prest VK, Grant DR, Rampton VN (1968) Glacial Map of Canada. Geological Survey of Canada, Map, p 1253A

Simon K, James TS, Forbes DL, Telka AM, Dyke AS, Henton JA (2014) A relative sea level history for Arviat, Nuanvut, and implications for Laurentide Ice Sheet thickness west of Hudson Bay. Quat Res 82:185–197

Shilts WW (1977) Geochemistry of till in perennially frozen terrain of the Canadian Shield—application to prospecting. Boreas 6:203–212

Shilts WW (1980) Flow patterns in the central North American Ice Sheet. Nature 286:213–218

Shilts WW (1984) Esker sedimentation models, Deep Rose Lake map-area, Distruict of Keewatin. Current Research in Canada, Part B. Geological Survey of Canada Paper 84-1B, pp 217–222

Shilts WW, Cunningham CM, Kaszycki CA (1979) Keewatin Ice Sheet—re-evaluation of the traditional concept of the Laurentide Ice Sheet. Geology 7(11):537–541

Shilts WW, Aylsworth JM, Kaszycki CA, Klassen RA (1987) Chapter 5: Canadian shield. In: Graf WL (ed) Geomorphic systems of North America. Boulder, Colorado, Geol Soc Am, Centennial Special Volume 2, pp 119–161

St-Onge DA (1984) Surficial deposits of the Redrock Lake area, District of Mackenzie. Curr Res Part A, Geological Survey of Canada, Paper 84-1A, 271–277

Stokes CR, Clark CD (1999) Geomorphological criteria for identifying Pleistocene ice streams. Ann Glaciol 28:67–74

Stokes CR, Clark CD (2001) Palaeo-ice streams. Quat Sci Rev 20:1437–1457

Stokes CR, Clark CD (2002) Are long subglacial bedforms indicative of fast ice flow? Boreas 31:239–249

Stokes CR, Clark CD (2003a) The Dubawnt Lake palaeo-ice stream: evidence for dynamic ice sheet behaviour on the Canadian Shield and insights regarding the controls on ice stream location and vigour. Boreas 32:263–279

Stokes CR, Clark CD (2003b) Laurentide ice streaming on the Canadian Shield: a conflict with the soft-bedded ice stream paradigm? Geology 31:347–350

Stokes CR, Clark CD (2004) Evolution of late glacial ice-marginal lakes on the northwestern Canadian Shield and their influence on the location of the Dubawnt Lake palaeo-ice stream. Palaeogeogr Palaeoecl 215:155–171

Stokes CR, Tarasov L (2010) Ice streaming in the Laurentide Ice Sheet: a first comparison between data-calibrated numerical model output and geological evidence. Geophys Res Lett. https://doi.org/10.1029/2009GL040990

Stokes CR, Clark CD, Lian O, Tulaczyk S (2006) Geomorphological map of ribbed moraine on the Dubawnt Lake Ice Stream bed: a signature of ice stream shut-down? J Maps 2006:1–9

Stokes CR, Lian OB, Tulaczyk S, Clark CD (2008) Superimposition of ribbed moraines on a palaeo-ice stream bed: implications for ice stream dynamics and shutdown. Earth Surf Process 33:593–609

Stokes CR, Spagnolo M, Clark CD, Ó Cofaigh C, Lian OB, Dunstone RB (2013) Formation of mega-scale glacial lineations on the Dubawnt Lake Ice Stream bed: 1. Size, shape and spacing from a large remote sensing dataset. Quat Sci Rev 77:190–209

Stokes CR, Margold M, Clark CD, Tarasov L (2016) Ice stream activity scaled to ice sheet volume during Laurentide ice sheet deglaciation. Nature 530:322–326

Storrar RD, Livingstone SJ (2017) Glacial geomorphology of the northern Kivalliq region, Nunavut, Canada, with an emphasis on meltwater drainage systems. J Maps 13(2):153–164

Storrar RD, Stokes CR, Evans DJA (2013) A map of Canadian eskers from Landsat satellite imagery. J Maps 9(3):456–473

Storrar RD, Stokes CR, Evans DJA (2014a) Morphometry and pattern of a large sample (>20,000) of Canadian eskers and implications for subglacial drainage beneath ice sheets. Quat Sci Rev 105:1–25

Storrar RD, Stokes CR, Evans DJA (2014b) Increased channelization of subglacial drainage during deglaciation of the Laurentide Ice Sheet. Geology 42:239–242

Tarasov L, Peltier WR (2004) A geophysically constrained large ensemble analysis of the deglacial history of the North American ice-sheet complex. Quat Sci Rev 23:359–388

Taylor RS (1956) Glacial geology or north-central Keewatin, Northwest Territories, Canada. Bull Geol Soc Am 67:943–956

Tyrell JB (1897) Report on the Dubawnt, Kazan and Ferguson Rivers. Canada Geological Survey Annual Report, 1986, Part E

Utting DJ, Ward BC, Little EC (2009) Genesis of hummocks in glaciofluvial corridors near the Keewatin Ice Divide, Canada. Boreas 38:471–481

Winsborrow MCM, Clark CD, Stokes CR (2010) What controls the location of ice streams? Earth-Sci Rev 103:45–59

Chris Stokes is a Professor in the Department of Geography at Durham University, UK. Following a Ph.D. (Geography) at the University of Sheffield (UK), he was appointed to a Lectureship at the University of Reading (UK). In 2007, he moved to take up a position as Reader in the Quaternary Environmental Change research group in the Department of Geography at Durham University, UK, where he was awarded a Personal Chair in 2014. His research focusses on glaciers and their interaction with the climate system, and ranges from monitoring small mountain glaciers over the last few decades to large-scale reconstructions of ice sheets over tens of thousands of years. His main research tool is satellite imagery, which affords a large-scale view of glaciers and glacial landscapes, but he has also visited the field in Arctic Canada, Norway, Svalbard, Russia and Patagonia. He has authored over 100 scientific papers and his work has been covered in both national and international media. He was awarded a prestigious Philip Leverhulme Prize (in 2009) in recognition of his research contributions, and has been awarded the British Society for Geomorphology's Gordon Warwick Medal (2013) and Wiley Award (2008). He currently serves on the Editorial Boards for two international journals and as an Editorial Advisor for a third. He is also a former President of the British Branch of the International Glaciological Society.

Deglacial to Paraglacial History of the Lake Saint-Jean Lowlands: A Geomorphological Perspective

5

Alexis Nutz, Patrick Lajeunesse, Jean-François Ghienne, Mathieu Schuster, Etienne Brouard, Pierre Dietrich, Frédéric Bouchette, Claude Roquin, and Pierre A. Cousineau

Abstract

The Lake Saint-Jean lowland is a particular area in eastern Canada as it constitutes an important Late Quaternary depocenter compared with other surrounding onshore regions. Here, the recent literature about the Late Quaternary history of the Lake Saint-Jean depocenter from the glaciated period to present-day is summarized; subsequently, we present some preserved landscape features that record such history.

Keywords

Late Quaternary • Deglacial • Paraglacial • Lake Saint-Jean • Depocenter

A. Nutz (✉)
Department of Geoscience, Aarhus University, Aarhus, Denmark
e-mail: nutz@cerege.fr

A. Nutz
Centre Européen de Recherche et d'Enseignement des Géosciences de l'Environnement, Aix-Marseille Université, CNRS, IRD, Collège de France, INRA, Aix-en-Provence, France

P. Lajeunesse · E. Brouard
Département de Géographie, Centre d'Études Nordiques, Université Laval, Quebec, Canada

J.-F. Ghienne · M. Schuster · C. Roquin
Institut de Physique du Globe de Strasbourg, UMR7516, CNRS, Strasbourg, France

P. Dietrich
Department of Geology, University of Johannesburg, Johannesburg, South Africa

P. Dietrich
Géosciences-Rennes, UMR6118, Université de Rennes 1, 35042 Rennes, France

F. Bouchette
GEOSCIENCES-M, Université Montpellier II et CNRS, Montpellier, France

P. A. Cousineau
Centre d'étude sur les Ressources minérales, Université du Québec à Chicoutimi, Chicoutimi, Canada

5.1 Introduction

The Lake Saint-Jean lowlands are an unusual area of eastern Canada where significant volumes of Late Quaternary sediments have been accumulated since deglaciation. As a restricted and relatively confined depocenter, the Lake Saint-Jean lowlands evolved individually and thus recorded the latest Pleistocene and the Holocene through an original morphostratigraphy. In recent years, onshore and offshore investigations have brought new insights on this evolution. Here, we show some of the emblematic landforms and features that illustrate particularly well the deglacial-to-paraglacial evolution of the area from both the onshore and offshore domains.

5.2 Lake Saint-Jean Lowlands

The Lake Saint-Jean lowlands are located in southern Québec, Canada (Fig. 5.1). They correspond to an 85 km long and 45 km wide low-relief area consisting of the western end of the Saguenay Graben (Kumarapelli 1985; Cousineau and Longuépée 2003). This low-relief is flanked by the 800 to 1000 m a.s.l. moderate- to high-relief of the Laurentian Shield. Rocky basement occasionally crops out in Lake Saint-Jean lowlands as ridges or extensive highs while downstream the Saguenay River incised the Saguenay Fjord within the graben structure. The modern Lake Saint-Jean is draining a 78,000 km^2 catchment area. It has mean and maximum depths of 11.6 and 68 m, respectively. The modern-day Lake Saint-Jean has a long axis of 42 km and a shorter one of 20 km. The three major tributaries (Péribonka, Mistassini and Ashuapmushuan rivers) enter the lake flow from the north and the northwest (Fig. 5.1). These rivers are subjected to strong seasonal discharge fluctuations inducing a rise of about 4 m, from the mean lowstand level in April (about 101 m a.s.l.) to the highstand level in June (about 105 m a.s.l.). The mean annual discharge is 1528 m^3 s^{-1} at

© Springer Nature Switzerland AG 2020
O. Slaymaker and N. Catto (eds.), *Landscapes and Landforms of Eastern Canada*,
World Geomorphological Landscapes, https://doi.org/10.1007/978-3-030-35137-3_5

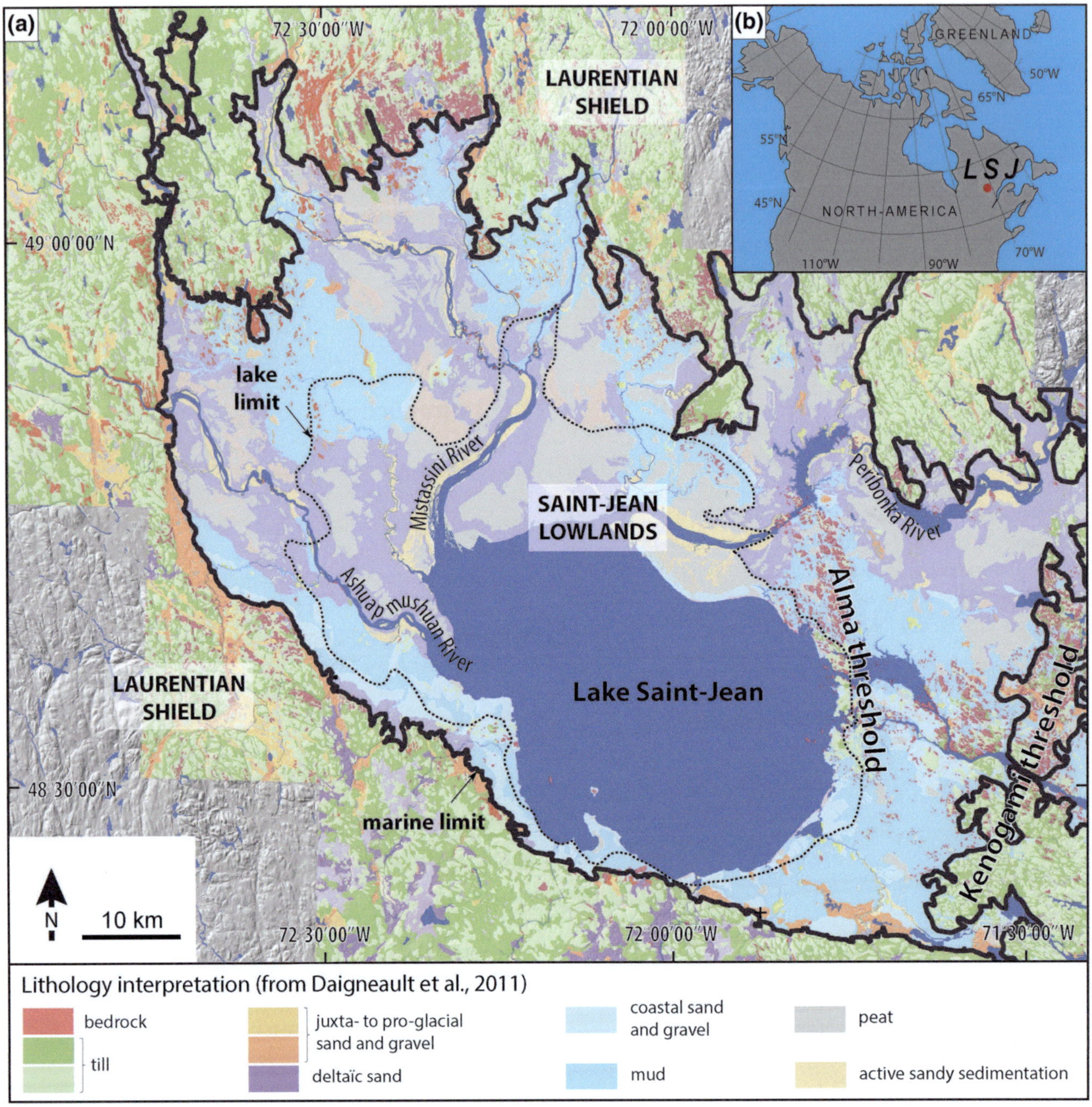

Fig. 5.1 Location maps (modified from Nutz et al. 2015a). **a** Quaternary deposits in the Lake Saint-Jean basin (modified from Roy et al. 2011). **b** Location of Lake Saint-Jean basin in North America

the outlet, in the southeastern part of the lake. The water renewal has been estimated to occur 4–5 times per year (Roy et al. 2011). Prevailing winds blow from northwest and west, but also from the southwest during summer (Meteorological Survey of Canada, http://www.climat-quebec.qc.ca).

5.2.1 Late Quaternary History

The retreat of the Laurentide Ice Sheet (LIS) margin is relatively well documented from the Last Glacial Maximum (LGM) to the Holocene (e.g. Fulton et al. 1986; Syvitski and

Praeg 1989; Shaw et al. 2002; Occhietti et al. 2001, 2004, 2011; Dyke 2004; Occhietti 2007; Nutz 2013; Dietrich et al. 2017). It is generally accepted that the LIS margin reached the northern shore of the present-day Saint-Lawrence Estuary between 13.6 and 11.55 cal ka BP. Afterwards, the LIS margin retreated northwestward up to the southeast edge of the Lake Saint-Jean lowlands (Fig. 5.2a) marking the onset of deglaciation in the area. The age of this stage is still a matter of debate. Previous studies suggest an age at c. 10.8 cal ka BP (Tremblay 1971a; Occhietti et al. 2011), whereas Nutz (2013) proposes an older age around c. 13.1 cal ka BP. Coevally with the progressive northwestward retreat of the LIS margin (Fig. 5.2b), marine waters invaded the newly deglaciated, glaci-isostatically flexured basement (Tremblay 1971a; Roy et al. 2011; Daigneault et al. 2011; Nutz et al. 2014, 2015a) forming a northwestern branch of the Goldthwait Sea (Dionne 1972) referred to as the Laflamme Gulf (Laverdière and Mailloux 1956). At its maximum extent, the Laflamme Gulf extended up to 50 km north of the present-day Lake Saint-Jean northern shoreline (Figs. 5.1 and 5.2c; Daigneault et al. 2011; Nutz et al. 2015a). At that time, the intense glaci-isostatic uplift led to significant rates of relative sea level (RSL) fall estimated around 3.5 cm y^{-1} in the northwest part of the area (Nutz 2013). The marine Laflamme Gulf became Lake Saint-Jean (Fig. 5.2d) once the connection with the Saguenay fjord ended due to the emergence of the Kenogami sill around 8.5 ± 0.6 cal ka BP (Roy et al. 2011). During the subsequent lacustrine period, the lake level continued to fall due to the differential glaci-isostatic uplift between the northwest and the southeast areas (Fig. 5.2e), but at slower rates estimated around 0.5 cm y^{-1} in the northwest part of the area. The LIS margin continued to retreat northward to leave the Saint-Jean lowlands catchment around 7.5 cal ka BP (Occhietti et al. 2011). After 7.5 cal ka BP, the lake level continuously dropped until 1926 C.E. to reach its lowest-stand during the Holocene at 2.5 m below the current lake level. Dam emplacement in 1926 raised and stabilized the lake level at its present-day configuration (Fig. 5.2f).

5.2.2 Stratigraphic Organization

The Lake Saint-Jean lowlands display a comprehensive stratigraphic record of the latest Quaternary, which appears very similar across the entire Lake Saint-Jean depocenter (Nutz et al. 2015a). In general, the sedimentary succession consists of lowermost sand and coarse-grained sediment bodies corresponding to glacigenic deposits such as basal till and ice-contact fans. These deposits are overlain by the Laflamme Mud Fm., covering extensive areas below the marine limit. This formation forms a sheet-like unit that grades vertically from proximal to distal ice-front-related fine-grained deposits to marine and then lacustrine prodeltaic fine-grained sediments, reflecting the progressive retreat of the ice-front and the subsequent transition to a continental ice-front. Above, the Laflamme Mud Fm. is overlain by diachronous deltaic and coastal sand wedges that fringe the basin and correspond to Gilbert deltas, mouth-bar deltas, spits, cuspates and beach ridges that have progressively prograded basinward following the falling base level.

5.3 Geomorphological Markers

The latest Quaternary history of the Lake Saint-Jean lowlands is documented by numerous remarkable onshore landforms and offshore features. The first group includes glacigenic features associated with the deglacial period. The second group comprises features associated with the subsequent paraglacial period.

5.3.1 Deglacial Period

The deglacial period is a relatively short period in the evolution of the Lake Saint-Jean lowlands corresponding to the period during which the LIS ice margin retreated over the area (Fig. 5.2a, b). Although eskers and ice-contact fans were identified in former studies and geomorphological maps, recent swath bathymetry imagery and seismic data now offer unprecedented images of well-defined, buried, ice-contact features preserved beneath the Laflamme Mud Fm. within Lake Saint-Jean.

5.3.1.1 Mega-Scale Glacial Lineations (MSGLs)

The swath bathymetry imagery of the Lake Saint-Jean floor reveals abundant elongate positive ridges (Fig. 5.3a, b) that are 1–3 km long, 100–200 m wide with heights ranging from 5 to 10 m. They represent subglacial mega-scale glacial lineations (MSGLs) indicating fast-flowing ice (Clark 1993; Stokes et al. 2013; Ó Cofaigh et al. 2013). MSGLs are broadly oriented NW–SE, parallel to the orientation of the graben structure, suggesting topographically controlled ice-flow. The occurrence of fast-flowing ice influenced by the topography supports the presence of an ice stream in the Lake Saint-Jean lowlands during the deglaciation, as proposed by Tremblay (1971a) and Nutz (2013). The reinterpretation of two groups of glacial striae previously identified by Tremblay (1971a) in the area suggests a transition from topographically independent to topographically controlled ice-flow, revealing that this ice stream evolved only during the deglaciation. A first group (black arrows, Fig. 5.2a), showing flow directions independent of topography, is interpreted to reflect the LIS climax

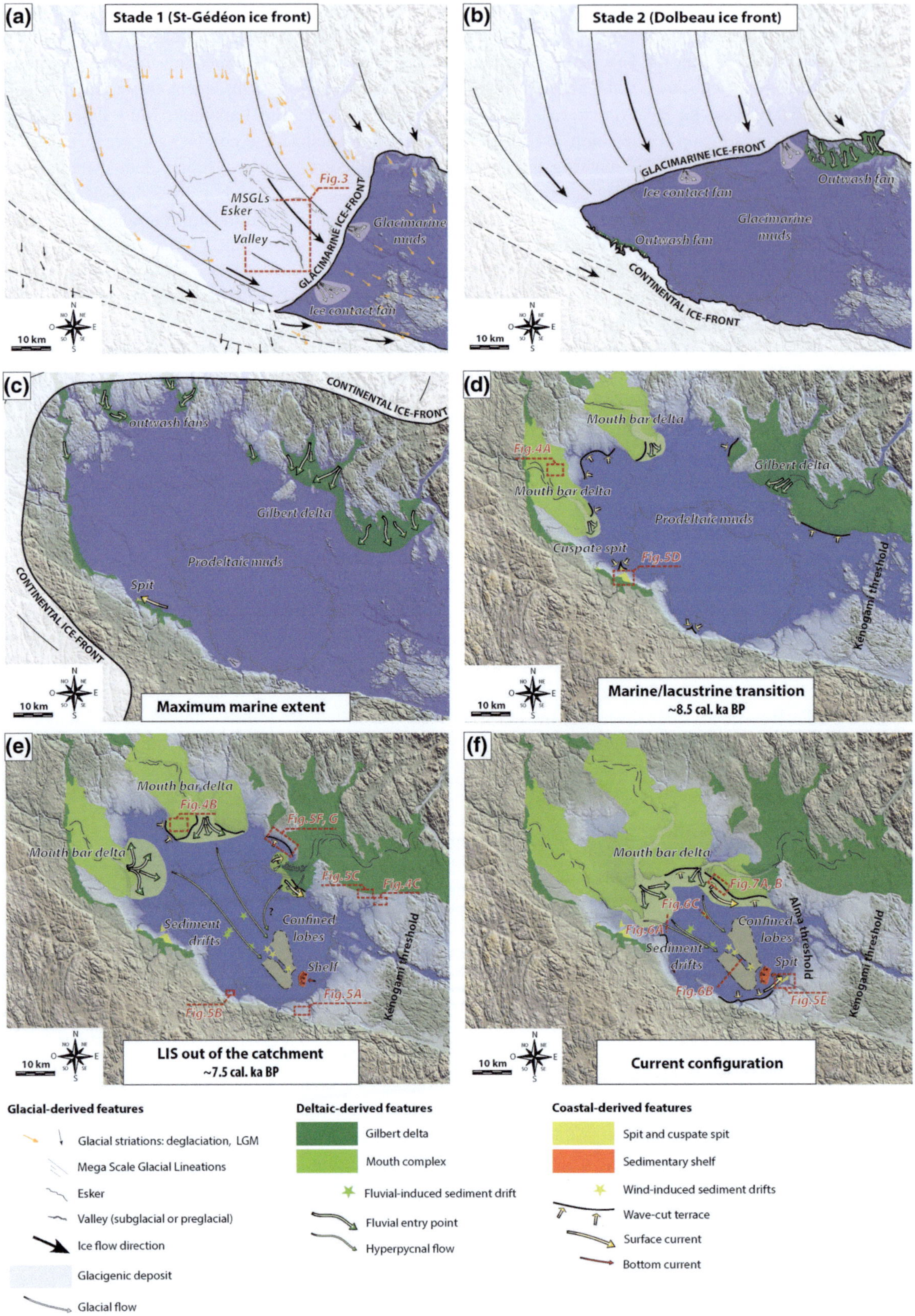

Fig. 5.2 Paleogeography and active sedimentary systems in the Lake Saint-Jean lowlands for six time steps from 13.1 cal ka to present-day (modified from Nutz 2013)

during the LGM when ice-sheet thickness overcame topographic controls on ice-flow patterns. A second group (orange arrows, Fig. 5.2a), displaying orientations parallel to the topography, is attributed to flow patterns associated with deglaciation when the ice thickness decreased enough to allow topographic control on ice-flow.

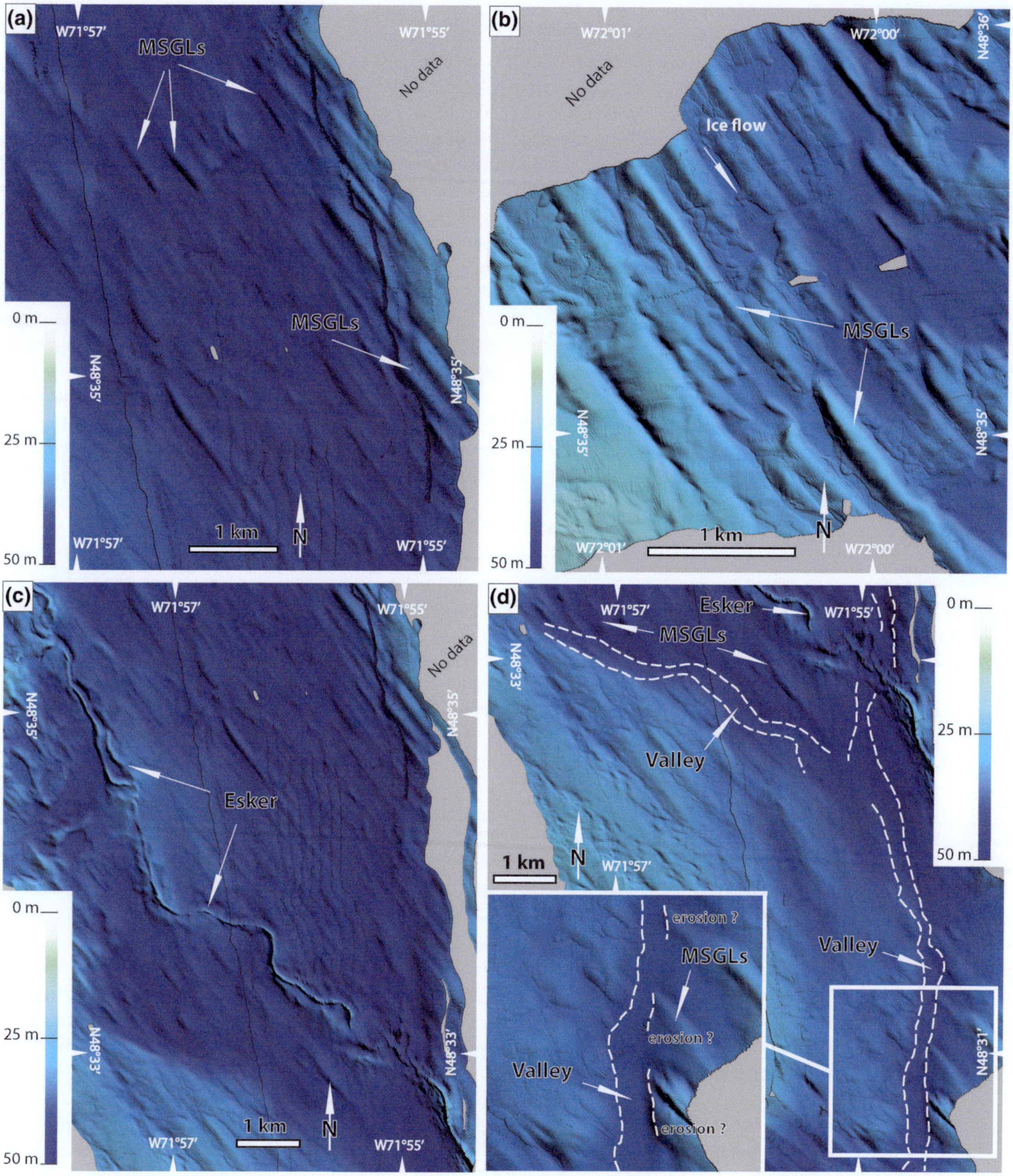

Fig. 5.3 Swath bathymetry imagery showing subglacial features in the Lake Saint-Jean lowlands deposited during the ice margin retreat (location Fig. 5.2a, b). **a** and **b** Mega-Scale Glacial Lineations (MSGLs). **c** Esker. **d** Preglacial fluvial or small tunnel valleys

5.3.1.2 Eskers

Several discontinuous relicts of eskers are identified onshore in the Lake Saint-Jean lowlands (Daigneault et al. 2011; Nutz 2013; Nutz et al. 2015a), but their preservation state is low. The swath bathymetry imagery reveals for the first time the presence of a well-preserved esker draped by the Laflamme Mud Fm. (Fig. 5.3c). This esker is oriented NNW–SSE to NW–SE; it is >10 km long, 80–100 m wide and 7–10 m high. Eskers are usually interpreted as the casts of subglacial R-channels (Brennand 2000; Cumming et al. 2011; Storrar et al. 2014). As characterized by a rounded crest, the esker most likely represents the infilling of a subglacial channel (Perkins et al. 2016). The short length and the relative straight morphology of the identified esker are typical of a type-II esker sensu Brennand (2000), indicating deposition and infill during ice-sheet retreat. Similar to MSGLs, this esker is parallel to the depression, suggesting topographically controlled ice-flow during deglaciation.

5.3.1.3 Valleys

Elongate negative morphologies are also identified in Lake Saint-Jean. Two segments, about 2–3 km long and 100 m wide, merge to form a single segment, more than 5 km long, 150–200 m wide and 5–10 m deep (Fig. 5.3d). These elongated negative morphologies can be interpreted either as preserved segments of small valleys cut into the subglacial sedimentary substrate (e.g. Kehew et al. 2012) by meltwater during deglaciation or as preglacial fluvial channels subsequently draped by glacigenic and glacimarine deposits. Valleys seem to truncate the MSGLs and are in places sealed by the esker described above. This might support a deglacial origin of these valleys. However, seismic data do not display clear evidence for incision in the underlying glacial deposits, supporting a preglacial fluvial origin. At present, it is difficult to determine the origin of these valleys.

5.3.2 Paraglacial Period

The paraglacial period is considered to have started after c. 12.6 cal ka, as soon as the LIS margin retreated beyond the marine limit (Fig. 5.2c–f). This period was characterized by the presence of a large marine water body grading to a lacustrine water body due to glaci-isostatic rebound. During that period, sedimentation is characterized by regressive coastal and deltaic landforms, rapidly perched after their deposition. In the central water body, sedimentation relates to low energy settling of fluvially derived turbid plumes, replaced after c. 7.5 cal ka by higher-energy conditions associated with recurrent bottom currents (Nutz et al. 2014, 2015a, b). Some of the most prominent landforms developed in terrestrial, coastal and subaqueous domains during this period are presented in the following sections.

5.3.2.1 Aeolian Dunes

Several aeolian dune fields are known in the Lake Saint-Jean lowlands (Côté et al. 1990; Daigneault et al. 2011). Dunes are parabolic, 50–600 m long, 50–400 m wide and 1–15 m high (Fig. 5.4). They are similar to those generally observed in high-latitude settings (Wolfe 2007). Aeolian parabolic dunes systematically developed over previous subaqueous deltaic deposits that were progressively raised due to glaci-isostatic rebound, revealing that aeolian processes constituted an important agent for the reworking of subaerial landscapes. Dune initiation likely started relatively quickly after emersion. Subsequently, the growth of vegetation interrupted the activity of dunes and led to fossilization. That is the reason why older are dune fields, higher in elevation they are. This is supported by both radiocarbon and OSL ages ranging between 9.7 and 4.3 ka BP (Côté et al. 1990; Nutz et al. 2015b). All dune fields indicate similar directions of migrations from northwesterly, westerly and southwesterly winds and suggest a relatively constant orientation of prevailing winds (Nutz et al. 2015b) in the area from deglaciation up to the present-day. Reconstructed palaeo-wind directions interestingly indicate the absence of important or recurrent northerly winds, which would be expected from katabatic air flows associated with the LIS.

5.3.2.2 Beach Ridges

Beach ridges represent the most abundant coastal landforms in the Lake Saint-Jean lowlands (Fig. 5.5a–c). They are especially well developed in the southwestern and southern parts of the basin. Beach ridges correspond to continuous, mound-shaped and linear accumulations of clastic sediments (sand, gravel, boulder and detritus such as shells or wood) built parallel to the shore. They frequently form strandplain systems in the form of undulating topography, reflecting the successive creation of beaches, rapidly abandoned due to the continuous water level fall (e.g. Billy et al. 2015; Schuster and Nutz 2017). Beach ridges indicate that the Laflamme Gulf and Lake Saint-Jean were impacted by significant wave action during the entire duration of the regression (Fig. 5.2d–f).

5.3.2.3 Spits

In addition to beach ridges, other coastal landforms developed in the form of spit systems during the paraglacial period. Spit systems are particularly well preserved in the western, southwestern and southern sectors (Nutz 2013; Nutz et al. 2014, 2015a). Two particularly well-preserved examples are discussed below. The first example is a cuspate spit (Ashton et al. 2001; Bouchette et al. 2014) emplaced in shallow waters and grew normal to the shoreline (Fig. 5.5d). The second example is a sandspit that developed during the lowest stand of the lake, before the artificial lake level rise of 1926 (Fig. 5.5e), in the form of a large asymmetric ridge

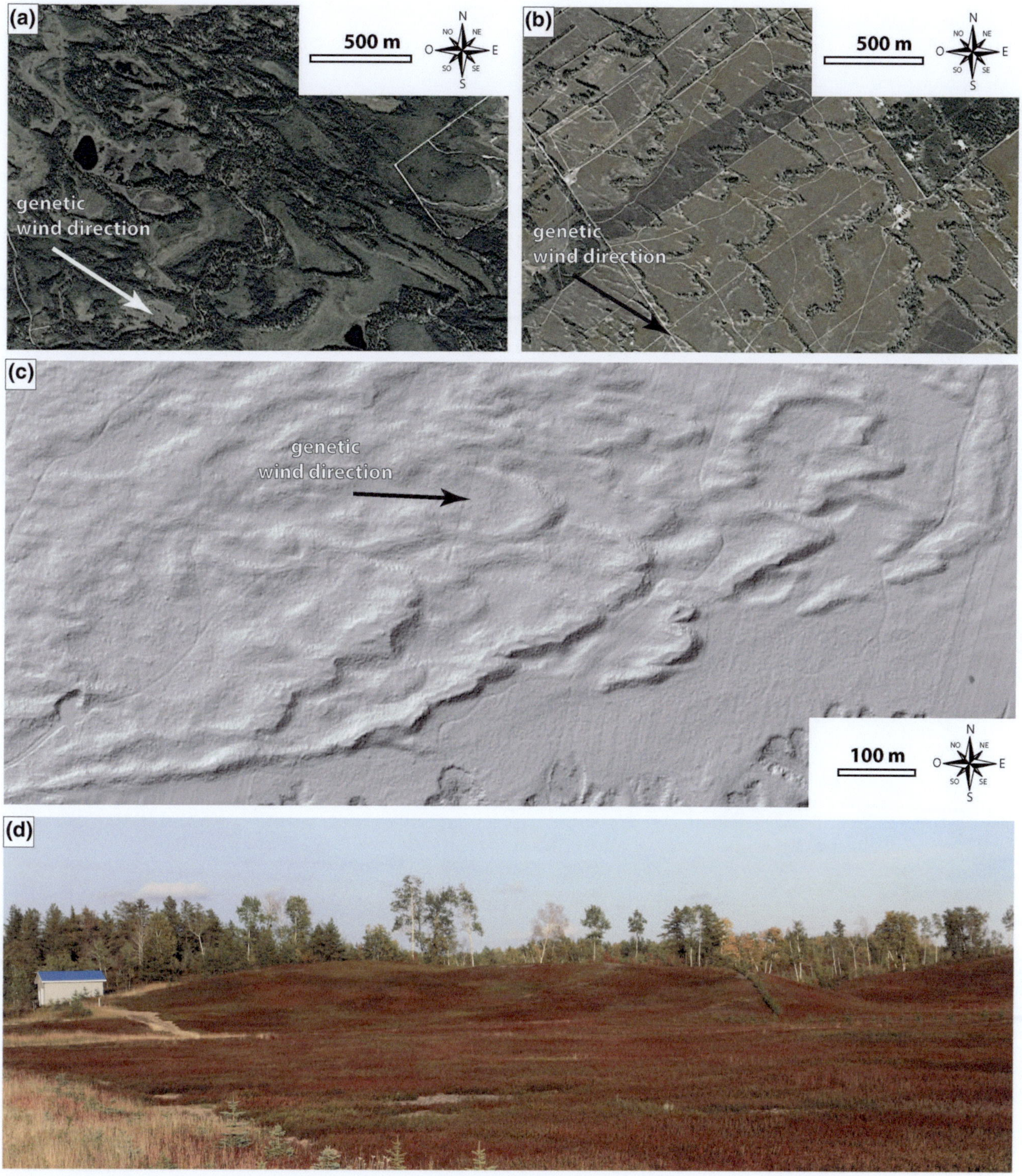

Fig. 5.4 Landforms related to aeolian dynamics in the Lake Saint-Jean lowlands emplaced during the paraglacial period of the basin (see location Fig. 5.2). **a** and **b** Satellite view (GoogleEarth 2017) of parabolic dunes, white arrows indicate the prevailing orientation of the genetic wind. **c** Lidar view of fossil parabolic dunes. **d** Ground view of a fossil aeolian dune

attached to the mainland at one end and growing obliquely to the coastline in the direction of the more recurrent longshore drift (Uda 2005; Bird 2008; Randazzo et al. 2015). Spits and cuspate spits are large-scale coastal landforms that are genetically related to waves and longshore drift. Thus, similar to beach ridges, sand spits and cuspate features

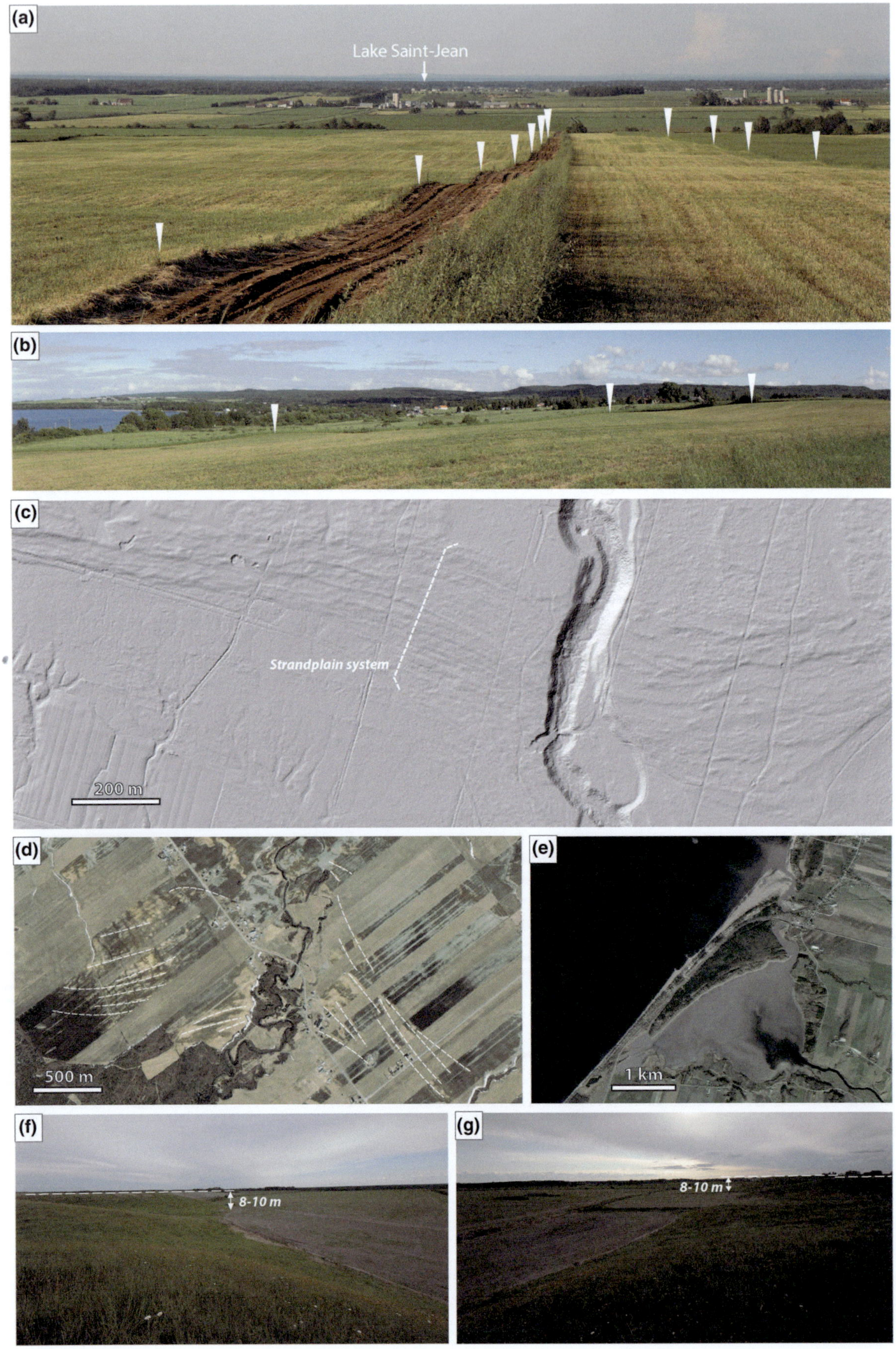
(a)
Lake Saint-Jean
(b)
(c)
Strandplain system
200 m
(d)
500 m
(e)
1 km
(f)
8-10 m
(g)
8-10 m

◀ **Fig. 5.5** Landforms related to coastal dynamics in the Lake Saint-Jean lowlands deposited during the paraglacial period of the basin. **a** and **b** Beach ridges (whites arrows) and swales in the western and southwestern parts of the depocenter (location Fig. 5.2e). **c** Lidar view showing successive beach ridges and swales. **d** Satellite image (GoogleEarth 2017) of a fossil cuspate foreland in the western part of the basin (location Fig. 5.2d). White dotted lines underline paleo-shorelines. **e** Satellite image (GoogleEarth 2017) of the St-Gédéon sandspit in the southern part of the lake (location Fig. 5.2f). **f** and **g** Wave-cut terrace in the eastern part of the depocenter (location Fig. 5.2)

suggest that waves and related longshore drift strongly influenced coastal sedimentation.

5.3.2.4 Wave-Cut Terraces

Several generations of wave-cut terraces have been identified in the Lake Saint-Jean lowlands (Tremblay 1971a, b). Daigneault et al. (2011) reported on the lateral extent of the wave-cut terraces, especially in the north and northeastern sectors. Some form more than 10 km long linear escarpments, with up to 10 m of erosion (Fig. 5.5f, g), and are partly dissected by recent drainage. Wave-cut terraces are coastal landforms resulting from erosional dynamics of wave impact along a shoreline (Trenhaile 1978, 2000). Locations of predominant wave-cut terraces in the northeastern and eastern areas reveal more efficient and energetic waves, attributed to the predominance of westerly winds in the basin and either support relative constant wind directions through time.

5.3.2.5 Hyperpycnal System

An important component in the origin of subaqueous features in Lake Saint-Jean is the fluvially derived sedimentation associated with hyperpycnal flows (Nutz et al. 2014). Hyperpycnal processes resulted in different landforms identified on the floor of Lake Saint-Jean. Erosional channels are observed at the Peribonka (Nutz et al. 2014) and Ashuapmushuan (Fig. 5.6a) river mouths. These channels show incisions ranging between 5 and 16 m (Nutz et al. 2014; Fig. 5.6a) compared to the surrounding lake floor. Laterally, erosional channels are associated with confined lobes (Nutz et al. 2014; here Fig. 5.6b) that represent the depositional areas of hyperpycnal flows. Recently collected acoustic stratigraphy profiles allowed documentation of the complete architecture of these lobes. The lobe on the western trough is 6 km wide, and reaches 10 m thick, draping the underlying Laflamme Mud Fm. and progressively attenuating the underlying topography (Fig. 5.6b). The deposition of confined lobes indicates that hyperpycnal flows transported a large amount of sediment into the lake during the Late Quaternary evolution of the depocenter.

5.3.2.6 Sediment Drifts

A specificity of Lake Saint-Jean consists of abundant features on the lake floor genetically related to wind-induced bottom currents (Fig. 5.6c; Nutz et al. 2014, 2015b). These are attributed to the emplacement of an internal wind-induced circulation due to the action of wind on the lake surface. Numerical modeling suggests that wind-induced circulations display a constant first-order pattern (Nutz et al. 2015b) whatever the orientation of the forcing wind (NW, W and SW). Pushed by wind, surface waters migrate from the upwind to the downwind side of the lake where water mass flows are forced to plunge initiating downwelling and reversed bottom currents that progressively flow toward the upwind side of the lake to reconnect with surface currents. Wind-induced circulations is supposed to develop during longer than 12 h storm events associated with wind intensity exceeding 10 m s^{-1}, circumstances currently occurring only a few days per decade. Velocities of generated currents range from 0.35 to 0.7 m s^{-1} at the surface and from 0.2 to 0.4 m s^{-1} along the lake floor. Among the resulting depositional features are positive mounds that are 100–500 m long and 5–10 m high, deposited on flat or inclined surfaces onto the lake floor referred to as sediment drifts. Some are typical elongated separate drifts (Fig. 5.6c) *sensus* Faugères et al. (1999) that traditionally develop in the marine realm. Indeed, in addition to those identified in Lake Saint-Jean, only few examples of sediment drifts are known in lake systems (Johnson et al. 1980; Tiercelin et al. 1992; Ceramicola et al. 2001; Gilli et al. 2005; Manley et al. 2012; Wagner et al. 2012). Combining sediment drifts with the abundant wave-related coastal landforms, Lake Saint-Jean has been among the pioneer examples used to recently develop the concept of wind-driven lake (Nutz et al. 2018). Wind-driven lakes share a sedimentation dominated by wave-related processes and wind-induced lake-scale water circulation. They are characterized by the construction of beach ridges, spits or cuspate spits along their shorelines, and by sediment drifts, sedimentary shelf progradation and erosional surfaces in their deeper, offshore domains. Wind-driven lakes are observed worldwide; they are relatively shallow regarding their sizes as expressed by the I_{WWB} index (Nutz et al. 2018), a ratio of the maximum representative fetch relative to mean basin depth. An index value greater than three is supposed to favor the evolution of a lake as a wind-driven lake.

5.4 Recent Geomorphological Processes

The anthropogenic lake level rise in 1926 is the last event that strongly impacted the evolution of the Lake Saint-Jean lowlands. The rise of the base level led to a change in the

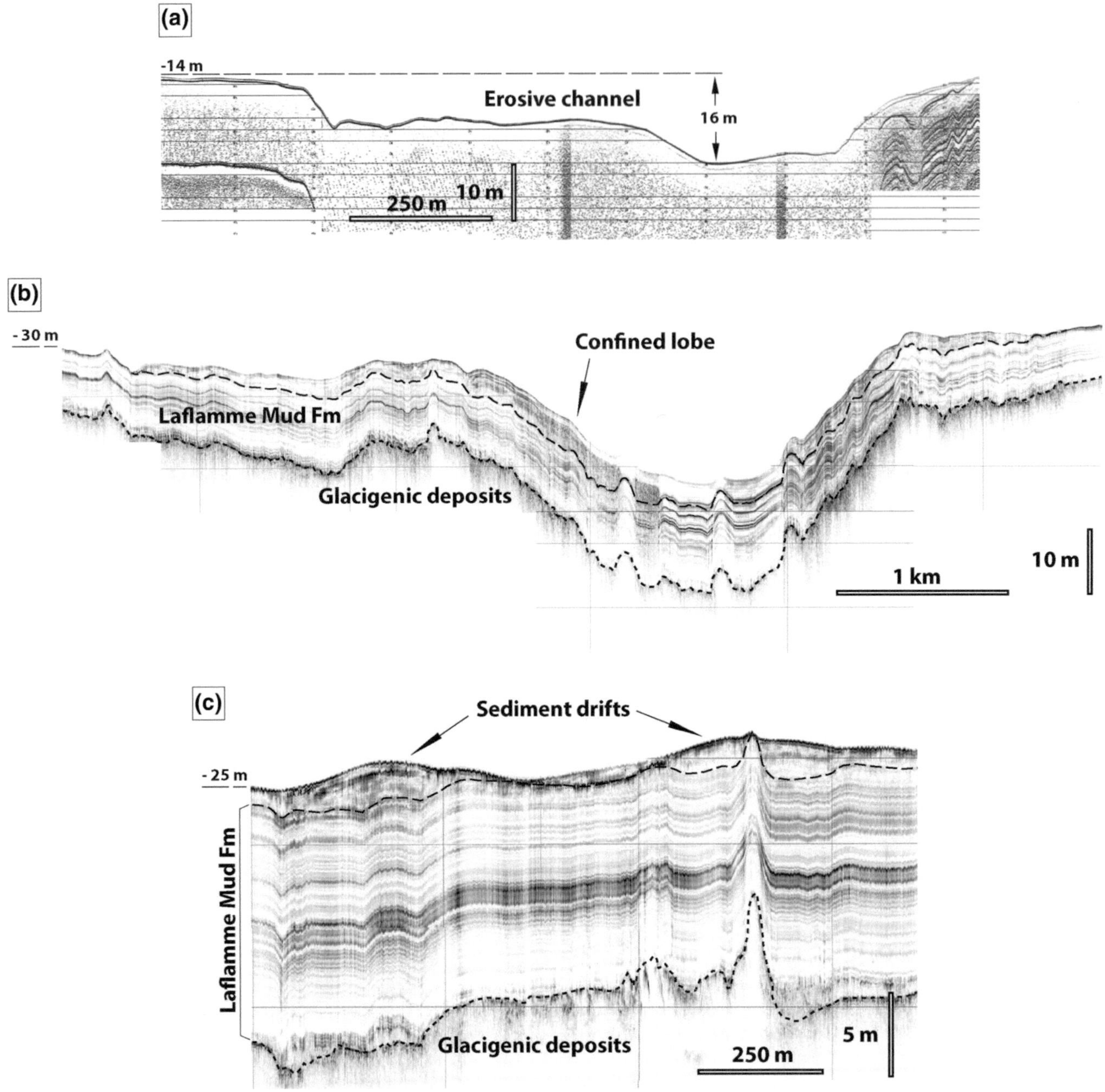

Fig. 5.6 Features related to hyperpycnal system and bottom currents in the Lake Saint-Jean developed during the paraglacial period of the basin (location Fig. 5.2f). **a** Erosional channel in front of the Ashuapmushaun river mouth. **b** Confined lobe in the western trough. **c** Mounds interpreted as sediment drift onto the lake floor. Note the three units constituting the common stratigraphic organization: glacigenic deposits at the base draped by the Laflamme Mud Fm. and bottom currents related features at the top

equilibrium profile of the lake margin, and enhanced erosion along the shore (Tremblay 1971b). Wave-cut terraces are now the predominant coastal landforms along the modern Lake Saint-Jean shoreline (Fig. 5.7a, b), in contrast to previous periods characterized by the coexistence of wave-cut terraces, constructive beach ridges and spit systems. Some parts of the lakeshore have been monitored since the lake level rise, and shoreline retreat has been measured. Between 1927 and 1966, erosion rates ranged between 1.2 and 3.2 m y^{-1} along the southeastern downwind sides of the lake (Tremblay 1971b). Today, mean erosion rates are around 0.2–0.3 m y^{-1} but maximum erosion rates reach 2–3 m y^{-1} in some places of the eastern part of the lake and despite the construction of shore protection measures, intense shoreline retreat has occurred. The second important geomorphic process in the Lake Saint-Jean lowlands is mass movement. These processes were

Fig. 5.7 Landforms associated with recent geomorphic processes. **a** and **b** Active wave-cut terraces as revealed by trees progressively falling in the lake along the Pointe-Taillon (eastern shore of Lake Saint-Jean). **c** and **d** Landslides in the Laflamme Mud Fm. (aerial photos from 1926, courtesy Rio Tinto Alcan)

identified since the 1960s (Chagnon 1968; La Rochelle et al. 1970, 1977), especially concerning localized places. Chagnon (1968) inventoried more than 20 landslides in the vicinity of Desbiens in the southern part of the depocenter among which the dramatic Saint-Jean-Vianney event of 1971 caused the death of 31 people. Mass movements are attributed to the exceptional sensitivity of muds comprising the Laflamme Mud Fm., as they meet all the criteria to behave as quick clay (e.g. Andersson-Sköld et al. 2005). Mass movements are observed both along the lake shore and along bluffs of incised rivers in the form of lobate or elongate scars extending over hundreds of meters (Fig. 5.7c, d). Finally, since the equilibration of the lake margin with the 1926 lake level will sustain erosion of the Lake Saint-Jean shoreline, and since the extensive subaerial exposure of the Laflamme Mud will likely continue to initiate mass movements, these two major geomorphic processes will probably continue to significantly impact landscapes and local communities of the Lake Saint-Jean lowlands in the near future.

5.5 Conclusion

The Lake Saint-Jean lowlands form a striking product of combined geomorphological processes that occurred since the last deglaciation. As a result, landscapes provide a valuable set of spatially diversified landforms which originated from different periods of the depocenter evolution. The acquisition of the high-resolution swath bathymetry imagery provided key elements dealing with the deglacial period of the area. Documenting sedimentation associated with the paraglacial period led to new paradigms concerning the evolution of restricted marine and lacustrine systems, especially about the role of wave-related deposition and erosion in the shaping of lacustrine systems. The development of wind-induced features onto the floor of the modern Lake Saint-Jean is an important characteristic of the area that makes it unique in eastern Canada. Finally, since 1926 the Lake Saint-Jean lowlands constitute an outstanding example of a geomorphological system controlled by both natural and anthropogenic forcings, combining mass movements in sensitive muds derived from the glacial history of the depocenter and intense lake shore erosion due to the anthropogenic lake-level rise.

References

Andersson-Sköld Y, Torrance JK, Lind B, Odén K, Stevens RL, Rankka K (2005) Quick clay-a case study of chemical perspective in Southwest Sweden. Eng Geol 82:107–111

Ashton AD, Murray AB, Arnoult O (2001) Formation of coastline features by large-scale instabilities induced by high-angle waves. Nature 414:296–300

Billy J, Robin N, Hein C, Certain R, FitzGerald D (2015) Insight into the late Holocene sea-level changes in the NW Atlantic from a paraglacial beach-ridge plain south of Newfoundland. Geomorphology 248:134–146

Bird E (ed) (2008) Coastal Geomorphology: an introduction, 2nd edn. Wiley, Chichester

Bouchette F, Manna M, Montalvo P, Nutz A, Schuster M, Ghienne J-F (2014) Growth of cuspate spits. J Coastal Res 70:47–52

Brennand TA (2000) Deglacial meltwater drainage and glaciodynamics: inferences from Laurentide eskers, Canada. Geomorphology 32:263–293

Ceramicola S, Rebesco M, De Batist M, Khlystov O (2001) Seismic evidence of small-scale lacustrine drifts in Lake Baïkal (Russia). Mar Geophys Res 22:445–464

Chagnon J-Y (1968) Les coulées d'argiles dans la province de Québec. Nat Can 95:1327–1343

Clark ChD (1993) Mega-scale glacial lineations and cross-cutting ice-flow landforms. Earth Surf Proc Land 18:1–29

Côté D, Dubois J-M, Nadeau L (1990) Les dunes du Québec méridionale: contribution à l'étude des vents dominants durant l'Holocène. Can Geogr 34:49–62

Cousineau PA, Longuépée H (2003) Lower Paleozoic configuration of the Quebec reentrant based on improved along-strike paleogeography. Can J Earth Sci 40:207–219

Cumming DI, Gorrell G, Guilbault J-P, Hunter JA, Logan Ch, Ponomarenko D, Pugin J-M, Pullan SE, Russell HAJ, Sharpe DR (2011) Sequence stratigraphy of a glaciated basin fill, with a focus on esker sedimentation. Geol Soc Am Bull 123:1478–1496

Daigneault R-A, Cousineau PA, Leduc E, Beaudoin G, Milette S, Horth N, Roy DW, Lamothe M, Allard G (2011) Rapport final sur les travaux de cartographie des formations superficielles réalisés dans le territoire municipalisé du Saguenay-Lac-Saint-Jean (Québec) entre 2009 et 2011. Ministère des Ressources naturelles et de la Faune, GM 65970

Dietrich P, Ghienne J-F, Lajeunesse P, Normandeau A (2017) Reconstructing ice-margin retreat using delta morphostratigraphy. Sci Rep 7:16936

Dionne J-C (1972) La dénomination des mers du postglaciaire au Québec. Cahiers de Géographie du Québec 16:483–487

Dyke AS (2004) An outline of North American deglaciation with emphasis on central and northern Canada. In: Ehlers J, Gibbard PL, Hughes PD (eds) Quaternary glaciations—extent and chronology, part II: North America. Dev Quat Sci 371–406

Faugères J-C, Stow DAV, Imbert P, Viana A (1999) Seismic features diagnostic of contourite drifts. Mar Geol 162:1–38

Fulton RJ, Karrow PF, LaSalle P, Grant DR (1986) Summary of quaternary stratigraphy and history, eastern Canada. Quat Sci Rev 5:211–228

Gilli A, Anselmetti FS, Ariztegui D, Beres M, Mckenzie JA, Markgraf V (2005) Seismic stratigraphy, buried beach ridges and contourite drifts: the Late Quaternary history of the closed Lago Cardiel basin, Argentina (49°S). Sedimentology 52:1–23

Johnson TC, Carlson TW, Evans JE (1980) Contourites in Lake Superior. Geology 8:437–441

Kehew AE, Piotrowski JA, Jorgensen F (2012) Tunnel valleys: concepts and controversies—a review. Earth Sci Rev 113:33–58

Kumarapeli PS (1985) Vestige of Iapetan Rifting in the Craton West of the Northern Appalachians. Geosci Can 12:54–59

La Rochelle P, Lefebvre G, Bilodeau PM (1977) The stabilization of a slide in Saint-Jérôme, Lac Saint-Jean. Can Geotech J 14:340–356

La Rochelle P, Chagnon JY, Lefebvre G (1970) Regional geology and landslides in the marine clay deposits of eastern Canada. Can Geotech J 7:145–156

Laverdière C, Mailloux A (1956) Etat de nos connaissances d'une transgression marine post-glaciaire dans les régions du Haut-Saguenay et du Lac Saint-Jean. Rev Géogr Montr 10:201–220

Manley PL, Manley TO, Hayo K, Cronin T (2012) Small-scale lacustrine drifts in Lake Champlain, Vermont. J Great Lakes Res 38:88–100

Nutz A, Schuster M, Ghienne J-F, Roquin C, Bouchette F (2018) Wind-driven waterbodies: a new category of lake within an alternative sedimentologically-based lake classification. J Paleolimnol 1–11 https://doi.org/10.1007/s10933-016-9894-2.

Nutz A, Ghienne JF, Schuster M, Roquin C, Bouchette F, Cousineau PA (2015a) Forced regressive deposits within a deglaciation sequence: example of the Late Quaternary succession in the Lake Saint-Jean basin (Québec, Canada). Sedimentology 62:1573–1610

Nutz A, Schuster M, Ghienne JF, Roquin C, Hay M, Rétif F, Certain R, Robin N, Cousineau PA, Bouchette F (2015b) Wind-driven bottom currents and related sedimentary bodies in the Lake Saint-Jean (Québec, Canada). Geol Soc Am Bull 127:1194–1208

Nutz A, Ghienne J-F, Schuster M, Certain R, Robin N, Roquin C, Raynal O, Bouchette F, Duringer Ph, Cousineau P (2014) Seismic-stratigraphic record of a deglaciation sequence: from the marine Laflamme Gulf to the Lake Saint-Jean (Late Quaternary, Québec Province, Canada). Boreas 43:309–329

Nutz A (2013) La déglaciation du bassin du Lac Saint-Jean (Wisconsinien/Holocène, Québec, Canada) enregistrement d'une régression forcée glacio-isostatique et de l'hydrodynamique d'un système fermé contrôlé par le vent. Dissertation, Université de Strasbourg

Ó Cofaigh C, Stokes CR, Lian OB, Clark CD, Tulacyzk S (2013) Formation of mega-scale glacial lineations on the Dubawnt Lake Ice Stream bed: 2. Sedimentology and stratigraphy. Quat Sci Rev 77:210–227

Occhietti S, Parent M, Lajeunesse P, Robert F, Govare E (2011) Late Pleistocene-early Holocene decay of the Laurentide ice sheet in Québec-Labrador. In: Ehlers J, Gibbard PL, Hughes PD (eds) Quaternary glaciations—extent and chronology, part II: a closer look. Dev Quat Sci 15:601–630

Occhietti S (2007) The Saint-Narcisse morainic complex and early Younger Dryas events on the southeastern margin of the Laurentide ice sheet. Géogr Phys Quat 61:89–117

Occhietti S, Govare E, Klassen R, Parent M, Vincent J-S (2004) Late Wisconsinan Early Holocene deglaciation of Québec-Labrador. In: Ehlers J, Gibbard PL, Hughes PD (eds) Quaternary glaciations—extent and chronology, part II North America. Dev Quat Sci 2:243–273

Occhietti S, Parent M, Shilts WW, Dionne J-C, Govare E, Harmand D (2001) Late Wisconsinan glacial dynamics, deglaciation, and marine invasion in southern Quebec. Geol Soc Am SP 351:243–273

Perkins AJ, Brennand TA, Burke MJ (2016) Towards a morphogenetic classification of eskers: Implications for modelling ice sheet hydrology. Quat Sci Rev 134:19–38

Randazzo G, Jackson DWT, Cooper JAG (2015) Sand and gravel spits. Coastal Research Library, vol 12. Springer, Dordrecht

Roy DW, Beaudoin G, Leduc É, Rouleau A, Walter J, Chesnaux R, Cousineau PA (2011) Isostasie postglaciaire différentielle au lac-Saint-Jean (Québec). GeoHydro 2011, Joint Meeting of Canadian Quaternary Association (CANQUA) and Canadian Chapter of the International Association of Hydrogeologists (IAH-CNC), Quebec City, 28–31 August 2011

Schuster M, Nutz A (2017) Lacustrine wave-dominated clastic shorelines: modern to ancient littoral landforms and deposits from the Lake Turkana Basin (East African Rift System. J Paleolimnol, Kenya). https://doi.org/10.1007/s10933-017-9960-4

Shaw J, Gareau P, Courtney RC (2002) Palaeogeography of Atlantic Canada 13-0 kyr. Quat Sci Rev 21:1861–1878

Stokes CR, Spagnolo M, Clark CD, O Cofaigh C, Lian OB, Dunstone RB (2013) Formation of mega-scale glacial lineations on the Dubawnt Lake Ice Stream bed: 1. Size, shape and spacing from a large remote sensing dataset. Quat Sci Rev 77:190–209

Storrar RD, Stokes ChR, Vans DJA (2014) Increased channelization of subglacial drainage during the deglaciation of the Laurentide Ice Sheet. Geology 42:239–242

Syvitski JPM, Praeg DB (1989) Quaternary Sedimentation in the St. Lawrence Estuary and Adjoining Areas, Eastern Canada: an overview based on high-resolution seismo-stratigraphy. Géogr Phys Quat 43:291–310

Tiercelin J-J, Soreghan M, Cohen AS, Lezzar K, Bouroullec J-L (1992) Sedimentation in large rift lakes: Example from the Middle Pleistocene Modern deposits of the Tanganyika Trough, East African Rift Systems. Bull Centres Rech Explo Prod Elf Aquitaine 16:83–111

Trenhaile AS (2000) Modeling the development of wave-cut shore platforms. Mar Geol 166:163–178

Trenhaile AS (1978) The shore platforms of Gaspé, Québec. Ann Assoc Am Geogr 68:95–114

Tremblay G (1971a) Géologie et Géomorphologie quaternaires à l'Est du Lac-Saint-Jean (Québec, Canada). Dissertation, Université de Paris 6

Tremblay G (1971b) Conséquences géomorphologiques du relèvement artificiel du niveau du Lac Saint-Jean. Cahiers de géographie du Québec 15:115–130

Uda T (2005) Spits. In: Schwartz ML (ed) Encyclopedia of coastal science. Springer, Dordrecht, pp 909–912

Wagner B, Aufgebauer A, Vogel H, Zanchetta G, Sulpizio R, Damaschke M (2012) Late Pleistocene and Holocene contourite drift in Lake Prespa (Albania/F.Y.R. of Macedonia/ Greece). Quat Int 274:112–121

Wolfe SA (2007) Dune fields/high latitudes. In: Elias SA (ed) Encyclopedia of quaternary science. Elsevier, Oxford

Alexis Nutz is an associate professor at Aix-Marseille Université (France) since 2018 in the Centre Européen de Recherche et d'Enseignement des Géosciences de l'Environnement. He completed his PhD in 2013 at the Université de Strasbourg (France). His PhD was awarded twice by both the Association des Sédimentologistes Français and the Université de Strasbourg. In 2014, he obtained a 2 years postdoc fellow at CNRS (Strasbourg, France) collaborating with industry. Then, he worked between 2016 and 2018 at the Department of Geosciences of the Aarhus University (Denmark). His PhD focused on the Late Quaternary deglacial and postglacial evolution of the Saint-Jean depocenter in northern Québec (eastern Canada). Subsequently, his research interests enlarged to the evolution of past fluvial-lacustrine dynamics and sedimentation in continental basins as proxies for past climate and tectonic forcings. He also has interests in modes of deglaciation during the last glacial cycle in Southern Alps (France).

Patrick Lajeunesse received his PhD in geomorphology from Université Laval (Canada) in 2000. He then undertook a postdoctoral project in glacial geology and geomorphology at the University of Alberta. He has previously worked at the Institut des sciences de la Mer de Rimouski (ISMER) at the University of Québec à Rimouski and at the Institut National de la Recherche Scientifique—Eau, Terre, Environnement (INRS-ETE). He now occupies a Professor (Full) position in the Geography Department at Université Laval. His research work focuses on the geomorphological and sedimentary record of glaciation, postglacial environmental changes and paleoseismicity on formerly glaciated coasts, continental margins and lakes, mainly of Eastern and Arctic Canada.

Jean-François Ghienne is a researcher of CNRS working at the 'Institut de Physique du Globe de Strasbourg', University of Strasbourg. He received his PhD in sedimentology from this university in 1998. Working mainly in western Europe, NW Africa and eastern Canada, his academic research focuses on the Lower Paleozoic Gondwana development, glacial sequence stratigraphy and tectonic/paleoclimatic signals in stratal records from the Cambrian to the Quaternary. He managed from 2013 to 2017 the SeqStrat-Ice project, founded by the French National Research Agency (ANR), dealing with the delineation of past glaciation events and related paleoclimatic scenarios.

Mathieu Schuster received his PhD in sedimentary geology from the University of Strasbourg (France) in 2002. He then worked at the universities of Cologne (Germany), of Brest and of Poitiers (France), as well as at the French geological survey (BRGM). He now works as a researcher at the CNRS, and he serves as vice-director of the Institut de Physique du Globe de Strasbourg, a multidisciplinary research lab on geophysics and geology hosted by the University of Strasbourg. His research concerns continental paleoenvironments, with a focus on clastic littoral deposits and landforms associated to lacustrine systems. His current main case studies are the Lake Chad Basin and the Turkana Depression (EARS). He was awarded the bronze medal of CNRS in 2008, and the scientific prize of Cercle Gutenberg and Fondation Université de Strasbourg. He was an Erskine Visiting Fellow at the University of Canterbury (Christchurch, New Zealand) in Spring 2016.

Etienne Brouard received his PhD in geographical sciences from L'Université Laval in 2018, which focused on the glacial geomorphology and paleo-ice dynamics records of the quaternary glaciations in northeastern Baffin Island shelf and fjords. He is now a postdoctoral fellow at the University of Québec in Montréal where he studies the potential impact of the drainage of the glacial lakes that formed during the demise of the Laurentide Ice Sheet. He is interested in global isostatic adjustment models, remote sensing mapping and cosmogenic dating of glacial sediments.

Pierre Dietrich is a Postdoctoral Fellow at the University of Johannesburg (South Africa) and received his PhD in sedimentary geology from the University of Strasbourg (France) in 2015. His thesis focused on the sedimentary, stratigraphic and geomorphological record of the glaci-isostatic adjustment and associated processes that accompanied the retreat of the Laurentide Ice Sheet along the North Shore of the St. Lawrence Estuary and Gulf (Québec, eastern Canada). His current research interest emphases the sedimentary record of both the Ordovician and Permo-Carboniferous ice ages in southern Africa. He also has interest in subglacial erosion, deltaic, fluvial, supercritical-flow-related processes as well as on the retreat of modern, high-latitude ice masses.

Frédéric Bouchette is an associate professor HDR (Habilitation à Diriger des Recherches) at the University of Montpellier in the laboratory 'Géosciences-Montpellier'. He is also an invited researcher at the Institute of

Mathematics Alexandre Grothendieck. His scientific activity concerns the development of concepts and methods in relation with the dynamics of shallow water environments. He studies the domain that extends from a few tens of meters of water depth at sea to the coastal watershed onshore, with a particular emphasis on the littoral area and the shoreline itself. He also focuses on the dynamics of ancient coastal geological systems in both marine and lacustrine systems.

Claude Roquin is a retired researcher from the French Research Center (CNRS), since 2016. He first worked as a geologist at the French Geological Survey (BRGM, Orléans), and received his PhD from the University of Orleans in 1984. Then, he joined the Center of sedimentology and surface geochemistry (Strasbourg) where he became a specialist of tropical soils and lateritic processes. Later he significantly contributed to the study of fossil littoral landforms associated to Megalake Chad (Africa) and Lake Saint-Jean (Québec). Claude Roquin is a gifted naturalist and photographer who currently contributes to several programs for the inventory and protection of birds and insects in Alsace.

Pierre Cousineau received his PhD at Laval University in Québec City (Canada). He has recently retired from Université du Québec à Chicoutimi (UQAC) where he has taught geomorphology, sedimentology and Earth systems science for more than 30 years. His early research has been focused on sedimentary processes, provenance studies and paleogeographic reconstruction of Archean and Paleozoic volcanic, volcaniclastic and siliciclastic sequences. In the last 10 years, he has been involved in regional mapping of superficial (quaternary) deposits in Québec and in paleogeographic evolution of these deposits.

Northwestern Ontario: The Thunder Bay Region

6

Kamil Zaniewski, Brian Phillips, and Fred Dean

Abstract

The scenery and landforms of the Thunder Bay region result from a varied bedrock geology, particularly notable being the flat-topped mountain features of the Sleeping Giant, Mt. McKay and others, resulting from the intrusion of the Logan diabase sill into the Precambrian shales and sandstones of the Animikie formation. There are no Palaeozoic or younger rocks other than Pleistocene glacial sediments. The region has a complex glacial history, having been affected by three ice lobes approaching from different directions in the most recent glaciation. The chapter reviews the geological and glacial history of the region and singles out five notable landforms for detailed discussion, each of which can be easily accessed by interested visitors.

Keywords

Precambrian • Diabase sills • Lava flows • Canyon • Ice lobes • Marquette Readvance • Moraine • Ice-contact delta • Raised shorelines • Bay-mouth bar

6.1 Introduction

By a coincidence of a varied bedrock geology, an interplay of three directions of ice movement in the latter stages of the MIS 2 glaciation and differential post-glacial isostatic recovery along an approximate southwest–northeast axis, the Thunder Bay region presents a particularly interesting array of solid rock, glacial and coastal landforms.

Dominating the present scenery are mesa-like, flat-topped, precipitous-walled features, often referred to locally as "mountains". The City of Thunder Bay is overlooked to the south by the 270 m high flanks of Mount McKay, the highest summit of the Nor'Westers (Fig. 6.1a, b), and looks out eastwards, 25 km across the Thunder Bay embayment, to the spectacular feature of the Sleeping Giant, its western facing walls rising from Lake Superior, 183 m, to the highest point, 563 m, the highest vertical rise in Ontario (380 m) (Fig. 6.1b).

Though the bulk of Mount McKay is composed of shale and greywackes of the Rove Formation, part of the Animikie Group, its flat top is composed of a 60 m thick diabase sill that forms a protective cap. Named the Logan diabase (1115 million years old), the cap is the erosional remnant of an intrusive sill that once extended over much of the region, the iconic outline of the Sleeping Giant also being formed by residual portions of the Sleeping Giant Sill, as it is locally known, also resting on rocks of the Rove Formation. Outcropping bedrock occurs across much of the region, sometimes covered by thin ablation moraine, and only deeply buried in localized accumulations of glacial, fluvial and lacustrine sediments or in ancient valleys such as those occupied by the present Kaministiquia (also spelled Kaministikwia) and Whitefish rivers.

6.2 Bedrock Geology

The oldest exposed rocks in the Thunder Bay region are granite, gneiss and greenstone volcanic rocks of the Precambrian Archaean Era, dated 3.6–2.7 billion years old, and part of the stable North American Craton. Separating these from later, unconformable Precambrian Proterozoic rocks was the Eparchaean Interval, estimated to have been a 200 million year interval of erosion during which both shallow sea sediments and glacial sediments were deposited in the northern parts of the shield.

It is early Proterozoic rocks that dominate the outcropping bedrock of the Thunder Bay region, particularly the platform sediments of the Animikie Basin. About 2.1 billion years ago a widening east–west basin opened up between a Northern (Superior) craton and a Southern craton, through

K. Zaniewski (✉) · B. Phillips · F. Dean
Department of Geography, Lakehead University, 955 Oliver Rd, Thunder Bay, ON P7B 5E1, Canada
e-mail: kzaniews@lakehead.ca

© Springer Nature Switzerland AG 2020

O. Slaymaker and N. Catto (eds.), *Landscapes and Landforms of Eastern Canada*, World Geomorphological Landscapes, https://doi.org/10.1007/978-3-030-35137-3_6

Fig. 6.1 **a** Location of features mentioned in the text. *Cartography* by E. Leinberger; **b** The iconic landforms of Mt. McKay and the Sleeping Giant. *Photograph* by B. Phillips

what is now Lake Superior and the northern parts of Wisconsin and Michigan. On its northern edge a basal conglomerate and quartzite was deposited on Archaean rock, locally known as the Kakabeka Formation and only a metre or so thick (Pye 1969). This same formation thickens southwards, the locally named Ajibic Quartzite of the Marquette, MI area, being several hundred metres thick (La Berge 1994).

As the Animikie Basin deepened, sedimentation of sandstones graded into mudstones, particularly iron-rich sediments which characterize this period of geological history in a manner which has not been repeated since. The Gunflint Formation comprises black shale, tuff, chert-carbonate and taconite, a more granular, shallower water variant of the iron-rich sediments containing stromatolites, microbial mats of a diverse group of microscopic organisms. The Gunflint Formation has yielded commercially viable grades of iron ore that played an important economic role in several locations within the Animikie Basin. Taconite, a highly iron-rich chert, occurs both as nodules and bands within the Gunflint and, because of its tendency to fracture in a conchoidal manner, producing razor sharp edges, it played a significant role in the occupation of the region by Palaeolithic peoples (Dawson 1984).

Overlying the Gunflint Formation is the Rove Formation, a thick blanket of greywackes and thin-bedded,

Fig. 6.1 (continued)

dark-coloured, fine-grained shales (argillite), much like the shales of the Gunflint, but accumulating to depths of several thousand metres in parts of the basin, outcrops being of a number of local names. Today, both the Gunflint and Rove Formations occur as flat lying or gently southeast dipping outcrops.

The Animikie was followed by the Keweenawan, beginning about 1.1 million years ago, but only after a five to six hundred million year period of erosion, during which the Superior region remained an undisturbed land surface. The Keweenawan period added rocks of sedimentary, volcanic and intrusive origin. The gently dipping group of Sibley sedimentary rocks lies unconformably on a variety of older rocks, its base marked by a conglomerate. The several hundred metres of sandstones and carbonate strata form the prominent cliffs of the present Sibley Peninsula, which continue the trend of the Sleeping Giant to the northeast. The carbonate rocks are characterized by the dark red colour of hematite, formed by oxidation in a warm, dry climate. A mottled, green and red Sibley sandstone has been used as a marker to indicate the direction of glacial transport across the region (Zoltai 1963).

Exposed on the islands off Thunder Bay and down the Minnesota north shore are between 2–3 km of extrusive lavas and associated conglomerates and sandstones, classified as the Osler Group (Pye 1969). The immense thickness of these lavas increases to over 6 km towards Duluth. Though most easily seen south of the US/Canada border at Pigeon River, the successive lava flows form a scenery of lakeward dipping slopes with steep inland facing cuestas (Fig. 6.2). However, the source of these huge volumes of lava lies deep below the area now occupied by Lake Superior, and being dominantly basaltic, amygdaloidal lavas, the vesicles formed by rising gases within them can clearly be seen curving northwest, opposing the direction of dip. Following the period of extrusion, it appears that foundering

Fig. 6.2 Keweenawan extrusive lava flows of the MN north shore. *Photograph* by B. Phillips

occurred over the source area, creating the lakeward dip apparent today. Often, the famed Lake Superior agates and, more rarely, thomsonite and other minerals can be found filling the vesicles in these lavas.

In the associated sedimentary rocks, particularly out on the Superior islands and on the Keweenawan Peninsula, MI, native copper occurs. Long before commercial mining began in the mid 1800s, this copper was exploited and worked by Palaeo-Indian people entering the region; indeed, at the time of publication, the dating of organics in a copper artefact, found near Thunder Bay, of 6800 cal yrs BP, was the earliest evidence of copper working in the world (Beukens et al. 1992).

At about the same time that lavas were being extruded, intrusions of basic magma, diabase, were injected as vertical or steeply angled dykes cutting through older rocks and as sills, injected between the layers of older rocks. In the present scenery the dykes form high, steep-sided, northeast trending ridges, though some occasionally cross-cut the trend to form box-like basins. Most visible as one approaches the US/Canada border on Hwy 61, the dykes form linear patterns of offshore islands (Fig. 6.3).

The sills, known as the Logan sills, are those that partly form, among other similar features, the Mt. McKay and the Sleeping Giant. The sills are characteristically well jointed and fissured, a characteristic exploited by later glacial erosion.

The Keweenawan period was followed by a very long erosion interval, lasting many hundreds of millions of years. Over most of the region, there was no deposition of Palaeozoic or younger rocks until the Pleistocene. There are, however, intermittent outcrops of Lake Superior sandstone, of Cambrian age, along the eastern shore of Lake Superior, towards Sault Ste. Marie, resting on Keweenawan and older rocks (La Berge 1994). These were deposited by a Cambrian sea encroaching from the direction of what is now eastern Michigan, but this never reached the Thunder Bay region.

6.3 Pleistocene Glaciation

The first ice sheet that ground over the Thunder Bay region, about one million years ago, met with a deeply weathered, relatively flat bedrock landscape. A shallow basin, of structural origin, existed in the area that is now Lake Superior, with higher ground existing both in northern Minnesota and northern Wisconsin and Michigan. Ancestral

Fig. 6.3 Keweenawan dyke swarms form linear island chains. *Source* Google Earth; annotated by B. Phillips

rivers flowed along the axes of what are now the Huron, Michigan and Superior basins, the latter apparently flowing northwards along the line of the present Nipigon basin.

During the Pleistocene, possibly ten or more major advances and retreats of ice took place, each with its own history of advancing and retreating lobes of ice, a repeated complexity about which little is known since each event roughly covered the same areas of the region, each successive event obliterating most of the evidence of the former. Undoubtedly, it took several glacial episodes to strip away the deeply weathered mantle of rock and transport it far south to the areas of present Wisconsin, Nebraska, Kansas and Illinois.

The glaciated landscape of the Thunder Bay region records only the most recent geomorphic processes because final advance and retreat of ice across the region has removed most of the evidence of earlier phases.

6.4 The Last Glacial Events in the Thunder Bay Region

The history of a glacial margin can become intricate in detail, and the present landscape of the Thunder Bay region is the product of just such interplay between three major ice lobes originating from three accumulation centres during the Wisconsin glaciation. Even then, it is only the final event, the Marquette Readvance and its subsequent retreat about which any true detail is known (Fig. 6.4).

The Patricia or Rainy River Lobe approached the region from the north; the Hudson Bay Lobe approached from the northeast, crossing the Nipigon basin and the Superior Lobe, an offshoot of the Hudson Bay Lobe, crossed the Superior north shore and flowed, relatively unimpeded, down the basin towards Duluth, MN, pushing inland against the Thunder Bay and Minnesota shores in a northwesterly direction, creating in places, features opposing the direction of those of the Patricia or Rainy River Lobe (Zoltai 1963).

The final Marquette event (circa 11,500 cal yrs BP) saw a resurgence of the Hudson Bay and Superior lobes. The Hudson Bay Lobe pushed again across the Nipigon basin, but halted to form the Dog Lake Moraine. The Superior Lobe once again headed down the axis of the Lake Superior basin, its margin climbing up the dip slope of the Sibley Peninsula, across where the city of Thunder Bay now lies and pushing up the Kaministiquia and Whitefish valleys. Known locally as the Murillo Lobe, it was initially thought to have formed the northward facing Marks Moraine, its sediments characterized by an appreciable content of the red, green-mottled Sibley sandstone. The Marks Moraine forms a prominent ridge on the skyline to the north of the city of Thunder Bay (Fig. 6.5).

Research has shown that the Superior Lobe merely plastered its deposits against the southern flanks of an older retreat moraine, the Brule Moraine, formed during the withdrawal of the last phase of the Patricia or Rainy River Lobe. In places, the Marks Moraine is a push-moraine

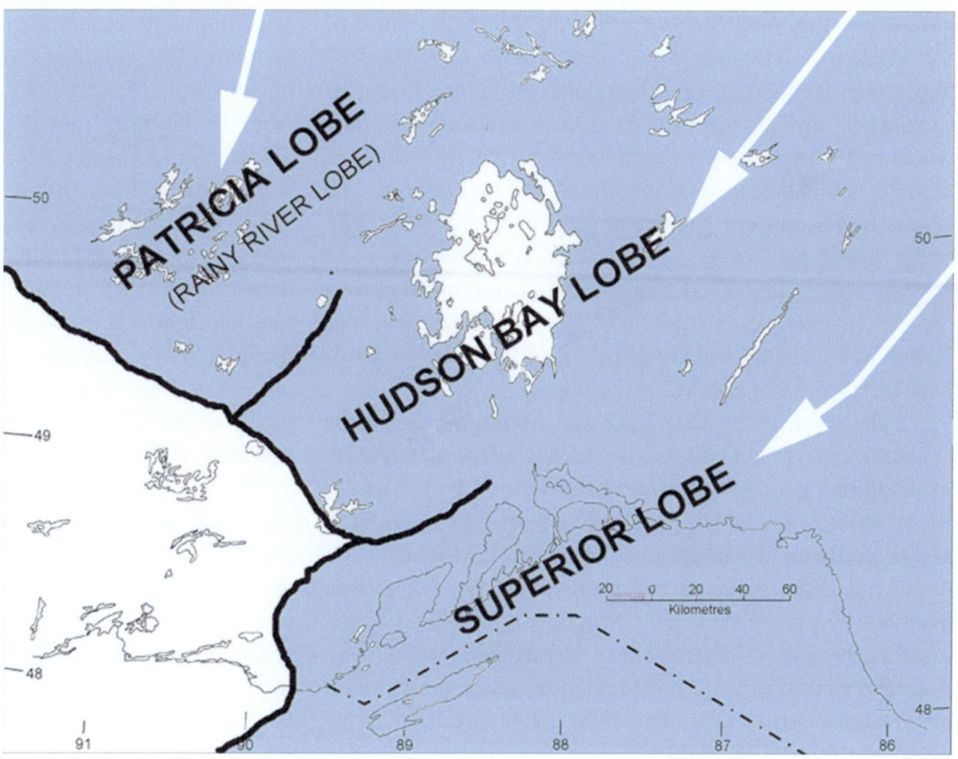

Fig. 6.4 The three ice lobes affecting the Thunder Bay region. Figure by B. Phillips

Fig. 6.5 The Marks moraine forms the skyline to the north of the city. *Photograph* by B. Phillips

(Phillips 2000). This is a fine example of "inheritance", where the younger event modifies a landform of an earlier event, a not uncommon complication in the interpretation of post-glacial landscapes (Phillips 1977).

That the Dog Lake and Marks Moraine were penecontemporaneous is shown by the presence of the Mackenzie Moraine, an interlobate moraine that runs to the east from the junction of the two moraines, but dwindles after a few kilometres (Fig. 6.6). Meltwater flowing along the course of the Mackenzie Moraine poured sediment towards the "Y" junction to form a large perched delta, the Lappe Delta, its apex at 463 m. This formed into glacial Lake Kaministiquia, which was temporarily trapped between the two ice lobe margins and extended west into the Shebandowan area. The history of glacial Lake Kaministiquia is more complex, there being interflows between it and the huge body of Lake Agassiz which inundated much of what is now Manitoba (Teller and Thorleifson 1983). Shorelines of several levels of former Lake Kaministiquia can be traced along the fronts of both the Dog Lake and Marks moraines.

To the west of the Dog Lake and Marks moraines, the landscape remained unaltered by the Marquette Readvance, its landforms dating from the retreat stages of the Patricia or Rainy River Lobe, which left a sequence of retreat moraines across northern Minnesota and into Canada (Fig. 6.7). Of these, the Brule Moraine has been noted as having been inherited by part of the Marks Moraine, its trend eastwards towards the city of Thunder Bay having been obliterated. A variety of landforms of Pre-Marquette age are preserved in the Whitefish/Arrow lakes area, most notably the East Arrow Delta, formed into the eastward extension of the high Herman level of Lake Agassiz from debris brought to that position on the Steep Rock Moraine by an esker/tunnel valley system crossing Whitefish Lake from the northeast. It is also in this area that the earliest evidence of Palaeo-Indian occupation of the region is found (Phillips and Hill 2004).

To the east of the Dog Lake and Marks moraines or closely associated with them are landforms of Marquette and post Marquette age. At its highest level, Lake Kaministiquia lay above low points of the Marks Moraine, lapping directly on Superior (Murillo Lobe) ice. As ice began to melt back from its terminal moraines, proglacial Dog Lake formed behind the Dog Lake Moraine (Fig. 6.8). More than once, water flowed from this unstable lake into Lake Kaministiquia, cutting a broad channel in the crest of the moraine. In turn, as Superior (Murillo Lobe) ice melted, pressure along parts of the Marks Moraine was such that catastrophic failure took place at a point where a much older rock channel was dammed by the moraine. The Cedar Creek outburst allowed waters of Lake Kaministiquia to pour into the Whitefish valley to form proglacial Lake Cedar Creek (442 m) (Phillips et al. 2012). This lake was prevented from flowing east by the still receding Superior (Murillo) Lobe and, instead, expanded west towards Whitefish Lake, escaping southwards down the Little Whitefish River valley to the valley of the Pigeon River, which was also prevented from flowing east by the presence of ice. The overflow found its way south, via the shallow Swamp River valley, to exit into a high level of Lake Superior some 244 m above the present lake level, near Hovland, MN (Dean and Phillips 2011).

The Wisconsin glaciation of the area, as those glacial events before it, depressed the earth's crust under its weight. Isostatic recovery since the Marquette Readvance has seen the Canadian north shore of Lake Superior rise more than the Duluth, MN end of the basin. As Superior ice withdrew from the Superior basin to the northeast, proglacial lakes, ever larger in area but lower in elevation, occupied the ice-free part of the basin until a phase known as Lake Minong (circa 10,500 cal yrs BP) occupied an area similar to that of present Lake Superior (Farrand 1960). Progressive tilt, coupled with alternating outlets to the basin, resulted in a series of shorelines now raised high above the present lake level. These, today, rise in elevation from Duluth towards the Canadian north shore, their trace being slightly curvilinear as isostatic recovery remains more rapid to the northeast than to the southwest (Fig. 6.9). The sequence is made more complex by a later rise in lake level, known as the Nipissing transgression (circa 5,500 cal years BP) which drowned earlier shorelines to a greater degree at the southwestern end of the basin than the northeastern end.

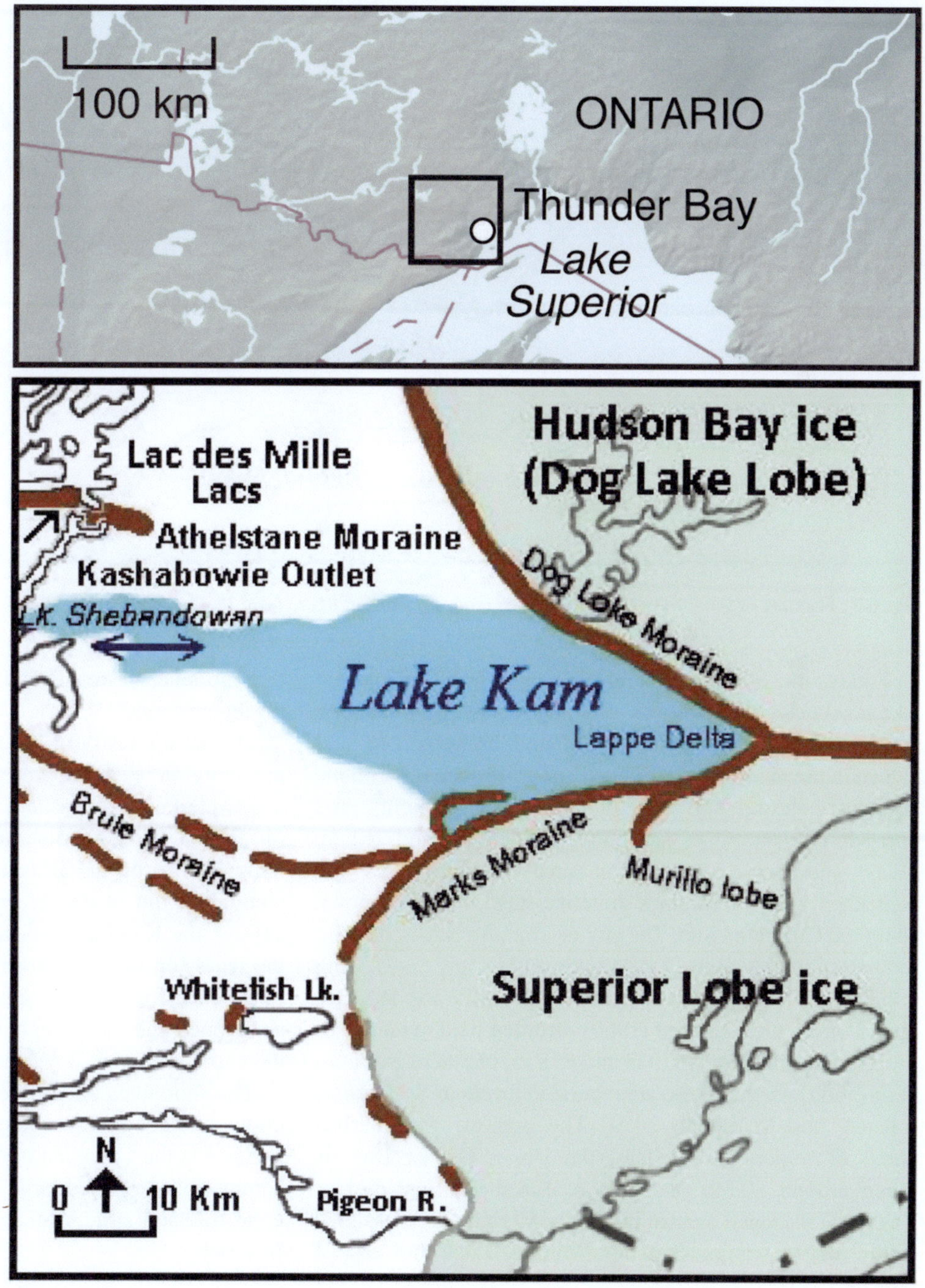

Fig. 6.6 The Marquette advance in the Thunder Bay region

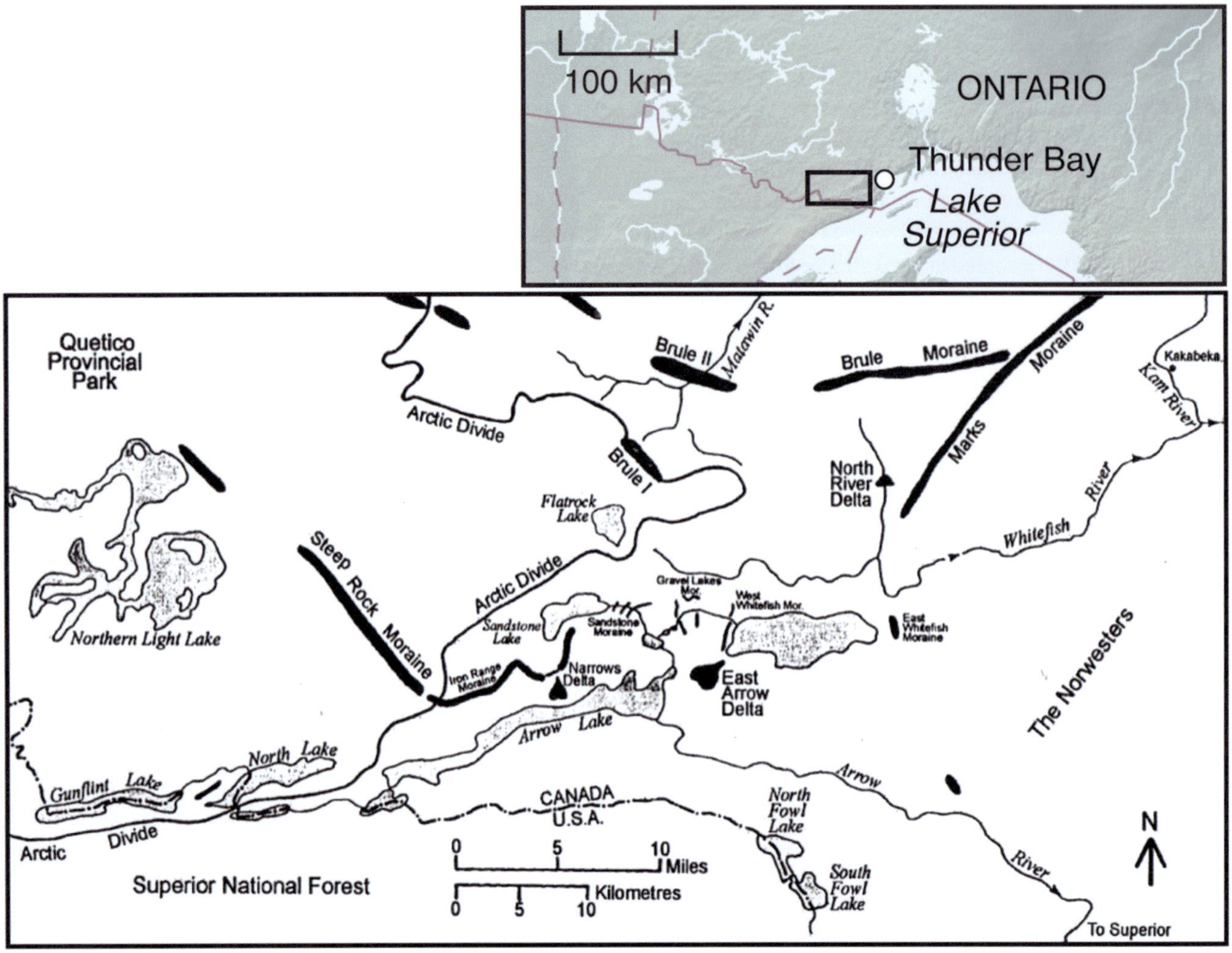

Fig. 6.7 Features of the arrow—Whitefish lakes corridor. Bottom figure from Phillips and Hill (2004; Fig. 10:12, p. 291)

Today, the Minong shoreline emerges from the present lake level (183 m) just south of Grand Marais, MN. Minong and post-Minong shorelines are found between 245 and 226 m in the city of Thunder Bay, and at Terrace Bay, on the northeast shore, six distinct levels of Minong and post-Minong shorelines are found between 305 and 261 m. Many shoreline Palaeo-Indian archaeological sites are associated with the Minong shoreline stage, the best known being the Cummins Site. The city of Thunder Bay is built on a staircase of shorelines, road alignment in many cases being influenced by the natural alignment of bluffs and terraces. Inland of the city, Farrand (1960) recorded field evidence of shoreline features at 358, 333 and 318 m, but there is reason to question whether these are related to levels of Superior or isolated post-Marquette lakes. The highest, well-marked phase of Superior shorelines, the Upper Beaver Bay, is found around 302 m just west of Kakabeka. Just east of Kakabeka it lapped against the Intola Moraine, a recessional feature of the waning Superior (Murillo) Lobe, but east of that moraine, nearer Thunder Bay, only lower Beaver Bay levels are found.

One scenic result of isostatic recovery is that most rivers in the region reach the present lake level in a series of rapids and waterfalls. One such feature is the Kakabeka Falls (40 m) on the Kaministiquia River. Another is the High Falls, (28 m) on the US/Canada border. Both their courses commonly run in deeply eroded gorges (Fig. 6.10). In the case of the Kaministiquia and other sediment laden rivers, the recession of water level has left a series of delta structures, each formed at a specific shoreline level, only to be abandoned and then deeply cut into by the river as it graded its course to a new, lower water level.

The shorelines themselves are marked either by erosional bluffs, often degraded slopes today, or by sub-parallel beach ridges. On the Canadian north shore, the raised beach ridges commonly comprised large cobbles (Fig. 6.11), one notable large foreland structure being found just south of Thunder Bay city.

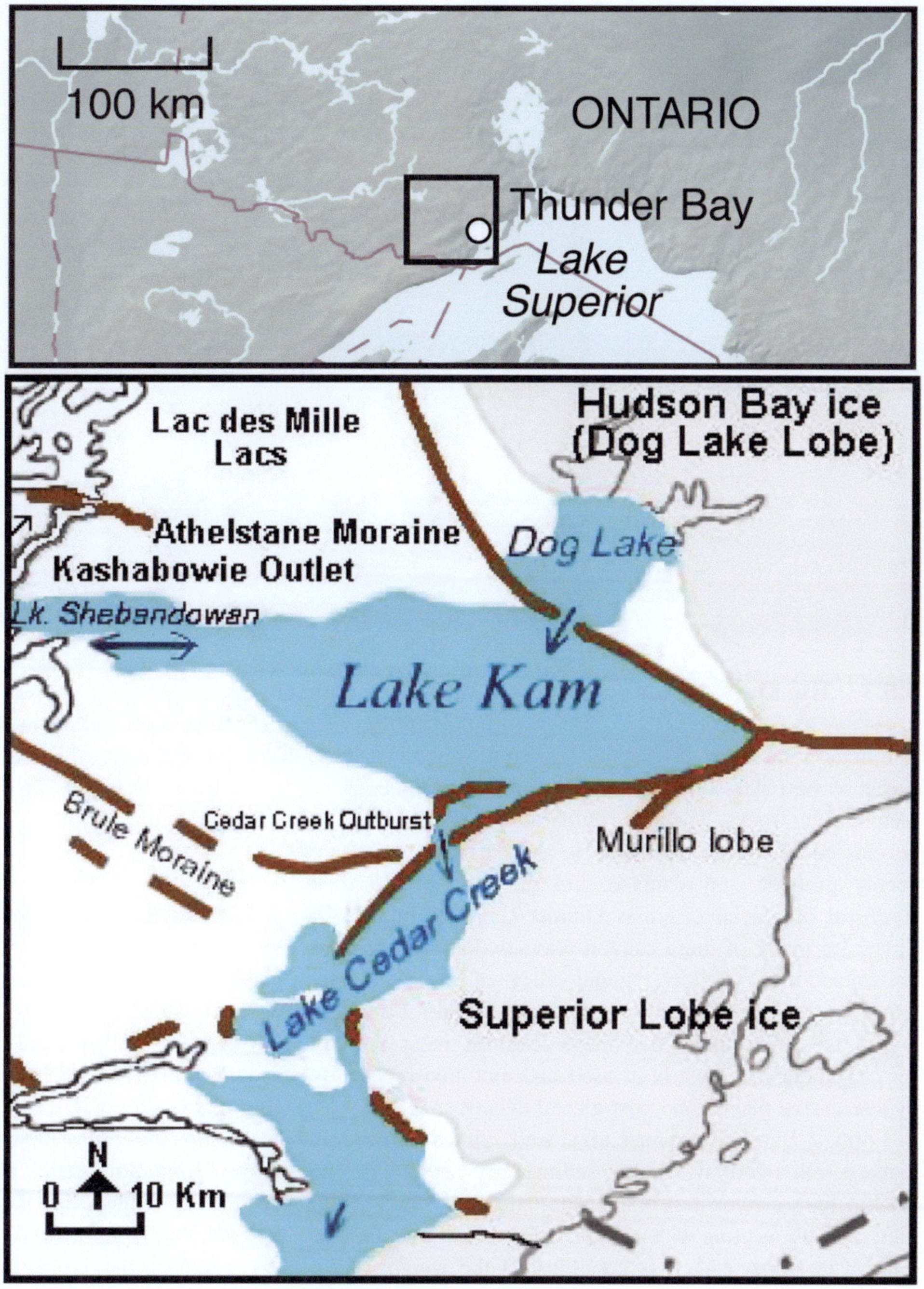

Fig. 6.8 Early post-Marquette events in the Thunder Bay region

6.5 Landforms of Particular Interest

Five landforms of the region have been chosen as worth exploring more closely for what they tell us of the palaeogeographic environment at the time of formation. They are all accessible to interested visitors and are presented in approximate chronological order.

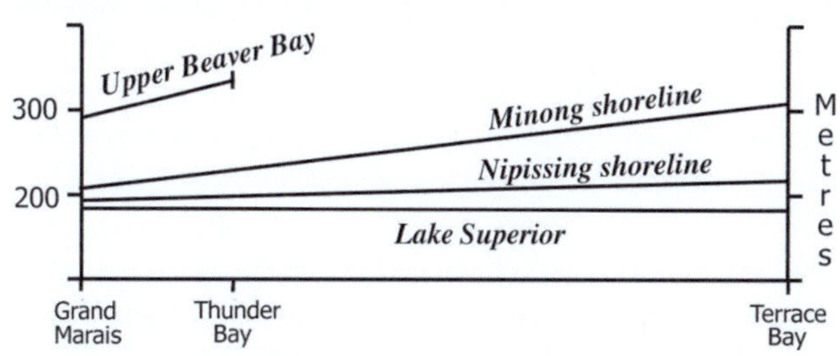

Fig. 6.9 Simplified diagram to illustrate differential isostatic recovery. Figure by B. Phillips

Fig. 6.10 Kakabeka Falls, the best known falls of many. *Photograph* by B. Phillips

6.5.1 The Ouimet Canyon

The region abounds in glacially striated, smoothed rock surfaces, best developed in granites and other igneous or metamorphic rocks. There are numerous features that could be classed as roches moutonnées, and surfaces which are deeply grooved and sculpted, but the best known rock landform of glacial origin is Ouimet Canyon (Fig. 6.12). There are many of these canyon features that are less well known and many shallower tunnel valleys of less spectacular appearance, especially between the Lake Nipigon basin and Dog Lake. It is likely that these features, most initially formed under ice, have been used and modified repeatedly by successive glacial and post-glacial events.

Formed across a remnant of a well jointed Palaeozoic diabase sill, overlying Sibley sedimentary sediments, the 2.4 km long canyon follows an approximately north–south fault or joint, meeting with an approximate east–west fault or joint at its upper end. It is possible that the weight of ice cambered the sill and widened an incipient fault or joint to the point at which flowing water could exploit it. Kor and Teller (1986) conclude that Ouimet Canyon has its origin as a sub-glacial feature, a tunnel valley, though likely further modified by sub-aerial meltwater flow. They also conclude that the location of the canyon is beyond the area in which Lake Agassiz discharges flowed. In fact, the canyon is a small segment of a sinuous line of channels, lakes and eskers that can be traced over 70 km northwards, through Cavern Lake, Wolf Lake and up towards Lake Nipigon.

No river runs in the present canyon. Since deglaciation, the columnar jointed, 100 m high canyon walls have receded by freeze-thaw and toppling and the original floor now lies deep below an unknown depth of large-scale rock debris. The microclimate of the floor is such that it acts as a refugium for arctic plants that more typically grow a thousand kilometres further north. The floor is closed to public access.

6.5.2 The East Arrow Delta

This is a large perched delta, with no present water source, approximately 2.5 km^2 in area, located at the east end of Arrow Lake (460 m), and its 503 m apex lying in a bedrock saddle above Whitefish Lake (403 m), directly to the east (Fig. 6.13).

Phillips and Hill (2004) demonstrated, using an isostatic retro-deformation model, that the lake occupying the Arrow basin, into which the East Arrow Delta formed, though known locally as glacial Lake Arrow, was the eastward extension of the Upper Herman level of Lake Agassiz. Contemporaneously, the margin of the Rainy River Lobe lay along what is known as the Steep Rock Moraine (circa 14,000–13,000 cal years BP), which can be traced as a chain of cross-valley morainic segments along the south side of Sandstone Lake, along the Gravel Lakes corridor to the West Whitefish Moraine, itself a kettled cross-valley moraine, and thence eastwards along the steep south shore of Whitefish Lake (Fig. 6.14). The materials of which the East Arrow Delta is built are of northern provenance, as are the materials in the moraine segments noted above. However, the volume of water and sediment issuing from the Steep Rock ice margin at the point of the rock saddle is linked to the

Fig. 6.11 Raised cobble beaches of Lake Superior *Source* Google Earth; *photograph* by B. Phillips

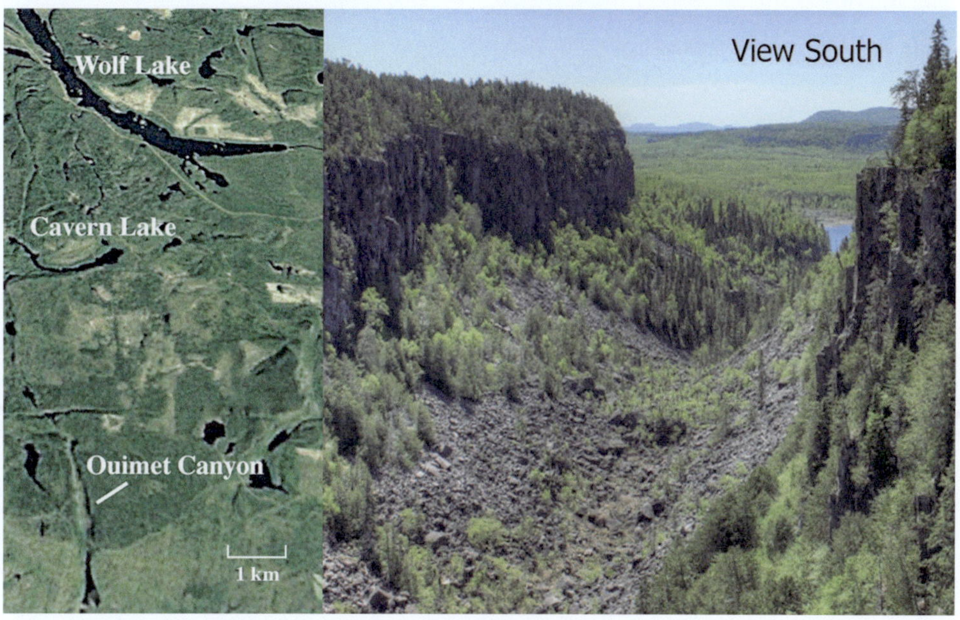

Fig. 6.12 Ouimet Canyon. *Source* Google Earth; annotated by B. Phillips; *photograph* by B. Phillips

Whitefish Esker that runs along the floor of Whitefish Lake, visible above water only as a chain of small islands. Furthermore, the trend of the esker can be traced northeast along a segment of tunnel valley which is oriented towards another tunnel valley segment at the mouth of which lies the North Delta, formed at the time of the succeeding Brule Moraine I phase (Fig. 6.7).

The foreset slope of the East Arrow Delta drops sharply, its base of 487 m with a marked shoreline that suggests that water level fell abruptly from the Upper Herman stage, exposing the delta structure to sub-aerial processes (Fig. 6.14). However, the delta is incised by a steep-sided channel, which runs from the rock saddle to the foreset margin, where three mega ripples show that it discharged a

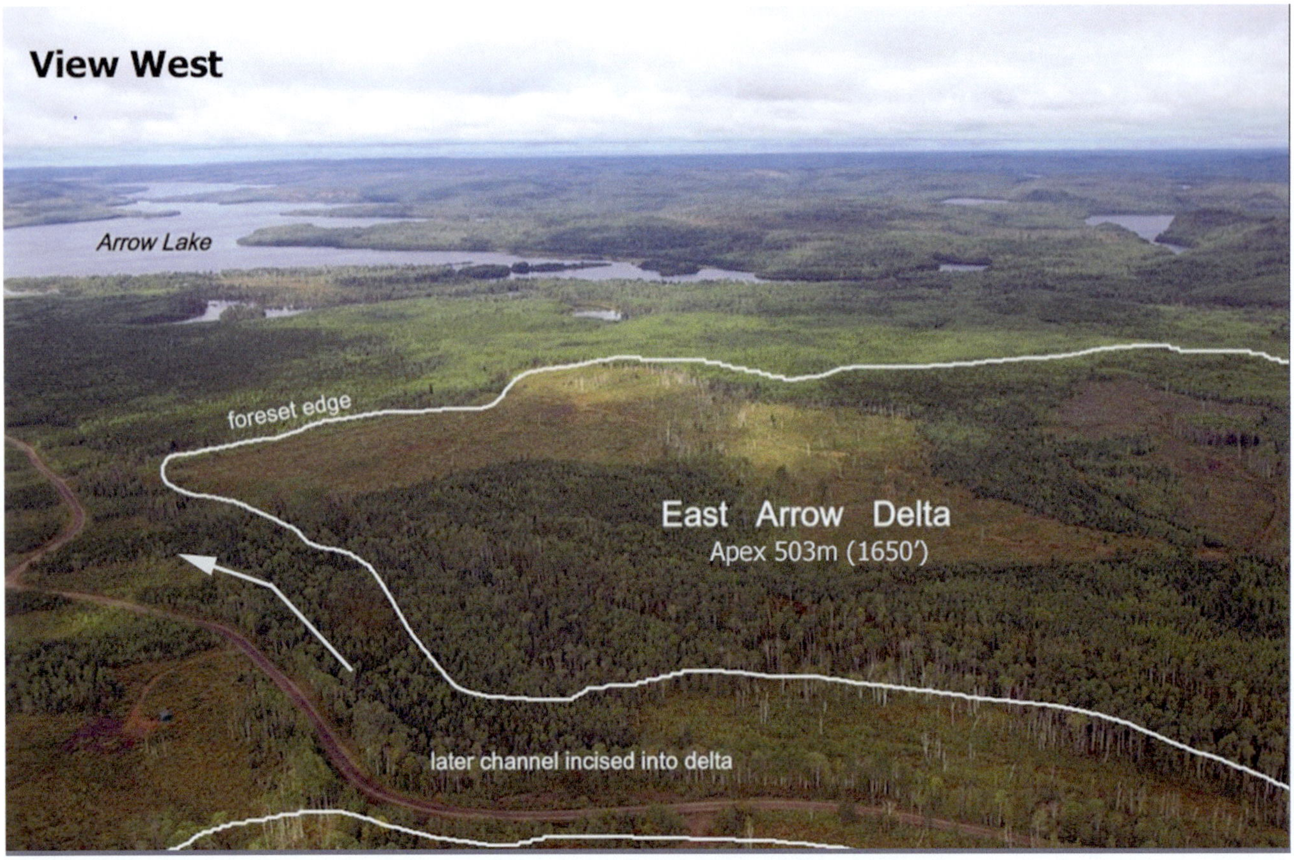

Fig. 6.13 East Arrow Delta. *Photograph* by F. Dean; annotated by B. Phillips

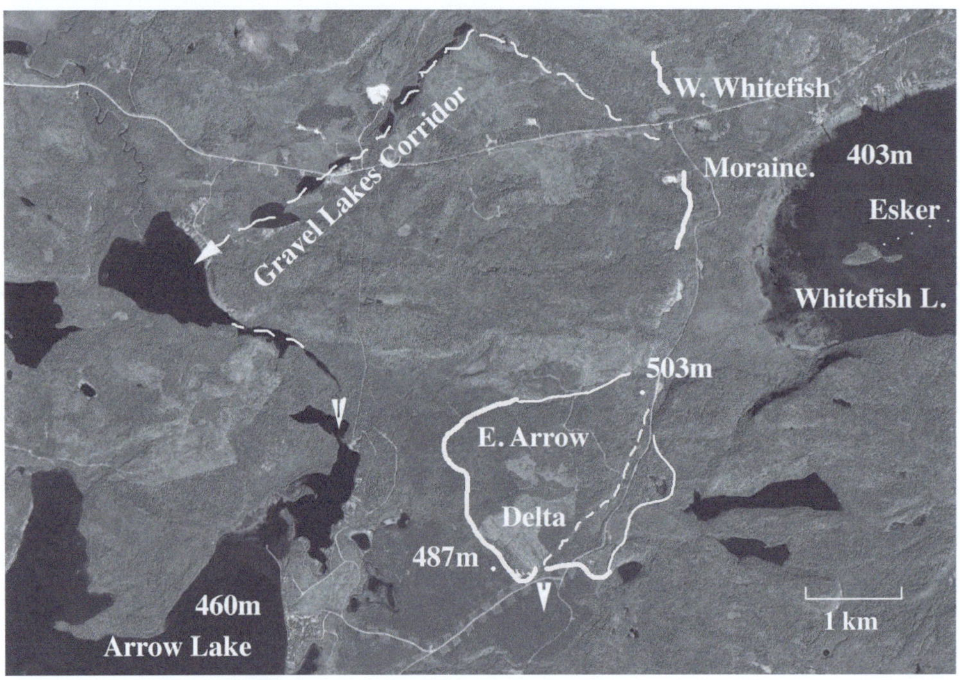

Fig. 6.14 East Arrow Delta and associated features. *Source* Google Earth; annotated by B. Phillips

high volume of water into the shallows of the 487 m water margin. No evidence of a 503 m lake level has been found against the rock walls of the west end of Whitefish Lake, but it is surmised that as ice melted back from the Steep Rock margin, a short-lived proglacial lake was trapped in the west end of the Whitefish basin, perhaps intermittently overflowing the rock saddle to erode the channel into the exposed delta surface.

That this lake, which was likely limited to the western end of the Whitefish basin, was "short-lived" is suggested by the large missing central portion of the West Whitefish Moraine. The huge spread of rock debris that carpets the floor of the Gravel Lakes corridor suggests that the pressure of this proglacial lake against the moraine, now acting as a dam, caused its sudden failure, draining the lake and flooding the Gravel Lakes area with debris which was carried southwest into the 487 m level of the Arrow basin.

As Rainy Lobe ice withdrew to the position of the Brule Moraine, a much larger and more stable proglacial lake at a level of 472 m was established, in both the Whitefish and Arrow Lake basins. The small perched delta at the Arrow Narrows and the much larger kettled structure of the North Delta are both graded to this level (Fig. 6.7). This lake is known locally as Glacial Lake Strange, after the Township in which the North River and the North Delta lie. As Rainy Lobe ice continued to withdraw to the north, this lake subsequently drained, and the lakes occupying the Arrow and Whitefish basins were likely reduced to the levels of their natural containment.

During the Marquette Readvance (circa 11,500 cal years BP), Superior Lobe ice pushed up the Kaministiquia/Whitefish valleys as far as the east end of Whitefish Lake, where a heavily kettled remnant, the East Whitefish Moraine, its material being of Sibley provenance, marked its limit. A proglacial lake once again occupied the Whitefish Basin but not at a level (460 m) sufficient to extend into Arrow Lake.

The archaeological significance of the events in the Whitefish/Arrow Lakes area is of particular note because the area remained unglaciated from circa 13,000 cal years BP. There is no environmental reason why people from the mid-west should not have penetrated the Thunder Bay region during the time between the retreat of the Rainy River Lobe and the Marquette Readvance, yet no definitive artefacts that would indicate such an early presence of Palaeo-Indian people have been found. That the Arrow/Whitefish corridor was used by people entering the region after the Marquette Readvance is well indicated by the style of lithic artefacts found, but the tantalizing question of whether earlier people reached the area remains unanswered.

6.5.3 The Lappe Delta and the Mackenzie Interlobate Moraine

During the Marquette Readvance, the Hudson Bay and Superior lobes only became distinguishable as separate entities as ice crossed the Superior northeast shore. Between the two diverging streams of ice, partly guided by high ground of outcropping rock, there lie a variety of ablation and fluviglacial landforms which together are referred to as the Mackenzie Interlobate Moraine. This moraine is not a single landform but a string of different features, many of which occupy areas of low topography, and which can only be identified as "interlobate" for about 45 km to the west of Pearl, on the Trans-Canada Highway between Thunder Bay and Nipigon. Thus, the Dog Lake and Marks moraines, both distinct ridges of morainic materials, diverge from the eastern end of the interlobate moraine near the community of Lappe (Fig. 6.15).

That the Dog Lake and Marks moraines were penecontemporaneous (Burwasser 1977) is confirmed by the presence of glacial Lake Kaministiquia which occupied the "V"-shaped area between them. Shorelines of Lake Kaministiquia occur on the leading flanks of both moraines and on the ice-contact Lappe Delta that formed at the apex where interlobate sediments were flushed westwards into the lake. The highest shoreline on the Lappe Delta lies at 469 m a.s.l., an elevation that is higher than the Marks Moraine at several points, and demonstrates that it was the Superior ice margin that held the water up to this maximum level, and not the moraine alone. The western end of glacial Lake Kaministiquia is connected with Lake Shebandowan, and it has been proposed that flow, both to and from Lake Agassiz, occurred through this lake (Zoltai 1963; Teller and Thorleifson 1983). At a minimum, 160 years of glacilacustrine sedimentation occurred in Lake Kaministiquia when the Superior Lobe was near its maximum extent at the Marks Moraine (Loope 2006).

Although clays were laid in Lake Kaministiquia, the topography within the basin is considerable and large portions of the inundated area show little or no sedimentation, especially the higher parts. The Lappe Delta comprised of uniformly stratified sands and gravels of early and late Precambrian era, built over a similar length of time, sloping at about three degrees to the edge of a sharp foreset slope at about 445 m, at which point there is a sudden 30 m drop down to the flat clay plain below. Today, surface aggregate mining has all but destroyed the original surface of the delta.

The Lappe Delta can be termed a "perched delta" or an "ice-contact delta", as was an isolated delta described by Phillips and Fralick (1994), this graded to a later, lower level of Lake Kaministiquia at 457 m, a level at which the Cedar

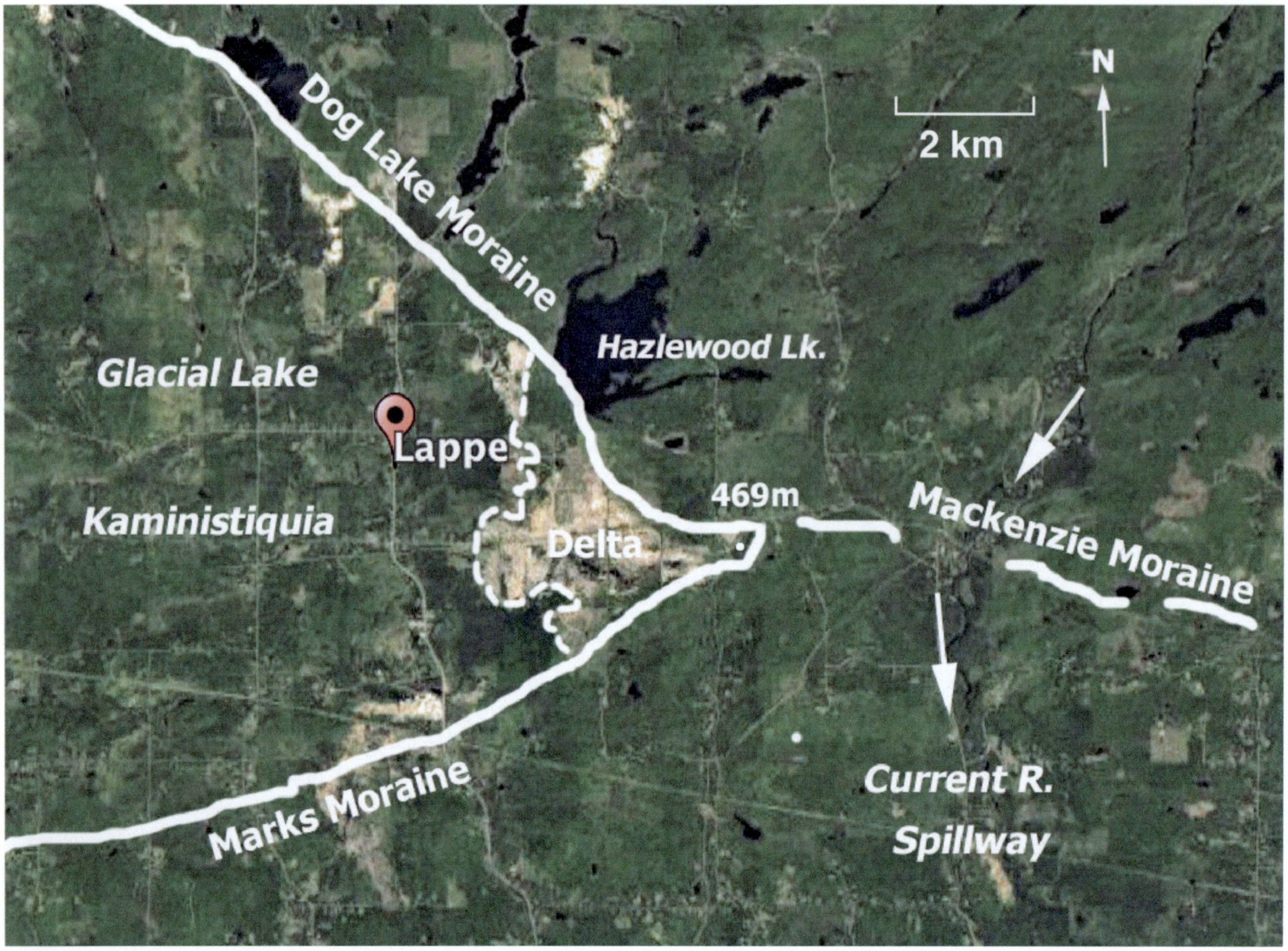

Fig. 6.15 The Lappe Delta and Mackenzie interlobate moraine. *Source* Google Earth; annotated by B. Phillips

Creek outburst may have occurred (Loope 2006; Phillips et al. 2012). Immediately east of the Lappe Delta apex, there is an eastward facing ice-contact slope, beyond which the land slopes steadily downhill towards the Current River (Fig. 6.15). There being no other possible source of water and sediment other than of ice, the term "perched" was applied. Segments of the Mackenzie Interlobate Moraine can be seen from nearby minor gravel roads, and much of it is characterized by kame and kettle topography and glacifluvial accumulations in which dropstones and severely collapsed, contorted sediment structures are common (Phillips and Fralick 1994).

A portion of the interlobate moraine is missing where the Current River valley cuts through it. At some point during the retreat of the Marquette ice from the Dog Lake and Mackenzie moraines, the Current River, the headwaters of which arise to the northeast, towards Lake Nipigon, acted as a meltwater spillway. Multiple channels are visible on Fig. 6.15. It is likely that the long, incised course of the Current River north of the Mackenzie Moraine, is in its origin, one of many tunnel valleys which radiate southwest from the Nipigon basin and may well have supplied sub-glacial meltwater flow to the interlobate area and a source of sediment for the Lappe Delta. The first Superior basin lake stage to inundate Thunder Bay immediately after the retreat of Marquette ice was the Lower Beaver Bay, found at around 268 m at the north edge of the city. A large delta structure of this elevation is found along the Current River below the Cascades Conservation Area, suggesting that the Current River was still carrying a heavy load of sediment from near its headwaters at that time.

6.5.4 The Murillo Drumlin Field

Located just north of Murillo, the drumlin field is fan shaped and spreads to the west and north to the foot of the Marks Moraine. The landforms were noted by Zoltai (1963) who described them as "long and narrow" and reported that an uneven, loose layer of ablation till masked the underlying moulded material which he described as "compacted, stony silty sand till" (p. 105). The Murillo Drumlin Field drumlins

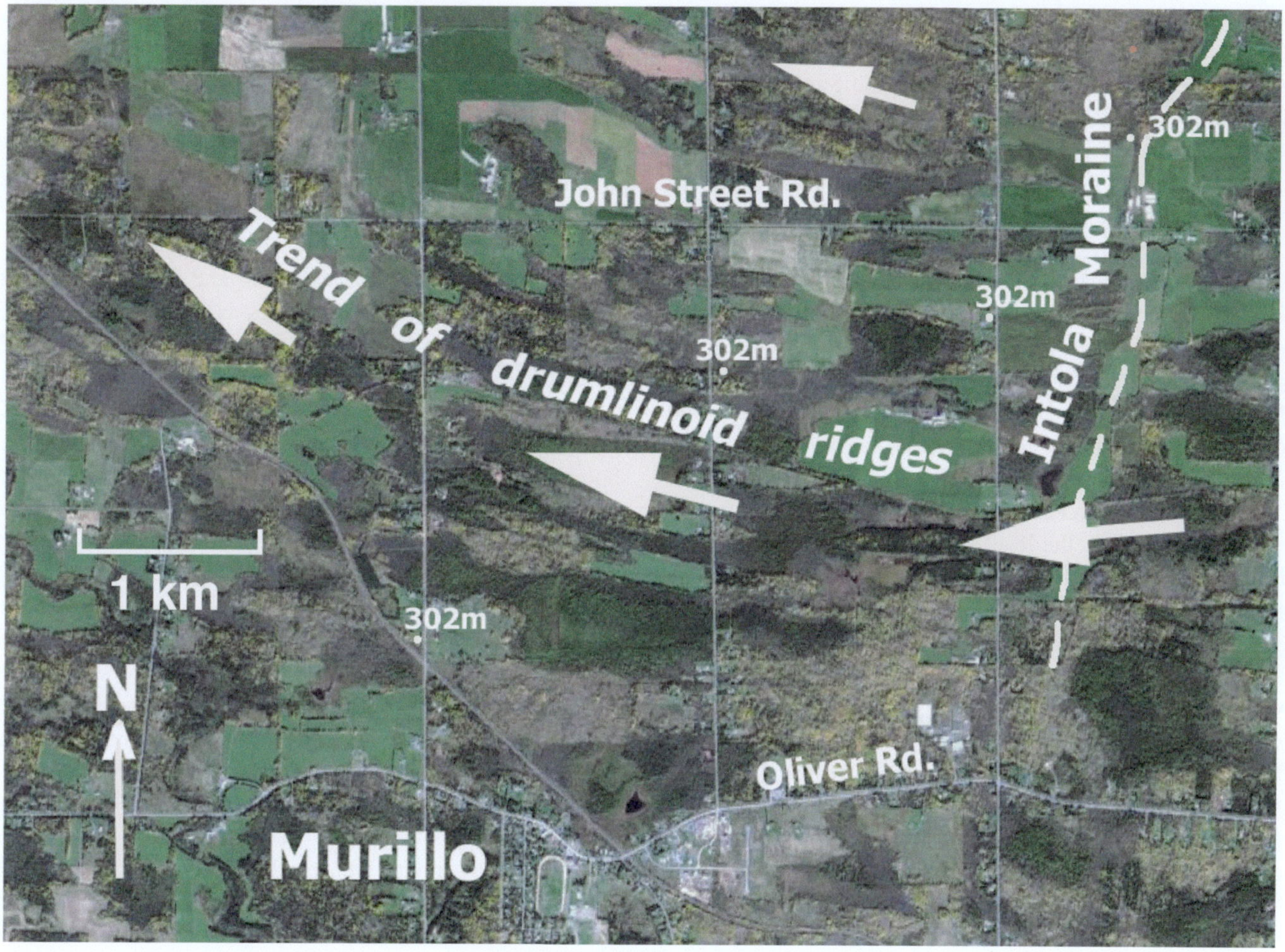

Fig. 6.16 The Murillo drumlin field. *Source* Google Earth; annotated by B. Phillips

can certainly not be compared to those of the Peterborough area in southern Ontario; hence the term "drumlinoid" may be a better choice to describe these landforms (Fig. 6.16). More thoroughly examined by Burwasser (1977), he described the compacted till as "gritty silt till" (p. 18).

Crossing the field in a north to south transect, as a number of country gravel roads do, one crosses a switchback of long, sub-parallel, rounded ridges, of an amplitude not exceeding 9 m, separated by swales in many of which wet, marshy ground persists. The clearing and agricultural use of the upper slopes of the ridges emphasizes the pattern as seen in air photography. The ridges are, for the most part, built over rock steps that occur on the long dip slope of the Rove formation. A gritty, silty till, noted by Burwasser (1977), directly overlies the rock. The till is of Superior provenance but is distinct from the redder, more clayey till that characterizes the Marquette Readvance. This younger till overlies the older, more compacted till, in places, but the area is mostly covered with a variable layer of looser ablation till which fills the swales to a greater depth, the initial amplitude of relief being greater than that seen at present. Though the drumlinoid features were initially interpreted as having been formed by Marquette ice, it appears that their origin was by an earlier phase of Superior Lobe ice, this till being found up to 488 m, north of what is now the Marks Moraine. This advance, more powerful than the Marquette, overrode the existing Brule Moraine (Fig. 6.7), whereas Marquette ice halted against it. The inheritance and further fashioning of the older drumlinoid features by Marquette ice is fully compatible with its inheritance of the already modified Brule Moraine, now named the Marks Moraine.

Superimposed on the drumlin field is the Intola Moraine, which extends southwards from the Marks Moraine and is a minor but important recessional/readvance, kettled moraine formed in the later stages of the melt back of the Superior (Murillo) Lobe ice. Interpreted as a stagnant ice feature, briefly overridden by a minor advance and then again stagnant (Burwasser 1977), the Intola Moraine would have extended across the Kaministiquia valley but for the fact that it came into contact with water and becomes indistinct when followed south into what would have been deeper water. Fronting part of the Superior (Murillo) Lobe ice at this time was the Upper Lake Beaver Bay stage of Superior water level, at 302 m, which flooded into the Murillo/Kakabeka part of the Kaministiquia valley from the Superior basin to the south, over a low saddle into the Slate River valley, once it became ice free. The approximate northern margin of the proglacial lake is shown by 302 m spot heights on Fig. 6.16, the lakeshore lying at an angle across the trend of the drumlinoid features. There would have been a transition from the final low stages of glacial Lake Cedar Creek to the Upper Beaver Bay level, though the high shoreline features observed by Farrand (1960), and referred to earlier, might possibly indicate incursions of earlier levels in the Superior basin, the Washburn and Manitou stages, these being more characteristically found on the Minnesota shore, south of the border.

The Upper Beaver Bay shoreline does not occur east of the Intola Moraine, the level of water in the Superior basin having fallen to the Lower Beaver Bay levels (277–268 m) by the time the site of the city of Thunder Bay was ice free. The stages of post-Marquette water level in the Superior basin were relatively brief still-stands and are not associated with layers of sediment but more by bluff and beach features where wave and sediment supply conditions so allowed. Thus, the inundation of parts of the Murillo drumlinoid features is marked primarily by shoreline features. However, at the base of the Intola Moraine, near John Street Rd., there are places where a thin apron of offshore clayey sediment marks the only remaining evidence of wave action on the moraine flank.

6.5.5 The Pass Lake Bay-Mouth Bar

The Lake Minong phase of lake level in the Superior basin (circa 10,500 cal yrs BP) is, in the Thunder Bay area, closely associated with numerous Palaeo-Indian archaeological sites, mostly found on gravel or cobble beaches, at river mouths or on lookout vantage points overlooking the shore. The Minong shore is thought to have provided a relatively easy route-way for both big game and early people making spring and summer forays from the mid-West into the new grounds relatively recently vacated by Marquette ice.

As noted earlier, because of differential isostatic recovery, the Minong and post-Minong shorelines are submerged by present Lake Superior at the Duluth, MN end of the basin, emerge just south of Grand Marais, MN, and are found between 245 and 226 m in the city of Thunder Bay. At Pass Lake, at the mainland base of the Sibley peninsula, the highest Minong shoreline is found 258 m, the main bay-mouth bar on which the Brohm Palaeo-Indian site was excavated, being at 250 m (Fig. 6.17).

Pass Lake is appropriately named as a pass between the Black Bay side of the Sibley Peninsula on the east and the Thunder Bay side on the west. Both rail and road squeeze along the narrow bench between Pass Lake and the steeply rising walls of the overlooking mesa-like forms of Sibley sediments. As seen just to the north and east of the small community, the origin of this pass is as a tunnel valley. Bathymetry of Pass Lake shows the trench continuing along its length. There are tunnel valleys running down the length of the Sibley Peninsula, but this east to west one is isolated and breaks through the west facing cuesta that forms the shoreline facing Thunder Bay. Like many of these features, it probably dates back several glaciations and has been used by sub-glacial flows several times.

During the Beaver Bay lake stage in the Superior basin, this break separated the island of Sibley from the mainland as a narrow strait, but as water level fell to the Minong stage, the pass began to emerge. On the east, dip slope side of the pass there are a succession of low amplitude beach ridges, but it is on the west side that there appears to have been both the wave action and the source of gravel and cobble sediments to build a succession of bars (Fig. 6.17) which span the narrow gap between the rising mesas on either side. These bars, bay-mouth bars by description, first connected the mainland to the island of Sibley, the largest and longest bar, at 250 m, forming a particularly useable route-way for large game to reach the former island. It seems no coincidence that during the relatively brief period of time this bar first provided access to the Sibley Peninsula, it formed an ideal natural trap for Palaeo-Indian hunters to use. The materials excavated at the Brohm site, on the top of this main bar, were all hunting related projectiles and scrapers, many made onsite from chunks of jasper taconite that they carried with them from quarry sources (MacNeish 1952). With a little imagination, one can see lookouts posted on the mesa bluffs above to give warning of animals approaching, a hushed waiting until the herd reached the central part of the bar, and the attack by armed hunters from either side. The site may have only served this purpose for a few decades, for as water level fell to post-Minong stages, the route way widened and lost its advantage.

Occasionally, geomorphic evidence tells us something of the palaeo environment of the time and at Pass Lake, a landform, thought to be unique at least within this region, provides just such evidence. Right beside Highway 587, at

Fig. 6.17 The Pass Lake bay-mouth bar and the Brohm site. *Source* Google Earth; annotated by B. Phillips

the Bak Rd. turnoff, is an almost circular pond which has the appearance of a kettle lake, with steep surrounding banks, though not at a location at which stagnant ice of a glacial retreat stage has stood. In the embayment to the north of the "Minong Headland" marked on Fig. 6.17, the equivalent of the bar on which the Brohm site was found, crosses the bay at 252 m a.s.l. Below it, a series of sub-parallel, concave beach bars fill the embayment, the lowest just on the road side of the pond. However, at this point there is a dramatic change of beach bar orientation, those beyond the pond being convex as if curved around an object that has deflected the incoming wave form (Fig. 6.18).

This explanation fits well with the use of the term "kettle" in that kettles are neither depositional nor erosional landforms but result from non-deposition. The most likely scenario is that of an iceberg drifting down the Thunder Bay embayment, when post-Minong lake level was around 236 m a.s.l., being pushed onshore, perhaps in Spring storm conditions, and becoming stranded. Subsequent longshore sediment movement was deflected around the front of the iceberg, creating a change of orientation in later beach ridges, even after the iceberg had eventually melted, leaving a hole of non-deposition. The equivalent elevation is shown in Fig. 6.17, below the Brohm site, the parallel post-Minong shoreline bars being shown in the texture of the forest cover.

The Minong lake level stage is found all round Lake Superior, indicating that the lake basin itself was free from glacial ice, yet the Pass Lake "Kettle" suggests that severe winter ice conditions prevailed, with Spring break-up producing sizeable icebergs, most eventually melting in open water, but one, at least, being driven onshore to form this unusual landform feature. The palaeo climatic glimpse of such winter ice conditions affirms the widely held belief among archaeologists that, at this time, Palaeo-Indian presence was strictly seasonal, perennial settlement coming some time later when climate conditions ameliorated.

In summary, the northwestern Ontario region, dominated by outcropping Precambrian rock and mostly thin soils, offers an interesting variety of landforms, due in part to the Keweenawan intrusion of diabase sills and dikes and then extrusion of lava flows, and in part to the interplay of three major ice lobes during Wisconsin glaciation.

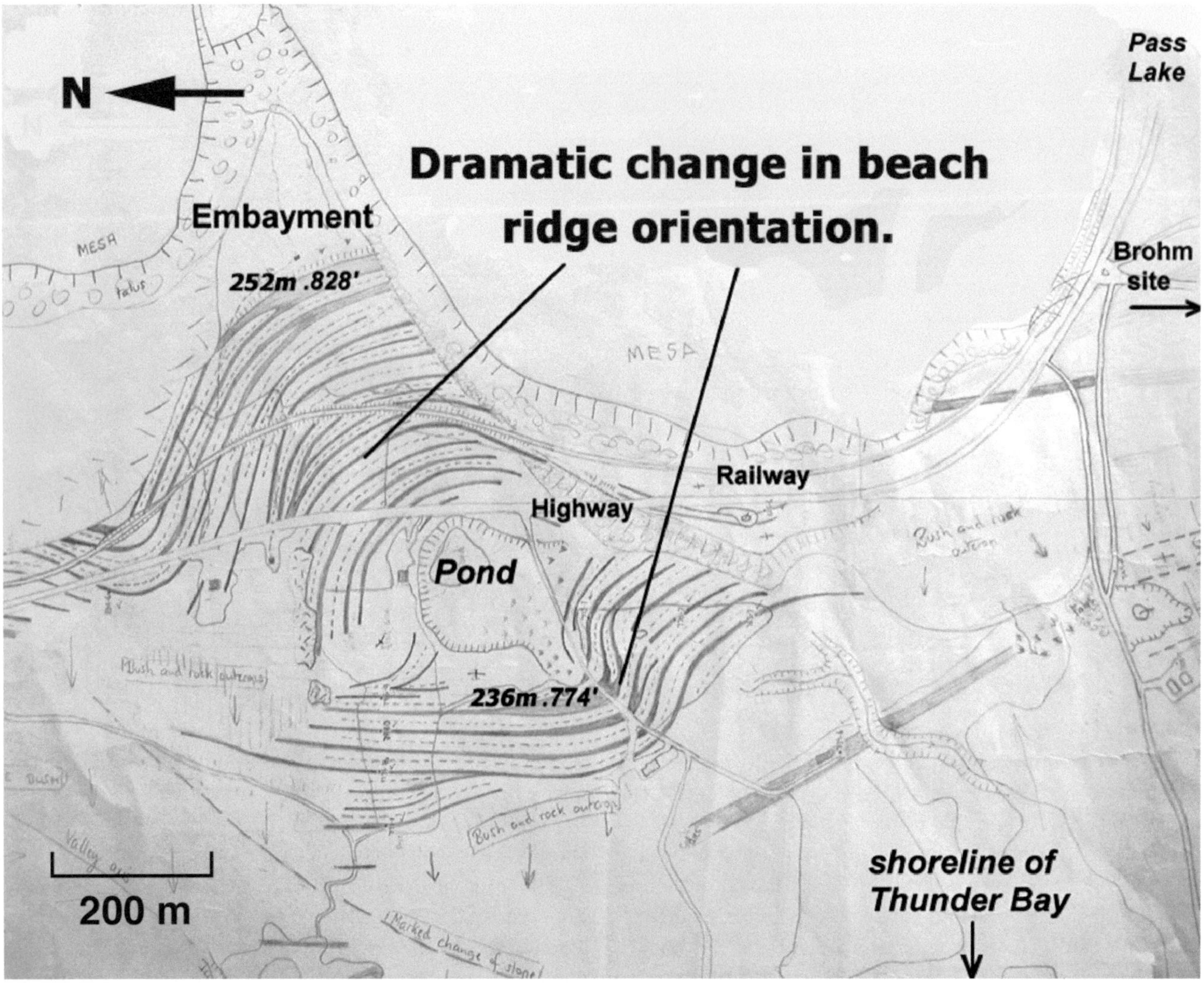

Fig. 6.18 The Pass Lake "Kettle". Unpublished Field Map; annotated by B. Phillips

References

Beukens RP, Pavlish LA, Hancock RG, Farquhar RM, Wilsons GC, Julig PJ, Ross W (1992) Radiocarbon dating of copper-preserved organics. Radiocarbon 34:890–897

Burwasser GJ (1977) Quaternary geology of the city of Thunder Bay and Vicinity, District of Thunder Bay, Ontario Geological Survey Report GR164, 70 p. Accompanied by Map 2372, scale 1:50,000

Dawson KCA (1984) A history of archaeology in Northern Ontario to 1893 with bibliographic contributions. Ont Archaeol 42:27–92

Dean JF, Phillips BAM (2011) Pre-and-Early Agassiz outlets to the Western Superior Basin. In: Miller JD, Hudak GJ, Witkop C, McLaughlin PI (eds) Archean to Anthropocene: field guides to the geology of the mid-continent of North America. Geol Soc Am Field Guide 24, pp 317–349

Farrand WR (1960) Former shorelines in Western and Northern Lake Superior Basin. PhD dissertation, University of Michigan

Kor PSG, Teller JT (1986) Canadian landform examples: Ouimet Canyon, Ontario—deep erosion by glacial meltwater. Can Geogr 30:273–276

La Berge GL (1994) Geology of the Lake Superior Region. Geoscience Press, Phoenix

Loope HM (2006) Deglacial chronology and glacial stratigraphy of the western Thunder Bay lowland, northwest Ontario, Canada. MSc thesis, University of Toledo

MacNeish RS (1952) A possible early site in the Thunder Bay district, Ontario, Annual Report of the National Museum of Canada for the Fiscal year 1950–51. National Museum of Canada Bull 126, p 25, pp 27–28

Pye EG (1969) Geology and scenery; North Shore of Lake Superior, Geological guidebook no. 2. Ontario Department of Mines, 144 p

Phillips BAM (1977) Shoreline inheritance in coastal histories. Science 195:11–16

Phillips BAM (2000) Geoarchaeology of the Thunder Bay area, pre-meeting field excursion #3. In: Proceedings of the 46th annual meeting, Institute on Lake Superior Geology, Thunder Bay, Ontario, v 46, Part 2. Field Trip Guide Books, 40 p

Phillips BAM, Dean JF, Zaniewski K (2012) A Catastrophic Breach and Inter-Lake Flow through the Marks Moraine via Brule Creek and Cedar Creek, Thunder Bay Region, Ontario: a small but significant detail of the Eastern Outflow History of Lake Agassiz.

In: Hollings P (ed) Institute on Lake Superior Geology Proceedings, 58th Annual Meeting, Thunder Bay, Ontario. Lakehead University, 16–20 May 2012

Phillips BAM, Fralick PW (1994) Interpretation of the sedimentology and morphology of perched glaciolacustrine deltas on the flanks of the Superior Basin, Thunder Bay, Ontario. J Great Lakes Res 20:390–406

Phillips BAM, Hill CL (2004) Deglaciation history and geomorphological character of the region between the Agassiz and Superior basins, associated with the 'Interlakes composite' of Minnesota and Ontario. In: Jackson LJ, Hinshelwood A (eds) The Late Palaeo-Indian Great Lakes: geological and archaeological investigations of late Pleistocene and early Holocene environments. University of Ottawa Press, pp 275–301

Teller JT, Thorleifson LH (1983) The Lake Agassi—Lake Superior connection. In: Teller JT, Clayton L (eds) Glacial Lake Agassiz. Geol Ass Can SP 26, pp 261–290

Zoltai SC (1963) Glacial features of the Canadian Lakehead Area. Can Geogr 7:101–115

Kamil Zaniewski has a PhD from the University of Amsterdam and is currently Associate Professor of Geography at Lakehead University in Thunder Bay, Ontario. His interests span a variety of research topics, ranging from microscopic analysis of glacial sediments (the subject of his doctoral dissertation), soil science and the geomorphic history, as well as current active geomorphology of northwestern Ontario. He has been known to extend his contributions to subjects such as lichenometry, lichen ecosystems and glacial history of southwestern Manitoba.

Brian A. M. Phillips (PhD University of Wales, Aberystwyth, Wales) is Emeritus Professor of Geography at Lakehead University, Ontario. He specializes in research related to coastal and glacial geomorphology, with a particular interest in the deglaciation history of northwestern Ontario. Hardly retired, he continues to pursue field research and publications in cooperation with colleagues in geomorphology, geology and archaeology, with the ultimate goal of helping to satisfactorily unravel the undoubtedly complex history of the region. His great pleasure remains in running field excursions for interested groups of students and adults and teaching people about the remarkable features found within only an hour or two's drive from the city of Thunder Bay. In cooperation with his wife Jenny (an amateur botanist), Brian had volunteered as a researcher for the U.S. Bureau of Land Management in the Mojave Desert, California, based in Needles, CA. He has completed a variety of field research reports ranging from studies of plant species and setting up of plots for future sequential studies, writing of OHV trail guides, locating and inventorying of abandoned mine sites and surveying the flow status and wildlife use of mountain springs. In discovering and interpreting the paleogeography of an archaeological site, Brian has been honoured by having the location named the "Phillips Site".

Fred Dean retired Research Technician, is a consummate "amateur" professional with an insatiable curiosity and tenacious drive for knowledge. His interests include a field of science he calls "geofisheries"—a field of study within biology/biogeography. Fred's interests specifically address the Brook Trout (Salvelinus fontinalis) aquatic habitat in northwestern Ontario: the groundwater needed for their existence and the geology that affects it. Fred's other major field of interest is, of course, glacial history of the Lakehead area. He has published a number of articles on the subject. His research led to a much more thorough understanding of the chronology of the deglaciation and the subsequent alterations to the hydrology of northwestern Ontario and northern Minnesota.

Labrador and the Northern Peninsula of Newfoundland: Foreshore and Flanking Geomorphology

7

Norm Catto

Abstract

The coastline of Labrador and the Northern Peninsula of Newfoundland displays a variety of marine, glacial, aeolian, karst, tectonic, and other geomorphic features. Structural geology and lithology underpin landscapes from Killiniq to the Strait of Belle Isle. In central and southern Labrador, extending to the northernmost Northern Peninsula, glacial features are related to MIS 2 Laurentide glaciation. Local glaciation influenced the distinctive landscapes of Torngat Mountains National Park, and Mealy Mountains (Akami-Uapishk^{U}; KakKasuak) National Park Reserve. Postglacial relative sea levels vary from less than 20 m to more than 170 m asl. Relative sea level history varies from falling through rising episodes, reflecting differing glaci-isostatic regimes. Currently, relative sea level is falling along Lake Melville, but rising in central and southern Labrador and along the Northern Peninsula. Since the initial occupation in the early Holocene, through indigenous settlement and Viking visits, to modern times, geomorphology has been a dominant influence on human activity along the coastline.

Keywords

Labrador • Northern Peninsula • Torngat Mountains • Mealy Mountains • L'anse-aux-Meadows • Coastal geomorphology • Glacial landscapes

7.1 Introduction

A north to south journey along the northeastern margin of North America passes more than 5000 km of shorelines of Labrador and northernmost Newfoundland (Fig. 7.1). Inuit residents and their ancestors have utilized the geomorphic resources of the shorelines and adjacent inland tundra flanks of Nunavik and Nunatsiavut since initial settlement in the mid-Holocene. The coastal taiga regions of Nitassinan and NunatuKavut supported the Innu and NunatuKavut peoples over the same period. In the far south, the Newfoundland coast of the Strait of Belle Isle was home to the Beothuk until the 1700s, flanking the Mi'kmaq traditional territory of Mi'kma'ki along the Gulf of St. Lawrence. These regions, identified with Markland and Vinland in the Grænlendinga saga, were visited by Leif Eirikson and Viking colleagues c. 1000 CE.

These regions of eastern Canada have distinct Indigenous cultures and histories, climates, vegetation and fauna, geologies, tectonics, and modern human communities. The northern Labrador tundra landscapes of the Eastern Shield are marked by deep fjords and coastal cliffs developed in Canadian Shield granites and gneisses. From Natuashish south to Port Hope Simpson, tectonic activity and glaciation have acted on the forested-covered Churchill and Grenville provinces' Precambrian rock to produce complex coastlines with myriads of islands and embayments in Nitassinan and NunatuKavut. Along the Strait of Belle Isle, Tommotian and Palaeozoic carbonate-clastic sequences form the coastal plains of the easternmost sector of the Great Lakes-St. Lawrence Lowlands. glacial, glacifluvial, and glacimarine landscapes contribute sediment to shorelines, with glacitectonic and relative sea level regimes varying from ongoing rebound to inundation. Seasonal ice and periglacial activity are important influences throughout. Tundra and taiga vegetation both responds to and influences geomorphology. All of these differences influence the geomorphologies seen from the ocean's edge.

7.2 Nunatsiavut and Natuashish

Nunatsiavut includes the coastal terrain of northern and central Labrador. The Nunatsiavut Government is an Inuit regional government, with offices in Nain and Makkovik.

N. Catto (✉)
Department of Geography, Memorial University of Newfoundland, St. John's, NL A1B3X9, Canada
e-mail: ncatto@mun.ca

© Springer Nature Switzerland AG 2020

O. Slaymaker and N. Catto (eds.), *Landscapes and Landforms of Eastern Canada*, World Geomorphological Landscapes, https://doi.org/10.1007/978-3-030-35137-3_7

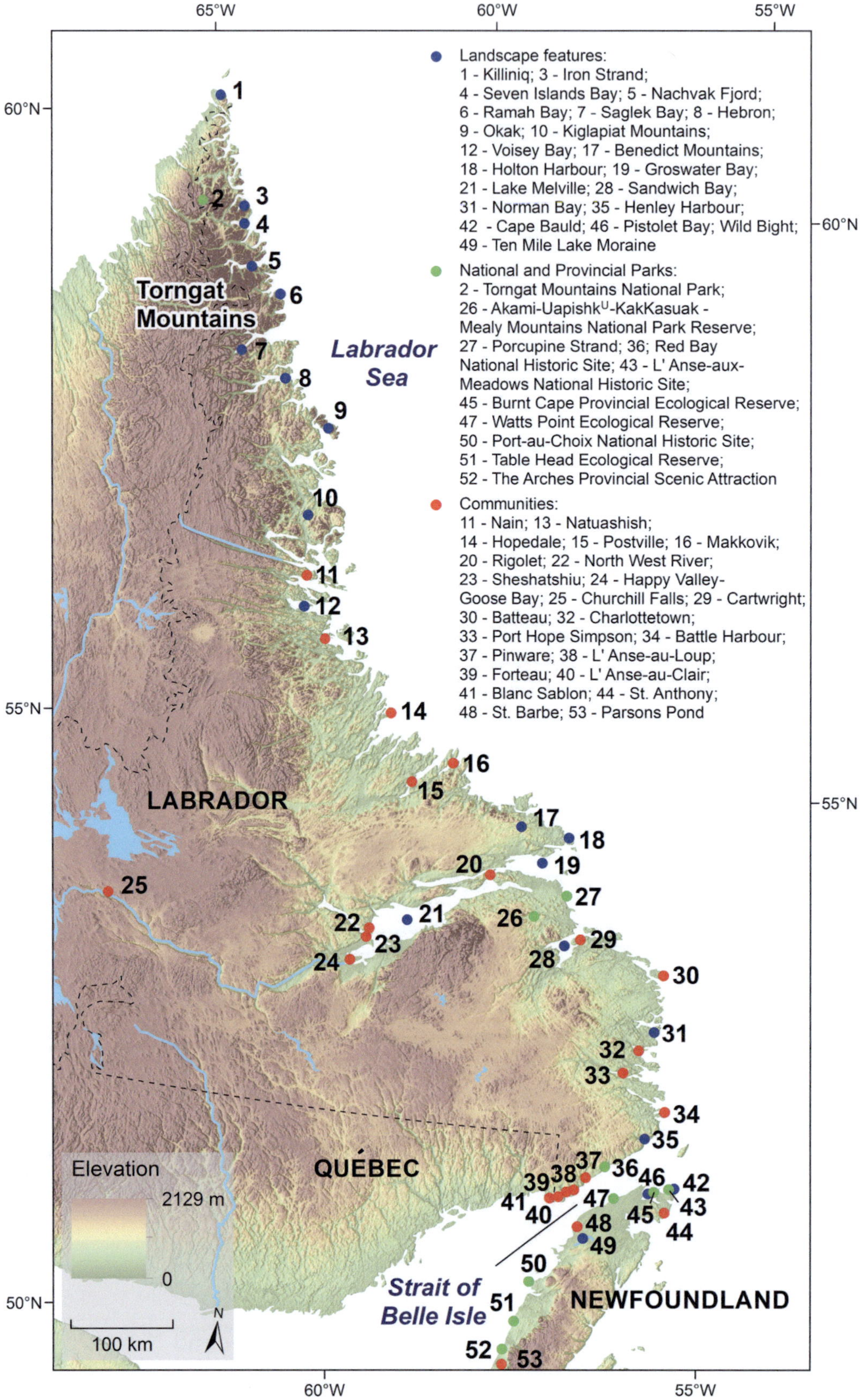

Fig. 7.1 Labrador and Northern Peninsula of Newfoundland, showing locations discussed in the text. *Cartography* D. Kovanen

Politically, Nunatsiavut is part of Newfoundland and Labrador, with the government having responsibility over many facets of health, education, culture and language, justice, and community matters. Torngat Mountains National Park encompasses the northern part of Nunatsiavut.

The Mushuau Innu community of Natuashish, located 75 km SE of Nain and 75 km NW of Hopedale, is part of the Innu Nation. Politically and culturally, the Innu Nation and Nunatsiavut are separate entities, with overlapping land claims. The Innu-aimun term "Nitassinan" includes the land of all boreal Labrador, and adjacent areas of Québec.

Nunavik (formerly Nouveau Québec; Ungava) includes all land and water north of 55° N latitude in Québec. Nunavik is an Inuit ancestral homeland, which currently is part of Québec. Ongoing negotiations are aimed at establishing Inuit self-government equivalent to that existing in adjacent Nunatsiavut.

7.2.1 Torngat Mountains

The Torngat Mountains form a series of rugged northwest–southeast ranges, cut by deep east–west fjords (Fig. 7.2). The northern part [west of Hebron north to Killiniq (Cape Chidley)] is more rugged than the southern (north from west of Nain). Fjords extend up to 15 km inland from the open coast, and main valleys dissect the ranges, extending across the drainage divide into Nunavik. Relief along the coastline is approximately 1200 m, and individual peaks rise above 1500 m. The highest point in the range is Mount Caubvick/Mont d'Iberville on the Labrador—Québec border, at an elevation of 1722 m asl. This peak has two "official" names, depending upon which side of the political border the observer stands on.

The Torngats lie within the Saglek Block of the Nain geological province (Wardle et al. 1997). Most of the Saglek Block rocks are Eoarchean to Paleoarchean gneisses (3800–3600 Ma). Additional metamorphism c. 2800 Ma further modified many of the Saglek Block rocks. The Saglek Block is cut by numerous faults, generally trending NNW–SSE. Most of the fjords cut across the structural grain of the fault systems.

Notable younger rock units within the Saglek Block include the Ramah sandstones, shales, and chert beds (approximately 2000 Ma); and layered intrusions of basic norite, gabbro, anorthosite, and other igneous rocks assigned to the Nain Plutonic Suite (1322–1305 Ma), exposed in the Kiglapiat Mountains and in scattered outcrops southwestward to the head of Voisey's Bay (Wardle et al. 1997). The chert units of the Ramah Group have been used for the manufacture of artefacts throughout the human occupation of Natuashish (Stopp 2016). The anorthosite units, formed almost exclusively of labradorite plagioclase feldspar, are used as facing stone and ornamental flooring in buildings, and are also valuable for jewellery production. The layered intrusions of the Nain Plutonic Suite contain nickel, actively exploited at Voisey's Bay.

The summits of most peaks in the Torngats are flat-topped or gently slope towards the coastline. North of Nachvak Fjord, four distinctive erosional surfaces can be recognized (Clark 1988): Torngat (highest and oldest), Komaktorvik, Koroksoak, and Saglek (lowest and youngest). South of Saglek Bay, the Torngat surface is not present, and the Komaktorvik surface is also absent south of Natuashish. The Torngat surface shows no signs of glaciation, and may have been subaerially exposed since the end of the Cretaceous.

Each of the three younger surfaces contains striated areas, erratic boulders derived from Nunavik, and glacial landforms, representing distinct glacial events. Ice accumulated in Nunavik and moved eastward towards the coast through the major valleys and fjords. The limited evidence available (Clark 1988) suggests that mountain glaciers did expand, but that their advances were not necessarily synchronous with those of the continental ice from Nunavik. The older two glaciations pre-date MIS 5.

The Saglek glaciation is considered to represent the MIS 2 Last Glacial Maximum advance. The Saglek glaciers extended to the coast along the major fjords, but the intervening coastlines were not ice-covered. Mountain glaciers were topographically confined (Fig. 7.3). The limited extent of the MIS 2 glaciers prevented significant fjordal erosion north of Saglek Bay. Former fjords are plugged on the seaward side by accumulations of glacial sediment, as at Iron Strand. Deglaciation began 14,000 years ago, and the last glacial ice had left the coastline no later than 11,000 years ago.

Currently, about 70 active glaciers are present in the Torngat Mountains (see Way et al. 2014). Most are less than 1 km long and are retreating (Way et al. 2015; Barrand et al. 2017). Rock glaciers are also present on the flanks of some of the higher summits.

Elevated sea levels were recorded in the fjords as the ice withdrew. The maximum limit reached by the sea declined from south to north, as the distance from the major Laurentide glacial sources increased. Maximum marine limit is approximately 90 m near Hopedale, 50 m at Saglek Fjord, and 17 m at Killiniq (Klassen et al. 1992).

As the glaciers withdrew from Nunatsiavut, Nunavik remained isostatically depressed. River systems drained to the west, following the isostatically controlled topographic slope, and were impounded against the retreating ice in two glacial lakes, Naskaupi and McLean. These lakes were short-lived, lasting for a few hundred years (Clark and Fitzhugh 1990; Dubé-Loubert et al. 2018). By 8,300 years ago, the glaciers had completely withdrawn from

Fig. 7.2 Torngat Mountains National Park. **a** West-east fjordal valleys, such as Nachvak Fjord, are carved into Eoarchaean to Paleoarchaean gneisses marked by flat-topped erosional surfaces. **b** Coastal cliffs display complex tectonic deformation of Saglek Block units

Nunatsiavut, and the George River drained north to Ungava Bay, into the postglacial d'Iberville Sea (Pienitz et al. 1991). Retreat continued rapidly throughout the early Holocene, and all of Labrador was deglaciated shortly after 7,500 years ago.

7.2.2 Vegetation

Northernmost Nunatsiavut (north of Okak) and summits above 900 m in the Torngat Mountains have tundra vegetation and terrain features (Fig. 7.4). The tundra area represents less than 5% of the land mass of Labrador.

Woody tundra shrubs include *Salix* (willow) and *Betula glandulosa* (dwarf birch), which commonly do not exceed 60 cm in height but may survive more than 100 years. *Salix* leaves were important supplements to traditional diets, as they are rich in Vitamin C. Other constituents of the tundra assemblage include *Empetrum nigrum* (crowberry), *Oxyria* (sorrel), *Polygonum* (knotweed), *Epilobium latifolium* (Broad-leaved fireweed), *Pyrola* (wintergreen), *Rhododendron decumbens* (Labrador tea), *Arctostaphylous* (bearberry), *Potentilla* (cinquefoil), *Eriophorum* (cottongrass), *Vaccinium vitis-idaea* (partridgeberry), *Dryas integrifolia* (mountain avens), *Azalea* (*loiseleuria*), *Saxifraga oppositifolia* (Purple saxifrage), and *Rhododendron lapponicum*

Fig. 7.3 Glacial features in Torngat Mountains National Park include **a** cubic parabolic troughs in trunk valleys, **b** truncated spurs and horns, with fan deltas at fjord mouths; and **c** small cirque glaciers currently undergoing rapid ablation

Fig. 7.4 Tundra vegetation developed on thin static and turbic cryosols overlying lithalsas and cryogenic mounds includes heath species, *Eriophorum*, *Dryas integrifolia*, and *Rhododendron lapponicum*, with isolated *Salix* and *Betula glandulosa* shrubs

(Lapland rosebay). Many of the tundra plants are edible, and some are of great medicinal value, especially sorrel, which prevents scurvy and was known to European visitors as "arctic salad". Soils in the tundra regions include both turbic and static cryosols.

The shore communities contain salt-tolerant plants which can endure repetitive freeze-thaw cycles, and prefer gravelly substrates. The assemblage includes *Diapensia lapponica, Armeria maritima* (thrift), *Arenaria* (sandwort), and *Cochlearia* (scurvy grass). The bright orange lichen *Caloplaca* is present in areas where birds nest and provide ample nourishment through their excrement to the lichens. Soils are poorly developed regosols.

7.2.3 Coastal Features and Communities

Garnet, ilmenite, magnetite, and sillimanite reworked from gneisses form beach placer deposits along Iron Strand and Seven Islands Bay (Fig. 7.5). At Iron Strand, promontories interfere with waves from the northeast. To the south of Helga River, shore-parallel transport is southeastward. To the north of Helga River, net littoral transport is northerly. Both segments of the beach are swash-aligned and shore-parallel. Iron Strand is a relatively low energy system, protected from winter storm events by seasonal ice, with abundant sediment flux. The system is transitional, tending towards dissipative, with shallow, gently sloping bathymetry, dominated by shore-parallel transport in independent cells. The geomorphology is controlled by littoral drift.

Seven Islands Beach is similar to Iron Strand in most respects, although sediment movement appears to be less confined, and offshore dispersion and southward movement of sediment towards Brownell Point is occurring. The transitional nature of the beach suggests that periodic reworking by shore-normal transport occurs, causing the beach to narrow over time. Taiga vegetation and northern boreal (Köppen-Geiger Dfc) climate marks the Labrador coastline south of Seven Islands Bay.

Hebron, established in 1830 by Moravian missionaries, consisted of approximately 20 houses along the southwest side of Kangershutsoak Bay, protected from southwesterly winds by hills up to 180 m high and from northeasterly winds by Cape Nuvotannak (Fig. 7.6). The population was entirely Inuit, aside from the Moravian missionaries and (later) government employees. The site was chosen by the Moravians because it offered low-lying, flat, sheltered ground for construction. The community was built on marine terraces, formed under higher sea levels (approximately 50 m) following deglaciation. Navigation and anchorage of sea-going vessels was facilitated by the location of the port in the sheltered lee of the Dog Islands, linked to the mainland by tidal flats and tombolos. Cape Nuvotannak acted to deflect marine ice from the community, limiting damage due to ice shove in the later autumn and spring. Hebron was abandoned in 1959, and has been declared a National Historic Site. Lithalsa is well-developed in the Hebron area.

Nain was established as a Moravian mission in 1771, the site chosen primarily for ease of access by marine vessels (Fig. 7.7). Located on the northwestern side of Unity Bay,

Fig. 7.5 Iron Strand is characterized by density-sorted concentrations of garnet and ilmenite, with associated quartz and feldspar. **a** Beach placer deposits are covered with patchy quartz-dominated aeolian sand. Net sediment transport is from south (background) to north. **b** Differential sorting is evident in both beach foreface (left) and channel-mouth (right) sediments. **c** Stone rings on elevated marine terrace, 8–10 m asl, indicate former locations of temporary hide tent structures

Fig. 7.6 Hebron Moravian mission site. **a** Buildings were constructed on thin marine gravel deposits overlying bedrock, interspersed with fibric peat. The site was protected from northeast winds by Cape Nuvotannak (background). **b** Silhouetted roche moutonées indicate glacial flow towards the northeast

and flanked to the south by a boulder tidal flat, Nain is directly open to the Labrador Sea through Strathcona Run. Although it is susceptible to avalanches from the >200 m high summits to the northwest, the southeasterly exposure was preferred for settlement.

At the western end of Sango Bay, Natuashish lies a broad marine terrace, with abundant level ground for construction. Freshwater sources are provided by Sango and several adjacent ponds, and by the wandering river to the south of the site. Sango Brook is a tidally dominated system, with brackish water extending inland to 2 m asl during high tides.

At Hopedale, relative sea level is falling by approximately 1 mm each year, exposing new coastal plains and scattered islets and skerries offshore. Kretschmer Island is now linked to the mainland by a tombolo exposed as a result of the gradual decline in relative sea level. To the south, a larger tombolo forms the north shore of Tooktoosner Bay. Hopedale is built on the flat ground formed by former marine terraces, extending up to 20 m above modern sea level. The community is located on the northern side of Hopedale Harbour, in the area least affected by tidal flats and seasonal ice shove.

Fig. 7.7 Nain. **a** Flanked by a boulder tidal flat, Nain is located on the northwestern side of Unity Bay. Avalanches are common in the area, claiming one life in January 2019. **b** Glacially eroded terrain indicates ice flow towards the Labrador Sea to the northeast

The area of Kaipokok Bay, an Inuktitut term meaning "frothy water", has attracted attention from people interested in marine fishery resources since the 1300s. Postville is located on the northeast side of Kaipokok Bay, built on a spit and raised marine beach.

Makkovik is situated on the west side of Makkovik Harbour. The dogleg configuration of the bay ensures that Makkovik is isolated from storm activity in the Labrador Sea.

7.2.4 Sea Ice and Icebergs

Sea ice is ubiquitous along the Labrador coast (Fig. 7.8). Off northern Labrador, the ice reaches its maximum thickness and its maximum extent seaward during late January to early February, extending more than 1000 km seaward of Nain and Hebron. For both terrestrial animals and humans, this expanse of sea ice provides additional terrain for movement.

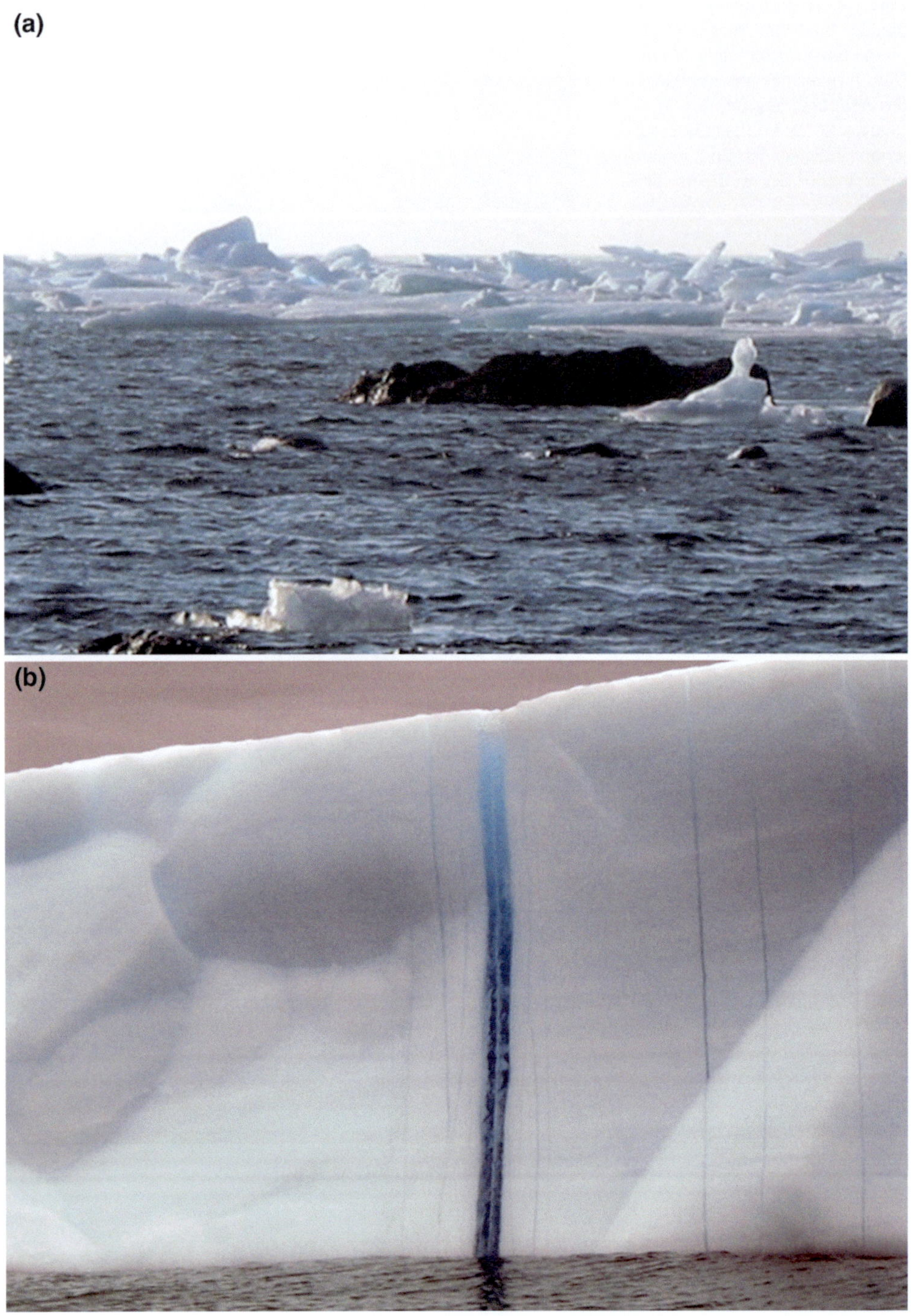

Fig. 7.8 Sea Ice and icebergs are geomorphic agents, tourist attractions, and influences on human life along the Labrador and northern Newfoundland coastline. **a** Wind-displaced pack and broken ice results in ice shove along the coastline, in addition to restricting coastal modification by winter storm waves. **b** Vertically oriented foliations indicate that this floating iceberg has flipped over 90°. Less than 10% of the icebergs which enter Labrador waters offshore of Killiniq reach the Strait of Belle Isle. *Photographs* courtesy of Gina Noordhof

During spring, recession of the seaward margins of ice in the north corresponds with the expansion of sea ice cover to the south and southeast. In mid- to late-March, ice commonly extends more than 800 km locally toward the east and northeast offshore of Hopedale and Makkovik. By early May, ice may extend farther offshore of Makkovik than it does to the east of Nain or Hebron.

Sea ice gradually recedes throughout the late spring, generally leaving the Strait of Belle Isle in late May or early June. In mid-May in a typical year, the ice extends 600 km seaward of Hebron, Nain, and Makkovik, 400 km seaward of Cartwright, and 200 km seaward of Battle Harbour. Open leads (*polyna; ashkui*) develop along the shoreline. Outflow of the spring freshet from the Churchill River and Lake Melville results in the formation of a tongue of sea ice extending southeastward from the main ice limit into the open Labrador Sea. Offshore sea surface temperatures vary from 1 to 3 °C.

By late June in most years, the sea ice has receded northwards from the coastline. Ice typically leaves the Killiniq area in early July, opening Hudson Strait for navigation by vessels plying the Churchill (Manitoba)—Europe grain shipping route. At this time, the sea surface temperature at Killiniq approximates 1 °C. Along the coastline, the warmest waters (>6 °C in July) are found from Groswater Bay south to Norman Bay. Sea surface temperatures generally decrease northwards, although the coldest surface water south of Killiniq is located offshore of Cape Kiglapit, averaging less than 2 °C in July. Sea surface temperatures further offshore (east of 58°W) commonly exceed 8 °C.

Ice expands southwards to close off Hudson Strait to navigation in September. By October, all areas north of Groswater Bay have sea surface temperatures less than 2 °C. The surface temperature in the Strait of Belle Isle is approximately 5–6 °C.

Icebergs are present throughout the year in the open Labrador Sea, and drift southward along the coast (depending on the fluctuations of the Labrador current) during the summer. Almost all icebergs are confined to the area of the Labrador Marginal Trough. Since 1990, iceberg numbers reaching Killiniq have varied between 1800 and 2000 annually. All of the icebergs are produced by calving along the margins of Baffin Bay and Davis Strait, predominantly from Kalaallit Nunaat; none of the Torngat glaciers reach tidewater. The icebergs begin to ablate rapidly once they move southward from Hudson Strait. During the peak iceberg transit season (typically mid-July), the sea surface temperature off Hopedale is 6 °C, and off Norman Bay it can reach 10 °C. Under these conditions, the icebergs ablate rapidly. An iceberg 40 m high above the water line and 100 m long (2,000,000 m^3, 10 million tonnes) will completely ablate in 24 days in 2 °C water, and requires only 15 days in 4 °C water. Such large icebergs make up approximately 20% of the iceberg population off Killiniq, about 10% off Makkovik, and about 5–6% at the mouth of the Strait of Belle Isle. Smaller fragments ("growlers") increase in proportion southwards as the larger icebergs disintegrate, but an average "growler" will endure for less than 5 days in 4 °C water. Less than 10% of the icebergs which enter Labrador waters offshore of Killiniq succeed in running the gauntlet to the Strait of Belle Isle.

7.3 Southern Labrador Coast: Southern Nunatsiavut and NunatuKavut

Southward from Makkovik, the southern Labrador coast extends to the mouth of the Strait of Belle Isle, including southern Nunatsiavut and NunatuKavut, the ancestral land of the NunatuKavut people. The Churchill (Grand) River discharges into the tidal Lake Melville estuary, which connects through The Narrows to Groswater Bay. Mealy Mountains (Akami-Uapishk^{U}; KakKasuak) National Park Reserve flanks Lake Melville to the southeast.

7.3.1 Lake Melville Estuary

Central Labrador and Lake Melville were glaciated by Laurentide ice from western Labrador, flowing northeastward along the Churchill and other major valleys (Klassen et al. 1992). As much as 2.5 km of ice covered Happy Valley-Goose Bay during the MIS 2 Last Glacial Maximum. Rögen moraines, drumlins, and flutings are preserved on the northern side of Lake Melville, and on the Churchill Plateau (Fig. 7.9).

Following deglaciation, the valley was flooded by marine waters (Figs. 7.10 and 7.11). Approximately 11,000 years ago, the ice front stood along a line of moraine ridges exposed from the Mulligan and Sebaskachu rivers southwestward to North West River and Sheshatshiu, forming the Sebaskachu-Little Drunken Moraine complex. South of the Churchill River, segments of moraine are preserved along the northwestern flank of the Mealy Mountains, and along the Little Drunken River. Near North West River, the moraine complex consists of up to four north–south aligned ridges. Individual ridges are up to 80 m high, and up to 2 km wide. The complex partially isolates Grand Lake from Lake Melville, and is more than 10 km wide. The Sebaskachu-Little Drunken moraine complex appears to represent a major pause in progressive retreat during late MIS 2, and may be correlated to the Younger Dryas readvance.

During subsequent recession, glaci-isostatic depression allowed marine waters to reach 135 m asl at Gull Island Rapids, along the Churchill River valley, c. 9 ka. Former marine beaches southwest of Goose Bay were reworked by southwest winds to form parabolic dunes, and other marine sediments were reworked by the Churchill and Goose Rivers. Ongoing rebound produced a series of marine beaches along the western shore of Lake Melville, with more than 30 documented positions. Progressively lower sea levels are chronologically constrained by dated archaeological materials, and by dated fragments of wood in deposits of river sediment. Approximately 3000 years ago, the sea stood at 18 m above its present position along the Churchill River, 12 m asl at Sheshatshiu, and 6 m asl near Rigolet. At present, sea level continues to fall in Lake Melville, approximately 1 mm per year.

The waters of Lake Melville are brackish (5–15 psu), due to the fresh water input from the 856 km-long Churchill River (Fig. 7.12). The mean discharge of the river at Happy Valley-Goose Bay is 1900 m^3/s, reflecting modifications resulting from regulation at the Churchill Falls dam. Freshet

Fig. 7.9 South of Hopedale, glacial features including flutings (**a**), ribbed moraines (**b**), and esker complexes (**c**) indicate MIS 2 Laurentide ice flow towards the northeast across central Labrador

Fig. 7.10 The Sebaskachu River meanders across brackish-water silt deposited in the glaci-isostatically depressed terrain north of Lake Melville. Following deglaciation c. 11.5 ka, relative sea level has successively fallen from 135 m asl, and continues to decline at ~1 mm/y. At least six generations of meanders have developed in the past 3000 years in the *Picea mariana*-dominated taiga forested valley

(spring runoff) values have been reduced as a result of dam construction and impoundment of flow in the Smallwood Reservoir, and the seasonal low flow values (usually in early September) have been increased. At present, turbid underflows during the spring runoff period bring cold, dense, sediment-laden waters to the bottom of the Lake Melville estuary.

Along the coastline of Lake Melville, the combination of low tidal ranges, limited influx from the open ocean, and dominant southwesterly winds largely preclude the development of many types of coastal features. In addition to the modern beaches, the most prominent features are tidal platforms and marshes. Marshes extend to elevations of 2 m asl, and inactive features are associated with higher strandlines to ±10 m asl.

7.3.2 Akami-Uapishk^{U}—KakKasuak—Mealy Mountains National Park Reserve

The terrain of Labrador's newest national park reserve is referred to as Akami-Uapishk^{U} in Innu-aimun, KakKasuak in Inuktitut, and Mealy Mountains in English. The park is flanked by Lake Melville on the northwest and the Trans-Labrador Highway on the west and south. Along Trumore Bay, Porcupine Strand forms the coastline of the National Park reserve.

The mountain landscape is rugged, marked by irregularly aligned short, high-gradient streams and steep cliffs (Fig. 7.13). The maximum elevations of the summits reach 1190 m. The Mealy Mountains are composed of resistant granites and gneisses, and extremely resistant anorthosites (Wardle et al. 1997). Irregular knobs of bedrock mark the terrain to the west of the Kenamu River, with local relief to 300 m. Streams are short and irregularly aligned, with gradients locally in excess of 100 m/km, and numerous small waterfalls and rapids. Meltwater channels are short and irregularly aligned.

The Kenamu River is the only stream that traverses the spine of the Mealy Mountains. It flows northward in a broad (up to 3 km wide) valley, with a maximum gradient of 0.006 in its steepest reach where it enters the Mealy Mountains. Tributaries to the Kenamu are short and steep. The valley of the Kenamu River is lined with glacimarine silt and clay below 100 m asl. Glacimarine sediments are very susceptible to disturbance and slope failure.

The eastern part of the Mealy Mountains is marked by the highest elevations, with local areas of permafrost (generally above 900 m asl). More than 20 cirques, carved by local glaciers, are present in the headwaters of the Etagulet and English Rivers, with floor elevations at 600 m or greater. Cirques facing northwards have linear or undulatory slopes, whereas tarns and riegels are common in cirques developed on the southern and eastern flanks of the Mealy Mountains. No active glaciers are present.

The summits of the Mealy Mountains support a tundra vegetation assemblage. Tree species are absent, and the largest plants are scrubby dwarf birches, willows, and green

Fig. 7.11 Brackish waters extended to 135 m asl at Gull Island Rapids along the Churchill River. **a** Silty clay deposits show sequences of deformations associated with initial deposition, submarine flow directed seaward downvalley, and recent slope failure from the north side of the modern Churchill River valley. White bar = 15 cm. **b** Although the Churchill valley deposits are not thixotropic, slope failures do occur. This moderate-viscosity debris flow in brackish-water silty clay was triggered by a combination of precipitation, undercutting by the Churchill River, and insect infestation of spruce. The resulting landform and disordered diamicton deposit indicates loss of internal cohesion during debris flow

alders. Northern species, such as *Rhododendron decumbens* (northern Labrador Tea) replace their southern equivalents. The tundra vegetation is dominated by Cyperaceae (sedges), *Sphagnum* moss, *Cladonia* lichens, and Poaceae (grasses). Other constituents of the tundra assemblage include *Azalea loiseleuria* (azalea), *Dryas integrifolia* (mountain avens), *Empetrum nigrum* (crowberry), *Rhododendron decumbens* (northern Labrador tea), *Lupinus sp*. (lupin), *Polygonum* (knotweed), *Potentilla* (cinquefoil), *Pyrola* (wintergreen), *Rhododendron lapponicum* (Lapland rosebay), *Silene* (campion), and *Vaccinium vitis-idaea* (partridgeberry). Static and turbic cryosols and regosols have developed in the tundra environment.

The southern limit of the Mealy Mountains is transitional to the gently rolling and undulatory terrain of the Churchill Upland. In contrast, the northern margin is abrupt, marked by a rampart up to 350 m high of steeply sloping anorthosite and other igneous rocks. With the exception of the Kenamu and Kenemich Rivers, streams have not cut valleys through this rampart, and small waterfalls cascade over the lip.

Fig. 7.12 Lake Melville. **a** Glacial sediments and overlying brackish-water sands and silts are exposed along the north shore, near Mokami Hill. Intervening coastal flats display ice-shove ridges. **b** Elevated strandlines are marked by alternating *Picea* (dry sediment, predominantly sand) and *Cladonia* lichen (pale yellow terrain, fibric peat). Low-lying areas support heathland and Cyperaceae assemblages, without *Picea* or *Cladonia*

The interior of the Mealy Mountains is dissected by streams that have developed along a well-defined jointing system. The longest stream reaches, such as the upper Kenemich, follow WSW–ENE joints. These joints are cut by north-to-south oriented joints, which also have been exploited by streams, generally shorter and steeper than the larger rivers. The drainage pattern in the core of the Mealy Mountains between the Kenamu and Etagaulet rivers is controlled entirely by bedrock morphology, with glacial erosion and deposition having no appreciable role.

Porcupine Strand, along the shore of Trumore Bay, may have been the original "Wunderstrand" site mentioned in the Norse sagas. The white medium-grained sand beaches extend over 35 km (Fig. 7.14).

Porcupine Strand is dominated by southeastwards littoral drift. Energy levels vary from moderate to high. Although the coastline is sheltered on the lee side from the prevailing westerly winds, strong coastal breezes blow onshore. Winter ice shove acts to move sand landward on the beach, forming ramparts which then gradually are lowered and flattened by

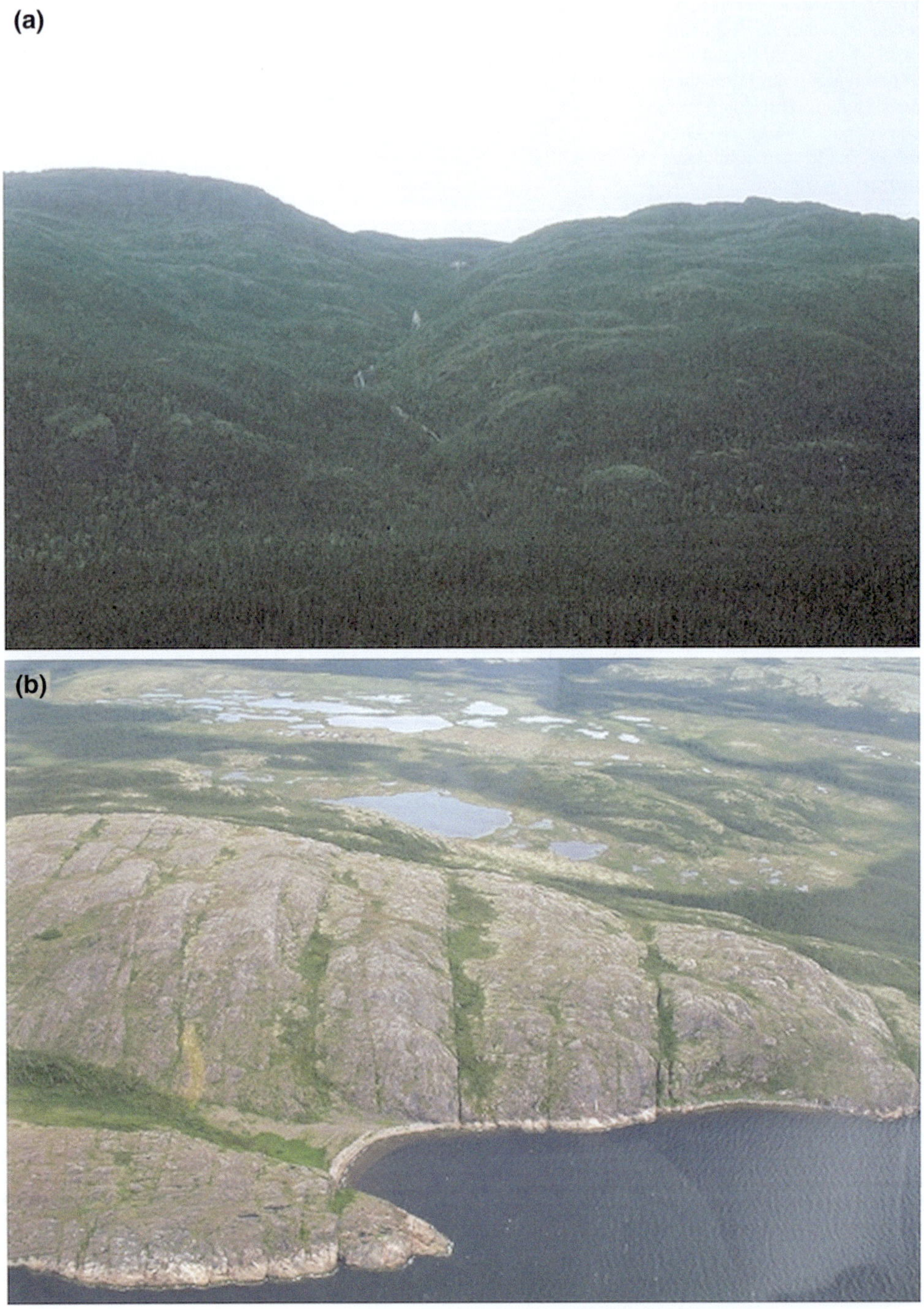

Fig. 7.13 Mealy Mountains (Akami-Uapishk[U]; KakKasuak) National Park Reserve. **a** Summits are flat to irregularly undulating, with scattered glacially transported boulders and thin regolith over anorthosite bedrock. **b** Steep cliffs form the southern shoreline of Lake Melville. *Photograph* b courtesy of Eric Watton

the prevalent summer littoral drift. Isolated bedrock knobs divide the strand into segments, and interrupt littoral drift during some wind conditions.

The strand is marked by broad, extremely gently sloping beaches composed of medium-grained, well-sorted sand. The beach slopes seaward with a planar to very gently concave profile during the summer months, with a modal slope of 2° or less. Swash and backwash features dominate, and cuspate features are generally not present. Transport of sediment during swash is both by suspension and by movement of diffuse sand sheets, and ripples are rarely produced except during phases of lower energy.

On the beach, contacts between individual sand strata are defined by magnetite, ilmenite, and garnet. Erosional scoured surfaces containing adhesion ripples and adhesion warts, micro-structures formed by wind blowing across the

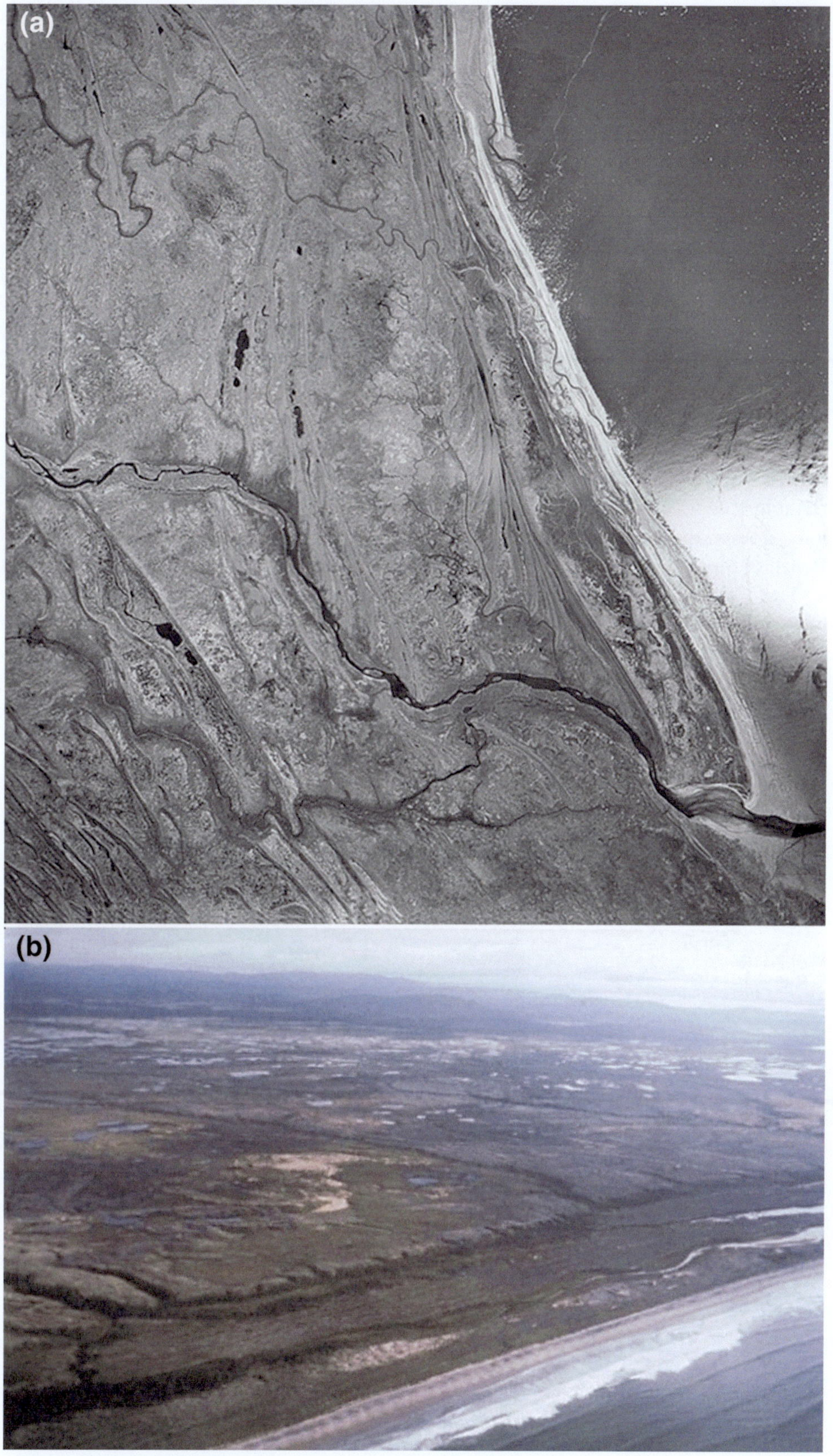

Fig. 7.14 Porcupine Strand, Trumore Bay. **a** Multiple marine strandlines indicate progressive fall in relative sea level from ~150 m asl. (National Air Photo Library). **b** The shoreline is dominated by a dissipative regime and southeastward littoral drift. Ephemeral foredunes develop during summers. *Photograph* b courtesy of Joyce MacPherson

wet sand surface, rain pits, and niveo-aeolian features, formed by the essentially simultaneous deposition of snow and aeolian sand, are present.

7.3.3 Southern Labrador Coast

The southern Labrador coast is dominated by the bedrock structure and lithology of the Grenville Province, modified by MIS 2 glaciation (Klassen et al. 1992; Wardle et al. 1997). In Labrador, the Grenville Front marks the position where the Grenville rocks were thrust northwards to collide with the pre-existing Makkovik, Nain, Churchill, and Superior structural provinces. The complex of thrusted and buckled terranes to the south of the Grenville Front, together with associated igneous orogenic plutons, is referred to as the Exterior Thrust Belt. The 1650 Ma Trans-Labrador Batholith, containing granite, granodiorite, and syenite plutons, forms the shoreline from Holton Harbour to the Benedict Mountains.

The Cartwright-Sandwich Bay coastline is marked by broad tidally influenced rock platforms, extending up to 1 kilometre offshore (Fig. 7.15). During the summer and autumn months (June through November), the rock platforms act to buffet waves. During the winter, the platforms are scoured by drifting pack ice and swish, pushed onshore by tides until freeze-up occurs. Renewed scouring accompanies spring break-up. In the mesotidal Cartwright area, sediment fluxes from either landward or seaward sources must remain low to keep the platforms exposed.

Seasonal ice activity acts to remove sediment and expose wide rock platforms. In some locations, the influence of seasonal ice shove results in the construction of a boulder rampart, or the emplacement of individual boulders perched on the rock substrate. Seasonal ice activity, however, cannot totally denude a rock platform that is routinely subjected to sediment influx from the adjacent land area. The impact of seasonal ice activity is directly related to tidal range, with greater effects evident in mesotidal areas, such as Cartwright. Sea ice must be driven aground, either by waves or tides, in order to be effective as an erosional agent.

The wide rock platform is most exposed during early to mid-spring (June). Small amounts of sediment cover can gradually accumulate throughout the summer months, and are removed during autumn storm events. The lateral extent of the exposed rock platforms has varied in accordance with the degree of ice shove activity since the 1950s. In years marked by large amounts of ice shove, the lower part of the platform is swept completely clean of sediment, and the boulders form several (discontinuous) lines across the upper

Fig. 7.15 Cartwright is developed on the northeast flank of Sandwich Bay. Broad tidally influenced rock platforms extend up to 1 km offshore (National Air Photo Library)

part representing different ice shove events. In years when ice shove activity was less or non-existent, no boulders may be present above mean high tide level.

In the areas above high tide level, the development of rock platforms is primarily of result the frost weathering during intervals when the rock is exposed. The rate of formation is controlled by the number of freeze-thaw cycles. Along Sandwich Bay, conditions are suitable for the formation of rock platforms at elevations ranging from mean low tidal position to several metres above present sea level. The amount of frost-induced erosion increases in the higher intertidal areas, where sea water only covers the rock for short periods each day at high tide. In contrast, abrasion by tidal action is most effective in the lower parts of the intertidal zone, where the rock is only exposed at low tide.

South of Cartwright, structural geology and metamorphic and igneous lithology dominate the shorelines from Norman Bay south to Henley Harbour. At Norman Bay, ponds, rivers, and coastal embayments form a grid-like pattern of northeast–southwest and southeast–northwest trending topography, the product of bedrock jointing and glacial erosion. Glacial erosional features indicate that the predominant ice flow direction was from the northwest towards the coastline. Local relief exceeds 150 m.

Mafic intrusive rocks associated with the Exterior Thrust Belt crop out in White Bear Arm, Charlottetown, and on Square Islands. At Charlottetown, the geological structure has created a "biscuit-pan" or "egg-box" topography, with intersecting east–west and NNE–SSW systems forming short valleys and diamond-shaped, isolated hills (Fig. 7.16). Seven Mile Pond is developed along the east–west structural trend. Glaciation has smoothed the northwestern flanks of the prominent summits. Rivers throughout the area are short cascade systems, largely developed directly on bedrock, and glacially modified "glint" lakes are common. Ultramafic and mafic rocks are less resistant and support diverse vegetation assemblages than do granite and granodiorite plutons (Port Hope Simpson, Williams Harbour) and granitic gneiss (Batteau).

Glaci-isostatic depression resulted in elevated maximum marine limits along the Labrador shore of 150 m asl west of Port Hope Simpson, located at the head of the fjordal Alexis estuary (McCuaig 2002). Influx of marine water is limited at the seaward margin by a bedrock obstruction, and is further slowed by the joint-controlled configuration. In addition to vertical gradients induced by salinity differences, transverse gradients are induced by bathymetry, with higher salinity along the northern side of the estuary. The prevailing southwesterly winds further accentuate the gradation, causing the southern side to be dominated by the Alexis River outflow and the northern side to be influenced by inflow from the Labrador Sea.

Fig. 7.16 Charlottetown, south shore of White Bear Arm. The southern Labrador Coast is dominated by intersecting east–west and NNE–SSW fault systems. Glacial flow towards the east-southeast has produced numerous roches mountonées and flyggbergs on the mafic intrusive Interior Thrust Belt rocks (National Air Photo Library)

The fringing marshes serve as sediment traps for sand and coarse silt. Much of the sediment impounded in these areas remains within the system, contributing to shoreline progradation. These areas provide important habitats for numerous species, including eels and crustaceans.

St. Lewis Inlet is the largest of a series of east-southeast to west-northwest aligned embayments, reflecting structural geology and glacial modification. Rögens mark the ragged shoreline of Fifty Islands Pond.

Battle Harbour was established on Battle Island, as a community aligned linearly from north to south along the eastern shore of Battle Harbour Tickle. Other houses were constructed on the western shore of the harbour, on Great Caribou Island. Great Caribou Island is cut by a series of south-southwest to north-northeast trending valleys developed along geological weaknesses, and subsequently eroded as sea level fell. The effect of falling sea level is apparent at Sand Cove and Cartridge Bight, where a tombolo of sand and gravel 500 m wide connects the two parts of Great Caribou Island. As sea level continues to fall along the southeastern Labrador coast, the tombolo is steadily becoming wider.

7.3.4 Vegetation

Coastal environments limit tree growth, with *Picea glauca* (white spruce) being commonly the only tree species. Krummholz (tuckamore) assemblages of low-lying white spruce occupy coastal settings where some soil development over glacimarine sediment or friable bedrock has occurred (Fig. 7.17). Tuckamore trees characteristically form a low, dense tangle of gnarled trunks and interlocking branches, aligned in response to the prevailing coastal wind. In many cases, the upper branches are shaped into waveforms. Although most white spruces in coastal Labrador do not exceed 3 m in height, they live up to 100 years. Many coastal areas are too exposed to permit any tree or shrub species to become established. Dwarf shrubs of *Salix* (willow), *Betula glandulosa* (tundra birch), *Betula borealis* (northern dwarf birch), *Betula pumila* (swamp birch), *Betula minor* (dwarf white birch), and *Alnus viridis crispa* (green alder) are also present.

The understory vegetation includes *Juniperus communis* (common juniper), *Empetrum nigrum* (crowberry), *Empetrum eamesii* (pink crowberry), *Cassiope hypnoides* (moss heather), *Linnaea borealis* (twinflower), *Diapensia lapponica, Viburnum edule* (squashberry), *Dryas integrifolia* (mountain avens), *Vaccinium vitis-idaea* (partridgeberry), *Vaccinium oxycoccus* (marshberry), *Rhododendron lapponicum* (lapland rosebay), *Arctostaphylous uva-ursi* (bearberry), *Rubus idaeus* (raspberry), *Rhododendron groenlandicum* (Labrador Tea), *Andromeda polifolia* (bog rosemary), *Myrica gale* (Sweet Gale), *Vaccinium boreale* (northern dwarf blueberry), *Vaccinium angustifolium* (blueberry), and *Vaccinium uliginosum* (bilberry).

7.4 Straits Region (Strait of Belle Isle)

The Straits region encompasses the terrain on the northern and southern shores of the Strait of Belle Isle. The Strait of Belle Isle is a northeastward extension of the Mecatina Trough and Esquiman Channel. It reaches a maximum depth in excess of 100 m off Point Amour and northeast of 56° W. The seaward extension of the trough is more than 200 m deep to the north of Belle Isle.

The Labrador shoreline extends from Henley Harbour to Blanc-Sablon (Québec). Small discontinuous patches of Ediacaran and Cambrian (Tommotian) sandstone and limestone overlie older rocks of the Pinware Terrane ($\sim$1500 Ma). Meso-Proterozoic igneous intrusive rocks of the Grenville province of the Canadian Shield are exposed along much of the north shore of the Strait of Belle Isle, including granite, syenite, monzonite, anorthosite, and gabbro plutons, with lesser amounts of gneiss.

In Newfoundland, the northern shoreline of the Northern Peninsula extends southwestward from Cape Bauld to Parsons Pond. The geology of the northernmost Northern Peninsula is dominated by mélanges of Iapetus Ocean rocks, including limestones, ophiolites, peridotites, and volcanic rocks (Knight 1977; Colman-Sadd et al. 1990). A major thrust fault parallels Highway 437, along a NW–SE orientation, although there is little surface topographic expression. Along the coastline to the west of the thrust fault, the terrain is underlain by the Ordovician Milan Arm mélange, consisting of limestone and shale with inclusions of volcanic rocks and peridotite. Offshore, notably at Trompe l'Oeil Point, Middle Ordovician shale and greywacke sandstone are exposed.

Initial human occupation of the area followed sometime after deglaciation, approximately 10 ka. The Straits region was initially occupied by "Maritime Archaic" people. A burial mound at L'Anse Amour containing the remains of a young Maritime Archaic male is dated to $\sim$7.5 ka. Approximately 500 years after the disappearance of the Maritime Archaic people from southern Labrador approximately 4,000 years ago, a second group of people referred to as the [Canadian] "Shield Archaic" occupied the area. These people, and their descendants, are the ancestors of today's Innu.

A second influx of people, the Sivullirmiut ("pre-Dorset" and "Dorset") also migrated approximately 3000 years ago into southern Labrador, spreading southward to occupy the entire Strait of Belle Isle coastline (Stopp 2016). Later, Taissumanialungmiut ("Thule") people moved

Fig. 7.17 Coastal krummholz conifers shaped by prevailing onshore winds are locally termed "tuckamore" in Newfoundland and Labrador. Tuckamore typically develop on gravel, coarse glacial diamicton, or directly over bedrock

southeastward from the area of Baffin Island c. 1350 CE. Although the Taissumanialungmiut did not occupy southernmost Labrador, they are the direct ancestors of the NunatuKavut, who were living in southernmost Labrador by 1600 CE.

7.4.1 Labrador Shoreline

7.4.1.1 Red Bay National Historic and UNESCO World Heritage Site

The topography surrounding Red Bay dictated the site's establishment on the eastern side of The Basin, protected from the open waters of the Strait of Belle Isle by the 30 m high Cox Hill on Saddle Island. Waves are further buffeted by the Twin Isles, located southeast of the community. The whale-processing operations required gently sloping terrain, provided by the bedrock-controlled platforms on the east side of Red Bay.

There is considerable debate amongst historians as to when Basque whalers reached Labrador: the earliest archival documents date to 1517. The Basques established a series of independent whaling stations along the coastline, the largest of which was the integrated factory operation at Red Bay ("Les Buttes"). The Basque whaling operations were expanded to Labrador in response to depletion of the whale stocks in the Bay of Biscay (beginning c. 1300), and the desire to harvest whales throughout the year. Gradual cooling of the northwestern North Atlantic during the Neoglacial (1400–1800 AD) acted to force whale stocks southward from Kalaallit Nunaat and Iceland, concentrating them in Labrador waters. Harvesting of predominantly bowhead whales greatly increased following 1545 (McLeod et al. 2008). More than 20 "tryworks" (oil and blubber processing stations) have been located, along with cooperages (barrel factories), and cutting-in stages (where the blubber was removed or "flensed" from the whales).

7.4.1.2 Pinware

Geomorphology influences the development of the communities of Pinware, L'Anse-au-Loup, Forteau, and L'Anse-au-Clair along the Straits. Pinware (for "le pied noir") is named for a foot-shaped 1500 Ma volcanic rock of the Pinware Terrane (Gower et al. 1997), at the mouth of Black Rock Brook. Ship Head, a 30 m high bedrock knob connected to the mainland by a gravel tombolo, protects the community from southeasterly and easterly winds, which are especially prevalent during autumn and winter and are enhanced during positive NAO years. The double tombolo connecting Ship Head to the mainland was formed by sediment transport from Pinware Harbour to the northwest and the southwest (Fig. 7.18).

Two cells of shore-parallel sediment movement are present. The southern cell, occupying most of the waterfront adjacent to the community of Pinware, is generally marked by southward movement of sediment from the harbour

Fig. 7.18 Pinware. **a** Ship Head (southwest) is connected to the mainland by a double tombolo. Directly to the north, southward movement of sediment feeds the tombolo. A northern cell moves sediment eastward to 1.5 km long barachoix at Pinware River Provincial Park (National Air Photo Library). **b** The north beach is sand-dominated, littoral-drift-aligned, generally low energy, and dissipative. *Photograph* b—courtesy Melanie Irvine, Geological Survey of Newfoundland and Labrador

towards Ship Head, feeding the tombolo. The northern cell, extending from the northern end of the community to the 1.5 km long barachoix (spit) at Pinware River Provincial Park, transports sediment northward to extend the barachoix to the east. The presence of the Pinware River barachoix protects the lagoon, lowering energy levels and minimizing storm exposure.

Immediately following deglaciation, the relative sea level was ~170 m higher than present (Grant 1989). The gradual decline from this height produced a succession of beach and aeolian deposits, which are the sediment sources facilitating the development of modern tombolos and barachoix. Under slowly changing relative sea level along a coastline with limited infrastructure and an abundant supply of sediment, barachoix and tombolos develop in synchronicity with the changing environment. If sediment influx to the coast is not curtailed, the coastal landforms will continue to develop in their present patterns. The position of the barachoix has developed in response to the prevalent storm wave, tidal, and ice regimes. The orientation of the barachoix is approximately parallel to the prevailing winds through the Strait of Belle Isle and to the net direction of ice drift through the strait. This orientation means that the barachoix is less subject to disruption by storm waves or ice shove directly at right angles to the beach. For the near future, the barachoix system at Pinware River Provincial Park will be similar to its present configuration, barring interruption of sediment influx.

The outlet of Pinware River from the lagoon is diverted to the far eastern bank as sediment is constantly supplied to the barachoix from the southwest. The outflow of the Pinware River maintains an open connection between the lagoon and the ocean, providing a suitable migration route for salmon, brook trout, and eels. The Pinware River is tidal in the lower reach, and longitudinal bars are shaped by both river and tidal currents.

7.4.2 Northern Peninsula

Striations, granite, and gneiss erratics derived from the Grenville Province indicate glaciation by Laurentide ice (Grant 1989, 1992). Laurentide ice merged with the Northern Peninsula glaciers to the south, near Ten Mile Lake, and extended south into the Gulf of St. Lawrence. Deglaciation of the coastline proceeded between c. 14 ka and c. 12.5 ka. Glaci-isostatic depression resulted in marine incursion, with elevations increasing from south to north, from 135 m above modern sea level at St. Barbe, 140 m at Watts Point, and approximately 150 m above modern sea level at Burnt Cape Ecological Reserve (Grant 1989). Isostatic recovery resulted in consistent sea level fall until c. 2.5 ka, followed by slow rise.

Relative rising of sea level is evident on the Great Northern Peninsula at Port-au-Choix (Bell et al. 2005; Smith et al. 2005); and along the Gulf of St. Lawrence coastline (Grant 1989; Daly et al. 2007). Currently, the rate of rise along Pistolet Bay and at L'Anse-aux-Meadows appears to be on the order of 1–2 mm per year (Catto 2006; Vasseur and Catto 2008), suggesting that relative sea level was approximately 1 m lower (shoreline 5–15 m seaward) during Leif Eirikson's visit. With increased storm activity evident since 2000, L'Anse-aux-Meadows could be vulnerable to storm surge, even under relatively stable relative sea level.

Saltwater intrusion into coastal aquifers is a local problem in the Northern Peninsula. At L'Anse-aux-Meadows, the influx of approximately 35,000 tourists during the Viking Millennium celebration between mid-July and mid-August 2000 caused the drilled well at the Norstead reenactment site to experience drawdown and salinization within one month. Ongoing salinization at L'Anse-aux-Meadows has effectively precluded establishment of other tourist-related businesses within the community of L'Anse-aux-Meadows. The increase in tourist usage of ground water has greatly accelerated the natural progress of salinization.

The modern Straits shoreline is dominated by erosion under the combined actions of sea ice and tidal action (Figs. 7.19 and 7.20). Tidal range along the central Northern Peninsula coast varies from 2.5 to 3.5 m. Offshore, a platform underlain by Ordovician rock and covered by gravel slopes towards the northwest at a gentle angle. The development of the rock platform offshore was primarily the result of frost weathering during intervals of lower sea level when the rock was exposed, rather than being the product of direct abrasion by waves, tides, or sea ice, comparable to the formation of the "strandflat" features along the Norwegian coast (Holtedahl 1998). The rock platform reflects incremental formation influenced by changing sea level. Good examples are present along the shoreline from St. Barbe south to Table Point Ecological Reserve and Bellburns.

Seasonal ice shove results in the formation of boulder ramparts along the shoreline of Pistolet Bay. Boulders pushed landward by sea ice form ridges up to 2 m above high tide level (Forbes and Taylor 1994). The development of the ramparts is most effective in areas with higher tidal ranges and relatively slow offshore currents, such as the Labrador current (estimated winter velocity ~0.5 km/h offshore of Belle Isle; Narayanan 1994).

Warmer winters in the southern Gulf of St. Lawrence, resulting in part from strong southwesterly winds, lead to restricted ice cover (Catto et al. 2002). However, southwesterly winds tend to push sea ice northeastwards along the Northern Peninsula, resulting in persistent ice cover north of St. Barbe, even during years when ice cover has been restricted or non-existent south of Bellburns.

Fig. 7.19 Ice shove features are common along the Gulf of St. Lawrence coast of the Northern Peninsula. **a** 5 m-high ice shove rampart, Port-au-Choix, March 2019. **b** Ice-shoved boulders in Pistolet Bay, exposed during low tide. Glacial faceting marks several granitic boulders derived from Grenville igneous units in southern Labrador, indicating transport by Laurentide MIS 2 glaciation. *Photographs* courtesy of Dominique Lavers

7.4.3 L'Anse-aux-Meadows

L'Anse aux Meadows (Fig. 7.21) is the only definitively authenticated Viking occupation site in North America, although many other candidates have been put forward. Archaeological evidence indicates that the site has been occupied previously on four separate occasions.

The palynological record from the area indicates that climatic deterioration commenced c. 2.5 ka. Patterned fen development began at this time. The late Holocene climatic deterioration coincides with the demise of the palaeo-Inuit/Innu culture at the site. The climate during the Norse occupation, as indicated by palynological data, was slightly milder than at present. Summer temperature was 1–2 °C higher (Henningsmoen 1985; Davis 1980, 1984, 1985; Bell et al. 2003; Bell and Renouf 2003), although the environment would have remained unsuitable for growing grapes. The palynological record shows no sign of human impact during the Norse occupation, suggesting that this was a short-lived event not marked by extensive modification of

Fig. 7.20 Coastal landscapes, Northern Peninsula.
a Ice-thrusted and wave-washed gravel-covered Ordovician limestone-shale rock platform, near St. Barbe. **b** *Ascophyllum* colonizes rock platforms in areas sheltered from pervasive ice shove. Associated sediment cover is sand to sandy gravel. Flowers Cove, 15 km north of St. Barbe. **c** Reflective shore-normal transport in more sheltered coves results in shallow cusps with alternating coarse-medium sand and granule-fine pebble gravel. Displaced kelp indicates wave limit ~12 h previously. Cook's Harbour, Pistolet Bay

Fig. 7.21 L'Anse-aux-Meadows National Historic Site, showing shoreline directly north of archaeological excavation. Relative sea level was approximately 1 m lower (shore 5–15 m seaward) c. 1000 CE. With increased storm activity evident since 2000, L'Anse-aux-Meadows could be vulnerable to storm surge, even under relatively stable relative sea level

the terrain around L'Anse-aux-Meadows, compatible with the saga record (e.g. Magnusson and Palsson 1982).

Identification of L'Anse-aux-Meadows as the spot of "Vinland" rests on its sailing distance from the long, gently sloping white sand beach of the "Wunderstrand" mentioned in the Grænlending saga. Porcupine Strand is considered by some archaeologists to be the "Wunderstrand", and hence L'Anse-aux-Meadows would be in the right geographical position.

"Vin" was (and is) commonly used in western Norway as a synonym for "farm" or "meadow". "Vinland" can be translated as "open meadow land" or "pastureland", in contrast to the boreal and taiga forest of coastal Labrador ("Markland"). Under this interpretation, "Vinland" carries no connotation, indicating what vegetation was growing at the site.

Today's coastal vegetation is dominated by tuckamore white spruce (*Picea glauca*). Exposed coastal bluffs and fringes of salt marshes are occupied by sea thrift (sea pink, *Armeria maritima*), Lapland rosebay (*Rhododendron lapponicum*), and *Rhacomitrium* moss.

7.4.4 Karst Topography

Burnt Cape Ecological Reserve and Watts Point Ecological Reserve were established to preserve the unique plant assemblages developed on the exposed Ordovician limestone and dolomite of the Strait of Belle Isle region. The Burnt Cape and Watts Point barrens, as well as other exposures at Flowers Cove and Eddies Cove, are developed on pavement karst.

Karren types are well developed at Burnt Cape and Flowers Cove, including evenly spaced gutter-like parallel troughs with sides rising to sharp cusps, developed on flat surfaces; step-like features on slopes; meandering channels; grooves which follow the joints in bedrock; and small, spiky points, where an individual part of the limestone is more resistant to dissolution (Fig. 7.22). Rates of dissolution of exposed pure limestone at Watts Point, estimated from meteorological conditions at St. Anthony (MAAT 2 °C, mean annual precipitation 1400 mm, 120 days frost-free period) (c.f. White 1988; Ford and Williams 1989), are ~40 mm/1000 y. The dolomitic limestone at Burnt Cape would be expected to dissolve at slightly slower rates than the purer calcium-rich limestone at Watts Point. Deglaciation occurred c. 12 ka, followed by marine inundation to 140 m asl at Watts Point and ~150 m asl at Burnt Cape (Grant 1989). Karst development recommenced c. 9 ka.

Frost wedging is responsible for the development of sorted and unsorted frost circles and frost polygons on the terrain at Burnt Cape. Development is enhanced in areas of exposed terrain, with minimal or no vegetation cover or snowdrifting. Frost geomorphic activity is accentuated where vegetation has been removed, as has occurred in other areas of Newfoundland (Henderson 1968; Taylor 1994). Coastal limestone areas with thin soil are particularly susceptible to damage by foot traffic, as has been documented under similar environmental and geological circumstances at Mingan Archipelago National Park Reserve (e.g. Couillard et al. 1987). Enhanced tourist usage is currently a greater potential problem at Burnt Cape, where tourist access is greater than at Watts Point. Erosion resulting from local residents' foot and vehicle traffic is evident at Flowers Cove.

Fig. 7.22 Karst pavement, Wild Bight. Limestone is directly exposed on the surface, with the exception of sedge and dwarf willow vegetation developed in grikes. **a** Undulations in the surface, initially caused by bedrock structure, are exploited by vegetation and retain snow and surface water, facilitating dissolution. **b** Limestone barrens vegetation must be tolerant of high pH, seaspray, and cold winters

Wedging by plant roots, and biochemical activity along the root surfaces, further accentuate modification. Coastal sites lacking vegetation are more susceptible to frost wedging than those with vegetation cover, as the combined result of exposure of the rock to the atmosphere and removal of any potential snow cover insulation during the winter and spring (Hicks 1995). Erosion is particularly accentuated when salt spray from storms or freezing rain damages or removes the vegetation, leaving wedged fractures exposed to the atmosphere and vulnerable to frost activity.

Climate change has the potential to adversely affect some of the rare plants present at Burnt Cape and Watts Point Ecological Reserves and other limestone barrens, such as Arctic bladderpod (*Lesquerella arctica*), alpine arnica (*Arnica alpina*), dwarf hawksbeard (*Crepis nana*), and Burnt Cape cinquefoil (*Potentilla usticapensis*). The endemic Fernald's braya (*Braya fernaldii*) is listed as "threatened", and Long's braya (*Braya longii*) and the barrens willow (*Salix jejuna*), both confined to Watts Point Ecological Reserve, are listed as "endangered" (COSEWIC 2000, 2001; Hermanutz et al. 2002).

Table Point Ecological Reserve protects the mid-Ordovician trilobite, orthocerid, and brachiopod fossil assemblage preserved in limestones of the Table Point Formation (Knight 1977, 1991; Colman-Sadd et al. 1990). Bryozoans, crinoids, gastropods, ostracods, and porifera fossils are also present.

Frost wedging and karst dissolution at Table Point appear to be less rapid than at Burnt Cape, possibly due to the less pronounced jointing. The difference in aspect, with Table Point being less subject to strong coastal winds (although still under their influence) than are Burnt Cape and Watts Point, could also reduce the amount of frost wedging.

The Arches Provincial Scenic Attraction preserves two sea arches eroded from Ordovician dolomitic limestone (Knight 1977, 1991; Colman-Sadd et al. 1990; Fig. 7.23). The configurations of the openings are controlled by the pattern of fractures and joints in the dolomitic limestone. The main fracture patterns, associated with thrust faulting (Williams and Cawood 1989), parallel the coast. Secondary joints are developed at approximately right angles to the main structural weaknesses, inclined between 30° and 50° SE. Fins of rock are loosened along the sides, particularly in the upper parts of the curved openings. The joint orientation also controls the shape of each arch, resulting in the asymmetrical form. Frost wedging is the primary process involved, with wave action responsible for removal of disaggregated material.

There is no evidence of glacial erosional features on the surfaces. The resistance of the dolomitic units, however, suggests that formation could have been initiated before MIS 2 glaciation, resuming after deglaciation and postglacial inundation.

7.4.5 Port-Au-Choix

At Port-au-Choix National Historic Park, excavations have revealed the presence of several phases of Maritime Archaic occupation, dating between 8 and 3 ka, punctuated by periods of climatic deterioration (Renouf 2011; Renouf et al. 2009; Anstey et al. 2016; Dussault et al. 2016). Sivullirmiut (Dorset) Inuit occupied the Cape Riche Peninsula between 1.9 and 1.4 ka.

Detailed studies on the fluctuations of sea level have been conducted at Port-au-Choix (Bell et al. 2005). After deglacial inundation to +125 m, relative sea level generally fell for almost 10,000 years, punctuated with periods of stability and slight rises (Bell et al. 2005; Smith et al. 2005). By about 3000 years ago, the sea had dropped to slightly below its modern level. At present, sea level is rising slowly, less than 2 mm/y (Fig. 7.24).

Port-au-Choix is marked by numerous shallow dolines and other surface karst features, formed in the Ordovician limestone and dolomite. In addition to climatic fluctuations, increases in fire frequency ~ 8, $\sim 5.5, 4.8-4.2$, and ~ 3.4 ka also contributed to modification of the vegetation assemblage. The relationship between fire frequency and human occupation suggests that climatic conditions were primarily responsible.

7.5 Conclusion

The North Atlantic seaboard of eastern Canada has been shaped by tectonic activity, glaciation, and marine processes. Structural geology and lithology underpin landscapes from Killiniq to the Strait of Belle Isle. Glacial features are dominantly related to extensive MIS 2 advances. Relative sea level history varies from falling through rising episodes, reflecting differing glaci-isostatic signals along more than 5000 km coastline.

The view from shipboard reveals human occupation and the use of coastal landscapes. Geomorphology has been a dominant influence on human activity since initial human settlement.

Fig. 7.23 The Arches Provincial Scenic Attraction. **a** Geomorphology of the Ordovician dolomitic limestone reflects fracture patterns related to thrust faulting, augmented by secondary jointing and frost wedging. **b** Mean high tide reaches the base of the supporting pinnacle. Beach gravels include local Ordovician carbonate and clastic pebbles, and granitic and gneissic clasts derived from the Grenville Province of the Canadian Shield

Fig. 7.24 Port-au-Choix. **a** The modern shoreline is a reflective gravel-dominated swash-aligned shore-normal transport system. Relative sea level currently is rising at less than 2 mm/y. **b** Grikes developed along bedding planes and tensional fractures, with karren and kamenitza. *Photograph* b—courtesy of Dominique Lavers

References

Anstey RJ, Guiry EJ, Renouf MAP, Deal M, Fuller BT (2016) Dating archaeobotanical remains: a cautionary tale from Port au Choix, Newfoundland. Veg Hist Archaeobot 25:75–84

Barrand NE, Way RG, Bell T, Sharp MJ (2017) Recent changes in area and thickness of Torngat Mountain glaciers (northern Labrador, Canada). Cryosphere 11:157–168

Bell T, Renouf MAP (2003) Prehistoric cultures, reconstructed coasts: Maritime Archaic Indian site distribution in Newfoundland. World Archaeol 35:350–370

Bell T, Macpherson JB, Renouf MAP (2003) "Wish you were here…" a thumbnail portrait of the Great Northern Peninsula Environment 1000 AD. In: Lewis-Simpson SM (ed) Vinland revisited: The Norse World at the turn of the first millennium. Selected Papers from the Viking Millennium International Symposium 15–24 September 2000, Newfoundland and Labrador. St. John's, Newfoundland: Historic Sites Association of Newfoundland and Labrador

Bell T, Smith IR, Renouf MAP (2005) Postglacial sea-level history and coastline change at Port au Choix, Great Northern Peninsula, Newfoundland. Newfoundland Labrador Stud 20:9–31

Catto NR (2006) Impacts of climate change and variation on provincial parks, NL. Newfoundland and Labrador Ministry of Environment and Conservation, 160 p

Catto NR, MacQuarrie K, Hermann M (2002) Geomorphic response to Late Holocene Climate variation and Anthropogenic pressure, northeastern Prince Edward Island, Canada. Quat Int 87:101–118

Clark PU (1988) Glacial geology of the Torngat Mountains, Labrador. Can J Earth Sci 25:1184–1198

Clark PU, Fitzhugh WW (1990) Late deglaciation of the Central Labrador coast and its implications for the Age of Glacial Lakes Naskaupi and McLean and for Prehistory. Quat Res 34:296–305

Colman-Sadd SP, Hayes JP, Knight I (1990) Geology of the island of Newfoundland. Map 90-01, Geological survey branch, Department of Mines and Energy, Government of Newfoundland and Labrador

COSEWIC (2000) Canadian species at risk, May 2000. Committee on the Status of Endangered Wildlife in Canada, 23 pp

Couillard L, Grondin P, Pelletier D (1987) Les plantes rares de l'Archipel de Mingan. Environment Canada

Daly JF, Belknap DF, Kelley JT, Bell T (2007) Late Holocene sea-level change around Newfoundland. Can J Earth Sci 44:1453–1465

Davis AM (1980) Modern pollen spectra from the tundra-boreal forest transition in northern Newfoundland, Canada. Boreas 9:89–100

Davis AM (1984) Ombrotrophic peatlands in Newfoundland, Canada: their origins, development, and trans-Atlantic affinities. Chem Geol 44:287–309

Davis AM (1985) Causes and character of paludification in Newfoundland. Can Geogr 29:361–384

Dubé-Loubert H, Roy M, Schaefer JM, Clark PU (2018) ^{10}Be dating of former glacial Lake Naskaupi (Québec-Labrador) and timing of its discharges during the last deglaciation. Quat Sci Rev 191:31–40

Dussault F, Bell TJ, Grimes V (2106) Archaeoentomological perspectives on dorset occupations in Newfoundland: a case study from the site of Phillip's Garden (EeBi-1). Arctic 69:1–15

Forbes DL, Taylor RB (1994) Ice in the shore zone and the geomorphology of cold coasts. Prog Phys Geog 18:59–89

Ford DC, Williams PW (1989) Karst geomorphology and hydrology. Unwin Hyman, London

Gower CF, Knight I, Perrault S (1997) Field guide to the geology of the Pinware terrane in southeasternmost Labrador and adjacent eastern Quebec. Newfoundland Department of Mines and Energy

Grant DR (1989) Quaternary geology of the Atlantic Appalachian region of Canada. In: Fulton RJ (ed) Quaternary Geology of Canada and Greenland. Geology of Canada Series, no. 1. Geological Survey of Canada, Ottawa, pp 393–440

Grant DR (1992) Quaternary geology of St. Anthony—Blanc-Sablon area, Newfoundland and Quebec. Geological Survey of Canada Memoir 427

Henderson EP (1968) Patterned ground in southeastern Newfoundland. Can J Earth Sci 5:1443–1453

Henningsmoen KE (1985) Pollen-analytical investigations of the L'Anse-aux-Meadows area. In: Ingstad AS (ed) The Norse Discovery of America, vol 1. Excavations of a Norse settlement at L'Anse-aux-Meadows, Newfoundland, 1961–1968. Universitetforlaget, Oslo, pp 309–362

Hermanutz L, Mann H, Anions MF, Ballam D, Bell T, Brazil J, Djan-Chékar N, Gibbons G, Maunder J, Meades SJ, Nicholls W, Smith N, Yetman G (2002) National recovery plan for Long's Braya (*Braya longii* Fernald) and Fernald's Braya (*Braya fernaldii* Abbe). National recovery plan no. 23. Recovery of Nationally Endangered Wildlife (RENEW), Ottawa, Ontario, 33 pp

Hicks D (1995) The morphology, composition, and characteristics of a coastal beach at Flower's Cove, the Great Northern Peninsula, Newfoundland: a study of strandflat coasts. Honours BA thesis, Department of Geography, Memorial University

Holtedahl H (1998) The Norwegian Strandflat—a geomorphological puzzle. Nor Geol Tidsskr 78:47–66

Klassen RA, Paradis S, Bolduc AM, Thomas RD (1992) Glacial landforms and deposits, Labrador (Newfoundland) and eastern Québec. Geological Survey of Canada Map 1814A

Knight I (1977) Cambro-Ordovician platformal rocks of the Northern Peninsula, Newfoundland (NTS 12I/NW; 12P/NE, SE). Geological Survey Newfoundland and Labrador Report 77-06

Knight I (1991) Geology of Cambro-Ordovician rocks in the Port Saunders (NTS 12I/11), Castors River (NTS 12I/5), St. John Island (NTS 12I/14), and Torrent River (NTS 12I/10) map areas. Geological Survey Newfoundland and Labrador Report 91-04

Magnusson M, Palsson H (1982) The Vinland sagas—the Norse Discovery of America. Penguin Books, London

McCuaig SJ (2002) Quaternary Geology of the Alexis River area, and the Blanc-Sablon to Mary's Harbour road corridor, southern Labrador. Current Research, Newfoundland Department of Mines and Energy Geological Survey, Report 02-1

McLeod BA, Brown MW, Moore MJ, Stevens W, Barkham SH, Barkham M, White BN (2008) Bowhead Whales, and not Right Whales, were the primary target of 16th- to 17th-century Basque Whalers in the Western North Atlantic. Arctic 6:61–75

Narayanan S (1994) Current meter observations from Hamilton Bank and NE Newfoundland Shelf, 1990 to 1993. St. John's, Newfoundland: Science Branch, Dept. of Fisheries and Oceans

Pienitz R, Lortie G, Allard M (1991) Isolation of Lacustrine Basins and Marine Regression in the Kuujjuaq Area, Northern Québec, as Inferred from Diatom Analysis. Géogr Phys Quat 45:155–174

Renouf MAP (2011) The cultural landscapes of Port au Choix: Precontact hunter-gatherers of northwestern Newfoundland. Springer, New York

Renouf MAP, Bell T, Macpherson J (2009) Hunter-gatherer impact on subarctic vegetation: Amerindian and Palaeoeskimo Occupations of Port au Choix, Northwestern Newfoundland. Arctic Anthropol 46:176–190

Smith IR, Bell T, Renouf MAP (2005) Testing a Proposed Late Holocene Sea-Level Oscillation using the 'Isolation Basin' Approach, Great Northern Peninsula, Newfoundland. Newfoundland Labrador Stud 20:33–55

Stopp MP (2016) Across the straits from Port au Choix: Mobility, Connection, and the Dorset of Southern Labrador. Arctic 69 (Supplement):1–10

Taylor C (1994) The formation of soil- frost related patterns along the Hawke Hills. BSc thesis, Department of Geography, Memorial University

Vasseur L, Catto NR (2008) Atlantic Canada. In: Lemmen DS, Warren FJ, Lacroix J, Bush E (eds) From impacts to adaptation: Canada in a changing climate 2007. Government of Canada, Ottawa, pp 119–170

Wardle RJ, Gower CF, Ryan B, Nunn GAG, James DT, Kerr A (1997) Geological Map of Labrador; 1:1 million scale. Government of Newfoundland and Labrador, Department of Mines and Energy, Geological Survey, Map 97-07

Way RG, Bell T, Barrand NE (2014) An inventory and topographic analysis of glaciers in the Torngat Mountains, northern Labrador, Canada. J Glaciol 60:945–956

Way RG, Bell T, Barrand NE (2015) Glacier change from the early Little Ice Age to 2005 in the Torngat Mountains, northern Labrador, Canada. Geomorphology 246:558–569

White WB (1988) Geomorphology and hydrology of karst terrains. Oxford University Press, New York

Williams H, Cawood PA (1989) Geology, Humber Arm Allochthon, Newfoundland. Geological Survey of Canada Map 1678A

Norm Catto joined Memorial University in July 1989 and is now the Head and Professor of Geography at that university. His research and teaching interests include coastal landforms, natural hazards, sea level change, and impacts in coastal environments; response of river systems to climate and weather events, and flood risk assessment; aeolian geomorphology and response to climate and human factors in boreal and costal environments; mass movements and slope failures; loess deposition; palaeosol formation; emergency measures policies and effectiveness human adaptations to climate and weather events; the impacts of climate and weather events on agriculture, transportation, fisheries, and communities; and the relationships among landscapes and people. His research has included projects and investigations in landscapes and environments in 16 countries, and in all Canadian provinces and territories, particularly in eastern Canada. For 18 years, he served as Editor-in-Chief for Quaternary International, a professional international journal dedicated to the study of landscape evolution and climate variation throughout the past 3 million years. He has taught more than 50 different undergraduate and graduate courses focused on numerous aspects of geomorphology, landscape analysis, Quaternary research, natural hazards, ocean sciences, environmental science, climate and climate change, soil science, geoarchaeology, and palaeontology. Study of landscapes forms his central focus for understanding of the various areas of Eastern Canada, the combination of physical and cultural constituents that contribute to its identities. For most species, and for all human occupants, life in Eastern Canada has been heavily influenced by geomorphology since deglaciation.

8 Landscapes and Landforms of the Hudson Bay Lowlands

L. A. Dredge and L. D. Dyke

Abstract

The Hudson Bay Lowlands and adjacent terrain form a vast wetland landscape with low relief and an abundance of organic terrain. Wetland landforms are the main features in the landscape. These include open and forested bogs and peat plateaus; flat fen meadows, and stringed and palsa fens; and swamps, marsh, and open water. Local relief is commonly <2 m. The landscape of the lowlands is the product of its geologic history and present conditions. The area is underlain by Precambrian and Palaeozoic rocks that were peneplaned to a gradient of <2 m/km. During the last glaciation, most of the area was covered by ice flowing out of, or across, the Palaeozoic limestone underlying and adjacent to Hudson Bay, although the western and eastern extremities of the lowlands were influenced, respectively, by Keewatin and Labradorean ice, both of shield provenance. Proglacial lakes Agassiz and Barlow-Ojibway followed the retreating ice sheet northward, deeply inundating the region. The southern margin of the Hudson ice sheet was unstable, and surged into the proglacial lakes. Following the break-up of Hudson ice about 8000 years ago, the lowlands were covered by a high postglacial sea (Tyrrell Sea), which regressed down to present sea level.

Keywords

Hudson Bay Lowlands • Wetlands • Peat • Glacial history • Landscape • Landforms

L. A. Dredge (✉)
Geological Survey of Canada, 601 Booth St., Ottawa, ON K1A 0E8, Canada
e-mail: lynda.dredge@gmail.com

L. D. Dyke
Nipissing University, 100 College Drive, North Bay, ON P15 8L7, Canada

8.1 Introduction

The Hudson Bay Lowlands, sensu stricto, refer to those lands surrounding Hudson Bay that are underlain by Palaeozoic rocks (Bostock 1970, 2014, and regions 9 and 10, Fig. 1.10, this volume). The 'lowlands' as used in this report are defined in a broader sense to include adjacent areas of flat terrain and wetlands. The Hudson Bay Lowlands (HBL) refer to that vast tract of land that rises from the tidal flats around the southern part of Hudson Bay and James Bay to form lowland plains 150–400 km wide (Fig. 8.1). The area is characterized by low local relief, poorly drained substrates, and an abundance of organic terrain. This region is one of the three largest wetlands in the world (Abraham and Keddy 2005), and although there are areas of inorganic surface sediments and bedrock, more than 80% of the lowlands are covered by wetlands. This chapter describes the landscape of the HBL, and geologic factors and climatic conditions that shaped its present form. It focuses on the distinctive wetland terrain that makes this region unique. Emphasis is on the Hudson Bay Lowlands in Manitoba, a representative region with conditions and features common to the larger area.

The area forms a gently sloping plain that rises inland monotonously from Hudson Bay at a rate typically of less than 1 m/km. Local relief over much of the area is less than 2 m. Exceptions occur in the extreme northwest and southeast where surface features are formed of glacial deposits, and where bedrock ridges rise above the plains. More subdued undulations in the terrain are buried glacial moraines, protruding glacial flutes, and flights of low beach ridges.

The plains are interrupted by incised valleys along major rivers, such as the Churchill, Nelson, Hayes, Severn, and Winisk, flowing into Hudson Bay, and Attawapiskat, Albany, Moose, and Harricana flowing into James Bay, where bedrock and stacks of glacial deposits form steep banks. The largest of these have ancient origins that predate the last glaciation and originate beyond the lowlands.

© Springer Nature Switzerland AG 2020
O. Slaymaker and N. Catto (eds.), *Landscapes and Landforms of Eastern Canada*,
World Geomorphological Landscapes, https://doi.org/10.1007/978-3-030-35137-3_8

Fig. 8.1 The Hudson Bay Lowlands: extent, major rivers, and places. *Cartography* D. Kovanen

Drainage over the rest of the land surface is extremely poor, and commonly chaotic in pattern. The rivers that originate within the lowlands have shallow channels, and become marshy at their mouths. The land surface is covered with a myriad of small, interconnected lakes and ponds. Some of these water bodies occupy depressions in glacial drift, but the majority lie completely within the peat that mantles the region (Fig. 8.2).

8.2 Wetland Landscape Ecotones

There are several geographically defined wetland landscape ecotones within the lowlands, although most areas are a composite of various wetland landforms. These tend to form concentric bands parallel to the coast (Fig. 8.3). Near the coast are subarctic tundra meadows of thin peat interspersed with non-organic substrates, and fens covered with grasses and sedges. Farther inland, a belt of thick bog with treeless surfaces of tundra lichen-heath and moss (Kuhry 1998) covers northeast Manitoba and other areas of continuous permafrost. To the south and inland, a subarctic/boreal tundra/forest transition zone is covered with bogs, fens, and swamps with open shrub, *alnus*, stunted *picea*, and *larix* communities (Rouse et al. 1995). A small outlier of this landscape occupies the vicinity of Churchill. Farther south and inland, away from Hudson Bay, and mainly in discontinuous permafrost, are boreal landscapes of forested bogs interspersed with sedge and grass fens. *Picea mariana* with *alnus*, *larix*, and *Betula glandulosa* are the main tree types,

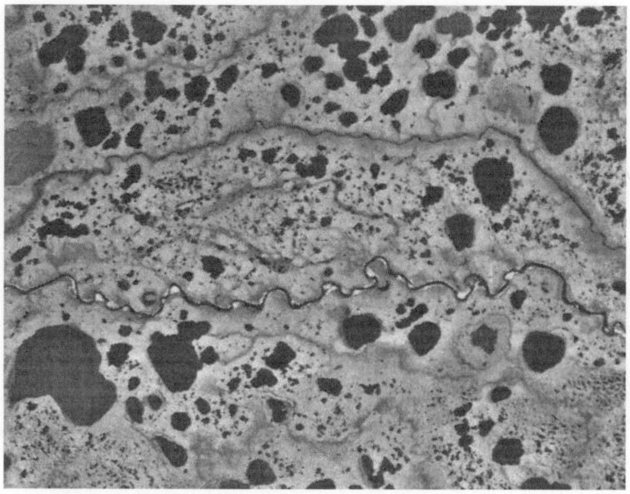

Fig. 8.2 Aerial view of the peatlands in the HBL with treeless bogs, shallow thermokarst ponds, and creeks lined with stunted trees. The mid-toned ovoid shapes are drained ponds. The photo covers an area of 25 km^2. *Source* NAPL 17404-14/Natural Resources Canada

and there is generally an understory of Labrador tea, blueberry, sphagnum moss, and caribou lichen (Dredge and Mott 2003). Boreal forest and wooded fens on non-organic substrate are present around the southern part of James Bay, where permafrost is sporadic (Fig. 1.6, this volume) and on the uplands on the Manitoba-Ontario border.

The main factors affecting the presence and distribution of wetlands are related to climatic conditions such as air temperature, ground temperatures, and precipitation; and to physiographic conditions.

8.3 Present Conditions Contributing to HBL Wetland Landscapes

8.3.1 Climate

The northern part of the lowlands lies within the marine subarctic climate zone, while areas to the south and inland experience a boreal climate. Sea ice in Hudson Bay has a significant cooling effect on the climate, which is more rigorous near the coast than in areas farther inland. The mean annual air temperature along the coast at Churchill is −6.5 °C (monthly averages are 12 °C for July and −26 °C for January), and −0.5 °C at Moosonee south of James Bay. There are numerous freeze-thaw cycles during any given year. Precipitation ranges from 452 mm annually at Churchill to 703 mm at Moosonee, with about half of that falling as snow. Prevailing winds are from the northwest (Environment Canada 2016).

Many of the thick organic deposits which are so extensive in the region are directly related to present climate conditions. The summers are sufficiently warm and humid to support healthy vegetative growth and biomass accumulation, but cold enough to prevent their decomposition; peat deposits are thick and widespread. Low annual temperatures and strong winds promote deep freezing.

Seasonal or diurnal freezing also promotes frost-churning where glacial materials form the surface, producing mudboils, and frost-shattered or frost-jacked bedrock.

8.3.2 Permafrost

Regional permafrost maps, such as displayed in the Atlas of Canada (Heginbottom et al. 1995; Smith et al. 2001), show permafrost to be continuous along a band extending inland from Hudson Bay for about 100 km from Manitoba to James Bay (Fig. 8.4). It is prevalent in bog peatlands, including peat plateaus and forested bogs, but tends to be discontinuous or sporadic beneath fens. Permafrost is also presently forming in land exposed by postglacial uplift at Churchill (Dyke 1988) and is inferred for the same situation on the Ontario Hudson Bay coast near the Manitoba-Ontario border (Kershaw 1976). The continuous permafrost designation is largely substantiated in Manitoba by several ground temperature-sensing boreholes placed in Wapusk National Park south of Churchill between 2007 and 2009. These installations indicate mean annual average ground temperatures to be −3 to −5 °C in peat plateaus and about −2 °C in adjacent fens (Dyke and Sladen 2010). Slightly inland near the mouth of Hayes River at York Factory several ground temperature installations show that permafrost is only present on the maintained grounds of the facility and is otherwise absent in surrounding fens (Sladen et al. 2009).

Few other ground temperature observations exist in the lowlands. An installation in a palsa in Polar Bear Provincial Park east of Winisk, Ontario shows a mean annual ground temperature of about −1 °C and installations in palsas inland of Attawapiskat show temperatures essentially constant at 0 °C (Ou et al. 2016a), indicating that only partial ice bonding exists. Permafrost modelling carried out in the lowlands adjacent to central James Bay indicates that only about 10–20% of the terrain is frozen (Ou et al. 2016b). Based on these observations and inferences, there is probably a narrowing zone of continuous permafrost from Churchill to Polar Bear Provincial Park. Discontinuous and sporadic permafrost, becoming isolated permafrost, is common farther inland and to the south, where mean annual air temperatures are 1–2 °C warmer.

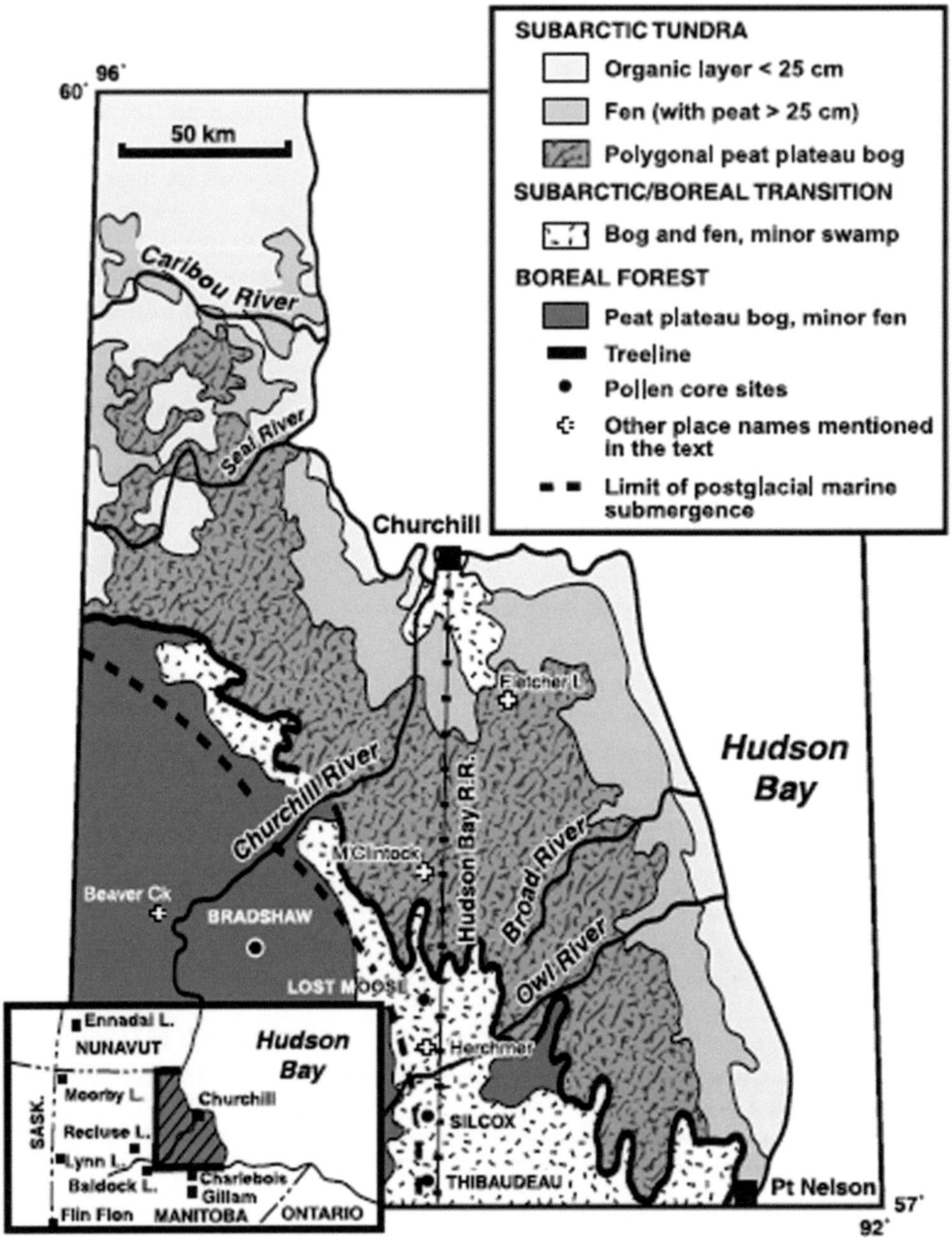

Fig. 8.3 Wetland ecotones in northeastern Manitoba. Treeline closely corresponds to the boundary between continuous and discontinuous permafrost. Tidal flats extend for a distance of 6 km beyond the coast. *Source* Dredge and Mott (2003)

8.4 Geologic Conditions Contributing to Wetland Landscapes

The physiographic factors leading to the creation of wetlands are those which affect land slope and relief, which controls the water flow gradient; and substrate type, which controls permeability and susceptibility to the growth of segregated ice. These factors are directly related to the geological evolution of the area.

8.4.1 Bedrock Geology

The lowlands contain the Hudson Bay Sedimentary Basin, a broad depression roughly coincident with Hudson Bay. Granitic or metasedimentary rocks of the Canadian Shield underlie the basin and extend beyond its edges. Palaeozoic rocks, mainly fine-grained carbonate strata, mantle the Precambrian rocks. Pockets of Cretaceous shale overlie the limestone in some places. The Palaeozoic cover rocks are either flat-lying, or dip gently towards the centre of Hudson

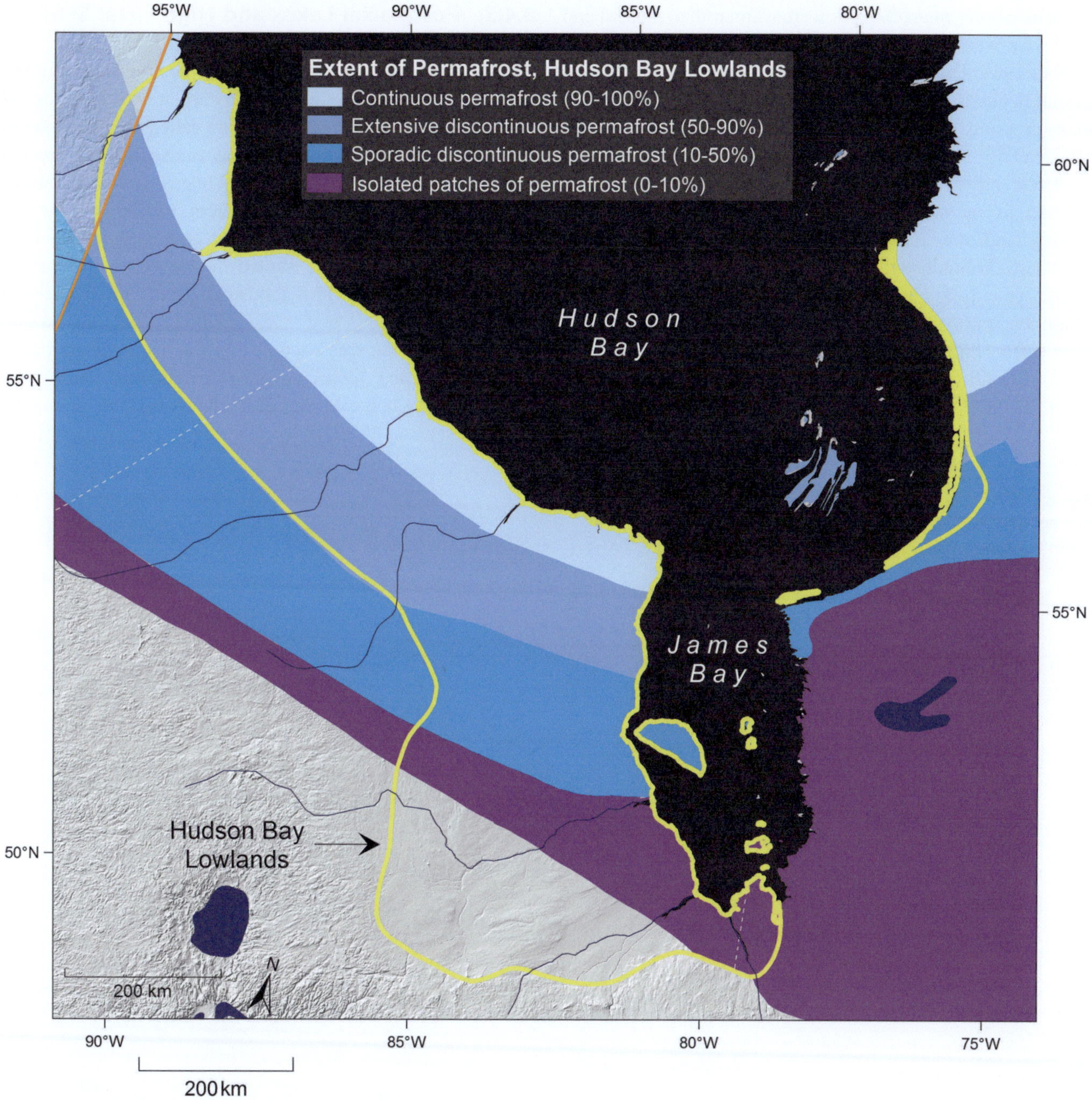

Fig. 8.4 Permafrost distribution in the Hudson Bay Lowlands. *Cartography* by D. Kovanen

Bay, and they are less resistant to erosion than the surrounding Precambrian Shield. Both the shield and the sedimentary basin have been peneplaned in post-Devonian times, forming a surface that slopes gently towards Hudson Bay (Dyke and Dredge 1989). The major regional elements of the Hudson Bay Lowlands landscape were shaped in pre-Quaternary time.

8.4.2 Glacial Geology

During the Quaternary era, glaciers centred in Keewatin, central Québec, Hudson Bay, and Northern Ontario crossed the region. A thick stack of sediments exposed in the cliffs along major rivers, such as the Nelson and the Moose, record a long record of multiple glaciations, and at least two interglaciations when climates were warmer than present and sea levels were somewhat higher than present (Dredge and Cowan 1989; Dredge et al. 1989). In the rest of the region, glacial deposits are generally thinner. Ice sheets originating in, or crossing, the Canadian Shield produced a stony sandy till, leading to permeable, well-drained substrates, and moderate relief. During the last glaciation, Keewatin ice reached as far south as the Churchill area in the HBL, leaving landforms that form surface features there.

Where Labradorean or Hudsonian ice sheets flowed across, or out of, Hudson Bay southwest through northern Manitoba, and south through northern Ontario, they crossed over softer, fine-grained bedrock, and produced calcareous stony silty till of relatively low permeability. These are the main tills underlying the wetlands in the HBL (Fig. 8.5).

8.4.3 Postglacial Lakes and High Level Seas

As the Hudsonian ice sheet retreated, deep proglacial lakes followed the receding glacier northward at least as far as the present coast. These water bodies are called Lake Agassiz in Manitoba and Lake Barlow-Ojibway in Ontario (Dredge and Cowan 1989; Vincent 1989). They began as separate water bodies, but joined in their northern extremity near the Manitoba-Ontario border during late phases of deglaciation. Although Lake Agassiz started to develop in southern Manitoba more than 11.8 thousand years ago (Dyke and Prest 1987a, b), it, and Lake Ojibway, did not extend northward into the area occupied by the lowlands until after 8400 radiocarbon years ago (9500 recalibrated years ago). Glacilacustrine deposits vary in thickness from 60 m in parts

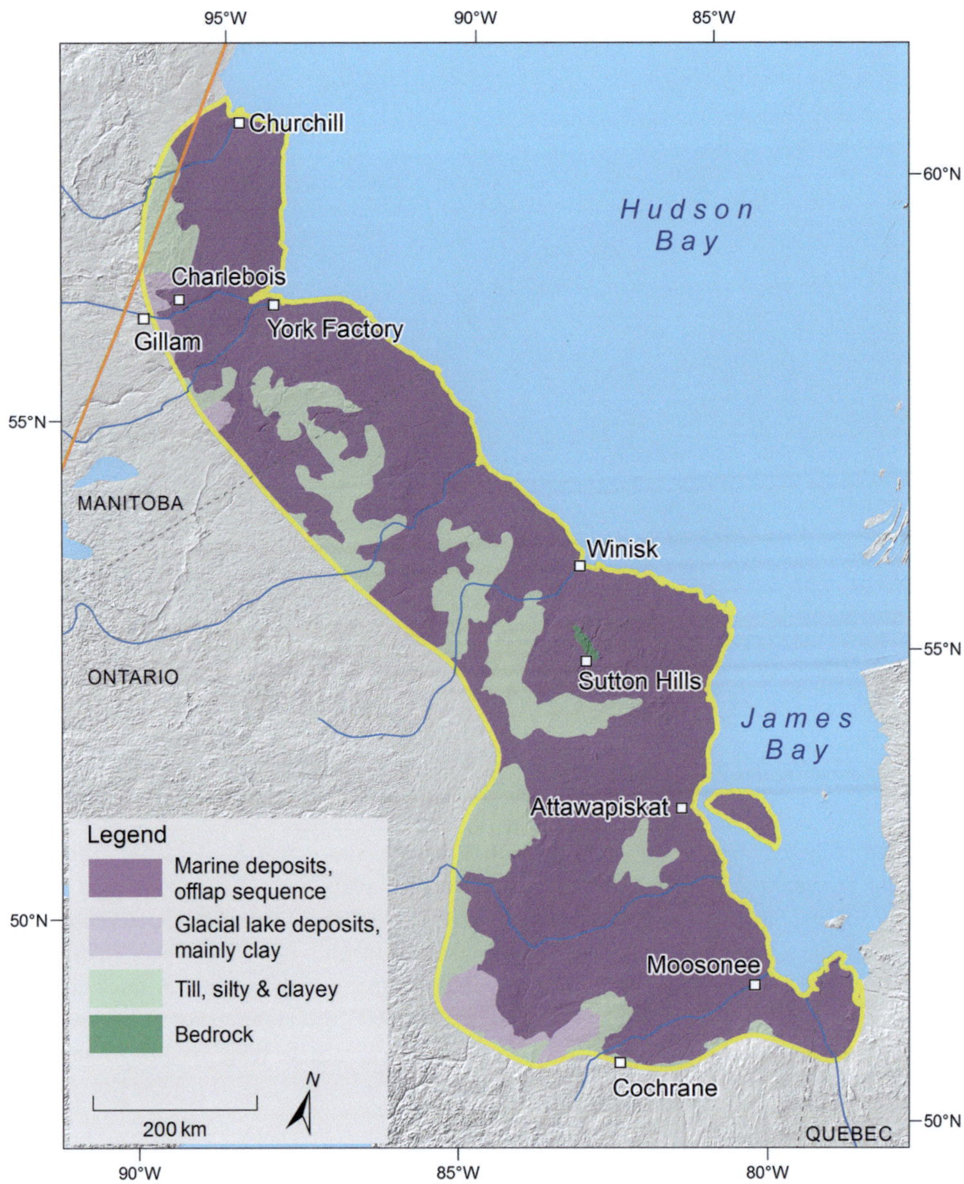

Fig. 8.5 Geological substrates that underlie the wetlands of the Hudson Bay Lowlands. *Cartography* D. Kovanen; original sketch by L. Dredge

of the James Bay Lowlands (Dredge and Cowan 1989) to about 5 m near the Manitoba-Ontario border. Glacial lake sediments consist primarily of massive clay, or varves made up of alternating beds and laminae of silty clay and clay (Fig. 8.6). Individual varves tend to be more than 1 m thick near the ice margin, but become thinner farther away. Where the retreating Hudsonian ice margin was unstable, parts of the ice front surged into the glacial lakes, producing clay till consisting of reworked glacilacustrine materials.

Radiocarbon dates on marine shells indicate that the Hudson ice sheet disintegrated about 7800 radiocarbon years ago (Dyke and Prest 1987a; Vincent 1989), or 8400 recalibrated calendar years ago (Dyke 2004). Marine waters, called the Tyrrell Sea, transgressed over the glaci-isostatically depressed land area that constitutes the HBL region, and extended as much as 300 km inland from the present coast. It reached a maximum elevation of 180 m asl in the HBL and greater elevations on the east side of James Bay. Glaci-isostatic rebound and regression of the sea has produced an offlap wedge of marine sediment that is thinnest at the marine limit, and thickens towards the present coast. The wedge consists of silty deposits overlain by littoral sand. The land is presently rebounding at a rate of about 40–60 cm/century at Churchill, and 100 cm/century around James Bay.

The geological events creating the HBL are responsible for the impermeable substrates, infilled topography, low relief, and poor drainage that characterize the region. The silty substrates have also provided materials prone to aggradation and segregation of ground ice. Almost all the landforms created by glacial and postglacial geologic events are buried by wetland terrain.

8.5 Bedrock and Glacial Landforms and Landscapes of the Lowlands

Although peat bogs, fens, and other wetlands create the distinctive landscape of the HBL, and have buried most of the glacial features, bedrock and glacial landforms are also present at the surface in some places.

Bedrock ridges poke out above surrounding wetlands, and create the greatest local relief in the region. The Sutton Hills form a conspicuous ridge made of Precambrian granite, metasediments, and gabbro southwest of Cape Henrietta Maria, Ontario (Bird 1972). They extend over a distance of 50 km and protrude up to 150 m above the surrounding lowland terrain. The ridge of greywacke and quartzite along the coast at Churchill is another prominent feature, creating about 10 m of relief. Its glaciated surface has been altered by present freeze-thaw processes, which have resulted in the frost jacking of huge joint-bound rock panels above the glaciated surface (Fig. 8.7). Dyke (1984) has attributed this movement to the trapping of water and build-up of pore pressure in joints during seasonal freeze-back, a form of natural fracking. Heaving of bedrock also results from the volumetric expansion of water as it turns into ice; the areas of felsenmeer at the edge of the lowlands northwest of Churchill may be the product of this process.

Fig. 8.6 Glacial Lake Agassiz varves of silt and clay overlain by marine sand in a 2-m exposure near Gillam, Manitoba. Varve couplets here are <2 cm thick. *Source* GSC-204314-X *photograph* by L Dredge/Natural Resources Canada

Fig. 8.7 The greywacke ridge at Churchill, grooved and striated by Keewatin ice. Large bedrock panels have been frost-jacked above the glaciated surface. *Source* GSC-205238-Q by L. Dredge/Natural Resources Canada

In the northern extremity of the lowlands north of Churchill, glacial landforms associated with the Keewatin ice regime form surface features, and have created local relief of up to 15 m. These include sandy till plains, drumlins, ribbed moraine (Fig. 8.8a), belts of small hummocky glacifluvial ridges, and the sandy Burntwood-Knife interlobate moraine (Dredge and Nixon1992), which are described in the companion volume on landscapes in Western Canada. Keewatin glaciers shaped, grooved, and striated the bedrock ridge that extends along the coast east of Churchill. The bouldery esker remnants (Fig. 8.8b) and pitted kame delta east of Churchill are also landforms created by Keewatin ice. At the southeast extremity of the lowlands is the Harricana interlobe moraine, which separates late-glacial Hudsonian ice deposits from the shield-derived deposits of the Labradorean ice regime.

Most of the glacial features associated with the Hudsonian ice regime underlie the lowlands but are buried by bog and fen; however, patches of glacial deposits form surface landforms, particularly at the outer limit of the lowlands where the land is highest. These include broad undulating moraines and plains of silty till with a relief of <5 m.

Glacial lake sediments, primarily massive and varved clays and silts have infilled topographic irregularities in the underlying till surface, producing flat clay plains. These plains underlay some of the organic terrain of the lowlands (Dredge and Cowan 1989; Dredge and Nixon 1992), form the ‘clay belt’ of northern Ontario, and are exposed in patches near the inner southern limit of the HBL in northwestern Ontario. Where these deposits are exposed, criss-crossing linear patterns are visible on aerial photographs (Fig. 8.9a), although they are rarely detectable at ground level. These are shallow grooves created where calving icebergs scoured the glacial lakebed (Dredge 1982). Ice-marginal glacial surges that advanced into the lake and reworked the lakebed have produced a clayey till characterized by long, low, narrow glacial streamlined forms of 1–2 m relief (Fig. 8.9b). These form surface deposits in parts of northern Ontario, particularly around James Bay.

Postglacial marine sediments form a plain that is generally covered by peatlands, although organics have not had time to accumulate near the present coast. Raised beaches that formed during the regression of the postglacial Tyrrell Sea are the main non-organic surface landforms. A large raised beach complex, 200–800 m wide, developed near the marine limit and can be traced across northern Manitoba and into Ontario as far as Winisk River, a distance of 800 km (Fig. 8.10a). At lower levels, individual ridges developed as the sea regressed. Flights of closely spaced beaches of 1–3 m relief form a distinctive landscape on a gently sloping upland facing Hudson Bay along the Manitoba-Ontario border. Widely spaced, low beach ridges of sand are found throughout the lowlands down to the present coast (Fig. 8.10b). Unvegetated, slightly raised, and modern tidal flats are exposed along the coast, forming a flat muddy plain, strewn with ice-rafted boulders, that extends for a distance of about 6 km into Hudson Bay.

Fig. 8.8 **a** Ribbed moraines of stony, sandy till deposited by Keewatin ice are present in the northwestern extremity of the Hudson Bay Lowlands. **b** Esker of Keewatin provenance east of Churchill has flanks consisting of granitic boulders. The surface has been winnowed by waves of the postglacial Tyrrell Sea. *Source* a—GSC-204036-G; b—GSC-204036-F *Photographs* L. Dredge/Natural Resources Canada

8.6 Wetland Landforms and Landscapes

Wetland landforms cover most of the bedrock, glacial, and postglacial geomorphic surfaces in the region. Bogs, fens, mineral soil-peat complexes, swamps, marshes, and open water bodies are the main wetland types forming the landscape within the HBL. Bogs and fens are the main peat-forming landforms.

8.6.1 Bogs

Bogs are organic accumulations that obtain their source of moisture from precipitation and generally lie above the water table (Zoltai et al. 1988; Halsey et al. 1997). They consist primarily of decomposed sphagnum moss. There are three main types of bogs in the HBL.

8.6.1.1 Open, Polygonized Peat Plateaus

These bogs form extensive, flat treeless plains in areas of continuous permafrost. The surface is vegetated with tundra lichen-heath communities with clumps of stunted *picea.* The peat is generally 2–3 m thick and is high in ice content (Dredge and Nixon 1992; Dredge and Mott 2003). The peat surface is wet in spring, but tends to dry out in summer. Surface relief is due to ice wedges making polygonal surface patterns, and to thaw ponds (Fig. 8.11). Networks of ice wedges typically measure about 15 m on a side, and are depressed about 60 cm, the seasonal depth of thaw, below the bog surface. The quasi-circular thermokarst ponds, generally <1 m deep, occupy 30–50% of the land surface, and do not penetrate into the mineral substrate. The size and location of these lakes vary from one year to another as a result of aggradation and degradation of ground ice. Their growth and evolution are sensitive to changes in both

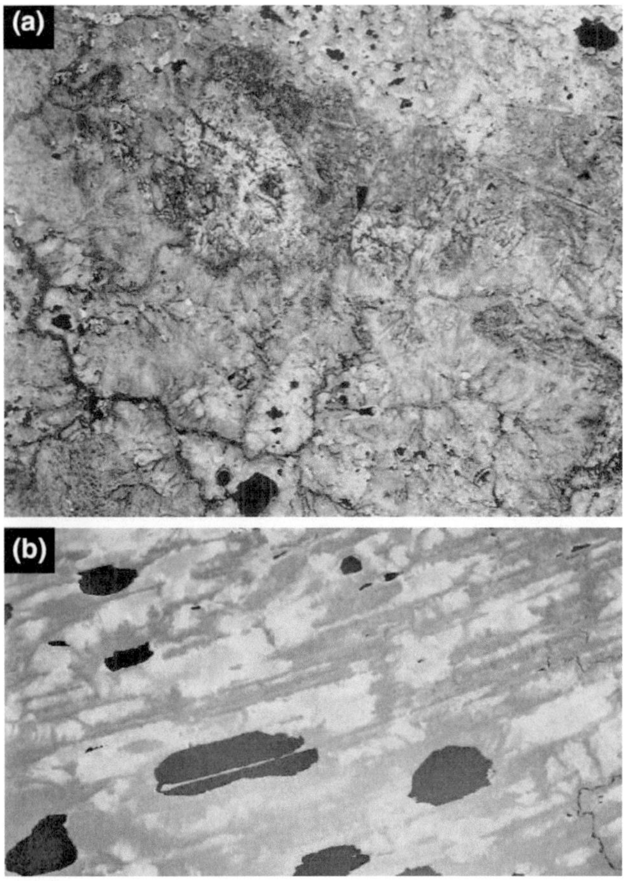

Fig. 8.9 **a** Criss-crossing linear iceberg scour marks on an area of exposed proglacial lakebed. The pitted terrain on either side of the lakebed sediments consists of forested peat bog with thermokarst ponds. **b** Multiple long narrow flutes of clay till were produced by glacial surges into proglacial Lake Ojibway. They are surrounded by fens and bogs. The lake in the lower left of the photo is 1 km across. *Source* NAPL A14128-21 and NAPL A13390-76/Natural Resources Canada

thermal and moisture regimes in the peat (Dredge 1978; Dredge and Nixon 1979; Dyke and Sladen 2010).

8.6.1.2 Forested Peat Plateaus

Forested peat plateau bogs cover extensive areas of boreal forest in areas of discontinuous permafrost. The surface is covered by spruce forest with a lichen-Labrador tea understory (Fig. 8.12a). The bog surface is almost flat, although there are spongy low peat hummocks. Peat varies in thickness from about 50 cm to 5 m (Dredge and Mott 2003) (Fig. 8.12b). Although peat thickness is highly variable regionally, there is some correlation with elevation, a measure of the time since the land emerged from the postglacial sea and became available for the establishment of vegetation. The forested bog expanses are interspersed with narrow fens, collapse scars, and thermokarst depressions that give the terrain a pitted appearance on air photos, and a local relief of 2–3 m. The depressions are waterfilled or contain sedge vegetation.

8.6.1.3 Coalesced Peat Plateaus

Coalesced peat bog plateaus consist of aligned or randomly patterned peat mounds that develop along the banks of creeks running through fens, or where ground ice aggradation has raised organic surfaces above the water table (Fig. 8.13). Individual peat plateaus south of Churchill are commonly about 200 m across and stand 2–3 m above surrounding fens. The small plateaus can have raised wooded or shrubby rims, and may have slightly depressed centres if ground ice is degrading. The mounds coalesce to form more extensive peat plateaus in areas north of regional treeline. Coalesced peat plateaus tend to maintain permafrost and support a cover of stunted trees, which form 'drunken forests' when individuals tip due to seasonal surface thawing of the peat deposits, or to changes in deeper ground ice.

8.6.2 Fens

Fens are wetland forms that obtain their moisture source from slow-flowing mineral soil waters (Zoltai et al. 1988; Halsey et al. 1997). Sedges are the main type of surface cover, although trees may also be present in shallow fens. On a regional scale, fen peat accumulations are thinner than bog peat, generally 1.5–2 m in maximum thickness. There are three types of fen in the HBL.

8.6.2.1 Horizontal Fens

Horizontal fens (terminology of Warner and Rubec 1997) are open areas that are characterized by flat homogeneous surfaces dominated by sedges. They occur in thermokarst collapse scars associated with forested bogs (Fig. 8.14a), or form featureless plains, particularly on the raised marine sediments nearer the coast that slope gently seawards. The peat is generally <1 m thick in this type of terrain, and most flat fens lie within discontinuous permafrost.

8.6.2.2 String Fens

String fens have surfaces characterized by open, shallow pools termed flarks, separated by drier, shrubby, or treed peaty ridges called strings, creating a local relief of 0.5–1 m (Fig. 8.14b). The strings are oriented perpendicular to the direction of water flow, forming sinuous ribs on gently sloping terrain and nets on flat terrain. They are prevalent around the James Bay Lowlands, and present throughout the rest of the HBL.

8.6.2.3 Palsa Fens

Palsa fens are common in areas adjacent to wooded peat plateaus where there is permafrost (Fig. 8.14c). The palsas develop with the aggradation of segregated ground ice, forming raised peat surfaces locally, which stand 1–3 m above the water table in surrounding flat fens. Coalesced palsas can evolve into peat plateau bogs.

Fig. 8.10 **a** The Great Beach. A marine complex of wide, sandy beach ridges supporting open woodlands extends across the HBL. **b** Widely spaced raised beach ridges near the coast of Hudson Bay, separated by grass and sedge meadows, thin peat, and pools of standing water. *Source* GSC-204860-C and GSC-204860-B *Photographs* L. Dredge/Natural Resources Canada

Fig. 8.11 A landscape of lichen-covered polygonized peat plateau bog, and shallow thermokarst ponds. *Source* GSC 205283-Q *Photograph* L. Dredge/Natural Resources Canada

8.6.3 Peat/Mineral Soil Complexes

These are hybrid wetland landforms consisting of a veneer of wet sedge and grass peat, forming meadows, through which mudboils have erupted. These meadows occur generally north of treeline, and near the coast where peat has had little time to accumulate. Peat deposits are generally <25 cm thick. The areas are extremely poorly drained, and ephemeral ponds are common. They occur in both discontinuous and continuous permafrost areas (Fig. 8.15).

8.6.4 Swamps

Swamps are treed wetlands located in areas of mineral substrate that have fluctuating water flow. Swamps are common in the James Bay area and in the forest/tundra transition zone in northern Manitoba, where the slope of the

Fig. 8.12 a Forested bog plateaus, covered with spruce trees and an understory of spongy lichen and Labrador tea, cover extensive areas of the Hudson Bay Lowlands. b A section through forested bog, showing trees growing on 4 m of sphagnum peat. *Source* GSC-204035-I and GSC-204035-S *Photographs* L. Dredge/Natural Resources Canada

land is greater than average, peat depths are shallow, and the mineral substrate is exposed (Fig. 8.16). The swamps in the HBL contain stunted *picea, alnus*, *salix*, and *larix*.

8.6.5 Marsh

Marshy areas are distinguished from other wetland landforms by their lack of trees and shrubs, predominance of sedges or grasses, and presence of mineral substrate. They are common along the coastline where streams enter the sea, particularly around James Bay.

8.6.6 Tidal Flats

Tidal flats form a strip of wetland terrain that extends up to 6 km into Hudson Bay. They consist of extensive mud flats dotted with ice-rafted boulders (Fig. 8.17).

8.6.7 Open Lakes and Ponds

The HBL are dotted with innumerable lakes, ponds, and waterfilled thermokarst depressions. Open water accounts for about 50% of the area of the HBL region (Parks Canada 2016).

8.7 Ground Ice in Peatlands

Ground ice is abundant in permafrost terrain, particularly in organic terrain or where exposed mineral substrates have a silty texture. Ground ice takes the form of interstitial

Fig. 8.13 Coalescing peat plateaus with rims of stunted trees, surrounded by fen. *Source* GSC-203992-Y *Photograph* L. Dredge/Natural Resources Canada

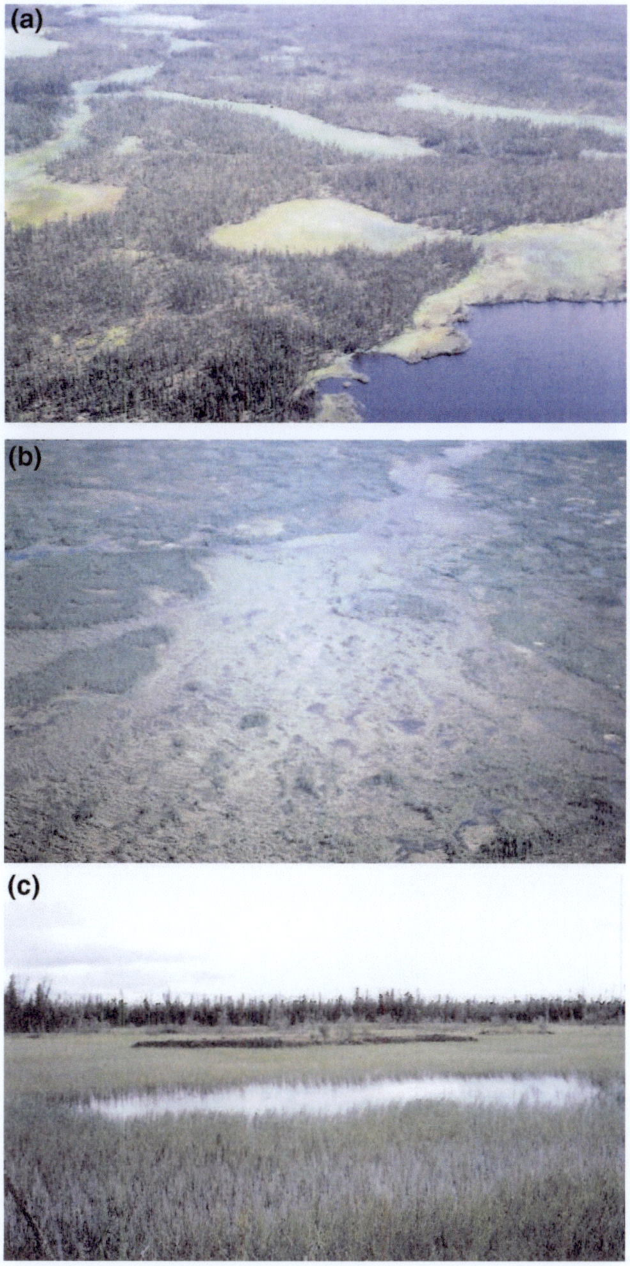

Fig. 8.14 **a** Open, horizontal fens and forested bog. *Source* GSC-204034-Q. **b** String fen, showing flarks and strings, with a 'creek' that has no defined bank in the distance. **c** A palsa mound in a sedge flat fen, surrounded by boreal forest. The mound is about 10 m long. Its surface has not yet been vegetated with bog species. *Source* GSC-203121-N *Photographs* L. Dredge/Natural Resources Canada

crystals, coatings on mineral particles and peat fibres, segregated laminae 2–10 cm thick, and ice wedges. In peatlands, a layer of white ice commonly separates the mineral substrate from overlying peat (Fig. 8.18a). Frost cracks, polygonal networks of ice wedge troughs (Fig. 8.18b), and emerging palsa mounds (Fig. 8.18c) are common surface features in the wetlands that are related to the repeated freezing and thawing of water in the peat soils, and to the segregation of ice into massive ice lenses. Ground ice commonly constitutes up to 22% of the volume of peat plateaus (Dredge and Nixon 1992). Thermokarst depressions are the surface expression of degrading ground ice in wetlands. Silty mineral substrates below the peat are also icy in the upper 3 m, but ice content tapers-off farther down. Ground ice is less prevalent in beach ridges, but can occur as interstitial ice in pore spaces.

8.8 Peatland Evolution

The earliest radiocarbon dates on basal peat in the lowlands are 6280 ± 80 (GSC-2760) from a site called Charlebois located 350 km south of Churchill at 100 m elevation, and 6280 ± 90 (GSC-5240) at Bradshaw Lake 125 km southwest of Churchill. The dates suggest that it took more than 500 years after exposure of the land surface for a measurable amount of peat to develop. Pollen analyses from four cores across the Manitoba lowlands (Dredge and Mott 2003) indicate a basal layer of woody peat or amorphous yellow sedge peat, overlain by fibrous brown sphagnum peat in most bogs, and interlayered sphagnum and sedge peat in the polygonized tundra peat plateaus. Although Nichols (1967) proposed that treeline was about 300 km farther north than present about 7000 years ago, the pollen in the peat cores postdating 6300 BP do not indicate major climatic changes in the HBL of Manitoba. The pollen shows that *picea* trees were abundant in the region for the last six millennia. Plant communities similar to those of today were present the entire time, with conditions shifting from bog to fen as hydrologic conditions changed locally. A change from fen to bog conditions in the last millennium is apparent at three sites. This could be due to natural vegetation succession (Klinger and Short 1996), climatic cooling (Nichols 1967), or to local changes in the position of the water table related to

Fig. 8.15 Meadows of grass and sedge, small ephemeral pools, and shallow peat characterize terrain near the Hudson Bay coast. *Photograph* L. Dredge

Fig. 8.16 Swampy wetland, with spruce and tamarack. *Source* GSC-205050-D/Natural Resources Canada

Fig. 8.17 Tidal flats consisting of mud and sea-ice-rafted boulders. Churchill radar station shows on the skyline. *Photograph* L. Dredge

Fig. 8.18 **a** A layer of white ground ice is insulated from summer thawing by a layer of peat at Churchill. **b** Ice wedge troughs form polygonal networks in peat in areas of permafrost. The ice wedge extends for several metres below the trough, which corresponds to the seasonal depth of thaw. *Source* GSC-203121-C; **c** Aggradation of laminae of ground ice has raised a small, recently formed palsa above the water table in a shallow fen that is surrounded by coalesced peat plateaus. *Source* GSC-203121-J, *Photographs* L. Dredge/Natural Resources Canada

aggradation of permafrost and ground ice at the sites examined (Hilbert et al. 2000). The interlayered sedge and sphagnum peat in the polygonized peat plateaus would support the latter conclusion, and relate to the local expansion or draining of tundra ponds resulting from permafrost degradation or aggradation.

8.9 Conclusion: The Future Landscape and Wetland Sensitivity

Climate change during the last century or so has resulted in the reduction of permafrost coverage, promoting the decomposition and oxidation of peat and releasing carbon to the atmosphere. Halsey et al. (1995) have documented a 30% decrease in permafrost in boreal wetland areas of the prairie provinces, including northern Manitoba, since the Little Ice Age. This decrease is expressed by the conversion of peat plateau to fen. The trend is also documented in Dyke and Sladen (2010) by the measured shrinkage of peat plateau borders that lie in the lee of snowbanks: snowbanks act as insulators from chilling winter air temperatures, permitting more summer thaw than occurs on snowfree surfaces.

In continuous permafrost, there seems to be a balance between thawing of peat accompanied by the expansion of tundra thaw ponds, and the aggradation of peat and permafrost. To some extent, this is part of a natural, self-reversing cycle (Dredge and Nixon 1979). However, the continually shifting of states of growth and thaw of peat plateaus south of Churchill indicates that this balance is delicate at present and tipping towards pond expansion (Fig. 8.19).

The peat terrain is out of equilibrium with the present climate. Northern peat plateau bogs have produced and maintained permafrost conditions in areas where air temperatures alone would have produced non-frozen conditions (i.e. areas south of the −3 °C air temperature isotherm) because dry peat is a good insulator from summer above-freezing air temperatures. Wet peat is not. Any change in temperature, precipitation, snow pack, or the water table can change the wetland/peat state (Klinger and Short 1996; Dyke and Sladen 2010). Much of the permafrost in the HBL wetlands is near 0 °C (Dyke and Sladen 2010; Sladen et al. 2009), so that small changes in air and ground temperature can lead to major changes in the peatland landscape. Cooling air temperatures or drying trends can lead to aggradation of permafrost, raising peatlands above the water table, drying bogs, and expanding peat plateaus. Conversely, warming conditions can cause melting of ground ice, subsidence of bogs to the water table, reversion of bogs to fens or thermokarst ponds, and expansion of open water bodies. In addition, since wet peat has a higher thermal conductivity than dry peat, temperature increases leading to thaw would likely accelerate the process of bog degradation as well as decomposition of organic matter. The Hudson Bay Lowlands are a sensitive ecosystem.

Fig. 8.19 Degrading polygonized bogs and expansion of thaw ponds has resulted from the melting of ground ice in permafrost. (Compare with Fig. 8.11). *Photograph* L. Dredge

References

Abraham KF, Keddy CJ (2005) The Hudson Bay Lowland. In: Fraser LH, Keddy PA (eds) The world's largest wetlands: ecology and conservation, Chapter 4. Cambridge University Press, New York, pp 118–148

Bird JB (1972) The natural landscapes of Canada. Wiley Publishers of Canada, Toronto

Bostock HS (1970) Physiographic subdivisions of Canada. In: Douglas RJW (ed) Geology and economic minerals of Canada. Geological Survey of Canada, Economic Geology Report 1, pp 10–30

Bostock HS (2014) Physiographic regions of Canada. Geological Survey of Canada Map 1254A (1st and 2nd edn), scale 1:5,000,000

Dredge LA (1978) Thaw depths and permafrost in polygonal peat terrain, Hudson Bay Lowlands, Manitoba. Geological Survey of Canada Paper 78–1C, p 27–30

Dredge LA (1982) Relict ice scour marks and late phases of Lake Agassiz in northernmost Manitoba. Can J Earth Sci 19:1079–1087

Dredge LA, Cowan WR (1989) Quaternary geology of the southwestern Canadian Shield. In Fulton RJ (ed) Quaternary Geology of Canada and Greenland. Geology of Canada Series, no. 1. Geological Survey of Canada, Ottawa, pp 214–249

Dredge LA, Morgan AV, Nielsen E (1989) Sangamon and preSangamon interglaciations in the Hudson Bay Lowlands of Manitoba. Geogr Physique et Quat 44:319–336

Dredge LA, Mott RJ (2003) Holocene pollen records and peatland development, northeastern Manitoba. Geogr Physi et Quat 7:7–19

Dredge LA, Nixon FM (1979) Thermal sensitivity and the development of tundra ponds and thermokarst lakes in the Manitoba portion of the Hudson Bay lowland. Geological Survey of Canada Paper 79-1C

Dredge LA, Nixon FM (1992) Glacial and environmental geology of northeastern Manitoba. Geological Survey of Canada Memoir 432

Dyke LD (1984) Frost heaving of bedrock in permafrost regions. Bull Ass Eng Geol 21:389–405

Dyke LD (1988) Permafrost aggradation along an emergent coast, Churchill, Manitoba. In: Proceedings of the 5th international permafrost conference, Trondheim, vol 1, pp 138–142

Dyke AS (2004) An outline of North American deglaciation with emphasis on central and northern Canada. In: Ehlers J, Gibbard PL (eds) Quaternary glaciations—extent and chronology, Part II. Elsevier, pp 373–424

Dyke AS, Dredge LA (1989) Quaternary geology of the northwestern Canadian Shield. In Fulton RJ (ed) Quaternary Geology of Canada and Greenland. Geology of Canada Series, no. 1. Geological Survey of Canada, Ottawa, pp 189–214

Dyke AS, Prest VK (1987a) Late Wisconsinan and Holocene history of the Laurentide ice sheet. Geogr Phys Quat 41:237–263

Dyke AS and Prest VK (1987b) Paleogeography of northern North America, 18,000–5000 years ago. Geological Survey of Canada Map, 1703A, scale 1:12,500,000

Dyke LD, Sladen WE (2010) Permafrost and peatland evolution in the northern Hudson Bay Lowland, Manitoba. Arctic 63:429–441

Environment Canada (2016) Canadian climate normals 1981–2010. http://climate.weather.gc.ca/climate_normals/index_e.html

Halsey L, Vitt D, Zoltai S (1995) Disequilibrium response of permafrost in boreal continental western Canada to climate change. Clim Change 30:57–73

Halsey L, Vitt D, Zoltai S (1997) Climatic and physiographic controls on wetland type and distribution in Manitoba, Canada. Wetlands 17:243–262

Heginbottom JA, Dubreuil MA, Harker PA (1995) Canada—Permafrost. In: National Atlas of Canada, 5th Edition. National Atlas Information Service, Natural Resources Canada, MCR 4177, scale 1:5 000 000

Hilbert DW, Roulet N, Moore T (2000) Modelling and analysis of peatlands as dynamical systems. J Ecol 88:230–242

Kershaw K (1976) The vegetational zonation of the east Pen Island salt marshes, Hudson Bay. Can J Botany 54:5–13

Klinger LF, Short SK (1996) Succession in the Hudson Bay Lowland, Northern Ontario. Arctic Alpine Res 28:172–183

Kuhry P (1998) Late Holocene permafrost dynamics in two subarctic peatlands of the Hudson Bay Lowlands. Eurasian Soil Sci 31:529–534

Nichols H (1967) The post-glacial history of vegetation and climate at Ennadai Lake, Keewatin, and Lynn Lake, Manitoba. Eiszeit Gegenw 18:176–197

Ou C, Leblona B, Zhangb Y, LaRocque A, Webster K, McLaughlin J (2016a) Modelling and mapping permafrost at high spatial resolution using Landsat and Radarsat images in northern Ontario, Canada: Part 1—model calibration. Int J Remote Sens 37:2727–2750

Ou C, LaRocque A, Leblon B, Zhang Y, Webster K, McLaughlin J (2016b) Modelling and mapping permafrost at high spatial resolution using Landsat and Radarsat-2 images in Northern Ontario, Canada: Part 2—regional mapping. Int J Remote Sen 37:2751–2779

Parks Canada. The Hudson Bay Lowlands. http://www.pc.gc.ca/eng/docs/v-g/nation/sec4.aspx. Accessed 2016

Rouse WR, Holland S, Moore T (1995) Variability in methane emissions from Wetlands at northern treeline near Churchill, Manitoba. Arctic Alpine Res 27:146–156

Sladen WE, Dyke LD, Smith SL (2009) Permafrost at York factory national historic site of Canada, Manitoba, Canada. Geological Survey of Canada, Current Research 2009–4, 10 pp

Smith SL, Burgess MM, Heginbottom JA (2001) Permafrost in Canada: a challenge to northern development. In: Brooks GR (ed) A synthesis of geological hazards in Canada. Geol Surv Can Bull 548

Vincent JS (1989) Quaternary geology of the southeastern Canadian Shield. In: Fulton RJ (ed) Quaternary Geology of Canada and Greenland. Geology of Canada Series, no. 1. Geological Survey of Canada, Ottawa, pp 249–275

Warner BG, Rubec CD (1997) The Canadian wetland classification system. Wetlands Research Centre, University of Waterloo, Waterloo Canada, p 68

Zoltai SC, Tarnocai C, Mills GF, Veldhuis H (1988) Wetlands of Subarctic Canada. In: Zoltai SC (ed) Wetlands of Canada. National Wetlands Working Group, Canada Committee on Ecological Land Classification, Environment Canada, Ecological Land Classification Series 24, pp 55–96

Lynda Dredge is an emeritus scientist with the Geological Survey of Canada in Ottawa. Her research has focused on Quaternary geology of northern Canada, including the reconstruction and dynamics of past ice sheets, and postglacial environmental changes. Her present work involves surficial geology mapping, dating, and interpretation of Quaternary deposits west of Hudson Bay.

Larry Dyke is an engineering geologist, formerly with the Geological Survey of Canada and the Department of Geological Sciences, Queen's University. His research has dealt with permafrost, hydrogeology, slope movement impacts on pipelines and environmental impacts of mining exploration. He is presently teaching part-time at Nippissing University, North Bay, Ontario.

9 Anticosti Island: Geomorphological Context of a Unique Karst Area

Michel Beaupré

Abstract

The Gulf of St. Lawrence Lowlands geomorphological landscape region includes Anticosti Island, the Mingan Islands and several small areas bordering the Gulf of St. Lawrence, the Strait of Belle Isle and the Newfoundland Coastal Lowland. The region is underlain by the St. Lawrence Platform geological province whose formations are made up of different types and proportions of Paleozoic limestones and detrital rocks. Anticosti Island itself covers an area of 7,943 km^2 and is located between the Jacques-Cartier and Honguedo straits. The island is a monoclinal SSW dipping cuesta structure of carbonate strata with wave-cut terraces up to c.120 m asl. Two cuesta fronts and canyons are distinctive features of the island. The island was glaciated during MIS 2 and MIS 4. The central plateau and interfluves above 76 m are covered with thin discontinuous till (ground moraine) or thin (≤ 1 m) organic deposits. Outcrops of preglacial weathered bedrock remain. Karst features are found everywhere on the island. In most areas, karst is of recent Holocene age with shallow underground flows. The Upper Salmon River karst is the most developed and seems somewhat older and is partly inherited from an earlier period of karst development. Newly discovered springs and their associated dry valleys appear definitively older. The holokarst covers an area of ±23.7 km^2 and is fringed to the west and south by fluvikarst areas with surface and underground flows.

Keywords

Anticosti Island • Gulf of St. Lawrence Lowlands • Karst • Cuesta • Canyon • Salmon River

9.1 Introduction

9.1.1 Historical Sketch

Anticosti Island is 222 km in maximum length, 55 km in maximum width and covers an area of ±7,943 km^2. It is located in the Gulf of St. Lawrence, between the Jacques-Cartier and Honguedo straits. It extends between latitudes 49°03′ N and 49°57′ N and longitudes 61°40′ W and 64°32′ W, between the Québec North Coast and Gaspé Peninsula (Fig. 9.1).

Anticosti Island, visited by people for 3500 years, was named *Notisquan* (where we hunt bears) by Innu (Montagnais) and *Natisgosteg* (foreland) by Mi'kmaq people. It was named *Isle d'Assumption* by Jacques Cartier in 1534, then *Île d'Enticosty* by Samuel de Champlain in 1625. The island was given to Louis Jolliet by Louis XIV in 1680. The famous French chocolatier Menier bought the island in 1895, built various infrastructures and in 1897 imported 220 deer. The deer population has now reached some 150,000. The island was bought by the Wayagamach Company, to become Consolidated Bathurst, in 1926. Finally, the Government of Québec bought the island in 1974, opening the door to cave and karst studies, among others. The 1974 caving report (Beaupré et al. 1974) suggested the creation of a thematic park for the Upper Salmon River karst.

Port-Menier, the only village of the island with its 240 residents, is in the western part of the island. It is equipped with an airport and a dock. Anticosti Island is a well-known fishing and hunting place as well as a recreational site, managed by the provincial parks department since 1980 which operates camps, lodges, pavilions and campgrounds. All are accessed using gravel roads. The Trans-Anticosti road, some 266 km long, runs from west to east on the northern side of the island. The Anticosti Parc national (572 km^2) has been operating since 2001 in the central part of the island. There are two ecological reserves: the eastern Heath Point and the southern Grand-lac-Salé. There are also

M. Beaupré (✉)
301 - 4540 Euclide-Brien St., Montréal, Québec H1X 3H4, Canada
e-mail: beaupremichel@outlook.fr

© Springer Nature Switzerland AG 2020
O. Slaymaker and N. Catto (eds.), *Landscapes and Landforms of Eastern Canada*, World Geomorphological Landscapes, https://doi.org/10.1007/978-3-030-35137-3_9

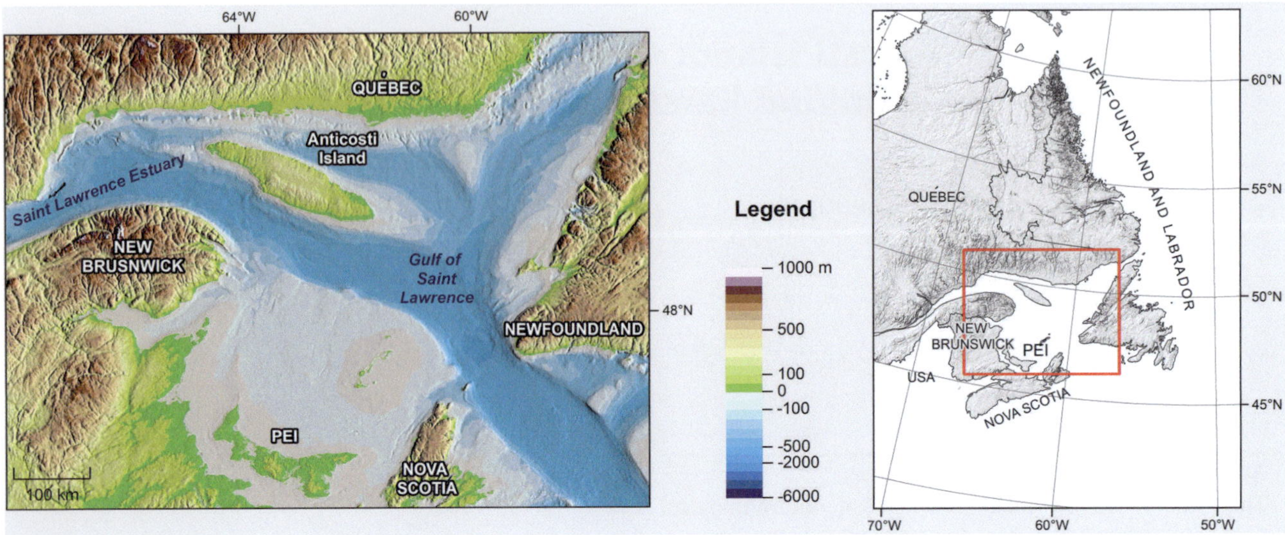

Fig. 9.1 Location of Anticosti Island. Background image adapted from Shaw and Courtney (2002). *Cartography* D. Kovanen

two outfitters, in the west and in the east, the rest of the territory being divided into numerous hunting sectors.

9.1.2 The Gulf of St. Lawrence Lowlands

The Gulf of St. Lawrence Lowlands are a part of the St. Lawrence Platform geological province that borders the Canadian Shield. It is a narrow, northeast extension of the stable interior of North America. The platform is underlain by relatively thin beds of, nearly flat-lying Cambrian to Lower Devonian orthoquartzites, carbonates and shales totaling 3500 m in thickness, that lie unconformably on the crystalline rocks of the Canadian Shield. The rocks of the Anticosti Basin consist of seven distinct formations, ranging from Upper Ordovician to Lower Silurian. These rocks are exposed on Anticosti Island, the Mingan Islands and several small areas bordering the Gulf of St. Lawrence, the Strait of Belle Isle and the Newfoundland Coastal Lowland (Fig. 9.2; see Chap. 7). Anticosti Island is a south dipping cuesta of carbonate strata. Wave-cut terraces to 120 m asl occur on both north and south sides, being generally lower on the south side. Most of the belt is less than 120 m asl, although the summit on Port au Port Peninsula, Newfoundland is at 360 m asl.

9.2 Climate, Hydrography and Vegetation of Anticosti Island

9.2.1 Climate

The island climate is temperate, with winter minimums higher than on the continent and cooler shorter summers, with frequent fogs. Climatic type is sub-polar, sub-humid (Köppen-Geiger Dfb) with an average growth season of ±171 days. Mean annual precipitation in Port-Menier is 1,005 mm, with 40% as snow. Mean annual temperature is 1.9 °C. July is the warmest month with a mean temperature of 14.8 °C, and February is the coldest one with an average temperature of −11.2 °C. Prevailing winds are from northwest and southwest (42% of the time) and from the northeast and southeast (33% of the time) (Boisclair 2004).

9.2.2 Hydrography

More than 10,000 km of water courses have been mapped on this island, 40% classified as intermittent. Rivers, lakes and ponds cover 214 km^2, 2.7% of the island area. There are 49 primary drainage basins covering an area of 5,800 km^2 (73% of the island), the remaining part being composed of residual basins (Fig. 9.3). Of the total number of water courses, 51 flow to the north and 52 to the south. Northern drainage basins are usually smaller and cover only one-third of the island. Basin areas range from 12 to 956 km^2, 24 being larger than 100 km^2. The most important ones are those of the Jupiter (956 km^2), Salmon (358 km^2), Chaloupe (209 km^2), Gallote (206 km^2), Otter (196 km^2), Vauréal (194 km^2), Oil (183 km^2) and Patate rivers (160 km^2) (Bazoge 2015). In the eastern and western lowlands rivers trend NNE–SSW and flow south or north, along or against the southern dip of the limestone strata. The dividing line between the northern and southern drainage basins follows the island center in the eastern part and passes more to the north on the western side. In the central plateau, rivers trend NNE–SSW and flow south or north, along or against the

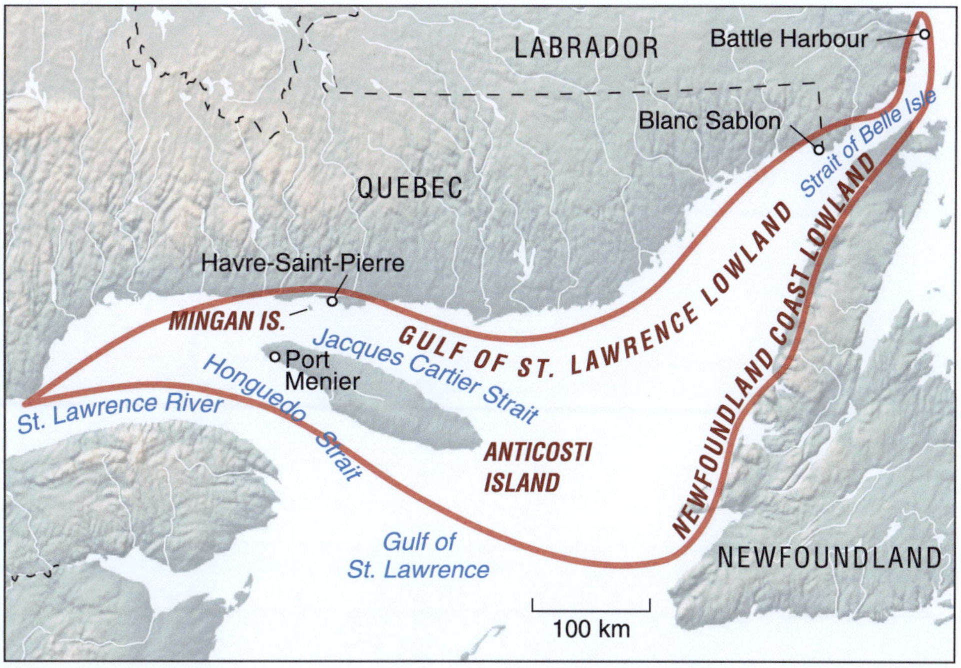

Fig. 9.2 Outer limits of the Gulf of St. Lawrence Lowland geomorphological landscape region. The Newfoundland Coast Lowland is one part of the Gulf of St. Lawrence Lowland. *Cartography* E. Leinberger

southern dip of the limestone strata. Salmon River and the upper part of the Jupiter River flow along the strike of strata (WNW–ESE).

Drainage networks are usually parallel on the southern side of the island, with rivers flowing along the natural slope. Networks are more rectangular on the northern side. Elsewhere networks are dendritic or rectangular with orientation mostly controlled by the main joint sets (NNE–SSW and ESE–WNW). Disordered networks prevail in the eastern part of the island and in its central part, at the head of Jupiter, Salmon and Vauréal rivers. Rivers are usually shallow with a flat rocky bed and crystalline waters. River entrenchment is small in the central plateau area and in the eastern lowlands. Entrenchment increases in the northern edge of the central plateau and on the dip side of the central cuesta, thus forming wide and deep canyons. On the northern side these canyons do not penetrate inland more than 10 km and rarely reach lands that are over 150 m (Fig. 9.4). On the dip side of the central cuesta they can reach 20 km in length, between the southern coast to the crest of the central cuesta. The change-over of non-entrenched rivers to canyons is usually associated with a rapid change in longitudinal slope.

9.2.3 Vegetation

The island is mainly covered with black spinet and balsam fir forests while the coastal area is mainly covered with white spinet and fir forests extending a few kilometers inland. Lumber activities have been going on for decades. Bogs and

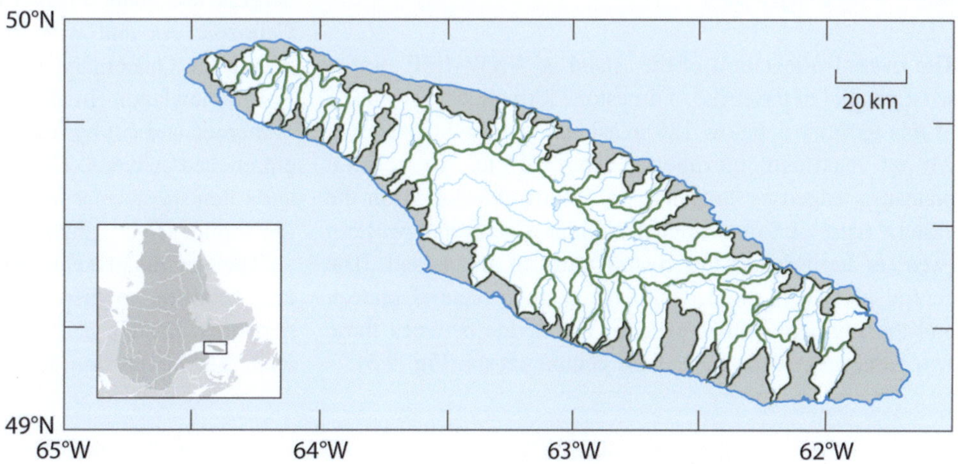

Fig. 9.3 Main drainage basins of Anticosti Island (after Bazoge 2015). *Cartography* E. Leinberger

Fig. 9.4 The upstream part of the Vauréal River canyon

fens cover some 25% of the island and are found mainly in the western and eastern lower regions and in the upstream part of Vauréal and Salmon rivers.

9.3 Physiography

The overall orientation of the island is WNW–ESE, somewhat parallel to the strike of limestone formations. Over 60% of this territory is below 120 m asl, while only 2.6% is over 240 m; maximum elevation reaches 313 m. The central plateau is edged by lowlands ($\leq$ 150 m) developed on the eastern (east of Salmon and Dauphiné rivers) and western (west of Jupiter and Oil rivers) parts of the island. The central plateau is divided into the northern central plateau and the southern central plateau. The divide between these two parts runs at the foot of the central cuesta (Fig. 9.5).

The overall physiography of the island is highly influenced by the monoclinal structure of the sedimentary formations dipping slightly to the SSW (0.5–2.0°). Two cuesta fronts have been mapped: the northern one along the north coast with height reaching 100 m and the central one with heights less than 30 m. Both have developed during the Cenozoic era and were partially modified by glaciations during the Quaternary. Cuestas are poorly developed in the eastern lowlands. In the western lowlands they are small, numerous and cut by numerous valleys. Valleys are open or entrenched canyons. These canyons are one of the important and often spectacular features of Anticosti Island. Some, like the Jupiter River canyon, can reach depths of 100 m.

The central plateau shows few glacial erosion features and dendritic to disordered, possibly preglacial, drainage networks. Eastern and western lowlands show more significant glacial erosion features such as giant grooves and

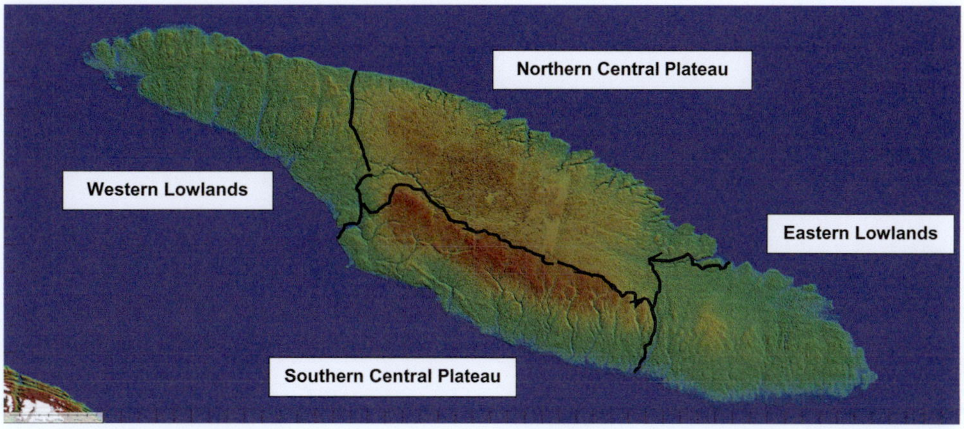

Fig. 9.5 Physiography and main divisions of Anticosti Island (SRTM)

"zig-zag" kilometric-shaped escarpments with funnel-shaped in-between valleys (Eyles and Putkinen 2014).

Many bedrock platforms, with width ranging from a few hundreds of meters to 2 km, can be seen along the island's 550 km coastline. They formed at different levels, most persistent ones having developed at 20.5 and 26 m levels. Some active cliffs can reach heights of 120 m in the middle-eastern part of the northern coast (Fig. 9.6). Fossil cliffs can locally reach more than 130 m in elevation in the northern coastal area. Active cliffs on the southern side of the island are usually lower (15–60 m) and less common.

9.4 Surface Materials

9.4.1 Studies

Surface materials on the island were mapped between 1980 and 1982 and a set of 1:50,000 maps has been published (Dubois et al. 1985). During this period numerous studies on glaciation/deglaciation of the island, marine invasion/regression, geomorphological and karst features and processes were completed. All these studies were completed by teachers and students at Sherbrooke University and a large set of related theses and papers was published.

9.4.2 Description

The island was glaciated during MIS 2 and MIS 4. Surface materials are usually scarce, especially in the central plateau and along interfluves above 76 m asl, 90% of the island area. These areas are covered with thin discontinuous till or thin (≤ 1 m) organic deposits. Four different till deposits have been mapped and are separated, in large valleys, by thick fluvial or marine sequences. Material from three of these tills was derived from the St. Lawrence North Shore, and the fourth one is related to a residual Anticosti cap. They are structureless, compact, homogeneous silty clayey sand with various amounts of cobbles and boulders. Most of the coarse fraction comes from the island, with only 4% from the North Shore. Thickness varies between 1 and 20 m. Glaciers, some 450–600 m thick, flowed southeast initially, then southwest. It seems that valleys did not influence glacier flow direction but served mostly as sediment traps. Some northern and southern fluviglacial deposits and frontal moraines were formed during the last central ice cap phase. Glaciers disappeared at ± 13 ka and marine regression started at ± 13.5 ka. Emergence rate of the island between 13 and 11 ka was around 2 m/100 years, but dropped to 0.25 m/100 years later ($11 - 6$ ka). Marine invasion reached a maximum at 80–85 m asl. Many terraces related to the marine regression have been mapped: at 70, 60, 40, 30, 25, 15 and 8–10 m asl. The largest accumulation of deposits is found along major rivers. Various types are present: glacial, fluviglacial, glacimarine, deltaic and alluvial. Frontal moraines have been mapped around the island, in its western half. Coastal deposits are frequent under 60 m asl in the eastern and western lowlands and on the island south side. They are also seen as terraces in the outlet area of northern rivers (Gratton et al. 1984; Painchaud et al. 1984; St-Pierre et al. 1987). Most karst features have developed in thin deposit areas.

Preglacial weathered bedrock does not appear to have been completely eroded away. Weathering debris up to 2.5 m thick is exposed in many areas and, in some places, regolith is covered by till. It seems that glacial erosion was not so intense in some parts of the island.

Fig. 9.6 Coastal active cliff in the Baie de la Tour

9.5 Geology of Anticosti

9.5.1 History of Studies

The Québec government bought Anticosti Island in 1974. Few geological studies were done before this event due to poor access. These studies mainly discussed lithological, paleontological and biostratigraphic aspects of this thick sedimentary sequence and, to a much lesser degree, geomorphological features and Quaternary deposits. The Québec government started a large geological mapping project at the end of the 1970s (Petryk 1981a, b, c). This new exploration phase coincided with the start of gas and petroleum exploration on the island. Since then, deep drilling and geophysical surveys have been done by the public and private entities. Many structural studies have been completed since to help understand the structural features and the tectonic framework of this island (Tanguay et al. 1986; Carboni 1988; Bordet 2007; Bordet et al. 2010; Brake and Pinet 2015; Pinet et al. 2015; SOQUIP 1985).

9.5.2 Limestone Sequence

Sedimentary rocks on the island range from Upper Ordovician to Lower Silurian and belong to the St. Lawrence Lowlands geological province and the Anticosti Platform of eastern Québec. They have been divided into seven formations (Vauréal, Ellis Bay, Becscie, Merrimack, Gun River, Jupiter, Chicotte, in decreasing age) based on their lithological characteristics (Desrochers and Gauthier 2009) (Fig. 9.7). Facies are more calcareous on the western side, with coarser clastic facies in the eastern part. These sediments were deposited in shallow marine environments. Thickness of formations ranges between 25 and 250+ m, and thickness of beds ranges between 5 and 20 cm. Most formations are made up of different lithological facies, ranging from coral reef rock, pure limestones of different grain sizes, and shaly, silty or sandy limestone interstratified with shale, siltstone or sandstone beds. The Merrimack formation (25 m) is made of very slightly soluble calcareous shales while the Chicotte formation (80 m) is composed of thick pervious limestone beds and coral reefs. The basal Vauréal and Ellis Bay formations tend to be more shaly, and are less prone to karst. Mapping of limits between each formation has been difficult because of gentle southern dips, the gradual facies changes, the lens-like nature of bedding, and difficult access. Primary permeability of these rocks is low, except in the Vauréal and Ellis Bay sandstones of the eastern part of the island.

The total thickness of this sedimentary sequence increases toward south and east, going from 915 m up north to more than 3,660 m in the south. Under this sequence are the black shales of the Macasty formation (gas schists), the sandstone and calcareous limestones of the Mingan formation, and the Romaine formation dolomites resting on the Precambrian rocks of the north coast. The last two formations outcrop in the Mingan Islands National Park.

Two vertical diabase dykes, 8 and 15 m thick, cut the Vauréal formation limestones in the central northern coast. They have been mapped along many kilometers and are related to the opening of the Atlantic during the Jurassic (Bedard 1992).

9.5.3 Structural Features

These limestone formations trend WNW–ESE, with gentle SSW dips. They are very slightly deformed by faults or folds. Four sets of vertical joints have been identified in the western part of the island (Bordet et al. 2010): a first orthogonal group with N–S and E–W joint sets and a second orthogonal one with ± NE–SW and NW–SE joint sets. A fifth irregular and random set has also been described. The first group is the most important one, and the second one has not been seen everywhere. Calcite fillings of joints are infrequent. The first joint set (N 082° − 110° E) is the most constant, with more regular and continuous joints. Strike of these joints tends to change from west (N 082° E) to east (N 110° E). The second set seems younger and shows two subsets (N 355° − 012° E and N 015° − 030° E). It is sometimes absent in the southern part of the island. These two sets are the most important ones and are usually found together. Karst features are mostly associated with these. The third and the fourth sets are minor and only seen locally.

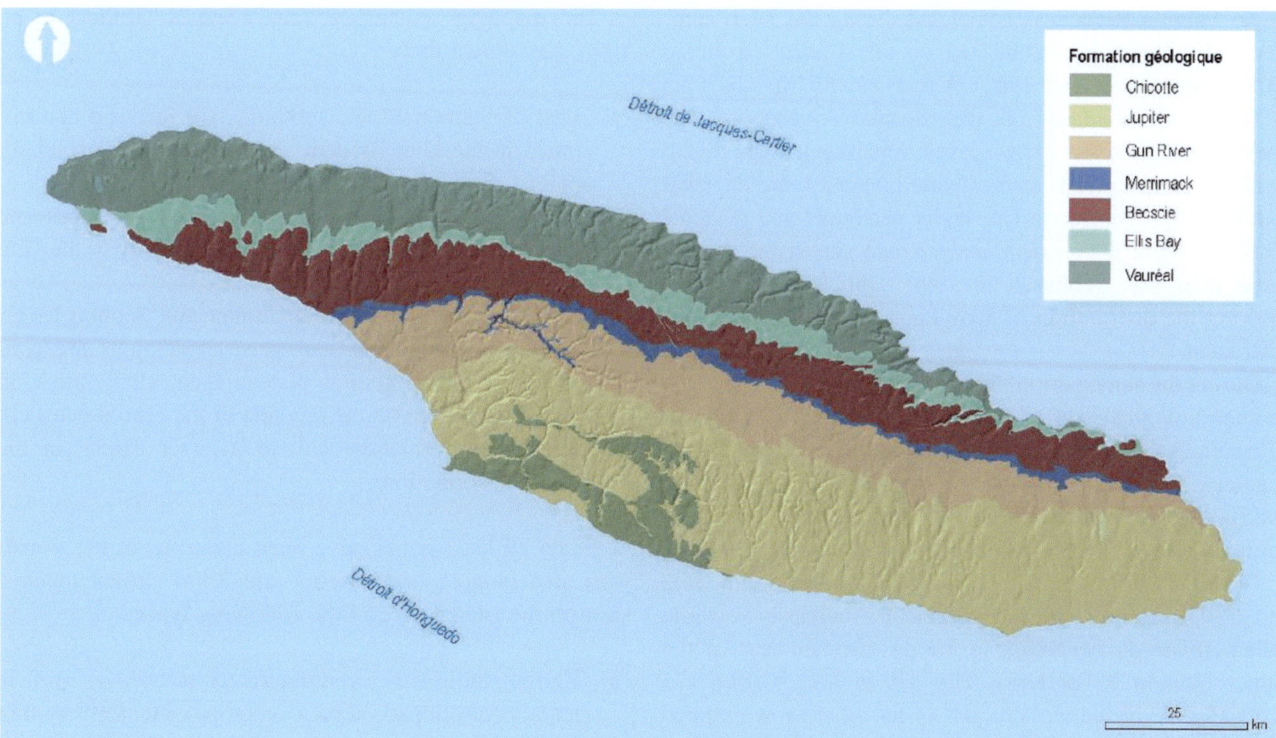

Fig. 9.7 Geological map of Anticosti Island (after Bazoge 2015; adapted from Desrochers and Gauthier 2009)

A few WSW–ENE trending open folds have been mapped in the southern Chicotte formation, elsewhere undulations being related mainly to sedimentary processes, not tectonic ones. Deep boreholes and seismic surveys have shown NW–SE trending gravity faults with 100–200 m drop of the southern block; some other gravity faults have been mapped with other trends. Faults are not easily seen on surface outcrops and displacements are small, between a few centimeters to a few meters (Bordet et al. 2010). Numerous sets of lineaments have been mapped on the island, some of them related to main joint sets or diabase dykes. Lengths range between 10 and 160 km (Brake and Pinet 2015; Pinet et al. 2015; SOQUIP 1985).

9.6 Karst

9.6.1 History of Studies

Karst studies began in 1974, following a first site visit to the island by members of the Société Québécoise de Spéléologie (Beaupré et al. 1974). The purpose of this reconnaissance was to evaluate and map as many features as possible based on local reporting. A second shorter visit was made in 1976 to acquire additional information on known features and explore new ones. Short caves and shallow pits were mapped. A study of the karst phenomena and drainage of the Upper Salmon River was completed during the 1976 summer and a thesis was published on this subject (Roberge 1979; Roberge and Ford 1983). A short caving trip was done in 1982 by members of the Société Québécoise de Spéléologie and a few known caves were mapped and dived.

Many surface karst features were mapped between 1980 and 1982 by Sherbrooke University personnel and students based on aerial photograph analysis and site visits. A synthetic map was published in 1990 (Dubois et al. 1990) showing the distribution of main karst features of the Anticosti karst.

During the same period, other studies concentrated on the geomorphological and hydrochemical processes associated with some karst lakes and drainage in the central plateau area (Cadieux 1983; Coté 1987; Coté et al. 2006; Lauriol et al. 1985). A few studies were done on specific caves, like the Grotte à la Patate (Roberge et al. 1985; Lauriol et al. 1987).

Clearing work was done, without results, during the 1992 and 1993 summers in the spectacular collapse sinkhole called *Abime du chevreuil*, in the downstream part of the Upper Salmon River karst. The 419 m long *Grotte des Trois-Plaines* in the western part of the island was mapped in 1996. The Québec Parks Department sponsored geomorphological and karst reconnaissances in 1996 and 2000 in the central and western parts of the island (Roberge 1996, 2000). A short visit was done in 2013 by the members of the Société Québécoise de Spéléologie to check changes in known karst features. A recent trip was done during the 2017 summer to revisit and reevaluate the main features of the Upper Salmon River karst (Beaupré et al. 2017).

Except for the significant work done by Roberge (Roberge 1979, 1996, 2000) and Sherbrooke University, most of the works completed on Anticosti karst features were quite specific (cave mapping and studies), usually reconnaissance or exploration work.

9.6.2 Karst Features

Even though karst studies have been fragmentary, incomplete and hindered by lack of good access, they suggest that karst features are found everywhere on the island, in all geological formations, in all areas and at all elevations (≤ 15 m − ≥ 220 m). The Upper Salmon River karst is, by far, the most developed karst, considering its area (38.5 km^2), its feeding drainage basin (143.6 km^2) and the length of some of its underground flows (≤ 14.1 km) (Roberge 1979). Elsewhere karst is in an embryonic stage with shallow underground flows (≤ 5 m) following surface drainage and fed by lakes, ponds, bogs or small brooks. The underground drainages are usually shorter than 1 km, always less than 2 km. Most of these phenomena seem to have developed recently, during the Holocene.

The mapping of surface karst features (Dubois et al. 1990) has shown that:

(a) East of Salmon River, sinkholes are few and concentrated in the Gun River formation and in the southern part of the Jupiter formation;
(b) West of the Jupiter River, sinkholes, karst areas and karst depressions have been mapped mainly in the Gun River and Becsie formations;
(c) In the central area between Jupiter and Salmon River, sinkholes, karst areas and karst depressions are concentrated in the central plateau area, along the Gun River formation, on the dip side of the central cuesta in the Jupiter formation and, to a lesser extent, in the Chicotte formation.

Some 79 karst zones have been identified in the island, with area ranging between 0.1 and 3 km^2. Karst features seen on the island were of the following types:

(a) Karren: millimetric, centimetric or metric-size surface features developed on rock outcrops or under very thin covers of surface materials or organic deposits;
(b) Open fractures: with widths ranging from a few centimeters to a few meters, lengths from a few meters to a few tens of meters and depths from a few meters

to ±18 m. This feature is the most frequent and well developed, mainly along E–W to ESE–WNW trending joints;

(c) Sinkholes: from a few meters to 1 km in diameter, with depths of a few meters to a few tens of meters, formed by solution, subsidence, collapse over large voids, or a combination of these processes;

(d) Sub-horizontal caves: active caves with lengths from a few meters to a few hundreds of meters, accessed from canyon sides, vertical shafts or springs. Only 8 are known, mainly along the Patate, Observation, Vauréal, Salmon rivers and the Malouin Lake, The longest one is Grotte à la Patate (625 m) (Fig. 9.8) followed by Grotte des Trois-Plaines (419 m) and Grotte de la Baie-de-la-Tour (264 m).

(e) Karst drainage: Many phenomena are related to underground flow: sinking streams in open fractures or sinkholes, surface or lake springs, blind valleys, dry valleys, karst windows, dry valleys and karst lakes. The longest underground flows are found in the Upper Salmon River karst (7–14 km). Elsewhere lengths range from a few hundred meters to less than 2 km. Of interest are those karst lakes found in the central plateau, inside a 60 km^2 area found west of Vauréal River (Fig. 9.9). The

Fig. 9.8 Entrance (spring) to *Grotte de la Patate*

hydrological behavior of these lakes is affected by open fractures, sinkholes and ponors. Lake-level variations depend mostly on the relation between local precipitation and karst seepage and can reach many meters. Levels gradually drop following snow melt and spring period, minimum observed levels being in winter. This progressive lowering of lake levels leads to a somewhat concentric zonation of littoral vegetation, intense ice dynamics and, in some cases, periglacial forms. The associated underground networks are small in extent, skin-deep and local. Most lakes are under 2 km^2 in area (Cadieux 1983; Lauriol et al. 1985; Coté et al. 2006).

(f) Residual hills: In the central plateau area, west of the Vauréal River, are hills with complex topography, usually less than 20 m in height but reaching locally 35 m, surrounded by flat surfaces. In some cases, they show rounded and profiled crests, in other bumpy or wrinkled veneers. They are densely punctuated with dry, often coalescing, depressions (Roberge 1996). These hills and depressions are highly variable in terms of dimensions, orientation, density, height, depth and area. In many cases they are located near embryonic karst or karst lakes and some seem derived from in situ bedrock. Surface is usually covered with cobbles and boulders, rarely by rock outcrops. Depressions show little relation with main joint sets and their form does not resemble glacial features. In some cases, these hills extend to rock surfaces with outcrops and karst pavements. Influence of structural features is clear north of Smith Lake where most depressions are trending N–S giving the impression of a labyrinth karst (Roberge 1996). Roberge (1996) suggests that they could be polygonal karst features partially eroded by glaciers while Cadieux (1983) sees them as regolith.

Fig. 9.9 Karst lakes in the Smith Lake area

9.6.3 Upper Salmon River Karst

This karst is, by far, the most important, developed and oldest one on the island. This is where surface karst features are spectacular. It has a WNW–ESE trend parallel to the strike of bedding and to the main joint set (N 110° E). Most features are found between elevations 170 and 190 m asl, with a main spring at 140 m asl. Maximum total drop is 60 m.

Karst features have developed mainly in the lower part of the Gun River formation and are limited to the north by the shale beds of the Merrimack formation and, to the south, by a more shaly section of the upper Gun River formation. Numerous peat bogs have formed south and west of the karst area and drainage network is disordered in those areas. Till thickness is usually less than 3 m.

The feeding basin (143.6 km^2) is divided into five non-karst sub-basins that are intercepted by open fractures or sinkholes when they enter the karst zone from the south (Fig. 9.10). Karst occupies some 27% of the basin and peat bogs 29% (Roberge 1979).

The holokarst, the area where there is no surface flow, covers an area of ±23.7 km^2 (2 × 12 km) and is fringed to the west and south by fluvikarst grounds with surface and underground flows.

The only known permanent spring is located at the eastern extremity of the karst, some 14.1 km from the Salmon River main sink (Fig. 9.11). Dye tracing tests have shown that it took less than 17 days ($\geq$0.85 km/day) for water to travel from sink to spring (Roberge 1979). Hydraulic gradients are small (0.2–1.0%) except near the main spring (6%). Karst seems to have developed from east to west.

Spring flow reacts more to heavy rainfall when the ground is soaked with water, that is when the spring outflow is large. Flood peaks were usually sharp and regular, with only one event per flood. Reaction time following a rainfall can be rapid, a few hours, suggesting pulse transfer in a saturated ground. Spring waters are under-saturated, suggesting rapid transit times and rapid flow. Chemical analysis of spring waters suggests water flow in a closed-circuit environment without access to CO_2 rich air.

Many karst features have been seen and mapped (Roberge 1979):

(a) Karst pavements where till is thin ($\leq$0.3 m) and forest is dry and open;
(b) Metric size open joints developed along the first and second joint sets where till is thicker (0.3–2.0 m);
(c) Infrequent subsidence sinkholes where till is thick ($\geq$2.0 m) (5–15 m wide, $\leq$4 m deep);
(d) Solution sinkholes developed along WNW–ESE joints and often associated with sinking streams. They are seen along water courses, dry beds or along peat bog edges. They can reach some 100 m in length and more than 10 m in depth. They can be active, intermittent or inactive, the last two types being often filled with till or alluvial material;
(e) Collapse sinkholes are the most spectacular ones. Five have been described by Roberge (1979), with volumes ranging from 300 m^3 to 6,000 m^3. Four are conical in shape while the last one, l'Abime du Chevreuil, is rectangular (44 × 4–7.5 × 18 m) (Fig. 9.12);

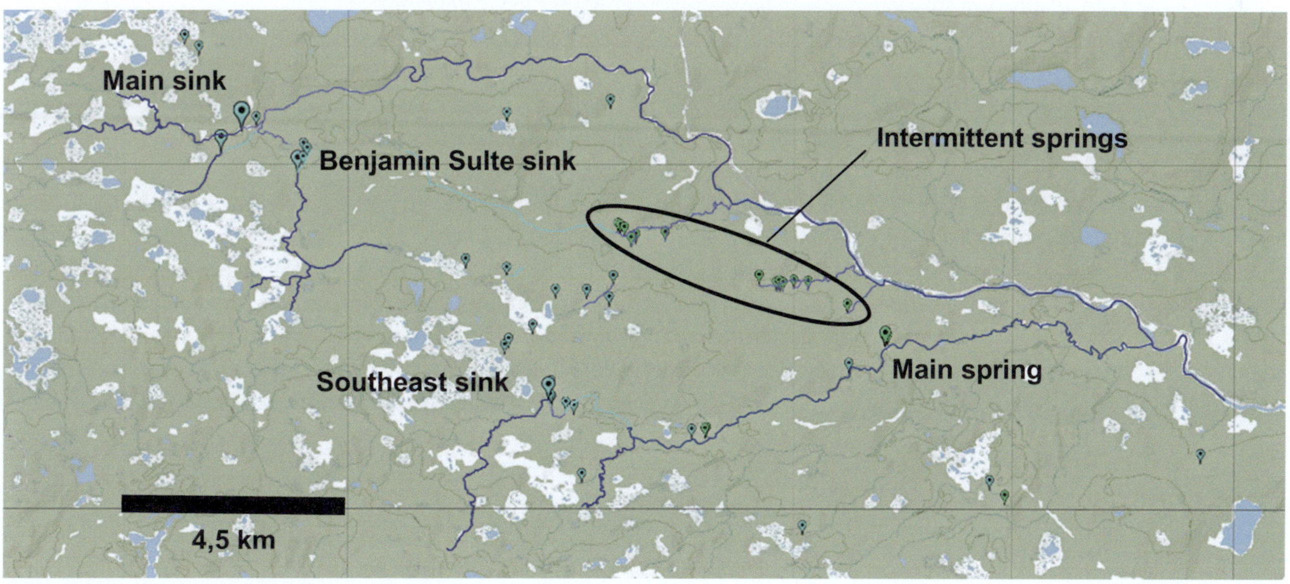

Fig. 9.10 The Upper Salmon River karst drainage (sinks in blue, spring in green)

Fig. 9.11 Main sink of Salmon River

(f) About ten deep sinkholes with steep sides (collapse, solution) are being used as creek sinks in the southern part of this karst;
(g) Some 20 km of dry, intermittent or permanent, stream beds have been mapped in this karst area. Many sinks filled with till, alluviums or organic debris are present (Fig. 9.13).

A recent visit to the Upper Salmon River karst has identified a series of intermittent springs on the north edge of the holokarst that seem to flow only in spring time or during large flood periods (Beaupré 2017; Fig. 9.14). They seem related to a westward old dry valley trending WNW–ESE, punctuated with large deep subsidence sinkholes. Head valleys associated with these springs are more developed and deeper than the main spring valley, suggesting that they are older.

Most karst features on the island are numerous, diversified and thought to be mainly Holocene in age. The Upper Salmon River karst with its large features and long drainage routes seems somewhat older and partly inherited from a previous karst development. The newly found springs and their associated dry valley appear definitively older.

9.7 Conclusion

This former huge private island seated in the Gulf of St. Lawrence has been, since its acquisition by the provincial government in 1974, the subject of many geological studies. Gas exploration, Quaternary studies to characterize glaciation and deglaciation history, and environmental studies to create a national park in a karst area, two ecological reserves and a set of more specific landmarks have been the subjects

Fig. 9.12 *L'Abime du Chevreuil*, a large collapse sinkhole

of these studies. Gas exploration has ceased and lumbering activity is on the decline. This island has become a salmon fishing and deer hunting paradise. Tourist activities are on the rise even though access by air or boat is not always easy and cheap, but those who visit the island are rewarded by something quite unique in the eastern part of Canada.

Karst studies have been somewhat slow moving since the 1970s, mainly because of the sheer size of the island, its remoteness, often difficult access to specific feature sites, and living conditions in the woods. Recently, the park service has been very helpful in facilitating access to remote and more interesting parts of the island, especially in the Upper Salmon River karst and in the southern parts of the national park. The size and variety of karst features and hydrogeological underground connections make this karst a quite unique feature in eastern Canada, more than worthy of future studies.

Fig. 9.13 The Benjamin Sulte River sink

Fig. 9.14 An intermittent spring on the northern edge of the holokarst

References

Bazoge A (2015) Étude AENV04, Caractérisation physique et biologique de l'île d'Anticosti, Étude ENV05, identification des zones de contraintes légales et réglementaires et d'autres zones de contraintes de l'île d'Anticosti, Étude AENV06, hydrologie et hydrométrie des bassins versants de l'Île d'Anticosti. Ministère du Développement durable, de l'Environnement et de la Lutte contre les changements climatiques, Direction générale de l'écologie et de la conservation, Report prepared on behalf of the Strategic Environmental Assessment of Anticosti Island

Beaupré M, Caron D, Roberge J (1974) Karstification de l'île d'Anticosti, reconnaissance 1974. Société Québécoise de Spéléologie, Montréal

Beaupré M, Caron D, Schroeder J (2017) Île d'Anticosti (Québec), reconnaissance spéléologique 2017. Mission report sent to the SÉPAQ and to the National park of Anticosti

Bedard JH (1992) Jurassic quartx-normative throleiite dikes from Anticosti island, Quebec. In: Puffer JH, Ragland PC (eds) Eastern America Mesozoic Magmatism. Geol Soc Am Spacial Papers 268, pp 161–168

Boisclair J (2004) Parc national d'Anticosti, plan directeur. Ministère de l'Environnement et de la Faune du Québec, Société de la Faune et des Parcs du Québec, Direction de la planification des parcs, Québec

Bordet E (2007) Analyse structurale de l'Île et de la plate-forme d'Anticosti, Québec. MSc thesis, Université du Québec

Bordet E, Malo M, Kirkwood D (2010) A structural study of Western Anticosti Island, St-Lawrence Platform, Quebec: a fracture analysis that integrates surface and subsurface structural data. Can Soc Petrol Geol Bull 58:36–55

Brake V, Pinet N (2015) Additional insights on fracture patterns in the northern Anticosti Basin from satellite, aerial and bathymetric images. Geological Survey of Canada Open File 7911

Cadieux R (1983) Contribution à l'hydrogéomorphologie du bassin de la rivière Jupiter, Île d'Anticosti, Québec. MSc thesis, Université de Sherbrooke

Carboni S (1988) Photointerprétation et analyses d'image des linéaments photogéologiques de l'île d'Anticosti, Québec. PhD thesis, Université de Montréal

Coté D (1987) Géomorphologie, hydrologie, hydrochimie des lacs du plateau central de l'Île d'Anticosti, Québec. MSc thesis, Université de Sherbrooke

Coté D, Dubois JM, Hétu B, Gwyn QHJ (2006) Les lacs karstiques de l'Île d'Anticosti: analyse hydrogéomorphologique. Université de Sherbrooke, départment de géomatique appliquée, Research Bull. 181, December

Desrochers A, Gauthier EL (2009) Carte géologique synthèse de l'Île d'Anticosti. Ministère des Ressources Naturelles et de la Faune du Québec, Quebec, DV 2009-03

Dubois JMM, Gwyn, QHJ, Bigras P, Gratton D, Perras S, St-Pierre L (1990) Géologie des formations en surface, Île d'Anticosti, Québec. Geological Survey of Canada Map 1660A

Dubois JM, Gwyn QHH, Gratton D, Painchaud A, Perras S, Cadieux D, St-Pierre L, Bigras P (1985) Géologie des formations en surface et géomorphologie de l'île d'Anticosti, Québec. Geological Survey of Canada Open File 1132

Eyles N, Putkinen N (2014) Glacially-megalineated limestone terrain of Anticosti Island, Gulf of St. Lawrence, Canada; onset zone of the Laurentian Channel Ice Stream. Quat Sci Rev 88:125–134

Gratton D, Dubois JM, Gwyn QHJ (1984) Les paléo-environnements sédimentaires au Wisconsinien moyen et supérieur, Île d'Anticosti, Golfe du Saint-Laurent. Géogr Phys Quat 38:229–242

Lauriol B, Hétu B, Coté D, Gwyn QHJ (1985) Phénomènes karstiques et périglaciaires dans un lac à niveau variable de l'Île d'Anticosti, Québec, Canada. Z Geomorphol 29:353–365

Lauriol B, Roberge L, Saint-Pierre L (1987) Les relations entre la géomorphologie, le climat et l'hydrochimie dans la caverne à la Patate, Île d'Anticosti, Canada. Z Geomorphol 31:419–431

Painchaud A, Dubois JM, Gwyn QHJ (1984) Déglaciation et émersion des terres de l'ouest de l'Île d'Anticosti, golfe du Saint-Laurent, Québec. Géogr Phys Quat 38:93–111

Petryk AA (1981a) Stratigraphy, sedimentology and paleogeography of the Upper Ordovician—Middle Silurian of Anticosti Island, Québec. In: l'Espérance, PJ (ed) Field meeting, Anticosti—Gaspé, 1981—Volume II: Stratigraphy and Paleontology, Université de Montréal, pp 11–40

Petryk AA (1981b) Carte géologique de l'ile d'Anticosti. Ministère de l'Énergie et des Ressources du Québec, Quebec, DPV-823

Petryk AA (1981c) Géologie de la partie ouest de l'Ile d'Anticosti. Ministère de l'Énergie et des ressources du Québec, Quebec, DPV-815

Pinet N, Brake V, Lavoie D (2015) Geometry and regional significance of joint sets in the Ordovician-Silurian Anticosti Basin: new insights from fracture mapping. Geological Survey of Canada Open File 7752

Roberge J (1979) Géomorphologie du karst de la Haute-Saumons, Île d'Anticosti, Québec. MSc thesis, McMaster University

Roberge J (1996). Géomorphologie de l'Île d'Anticosti et de la région de la rivière Vauréal: état des connaissances. Ministère de l'Environnement et de la Faune du Québec, Direction du Plein-Air et des Parcs, March, 214 p

Roberge J (2000) Spécificité de la géomorphologie de la pointe Ouest de l'île d'Anticosti. Ministère de l'Environnement et de la Faune du Québec, Direction du Plein-Air et des Parcs, unpublished report

Roberge J, Lauriol B, St-Pierre L (1985) La morphogénèse de la caverne à la Patate Île d'Anticosti. Québec. Géogr Phys Quat 39:67–75

Roberge J, Ford DC (1983) The Upper Salmon River karst, Anticosti Island, Québec, Canada. J Hydrol 61:159–162

Shaw J, Courtney RC (2002) Postglacial coastlines of Atlantic Canada: digital images. Geological Survey of Canada Open File Report 4302

St-Pierre L, Gwyn QHJ, Dubois JM (1987) Lithostratigraphie et dynamique glaciaires au Wisconsinien, Île d'Anticosti, Golfe du Saint-Laurent. Can J Earth Sci 24:1847–1858

SOQUIP (1985) Interprétation géologique d'images LANDSAT de l'Île d'Anticosti. Ministère de l'Énergie et des Ressources du Québec, GM-43323, October

Tanguay MG et al (1986) Étude des linéaments de l'île d'Anticosti sur les photos aériennes noir et blanc, au 1:40 000. Institut de Recherche en Exploration Minérale/ Mineral Exploration Research Institute

Michel Beaupré is a senior engineering geologist working for Hatch. After an M.Sc. in geology from Montréal University, he studied karst geomorphology at PhD level at MacMaster University for one year and he completed his B.ScA. at L'École Polytechnique de Montréal in engineering geology. He worked for Hydro-Québec for 29 years, first as an engineering geologist and head of the geology department, then as a project engineer. Since then he has been working for 12 years as a senior engineering geologist for different engineering consulting companies mainly in Canada, Central and South Americas, Africa, Southeast Asia and China. The main fields of interest are hydroelectric projects and karst and caves studies.

Deltaic Complexes of the Québec North Shore

10

Pierre Dietrich, Alexandre Normandeau, Patrick Lajeunesse, Jean-François Ghienne, Mathieu Schuster, and Alexis Nutz

Abstract

One of the most particular morphological features of the Québec North Shore (North Shore of the Estuary and Gulf of the St. Lawrence, eastern Canada) is the occurrence of deltaic complexes that form thick and extensive sediment bodies (gravel, sand and mud) along the modern coastline. The deltaic complexes were emplaced during the retreat of the Laurentide Ice Sheet from the region that commenced about 11 ka in the context of falling relative sea level. Deltaic complexes are made up of three distinct, superimposed depositional systems, consisting of: (1) ice-contact subaqueous fans and deltas, (2) ice-distal glacifluvial deltas, and (3) coastal suites and meandering channel belts. Ice-contact systems were emplaced immediately after deglaciation of the region at the front of the ice-margins. Subsequent ice retreat fed glacifluvial deltas in meltwater and glacigenic sediments until the complete retreat of the ice-margin from the drainage basin a few thousand years ago. The resulting shutdown in sediment supply led to the reworking of the now-inactive glacifluvial deltas by shore-related and fluvial processes that resulted in the deposition of coastal suites and meander channel belts. Submarine sediment accumulations and related morphologies revealed by high-resolution swath bathymetric and seismostratigraphic data collected off these deltaic complexes are the subaqueous counterparts of these deltas. The modern evolution of these deltaic complexes is controlled by the reworking, transport, and deposition of sediments by shore-related processes along the coast, on the shallow shelf, and through submarine channels and canyons.

Keywords

Deltaic complexes • Laurentide Ice Sheet • Deglaciation • Ice-contact deposits • Glacifluvial deltas • Submarine fans

P. Dietrich (✉)
Univ Rennes, CNRS, Géosciences Rennes, UMR 6118, 35000 Rennes, France
e-mail: pierre.dietrich@univ-rennes1.fr

P. Dietrich
Department of Geology, University of Johannesburg, Johannesburg, South Africa

A. Normandeau
Geological Survey of Canada, Atlantic, Dartmouth, NS, Canada

P. Lajeunesse
Département de Géographie, Université Laval, Québec, Canada

J.-F. Ghienne · M. Schuster
Institut de Physique du Globe de Strasbourg, UMR 7516, CNRS, Université de Strasbourg, Strasbourg, France

A. Nutz
Centre Européen de Recherche et d'Enseignement des Géosciences de l'Environnement (CEREGE), Aix-Marseille University, 13545 Aix-en-Provence Cedex 4, France

A. Nutz
Department of Geoscience, Aarhus University, Aarhus, Denmark

10.1 Introduction

Thick and extensive (tens to hundreds of km^2) sediment accumulations consisting of gravel, sand and mud form prominent deltaic complexes along the Québec North Shore (QNS, North Shore of the Estuary and Gulf of the St. Lawrence, Québec, eastern Canada (Figs. 10.1 and 10.2). These deltaic complexes lie at the outlet of deep and steep-sided valleys carved into the hinterland bedrock in which modern rivers flow. Although some of these deltaic complexes are conspicuously protruding in the Estuary and Gulf of St. Lawrence (e.g. Fig. 10.2), most are characterized by straight shorelines and are hence recognized by their smooth aspect on digital elevation models (DEM) and LiDAR compared to the rough aspect of the surrounding bedrock.

These deltaic complexes essentially consist of relict sedimentary bodies raised above the modern sea level (Normandeau et al. 2015; Dietrich et al. 2017a) as they were emplaced immediately after the retreat of the Laurentide Ice

© Springer Nature Switzerland AG 2020
O. Slaymaker and N. Catto (eds.), *Landscapes and Landforms of Eastern Canada*, World Geomorphological Landscapes, https://doi.org/10.1007/978-3-030-35137-3_10

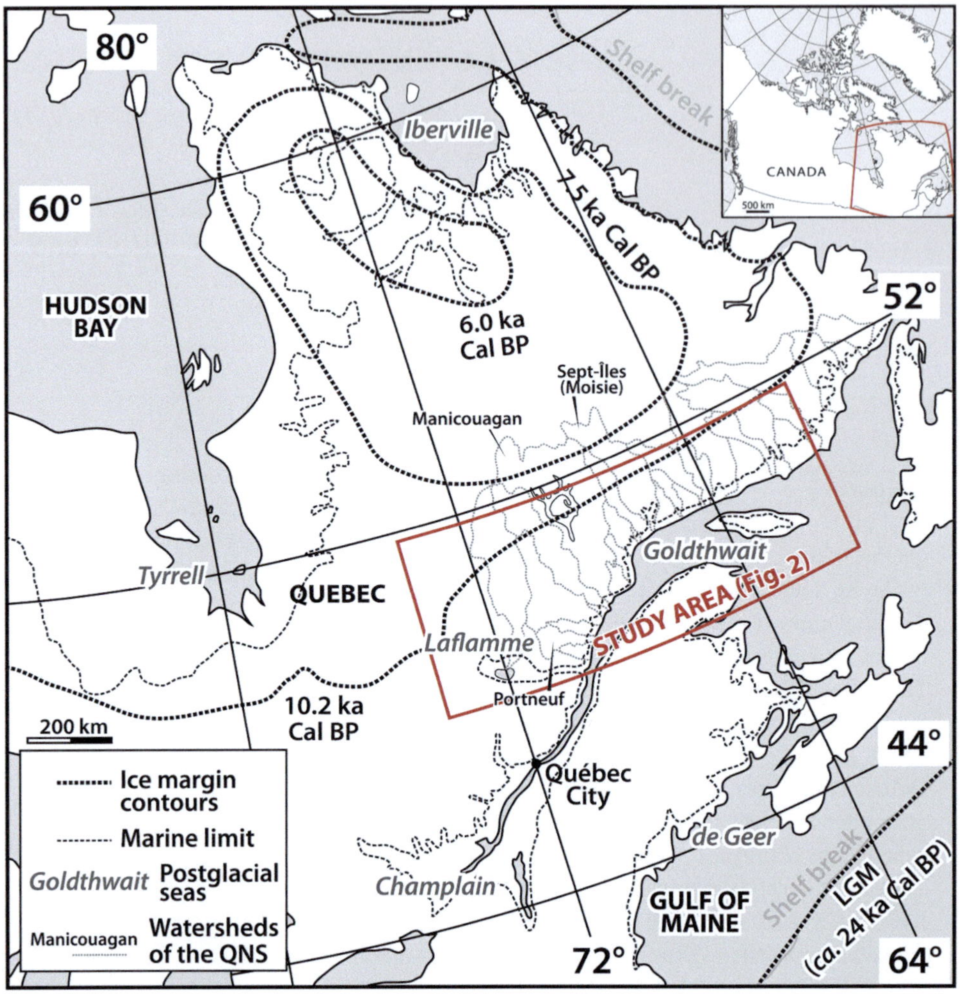

Fig. 10.1 Location map of the study area (Québec North Shore, Estuary and Gulf of St. Lawrence) within the frame of the deglaciated eastern Canada. Bold dashed lines represent successive contours of the retreating Laurentide Ice Sheet. Maximal extent of postglacial seas (the Goldthwait Sea in the study area) is symbolized by the marine limit. Watersheds of the Estuary and Gulf of St. Lawrence are also represented (gray dotted lines)

Sheet (LIS) around 11 ka in the glacimarine and postglacial Goldthwait Sea (Fig. 10.1), which subsequently experienced relative sea level (RSL) fall (Fig. 10.3). Prominent morphosedimentary structures such as relict-braided fluvial patterns, kettle lakes, raised beach ridges, and sandspits characterize the surface of these sedimentary systems (Dredge 1983; Drapeau 1992; Hein et al. 1993; Dionne et al. 2004).

High-resolution swath bathymetric and seismostratigraphic surveys conducted off the deltaic complexes revealed the presence of subaqueous sediment accumulations characterized by key morphological features such as bedforms and submarine channels (e.g. Normandeau et al. 2013, 2017). These subaqueous sediment accumulations and their correlative morphologies are the marine counterpart of the deltas observed on land. Both settings, subaqueous and subaerial, are part of the same system and as such have been studied as a continuum.

10.2 Setting

10.2.1 Geographic Setting

The St. Lawrence Estuary and Gulf form a deep (mean depth 150 m) semi-enclosed sea lying on the continental platform and supplied with fresh water from the St. Lawrence River to the southwest. The St. Lawrence Estuary and Gulf are bordered by the moderate to high relief forested Shield geomorphological region (Grenville geological province) and the Appalachian mountain geomorphological region (Appalachian geological province) on its western and south-eastern sides, respectively, and open to the Atlantic Ocean to the east (Fig. 10.1). Palaeozoic sedimentary rocks of the eastern part of the St. Lawrence Lowland (St. Lawrence platform) are also found from place to place along the QNS (Lajeunesse et al. 2013). The Estuary and Gulf of St. Lawrence are characterized by the prominent Laurentian

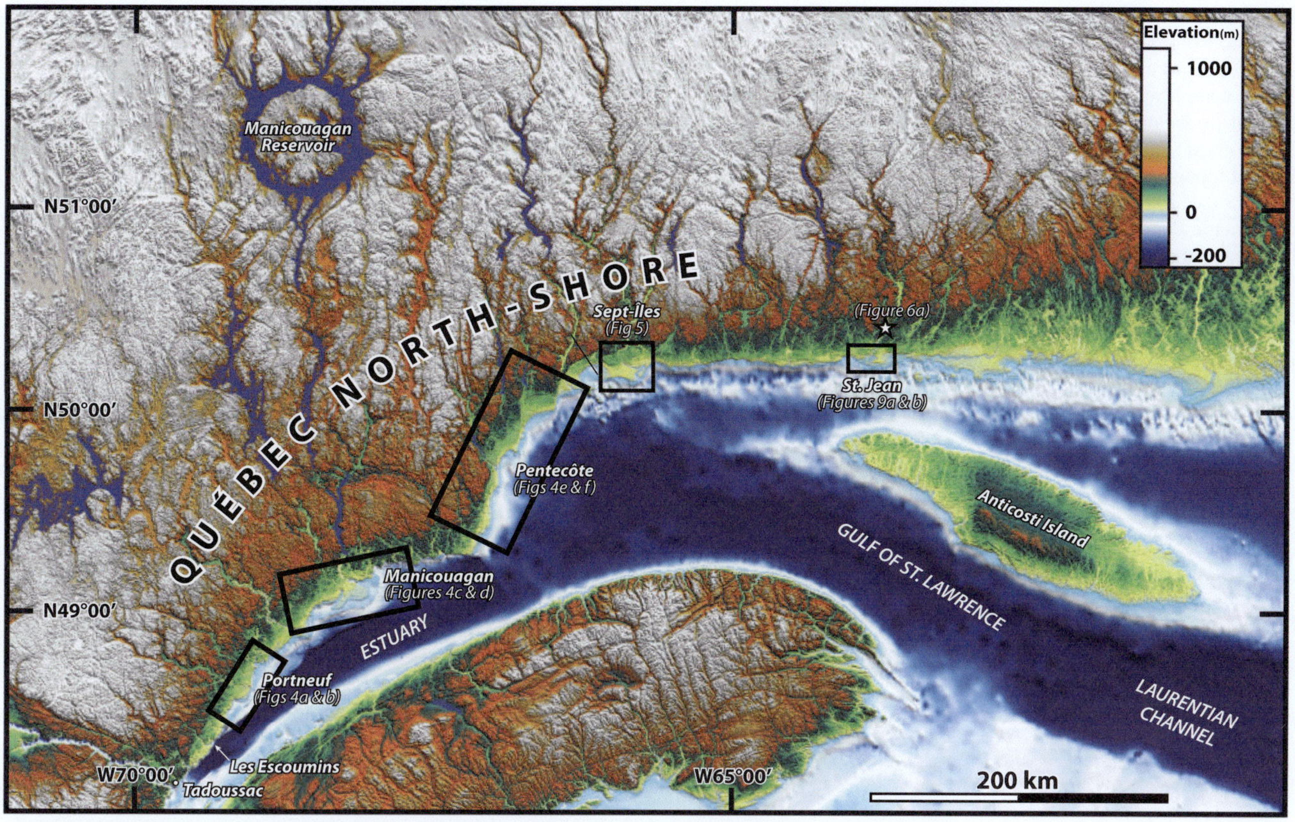

Fig. 10.2 Map of the Québec North Shore (digital elevation model, SRTM3, available at https://www2.jpl.nasa.gov/srtm/) showing locations of deltaic complexes mentioned in the text. See Fig. 10.1 for location in eastern Canada

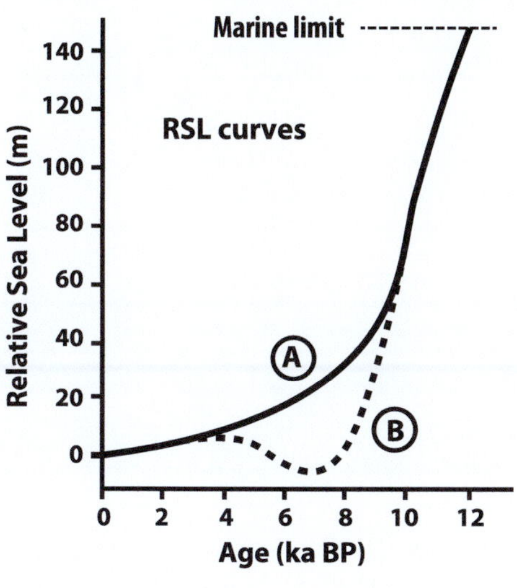

Fig. 10.3 Relative sea level curves for the Québec North Shore, where A is from Dietrich et al. (2017a) and B from Dionne (2001). The latter inferred the presence of a lowstand between 7 and 6 ka followed by a 10 m RSL rise (the so-called *Laurentian transgression*) for which there is evidence on the Québec South Shore (Dionne 2001) but none observed on the Québec North Shore by the present authors

Channel, a 1000 km-long and 80–90 km-wide U-shaped cross-shelf trough that extends from Tadoussac to the edge of the continental platform (Fig. 10.2; Todd 2016; Pinet et al. 2016). The Laurentian Channel is separated from the QNS by a shallow 0–20 km-wide structural shelf that extends from Tadoussac to Sept-Iles (Fig. 10.2; Pinet et al. 2011). The slopes linking the Laurentian Channel to the shallow shelf vary in angle from 2 to 20°. The terrestrial landscape of the QNS is characterized by a series of gently sloping hills reaching hundreds of meters in elevation (Figs. 10.2 and 10.4). These hills, consisting of glacially carved crystalline bedrock of the Grenville Province, are deeply incised (100–400 m) by steep-flanked structural valleys, 0.5–3 km in width, oriented at 90° to the axis of the Estuary and with their bases lying between 50 m above and 300 m below the sea level (Figs. 10.4; Lajeunesse 2014).

10.2.2 Deglaciation and RSL History

After the last glacial maximum (LGM) that occurred approximately 23 ka, the Laurentide Ice Sheet (LIS) began

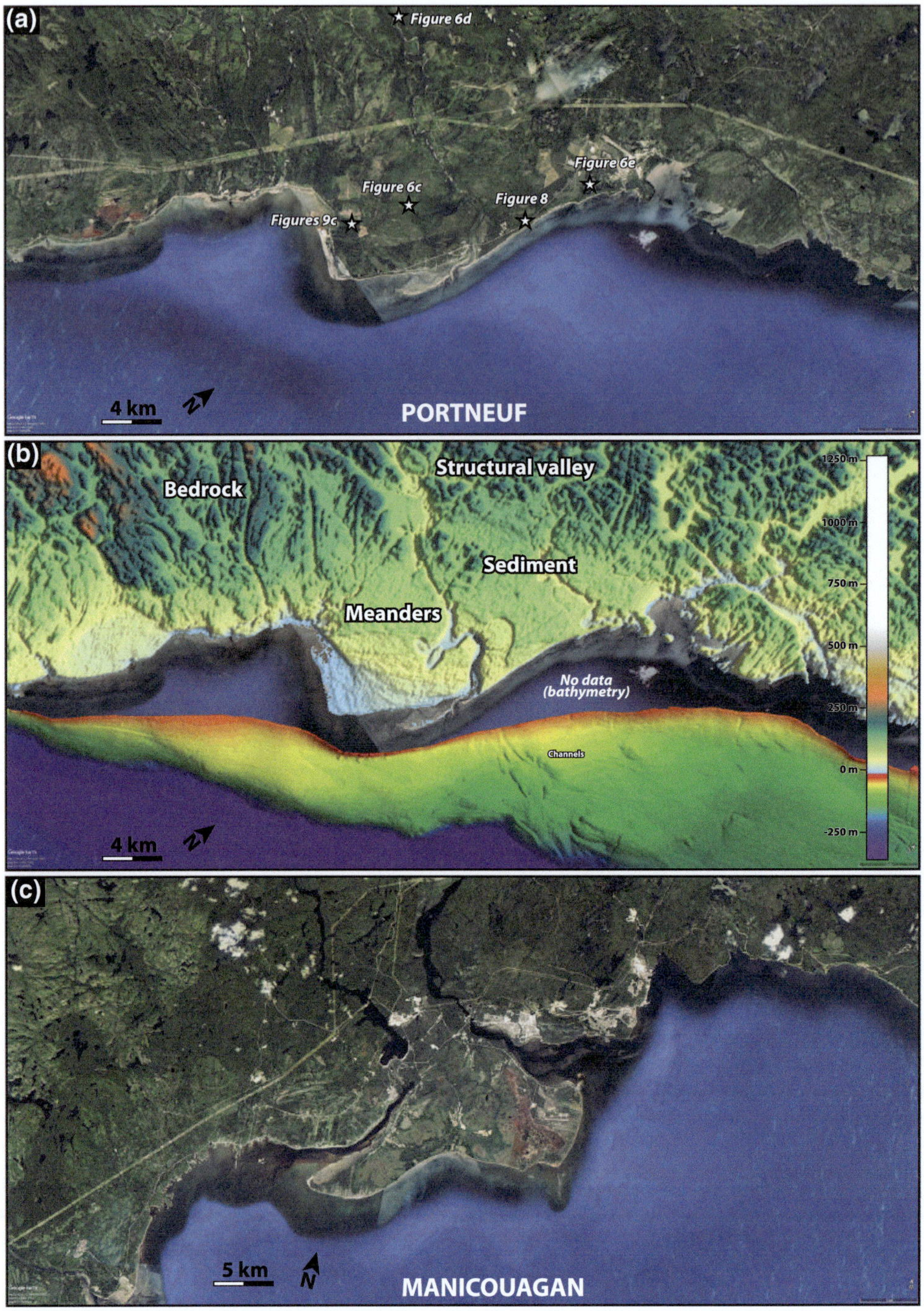

Fig. 10.4 Google Earth and correlative digital elevation model (DEM) images of some of the deltaic complexes mentioned in the text (**a**, **b** Portneuf, **c**, **d** Manicouagan, and **e**, **f** Pentecôte). Onland topography is from SRTM3 (see Fig. 10.1) and bathymetric data derive from Normandeau et al. (2015)

retreating by rapid collapse, generating massive iceberg calving in the Gulf of St. Lawrence (Shaw et al. 2002, 2006). The QNS was essentially deglaciated by c. 11 ka. Retreat of the LIS was subsequently achieved by ablation, deglaciating the northern extent of the Québec North Shore drainage basins by c. 7 ka (Fig. 10.1; Occhietti et al. 2011; Nutz et al. 2015). Along the QNS, immediately after the ice retreat, the invasion of the Goldthwait Sea on glaci-isostatically flexured lowlands allowed the relative sea level (RSL) to reach an elevation of 130–150 m above the modern sea level, forming the marine limit (Figs. 10.1 and 10.3). This 'postglacial' sea also occupied the bottom of some of the structural valleys that formed fjords or fjord-lakes at that time. Glaci-isostatic adjustment (GIA) led to a continuous RSL fall with rates reaching several centimeters per year, in spite of the concomitant global eustatic sea level rise (Fig. 10.3;

Fig. 10.4 (continued)

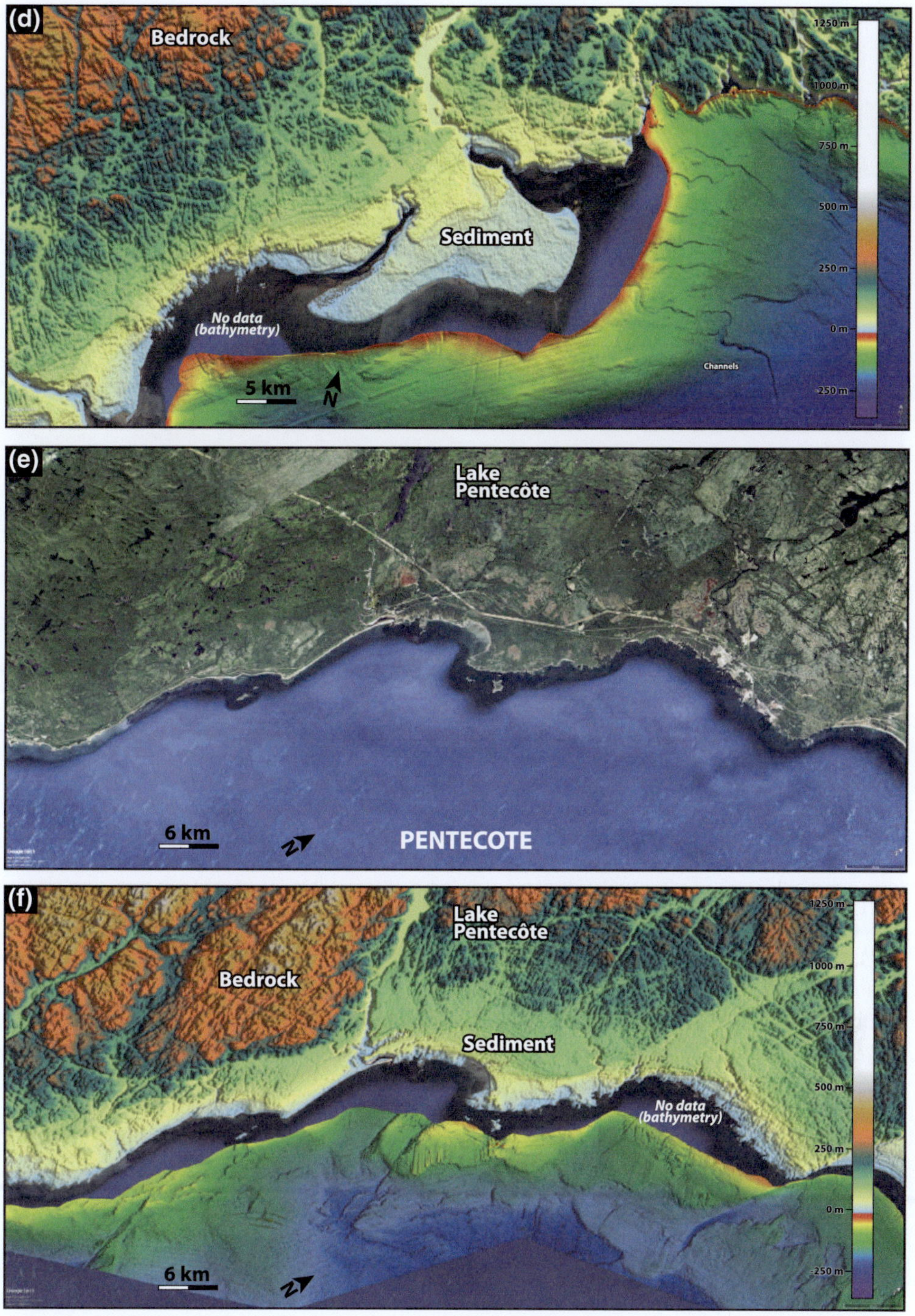

Boulton 1990; Peltier et al. 2015). The deltaic complexes were thus emplaced in a context of falling RSL forced by the GIA and fed by meltwater and glacigenic sediments from the retreating LIS margin. Nowadays, RSL is stable in this region, as the still ongoing GIA is essentially compensated by modern global eustatic rise (Sella et al. 2007; Woodworth et al. 2011). The modern evolution of deltaic complexes is characterized by a generalized retreat of the shoreline (Bernatchez and Dubois 2008).

10.3 Deltaic Complexes

The deltaic complexes lying on the QNS are made up of three distinct depositional systems, each one deposited in accordance with the pattern of ice-margin retreat and associated glaci-isostatically-forced RSL fall (Dietrich et al. 2017a). The three depositional systems, either vertically superimposed or laterally juxtaposed, consist of (1) ice-contact

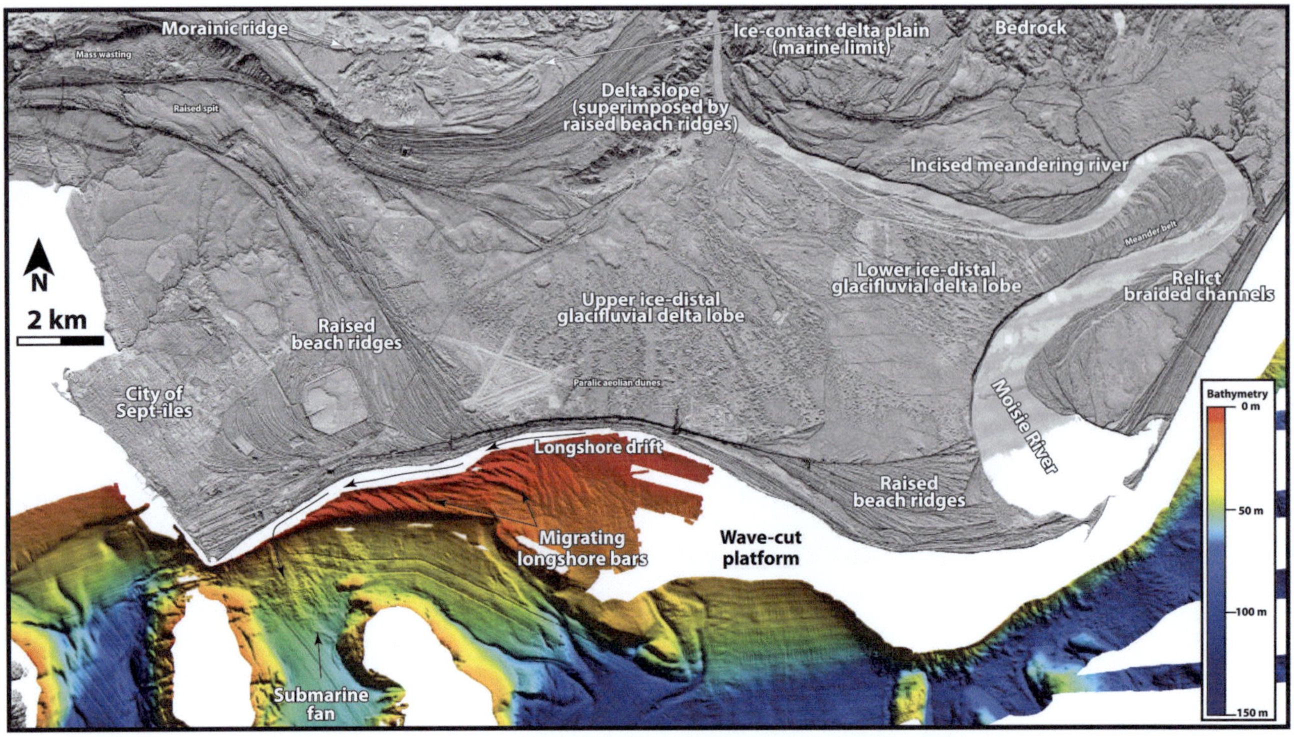

Fig. 10.5 LiDAR cover and multibeam bathymetry of the Sept-Iles deltaic complex. Note well-expressed morphologies (morainic ridges, braided channels, paralic aeolian dunes, etc.) on the LiDAR. The bathymetric cover illustrates the modern sediment dynamics on delta fronts. The delta front is now being eroded and sediment is transported by longshore drift. Where the shelf narrows, sediment overspills on the slope and generates turbidity currents, that in turn accumulate as submarine fans. See Fig. 10.2 for location. LiDAR available at https://geoegl.msp.gouv.qc.ca/igo/mffpecofor/; bathymetry derives from Normandeau et al. (2013)

subaqueous fans and ice-contact deltas, together referred to as ice-contact depositional systems; (2) ice-distal glacifluvial deltas and submarine counterparts; and (3) coastal suites and meandering channel belts. As these depositional systems have specific and unique morphological and surficial expressions, they can confidently be identified on satellite and aerial imagery (Figs. 10.2, 10.4, 10.5, and 10.6) which renders their identification straightforward. Submarine channels and canyons revealed by high-resolution swath bathymetry and seismostratigraphic surveys intrinsically belong to one or more of these three architectural elements (Figs. 10.4 and 10.5; Normandeau et al. 2017).

10.3.1 Subaqueous Ice-Contact Fans and Ice-Contact Deltas

Ice-contact depositional systems form sediment wedges that were deposited at the direct vicinity of a stillstanding ice-margin, immediately following deglaciation of the area. They, therefore, most often lie directly on the bedrock (Fig. 10.7). Although grounding zone wedges and morainal banks have recently been mapped on the shallow structural shelf (Lajeunesse 2016; Lajeunesse et al. 2018), subaqueous ice-contact fans and ice-contact deltas are present on land, at the outlet of structural valleys (Dietrich et al. 2017a, b; Gagnon-Poiré et al. 2018). Ice-contact systems generally constitute the core of the deltaic complexes. Ice-contact deltas are characterized by flat-lying topsets (delta plain) and basinward-dipping clinothems (delta slope), regularly superimposed by raised beach ridges (Fig. 10.5). As deposition of these specific ice-contact systems occurred during early stages of deglaciation, in the context of high RSL, their topsets lie directly at, or slightly below, marine limit (Figs. 10.5 and 10.7). As revealed by aerial and satellite imagery (Fig. 10.5), their topsets are characterized by relict kettle holes occupied today by circular lakes, inactive braided channels, and more rarely by morainic ridges, indicative of the direct vicinity of the ice-margin during deposition (Dubois and Dionne 1985; Dietrich et al. 2016).

The ice-margin stillstands that punctuated the general retreat of the LIS over the study area and permitted the deposition of ice-contact subaqueous fans and deltas generally occurred at the outlet of the structural valleys. After the retreat of the ice-margin from its stillstanding position, the ice-contacts systems may have dammed fjord-lakes, some of which still exist, in the axes of the structural valleys (Normandeau et al. 2016; Dietrich et al. 2017b; Gagnon-Poiré et al. 2018).

Fig. 10.6 **a** Steep-flanked structural valley carved into the bedrock of the Québec North Shore in which the modern St. Jean River flows. Note that the bottom of the valley lies below marine limit (highest elevation reached by the sea immediately after the deglaciation of the area). The valley was invaded at times by the sea and thus formed a fjord (see Fig. 10.7). Ice-distal glacifluvial sediments are preserved in the valley in which the modern meandering river is incised. **b** Relict-braided fluvial pattern characterizing the top surface of an ice-contact depositional system. **c** Abandoned meandering channel of the proto-Portneuf River incised of 20–30 m within the glacifluvial delta. Abandoned channels are occupied by lakes and peat bogs. **d** The modern meandering Portneuf River deeply incised (40 m) within the ice-distal glacifluvial delta sediments. **e** The meandering Sault-au-Cochon River. Although standing very close to the modern shoreline, the river lies at 15 m above the modern sea level because a bedrock sill on the river path prevents any further adjustment of the river long profile to the sea level. See Figs. 10.2 and 10.4 for location

10.3.2 Ice-Distal Glacifluvial Delta

Ice-distal glacifluvial deltas relate to spatially extensive (several tens of km^2; Fig. 10.5) and thick (up to 70 m, Fig. 10.8) sediment bodies that were deposited over or next to ice-contact systems. Glacifluvial deltas were deposited when the ice-margin retreated throughout the drainage basins, and fed meltwater and glacigenic sediment into the coastal delta systems (Fig. 10.7). Glacifluvial deltas are identified by their well-defined stratigraphic architecture formed by topset, foreset (Fig. 10.8b), and bottomset beds, respectively, corresponding to delta-plain, delta-slope, and prodelta deposits. As the deposition of glacifluvial deltas occurred following the ice-contact systems during a period of RSL fall, the elevation of the sea level was at that time lower than marine limit by tens of meters, although much higher than today (Figs. 10.3 and 10.7). The top surface (sandy delta plain) of glacifluvial deltas hence lies at

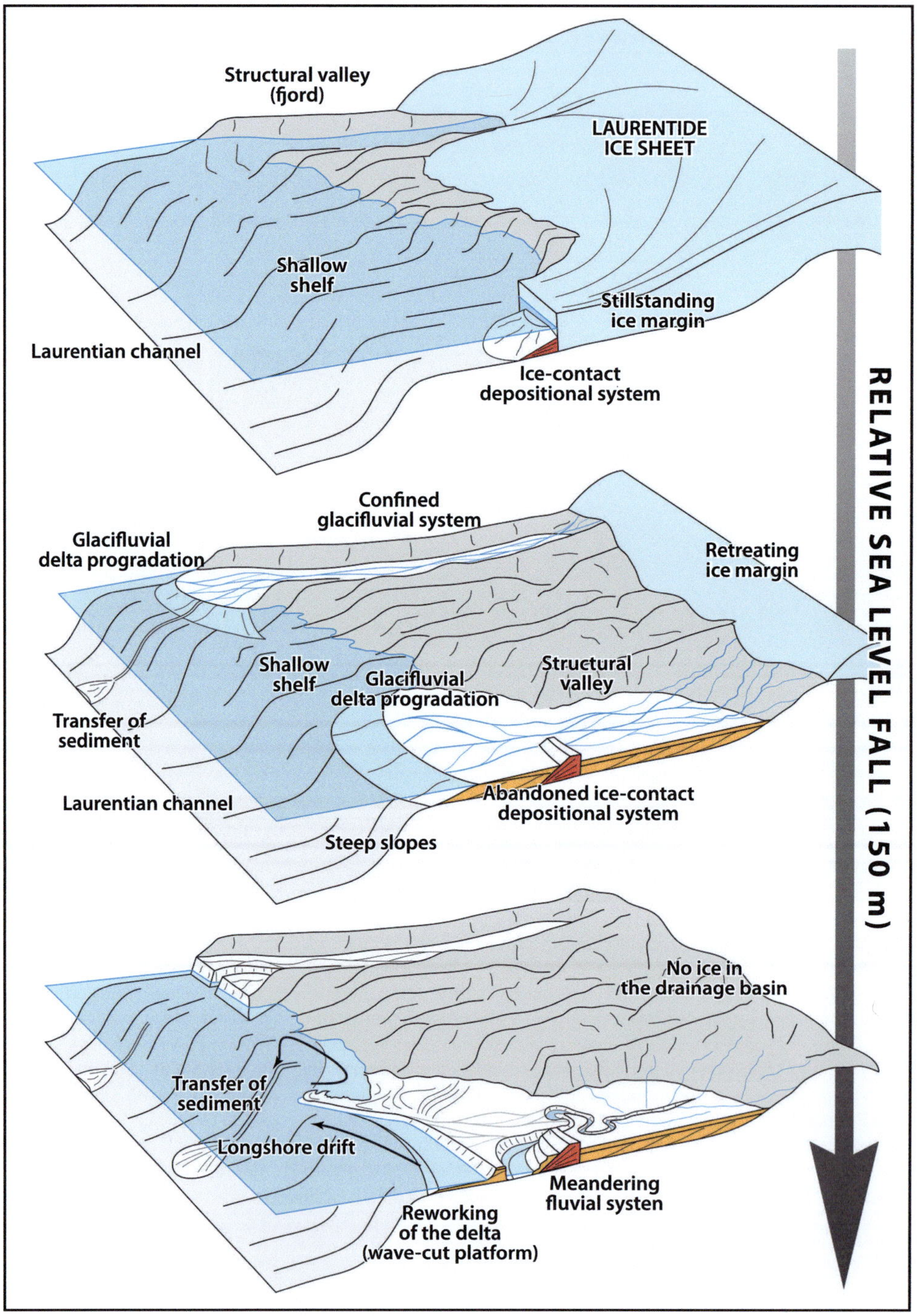

Fig. 10.7 Three-staged conceptual model of the deposition of the deltaic complexes of the Québec North Shore

Fig. 10.8 Aerial (drone) pictures of the Portneuf deltaic complex. **a** Panoramic view of the whole Portneuf deltaic complex. Ice-contact and ice-distal glacifluvial sediments are well-exposed in the cliff that experience active retreat due to wave erosion processes. Note that the extensive top surface of the deltaic complex gently dips basinward due to its emplacement throughout RSL fall. **b** Panoramic view of the exposed foreset beds of the ice-distal glacifluvial delta truncated by erosive-based coastal suites (dashed black lines). Note the dipping of the foreset beds toward the left of the picture. **c** Panoramic view of the spit platform (belonging to coastal suites) lying at an extremity of the deltaic complex. Note the steeply dipping foresets, the modern beach, and the currently active longshore bars fed by the erosion of the deltaic complex. The cliff is 20 m high. See Fig. 10.2 for location

elevations between 120 and 40 m asl and dips gently toward the sea (Dredge 1983; Dionne et al. 2004). In addition, the top surfaces of the glacifluvial deltas are characterized by rare and poorly expressed relict-braided fluvial patterns and isolated raised beach ridges lying on otherwise featureless flat surfaces (Fig. 10.5; Dietrich et al. 2017a).

After the retreat of the stillstanding ice-margin that deposited ice-contact systems, ice-distal glacifluvial deltas supplied by meltwater and glacigenic sediments from the ablation-dominated ice-margin were initially confined in structural valleys, either occupied by fjord-lakes, or by the sea (fjords; Fig. 10.7). Restricted in-valley accumulation space combined with high rates of sediment supply permitted high rates of delta progradation. Subsequently, as the entire in-valley space was filled, glacifluvial deltas emerged on the St. Lawrence shallow, structural shelf, where the increase of accommodation space caused sudden drops in progradation rates (Dietrich et al. 2017b). In some cases, depending on the width of the shallow shelf and the duration of the glacifluvial dynamic, coarse-grained materials were supplied to the deep realm of the St. Lawrence (Laurentian Channel, see below; Normandeau et al. 2017).

10.3.3 Submarine Channels

The ice-distal glacifluvial deltas described above prograded into the Goldthwait Sea as long as glacial meltwaters were present in their glaci-hydrological drainage basins. The inherited morphology of these deltas in some cases can still be observed in the St. Lawrence Estuary (Fig. 10.4). Multibeam bathymetric data shows numerous channels at river mouths, on the shallow shelf and on the slopes of the Laurentian Channel (Fig. 10.4). The channels may reach lengths of tens of kilometers offshore and may be as wide as 1 km (Fig. 10.4). Channel fills consist of an alternation of hemipelagic mud, deposited from normal settling of sediment in the water column, and turbidites, resulting from hyperpycnal flows at the mouths of the rivers. The generation of hyperpycnal flows, formed by the direct plunging of a mixture of sediment and water at the mouth of the river, was made possible due to the high content of suspended and bedload sediments carried by glacifluvial rivers entering the Goldthwait Sea (Mulder and Syvitski 1995). Hyperpycnal flows were responsible for the formation of the channels observed offshore. High sedimentation rates at river mouths also favored the formation of slope failures on delta fronts (Duchesne et al. 2003). Seismic reflection data show that the channels have bedforms both on their levees and in their axes, interpreted as sediment waves formed during the passage of turbidity currents. The morphology of these bedforms and channels is today much more attenuated than it was during their formation. Today, they are either completely or partly buried by tens of meters of mud, because sediment concentration in the rivers is no longer high enough to overcome the density of the seawater, thus not generating hyperpycnal flows. Related channels are hence relict forms.

Some of the submarine components of the deltas do not reveal any type of channels or bedforms that would provide evidence for the activity of hyperpycnal flows during their formation. One example is the Pentecôte River, where the offshore environment is devoid of delta-related morphologies (Fig. 10.4f). For those deltas, it is assumed that sediments were trapped inland in lakes, such as Lake Pentecôte (Fig. 10.4f). Sediment supply at the mouth of the river was greatly reduced, precluding the formation of large ice-distal glacifluvial deltas. This lake trapping can be observed for the Aux-Rochers and Magpie rivers as well (Gagnon-Poiré et al. 2018).

10.3.4 Coastal Suites and Meandering Channel Belts

After the retreat of the LIS margin from individual drainage basins and the resulting shutdown in meltwater and glacigenic sediment supply, deltaic progradation was no longer possible. Therefore, reworking processes started to dominate in a setting characterized by a still ongoing RSL fall (Fig. 10.7). Shallow-marine processes (wave ravinement) allowed the development of well-expressed shorelines, beach ridges, spits and spit platforms (Fig. 10.9c), coastal cliffs, and backshore systems, referred to as coastal suites (Fig. 10.9; Dionne et al. 2004). These coastal suites were successively formed, abandoned, and then raised as the RSL was falling (Fig. 10.3). The pattern of deposition and successive abandonment of shorelines was also likely controlled by the availability of local finite sediment sources (erosion of inactive ice-contact and/or glacifluvial depocenters; Hein et al. 2014), progressive emersion of bedrock exposures (Fraser et al. 2005), inherited delta morphology, and patterns of local longshore drift and coast morphology, among other autogenic or local processes. The resulting modern surface expression of the most basinward side of the deltaic complexes is characterized by stepped, relict and raised shorelines and associated morphologies, from 100 m asl down to the modern sea level (Figs. 10.5 and 10.9). On the other hand, the abrupt decrease in supply of glacigenic sediments also forced the transition from braided to meandering fluvial systems (Fig. 10.6). In a similar manner as for the coastal suites, the successive incisions, avulsions, and abandonments of meandering channels throughout the ongoing RSL fall left behind stepped and relict meander belts within the ancient and inactive glacifluvial deltas (Fig. 10.6). These abandoned and stepped meander belts are prominently

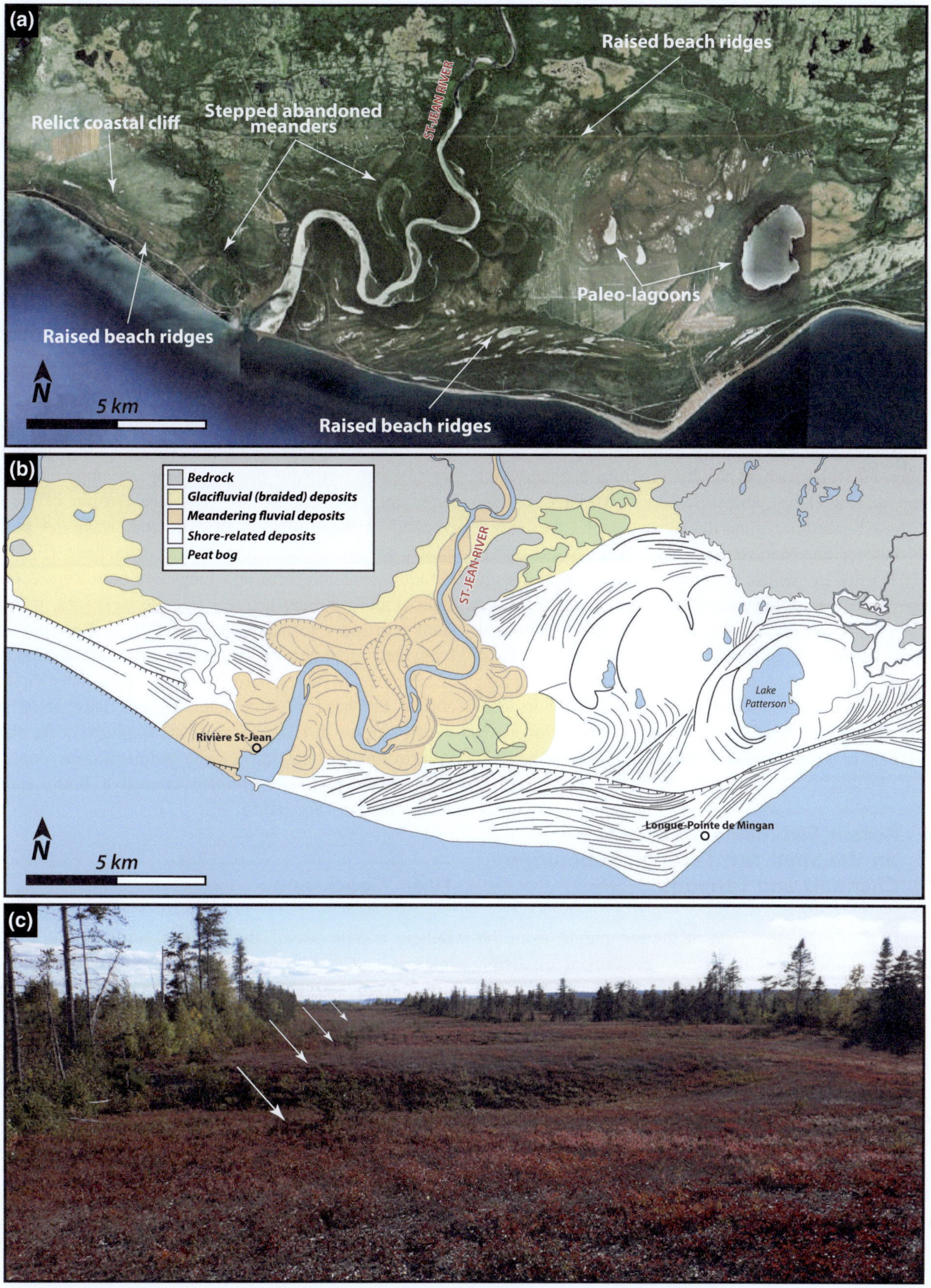

Fig. 10.9 **a** and **b** Satellite and correlative line-drawing of the St. Jean deltaic complex evidencing the large array of coastal structures (beach ridges, coastal cliffs, raised spits) characterizing the top surface of the deltaic complex imputed to coastal processes. Also note the presence of the abandoned meandering channels. **c** Stepped beach ridges raised well above the modern sea level (altitude: ~50 m). See Figs. 10.2 and 10.4 for location

exposed on aerial and satellite imagery as well as on LiDAR and DEM (Figs. 10.4 and 10.5). Although modern rivers are still meandering and encased within the glacifluvial deltas they are eroding (Figs. 10.6d, e), RSL fall has now mainly stopped. Presently, meandering, meander amplification, and avulsion processes are thought to be partly controlled by bedrock threshold exhumed along the path of the rivers (Fig. 10.6e; Tooth et al. 2004), or in places by retreat of the shoreline that forces constant readjustment of the fluvial longitudinal profiles.

Drainage basins of the QNS have different spatial extents and shapes, ranging from a few hundreds to more than 45,000 km^2 (Manicouagan; Fig. 10.1) in area. Therefore, the retreat of the ice-margin from the respective drainage basins, the resulting shutdown of glacifluvial delta progradation, and the onset of reworking processes did not occur at the same time in each drainage basin of the QNS. The smallest drainage basins experienced an early transition (10.5 ka; Portneuf; Fig. 10.1; Dietrich et al. 2017a) from a glacifluvial dynamic to reworking processes. More extensive basins, and/or those located in regions deglaciated later, experienced a later shift (7 ka; Manicouagan; Fig. 10.1; Normandeau et al. 2017). Hence, deltaic complexes with drainage basins that have been deglaciated earlier were submitted to a greater RSL fall (up to 100 m) after this transition, which permitted enhanced deposition of prominent stepped shore-related landforms on deltaic complexes. Conversely, drainage basins that were deglaciated later only endured a short RSL fall after glacial retreat, and hence have poorly developed, or at least weakly stepped, coastal suites and meander belts.

10.4 Recent Sedimentary Processes on the Shelf and Through Submarine Channels and Canyons

Following the complete retreat of the ice-margin from the watersheds of the respective rivers, glacifluvial discharge was greatly reduced. Deltas ceased being dominated by river discharge (fluvial-dominated) and became dominated by wave action. Nearshore, as erosion processes were now dominant, delta fronts began to be cannibalized by shore-related processes, which made the deltas more linear (Figs. 10.7 and 10.8; Bernatchez and Dubois 2008). The erosion of the deltas, which began in some cases when RSL was still falling, led to the transfer of sediment along the coast instead of directly offshore (Normandeau et al. 2017). Eroded sediments were transferred in adjacent bays by longshore drift, or reached the edge of the narrow, shallow shelf (Fig. 10.7). At these locations, sediments were transferred toward the Laurentian Channel. Accumulation of sediment on steep slope eventually led to slope failures, either due to wave action during storms or oversteepening of the slope. In some cases, as along the Les Escoumins region (Fig. 10.2), the alongshore sediment transfer occurred over long distances, until the shelf narrows enough to amplify offshore transfer (Gagné et al. 2009). Here, three small submarine canyons formed in response to the activity of longshore drift and resulting slope failures and turbidity currents on the Laurentian Channel slope. These three canyons incise the shelf from less than 300 m from the coast and extend down to 300 m water depth. At the base of the canyons, submarine fans form, one of which is characterized by crescent-shaped sediment waves, termed cyclic steps. In other cases, as in Sept-Îles (Fig. 10.5), longshore-transport processes occur on the delta itself. Sediments supplied at the river mouth are transferred almost immediately westward along the coast through longshore-drift instead of being exported offshore. Combined with RSL fall, the longshore-drift transport created beach ridges west of the river mouth. In addition to the sediments coming from the river, coastal erosion began playing an important role in supplying sediment to the west. The erosion of the delta resulted in the formation of a coastal, wave-cut platform which was previously part of the delta front (Fig. 10.7). Oblique nearshore bars and longshore bars formed on this shelf (Fig. 10.5) and actively migrate westward, where the shelf narrows (Normandeau et al. 2013). Repeated swath bathymetry mapping in the area showed that sediment also accumulates at the western extremity of the narrowing shelf and at the head of a channel. Occasionally, most likely during storms, turbidity currents are triggered at the shallow head of this channel and transfer sediment downslope. This offshore transport of sediment created a bird's foot-like submarine fan (Fig. 10.5).

10.5 Conclusions

Deltaic complexes of the Québec North Shore form the thickest and most extensive accumulation of coarse and fine-grained sediments in eastern Canada. The various morphologies of the drainage basins led to different morphostratigraphies and geomorphological characteristics of the deltaic complexes. Some deltas protruded into the St. Lawrence Estuary and Gulf and are conspicuously characterized by extensive submarine channels on the seafloor. In other cases, the presence of lakes in the watershed trapped most glacigenic sediments, and resulted in an attenuated geomorphology of the deltas in the St. Lawrence. Nonetheless, the deltaic complexes of the QNS are composed of a lower ice-contact depositional system, superimposed by ice-distal glacifluvial deltas that are currently being reworked by coastal and fluvial processes. These three components of the deltas are attributed to the progressive north-westward retreat of the ice-margin on the QNS.

Ice-contact deposits were emplaced at the front of the LIS margin. Glacifluvial deltas were formed as the ice-margin retreated on land, while supplying large volumes of meltwaters and glacigenic sediments to the rivers. The retreat of the LIS from the watersheds forced a transition from previously fluvial- to wave-dominated deltaic conditions. Deltas subsequently began to be eroded, forming coastal suites. The erosion of the deltas also led to the transport of sediment alongshore and direct transfer offshore ceased. Instead, offshore sediment transport was focused where the shallow shelf narrows. Sediment accumulation leads to slope failures and turbidity currents that can reach deeper marine basins.

References

Bernatchez P, Dubois J-MM (2008) Seasonal quantification of coastal processes and cliff erosion on fine sediment shorelines in a cold temperate climate, North Shore of the St. Lawrence Maritime Estuary, Québec. J Coastal Res 24:169–180

Boulton GS (1990) Sedimentary and sea level changes during glacial cycles and their control on glacimarine facies architecture. In: Dowdeswell JA, Scourse JD (eds) Glacimarine environments: processes and sediments. Geological Society of London SP 53: 15–52

Dietrich P, Ghienne J-F, Normandeau A, Lajeunesse P (2016) Upslope-migrating bedforms in a proglacial sandur delta: cyclic steps from river-derived underflows? J Sediment Res 86:113–123

Dietrich P, Ghienne J-F, Schuster M, Lajeunesse P, Nutz A, Deschamps R, Roquin C, Duringer P (2017a) From outwash to coastal systems in the Portneuf-Forestville deltaic complex (Québec North Shore): Anatomy of a forced regressive deglacial sequence. Sedimentology 64:1044–1078

Dietrich P, Ghienne J-F, Normandeau A, Lajeunesse P (2017b) Reconstructing ice-margin retreat using delta morphostratigraphy. Sci Rep 16936

Dionne JC (2001) Relative sea-level changes in the St. Lawrence Estuary from deglaciation to present day. In: Weddle TK, Retelle MJ (eds) Deglacial history and relative sea-level changes, Northern New England and Adjacent Canada. Geological Society America Special Paper 351, pp 271–284

Dionne J-C, Dubois J-MM, Bernatchez P (2004) La terrasse Mitis à la pointe de Mille-Vaches (péninsule de Portneuf), rive nord de l'estuaire maritime du Saint-Laurent: nature des dépôts et évolution du niveau marin relatif à l'Holocène. Géogr Phy Quat 58:281–295

Drapeau G (1992) Dynamique sédimentaire des littoraux de l'estuaire du Saint-Laurent. Géogr Phys Quat 46:233–242

Dredge LA (1983) Surficial geology of the Sept-Îles area, Quebec North Shore. Geological Survey of Canada Memoir 408, 40 p

Dubois JMM, Dionne JC (1985). The Québec North Shore moraine system: A major feature of Late Wisconsin deglaciation. Geological Society America Special Papers 197, pp 125–134

Duchesne MJ, Long B, Urgeles R, Locat J (2003) New evidence of slope instability in the Outardes Bay delta area, Quebec, Canada. Geo-Mar Lett 22:233–242

Fraser C, Hill PR, Allard M (2005) Morphology and facies architecture of a falling sea level strandplain, Umiujaq, Hudson Bay, Canada. Sedimentology 52:141–160

Gagné H, Lajeunesse P, St-Onge G, Bolduc A (2009) recent transfer of coastal sediments to the Laurentian Channel, Lower St. Lawrence Estuary (Eastern Canada), through submarine canyon and fan systems. Geo-Mar Lett 29:191–200

Gagnon-Poiré A, Lajeunesse P, Normandeau A, Francus P, St-Onge G, Nzekwe OP (2018) Late-Quaternary glacial to postglacial sedimentation in three adjacent fjord-lakes of the Quebec North Shore (eastern Canadian Shield). Quat Sci Rev 186:91–110

Hein FJ, Syvitski JPM, Dredge LA, Long BF (1993) Quaternary sedimentation and marine placers along the North Shore, Gulf of St-Lawrence. Can J Earth Sci 30:553–574

Hein CJ, FitzGerald DM, Buynevich IV, Van Heteren S, Kelley JT (2014) Evolution of paraglacial coasts in response to changes in fluvial sediment supply. In: Martini IP, Wanless HR (eds) Sedimentary coastal zones from high to low latitudes: similarities and differences. Geol Soc London SP 388:247–280

Lajeunesse P, St-Onge G, Locat J, Duchesne MJ, Higgins MD, Sanfaçon R, Ortiz J (2013) The Corossol structure: a possible impact crater on the seafloor of the northwestern Gulf of St. Lawrence, Eastern Canada. Meteorit Planet Sci 48:2542–2558

Lajeunesse P (2014) Buried preglacial fluvial gorges and valleys preserved through quaternary glaciations beneath the eastern Laurentide Ice Sheet. Geol Soc Am Bull 126:447–458

Lajeunesse P (2016) Late-Wisconsinan grounding-zone wedges, northwestern Gulf of St. Lawrence (eastern Canada). In: Dowdeswell J, Canals M, Jakobsson M, Todd BJ, Dowdeswell EK, Hogan KA (eds) Atlas of submarine glacial landforms: modern, quaternary and ancient. Geol Soc London Mem 46:227–228

Lajeunesse P, Dietrich P, Ghienne J-F (2018) Late Wisconsinan grounding zones of the Laurentide Ice Sheet margin off the Québec North Shore (NW Gulf of St. Lawrence). In: Le Heron DP, Hogan K, Phillips E, Huuse M, Busfield M, Graham A, Kouwe W (eds) Glaciated margins: the sedimentary and geophysical archive. Geol Soc London SP 475-10

Tooth S, Brandt D, Hancox PJ, McCarthy TS (2004) Geological controls on alluvial river behavior: a comparative study of three rivers on the South African Highveld. J Afr Earth Sci 38:79–97

Mulder T, Syvitski JP (1995) Turbidity currents generated at river mouths during exceptional discharges to the world oceans. J Geol 103:285–299

Normandeau A, Lajeunesse P, St-Onge G (2013) Shallow-water longshore drift-fed submarine fan deposition (Moisie River Delta, Eastern Canada). Geo-Mar Lett 33:391–403

Normandeau A, Lajeunesse P, St-Onge G (2015) Submarine canyons and channels in the Lower St. Lawrence Estuary (Eastern Canada): Morphology, classification and recent sediment dynamics. Geomorphology 241:1–18

Normandeau A, Lajeunesse P, Poiré AG, Francus P (2016) Morphological expression of bedforms formed by supercritical sediment density flows on four fjord-lake deltas of the south-eastern Canadian Shield (Eastern Canada). Sedimentology 63:2106–2129

Normandeau A, Dietrich P, Lajeunesse P, St-Onge G, Ghienne J-F, Duchesne M, Francus P (2017) Timing and controls on the delivery of coarse sediment to deltas and submarine fans on a formerly glaciated coast and shelf. Geol Soc Am Bull 129:1424–1441

Nutz A, Ghienne J-F, Schuster M, Dietrich P, Roquin C, Hay MB, Bouchette F, Cousineau P (2015) Forced regressive deposits of a deglaciation sequence: Example from the Late Quaternary succession in the Lake Saint-Jean basin (Québec, Canada). Sedimentology 62:1573–1610

Occhietti S, Parent M, Lajeunesse P, Robert F, Govare E (2011) Late Pleistocene-early Holocene decay of the Laurentide Ice Sheet. In: Ehlers J, Gibbards PL, Hughes PD (eds) Quaternary glaciations—extent and chronology. A closer look. Developments in Quaternary Science, 15, 601–630, Amsterdam, The Netherlands

Peltier WR, Argus DF, Drummond R (2015) Space geodesy constrains ice age terminal deglaciation: the global ICE-6G_C (VM5a) model. J Geophys Res 120:450–487
Pinet N, Brake V, Campbell DC, Duchesne MJ (2011) Seafloor and shallow subsurface of the St. Lawrence River Estuary. Geosci Can 38:31–40
Pinet N, Brake V, Duchesne MJ (2016) Partly-filled U-shaped morphology of the Laurentian Channel, St Lawrence Estuary, Canada. In: Dowdeswell J, Canals M, Jakobsson M, Todd BJ, Dowdeswell EK, Hogan KA (eds) Atlas of submarine glacial landforms: modern, quaternary and ancient. Geol Soc London Mem 46:163–164
Sella GF, Stein S, Dixon TH, Craymer M, James TS, Mazzotti S, Dokka RK (2007) Observation of glacial isostatic adjustment in "stable" North America with GPS. Geophys Res Lett 34
Shaw J, Gareau P, Courtney RC (2002) Palaeogeography of Atlantic Canada 13-0kyr. Quat Sci Rev 21:1861–1878
Shaw J, Piper DJW, Fader GBJ, King EL, Todd BJ, Bell T, Batterson MJ, Liverman DGE (2006) A conceptual model of the deglaciation of Atlantic Canada. Quat Sci Rev 25:2059–2081
Todd BJ (2016) The Laurentian Channel: a major cross-shelf trough in Atlantic Canada. In: Dowdeswell J, Canals M, Jakobsson M, Todd BJ, Dowdeswell EK, Hogan KA (eds) Atlas of submarine glacial landforms: modern, quaternary and ancient. Geol Soc London Mem 46:161–162
Woodworth PL, Gehrels WR, Nerem RS (2011) Nineteenth and twentieth century changes in sea level. Oceanography 24:80–93

Pierre Dietrich is a Postdoctoral Fellow at the University of Johannesburg (South Africa) and received his Ph.D. in sedimentary geology from the University of Strasbourg (France) in 2015. His thesis focused on the sedimentary, stratigraphic and geomorphological record of the glaci-isostatic adjustment and associated processes that accompanied the retreat of the Laurentide Ice Sheet along the North Shore of the St. Lawrence Estuary and Gulf (Québec, eastern Canada). His current research interest emphases the sedimentary record of both the Ordovician and Permo-Carboniferous ice ages in southern Africa. He also has interest in subglacial erosion, deltaic, fluvial, supercritical-flow related processes as well as on the retreat of modern, high-latitude ice masses.

Alexandre Normandeau holds a Ph.D. from Université Laval and did a 1-year post-doc at Queen's University studying High Arctic lakes on Ellesmere and Melville Islands. In 2016, he joined the Marine Environmental Geoscience group at the Geological Survey of Canada (Atlantic). His research focuses on sediment transport processes in marine and lacustrine environments across a broad range of temporal and spatial scales. Specifically, his goal is to understand the triggers and the frequency of extreme events such as submarine landslides and turbidity currents that can severely damage submarine infrastructures. At the GSC, he is involved in the study of marine geohazards along the Eastern Canadian margin, the Western and Eastern Canadian Arctic and the Gulf of St. Lawrence. He uses a wide variety of methods, including high-resolution seafloor mapping techniques (multibeam bathymetry) and geophysics, water column measurements using current profilers and the collection of sediment cores that are analyzed for physical and geochemical properties.

Patrick Lajeunesse received his Ph.D. in geomorphology from Université Laval (Canada) in 2000. He then undertook a postdoctoral project in glacial geology and geomorphology at the University of Alberta. He has previously worked at the Institut des sciences de la Mer de Rimouski (ISMER) at the University of Québec à Rimouski and the Institut National de la Recherche Scientifique—Eau, Terre, Environnement (INRS-ETE). He now occupies a Professor (Full) position in the Geography Department at Université Laval. His research work focuses on the geomorphological and sedimentary record of glaciation, postglacial environmental changes and paleoseismicity on formerly glaciated coasts, continental margins and lakes, mainly of Eastern and Arctic Canada.

Jean-François Ghienne is a researcher of CNRS working at the 'Institut de Physique du Globe de Strasbourg', University of Strasbourg. He received his Ph.D. in sedimentology from this university in 1998. Working mainly in Western Europe, NW Africa and eastern Canada, his academic research focuses on the Lower Paleozoic Gondwana development, glacial sequence stratigraphy and tectonic/paleoclimatic signals in stratal records from the Cambrian to the Quaternary. He managed from 2013 to 2017 the SeqStrat-Ice project, founded by the French National Research Agency (ANR), dealing with the delineation of past glaciation events and related paleoclimatic scenarios.

Mathieu Schuster received his Ph.D. in sedimentary geology from the University of Strasbourg (France) in 2002. He then worked at the universities of Cologne (Germany), of Brest and of Poitiers (France), as well as at the French geological survey (BRGM). He now works as a researcher at the CNRS and serves as Vice-director of the Institut de Physique du Globe de Strasbourg, a multidisciplinary research lab on geophysics and geology hosted by the University of Strasbourg. His research concerns continental paleoenvironments, with a focus on clastic littoral deposits and landforms associated to lacustrine systems. His current main case studies are the Lake Chad Basin and the Turkana Depression (EARS). He was awarded the bronze medal of CNRS in 2008, and the scientific prize of Cercle Gutenberg and Fondation Université de Strasbourg. He was an Erskine Visiting Fellow at the University of Canterbury (Christchurch, New Zealand) in Spring 2016.

Alexis Nutz is an associate professor at Aix-Marseille Université (France) since 2018 in the Centre Européen de Recherche et d'Enseignement des Géosciences de l'Environnement. He completed his Ph.D. in 2013 at the Université de Strasbourg (France). His Ph.D. was awarded twice by both the Association des Sédimentologistes Français and the Université de Strasbourg. In 2014, he obtained a 2 years post-doc fellow at CNRS (Strasbourg, France) collaborating with industry. Then, he worked between 2016 and 2018 at the Department of Geosciences of the Aarhus University (Denmark). His Ph.D. focused on the Late Quaternary deglacial and postglacial evolution of the Saint-Jean depocenter in northern Québec (eastern Canada). Subsequently, his research interests enlarged to the evolution of past fluvial-lacustrine dynamics and sedimentation in continental basins as proxies for past climate and tectonic forcings. He has also interests in modes of deglaciation during the last glacial cycle in Southern Alps (France).

11 Geomorphology of the Great Lakes Lowlands of Eastern Canada

Joseph R. Desloges, Roger T. J. Phillips, Mary-Louise Byrne, and Jaclyn M. H. Cockburn

Abstract

The Great Lakes Lowlands is Canada's most populated region. While landform distributions are mainly related to a legacy of Quaternary glaciation and pre-Quaternary bedrock erosion, many modern processes of sediment erosion and deposition are significantly influenced by human modifications to the landscape, including deforestation, agriculture, aggregate extraction and urbanization. Ice lobes and/or streams of the Laurentide Ice Sheet Complex have resulted in a complex arrangement of ice-contact, ice-marginal and pro-glacial sediment sequences infilling pre-glacial bedrock channels. End moraines and interlobate kame moraines are thick and their spatial arrangement records multiple ice-flow directions around the eroded Great Lakes. The "peninsula" of southern Ontario undergoes significant coastal modifications as wave energy from three of the lakes erodes and redistributes glacilacustrine and glacifluvial sediments along shorelines. Post-glacial isostatic uplift has resulted in entrenchment of both subsequent and consequent river systems exposing mixed channel boundaries of glacial sediments and bedrock. Modern infrastructure has developed on these surfaces that are still undergoing post-glacial adjustment.

Keywords

Lower Great Lakes • Landforms • Glacial • Coastal • Fluvial • Human impact

11.1 Introduction

The Great Lakes Lowlands, bounded to the south and west by lakes Ontario, Erie and Huron, and to the north and east by the Canadian Shield (including the Frontenac Axis of the Canadian Shield at Gananoque), is an area of relatively low relief (Fig. 11.1) that is heavily influenced by its bedrock origins, extensive glaciations, post-glacial fluvial and coastal activity and historical land use. Figure 11.1 shows the contemporary topography and river drainage. The region is densely populated with close to 9.5 million people. This is more than one quarter of Canada's entire population living in approximately 80,000 km^2, only 0.8% of the country's total surface land area. Southern Ontario is a rapidly growing region with a forecasted population of 13.5 million by 2041 (Ontario Ministry of Finance 2017). Therefore, the landscape seen today is primarily a composite of its geologic and geomorphic origins with substantial human impact on processes and landform changes over the past 200 years.

Since deglaciation, water and wind forces along river valleys and shorelines have shaped the surface morphology of the Great Lakes Lowlands. The temperate, mid-latitude, climate has not changed significantly over the Holocene (Edwards et al. 1996; McFadden et al. 2005) and thus the hydrology continues to be dominated by a modest precipitation regime with seasonal temperature contrasts (Colombo et al. 2007). Unlike other continental climate zones of central Canada, the Lower Great Lakes have significantly moderate seasonal temperatures and enhanced winter snowfall with prevailing westerly winds off the Great Lakes. A strong nival flood regime is dominant (Gingras et al. 1994; Javelle et al. 2003).

J. R. Desloges (✉)
Departments of Geography and Earth Sciences, University of Toronto, Toronto, ON, Canada
e-mail: joseph.desloges@utoronto.ca

R. T. J. Phillips
Department of Geography, University of Toronto, Toronto, ON, Canada

M.-L. Byrne
Department of Geography and Environmental Studies, Wilfrid Laurier University, Waterloo, ON, Canada

J. M. H. Cockburn
Department of Geography, Environment & Geomatics, University of Guelph, Guelph, ON, Canada

© Springer Nature Switzerland AG 2020
O. Slaymaker and N. Catto (eds.), *Landscapes and Landforms of Eastern Canada*,
World Geomorphological Landscapes, https://doi.org/10.1007/978-3-030-35137-3_11

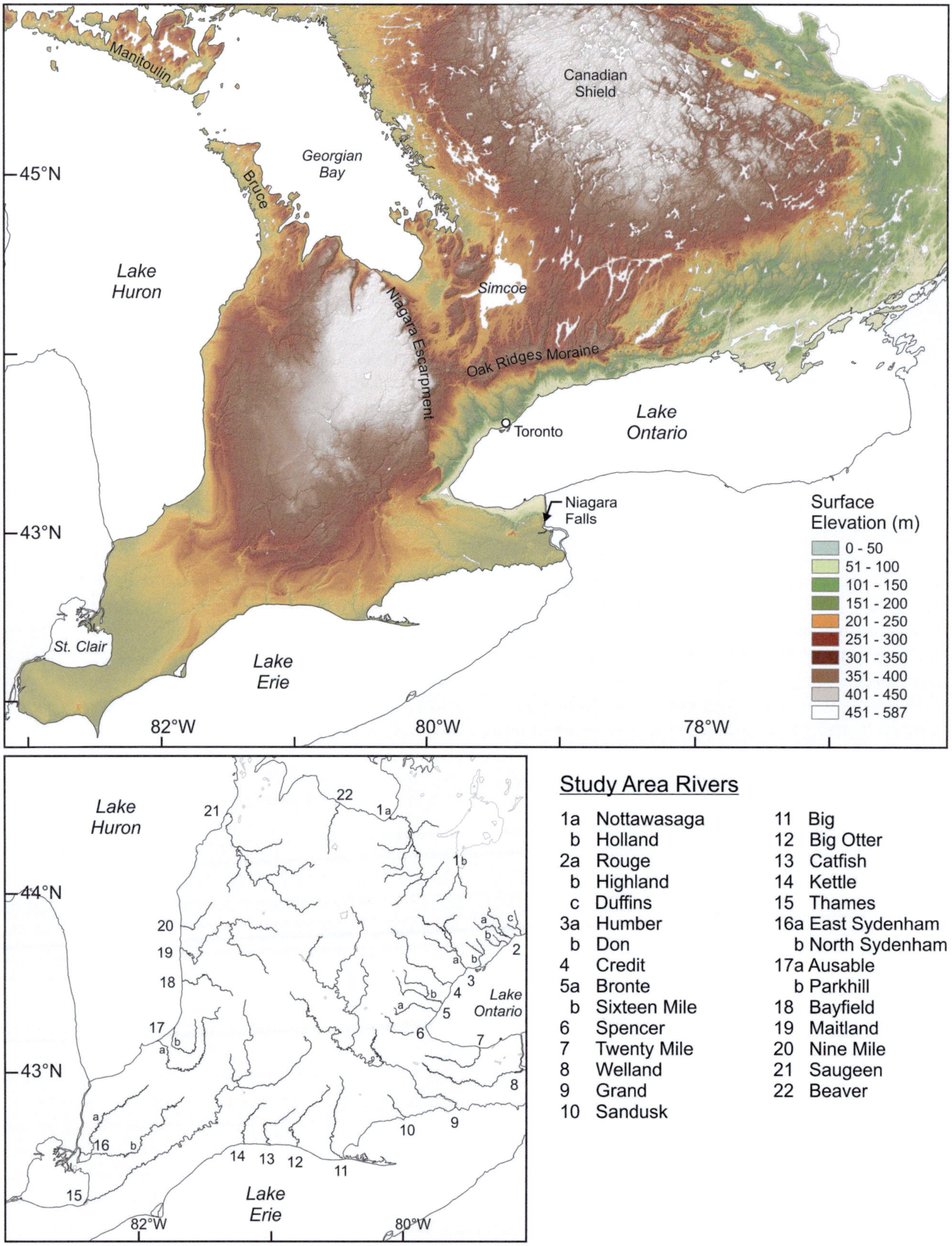

Fig. 11.1 Digital Elevation Model of the surface topography (m a.s.l) of southern Ontario with study streams used for classification of floodplains (Fig. 11.10). *Source* Ontario Geological Survey: https://www.mndm.gov.on.ca/en/mines-and-minerals/applications/ogsearth

11.2 Bedrock

Geomorphic activity is, and continues to be, influenced by the lithological and structural make-up of basement and surface bedrock features. Extending outward from the margin of the Grenville-aged Canadian Shield (1.1 billion years), the Algonquin Arch (Fig. 11.2) forms a buried NE–SW trending topographic high of basement crystalline rocks (Carter et al. 1996). Granitic rocks are draped by a thick sequence of early Palaeozoic sedimentary units predominately comprising limestone, shales, dolostone and sandstones deposited in shallow marine and deltaic environments of the Michigan Basin between 450 and 375 million years ago (Catacosinos and Daniels 1991). Epeirogenic deformation of the surface crust resulted in shallow tilting of the sedimentary sequence to the southwest such that most formations of the eastern outer rim of the Michigan Basin were exposed to differential weathering over the Mesozoic and Cenozoic, ultimately forming the Niagara Escarpment. The escarpment, extending NW to form the Manitoulin Islands, is a major topographic feature separating Georgian Bay from Lake Huron (Tovell 1992). Along the southern margin, the Niagara River cuts across the escarpment between Lake Erie and Lake Ontario forming the largest waterfall (by volume) in Canada (Niagara Falls). The overlying Palaeozoic strata in southern Ontario have eroded to form a network of bedrock channels now buried by Pleistocene sediments (Fig. 11.2). Eyles et al. (1997) demonstrated that the bedrock channel orientation is correlated with the regional bedrock joint sets formed from mid-continent tectonic stresses and that these conditions have helped "pre-determine" post-glacial river alignment. Bedrock deformation in the form of pop-up structures and joint sets relate to glacial rebound processes (Adams 1989; Godin et al. 2002).

The spatial distribution and style of associated earthquake activity in eastern Canada is related to both tectonic and ice rebound stresses (Stewart et al. 2000). The southern Great Lakes (SGL) seismic zone has low to moderate levels of seismic activity compared to the rest of eastern Canada (Dineva et al. 2004). Evidence of past seismic activity, including small vertical and horizontal offset faults, can be found in the Late Glacial and Holocene sediments of these seismically active zones (Doughty et al. 2014).

The bedrock surface topography was largely in place until the beginning of the Quaternary period (Fig. 11.2). Highest bedrock terrains are over the Algonquin Arch and Niagara Escarpment exposures forming a bedrock island surrounded by lower bedrock terrain. Connecting present-day Georgian Bay of Lake Huron with Lake Ontario is the Laurentian bedrock channel that would have been a significant part of the pre-glacial drainage network in the region (Fig. 11.2).

11.3 Glacial Landforms and History

Quaternary continental-scale glaciations resulted in extensive sedimentary sequences (glacial drift) in the region, including a broad range of glacial landforms and a post-glacial legacy of isostatic rebound. The glacial drift covering much of southern Ontario averages about 40 m in thickness (median thickness ~30 m) and is locally up to 262 m deep (Fig. 11.3). Greatest fill depths to bedrock are at the crest of the Oak Ridges Moraine (ORM) and under the Simcoe Uplands (Fig. 11.3) associated with infill of bedrock channels (Mulligan et al. 2016). The total volume of overburden in southern Ontario is approximately 2500 km^3 (estimated from Gao et al. 2006), which is roughly the combined volume of the Lake Ontario and Lake Erie basins.

Major surface landforms and glacial deposits with local names are shown in Fig. 11.4 based on the original physiographic mapping of Chapman and Putnam (1951, 1984) and digital data from the Ontario Geological Survey (Chapman and Putnam 2007). Differential rates of glacial erosion into less resistant shale formations left residual outcrops of dolomitic limestone as topographic highs—namely the Niagara Escarpment (see 1 in Fig. 11.4a)—and as alvar limestone plains with thin to no overburden along the Niagara Escarpment and to the east along the fringes of the Michigan Basin. MIS 2 (Late Wisconsinan) lodgement till, drumlinized till plains and other mega-scale glacial lineations (Fig. 11.5) blanket large areas of the region. Older glacial drift recording ice advance, interglacial and Illionian-age sediments are documented in outcrops and subsurface borehole and seismic data (Karrow 1967; Sharpe et al. 1997). Subglacial eskers and drumlins (Fig. 11.4b), formed under separated lobes of the Laurentide Ice Sheet, record evidence of ice flow directions (Boyce and Eyles 1991; Brennand 1994, 2000; Eyles and Doughty 2016), with major streamlines associated with ice moving out of Lake Huron (Teeswater (see 14 in Fig. 11.4b) and Arran (see 16 in Fig. 11.4b) drumlin fields) and Lake Ontario (e.g. Peterborough Drumlin Field; see 18 in Fig. 11.4b and 11.5). With respect to the dominant processes of drumlin formation, the role of glacial erosion and bed deformation by fast-moving ice-streams (Eyles et al. 2016) and the role of deposition by subglacial meltwater floods (Shaw 2002) have been presented in the literature.

With the onset of MIS 2 (Late Wisconsinan) deglaciation, a succession of kame moraines, recessional till moraines (the "Horseshoe Moraines") and associated outwash channels were formed roughly surrounding the central uplands (Fig. 11.4c)—the central upland area has been aptly referred to as the *Ontario Island* as it was briefly surrounded by glacial ice lobes and proglacial lakes during the early stages of deglaciation around 13 ka BP (Chapman and Putnam 1951). To the east of the Niagara Escarpment is the

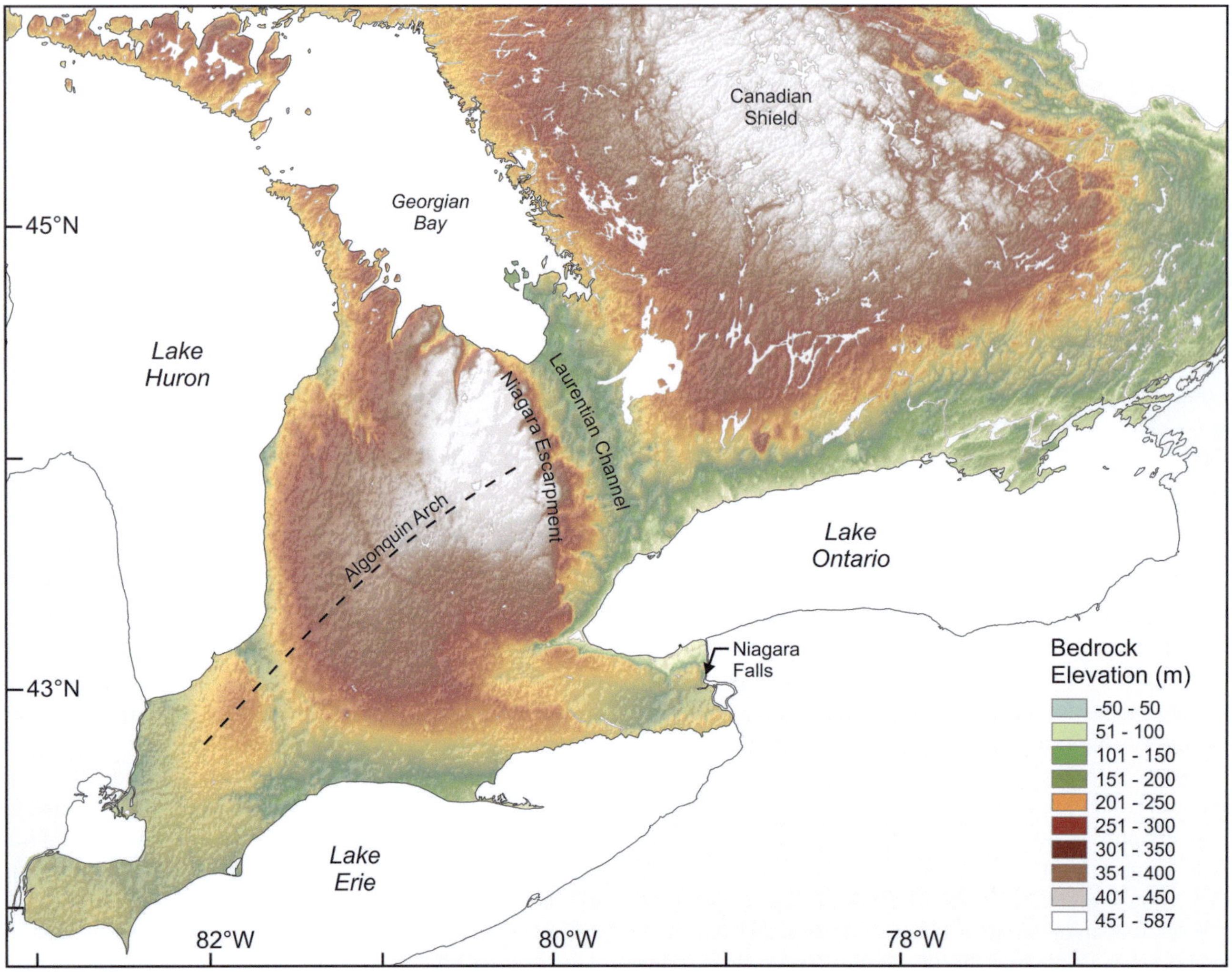

Fig. 11.2 Digital Elevation Model of the bedrock topography of southern Ontario, highlighting key features such as the Algonquin Arch, Niagara Escarpment, Laurentian Channel and Canadian Shield. *Source* Ontario Geological Survey: https://www.mndm.gov.on.ca/en/mines-and-minerals/applications/ogsearth

prominent ORM, a stratified interlobate moraine (dominantly sand and gravel, kame-type) formed within a complex sequence of subglacial, ice-contact and proglacial depositional environments (Gilbert 1997; Barnett et al. 1998). The ORM is a major groundwater reservoir discharging into drainages that flow north and south off the moraine crest (Gerber and Howard 2000). The Dummer Moraine north of the Peterborough Drumlin Field is a hummocky ridge of limestone-clast debris along the transition between the surface Palaeozoic limestone and Precambrian crystalline bedrock of the Canadian Shield (Eyles and Doughty 2016).

Within the greater Great Lakes Basin, meltwater during deglaciation ponded in a series of large proglacial lakes leaving behind beach deposits of gravel and glacilacustrine plains of sand, silt and clay (Fig. 11.6; Larsen and Schaetzl 2001; Lewis et al. 2008). The proglacial lake shorelines tend to be tilted across each basin due to greater glacial isostatic rebound to the northeast of the Great Lakes Basin (Fig. 11.7; Lewis et al. 2005).

Given the low-relief topography and large-scale ice-sheet dynamics of the region, the geomorphic expression of glacial landforms viewed from the ground is subtle. However, the extensive glacial sedimentary sequences are frequently cut and exposed along river valleys (Fig. 11.6) and coastal bluffs as the post-glacial surface has been reworked by Holocene fluvial, coastal and aeolian processes.

11.4 Coastal Geomorphology in Southern Ontario

Glacial erosion, extensive late Pleistocene glacial lakes and post-glacial lake level adjustment (including isostatic tilting, Fig. 11.7) have led to the modern configuration of the lower

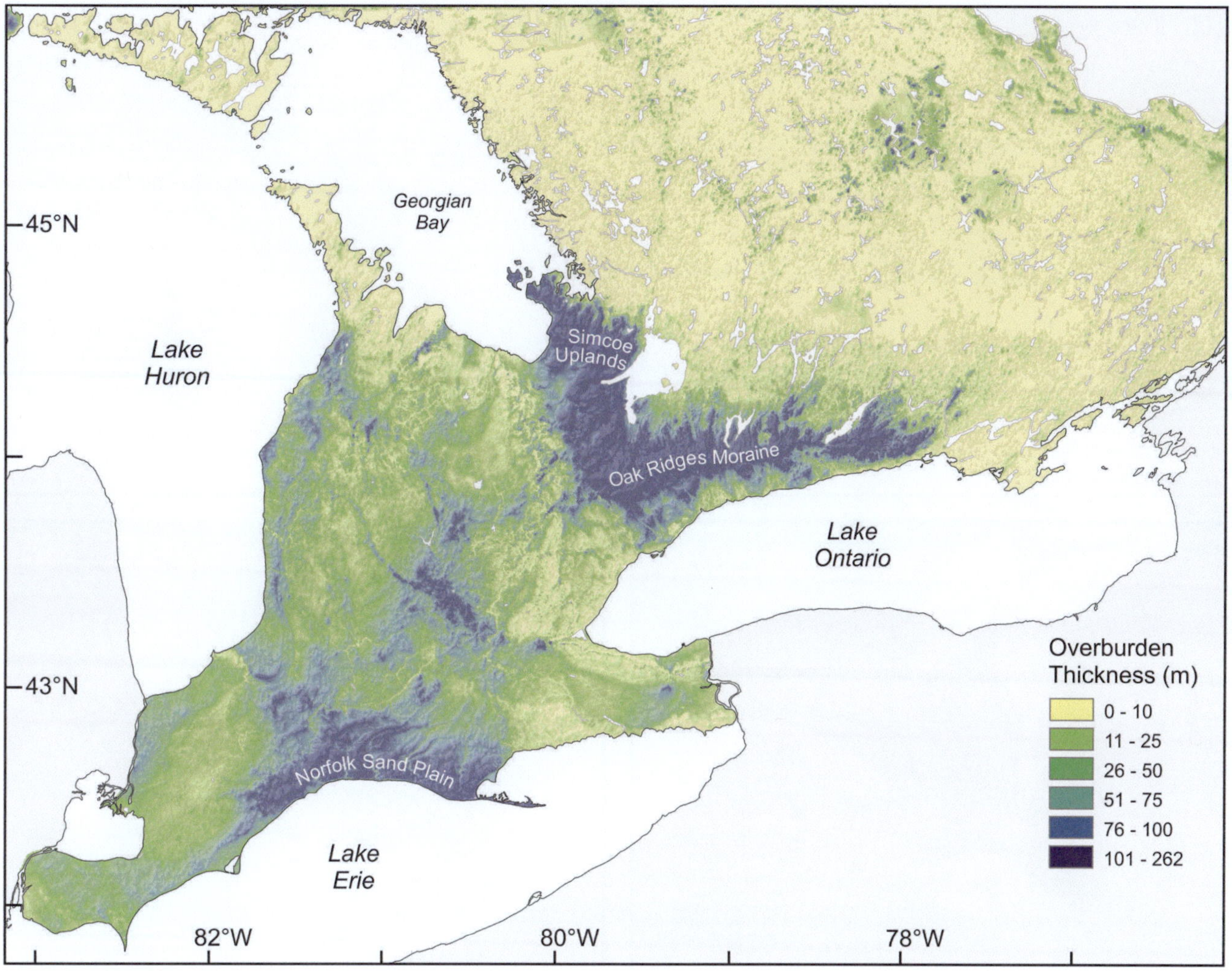

Fig. 11.3 Sediment overburden thickness in the Lower Great Lakes, with the thickest deposits within the Oak Ridges Moraine and Simcoe Uplands. *Source* Ontario Geological Survey: https://www.mndm.gov.on.ca/en/mines-and-minerals/applications/ogsearth

Great Lakes shorelines (Fig. 11.8). Shoreline length of lakes Erie, Ontario and southern Huron on the Canadian side exceeds 1400 km. Contemporary wave and current processes are strongly influenced by variations in lake water levels on a seasonal and long-term basis (Quinn 2002; Rao and Schwab 2007; Department of Fisheries and Oceans 2016). Seasonal signatures of high water are associated with spring melt and low water with late summer/fall temperature induced evaporation. Long-term high lake levels result in higher rates of shoreline erosion and higher rates of sedimentation within the lakes.

11.4.1 Lake Erie

Lake Erie is comprised of three basins—Western, Central and Eastern—with shoreline features reflecting the configuration of each. The Western Basin is associated with bluffs bordering the Essex Clay Plain, with narrow beaches and a depositional sink that makes up Point Pélée. It is shallowest and receives inflow from the Detroit River. The land bordering the north side of the Western Basin is characterized by clay-rich till with bluffs along the north shore that range in height from a few metres above the lake level in the west to over 20 m above the lake level in the east. Narrow beaches border the bluff toe in an overall erosional environment that has resulted in various approaches to shoreline protection (e.g. Lawrence 1994). The national park at Point Pélée is a large cuspate foreland composed of beach ridges, dunes and marshland (Coakley 1976; Trenhaile et al. 2000), and marks the eastern end of this basin. The pre-European contact processes active in this area resulted in a tenuous balance of erosion and deposition along the shoreline and the development of Point Pélée. Post-European contact settlement and development resulted in sediment starvation of the coastal system and a resulting increase in erosion

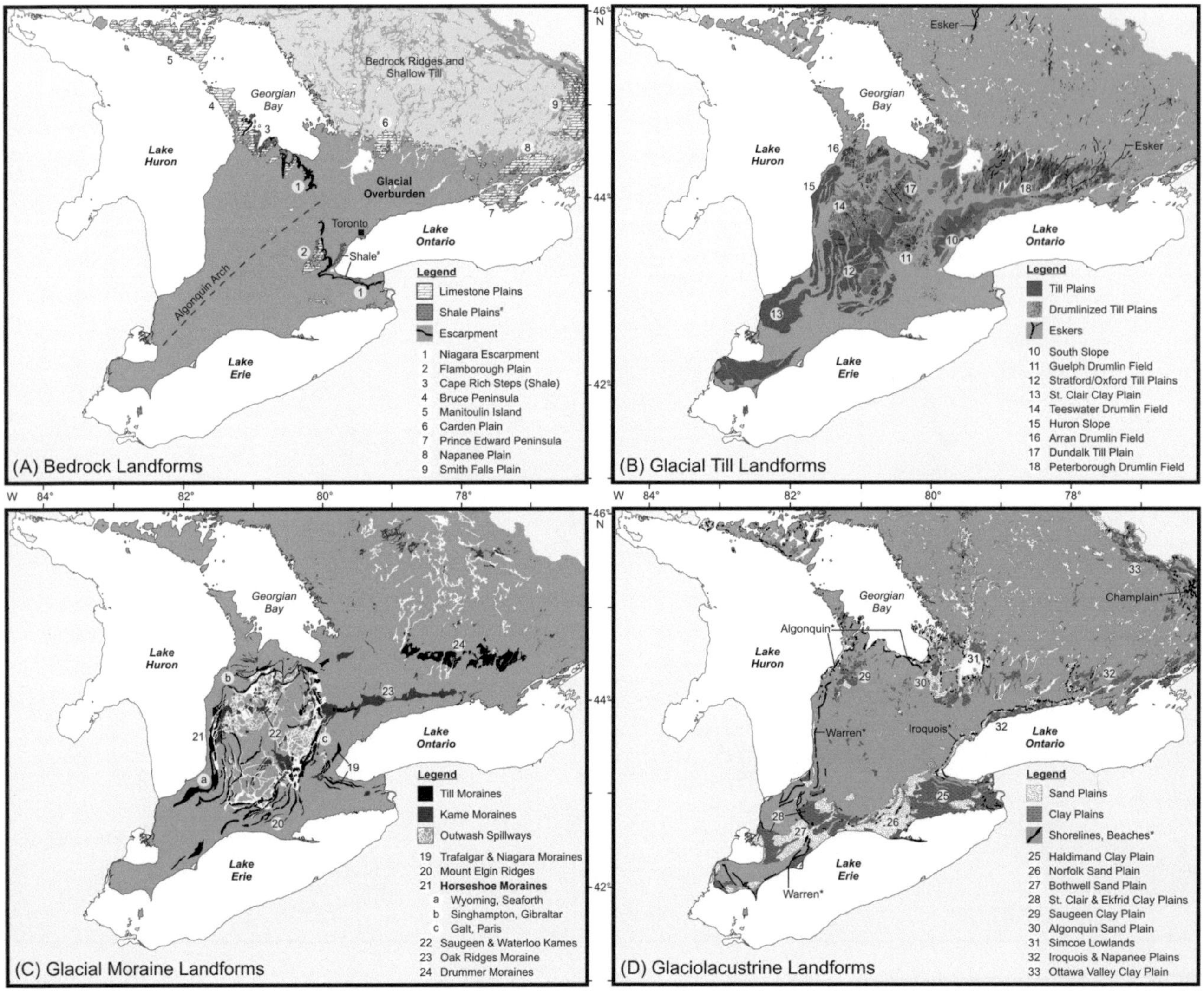

Fig. 11.4 Glacial landforms of southern Ontario. **a** Bedrock landforms; **b** glacial landforms—till plains drumlin fields, eskers; **c** glacial landforms—moraines; and **d** glacilacustrine landforms with post-glacial lakeshores identified with an asterix (Warren (Erie and southern Huron), Iroquois (Ontario), Algonquin (Bruce Peninsula), Champlain (eastern edge of the map)). After Chapman and Putnam (2007)

(BaMasoud and Byrne 2011) and thus the need for shoreline protection or sediment nourishment to maintain beaches. Point Pélée National Park provides a good example of both shoreline protections with the armouring of some beaches to arrest the rate of erosion and recent sediment nourishment on others (Dobbie, pers.com).

The Central Basin, extending eastward and terminating in Long Point and the Long Point Spit (Fig. 11.8b), is bordered to the north by sand plains overlying till. Several small streams drain the Norfolk and Bothwell Sand Plains (see 26, 27 in Fig. 11.4d) and the Ekfrid Clay Plain emptying into the Central Basin. Two sediment depositional sinks mark the shoreline—first is Long Point and the second is the Pointe aux Pins complex, composed of the Erieau and Rondeau spits. Between these two sinks are erosional bluffs that range in height from just a few metres to over 30 m. Sections of thick sand in the Norfolk Sand Plain are highly erosive (Fig. 11.8e) with shoreline retreat rates varying from 0.2 to 4.9 m per year (Boyd 1981). Rukavina and Zeman (1987) noted that most of this eroded material is removed from the coastal system and lost to offshore basins.

The Eastern Basin is the deepest and is bordered on the north shore by a low sloping coast associated with the Haldimand Clay Plain (25, Fig. 11.4d) with wetlands and pocket beaches between limestone headlands. A present-day delta to the Grand River (9, Fig. 11.1) is marked by sand accumulation and a minor depositional sink with active sand dunes.

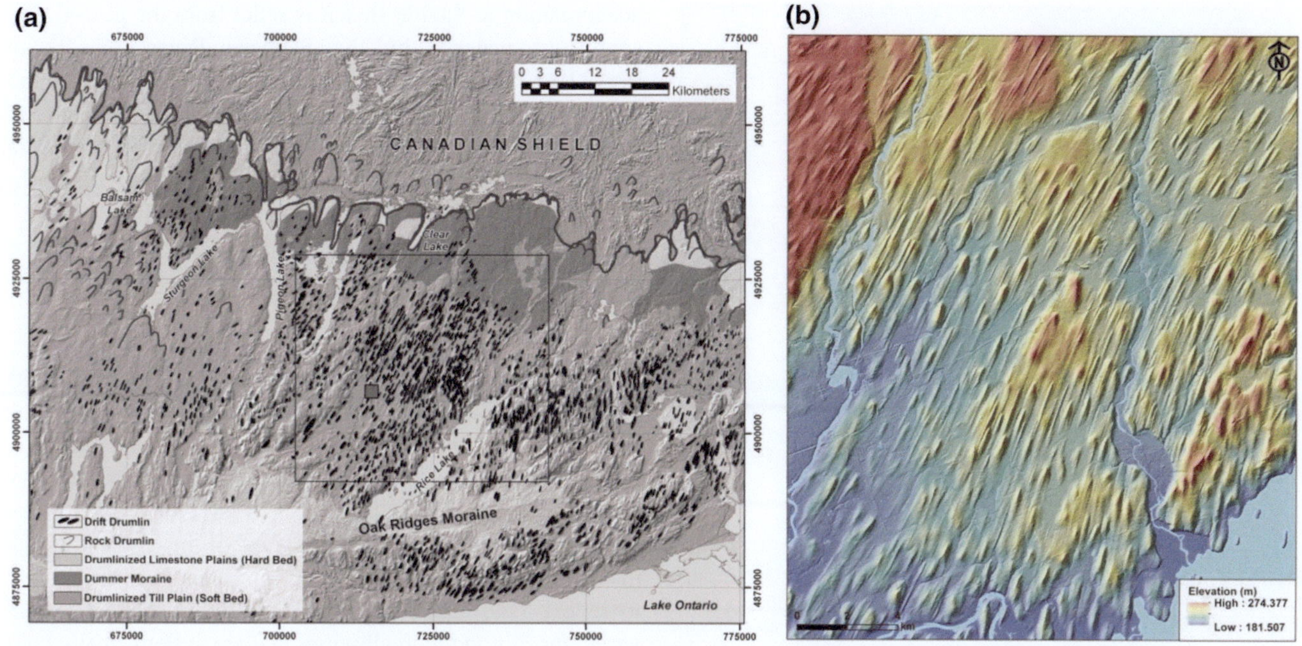

Fig. 11.5 Drumlins and megaridges: **a** centred on City of Peterborough; **b** drumlins and mega-scale glacial lineations of dissected drumlins near Rice Lake. From Eyles and Doughty (2016)

Fig. 11.6 **a** Glacial till bluff along Little Rouge River (43°48′29″ N′ 79°08′09″ W) showing contribution of cobble and gravel till clasts supplied to the fluvial channel. *Photograph* James Thayer, with permission. **b** 45 m high valley bluff of exposed glacilacustrine sand over till along Saugeen River (see 21 in Fig. 11.1) near Walkerton, Ontario (44°10′33″ N′ 81°10′51″ W)

11.4.2 Lake Ontario

This lake is characterized by the Iroquois Plain (see 32 in Fig. 11.4d), a relatively flat, low-lying former lake bottom that contrasts with the undulating surface of the till plains that it borders. The Iroquois Plain extends over 300 km from the Niagara River around the western end of Lake Ontario and then eastward to the Trent River. The Iroquois Plain contains the Niagara fruit belt, Lake Ontario's lake-head including a large and productive wetland complex—Cootes Paradise, and the sands, gravels and shale plains of the Hamilton to Toronto shoreline. This landscape was so

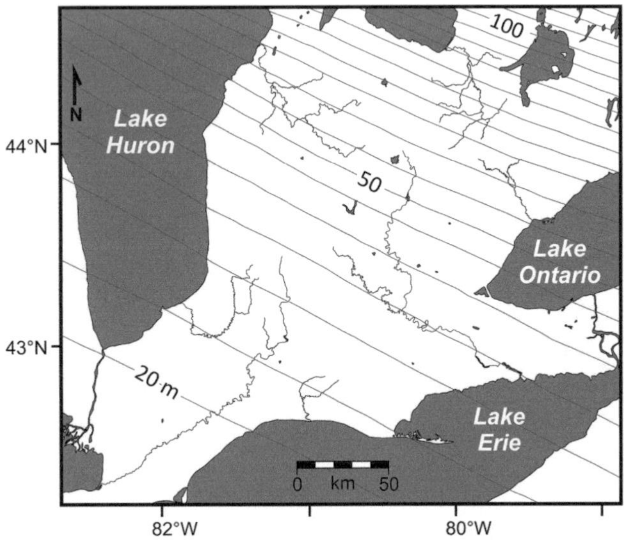

Fig. 11.7 Isoline map of southern Ontario for isostatic rebound since 10.6 ka BP (uncalibrated ^{14}C, glacial Lake Algonquin phase) (after Lewis et al. 2005). Contour interval is 5 m with reference area of zero uplift southwest of Lake Michigan beyond the limit of the last glaciation

desirable for settlement that it is now almost completely developed. The Iroquois Plain is very narrow in the eastern part of Toronto (Fig. 11.1). Here the Scarborough Bluffs tower over 100 m above the lake level. Eyles and Eyles (1983) interpreted these deposits to be the preserved stratigraphy of a large Pleistocene lake (Fig. 11.8f). Eastward, the Iroquois Plain continues with former shoreline sand and gravel complexes now nearly invisible beneath modern urban and rural development. Other notable features of Lake Ontario's coastal geomorphology include Toronto Islands, the Limestone Plains of Prince Edward County with the sand spits and the drumlinized clay plains at the east end of the lake.

11.4.3 Lake Huron (South)

The southern part of Lake Huron is marked by a series of moraines that parallel the present-day lakeshore. The Huron Slope (15, Fig. 11.4b) marks the recessional shorelines of glacial Lake Warren (Fig. 11.4d) and glacial Lake Algonquin (Fig. 11.4d). Bordering this slope is the Huron Fringe that is comprised of glacial Lake Algonquin and Lake Nipissing shorelines and the modern Lake Huron. On the Bruce Peninsula, the fringe is characterized by scoured limestone plains just metres above present-day lake level, interspersed with small beaches and wetlands. A large sand accumulation at Sauble Beach is relict from the post-glacial supply of material by the Saugeen River (see 21 in Fig. 11.1). Southward from the Bruce Peninsula to Grand Bend, the shoreline is marked by shore bluffs 15–30 m tall that are highly erosive (Fig. 11.8c). Sediment is carried in the littoral system to be deposited from Grand Bend northward as active beaches and dune complexes (Fig. 11.8d).

11.5 Fluvial

Holocene fluvial systems have carved river drainage networks into the deglaciated surfaces forming valleys typically in the range of 10–50 m deep—30 m being a reliable average. Longitudinal river profiles retain topographic and sedimentary signatures from glacial landforms (Fig. 11.9), with fluvial systems incorporating a large range of materials through incision (Phillips and Desloges 2014). Channel materials tend to reflect local sediment sources, resulting in longitudinal variations in floodplain alluvium that are systematically coupled with the surrounding glacial landforms within each watershed (Phillips and Desloges 2015a, b).

Given the dominance of silt and clay from till and glacilacustrine deposits in the region, variations in the alluvial floodplain sequences and channel morphologies tend to be most strongly influenced by the local sources of cobble-gravel (from moraines and outwash) and of sand (from lake bottom, delta and beach deposits). The continuum of alluvial floodplain types (Fig. 11.10) tend to statistically cluster—providing a useful classification tool (Phillips and Desloges 2015a, b)—based on stream power (correlated with bed grain size and boundary resistance) and floodplain sedimentology (based on sand availability and floodplain thickness).

The glacial conditioning of fluvial systems in southern Ontario is also revealed in the comparison of specific stream power mapped over the glacial landforms (Fig. 11.11). However, the correlation between stream energy and bed grain size may be mismatched in some cases resulting in oversteepened or understeepened reaches (Phillips and Desloges 2014). For example, slope-driven changes in grain size are obscured by the continual recruitment of coarse sediment from incision into glacial moraine sediments that supply a relatively constant grain size even as slope increases or decreases downstream (Thayer et al. 2016). Not only is the typical Holocene response of the fluvial systems confounded by numerous and distinct glacial landform assemblages, rivers in the lower Great Lakes region are also affected by base-level changes associated with water-level fluctuations (±100 m) and post-glacial isostatic rebound (Phillips and Desloges 2015b).

Fig. 11.8 **a** Beach strandlines due to Holocene isostatic rebound and tilting at Michael's Bay, Manitoulin Island (see 5 in Fig. 11.4a); **b** Long Point in Lake Erie developed by prevailing longshore currents; **c** Glacial Lake Warren shoreline above modern day Lake Huron (Fig. 11.4d); **d** sand dunes at Pinery Provincial Park—This depositional sink has been receiving sediment since the Lake Algonquin stage. Modern coastal processes still deliver sand to this location; **e** eroding bluffs of the Central Basin, north shore Lake Erie—Sediment removed by subaerial and coastal processes is removed from the coastal system and carried to offshore basins; and **f** eroding Scarborough Bluffs east of Toronto on the north shore of Lake Ontario in 2008—the bluffs have since been protected from further erosion with engineered rip-rap at the base of the slopes

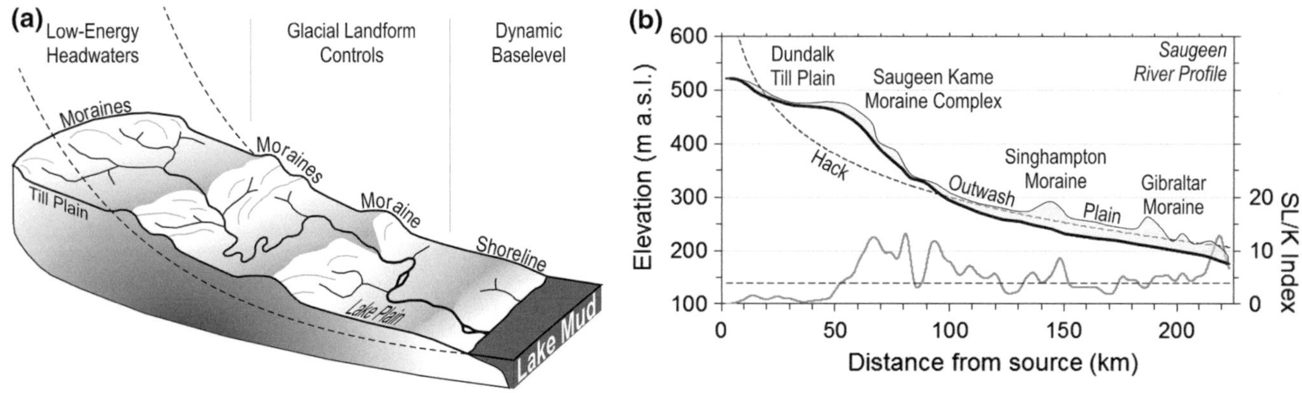

Fig. 11.9 Longitudinal river profiles retain topographic and sedimentary signatures from glacial landforms. **a** Generalized fluvial landscape model for southern Ontario; and **b** example longitudinal profile for Saugeen River (see 21 in Fig. 11.1; Phillips and Desloges 2014, 2015b), including local names for glacial landforms and the stream length-gradient (*SL/K*) index relative to an exponential best-fit Hack profile

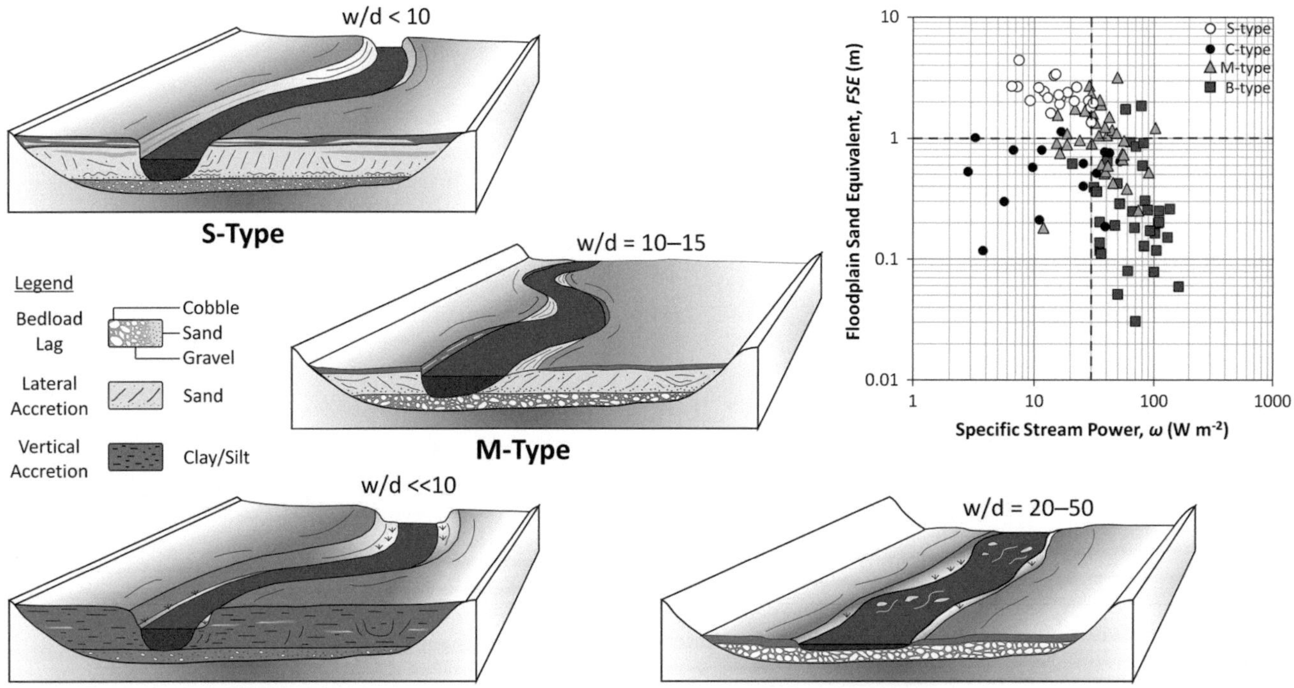

Fig. 11.10 Alluvial floodplain types for low-relief, glacially conditioned rivers (labelled in Fig. 11.1 lower map) as end-members in a continuum of stream power and floodplain sedimentology, with graph in upper right corner showing clustering of the four types by specific stream power and sand availability (defined as floodplain sand equivalent, *FSE*) as per Phillips and Desloges (2015a, b)

11.6 Human Influences in the Lower Great Lakes

European settlement in the seventeenth and eighteenth centuries dramatically altered the lower Great Lakes region, but early horticulture started more than a few millennia before European settlement (Hart and Lovis 2013) including on fertile river floodplains (Crawford et al. 1998). Extensive deforestation for agriculture across southern Ontario, and subsequent land development in urban areas, has resulted in significant changes in land cover and watershed hydrology. Cleared agricultural and urban land cover currently account for about 62 and 9%, respectively, with the balance of 29% making up the remaining forests, wetlands, and shorelines (from SOLRIS 2000 v1.2 dataset, Ontario Ministry of

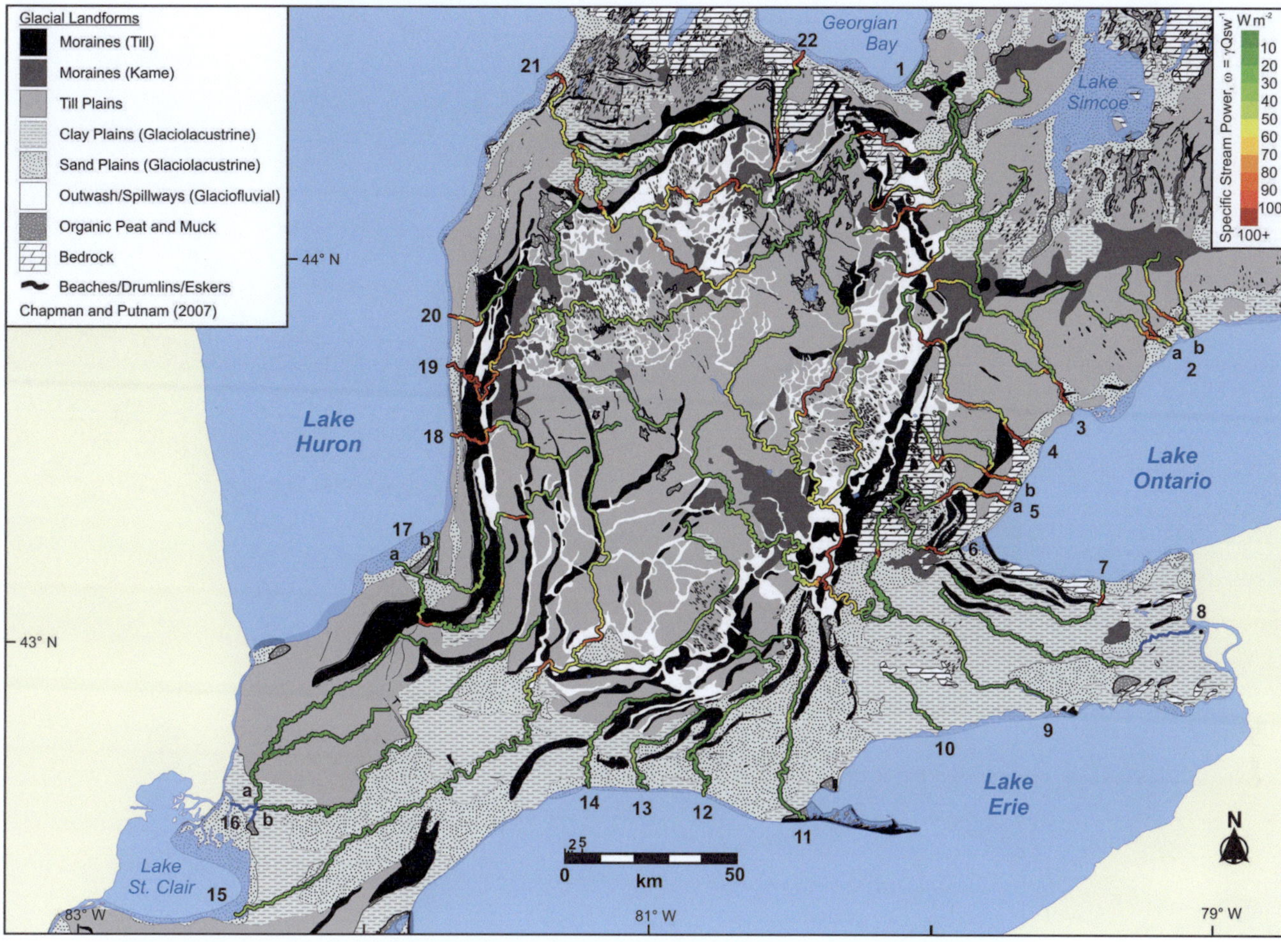

Fig. 11.11 Specific stream power conditioned by glacial landforms, numbers correspond to streams labelled in Fig. 11.1 (Phillips and Desloges 2014)

Natural Resources 2008). Land development and road construction have also driven aggregate resource extraction across the landscape, with provincial records for more than 8,000 open pits and quarries (Ontario Ministry of Natural Resources 2013), covering over 800 km^2 and accounting for about 1% of the land area in southern Ontario (Fig. 11.12). Aggregate mining opportunities have tended to focus on glacifluvial gravel and sand deposits or near-surface limestone formations in the region. Sand mining of the Great Lakes shorelines also occurred in the past. On average, about 130 Mt of aggregate (sand and gravel; crushed stone) are mined and moved in this region each year (Altus Group Economic Consulting 2009). This is equal to $\sim$1600 t/km^2/year. Clubine et al. (2010) showed that the average suspended sediment yield for southern Ontario drainages are in the range of 200–300 t/km^2/year, indicating that human movement of aggregates is 5–8 times river sediment discharge from the region.

The hydrological cycle is modified by anthropogenic activities, which has led to changes in watershed and fluvial processes (e.g. Campo and Desloges 1994; Allan 2004; Chin 2006; Gregory 2006). With European settlement and intensified agricultural practices, stream channels were moved and in some cases filled, while subsurface drainage was augmented through tile-drain networks (e.g. Macrae et al. 2010; Molder et al. 2015). Over 68% of wetlands were drained (Ducks Unlimited Canada 2010) and the overall watershed hydrology has changed because of water use and extraction (Mao and Cherkauer 2009). As agricultural land use around urban centres evolved to allow urban expansion (e.g. in the form of suburban communities, road networks, aggregate extraction, etc.), impervious surface area increased and further modified hydrological processes and aquatic ecosystems (Paul and Meyer 2001; Conway and Hackworth 2007). River engineering practices have also directly impacted the stream channels with a long history of dam and

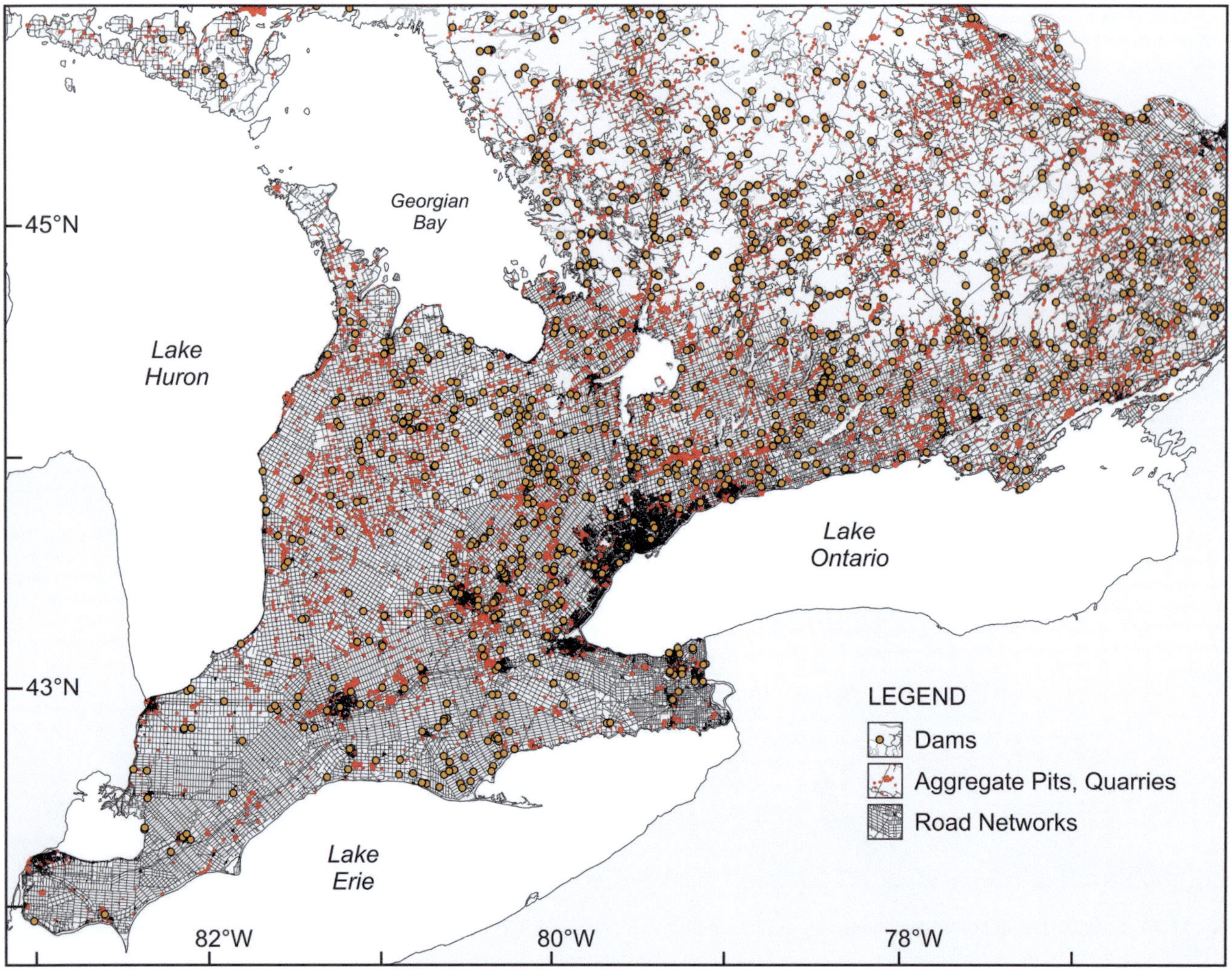

Fig. 11.12 Regional human impacts illustrated with location of dams, aggregate pits and quarries, and road networks. The increasing density of road networks provides a graphical representation of development intensity from low density in the northern portion of the figure (Canadian Shield), to moderate density in southern Ontario (rural agriculture), to high density urban areas. Dams tend to cluster in areas of higher relief and aggregate activities cluster based on source deposits and development intensity and history

road construction, including generations of erosion and flood mitigation measures that tend to deteriorate within a few decades resulting in further degradation of "natural" stream functions.

For example, the provincial dam inventory has in the order of 1000 medium and large dams on record in southern Ontario (Fig. 11.12), about 100 km of stream per dam, with many more small dams not in the current database (Ontario Ministry of Natural Resources and Forestry 2015). Given the approximately 100,000 km of existing stream channels, and a drainage density of about 1.2 km/km^2 in the region, there is on average a road crossing for each 1.8 km of stream ($\sim$0.6 crossings/km), which is just slightly less than the distance of 2 km between the concession roads of the 1,000 acre sectional system used in the survey layout in many counties of southern Ontario. To highlight river engineering for erosion and flood control, Highland Creek—one of Toronto's most urbanized watersheds—has had over 50% of the stream banks hardened including treatments like concrete and rip-rap stone, with some reaches as high as 89% (Aquafor Beech Ltd. 2010). Dams, road crossings and river engineering practices in southern Ontario have contributed significantly to fluvial discontinuity and habitat fragmentation in hydrological, geomorphological and ecological terms.

The results of the anthropogenically modified hydrological cycle and history of channel engineering are evident in the degraded forms and functions of freshwater streams throughout the lower Great Lakes (Stanfield 2012). Baseflow dynamics, runoff-response relationships and fluvial processes have changed, in turn impacting sediment transport

Fig. 11.13 Stream restoration and river engineering (before and after photos); a–a′ stream restoration within formerly agricultural lands in the headwaters of Fletcher's Creek, Brampton a tributary to Credit River (see 4 in Fig. 11.1); b–b′ river engineering to protect sanitary sewer pipe within Highland Creek (see 2b in Fig. 11.1), Toronto (white arrows for common reference point; black arrow and dashed line indicate exposed trunk sewer pipe in before photo); c–c′ slope restoration and channel realignment within Amberlea Creek, Pickering

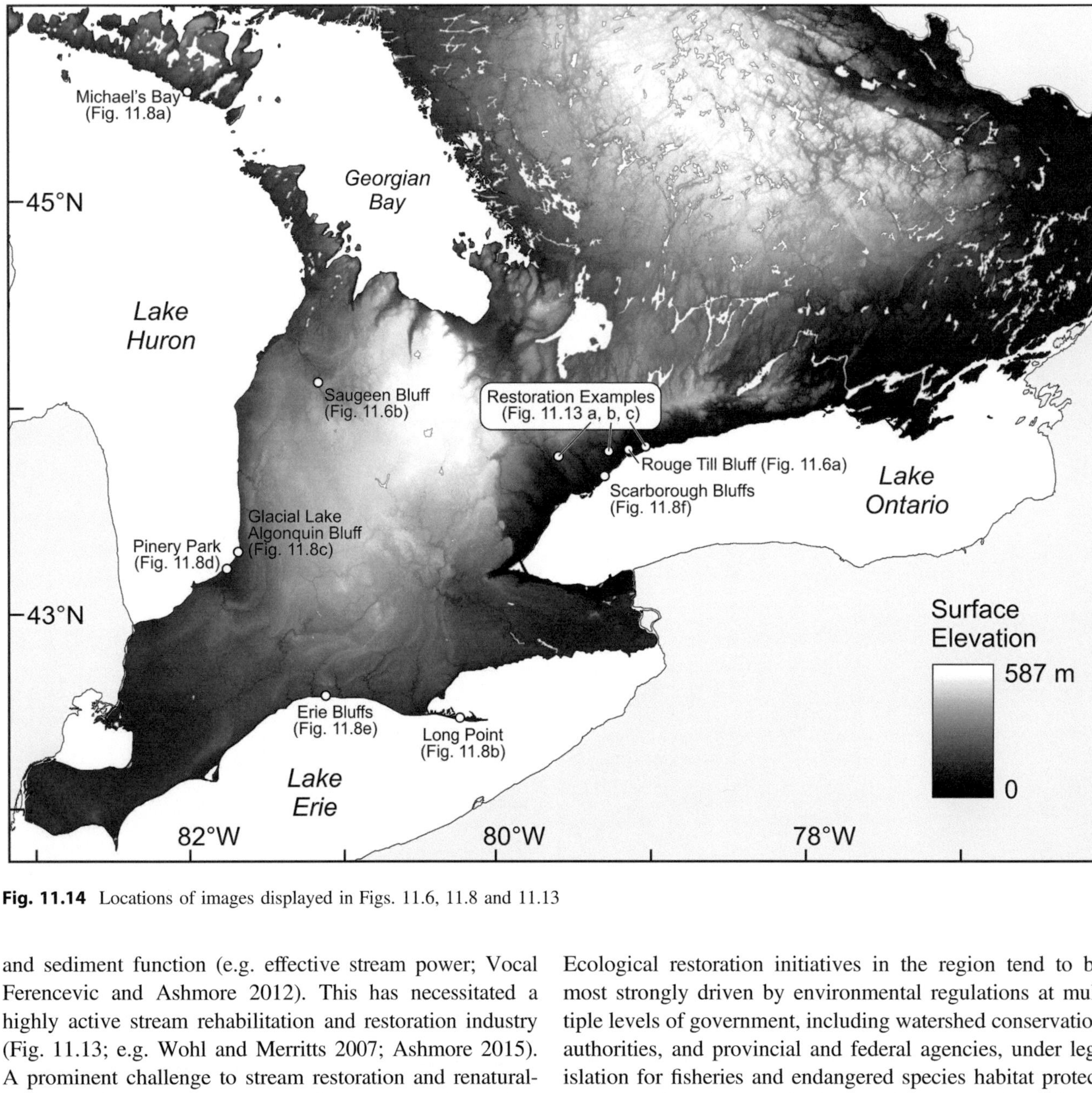

Fig. 11.14 Locations of images displayed in Figs. 11.6, 11.8 and 11.13

and sediment function (e.g. effective stream power; Vocal Ferencevic and Ashmore 2012). This has necessitated a highly active stream rehabilitation and restoration industry (Fig. 11.13; e.g. Wohl and Merritts 2007; Ashmore 2015). A prominent challenge to stream restoration and renaturalization in the region is existing infrastructure and therefore space constraints. Designed stream channels are often engineered within confined corridors, but with planning objectives that still expect flexibility for flood attenuation and aquatic habitat (Walsh et al. 2005). With these design challenges for restoring urban watercourses, projects often attempt to implement "hybrid" approaches that mimic stream functions with static features rather than restore dynamic fluvial processes that addresses sediment supply connectivity within the watershed (Cockburn et al. 2016; Thayer et al. 2016; Wohl et al. 2017). Sustainable long-term management policies for stream restoration are largely restrained by practical, social and economic limitations. Ecological restoration initiatives in the region tend to be most strongly driven by environmental regulations at multiple levels of government, including watershed conservation authorities, and provincial and federal agencies, under legislation for fisheries and endangered species habitat protection (e.g. Committee on the Status of Endangered Species in Canada (COSEWIC 2007)) (Fig. 11.14).

11.7 Conclusion

Geomorphic processes and surface landforms in the lower Great Lakes region have undergone significant change over a relatively short time in Earth's history (<25 ka). These changes have been accelerated substantially in an even shorter time span of less than 200 years. Predicting landscape response to regionally and globally influenced climate change in a context of significant human impacts on surface form will be a critical challenge for future geomorphic research in this region.

References

Adams J (1989) Postglacial faulting in eastern Canada: nature, origin and seismic hazard implications. Tectonophysics 163:321–331

Allan JD (2004) Landscapes and riverscapes: the influence of land use on stream ecosystems. Ann Rev Ecol Evol S 35:257–284

Aquafor Beech Ltd. (2010) Highland Creek Geomorphic systems master plan. Prepared for the City of Toronto (ABL Project Reference: 64473)

Ashmore P (2015) Towards a sociogeomorphology of rivers. Geomorphology 251:149–156

Altus Group Economic Consulting (2009) State of the aggregate resource in Ontario study paper 1—aggregate consumption and demand, p 69

BaMasoud A, Byrne ML (2011) Analysis of shoreline changes (1959–2004) in Point Pelee National Park, Canada. J Coastal Res 27:839–846

Barnett PJ, Sharpe DR, Russell HAJ, Brennand TA, Gorrell G, Kenny F, Pugin A (1998) On the origins of the Oak Ridges Moraine. Can J Earth Sci 67:1152–1167

Boyce JI, Eyles N (1991) Drumlins carved by deforming till streams below the Laurentide Ice Sheet. Geology 19:787–790

Boyd G (1981) Great Lakes Erosion Monitoring Programme 1973–1980. Final Report. Unpublished Manuscript. Bayfield Laboratory for Marine Science and Surveys, Ocean Science and Surveys, and Department of Fisheries and Oceans, Burlington, Ontario

Brennand TA (1994) Macroforms, large bedforms and rhythmic sedimentary sequences in subglacial eskers, south-central Ontario: implications for esker genesis and meltwater regime. Sed Geol 91:9–55

Brennand TA (2000) Deglacial meltwater drainage and glaciodynamics: inferences from Laurentide eskers, Canada. Geomorphology 32:263–293

Carter TR, Trevail RA, Easton RM (1996) Basement controls on some hydrocarbon traps of southern Ontario, Canada. In: van der Pluijm BA, Catacosinos PA (eds) Basement and basins of Eastern North America, Geological Society America Special Paper 308, pp 96–211

Campo SH, Desloges JR (1994) Sediment yield conditioned by glaciation in a rural agricultural basin of southern Ontario, Canada. Phys Geog 15:495–515

Catacosinos PA, Daniels PA Jr (1991) Early sedimentary evolution of the Michigan Basin. Geological Society America, Boulder Colorado

Chapman LJ, Putnam DF (1951) Physiography of Southern Ontario. University of Toronto Press, Toronto

Chapman LJ, Putnam DF (1984) The physiography of Southern Ontario, 3rd ed. Ontario Geological Survey Special Volume 2, p 270

Chapman LJ, Putnam DF (2007) Physiography of Southern Ontario [computer file]. Miscellaneous Release-Data 228. Ontario Geological Survey

Chin A (2006) Urban transformation of river landscapes in a global context. Geomorphology 79:460–487

Clubine NG, Desloges JR and Ashmore P (2010) A quarter century of seasonal and annual sediment yield variations into Lake Huron from Ausable River, Ontario. International Association for Great Lakes Research. 53rd Annual conference on Great Lakes research, May 15–19, Toronto, Ontario. Abstract p 53

Coakley JP (1976) The formation and evolution of Point Pelee, western Lake Erie. Can J Earth Sci 13:136–144

Cockburn JMH, Villard PV, Hutton C (2016) Assessing instream habitat suitability and hydraulic signatures of geomorphic units in a reconstructed single thread meandering channel. Ecohydrology 9:1094–1104

Colombo SJ, McKenney DW, Lawrence KM, Gray PA (2007) Climate changep projections for Ontario: practical information for policymakers and planners. Ontario Ministry of Natural Resources. Science and Information Resources Division, 38 p

Committee on the Status of Endangered Species in Canada (COSEWIC) (2007) COSEWIC assessment and update status report on the redside dace (Clinostomus elongatus) in Canada. Ottawa, 59 pp

Conway T, Hackworth J (2007) Urban pattern and land cover variation in the greater Toronto area. Can Geogr 51:43–57

Crawford GW, Smith DG, Desloges JR, Davis AM (1998) Floodplains and agricultural origins: a case study in south-central Ontario. Can J Field Arch 25:23–137

Department of Fisheries and Oceans (2016) Natural factors affecting lake levels. http://www.tides-marees.gc.ca/C&A/natural-factors-eng.html. Accessed 10 July 2018

Dineva S, Mereu R, Eaton D (2004) Seismicity of the Southern Great Lakes: revised earthquake hypocenters and possible tectonic controls. Bull Seism Soc Am 94:1902–1918

Dobbie T, Ecologist Team Leader, Point Pelee National Park, personal communication, June 27, 2018

Doughty M, Boyce JI, Wallace K, Eyles N, Eyles CH (2014) Lake sediments as natural seismographs: earthquake-related deformations (seismites) in central Canadian lakes. Sed Geol 313:45–67

Ducks Unlimited Canada (2010) Southern Ontario wetland conversion analysis. http://www.ducks.ca/assets/2010/10/duc_ontariowca_optimized.pdf?9d7bd4

Edwards TWD, Wolfe BB, MacDonald GM (1996) Influence of changing atmospheric circulation on precipitation δ18O-temperature relations in Canada during the Holocene. Quat Res 46:211–218

Eyles CH, Eyles N (1983) Sedimentation in a large lake: a reinterpretation of the late Pleistocene stratigraphy at Scarborough Bluffs, Ontario, Canada. Geology 11:146–152

Eyles N, Arnaud E, Scheidegger AE, Eyles CH (1997) Bedrock jointing and geomorphology in southwestern Ontario, Canada: an example of tectonic predesign. Geomorphology 19:17–34

Eyles N, Putkinen N, Sookhan S, Arbelaez-Moreno L (2016) Erosional origin of drumlins and megaridges. Sed Geol 338:2–23

Eyles N, Doughty D (2016) Glacially-streamlined hard and soft beds of the paleo-Ontario ice stream in Southern Ontario and New York state. Sed Geol 338:51–71

Gao C, Shirota J, Kelly RI, Brunton FR, van Haaften S (2006) Bedrock topography and overburden thickness mapping, southern Ontario; Ontario Geological Survey Miscellaneous Release–Data 207

Gerber R, Howard K (2000) Recharge through a regional till aquitard: three dimensional flow model water balance approach. Groundwater 38:410–422

Gilbert R (1997) Glaciolacustrine environment of part of the Oak Ridges Moraine, Southern Ontario. Geogr Phys Quat 51:55–66

Gingras D, Adamowski K, Pilon PJ (1994) Regional flood equations for the provinces of Ontario and Quebec. J Am Water Res Assoc 30:55–67

Godin L, Brown RL, Dreimanis A, Atkinson GM Armstrong DK (2002) Analysis and reinterpretation of deformation features in the Rouge River Valley, Scarborough, Ontario: Can J Earth Sci 39:1373–1391

Gregory KJ (2006) The human role in changing river channels. Geomorphology 79:172–191

Hart JP, Lovis WA (2013) Revaluating what we know about the histories of maize in northeastern North America: a review of current evidence. J Arch Res 21:175–216

Javelle P, Ouarda TBMJ, Bobée B (2003) Spring flood analysis using the flood-duration-frequency approach: application to the provinces of Quebec and Ontario, Canada. Hydro Proc 17:3717–3736

Karrow PF (1967) Pliestocene geology of the Scarborough area; Ontario Department of Mines, Geological Report, p 46

Larson G, Schaetzl R (2001) Origin and evolution of the great lakes. J Great Lakes Res 27:518–546

Lewis CFM, Blasco SM, Gareau PL (2005) Glacial isostatic adjustments of the Laurentian Great Lakes basin: Using empirical record of strandline deformation for reconstruction of early Holocene paleo-lakes and discovery of a hydrologically closed phase. Géogr Phys Quat 59:187–210

Lewis CFM, Karrow PF, Blasco SM, McCarthy FMG, King JW, Moore TC, Rea DK (2008) Evolution of lakes in the Huron basin: deglaciation to present. Aqu Eco Health Manage 11:127–136

Macrae ML, English MC, Schiff SL, Stone M (2010) Influence of antecedent hydrologic conditions on patterns of hydrochemical export from a first-order agricultural watershed in Southern Ontario, Canada. J Hydro 389:101–110

Mao D, Cherkauer KA (2009) Impacts of land-use change on hydrologic responses in the Great Lakes region. J Hydro 374:71–82

McFadden MA, Patterson WP, Mullins HT, Anderson WT (2005) Multi-proxy approach to long- and short-term Holocene climate-change: evidence from eastern Lake Ontario. J Paleolim 33:371–391

Molder B, Cockburn JMH, Berg A, Lindsay JB, Woodrow K (2015) Sediment-assisted nutrient transfer from a small, no-till, tile drained watershed in Southwestern Ontario, Canada. Agri Water Manag 152:31–40

Mulligan RPM, Bajc AF, Eyles CH (2016) One piece of the puzzle: uncovering the genesis of large valleys cut into Quaternary sediment and Paleozoic bedrock in south-central Ontario, Canada. Geological Society America—N-Central Region Conference. Abstracts with Programs, vol 48, no 5

Ontario Ministry of Finance (2017) Population Projections. http://www.fin.gov.on.ca/en/economy/demographics/projections/#s3cc. Accessed 20 July 2017

Ontario Ministry of Natural Resources (2008) Southern Ontario land resource information system (SOLRIS): Phase 2—data specifications version 1.2. Inventory, Monitoring and Assessment Section Technical Report

Ontario Ministry of Natural Resources (2013) Aggregate data classes users guide

Ontario Ministry of Natural Resources and Forestry (2015) User guide for Ontario dam inventory: LIO data class

Paul MJ, Meyer JL (2001) Streams in the urban landscape. Ann Rev Eco Syst 32:333–365

Phillips RTJ, Desloges JR (2014) Glacially conditioned specific stream powers in low-relief river catchments of the southern Laurentian Great Lakes. Geomorphology 206:271–287

Phillips RTJ, Desloges JR (2015a) Glacial legacy effects on river landforms of the southern Laurentian Great Lakes. J Great Lakes Res 41:951–964

Phillips RTJ, Desloges JR (2015b) Alluvial floodplain classification by multivariate discriminant analysis for low-relief glacially conditioned catchments. Earth Sur Proc Land 40:756–770

Quinn FH (2002) Secular changes in Great Lakes water level seasonal cycles. J Great Lakes Res 28:451–465

Rao VR, Schwab DJ (2007) Transport and mixing between the coastal and offshore waters in the Great Lakes: a Review. J Great Lakes Res 33:202–218

Rukavina NA, Zeman AJ (1987) Erosion and sedimentation along a cohesive shoreline—the north-central shore of Lake Erie. J Great Lakes Res 13:202–217

Sharpe DR, Barnett PJ, Brennand TA, Finley, D, Gorrell, G, Russel HAJ, Stacey P (1997) Surface geology of the Greater Toronto and Oak Ridges Moraine Area, Southern Ontario. Geological Survey of Canada Open File 3062

Shaw J (2002) The meltwater hypothesis for subglacial bedforms. Quat Int 90:15–22

Stanfield LW (2012) Reporting on the condition of stream fish communities in the Canadian tributaries of Lake Ontario at various spatial scales. J Great Lakes Res 38:196–205

Stewart IS, Sauber J, Rose J (2000) Glacio-seismotectonics: ice sheets, crustal deformation and seismicity. Quat Sci Rev 19:1367–1389

Thayer JB, Phillips RTJ, Desloges JR (2016) Downstream channel adjustment in a low-relief, glacially conditioned watershed. Geomorphology 262:101–111

Tovell WM (1992) Guide to the geology of the Niagara Escarpment: with field trips. In: Brown L (ed) Niagara Escarpment Commission with the assistance of the Ontario Heritage Foundation, ON

Trenhaile AS, Lavalle PD, Lakhan VC (2000) Point Pelee—a large cuspate foreland on Lake Erie, Canadian landform examples: 39. Can Geogr 44:191–193

Vocal Ferencevic M, Ashmore P (2012) Creating and evaluating digital elevation model-based stream-power map as a stream assessment tool. River Res App 28:1394–1416

Walsh CJ, Roy AH, Feminella JW, Cottingham PD, Groffman PM, Morgan RP (2005) The urban stream syndrome: current knowledge and the search for a cure. J North Am Benthol Soc 24:706–723

Wohl E, Merritts D (2007) What is a natural river? Geog Comp 1:871–900

Wohl E, Magilligan FJ, Rathburn SL (2017) Introduction to the special issue: connectivity in geomorphology. Geomorphology 277:1–5

Joe Desloges is a professor in the Department of Geography and Planning and the Department of Earth Sciences at the University of Toronto. He has a Ph.D. from the University of British Columbia, and M.Sc. from the University of Wisconsin-Madison. Along with dedicated graduate students and Canadian and international colleagues, research has been dedicated to improving knowledge about the influence of late glacial and Holocene climate change on fluvial, glacilacustrine and glacimarine systems in the Great Lakes and Cordillera regions of Canada, in the fiord lands of western Greenland, and glacier-fed lakes of central and eastern Iceland. Floodplain, lake and fiord sedimentary deposits have been extensively explored as archives of sediment flux often with high temporal resolution (e.g. annual, event-based, etc.). Recent geoarchaeological work has been undertaken with archaeology colleagues in Canada, China and Greece examining human–environment interactions in cultural settings of the early- to mid-Holocene. He is a registered geoscientist (PGeo), a former president of the Canadian Geomorphology Research Group and is a Fellow of the Royal Canadian Geographic Society.

Roger T. J. Phillips is a professional geoscientist (P.Geo.) and lecturer at the University of Toronto, with graduate degrees from the University of Toronto (Ph.D.) and York University (M.Sc.) His doctoral research in geomorphology has contributed to understanding Quaternary glacial legacy effects on rivers and stream systems in Canada, with a focus on stream energy and sediment supply in the Great Lakes lowlands of Ontario. As an environmental geoscience consultant working in the water resources engineering industry, he specializes in watershed impacts, natural hazards, stormwater management and ecological restoration in urban and agricultural settings. Combining an environmental and Earth science background with geoscience and engineering, his demonstrated career ambition is to bridge gaps between natural science and applied practice. As an active member of Professional Geoscientists of Ontario, Roger is a strong advocate for professional standards in environmental geoscience and applied geomorphology in Canada.

Mary-Louise Byrne is a professor of geography in the Department of Geography and Environmental Studies at Wilfrid Laurier University in Waterloo, Ontario. She earned her M.Sc. and Ph.D. degrees at McMaster University. Her research focus is Great Lakes coastal geomorphology and coastal management, typically in partnership with national and provincial parks. Her work examines direct and indirect human impacts on coastal processes in protected areas concentrating on beach and dune interactions and restoration of impacted coastal environments. She is the first Canadian author of a widely adopted, best-selling introductory physical geography textbook and an award winning undergraduate instructor. Dr. Byrne has a wide range of experience in administrative service to the University and to the external community. In addition to her Board and Senate experience on campus, she was Financial Director for Canadian Federation of Earth Sciences from 2011 to April of 2016 and Treasurer for the Canadian Association of Geographers from 2005 to 2011. She is currently a board member of the Canadian Federation of Earth Sciences as the representative of the Canadian Association of Geographers.

Jackie Cockburn grew up in the interior of British Columbia, and before she even knew what they were, was fascinated by the glacilacustrine deposits in her backyard. It would all come together while completing her degrees at Queen's University, with the help of Bob Gilbert and Scott Lamoureux; the secret is in the sediments. The work at Queen's was focused on palaeoclimatology. Her interest in the complex story between water and sediment developed further with a keen lens to human modification within the landscape, focused on disruptions to sediment cascades (dams, landslides, agriculture). In 2011, she moved to the University of Guelph, and has continued the work started in the USA as well as working in areas closer to home. Her research interests are centred on understanding the impacts of environmental and climate changes on surface processes. Specifically, issues related to sediment mobilization, transport and deposition as a result of these changes. In addition to working with excellent colleagues in the Department of Geography, Environment and Geomatics at the University of Guelph, her research also involves environmental consultants at GEO Morphix Ltd in Milton, ON, in particular Paul Villard.

12 Karst Geomorphology of the Saugeen-Bruce Peninsula, Ontario

Daryl W. Cowell

Abstract

The Saugeen-Bruce peninsula is distinguished by the widespread surface exposure of dolostone and its association with glacial landforms and processes. Given that it has formed within late glacial and postglacial times, it is considered as a young karst that is actively expanding. But palaeokarst forms have also been identified. The Lockport Group carbonates (Early Silurian) are the most significant karst forming rocks on the peninsula. The release of stresses created by over 40 cycles of glacial loading and unloading has led to mechanical separation along major joint surfaces and bedding planes. In addition, the last deglaciation was accompanied by massive flows of sub-glacial meltwater that penetrated these structural openings. High hydraulic gradients associated with the Niagara Escarpment have created a holokarst zone (13% of the peninsula) where surface channels are largely absent and all precipitation is taken underground via sinkholes and bedrock fractures. Four types of dolostone pavements have been identified: intensely pitted littoral zone (type 1), vuggy pavements on reefal rocks (types 2a and 2b) and pavements on non-reefal rocks (type 3) which have the greatest diversity of karren forms. Fluvïkarst systems occur throughout the peninsula, with sinkholes, springs, suffusion dolines and interconnecting conduits; five such systems are described. The role of sub-glacial meltwater systems is discussed.

Keywords

Karstification • Holokarst • Dolostone pavements • Karren • Geomorphology • Hydrogeology • Niagara escarpment • Georgian Bay

D. W. Cowell (✉)
27 Rita Crescent, RR1, Tobermory, ON N0H 2R0, Canada
e-mail: dcowell@amtelecom.net

© Springer Nature Switzerland AG 2020
O. Slaymaker and N. Catto (eds.), *Landscapes and Landforms of Eastern Canada*,
World Geomorphological Landscapes, https://doi.org/10.1007/978-3-030-35137-3_12

12.1 Introduction

12.1.1 Study Area

The Saugeen-Bruce Peninsula forms the northernmost extension of southwestern Ontario into Lake Huron. The peninsula is a part of the Great Lakes Lowland geomorphological region, identified as region 11c on Fig. 1.10 (this volume). The area of interest includes the entire peninsula from Tobermory in the north to Wiarton on Colpoy Bay (Fig. 12.1). This incorporates a 1060 km^2 landmass that separates the main body of Lake Huron on the west from Georgian Bay to the east. As defined, the peninsula is about 70 km long and varies in width from 35 km across the top between Tobermory and Cabot Head to only 7.9 km at Lions Head (Cowell 1976). Originally referred to as the Saugeen Peninsula after a nearby river of the same name as well as the name given to the local indigenous people, it became more commonly known as the Bruce Peninsula after about 1854 when most of it was opened for European settlement (Koenig 2000). The use of the word 'Saugeen' has been gaining modern acceptance and this paper will use the term 'Saugeen-Bruce Peninsula' following Koenig (2000).

12.1.2 Purpose

The purpose of this chapter is to provide an understanding of the geology and geomorphological processes on the Saugeen-Bruce Peninsula that have resulted in the peninsula's unique landscape. Although connected to southern Ontario by the Niagara Escarpment, the geomorphology of the peninsula is very much different due to the widespread exposure of dolostone bedrock which is soluble in natural water. Dolostone is a carbonate rock composed of the mineral dolomite which consists of both calcium and magnesium ions bonded to carbonate ions: $CaMg(CO_3)_2$. It has similar properties to limestone which is composed of the calcite

Fig. 12.1 Location of the Saugeen-Bruce Peninsula in Southern Ontario

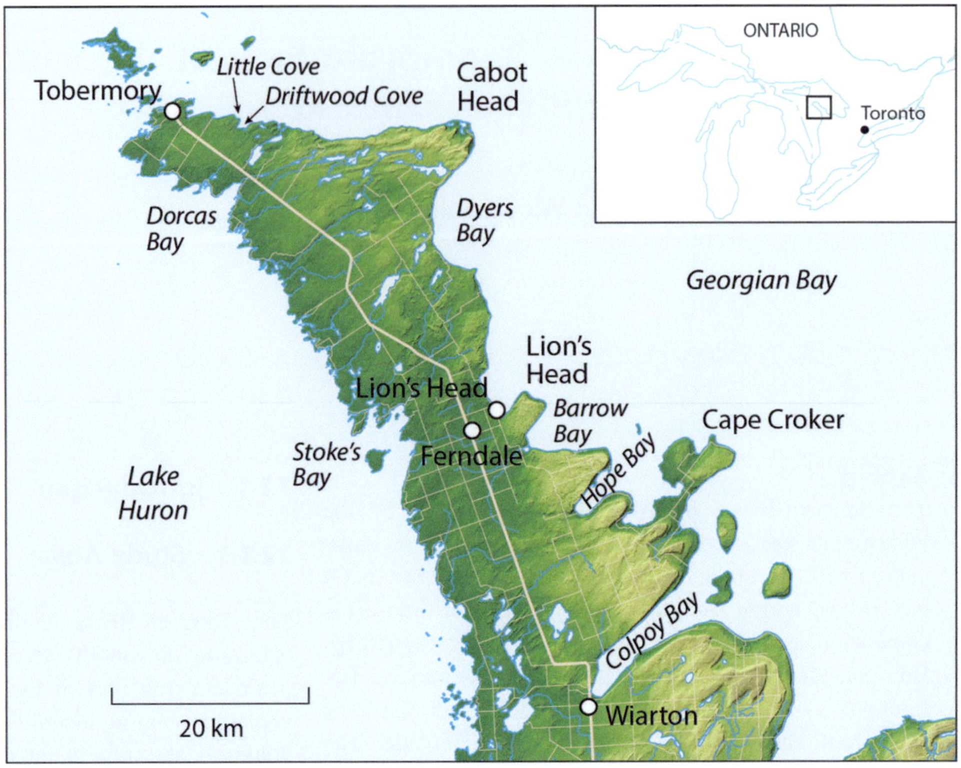

mineral ($CaCO_3$). Most dolostones are typically formed by the postdepositional alteration of lime sediments, whereby some of the calcium is replaced by magnesium from deep circulating meteoric water (water derived from the atmosphere). Calcite is slightly more soluble than dolomite (Ford and Williams 2007) but solution rates are similar under temperate, forested conditions (Szramek and Walter 2004).

The solution of dolostone, the widespread surface exposure of these rocks and their intimate association with glacial landforms and processes has created a complex glacikarst geomorphology that gives the Saugeen-Bruce Peninsula its unique character. Normal surface channel systems compete with underground karst capture into the groundwater system resulting in the formation of karst landforms, including sinkholes, conduits, caves and springs. Karstification is most intense within approximately 1–3 km of the escarpment brow (Fig. 12.2) where high hydraulic gradients dominate the groundwater flow regime (Cowell and Ford 1983).

Karst processes control the modern surface hydrology, including the distribution and size of surface water features and their efficiencies. The ongoing karstification will continue to alter this pattern, evolving towards a drying-out of the peninsula's surface. In the short term, the control of surface water flows and levels by underground conduits has significant implications for stormwater management, water wells, septic beds and associated water quality in built-up areas. This is of particular importance given the dramatic increase in tourism in recent years and the ongoing development.

12.1.3 Previous Work

The karst geomorphology of the Saugeen-Bruce Peninsula was first investigated more than 40 years ago (Cowell 1976). This work covered a wide range of karst features and processes, including dolostone pavements, small-scale surface etchings (karren forms), surface geomorphology, karst hydrogeology and hydrochemistry. The work also formed the basis of a provincial research paper (Cowell 1974) and resulted in two published papers (Cowell and Ford 1980, 1983). Additional karst research conducted out of McMaster University on the peninsula focused on small-scale karst forms, including karst pavements and solution pitting (pit karren) in the Tobermory area.

Federal and provincial research and management programs have contributed significantly to the karst knowledge of the peninsula. Parks Canada commissioned cave studies of the Tobermory Islands (Goodchild 1978, 1984) and an extensive geomorphological investigation of Bruce Peninsula National Park, Fathom Fiver National Marine Park and all the islands between Tobermory and Manitoulin Island, to the northwest of the study area (Cowell et al. 1994 and Geomatics International Inc. 1994).

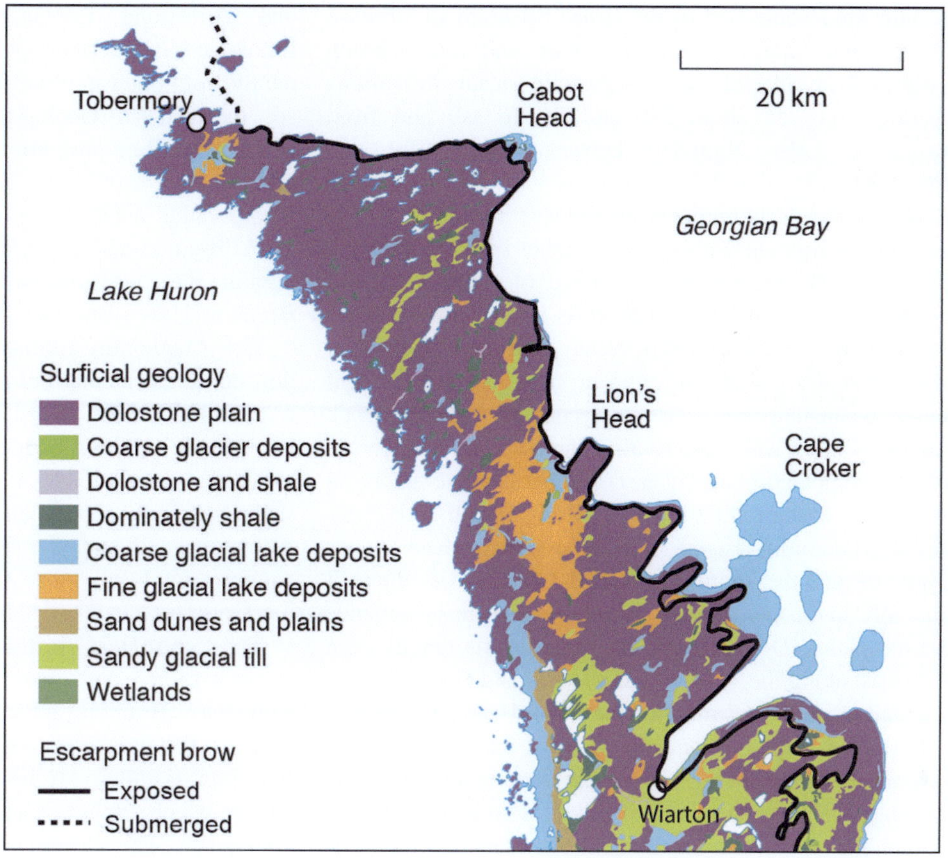

Fig. 12.2 Physiography of the Saugeen-Bruce Peninsula (modified from Ontario Geological Survey 2010)

The Ontario Ministry of Natural Resources and Forestry commissioned several protected areas inventories that included karst features and processes of the Saugeen-Bruce Peninsula. These included an extensive unpublished Nature Reserve inventory of the entire Niagara Escarpment Planning Area as well as a series of update reports (e.g. Buck and Cowell 2002; Cowell 2002a, b). This information was largely summarized in a geology book by the Saugeen Field Naturalists (Bruce-Grey Geology Committee 2006).

The Ontario Geological Survey compiled the locations of all known karst features and areas for southern Ontario in a map and report by Brunton and Dodge (2008a, b). Worthington et al. (2012) summarized much of the early karst work on the peninsula in a field trip guide in support of a Niagara Escarpment karst field trip undertaken for the 39th Congress of the International Association of Hydrogeologists in September 2012. The Ontario Geological Survey has been undertaking an extensive drilling programme over the past 10 years or so to update the Silurian stratigraphy of southwestern Ontario and better understand groundwater flow systems in the carbonate rocks. This work has direct implications on the karst hydrogeology of the study area (e.g. Banks and Brunton 2017; Brunton et al. 2012; Priebe et al. 2012, 2014; Priebe and Brunton 2016).

Geomorphology specific to the Tobermory Islands within Fathom Five National Marine Park was also described and mapped by Cowell (2001). Karst features including pavements and caves were described along with sub-glacial meltwater features, raised shorelines, wave-cut features and other landforms. The most recent karst investigation on the peninsula was commissioned by the Bruce Peninsula Biosphere Association for the Judges Creek Watershed (see Sect. 12.4.4). An important agricultural watershed on the peninsula, Judges Creek is fed by one of the largest karst headwater systems on the Niagara Escarpment.

12.2 Setting

12.2.1 Physiography

The Saugeen-Bruce Peninsula is a glacially scoured dolostone plain partially covered with thin, discontinuous drift (Fig. 12.2). Large areas of glacial deposits are restricted to four principal locations: sub-glacial meltwater outwash fans and associated sand plains near Tobermory and Lion's Head; a large glacilacustrine silt deposits up to several meters thick covering the central flats around Ferndale; and the Mar drumlin field composed of the Bruce till north of Wiarton (Cowan and Sharpe 2007). Nearshore and raised beach deposits from glacial Lake Algonquin occur in various locations between about 240 and 250 m asl, and from postglacial Lake Nipissing between about 190 and 196 m asl.

The peninsula is bordered on the east and north by the Niagara Escarpment with a relative relief of 60 m or more (Fig. 12.2). The escarpment has a range of morphologies on the peninsula including vertical cliffs rising abruptly from Georgian Bay (west of Cabot Head), vertical cliffs with a talus apron above shale and siltstone steps (Cabot Head and central peninsula) and a distinct secondary carbonate rock scarp (at Cape Croker and south). The escarpment becomes submerged in Georgian Bay east of Tobermory, re-emerging to form the eastern cliffs of Bear's Rump Island and, eventually, Manitoulin Island. The dolostone surface slopes gently toward the southwest toward Lake Huron where it gradually intercepts the lake forming widespread wetlands and flat plains. Overall relief on the peninsula ranges from 177 m asl along the shore of Lake Huron and Georgian Bay to a high of about 267 m asl on the Cabot Head promontory.

Surface drainage principally follows the surface slope toward the southwest but most streams are undergoing various stages of karst capture. Streamflow is limited due to this capture and due to the small extent of watersheds. The largest surface systems, from north to south, include Willow Creek, the Crane River, Spring Creek, Judges Creek, the Rankin River and Albermarle Brook. Although most of these flow into Lake Huron, Judges Creek drains much of the central clay plain flowing into Georgian Bay at Barrow Bay. Although the Rankin River drains into Lake Huron (via the Sable River), it mostly flows southward through a series of small lakes that are dammed on their western shores by the presence of a low bedrock escarpment.

12.2.2 Bedrock Geology

The bedrock geology of the Saugeen-Bruce Peninsula consists of basal Late Ordovician shales and sandstones overlain by Early Silurian shales, siltstones and carbonates located near the northeast edge of the Michigan structural basin. The Early Silurian Lockport Group (Table 12.1) consists of stacked dolostone carbonates that form the vertical cliffs of the Niagara Escarpment. Within this group, the Guelph, Eramosa, Goat Island and Gasport formations form the caprock of the escarpment as well as forming most of the peninsula's surface and karst aquifers.

Detailed bedrock geology mapping was prepared by Liberty and Bolton (1971), Armstrong and Dubord (1992) and Armstrong (1993a, b). These authors used a lithologic-based stratigraphic nomenclature derived dominantly from outcrop descriptions. Over the past 20 years or so, the Ontario Geological Survey has undertaken an extensive drilling program throughout southwestern Ontario and has significantly revised the stratigraphic nomenclature incorporating a biostratigraphic framework (Bancroft et al. 2016; Brunton and Britnell 2017). Revised mapping incorporating these revisions have been prepared by Brintnell and Brunton (2009), Brunton (2009) and Brunton et al. (2009).

The Manitoulin formation is a thin- to thick-bedded dolostone, calcareous dolostone and minor limestone unit (Armstrong 1993b). It has a thickness of 5.5–7.6 m in drill holes in the northern Saugeen-Bruce Peninsula. Where it is exposed, it characteristically forms a prominent secondary scarp below the upper cliffs of the Niagara Escarpment. Although a carbonate this formation is not a significant karst aquifer. Only locally does it develop small sinkhole to spring karst systems.

The Cabot Head formation overlies the Manitoulin and consists of red and green shales and minor siltstone and carbonates. Brunton (2009) and Banks and Brunton (2017) note that this unit forms a regional aquitard from Manitoulin Island to Hamilton. It immediately underlies the Clinton Group which includes interbedded shales, siltstones and dolostones. The lowermost units of the Clinton Group comprise the Dyer Bay, Wingfield and St. Edmund formations which are dominantly shales and siltstones with minor dolostone. These three units along with the Cabot Head Formation are best exposed on the west side of Cabot Head where they form a 35 m high section referred to as the 'Clay Cliffs' by Liberty and Bolton (1971).

The upper portion of the Clinton Group on the peninsula consists of four dolostone units, including the Fossil Hill, Rockway, Irondequoit and Lions Head formations. Although carbonates, they typically have low transmissivities and Banks and Brunton (2017) consider the finely crystalline dolostone of the Lions Head formation to form a sub-regional aquitard at the base of the higher transmissive Lockport Group dolostones.

The Lockport Group carbonates are the most significant karst forming rocks on the peninsula and Niagara Escarpment. The Gasport formation (formerly the Wiarton and Colpoy Bay Members of the Amabel, Table 12.1) consists of coarse crystalline encrinites (Brunton et al. 2012). It has a complex lithofacies including cross-bedded crinoidal grainstones which are pure (low terrigenous mud), coarsely-crystalline fossiliferous carbonates and microbial reef mounds (Brunton et al. 2012). The overlying and basal Goat Island formation is a finer crystalline encrinitic dolostone (crinoidal) with a distinctive blue-grey 'pin-striped

Table 12.1 Stratigraphic nomenclature for the Saugeen-Bruce Peninsula (modified after Brunton and Britnell 2017)

Age	Series/stage	Liberty and Bolton (1971)				Brunton (2009), Brunton and Brintnell (2017)	
		Group	Formation/member			Group	Formation
Wenlock ?	Niagaran		Guelph Fm			Lockport	Guelph
			Lockport Fm	Mbr 3/Amabel Fm	Eramosa Mbr		Eramosa
					Wiarton/Colpoy Bay Mbr		Goat Island
					Lions Head Mbr		Gasport
				Fossil Hill (Mbr 2)		Clinton	Lions Head
				Unnamed (Mbr 1)			Irondequoit
Llandovery ?		Cataract	Cabot Head Fm	St. Edmund Mbr			Rockway
				Wingfield Mbr			Fossil Hill
				Dyer Bay Mbr			St. Edmund
				Cabot Head Mbr (restricted)			Wingfield
							Dyer Bay
						Medina	Cabot Head
				Manitoulin			Manitoulin

appearance' (Brunton et al. 2012). Armstrong (1993b) reports a thickness of from 25.7 to 31.9 m for the Lions Head through Amabel formations in drill holes on the peninsula.

The Eramosa formation is a thin-bedded bituminous dolostone that Armstrong (1993b) subdivided into two lithofacies forming three 'informal units'. Brunton (2009) and Brunton et al. (2012) formally recognized three members and Brunton and Brintnell (2011) and Banks and Brunton (2017) consider the lowermost unit (Vinemount Member) to be a sub-regional aquitard, particularly in sub-crop. Karst evidence from immediately south of Wiarton at Hepworth confirms this with a well-developed karst system feeding Spring Creek that emerges near the contact with the lower member. The lower member appears to confine the underlying Gasport/Goat Island aquifer creating artesian conditions. Armstrong (1993b) gives an average thickness of 33 m for the Eramosa formation in boreholes in the southern peninsula.

The uppermost bedrock on the Saugeen-Bruce Peninsula is the Guelph formation, which is a light-tan to brown, fine- to medium-crystalline, thin- to thick- and massively-bedded (reefal), sparsely to very fossiliferous dolostone (Armstrong and Dubord 1992). Brunton et al. (2012) describe a lower reef mound-bearing member and an upper microbial lagoonal member. Overall, they note that eight major facies comprise the Guelph formation. Armstrong (1993b) gives a thickness of 26.5 m from one borehole on the peninsula.

12.2.3 Climate

The Saugeen-Bruce Peninsula has a cool temperate climate (Köppen Classification Dfb) with the warmest mean monthly temperature greater than 10 °C and less than 22 °C and at least four months with means above 10 °C. Precipitation may occur at any time with snow cover common between mid-November and April. Short sudden thaws may occur during this period. Local lake effects moderate temperatures resulting in warmer falls and cooler spring periods than inland or south of the peninsula. Significant short-term winter squalls are common when cool air passes over the relatively warmer lake during ice-free periods. Canadian Climate Normals at the Wiarton Environment Canada Station from 1981 to 2010 provide an average annual daily temperature of 6.6 °C with an average daily maximum of 11.2 °C and minimum of 1.9 °C (Government of Canada). Average annual precipitation is 1047.9 mm for the period including averaged snowfall amounts of 404.8 mm and rainfall amounts of 750.6 mm. Heaviest snowfall occurs in December and January (108.9 and 111.7 mm, respectively) with average rainfall highest in May (83 mm), September (103.1) and October (97.1).

12.3 Karst Features and Processes

According to Ford and Williams (2007) the word "karst" has a long history going back to pre-Indoeuropean time but its modern usage stems from the Slovenian word "kar(r)a" after the name of a region in Slovenia where all precipitation is taken underground. In this regard it is used to describe a unique suite of landforms created by specialized hydrogeological and hydrogeochemical processes.

Ford and Williams (2007) describe karst as:

> *...the term used to describe a special style of landscape containing caves and extensive underground water systems that is developed on especially soluble rocks such as limestone, marble, and gypsum.and, ...as comprising terrain with distinctive hydrology and landforms that arise from a combination of high rock solubility and well developed secondary (fracture) porosity. Such areas are characterized by sinking streams, caves, enclosed depressions, fluted rock outcrops, and large springs. Considerable rock solubility alone is insufficient to produce karst. Rock structure and lithology are also important...*

The solution of carbonate rocks operates primarily in a vertical direction creating landforms that appear isolated, unorganized and disparate (Sweeting 1972). This and the fact that many underground karst landforms are not readily accessible because of small size or isolation requires the use of geochemical and geophysical investigative methods such as dye tracing, carbonate chemistry analysis, electromagnetic surveys, and so on to investigate karst systems.

Carbonate solution involves a three-phase system of rock, water, and gas with the gas being carbon dioxide (CO_2). The solution of CO_2 in water forms carbonic acid:

$$CO_2 + H_2O \Longleftrightarrow H_2CO_3 \qquad (1)$$

The carbonic acid further dissociates into free hydrogen ions that lowers the pH of the solution making it 'aggressive' to carbonate rocks including limestone (calcium carbonate—$CaCO_3$) and dolostone (calcium, magnesium carbonate—$CaMgCO_3$). This, in turn, results in the dissociation of the relatively weak bonds between calcium and carbonate ions and magnesium and carbonate in dolostone:

$$CaMg(CO_3)_2 \Longleftrightarrow Ca^{++} + Mg^{++} + 2CO_3^- \qquad (2)$$

Hence, one of the most important controls on the solution of carbonates in the concentration of CO_2. The greater the partial pressure of this gas, the more Equation 2 will react to the right and the greater the amount of rock that will be dissolved (Szramek and Walter 2004).

The primary source of carbon dioxide is the atmosphere, currently at about 0.04% of air. However, the greatest source of carbon dioxide with regard to bedrock solution occurs in overlying soils as a result of the decomposition of forest litter and respiration from vegetation. Cowell (1976) measured partial pressures of carbon dioxide up to 8 times atmospheric in litter within open joints on the peninsula (Cowell and Ford 1980).

12.4 Karst Groundwater Setting

The hydrogeological setting of the Saugeen-Bruce Peninsula can be separated into two distinct flow regimes: (1) a high hydraulic gradient induced flow regime along the Niagara Escarpment and extending some distance down the backslope; and (2) a shallower bedrock dip-slope gradient regime (Fig. 12.3). Flows in the first are oriented northerly and easterly from the backslope of the scarp towards the base of the escarpment where Georgian Bay or the Cabot Head aquiclude form the base level. The second is controlled by the stratal dip directed predominantly to the southwest where Lake Huron forms the base level. The Niagara Escarpment rises to a maximum of almost 90 m at Cabot Head which exerts a strong sub-regional hydraulic gradient on waters entering the carbonate strata above the cliffs. Highest gradients are found closest to the escarpment brow and decline with distance from the brow (Fig. 12.3). Along the Northern Peninsula between Tobermory and west of Cabot Head, the Lockport carbonate rock sequence extends below the level of Georgian Bay which forms the modern base level for karstification in this area. However, throughout most of the escarpment length (i.e. Cabot Head and south) significant karstification does not extend below the top of the Fossil Hill Formation.

As hydraulic gradients decrease back of the escarpment, the stratal dip becomes more prominent in controlling the direction of groundwater flow. The dip slope has a gradient of only about 5.6 m/km (Liberty and Bolton 1971) creating hydraulic gradients generally less than 1:100 resulting in more restricted and localized karst flows. Regardless of the particular flow regime, geological structure plays a critical role in controlling the specific pathways at any given location. In the case of the escarpment induced karst, this is predominantly via joint and bedding plane fractures although some lithofacies may provide local controls as well. The dominant joint set is oriented between 70° and 100°, roughly paralleling the face of the escarpment between Tobermory and Cabot Head. There is also an approximately orthogonal secondary set and a minor third set (Cowell 1976). The primary and secondary joint sets are typically fairly continuous and can completely penetrate individual carbonate rock units. In locations dominated by reefal structures, reef margins and major joints provide pathways for the movement of meteoric water.

The release of horizontal and vertical stresses created by over 40 cycles of glacial loading and unloading (e.g. Eyles

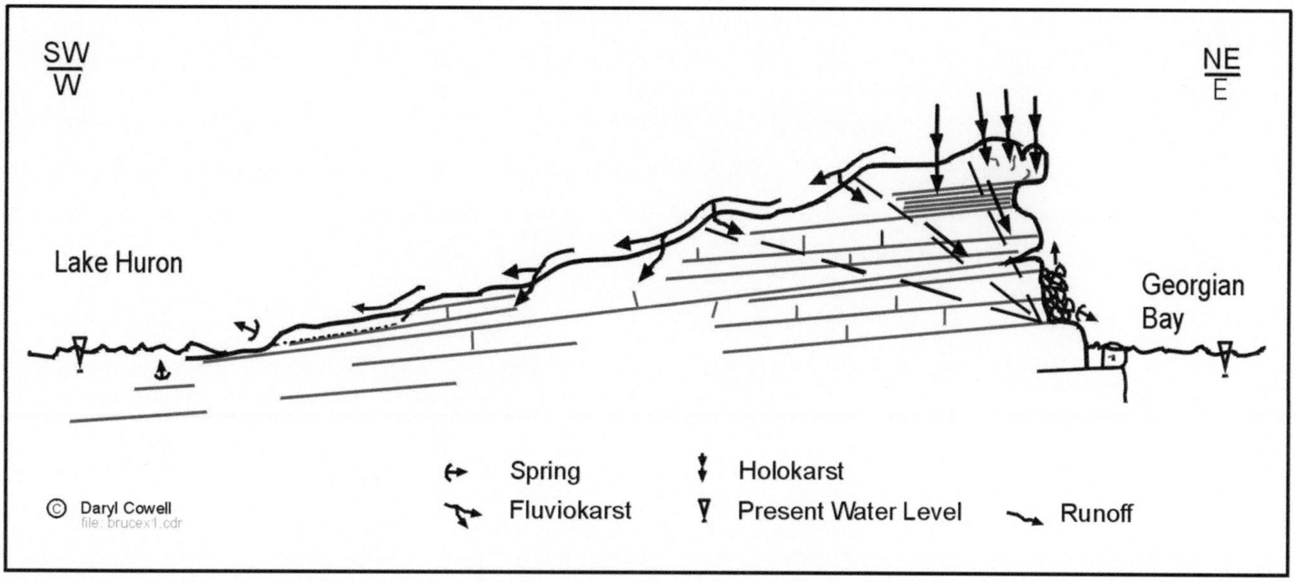

Fig. 12.3 Schematic cross-section of the northern Saugeen-Bruce Peninsula showing competing hydrogeological flow regimes controlled by high hydraulic gradients at the Niagara Escarpment adjacent to Georgian Bay and the stratal dip towards Lake Huron

2002) have allowed differential mechanical separation along major joint surfaces and bedding planes. Further, the last deglacial event (beginning approximately 14 ka) was accompanied by massive flows of sub-glacial meltwater that would have penetrated structural openings (Kor and Cowell 1998). The combination of these events along with high hydraulic gradients created ideal conditions for the capture of surface water and development of karst flow systems. Mechanical opening of fractures is very important to the formation of karst as it is the primary precursor to surface capture. Banks and Brunton (2017) refer to this as a 'stress relief zone' that abets groundwater movement along the escarpment edge.

This zone corresponds to the holokarst zone of Cowell (1976) and Cowell and Ford (1983). Holokarst conditions occur where normal surface channels are largely absent and all precipitation is taken underground via sinkholes and bedrock fractures. The holokarst zone occupies roughly 13% of the peninsula lying within 1–3 km of the scarp face (Fig. 12.4). Given that it has formed within late glacial and postglacial times, it is considered a young karst that is actively expanding inland with holokarst headwater areas competing with adjacent surface water headwaters over time. It is within this zone where most of the largest surface and subsurface karst forms can be found.

The dip slope flow regime has not been significantly investigated on the Saugeen-Bruce Peninsula. It is likely that joints and bedding planes also play a significant role in this zone, particularly in regards to the underground capture of surface water in vadose zones (air-filled pore spaces located above the water table). Sinkholes in this zone tend to be small and karst drainage is not efficient, resulting in local flooding and ponding except during low flow conditions.

Flow in deep bedrock below the top of the water table may have both structural and lithofacies controls. This is referred to as the phreatic zone and Banks and Brunton (2017) note that preferred groundwater pathways in the phreatic zone of the Lockport Group are influenced by stratigraphic boundaries and lithological contrasts.

12.5 Karst Geomorphology

The Saugeen-Bruce Peninsula is an excellent example of glacikarst development where all active karst features are directly associated with glacial landforms and processes. Older pre- and inter-glacial surface karst features which may have existed were removed during the MIS 2 (Cowell 1976). The predominance of an exposed bedrock surface is the product of a lack of overburden deposits and the result of overwash by glacial Lake Algonquin and postglacial Lake Nipissing. Cowell and Ford (1983) note that this makes the peninsula unlike most other Canadian lowland carbonate surfaces. The glacikarst association includes:

(a) Expansion of pre-existing fractures by stress relief during isostatic loading and unloading contribute to enhanced groundwater circulation (Sect. 12.3);
(b) Direct glacial abrasion of the bedrock surface created an extensive flat to ridged surface that has been subject to chemical solution for 10–12 ka and is thought to be the largest dolostone pavement in the world (Sect. 12.4.1);

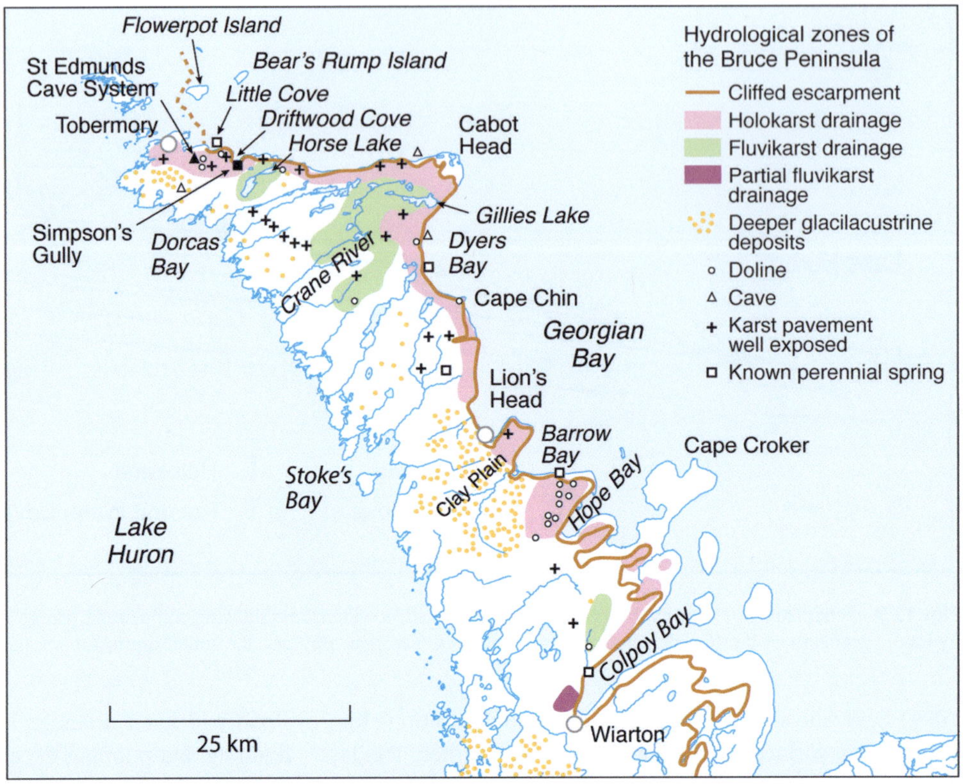

Fig. 12.4 Karst hydrological conditions and pavement exposures on the Saugeen-Bruce Peninsula (from Cowell 1976)

(c) Residual drift deposits allow for the accumulation of surface water that is subject to underground capture (Sect. 12.4.2); and
(d) Pressurized injection of sub-glacial meltwater during sheetflood events result in conduit and cave formation or enhancement along fractures (Sect. 12.4.3).

Relict karst features that may not have an association with glacial landforms or processes are also known to exist. These are referred to as palaeokarst forms (Ford and Williams 2007) being the product of an earlier hydrological regime, now abandoned. Palaeokarst features can become rejuvenated within the modern hydrological regime (Sect. 12.4.4).

12.5.1 Pavements

Small-scale karst surface forms are referred to as karren (Williams 1966; Jennings 1971; Sweeting 1972). These surface etchings have a variety of shapes and sizes, both gravity-assisted (e.g. runnels) and non-gravity (e.g. pitting). Collectively, they occur as part of a limestone or dolostone pavement (Fig. 12.5), which is characterized by an etched bedrock expanse cut into blocks resembling man-made pavements. The pavement forms due to solution are concentrated along weaknesses such as joints or small faults resulting in the opening of these structures forming a semi-rectangular pattern of isolated blocks like paving stones. The residual blocks are referred to as clints and the solution-widened joints are referred to as grikes (Fig. 12.6).

Fox (1952), the first writer to bring attention to the beauty and history of the peninsula, best described the striking visual impact of these pavements as sterile flat slabs of limestone, separated at intervals of a meter or more by vertical cracks, 4–5 cm in width and virtually without vegetation. There is a great deal of literature dealing with limestone pavements (e.g. Williams 1966; Sweeting 1972) but very little has been presented on dolostone pavements. Pluhar and Ford (1970) provide a detailed discussion of karren forms (*fr.* lapiés) on the dolostone pavements in the Hamilton, Ontario area of the Niagara Escarpment. They define karren as structurally controlled, selective solutional or solution-erosional pits developed in carbonate rocks evolving into rectangular blocks and ridges (Pluhar and Ford 1970). The pavements are visible throughout the peninsula where soils are absent or discontinuous (Fig. 12.2). They are particularly well displayed in the northern portion of the peninsula and good exposures can be found all along Highway #6 between Cabot Head and Tobermory (Fig. 12.4). Cowell (1976) distinguishes three types of pavements, two based on bedrock formation (Guelph vs Gasport/Goat Island) and one based largely on its location within the emergent shore zone of Lake Huron. Each was described according to overall surface form and

Fig. 12.5 **a** View of Malcolm Bluff as seen across Colpoy's Bay. At the horizon is the dolostone pavement. See Fig. 12.1 for location. *Photograph* P199/Wikimedia Commons, CC BY-SA 3.0. **b** Exposed limestone along Georgian Bay. *Photograph* E. Lung/Flickr, CC BY-SA 2.0

characteristic karren assemblages. Cowell later (in Geomatics International Inc. 1994) subdivided the pavements into four types incorporating local relief and based on karren assemblages.

Type 1 pavements correspond to the intensely pitted littoral zone pavements found along most of the Lake Huron shore on the Guelph formation (Fig. 12.7). They occur within the active wave zone, on the recently emerged shore and extending below the modern water level. They are also found on some inland lakes (e.g. Marr Lake in Bruce Peninsula National Park) where they also occur along the shore in seasonally inundated areas. Lithologically, they form on flat to slightly undulating thin-bedded strata.

The pitting (pit karren) covers much of the bedrock surface with pits being very regular in form having an egg-cupped shape and flat interfluves (Cowell 1976). Average depths and widths of the pits are 14.3 and 18.1 mm, respectively, at one location measured by Cowell (1976). They tend to increase in depth and diameter with increasing depth in Lake Huron (Cowell 1976). This can lead to coalescing of adjacent pits leaving sharp finger-like protrusions as in the case of one sample found at a depth of 18 m near Tobermory (Fig. 12.8).

The origin of the pits is not certain but they tend to occur on non-reefal rocks, often in association with former glacial striae. The striae may be visible or highlighted only by the perfect alignment of pits in the direction of glacial flow (Fig. 12.7). The reason for the initiation of a pit at a specific location is not known but could be random or related to primary pores, being initiated either by rain drop

Fig. 12.6 Photograph of dolostone pavement in the Fossil Hill Formation composed of clint blocks separated by solution-widened joints (grikes). *Photograph* D. Cowell

impingement and/or from surface lichen. Once a depression begins to form, the pit becomes a small reservoir that remains wetted longer than the flat interfluves, hence providing a micro habitat for lichens and mosses. The presence of vegetation enhances the development of the pit through excretion of carbon dioxide which enhances localized solution. Wave action and freeze and thaw eventually fracture the surface along bedding planes leaving a fresh bedrock surface which typically displays micro-pits as the process starts over on the new surface. The underwater pits must have originated above water (e.g. at lower postglacial lake levels) due to a lack of micro-pitting observed on subaqueous surfaces. The greater width and depth of the subaqueous pitting is not known but probably relates to protection from direct wave action.

Type 2 pavements are referred to as vuggy pavements due to the pocketed character of the rock surface. They occur only on reefal rocks, most commonly in the Guelph formation with solutional etching focused on coarse primary porosity typical of biohermal environments. The only 'true' karren forms on these pavements are clint blocks and grikes.

These pavements can be subdivided into two subtypes based on degree of relief. Type 2a pavements are flat to gently ridged with less than 1 m of relief (biostromal reef formation) and Type 2b form where the surface has been glacially moulded into ridges of 1 m or greater in relief (biohermal). Although both types are typically vuggy, the flat pavements have a more regular and dense pattern of clints and grikes. Type 2a pavements are the most extensive of all four types and are particularly well displayed along Highway #6 between Cabot Head and Tobermory. Type 2b reach their best expression in a large ridge complex located north of Highway #6 and west of Cabot Head.

Type 3 pavements are formed on non-reefal rocks, principally on the Gasport/Goat Island formations; however, some examples were found on Guelph formation rocks west of Cabot Head. They have the greatest diversity of true karren forms including sharp-crested pit karren, runnel karren (slope controlled linear rills) and small bedrock basins referred to as kamenitzas (Cowell 1976). They typically occur on sloping surfaces and where the bedrock is lithologically more homogeneous than in the case of the Type 2 forms. These pavements are fairly restricted in extent, most commonly found on Gasport/Goat Island exposures on the immediate backslope of the Niagara Escarpment (Fig. 12.9).

12.5.2 Fluvïkarst

Exposed pavements generally prevent the accumulation of surface water in the form of streams or lakes because rainfall and snowmelt infiltrate everywhere through the extensive network of grikes. This is considered a type of diffuse recharge which restricts the formation of large surface karst landforms, particularly sinkholes (Cowell and Ford 1983). Infiltrating water is directed by prominent joints and bedding planes down the hydraulic gradient, integrating into larger

Fig. 12.7 Littoral zone pitting in the Guelph formation at Marr Lake. Note former glacial striae controlling pit alignment. *Photograph* D. Cowell

Fig. 12.8 Finger-like protrusions on a deeply pitted sample from 18 m depth in Lake Huron. *Photograph* D. Cowell

and fewer karst conduit and cave systems at depth. The presence of residual drift deposits permits the formation of streams, lakes, ponds and wetlands on the surface, which can be captured underground via the formation of sinkholes within streams (streamsinks) or in lake bottoms. This creates a surface karst characterized as fluvīkarst, whereby the surface and subsurface hydrology is the product of both chemical solution and physical erosion (Ford and Williams 2007). Fluvīkarst systems occur throughout the peninsula and range from large stream and lake capture to small local ephemeral creeks. The latter case includes sinkholes fed by ditches dug by landowners to improve local field drainage. All fluvīkarst systems are postglacial reflecting the modern hydrological setting, however, they may incorporate sub-glacial injection caves (e.g. Leopard Frog cave, Sect. 12.4.3) or palaeokarst voids. The largest fluvīkarst systems are found within or adjacent to the holokarst zone associated with the high hydraulic gradients provided by the Niagara Escarpment. Fluvīkarst systems are characterized by sinkholes, springs at lower elevations and interconnecting caves and conduits.

Sinkholes (also called dolines) include stream sinks and suffosion dolines. The former is formed by the downward movement of water into the upper rock mass, gradually enlarging the opening by solution from the precipitation and stream flows. Suffosion dolines form where the bedrock is mantled with sediment and solution is initiated within the bedrock beneath the mantle, gradually piping the sediment downward into the enlarging conduit. A third type of sinkhole is referred to as collapse doline and forms where ongoing karstification in the upper bedrock eventually weakens the rock resulting in the sudden collapse of surface materials into the karst void. This type is rare on the Silurian dolostones in Ontario.

Streamsinks typically occur in series within channels with the largest and oldest typically located furthest downstream. As the karst matures, the stream gradually becomes captured further upstream with the downstream sinks are activated only during spring freshet. The original capture occurs where the stream channel contacts bedrock either at the surface or beneath the channel at bedrock rises or in zones of higher permeability.

Fig. 12.9 Gravity controlled runnel karren on outcrop of Gasport/Goat Island dolostone. *Photograph* D. Cowell

Suffosion dolines are initiated at the bedrock-drift interface gradually punching up to the surface as expanding conduits enlarge and remove overburden. They can be found in isolated areas but typically are found in combination with streamsinks but outside the channel. Although suffosion dolines are not formed by a flowing stream, they can become stream sinks over time as they expand their local basins (Cowell and Ford 1975; Worthington et al. 2012).

Cowell 1976 described five fluvïkarst systems within three of which he performed tracer studies using a fluorescent dye (Rhodamine WT). Traces were undertaken at perennial flow systems including the outlet from Gillies Lake; an in-lake sinkhole connecting Horse Lake to Marr Lake; and the St. Edmunds karst system. Other systems investigated included a small stream flowing into Colpoy Bay and the outlet of a large wetland complex near Dyer Bay (Fig. 12.4).

Table 12.2 summarizes key morphological information pertaining to these systems. Basin sizes range from about 2 km^2 to over 15 km^2, however, precise measurements are not easily determined because surface watershed boundaries do not usually coincide with groundwater watersheds in karst systems. Each of these systems is unique and they have few common characteristics. This is reflected in the wide range of hydraulic gradients and nature of sink points and resurgences.

12.5.2.1 Colpoy Creek

Colpoy Creek system is a small creek that drains small pastures and forested areas on the north flank of Colpoy Bay. The karst is very immature without distinct sinkholes or springs. Flow is captured gradually within the stream channel by shallow bedding planes (on-strike) within the Gasport/Goat Island formation. Subsurface flow appears to parallel the surface channel re-emerging in small springs within the channel. The length (and gradient) varies throughout the year as the capture and emergence locations appear to change with flow regime.

12.5.2.2 Dyers Creek

Dyers Creek has been enlarged by ditching to drain a large wetland area. The entire till plain above the village is drained

Table 12.2 Summary of five fluvikarst systems, Saugeen-Bruce Peninsula (Cowell 1976; Cowell and Ford 1983)

System	Location	Approximate basin (km^2)[a]	Lake/stream capture	Minimum gradient	Tracer arrival	Resurgence type
Colpoy Creek	Flows into Colpoy Bay at Colpoy Bay Village	7.5	In-stream	176:1	N/A	In-stream
Dyers Creek	1 km SW Dyer Bay Village	3	Streamsink (ditched?)	39:1	N/A	Small spring
Gillies Lake	5 km N of Dyer Bay Village	7	Stream base and In-lake	5:1	45 min	Large springs at top of Cabot Head Fm.
Horse Lake	Cyprus Lake Campground of BPNP	15.5	Sinkholes and streamsink	42:1	90 min	Four small springs
St. Edmunds Karst	4 km SE of Tobermory	2.2	Two large streamsinks	139:1 200:1 (cave)	57–72 h	Little Stream Cave

[a]Areas are very approximate because basins are partly spring fed and may be drained via other sinkholes

via karst. The stream is captured by a prominent N-S joint (secondary) within the Gasport/Goat Island formation. Immediately north (downstream) of the streamsink is a large pavement complex with evident slumping of clint blocks evidencing significant bedding plane solution.

12.5.2.3 Gillies Lake

Gillies Lake is the deepest lake on the peninsula having a depth of at least 36.5 m (Cowell 1976). Interestingly, it is deepest where it is closest to the brow of the escarpment, only approximately 250 m distant. Although thought to be a karst collapse feature (Fox 1952), it is more likely a sub-glacial meltwater erosional pothole (Kor and Cowell 1998). The lake does not have a surface channel connection to Georgian Bay but is drained entirely through karst within the Gasport/Goat Island formations. Capture occurs within the base of a dug outlet channel that was excavated to provide water to a wooden flume constructed in 1882. The flume was used to conduct logs to a former sawmill at the base of the escarpment (Folkes 2008). The original and primary karst drain, however, is a major E–W joint. Flow into the joint occurs within the lake body and is not visible as an individual sink point.

12.5.2.4 Horse Lake

Horse Lake lies downstream of two larger lakes (Cyprus and Cameron) within the Cyprus Lake Campground Recreation Area in Bruce Peninsula National Park located about 12 km SE of Tobermory. Horse Lake currently drains into a large sinkhole on its northern edge and is separated from Marr Lake to the north by a low bedrock ridge of Guelph formation dolostone. The two lakes are approximately half a kilometre apart. It is not certain whether they were ever connected at the surface but a compound sinkhole lies at the base of the ridge at the end of a former embayment of Horse Lake. This sinkhole probably marks the original sinking point with capture gradually advancing towards the lake where most of the sinking currently takes place. This occurs at a large sinkhole that lies within the lake during spring freshet gradually becoming exposed as water levels fall. At one time, beavers living in Horse Lake constructed a dam around the sinkhole in an attempt to augment late season water levels (Cowell 1984). The karst follows the major E–W joint set.

12.5.2.5 The St. Edmunds Karst

The St. Edmunds karst is the best example of a fluvïkarst system on the peninsula and consists of what may be the most diverse complex of karst and glacial landforms in Ontario (Worthington et al. 2012). The karst lies entirely within the Guelph formation. Museum Cave and Leopard Frog Cave serve to drain a small lowland area that has no surface outlet to Georgian Bay. Leopard Frog Cave parallels the escarpment for a distance of 2.5–3.5 km, which is one of the longest straight-line passages of known Ontario caves. Flow in the system has been traced from Museum Cave towards the northeast into Little Cove (Cowell 1976). Having a straight-line distance of about 2.5 km and a total fall of about 18 m, the system's overall gradient is only 139:1 (Table 12.2). At least half of this fall occurs at the cave entrances and thus the cave itself is virtually flat (200:1) and, as a result, it remains flooded with little or no throughflow during most of the season. Surface ponding of the entrances during spring freshet results in significant flooding of overlying fields and a local road. This effectively increases the hydraulic gradient to a point where the cave system becomes activated and local landowners report that the entire surface drains within about 24 h (Cowell 1976). This suggests a sump type system whereby the rising head eventually reaches a level that literally 'flushes' the system. Leopard Frog and Museum caves are interpreted to be the reactivation of older karst voids that were created during an earlier, now relict hydrological system. These are referred to as palaeokarst features. They were likely to be reactivated by sub-glacial meltwater injection (Sect. 12.4.3) which is effectively a relict event. Evidence for this includes the fact that the cave is not directed by the steep hydraulic gradient provided by the nearby escarpment face as well as by the relatively large size of the passages. The scale of these passages is much greater than other postglacial fluvïkarst systems observed on the peninsula or in association with the escarpment.

12.5.3 Sub-glacial Meltwater Injection

The role of catastrophic sub-glacial meltwater erosion in sculpting bedrock surfaces over large areas has been interpreted by Sharpe and Shaw (1989), Gilbert and Shaw (1994) and Shaw and Gilbert (1990). Collectively referred to as 's-forms' (sculpted forms), they include the formation of re-entrant valleys, lake basins, rock drumlins, bedrock scours and many other landforms, and are typically associated with the presence of escarpments (Gilbert and Shaw 1994). The mechanism involves long-term accumulation of meltwater at the base of a continental ice sheet sufficient to abruptly raise the ice sheet resulting in the sudden short-term release of basal water under tremendous pressure. Sculpting occurs via widespread sheetflood flows as well as by the localized separation of these flows into vortices. Entrained sediment, cobbles and boulders enhance erosion. Whilst the effect of sheetfloods in shaping surface geomorphology can be readily observed, there is also a less well-known subsurface role that is especially important in karst terrains. Ford and Williams (2007) refer to the 'direct injection of meltwater into karst aquifers'. Erosion is both mechanical and chemical and can create conduits and caves along fractures as well as rejuvenate palaeokarst. Both have been documented on the Saugeen-Bruce Peninsula.

Kor and Cowell (1998) describe landforms created during catastrophic sub-glacial sheetflood events on the peninsula during the MIS 2. Most of these are surface forms but they also note the presence of caves in bedrock ridges that they refer to as a 'new type of s-form'. Although certainly involving direct abrasion, solution of the bedrock also occurred as shown by the presence of phreatic cave forms in some locations. Several isolated caves located in bedrock ridges near Berford Lake and George Lake, as described by Kor and Cowell (1998), are considered to have been formed during one or more sub-glacial meltwater events. At George Lake on the Northern Peninsula, at least four narrow solution caves up to 40 m long occur completely within a rock drumlin on the northeast side of the lake (Geomatics International Inc. 1994). The caves have been formed along joints that are mostly sub-parallel to the ridge orientation which also marks the glacial surface grain orientation (NE–SW). The caves are not connected to any existing surface water system but are totally isolated from surrounding streams and lakes. Although they could be older palaeokarst cavities isolated from an earlier karst aquifer by glacial erosion, the presence of distinct abrasional s-form features on some cave openings such as cavettos and furrows indicate at least partial formation by sub-glacial meltwater.

Leopard Frog Cave, located east of Tobermory, is a complicated karst incorporating palaeokarst and active features but appears to also have experienced a sub-glacial meltwater injection event (Sect. 12.4.2). One of the most remarkable features of Leopard Frog Cave is the presence of the largest cavetto s-form feature known from the peninsula on the wall of a deep glacial scour above the cave entrance (Fig. 12.10). Cavettos are horizontally oriented u-shaped landforms mostly found in the walls of sub-glacial bedrock channels (Kor and Cowell 1998). The cavetto is located at the upstream end of the cave but at the downstream end of the glacial flow direction presenting compelling evidence of at least one glacial injection processes. It is unlikely that the entire cave system was formed during the MIS 2 sub-glacial meltwater event(s), however, it would have been a significant factor in the evolution of the system.

Fig. 12.10 Large sub-glacial meltwater cavetto form sculpted into the wall of a glacial scour above the entrance to Leopard Frog Cave in the Guelph formation. The cave descends vertically from the base of the cavetto near centre of photo. *Photograph* D. Cowell

12.5.4 Palaeokarst

The role of palaeokarst landforms within the modern hydrological setting is an important component in our understanding of karst geomorphology and hydrogeology on the Saugeen-Bruce Peninsula. Cowell (1976) first postulated the presence of palaeokarst in his investigation of the St. Edmunds karst system, however, the significance and extent of palaeokarst features on the peninsula (or Niagara Escarpment in general) was not understood. It is becoming clear that the palaeokarst landscape is very significant to the karst story. Some additional palaeokarst features known to occur on the peninsula include the Judges Creek headwater aquifer, Root Cave and Ontario's largest known sinkhole at Simpson's Gully.

12.5.4.1 Judges Creek

Judges Creek is currently being studied by the Bruce Peninsula Biosphere Association (BPBA) with the objective of protecting and enhancing water quantity and quality within this agricultural watershed. The watershed lies in the central peninsula in a major livestock farming area. As part of their studies, the BPBA became aware of the presence of several large karst springs in the eastern portion of the watershed. These springs were recognized as critical components of the base flow to the creek. The Judges Creek Watershed consists of a central low glacilacustrine clay plain

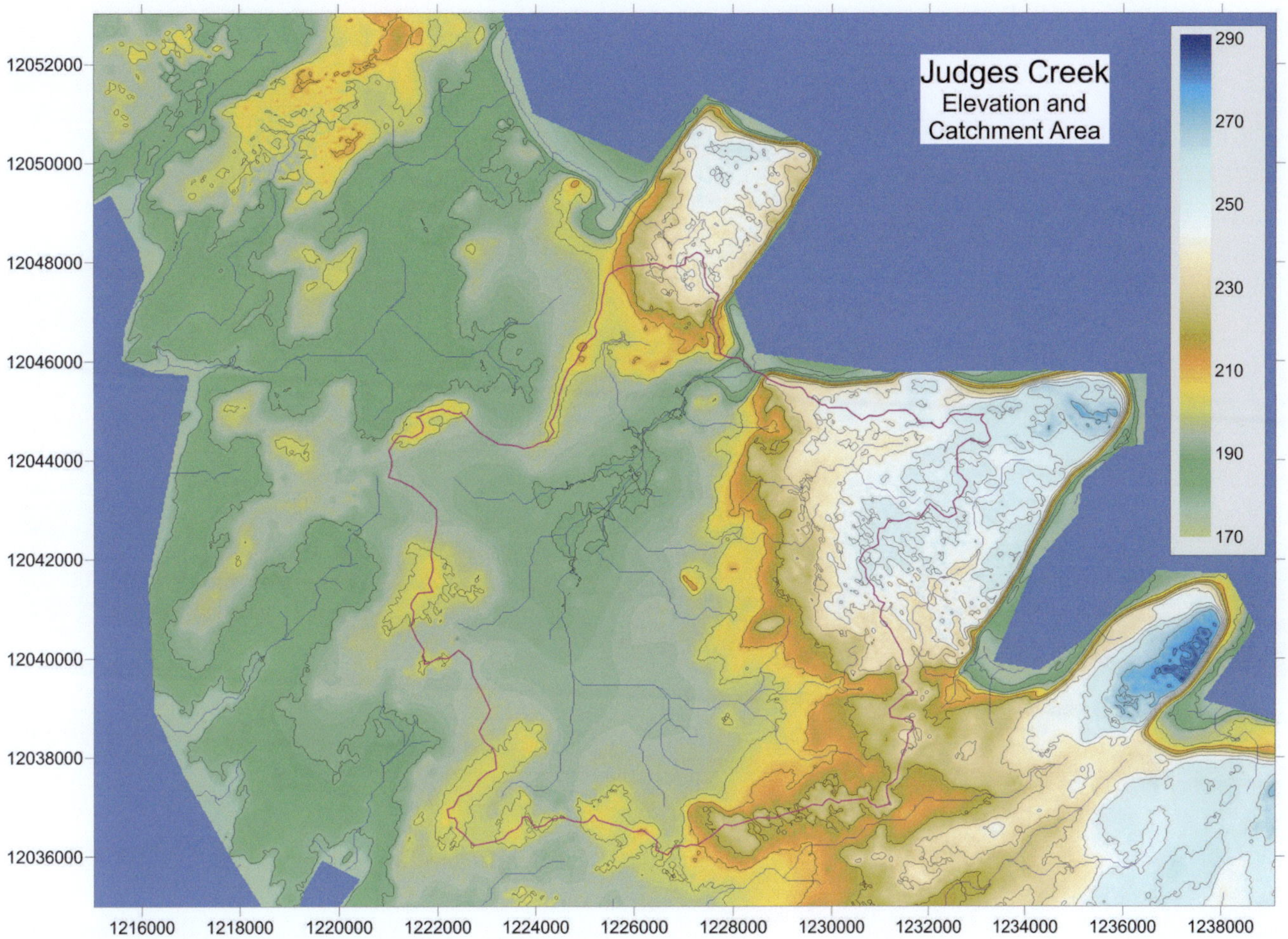

Fig. 12.11 The Judges Creek Watershed (red line) south of Ferndale on the Saugeen-Bruce Peninsula. The primary karst aquifer lies in the eastern highlands and the springs appear at about 193 m asl (brown-green transition). The area in green and beige outlines the extent of the clay plain within the watershed (unpublished report by Cowell et al. from 2016)

lying between 190 and 200 m asl. The main channel flows to the north to outlet at Barrow Bay in Georgian Bay (Fig. 12.11). A series of bedrock ridges having in the order of 50–80 m of relief rise to the east above the clay plain towards the Niagara Escarpment. Surface bedrock in the uplands consists of the Guelph and Eramosa formations but the upper Gasport/Goat Island formations immediately underlie the surrounding clay plain (Armstrong 1993a).

The most striking karst landforms of the study area include several large springs located in an almost north–south alignment along the eastern edge of the clay plain at an elevation of about 191–193 m asl along the foot of the bedrock uplands. Figure 12.12 shows one of the large springs located immediately adjacent to the uplands. The bedrock uplands are mostly forested and are marked by the absence of surface streams, ponds or lakes of significant size. Short ephemeral channels and occasional small ponds are known to exist on pockets of surficial materials that occupy low areas between bedrock ridges. All surface waters in this upland area recharge the bedrock via grikes or via small fluvïkarst stream sinks on localized till deposits. Cowell (1976) considered most of the promontory between Hope Bay and Barrow Bay as a holokarst zone (Fig. 12.4). Although the karst aquifer in upper Judges Creek has not been defined in detail, preliminary analyses of static well levels suggest that the groundwater watershed extending onto the promontory between Barrow Bay and Hope Bay roughly parallels the surface water watershed.

Figure 12.13 provides a conceptual groundwater model of the Judges Creek karst. A groundwater divide occurs in the east within the promontory separating northerly and easterly karst flow to the base of the escarpment from dominantly westerly flows to the springs feeding Judges Creek. Confining lake sediments in the Judges Creek basin force groundwater upwards to the springs which lie along the bedrock/clay interface (Fig. 12.11). A couple of springs occur at slightly lower elevation further out on the clay plain but it is probable that water levels in all springs are within a

Fig. 12.12 Large spring at the boundary between the clay plain (to right) and bedrock upland. This spring only flows when the aquifer is fully charged. *Photograph* D. Cowell

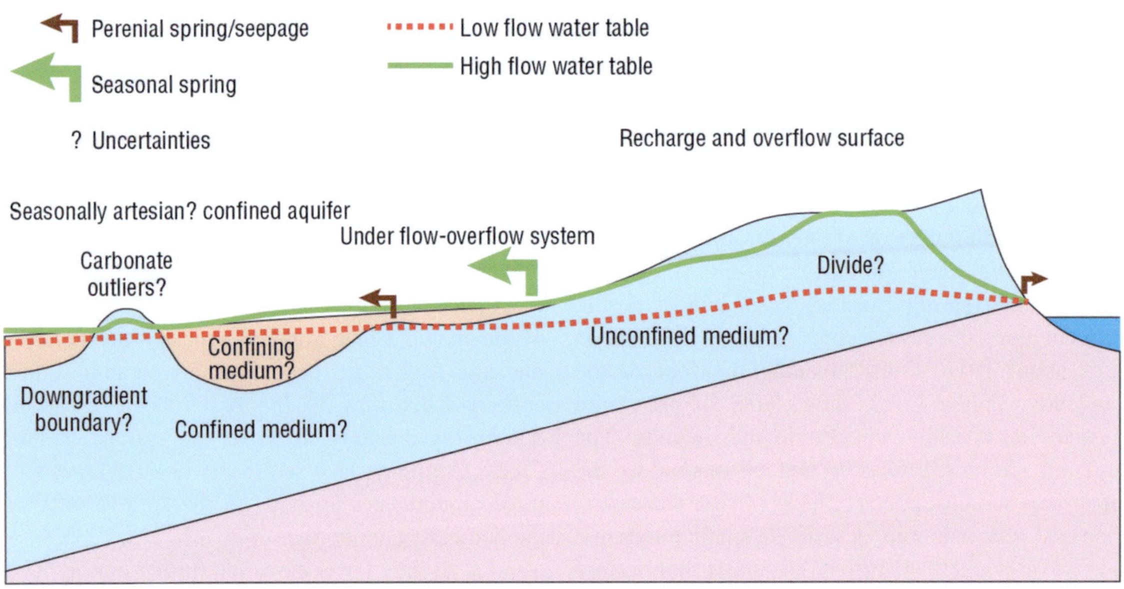

Fig. 12.13 Conceptual groundwater model of the Judges Creek Watershed. The Niagara Escarpment is on the right

Fig. 12.14 Large stalactites in the middle room of Root Cave. *Photograph* Willie Waterton

narrow elevational range, particularly during low aquifer levels. The springs furthest from bedrock interface flow throughout the year even during very dry years indicating a significant storage capacity within the aquifer. Higher springs located closer to the bedrock upland (Fig. 12.12) are activated only when the aquifer level rises such as during spring freshet.

In an unpublished report we interpret the aquifer feeding these springs as representing the reactivation of a relict karst aquifer. They note that the information available indicate the presence of a well-developed integrated karst that if fully connected would represent the best developed and largest karst system so far described for the Bruce Peninsula.

The role of palaeokarst is suggested by the size and number of large springs which are not typical of known postglacial karst. The presence of springs on the clay plain and separated from the bedrock upland suggests the reactivation of older features because the bedrock is covered by substantial lake clays. Further, direct evidence of palaeokarst was revealed in boreholes drilled in the upland area for a proposed quarry development. Six of 14 boreholes had sediment-filled cavities between 0.5 and 1.8 m in thickness within the bedrock sequence (Genivar Inc. 2011). The cavities indicate a period of solution followed by in-filling of sediment. The sediment source is unknown but the surface in this area has virtually no sediment, however, even if the sediment were derived from MIS 2 events, the cavities must have pre-existed the modern groundwater system.

12.5.4.2 Root Cave

Root Cave has the largest, most abundant and diverse range of speleothems of any cave in Ontario (Worthington et al. 2012). The cave entrance lies at the base of the Gasport/Goat Island cliffs on the escarpment west of Cabot Head. It has a low (50–75 cm) but wide (4–5 m) entrance. This passage follows a bedding plane to the south for about 30 m and ends at a vertical rift that parallels the face of the escarpment, following the major E–W joint set (Bruce-Grey Geology Committee 2006 and Beddows et al. 2008).

Thin short stalactites occupy the walls of the entrance passage and a rimstone dam has formed from standing water on the east wall at the end of the passage. The rift rises about 17 m in two steps and terminates within a meter or so of the ground surface. The rift is entirely lined with cave deposits including flowstone covering the walls of the lowest step. The upper step contains a phenomenal collection of stalactites, stalagmites and helictites, including large columnar and drapery stalactites that rise 7–8 m on the north wall (Fig. 12.14). Stalactites grow downward from the ceiling and stalagmites grow upward from the floor, both in the vertical orientation. Helictites are curving or angular, often growing sub-horizontally from cave walls. The large columnar and drapery stalactites terminate upward in a flow stone mass intercepted by live cedar roots that have descended from the surface. The root mass continues downward on the flowstone and become encased in calcite forming a portion of one of the long stalactites. The source of the calcite is drip water that enters the bedrock in the vicinity of the roots. The water fills a shallow flowstone basin in the uppermost part of the cave. As the basin overflows, a thin sheet of water drains over the roots and stalactites into the main rift (Bruce-Grey Geology Committee 2006; Beddows et al. 2008; Worthington et al. 2012).

The cave is clearly a palaeokarst feature given its complex history of formation, abandonment and in-filling. After the cave was first abandoned there have been at least two stages of speleothem development between which there was at least one period where water-laid deposits filled all or part of the cave with laminated sands, fine gravels and clays. The current stage is one of speleothem development with the active formation of the large stalactites and calcite capping some sediment. The age of cave formation is not known but would be at least inter-glacial if not pre-glacial.

12.5.4.3 Simpson's Gully

Simpson's Gully is a large sinkhole situated in a relatively small ephemeral channel that drains a glacilacustrine sand plain immediately above the escarpment at Driftwood Cove. The channel flows towards the escarpment brow but is captured in a series of three streamsinks located immediately back of the escarpment brow. The sinkholes increase in size in the downstream direction with the largest being 11.8 m deep and 31 m across (Geomatics International Inc. 1994). This is the largest sinkhole known to occur in Ontario. The walls of the sinkhole are mostly composed of sand but bedrock is exposed at its lowest point. The streamsinks and stream itself align with the major joint set, approximately E–W (80°–260°). The scale of the largest sinkhole cannot be explained as having formed in the current hydrological setting given its location within a small intermittent creek. The major development of this sinkhole is interpreted to be as a rejuvenated palaeokarst. Further, its proximity to the St. Edmunds karst (approximately 3 km), the use of the same joint set by both systems, and the comparable scale of the two systems provides convincing evidence that Simpson's Gully and the St. Edmunds karst may have been part of a much larger palaeokarst (Geomatics International Inc. 1994). They are not connected within the modern hydrological environment but likely represent local reactivation of a pre- or inter-glacial karst.

12.6 Conclusion

The Saugeen-Bruce Peninsula is a glacially scoured dolostone plain exhibiting a complex geomorphology with a wide diversity of landforms. Glacial landforms have had tremendous influence on the presence and development of karst features and processes. Of particular significance are the extensive dolostone pavements whose surfaces were prepared during the MIS 2 or 4 by the removal of surface deposits and the stripping of any previous pavements. The pavement complex on the peninsula and continuing northward onto Manitoulin Island is one of the best examples of a dolostone pavement described in the literature.

The interaction between glacial landforms and karst solution has created a complex hydrological setting. Isostatically opened fractures, in combination with high hydraulic gradients created by the Niagara Escarpment, has resulted in the development of a holokarst where all surface waters are taken underground. Drift deposits permit the formation of streams and lakes whose flows eventually become captured creating a diverse assemblage of surface and subsurface karst landforms such as sinkholes, caves and springs. These fluvïkarst systems occur throughout the peninsula and display a wide range of conditions from small in-stream capture and resurgence to having the longest straight-line cave in Ontario.

Karst landforms on the peninsula also have a complicated history ranging from postglacial (e.g. pavements) through late-glacial (sub-glacial meltwater injection) to inter- and pre-glacial (palaeokarst). The Early Silurian carbonates hosting the karst are in the order of 400 million years old and consequently have experienced many varying geological, morphological and hydrogeological settings. Glimpses of these are provided by the palaeokarst landforms of the present study area including explorable cave passages of the St. Edmunds—Simpson's Gully Karst, an incredibly well-decorated cave along the escarpment (Root Cave), and an efficient aquifer feeding the Judges Creek watershed. This karst complex defines the modern hydrological setting in terms of the distribution of water at the surface and in the subsurface. Importantly, the current hydrological setting is not static but is evolving leading to complications for infrastructure development and water quality issues on the Saugeen-Bruce Peninsula.

References

Armstrong DK, Dubord MP (1992) Paleozoic geology, northern Bruce Peninsula, southern Ontario. Ont. Geol. Surv. Open File Map 198

Armstrong DK (1993a) Paleozoic geology of the central Bruce Peninsula, southern Ontario. Ont. Geol. Surv. Open File Report 5856

Armstrong DK (1993b) Paleozoic geology of southern Bruce Peninsula, southern Ontario. Ont. Geol. Surv. Open File Report 5875

Banks WD, Brunton FR (2017) Collaboration between Ontario Geological Survey, consultants and municipal staff results in discovery and development of a safe and sustainable bedrock groundwater supply for the Town of Shelburne, Southern Ontario, Canada. Proceedings of the GEO OTTAWA 2017 conference 6

Bancroft AM, Kleffner MA, Brunton FR (2016) Biochemistry of the Eramosa Formation in southwestern Ontario, Canada. Can J Earth Sci 53:749–762

Beddows PA, Brunton FR, Buck MJ, Cowell DW, Parker S (2008) Root cave project, Southern Ontario. In: Summary of field work and other activities 2008. Ont. Geol. Surv. Open File Report

Bell JR (1985) Solution pits. A research paper submitted to the Department of Geography in fulfilment of the requirements of geography 4C6, McMaster University, Hamilton, Ontario

Brintnell C, Brunton F (2009) Characterization of the Fossil Hill-Cabot Head formational disconformity between Tobermory and Guelph, Niagara Escarpment region, Southern Ontario. In: Summary of Field Work and Other Activities 2009, Project Unit 09-026. Ont. Geol. Surv. Open File Report 6240

Bruce-Grey Geology Committee (2006) A guide to the geology and landforms of Grey & Bruce Counties. The Owen Sound Field Naturalists, Stan Brown Printers, pp 1–192

Brunton F, Dodge JEP (2008a) Karst map of Southern Ontario, including Manitoulin Island. Ont. Geol. Surv. Groundwater Resource Study 5

Brunton F, Dodge JEP (2008b) Karst of Southern Ontario and Manitoulin Island. Ont. Geol. Surv. Groundwater Resource Study 5, pp 1–99

Brunton F, Brintnell C (2011) Final update of Early Silurian stratigraphy of the Niagara Escarpment and correlation with subsurface units across southwestern Ontario and the Great Lakes Basin. In: Summary of field work and other activities 2011, Project Unit 08-004. Ont. Geol. Surv. Open File Report 6270, pp 30–1 to 30–11

Brunton F, Brintnell C (2017) Early Silurian stratigraphy and geological controls on karstic bedrock groundwater flow zones, Niagara Escarpment region and subsurface of southwestern Ontario. Ont. Geol. Surv. Groundwater Resource Study 13, pp 1–200

Brunton F, Brintnell C, Brett CE, Jin J, Bancroft A, Kleffner, M (2009) Update on Early Silurian stratigraphy of eastern Michigan Basin, Niagara Escarpment, southern Ontario. In: Canadian Paleontology conference proceedings no. 7, Sudbury, Ontario, pp 1–13

Brunton F, Brintnell C, Jin J, Bancroft A (2012) Stratigraphic architecture of the Lockport Group in Ontario and Michigan—a new interpretation of Early Silurian 'Basin Geometries' & 'Guelph Pinnacle Reefs'. In: Proceedings of the 51st annual conference—Ontario—New York oil & gas conference, Oct 23–25th, Niagara Falls, Ontario, pp 1–37

Buck M, Cowell DW (2002) The St. Edmunds Karst. Earth science ANSI checklist report, Ontario Ministry of Natural Resources, Owen Sound, Ontario

Cowan WR, Sharpe WR (2007) Surficial geology of the Bruce Peninsula, southern Ontario. Ont. Geol. Surv. Open File Report 6211

Cowell DW (1974) Karst geomorphology of the northern Bruce Peninsula. Ontario Ministry of Natural Resources, Environmental Planning Series, Miscellaneous Report, Peterborough, Ontario, pp 1–152

Cowell DW (1976) Karst geomorphology of the Bruce Peninsula, Ontario. Master of Science thesis, McMaster University, Hamilton, Ontario

Cowell DW (1984) The Canadian beaver (Castor canadensis) as a geomorphic agent in karst terrain. Can Field Nat 98:169–175

Cowell DW (2001) Island geology and landforms. In: Parker S, Munawar M (eds) Ecology, culture, and conservation of a protected area: Fathom Five National Marine Park, Canada. Ecovision World Monograph Series, Backhuys Publishers, Leiden, The Netherlands, pp 63–79

Cowell DW (2002a) Cabot Head (west). Earth science ANSI checklist report and map, Ontario Ministry of Natural Resources, Owen Sound, Ontario

Cowell DW (2002b) Little cove nature reserve. Earth Science ANSI checklist report, Ontario Ministry of Natural Resources, Owen Sound, Ontario

Cowell DW, Ford DC (1975) The Wodehouse Creek karst, Grey County, Ontario. Can Geogr xix:196–205

Cowell DW, Ford DC (1980) Hydrochemistry of a dolomite karst: the Bruce Peninsula of Ontario. Can J Earth Sci 17:520–526

Cowell DW, Ford DC (1983) Karst hydrogeology of the Niagara Escarpment, Bruce Peninsula, Ontario. In: Proceedings of V.T. Stringfield Symposium. Karst Hydrology. J Hydrol 16:163–168

Cowell DW, Fisher J, Delorme RJ, Stewart CJ (1994) Geomorphological research on the northern Bruce Peninsula and Tobermory Islands. In: Proceedings leading edge '94 conference, Hockley Valley, Ontario Ministry of Environment, Toronto, ON, pp 336–349

Eyles N (2002) Ontario Rocks: three billion years of environmental change, 1st edn. University of Toronto

Folkes P (2008) Cabot Head: a history of its lighthouse, shipwrecks, fishery and timber industry. Copyright, Friends of Cabot Head, pp 1–38

Ford DC, Williams P (2007) Karst hydrogeology and geomorphology. Wiley, The Atrium, Southern Gate, Chichester, West Sussex, England, pp 1–562

Fox WS (1952) The Bruce Beckons. University of Toronto Press, Toronto, Ontario

Genivar Inc. (2011) Groundwater investigation Albemarle Quarry, Lots 1, 2 & 3, Concession 7EBR (former township of Albemarle), Municipality of South Bruce Peninsula. Prepared for Drysdale Aggregate Consulting, Owen Sound, Ontario

Geomatics International Inc. (1994) Surface geomorphological inventory: St. Edmunds Township, Cabot Head and Islands between Tobermory and Manitoulin. Canadian Parks Service, Ontario Region, Cornwall Ontario

Gilbert R, Shaw J (1994) Inferred subglacial meltwater origin of lakes on the southern border of the Canadian Shield. Can J Earth Sci 31:1630–1637

Goodchild MF (1978) Final report: cave safety study, Flowerpot Island, Georgian Bay Islands National Park. Parks Canada, Cornwall Ontario

Goodchild MF (1984) Final report: resource management study of caves, Tobermory Islands Unit, Georgian Bay Islands National Park. Parks Canada, Cornwall Ontario

Government of Canada 1981–2010. Climate Normals for Wiarton, Ontario. http://climate.weather.gc.ca/climate_normals/results_1981_2010_e.html?searchType=stnName&txtStationName=wiarton+a&searchMethod=contains&txtCentralLatMin=0&txtCentralLatSec=0&txtCentralLongMin=0&txtCentralLongSec=0&stnID=4538&dispBack=1&month1=0&month2=12

Jennings JN (1971) Karst. M.I.T. Press, London

Koenig EC (2000) Native fishing conflicts on the Saugeen-Bruce Peninsula: perspectives on resource relations past and present. Dissertation, McMaster University, Hamilton, Ontario

Kor PSG, Cowell DW (1998) Evidence for catastrophic subglacial meltwater sheetflood events on the Bruce Peninsula, Ontario. Can J Earth Sci 35:1180–1202

Liberty BA, Bolton TE (1971) Paleozoic geology of the Bruce Peninsula area. Ontario. Geological Survey of Canada Memoir 360, Map 1194A

Ontario Geological Survey (2010) Surficial geology of southern Ontario. Miscellaneous Release-Data, MRD128-123 (DVD)

Pluhar A (1966) Lapies and related small karst features on the Niagara Escarpment. Master of Science thesis, McMaster University, Hamilton, Ontario

Pluhar A, Ford DC (1970) Dolomite karren of the Niagara Escarpment, Hamilton, Ontario. Z Geomorphol 14:392–410

Priebe EH, Brunton FR (2016) Regional-scale groundwater mapping in the Early Silurian Carbonates of the Niagara Escarpment: final update. In: Summary of Field Work and Other Activities 2016, Project Unit 11-032. Ontario Geological Survey Open File Report 6323

Priebe EH, Brunton F, Lee VL (2012) Regional-scale groundwater mapping in Early Silurian carbonate Rocks of the Niagara Escarpment Cuesta: multi-level monitoring well sampling. In: Summary of field work and other activities 2012, Project Unit 11-032. Ontario Geological Survey Open File Report 6280

Priebe EH, Neville CJ, Brunton FR (2014) Evaluating the influence of geological features on hydraulic conductivity variability in Early Silurian carbonate rock aquifers of the Guelph Region. In: Summary of field work and other activities 2014, Project Unit 08-004. Ontario Geological Survey Open File Report 6300

Sharpe DR, Shaw J (1989) Erosion of bedrock by subglacial meltwater, Cantley, Quebec. Geol Soc Am Bull 101:1011–1020

Shaw J, Gilbert R (1990) Evidence for large-scale subglacial meltwater flood events in southern Ontario and northern New York State. Geology 18:1169–1172

Sweeting MM (1972) Karst Landforms. Macmillan, London

Szramek K, Walter LM (2004) Impact of carbonate precipitation on riverine inorganic carbon mass transport from a mid-continent, forested watershed. Aquat Geochem 10:99–137

Williams PW (1966) Limestone pavements with special reference to western Ireland. Trans Inst Br Geogr 40:155–172

Worthington SRH, Cowell DW, Ford DC (2012) Problems of karst hydrogeology, conservation, and economic development on the Niagara Escarpment. In: Field guidebook for the 39th congress of the international association of hydrogeologists, Niagara Falls, Ontario, September 16–21, 2012

Ice Masses of the Eastern Canadian Arctic Archipelago

13

Wesley Van Wychen, Luke Copland, and David Burgess

Abstract

The islands of the Canadian Arctic, known as the Canadian Arctic Archipelago (CAA), contain the largest area of glacierized terrain outside of the ice sheets of Greenland and Antarctica. The ice masses are focused in the eastern part of the CAA, and stretch from the ice shelves of northernmost Ellesmere Island to glaciers in the southernmost parts of Baffin Island, together with a few mountain glaciers in northern Labrador. The majority of glacier ice (roughly 70%) is contained within the large ice masses located on Baffin Island (Penny and Barnes Ice Caps), Bylot Island (Bylot Island Ice Cap), Devon Island (Devon Ice Cap), Ellesmere Island (Prince of Wales Icefield, Manson Icefield, Sydkap Ice Cap, Agassiz Ice Cap, Northern Ellesmere Icefield) and Axel Heiberg Island (Steacie and Müller Ice Caps). The remaining ice is stored mainly in smaller ice caps and valley glaciers that skirt the coastline. The region's major ice caps typically have maximum elevations around 2000 m asl and descend to sea level where outlet glaciers meet the ocean. Scientific studies of glaciers in the Canadian Arctic began in earnest in the 1950s, with previous knowledge largely gleaned from historical materials and expedition reports. Systematic in situ surface mass balance measurements initiated in the late 1950s and continue to present day, providing one of the longest continuous records of glacier mass balance within the Arctic. The Canadian Arctic also contains some of the very few remaining northern hemisphere ice shelves. Recent analysis indicates that since ~2000 these ice shelves have been undergoing significant reductions in their area and volume, often caused by episodic calving events that can produce ice islands with diameters of 10 km or more. Once detached, these ice islands can drift within the waters of the CAA and Arctic Ocean for years to decades and may pose a threat to Arctic shipping and offshore oil developments.

Keywords

Glaciers • Mass balance • Glacier motion • Glacier dynamics • Canadian Arctic Archipelago

13.1 Physical Geography of Ice Masses in the Canadian Arctic

The islands of the eastern Canadian Arctic Archipelago (CAA) contain ~150,000 km^2 of glacierized terrain, representing 28% of the worlds glacierized area outside of the ice sheets (Sharp et al. 2014) (Fig. 13.1). Approximately 107,000 km^2 of the glacier area is located in the northern CAA, in the Queen Elizabeth Islands (QEI: Devon, Axel Heiberg and Ellesmere Islands), while the remaining ice is located in the southern CAA, on Baffin Island (~37,500 km^2) and Bylot Island (~5000 km^2). The region's major ice masses are classified either as "ice caps" (broad dome-like ice masses less than 50,000 km^2 in area that cover much of the underlying topography), or "icefields" (differentiated from ice caps by the presence of bedrock nunataks that protrude through the ice in their interior regions) (Copland 2013). The remaining ~10% terrestrial ice cover in this region is composed of stagnant plateau ice caps in the QEI less than ~1200 km^2 in size, and standalone Alpine cirque glaciers, many of which fringe the coastlines of Baffin Bay, Nares Strait and the Arctic Ocean. Figures 13.2, 13.3, and 13.4 provide field photos of various glacier features in the Canadian Arctic.

W. Van Wychen (✉)
Department of Geography and Environment Management, University of Waterloo, Waterloo, ON, Canada
e-mail: wvanwychen@uwaterloo.ca

W. Van Wychen · L. Copland
Department of Geography, Environment and Geomatics, University of Ottawa, Ottawa, ON, Canada

D. Burgess
Natural Resources Canada, 601 Booth Street, Ottawa, ON K1A 0E8, Canada

© Springer Nature Switzerland AG 2020

O. Slaymaker and N. Catto (eds.), *Landscapes and Landforms of Eastern Canada*, World Geomorphological Landscapes, https://doi.org/10.1007/978-3-030-35137-3_13

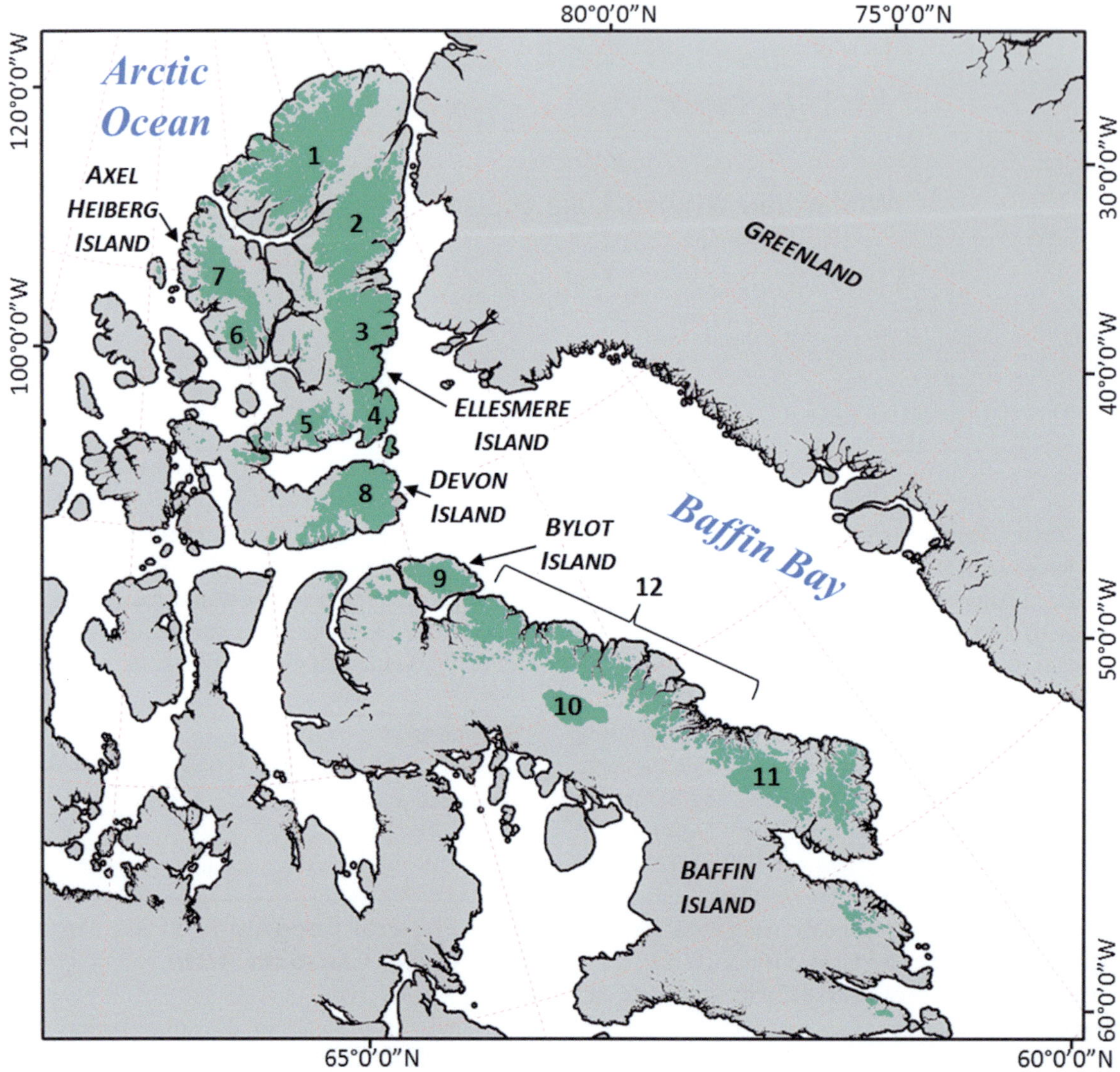

Fig. 13.1 Overview of the major ice masses of the eastern Canadian High Arctic (denoted by green shading). 1 = Northern Ellesmere Icefield; 2 = Agassiz Ice Cap; 3 = Prince of Wales Icefield; 4 = Manson Icefield; 5 = Sydkap Ice Cap; 6 = Steacie Ice Cap, 7 = Müller Ice Cap; 8 = Devon Ice Cap; 9 = Bylot Island Ice Cap, 10 = Barnes Ice Cap; 11 = Penny Ice Cap; 12 = Coastal Glaciers and Ice Caps of Baffin Island

To understand the distribution of glaciers within the CAA, it is necessary to understand the regional climatology of the Arctic Islands. Mean annual air temperatures within the Canadian Arctic decrease as a function of latitude and elevation, such that areas located at high elevations and more northerly latitudes are colder than those located at low elevations and latitudes (Sharp et al. 2014). In general, the climate of the CAA is characterized by relatively short, cool summers followed by longer, colder winters, which provides the climatic conditions necessary for glacier formation and survival. In terms of precipitation, there is a southeast to northwest gradient across the Canadian Arctic Archipelago, with annual average precipitation of 420 mm in Iqaluit and 70 mm in Eureka, with most of the region receiving less than 200 mm annually. Baffin Bay is the dominant moisture source, with the Arctic Ocean a secondary source (Koerner 1979). As a consequence, the distribution of glaciers within the region follows these climatic conditions, with glacierized terrain concentrated in high latitude, mountainous areas adjacent to moisture sources, such as along the east coast of Ellesmere Island and west coast of Axel Heiberg Island (Sharp et al. 2014). The major exception to this distribution is Barnes Ice Cap, which is located on a plateau in the interior of Baffin Island and is thought to exist as a remnant of the Laurentide Ice Sheet Complex (Briner et al. 2009).

Topographically, the major ice masses of the CAA generally rise to elevations of 1800–2000 m asl at their summits, with their inland ice descending gradually to margins at 400–500 m asl, and their large outlet glaciers often reaching the ocean. Ice cap sectors flowing towards coastlines

Fig. 13.2 The snout of a glacier on southeastern Penny Ice Cap, July 2016. *Photograph* L. Copland

Fig. 13.3 A land-terminating outlet glacier located on western Axel Heiberg Island, July 2015. *Photograph* L. Copland

adjacent to the Baffin Bay moisture source tend to be thicker than their inland counterparts (Koerner 1979; Leuschen et al. 2017) and are drained primarily by large fast-flowing outlet glaciers which terminate at sea-level. Locations of maximum ice thickness (600–800 m) exist under the head of major outlet glaciers, where the onset of channelization funnels ice from high elevation accumulation areas to the ocean and has eroded the subglacial topography (Dowdeswell et al. 2004). The most recent ice thickness surveys conducted in 2014 under NASA's Operation IceBridge program mapped the bed elevation profiles (3–4 km in width) along most tidewater glaciers in the Canadian Arctic (https://nsidc.org/data/icebridge). An example from these datasets, combined with high resolution surface elevation data from the ArcticDEM, reveals detailed surface (Fig. 13.5a) and glacier bed (Fig. 13.5b) topography of the Agassiz Ice Cap, NE Ellesmere Island.

Longitudinal profiles extracted along Cañon Glacier highlight the contrasting geometry between the surface and bed geometry, which is characteristic of many glaciers

Fig. 13.4 Photo of the calving front of Trinity Glacier, Prince of Wales Icefield, August 2016. *Photograph* L. Copland

across the CAA. The relatively smooth descent in surface elevation of Cañon Glacier as it extends from >2000 m asl to sea level along its 80 km length (Fig. 13.5c) contrasts with the ice-bed interface, which is controlled by the underlying bedrock and drops sharply to 300–400 m below sea-level for the lowermost 50 km of the glacier (Fig. 13.5d). The fact that many tidewater glaciers of the northern CAA are grounded well below sea level for tens of kilometers inland from their termini make them susceptible to thinning and acceleration driven by changes in sea level and the penetration of warm ocean water at their beds. This can result in enhanced subaqueous melting, floatation, and the likely presence of highly deformable marine sediments (Burgess et al. 2005; Dowdeswell et al. 2004). Ongoing analysis of high resolution geophysical datasets from airborne and satellite measurements are critical for improving current estimates of mass loss through iceberg calving, changing glacier dynamics, and the potential for catastrophic collapse of glaciers in response to measured and predicted climate and ocean warming (Van Wychen et al. 2016).

13.2 Ice Flow of the Eastern CAA

13.2.1 Spatial Pattern of Ice Flow in the Eastern CAA

Regional scale velocity mapping has provided significant insight into the spatial variability in motion for the ice masses of the eastern CAA (Figs. 13.6 and 13.7) (Burgess et al. 2005; Van Wychen et al. 2012, 2014). Typically, land-terminating glaciers in the Canadian Arctic attain their highest velocities (25–75 m y^{-1}) near the equilibrium line altitude (ELA) with velocities decreasing to stagnation in both an up-glacier and down-glacier direction from this point. This pattern also generally holds true for the small tidewater terminating glaciers of the southern CAA, where speeds of 40–100 m y^{-1} are common near the ELA with ice motion often decreasing to near stagnation both up-glacier and down-glacier from there (Van Wychen et al. 2015). On the other hand, velocities of 30–90 m y^{-1} are commonly found near the ELA of large tidewater terminating glaciers of the QEI, with velocities increasing in a down-glacier direction typically to 100–250 m y^{-1} at their terminus (Van Wychen et al. 2014). Of particular note are the Trinity and Wykeham Glaciers of Prince of Wales Icefield, which have recently attained flow speeds of >1200 m y^{-1} and >600 m y^{-1} respectively. These maximum velocities attained by both glaciers make them currently the fastest flowing glaciers in the Canadian Arctic (Fig. 13.6a).

The velocity structure of Belcher Glacier on Devon Ice Cap can be used as an example of the general patterns of ice motion of large outlet glaciers in the northern CAA (Fig. 13.8a, b). Within the upper portion of the ice mass (40–60 km from the terminus), velocities of 5–15 m y^{-1} are common and reflects the fact that ice motion within these areas is likely limited to "internal deformation", that is, ice that is frozen to its bed and flowing under its own weight. As the glacier descends to lower elevations and begins to converge towards channelized flow where it enters confined valleys (15–40 km from the terminus), the convergent flow tends to thicken and accelerate. As a result of increasing overburden pressure and longitudinal stress, liquid water is generated at the glacier bed which facilitates a transition towards glacier movement dominated by basal sliding.

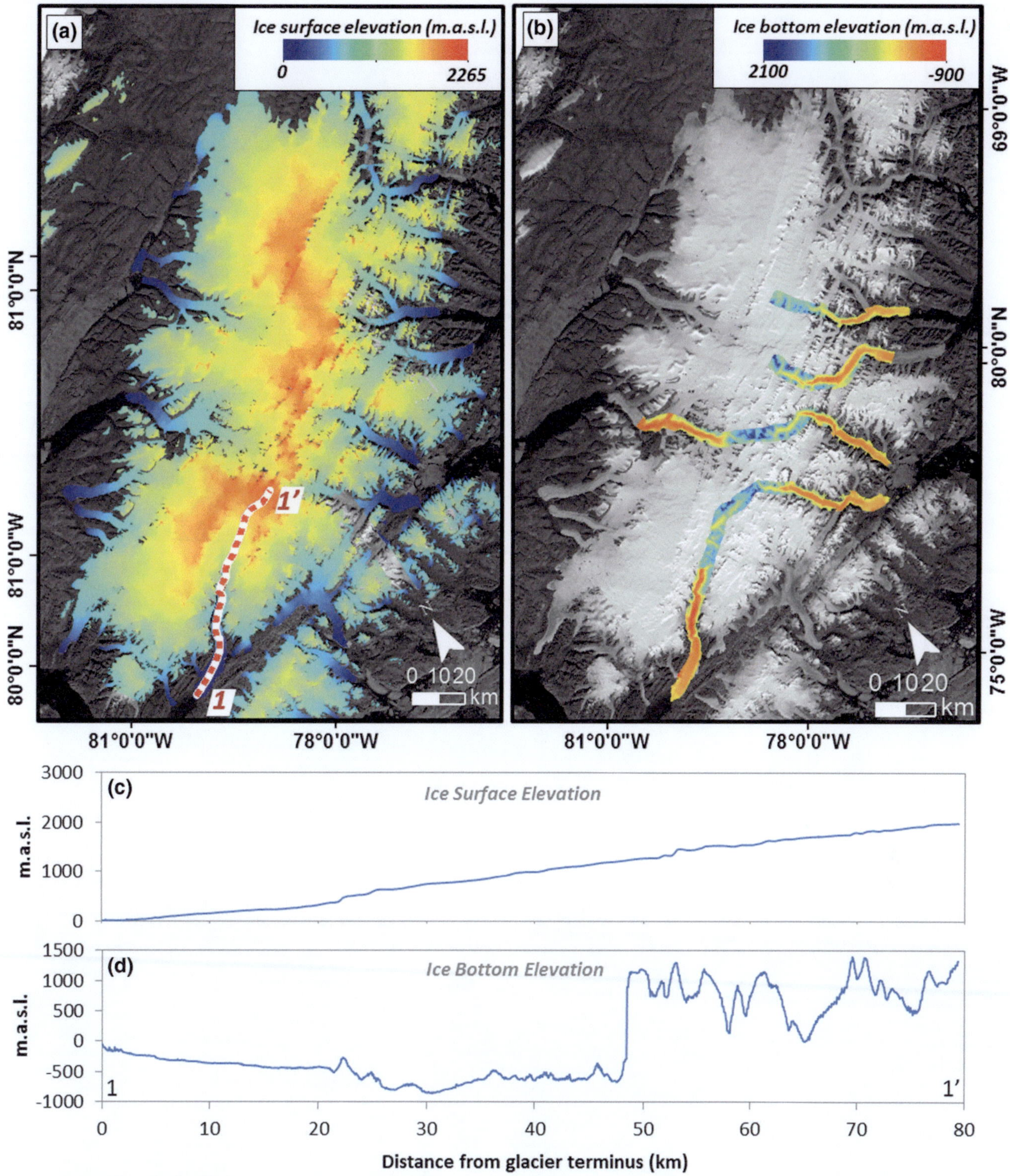

Fig. 13.5 **a** Ice surface elevation of Agassiz Ice Cap (surface elevations provided by the ArcticDEM project available online https://www.pgc.umn.edu/data/arcticdem/, dashed red line indicates the centreline of Cañon Glacier. **b** Ice bottom elevations for select glaciers on Agassiz Ice Cap. Data available via NASA's Operation IceBridge. **c** Extracted centreline ice surface elevations of Cañon Glacier. Profile location shown in (**a**). **d** Extracted centreline ice bottom elevation for Cañon Glacier. Profile delineated in (**a**)

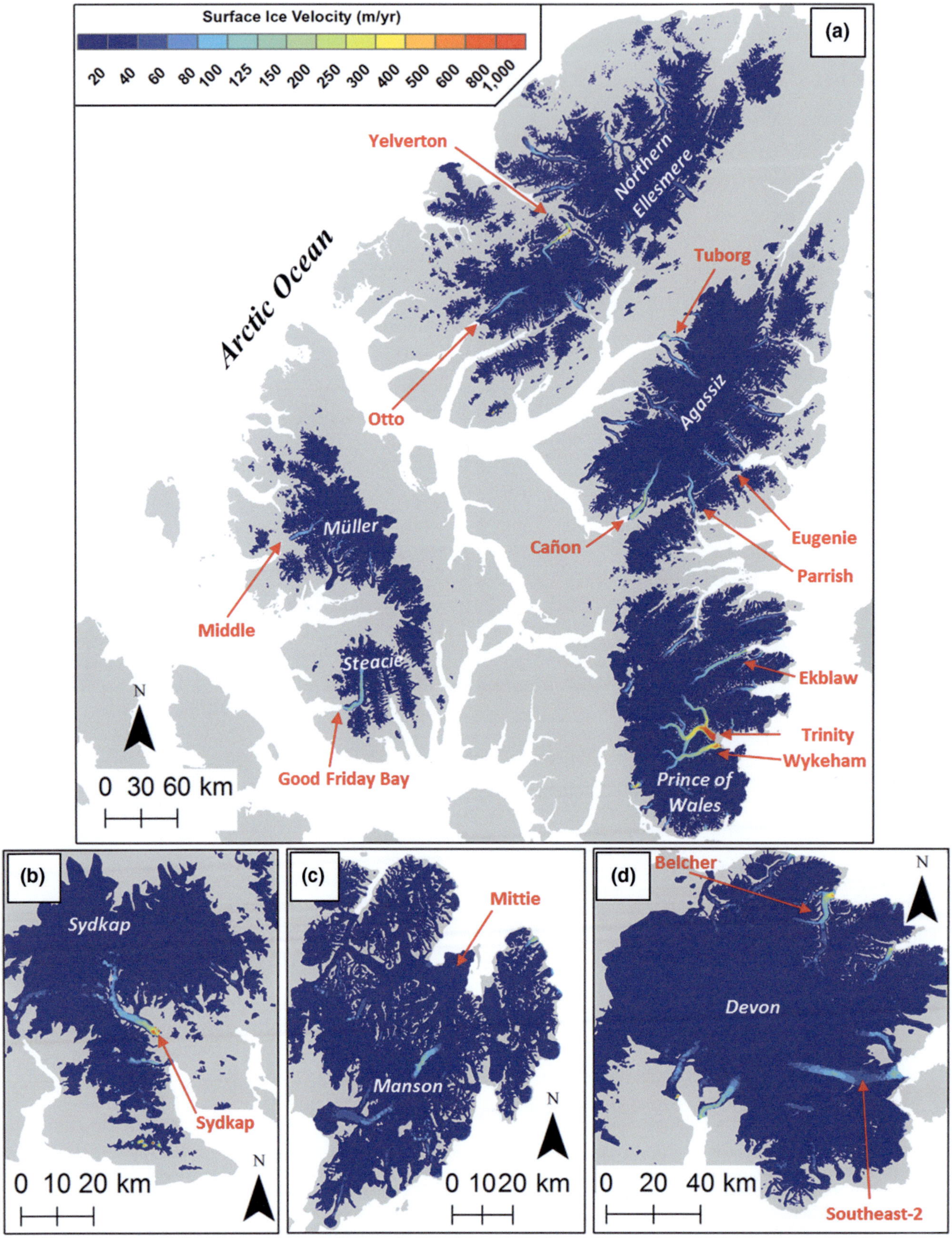

Fig. 13.6 Overview of surface ice motion for the main ice masses of the northern Canadian Arctic Archipelago derived from speckle tracking of RADARSAT-2 imagery acquired in winter 2016–17 for: **a** Northern Ellesmere Icefield, Agassiz Ice Cap, Prince of Wales Icefield, Manson Icefield, Müller Ice Cap and Steacie Ice Cap. **b** Sydkap Ice Cap. **c** Manson Icefield. **d** Devon Ice Cap. Red labels indicate names of major outlet glaciers

Ice velocities in this region increase to 25–100 m y^{-1}, and as the glacier continues its descent towards the ocean, basal sliding becomes an increasingly important component of overall motion. Ice speeds continue to increase to the very lowermost region of the terminus where velocities commonly exceed 200 m y^{-1} (Fig. 13.8b). In many cases, maximum velocities of 300–500 m y^{-1} usually occur over low sloping or flat surfaces in the terminus region. This likely indicates that the glacier is overriding deformable sediments that offer little-to-no resistance to ice flow (Burgess et al. 2005).

In contrast, the large outlet glaciers of the southern Canadian Arctic generally display a different velocity structure than their northern counterparts, with the velocity structure of Coronation Glacier on Penny Ice Cap providing a good example (Fig. 13.8c, d). In the upper most portions of Coronation Glacier, ice motion is low and indicative of rates of motion expected by internal deformation alone. However, ice velocities peak ($\sim$150 m y^{-1}) 25–35 km from the terminus, as ice funnels through relatively narrow channels and glacier motion (as indicated above) is likely dominated by basal sliding. However, unlike large outlet glaciers in the north, velocities decrease to 25–50 m y^{-1} in the lowermost lengths of the glacier indicative of increased basal friction and a reduced importance of sliding as a control on glacier motion. Van Wychen et al. (2015) attributed the asymmetric velocity pattern between the NCAA and SCAA to the fact that outlet glaciers in the NCAA are typically fed by much larger source (i.e. accumulation) areas which help maintain driving stresses that perpetuate glacier flow towards the terminus. In addition, reduced friction of a tidewater glacier as it flows over a subaqueous grounding line into the ocean, compared to a terrestrial terminus, is likely also a factor contributing to the general differences in flow patterns observed between NCAA and SCAA outlet glaciers. It should be noted, however, that these patterns of ice motion are broad generalizations only, as glacier surging and pulsing can cause significant deviations from these patterns.

13.2.2 Temporal Variability in Ice Flow

Inter- and intra-annual variability in glacier motion is widespread across the Canadian Arctic, with research as early as the 1960s linking seasonal changes in surface hydrology with accelerated ice motion (Cress and Wyness 1961; Müller and Iken 1973), and identifying surge-type glacier behaviour (Hattersley-Smith 1969; Müller 1969; Løken 1969; Holdsworth 1973, 1977). Beginning in the early 2000s, a renewed focus was placed on understanding ice motion and its variability across the CAA, including further investigation of the connections between surface

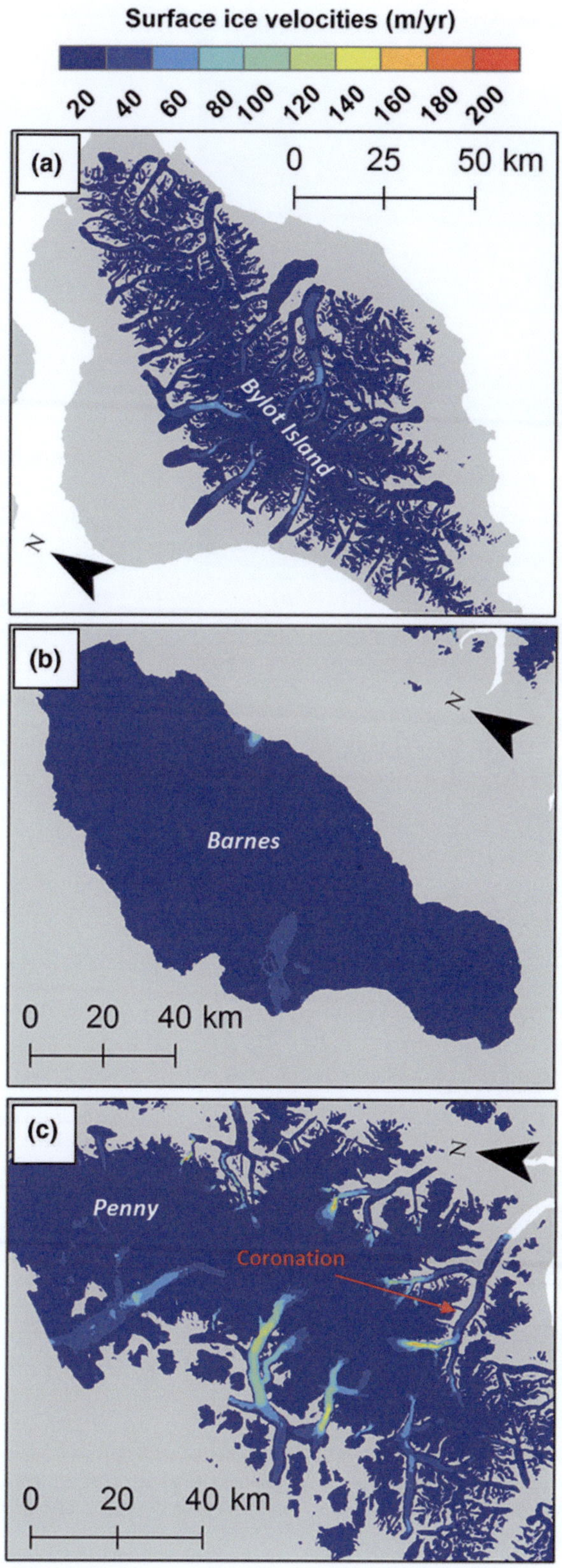

Fig. 13.7 Overview of surface ice motion for the ice masses of the southern Canadian Arctic Archipelago derived from speckle tracking of ALOS PALSAR imagery acquired in winter 2007–08 and 2010–11, for: **a** Bylot Island Ice Cap; **b** Barnes Ice Cap; and **c** Penny Ice Cap. Red labels indicate names of major outlet glaciers

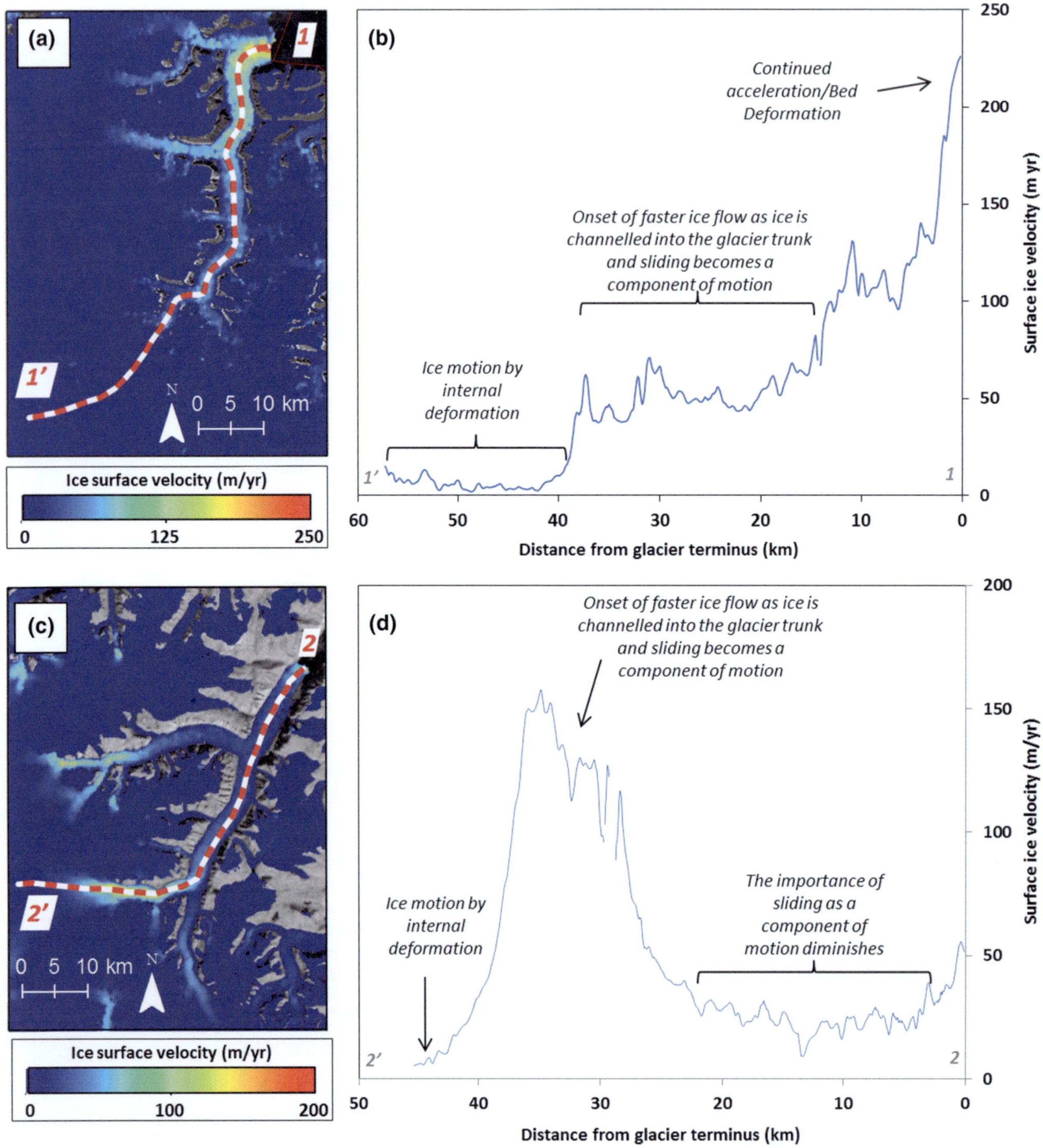

Fig. 13.8 **a** Overview of surface ice velocities of Belcher Glacier, and **b** extracted centreline surface ice velocities of Belcher Glacier. Transect shown as dashed red line from 1 to 1′ in (**a**). **c** Overview of extracted centreline surface ice velocities of Coronation Glacier, Penny Ice Cap, and **d** extracted centreline surface velocities of Coronation Glacier. Transect shown as dashed red line from 2 to 2′ in (**c**)

hydrology and seasonal acceleration (Copland et al. 2003a, b, c; Dowdeswell et al. 2004; Short and Gray 2005; Burgess et al. 2005; Bingham et al. 2003, 2006, 2008), as well as systematic cataloguing of surge-type glaciers (Copland et al. 2003a). More recently, repeated regional velocity mapping has helped reveal how large-scale glacier dynamics have evolved over time (Williamson et al. 2008; Van Wychen et al. 2012, 2016, 2017; Millian et al. 2017; Schaffer et al. 2017; Strozzi et al. 2017), as well as on individual valley glaciers (Thomson and Copland 2017, 2018).

From ~2000–2015 most glaciers across the Queen Elizabeth Islands showed no detectable change in their dynamics. For example, for the ice masses of Axel Heiberg and Ellesmere Islands, Van Wychen et al. (2017) found that 101 of the 117 (~86%) sampled glaciers showed no significant (>20 m y^{-1}) change in velocity based on the available data. Many of these glaciers were small and land-terminating, while significant velocity variability was determined on larger outlet glaciers that terminated in the ocean. Eight of these larger outlet glaciers slowed down in the period 2000–2015, while six underwent acceleration followed by deceleration (Van Wychen et al. 2016). Mittie Glacier, which slowed throughout this period, provides a good illustration of the magnitude of velocity change that is possible within the Canadian Arctic. In the early 2000s, the main trunk of the glacier was flowing at speeds of >1000 m y^{-1}, but slowed to stagnation (<20 m y^{-1}) along the entirety of its trunk by 2015. Van Wychen et al. (2016) attributed the observed deceleration of these eight outlet glaciers to the end of their surge cycles. For the glaciers that underwent speed-up followed by slowdown, Van Wychen et al. (2016) determined that based on the scale, pattern and configuration of the glacier bed topography, these glaciers were responding to either surging or pulsing mechanisms. This means that the majority of the observed changes in glacier dynamics over the study period were due to regular processes inherent to the glaciers of the region, rather than attributable to external drivers (for example, increases in surface melt). The two major exceptions to these trends were the Trinity and Wykeham glaciers, which both at least doubled in surface velocity over the period 2000–2015 (Van Wychen et al. 2016). Rapid acceleration of these glaciers accompanied by significant terminus retreat and substantial thinning, led Van Wychen et al. (2016) to speculate that either ocean or climate warming was responsible for the acceleration and retreat of these adjacent ice masses. In addition, because the main trunks of these glaciers are grounded below sea level for up to 45 km up-glacier from their termini (Leuschen et al. 2017), they may be prone to continued acceleration into the future and catastrophic collapse.

Although repeat velocity mapping at the annual scale has not been systematically undertaken in the southern Canadian Arctic, recent research has revealed multi-decadal scale ice motion variability in this region (Schaffer et al. 2017; Strozzi et al. 2017; Heid and Kääb 2012). For example, Schaffer et al. (2017) used optical and RADAR remote sensing imagery to determine that ice velocities on six major outlet glaciers on Penny Ice Cap increased between the mid-1980s and mid-1990s and then subsequently decreased to 2011/2014. Schaffer et al. (2017) attributed the deceleration of these six glaciers as a response to increasingly negative surface mass balances since the mid-1980s, and postulated that the higher melt rates have resulted in the basal drainage network of these glaciers becoming more efficient, producing lower basal water pressures and a reduction in basal sliding. Multi-annual velocity variability, perhaps due to surging, has also previously been reported for the glaciers of Bylot Island Ice Cap (Dowdeswell et al. 2007) and Barnes Ice Cap (Løken 1969; Holdsworth 1973, 1977), located in the southern CAA. As the potential of repeated annual and sub-annual velocity mapping using satellite imagery becomes increasingly feasible, the knowledge of velocity variability in this region is likely to improve.

13.2.3 Dynamic Discharge: Mass Loss via Ice Flow

Dynamic discharge, or mass loss via iceberg calving from the termini of tidewater glaciers, can be calculated as a function of the surface velocities and measured (or modelled) ice thicknesses. When combined with the surface mass balance (discussed below in Sect. 13.3), dynamic discharge allows scientists to refine estimates of total glacier contributions to global sea level rise and associated freshwater fluxes to the ocean, while providing the information necessary for determining the dominant processes responsible for regional mass loss. For the ice masses of the southern CAA, Van Wychen et al. (2015) determined a dynamic discharge estimate of 0.017–0.108 Gt y^{-1}. By comparison, mean dynamic discharge (2011–2015) from the northern CAA was ~2.47 ± 0.88 Gt y^{-1} indicating that the northern ice masses produce ~95% of icebergs within Canada. The data presented in Van Wychen et al. (2016, 2017) indicates that only 8 glaciers accounted for ~75% of total discharge, with Trinity and Wykeham Glaciers alone accounting for ~50% of dynamic discharge (and up to ~62% in recent years). Although many of the tidewater terminating glaciers in the Canadian High Arctic had reduced discharge between 2000 and 2015, these reductions have been more than compensated by increases in velocity and discharge from the Trinity and Wykeham Glaciers. These patterns highlight the importance of routine glacier velocity monitoring as total dynamic discharge from the CAA is sensitive to the variations in dynamics of just a few select glaciers in this region.

Millan et al. (2017) compared rates of glacier discharge to surface mass balance for the northern CAA and found that during the 1991–2005 period dynamic discharge accounted for ~52% of mass loss, while surface melt accounted for the remainder. However, during the 2005–2014 period, ice discharge only accounted for ~10% of total mass loss while surface melt accounted for ~90% of total mass loss. These results are similar to those found by previous studies (Gardner et al. 2011; Van Wychen et al. 2014) and highlight the fact that surface melt and runoff is increasingly dominating glacier mass losses on a regional scale, although dynamic discharge can still be important for specific basins that are drained by large dynamic tidewater glaciers.

13.3 Glacier Mass Balance Within the CAA and Recent Trends

Total mass balance of glaciers in the CAA is estimated through combined knowledge of the mass lost through dynamic discharge (as described above) and surface mass balance, where surface mass balance is the difference between mass gained through snowfall, and mass loss due to melt runoff. Surface mass balance is calculated annually over a hydrological year referenced to the end of the summer melt season which, on the remote High Arctic glaciers, occurs around the end of August or early September. When mass gain, or accumulation, exceeds total ablation (i.e. melt runoff plus dynamic discharge), glaciers will grow and advance, while glaciers will shrink and recede when total ablation exceeds accumulation.

In situ measurements of glacier surface mass balance over long-term reference glaciers in the Canadian High Arctic (White Glacier, Meighen Ice Cap, and Devon Ice Cap (NW)), indicate that the monitored glaciers in this region have thinned by 8–12 m since the early-1960s (Fig. 13.9). Winter accumulation over the monitored glaciers in the Canadian high Arctic has remained relatively consistent since the ~1960s, which means that total glacier mass balances in the Canadian Arctic have largely been driven by increased summer melt (Koerner 2005; Sharp et al. 2011). Since the late 1980s, and especially since 2005, changing summer atmospheric circulation patterns have increased advection of warm air from the northwest Atlantic to the Canadian High Arctic and increased anti-cyclonic air circulation over the QEI in summer (Gardner et al. 2011; Sharp et al. 2011; Lenaerts et al. 2013). These changes in atmospheric circulation have accounted for the longer and more intense melt seasons within the Canadian Arctic in recent years (Wolken et al. 2009; Dupont et al. 2012; Zdanowicz et al. 2012; Fisher et al. 2012; Mortimer et al. 2016). As a result, glaciers and ice caps in the CAA were largely in balance until the late 1980s, but became significantly negative by the mid-1990s. Since the mid-2000s however, melt runoff has increased three-fold relative to the 1960–2004 average (Sharp et al. 2011) accounting for 50% of the total mass loss that has occurred over the 57-year period of observation.

In the ~mid-1990s, airborne campaigns to map the surface elevation (and surface elevation change) of Canada's High Arctic glaciers and ice caps have allowed for more regionally expansive inferences of changes in glacier mass to be made augmenting results from the in situ monitoring programme (Abdalati et al. 2004). Since the early 2000s, space-borne measurement techniques (such as gravimetric measurements provided by the GRACE (2002–2017) satellite system and repeated measurements of glacier surface elevation provided by ICESAT (2003–2009)) have provided insight into regional changes in mass of glaciers and ice caps across the eastern Canadian Arctic. Mass loss from glaciers in the CAA increased more than two-fold from ~25 Gt y^{-1} for the 1995–2000 epoch (Abdalati et al. 2004) to 60 ± 8 Gt y^{-1} between 2004 and 2009 (Gardner et al. 2004). Jacob et al. (2012) documented similar mass loss estimates of 67 ± 6 Gt y^{-1} between 2003 and 2010, with mass losses

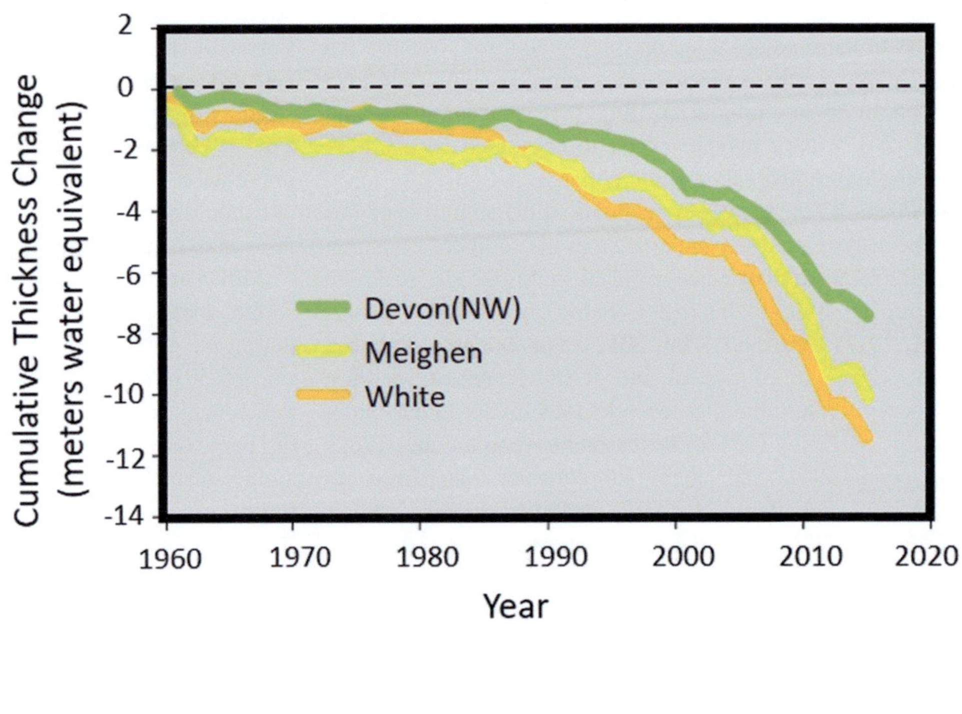

Fig. 13.9 Cumulative thickness change of reference glaciers in the northern Canadian Arctic Archipelago. Mass balance data for Devon (NW) and Meighen Ice Cap is courtesy of the Geological Survey of Canada, mass balance data for White Glacier (Müller Ice Cap) from Trent University (1960–2012; G. Cogley) and University of Ottawa (2013–2015; L. Copland & L. Thomson). See Fig. 13.1 for site locations

continuing to accelerate to 2015 (Harig and Simons 2016). Notwithstanding contributions from the continental ice sheets, Canadian glaciers and ice caps have become the largest contributors to eustatic sea-level rise (i.e. the component of sea-level change by changes in quantity of global ocean water) on Earth. Canada's glaciers and ice caps currently contribute 0.16 mm a^{-1} to global sea-level rise, which is 23% and 75% of net contributions from the Greenland and Antarctic ice sheets, respectively (Gardner et al. 2013; Shepherd et al. 2012).

Nearly all model projections indicate that glacier surface mass balance will continue to be negative for the foreseeable future across the Canadian Arctic Archipelago. Lenaerts et al. (2013) and Marzeion et al. (2012) predict that under moderate warming scenarios published in the Intergovernmental Panel on Climate Change AR5 report, glaciers and ice caps across this region are expected to lose 18% (or 35 mm sea-level equivalent) of their total mass by 2100. Similarly, Radić et al. (2014) estimate that by 2100 mass loss from Canada's Arctic glaciers and ice caps will contribute 41–57 $\pm$ 18 mm of SLE under moderate (RCP4.5) to high (RCP8.5) warming scenarios. Canadian land ice will thus continue to lose mass and be an important contributor to global sea-level rise into the twenty-second century, with significant implications for ocean circulation and the health and sustainability of marine ecosystems (Böning et al. 2016).

13.4 Ice Shelves and Ice Islands

13.4.1 Definition and Origin of Arctic Ice Shelves

The largest concentration of ice shelves in the Arctic is found along the northernmost coast of the CAA. Ice shelves comprise permanent floating ice masses that can fill fiords and extend many kilometres into the ocean, and are defined by Dowdeswell and Jeffries (2017) as: "*an ice mass of considerable thickness ($\geq$20 m) that is afloat on the ocean but attached to the coast. It is often of great horizontal extent (many km) and has, typically but not exclusively, a regularly undulating surface. An ice shelf can form by the seaward extension of a glacier or glaciers, or by the formation of multiyear sea ice, or both and thicken further by the accumulation of snow at the top surface and the accretion of ice from water at the bottom surface*". Although they are commonly associated with the Antarctic, where they make up about 75% of the coast, ice shelves also exist in the Northern Hemisphere where they are primarily concentrated on the northern coastline of Ellesmere Island (Jeffries 2017), some Eurasian fjords that fringe the Arctic Ocean (Dowdeswell 2017) and the floating ice tongues of northernmost Greenland where fast-flowing glaciers reach the ocean (Reeh 2017). Arctic ice shelves are further differentiated from their Southern Hemisphere counterparts by their size, with the largest individual arctic ice shelf, the Ward Hunt Ice Shelf of Northern Ellesmere Island, being three orders of magnitude smaller than the large ice shelves of Antarctica (Dowdeswell and Jeffries 2017).

Evidence based on radiocarbon data indicates that the ice shelves of northern Ellesmere Island developed ~5500 years ago in response to Holocene cooling (England et al. 2017), which allowed for the advance of glaciers beyond their current positions and the formation of a large contiguous ice shelf termed the "Ellesmere Ice Shelf" (Vincent et al. 2001; Dowdeswell and Jeffries 2017). In 1906, the "Ellesmere Ice Shelf" had an estimated area of 8597 km^2, but as of 2015 only smaller remnants were left in individual fiords, totalling 535 km^2 (Mueller et al. 2017). The main remaining pieces comprise the Ward Hunt Ice Shelf (total 224.3 km^2, now split into separate eastern and western portions), the Milne Ice Shelf (187.4 km^2) and the Petersen Ice Shelf (19.6 km^2). Small pieces <10 km^2 in area remain in a few other fiords, but almost all of these have been undergoing rapid reductions in extent over the past decade (Mueller et al. 2017).

13.4.2 Physical Structure of Arctic Ice Shelves

The ice shelves of Northern Ellesmere Island range in thickness from ~20–100 m (Mortimer et al. 2012; White et al. 2015), compared to antarctic ice shelves that can approach 1 km in ice thickness (Griggs and Bamber 2011). They are characterized by an undulating surface topography punctuated by troughs and ridges with wavelengths of ~30–300 m (Jeffries 2017) and amplitudes up to 7.5 m (Mortimer et al. 2012) (Fig. 13.10). The origin of the rolling topography of these ice shelves is somewhat unclear, but Holdsworth (1987) and Jeffries (2017) offer a number of possible explanations including: thermal stress, sea-ice pack pressure, stacking of sea ice pressure ridges, glacier compression, tidal action, snow dune formation due to prevailing wind patterns and surface meltwater that pools into elongated lakes due to the prevailing wind patterns. Although the exact process (or combination of processes) responsible for these features is unclear, they nevertheless provide a distinctive surface topography that aids in distinguishing them from adjacent sea ice, glaciers and icebergs.

Dowdeswell and Jeffries (2017) describe four types of ice shelves based on their method of formation, which provides a useful way to classify these features:

(a) *Classical Ice Shelf*: this type is commonly associated with those found in Antarctica, and is comprised of floating margins of glaciers or an ice sheet, where

Fig. 13.10 View across the central portion of the Milne Ice Shelf, July 17, 2014. *Photograph* L. Copland

glacial ice reaches tidewater and remains intact on the ocean surface. The Milne Ice Shelf (Fig. 13.10) of northern Ellesmere Island, which is comprised of a number of glacier outlets that have coalesced within the confines of the steep-sided Milne Fiord, provides the closest approximation of this type of ice shelf in the Northern Hemisphere. Ground-penetrating radar measurements in 2008/9 indicate that it reaches a maximum thickness of ~94 m (Mortimer et al. 2012).

(b) *Sea-ice Ice Shelf*: this describes ice shelves which form due to the thickening of multi-year sea ice which has remained in place in protected inlets and embayments for long periods of time (centuries to millennia). These ice shelves act as a platform for new ice to accumulate on their surface from snowfall, and on their bottom from basal freeze-on of sea water, which allows them to thicken substantially over time (Lemmen et al. 1988; Jeffries et al. 1989; Dowdeswell and Jeffries 2017). By definition, they are at least 20 m thick, but can be substantially thicker than this. The Ward Hunt Ice Shelf is the best current example of this type of ice shelf in the Northern Hemisphere, and was estimated to be 44.5 m thick from a borehole drilled in its western portion in 1983 (Jeffries 1991). The entire contiguous "Ellesmere Ice Shelf" is thought to have primarily been of this type when it existed at the start of the 20th century.

(c) *Composite Ice Shelf*: this refers to an ice shelf that is comprised of both glacier ice and sea ice, sometimes also with the inclusion of icebergs. The Serson Ice Shelf on northern Ellesmere Island provided a good example of this type of ice shelf prior to its almost complete break-up in 2008 (Lemmen et al. 1988; Dowdeswell and Jeffries 2017; Mueller et al. 2017). This ice shelf was made up of two main glacier tongues which extended many kilometres into the ocean, with intervening areas infilled with ice that had originated as very old multi-year landfast sea ice (Dowdeswell and Jeffries 2017).

(d) *Glacier Tongues*: this refers to the floating margins of glaciers that are narrow relative to their overall lengths, confined by either slow moving ice (typically sea ice) or bedrock on their sides and primarily fed by fast-flowing outlet glaciers (Dowdeswell and Jeffries 2017; Hambrey 1994). These features are common on the northern coastline of Greenland, and a number are found in the protected fjords of northern Ellesmere Island. Mueller et al. (2017) identified eight ice tongues with a total area of 94 km^2 in 1959, compared to 9 ice tongues with a combined area of 23 km^2 in 2015. The number of tongues increased over this period, even though their area declined by >75%, due to the reclassification of some ice shelves to ice tongues as the surrounding ice shelf was lost. The largest remaining feature of this type in the Canadian Arctic is the Milne Glacier Tongue, which comprises the northernmost ~16 km of the Milne Glacier (Hamilton et al. 2017).

In terms of mass balance processes, ice shelves gain mass through surface accumulation of snow and superimposed ice, sea ice thickening, water freezing to their base, and glacier input (Mueller et al. 2008; Jeffries 2002; Dowdeswell and Jeffries 2017). The relative importance of each of these accumulation processes varies by ice shelf type. Ice shelves lose mass via surface melt and runoff, sublimation, basal melt at the ice-ocean interface and the calving of icebergs or ice islands.

13.4.3 Ice Islands

A unique feature of the ice shelves of northern Ellesmere Island, compared to any other ice mass in the CAA, is that they can produce very large ice islands when they calve. Ice islands refer to large tabular icebergs, which typically have a freeboard of at least ~2–5 m and can reach >500 km^2 in area (Jeffries 1992; Van Wychen and Copland 2017). They are frequently distinguishable by their ribbed surface topography, inherited from the ice shelf from which they formed. The largest recorded ice island calved from northern Ellesmere Island in ~1950 and was named T-2, and had dimensions of ~31 × 33 km (Van Wychen and Copland 2017; Jeffries 1992). As these ice islands primarily calve from free-floating ice shelves, their ice thicknesses are similar to their parent source and can reach values of 20–100 m.

Given the large dimensions of ice islands they can pose particular threats to shipping and offshore infrastructure, such as oil production platforms, as they circulate within Arctic waters. Van Wychen and Copland (2017) catalogued historical drift tracks of ice islands that originated from the northern coastline of Ellesmere Island and found that these ice features undertook three general routes (Fig. 13.11).

(1) *Arctic Ocean Drift* occurs when an ice island drifts westward after detaching from the coastline and becomes entrained in the Beaufort Gyre, allowing it to circulate within the Arctic Ocean. Eventually the ice island either undertakes Archipelago Drift (described next), or is evacuated from the Arctic Ocean along the eastern coastline of Greenland via the Transpolar Drift ocean current.
(2) *Archipelago Drift* occurs when an ice island drifts into the channels between the islands of the Canadian Arctic Archipelago and eventually disintegrates within these waterways. For example, Hobson's Choice Island drifted into Queen's Channel and broke up there in 1992 (Jeffries 1992) (Fig. 13.11).
(3) *Nares Strait Drift* describes the situation when ice islands drift eastward immediately after calving and become drawn into Nares Strait, the channel located between eastern Ellesmere Island and northwest Greenland. These ice islands eventually drift southward through Baffin Bay to the east coast of Newfoundland and Labrador, where they melt and break-up. Ice island WH-5 is the only ice island from northern Ellesmere Island known to have made this route, between approximately 1961 and 1964 (Nutt 1966).

Depending on the drift path that an ice island undertakes it can potentially have a significant lifespan before it breaks up. This is especially true for ice islands that undertake Arctic Ocean Drift. For example, Ice Island T-3 circulated three times around the Arctic Ocean over a period of ~40 years, before eventually exiting the Arctic Ocean through Fram Strait and breaking up off southern Greenland in 1984 (Jeffries 1992; Van Wychen and Copland 2017). Additionally, as ice islands disintegrate they typically become increasingly difficult to track, and may spawn many smaller ice islands from one large original feature. For these reasons, ice islands can pose a particular threat to offshore activities and infrastructure as a single original ice island may threaten the same region multiple times as it breaks apart.

13.4.4 Recent Changes of the Northern Ellesmere Ice Shelves

Between 1906 and 1982, ~90% of the ice shelf area that skirted the northern coastline of Ellesmere Island at the start of the 20th century was lost. Over this time, losses mainly occurred during large calving events prior to the 1950s, which released many ice islands into the Arctic Ocean and left intact ice shelves in protected fiords of northern Ellesmere Island (Copland et al. 2017; Koenig et al. 1952). A second period of large calving events started in ~2005, with the remaining ice shelves losing ~50% of their 2005 area by 2012 (Copland et al. 2017). Of the three large remaining ice shelves (Petersen, Milne and Ward Hunt), the Milne thinned by an average of 8.1 ± 2.8 m between 1981 and 2008/2009 and lost 13% of its 1981 volume (Mortimer

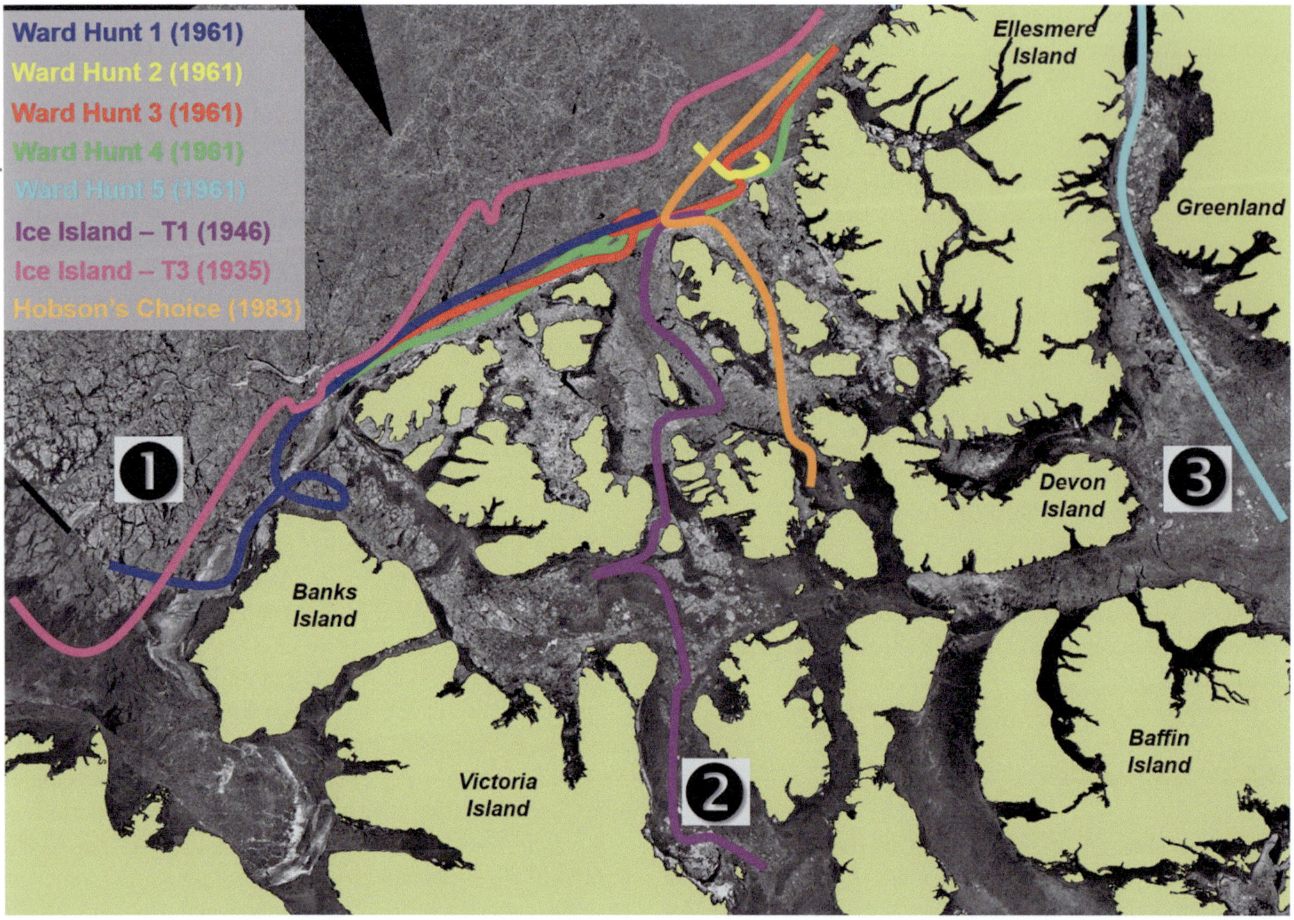

Fig. 13.11 Drift paths taken by large ice islands over the past ~80 years across the CAA: (1) Arctic Ocean Drift; (2) Archipelago Drift; (3) Nares Strait Drift. Dates in legend refer to earliest possible date of calving. Modified from Van Wychen and Copland (2017)

et al. 2012). Point measurements of ice thickness on the Ward Hunt Ice Shelf indicate that it thinned from ~40–50 m in the 1950s to ~25–30 m in the early 2000s, with most of the thinning taking place since the 1990s (Braun 2017). Although repeated ice thickness mapping has not been completed on the Petersen Ice Shelf (Fig. 13.12), mass balance measurements indicate that it is losing mass at a rate of 28.45 Mt y^{-1} at the surface, with little evidence that ice is being replaced via basal freeze-on, and will be entirely lost by the 2040s or earlier (White et al. 2015). As such, it is important to understand the processes driving the current loss of these ice shelves.

Copland et al. (2017) identified several factors that have likely led to the recent loss of the northern Ellesmere Ice Shelves, including.

(1) *Rising Regional Air Temperatures*, particularly in the autumn, winter and spring seasons, which have reduced the overall number of freezing degree days (FDD) that are important for limiting the impacts of summer melt and for ice shelf recovery via processes such as basal freeze-on. The decrease in FDD was identified as particularly important leading up to the loss of the Ayles Ice Shelf in 2007 (Copland et al. 2007). Although changes in summer air temperature have been less apparent in the region than changes in other seasons, White et al. (2015) showed that losses of the Petersen Ice Shelf occurred coincident with record-breaking summer air temperatures in 2005, 2008, 2011 and 2012.

(2) *Reductions in Glacier Input into Ice Shelves*: for Classical and Composite ice shelves, there is evidence of long-term reductions, or the complete loss, of glacier input to the Ayles, Milne and Petersen ice shelves since ~2000 (Mortimer et al. 2012; White et al. 2015; Copland et al. 2017). For example, five tributary glaciers provided input to the Milne Ice Shelf in 1959, but this was reduced to only one by 2011. These reductions in ice input reduce the mass delivered to these systems and prime them for break-up and disintegration.

(3) *Long-Term Negative Mass Balances*: in situ surface mass balance stakes surveyed on the Ward Hunt Ice Shelf indicate that winter accumulation rates have

Fig. 13.12 View across the Petersen Ice Shelf, July 13, 2014. *Photograph* L. Copland

remained relatively constant since the 1950s, while summer surface melt has been much more variable, resulting in generally negative surface mass balances over this period (Braun 2017). This trend is similar to all glaciers and ice caps in the Canadian Arctic (see Sect. 13.3), as well as for the other ice shelves within the region (Mortimer et al. 2012). Incursions of warm Arctic Ocean waters are likely to have promoted increased basal melt in the 1990s and again from 2005 to 2007 (Beszczynska-Moller et al. 2012; Copland et al. 2017) which would have further promoted mass loss from these ice shelves.

(4) *Changes in Arctic Sea Ice*: episodic losses of ice shelves have tended to occur when the northern coastline of Ellesmere Island is free of sea ice, suggesting that sea ice plays an important role in buttressing ice shelves and aiding their stability and long-term survival (Copland et al. 2017).

Based on the above review, and that of Copland et al. (2017), there is not a single factor that can be attributed as the mechanism driving recent ice shelf losses. Rather, it is likely the combination of all four of these factors, and perhaps others, that has led to the loss of ice shelves in the northern CAA and the reductions in mass of the remaining intact ice shelves there. Taken together, the rapid loss of ice shelf extent since 2005, indications of continued mass loss from those that remain, and projections of continued climate warming and sea ice loss in the Arctic, implies that the long-term survival of these features is doubtful.

13.5 Conclusions

Knowledge of the mass balance and characteristics of glaciers and ice masses of the eastern Canadian Arctic Islands has been primarily based on in situ observations initiated in the 1960s. Since ~2000, the increased availability and resolution of remote sensing imagery and methodologies to processes these data have helped to further expand our collective understanding of these ice masses on a regional basis. All of these observations converge to reveal a rapidly changing landscape. In terms of glacier dynamics, regional velocity mapping has shown that dynamic variability within the region is now much more widespread than previously thought. Glacier mass balances indicate that snowfall within the region has remained relatively steady since the 1960s, but that surface melt and runoff have increased substantially, beginning in the 1980s and especially since 2005. Additionally, nearly all models project that the Canadian arctic glaciers will continue to lose mass and be a major contributor to sea level rise over the next century. Within this context, the ice shelves of Northern Ellesmere Island also continue to lose mass, punctuated by the episodic detachment of large pieces of ice shelf and the creation of ice islands. In light of this, continued monitoring of the ice features of the eastern CAA is essential for furthering our understanding of how the region will continue to respond to climate change.

References

Abdalati W, Krabill W, Frederick E, Manizade S, Martin C, Sonntag J, Swift R, Thomas R, Yungel J, Koerner R (2004) Elevation changes of ice caps in the Canadian Arctic Archipelago. J Geophys Res 109. https://doi.org/10.1029/2003jf000045

Beszczynska-Moller A, Fahrbach E, Schauer U, Hansen E (2012) Variability in Atlantic water temperature and transport at the entrance to the Arctic Ocean, 1997–2010. ICES J Mar Sci 69:852–863

Bingham R, Nienow P, Sharp M (2003) Intra-annual and intra-seasonal flow dynamics of a High Arctic polythermal valley glacier. Ann Glaciol 37:181–188

Bingham RG, Nienow PW, Sharp MJ, Copland L (2006) Hydrology and dynamics of a polythermal (mostly cold) High Arctic glacier. Earth Surf Processes 31:1463–1479

Bingham RG, Hubbard AL, Nienow PW, Sharp MJ (2008) An investigation into the mechanisms controlling seasonal speed-up events at a High Arctic glacier. J Geophys Res 113. https://doi.org/10.1029/2007/jf000832

Böning CW, Behrens E, Biastoch A, Getzlaff K, Bamber JL (2016) Emerging impact of Greenland meltwater on deepwater formation in the North Atlantic Ocean. Nat Geosci 9:523–527

Braun C (2017) The surface mass balance of the Ward Hunt Ice Shelf and Ward Hunt Ice Rise, Ellesmere Island, Nunavut, Canada. In: Copland L, Mueller D (eds) Arctic Ice Shelves and Ice Islands, Ch. 6. Springer Nature, Dordrecht

Briner JP, Davis PT, Miler GH (2009) Latest Pleistocene and Holocen glaciation of Baffin Island, Arctic Canada: key patterns and chronologies. Quat Sci Rev 28:21–22

Burgess D, Sharp M, Mair D, Dowdeswell JA, Benham TJ (2005) Flow dynamics and iceberg calving rates of Devon Ice Cap, Nunavut, Canada. J Glaciol 51:219–230

Copland L, Sharp M, Dowdeswell JA (2003a) The distribution and flow characteristics of surge-type glaciers in the Canadian High Arctic. Ann Glaciol 36:73–81

Copland L, Sharp M, Nienow P (2003b) Links between short-term velocity variations and the subglacial hydrology of a predominantly cold polythermal glacier. J Glaciol 49:337–348

Copland L, Sharp M, Nienow P, Bingham R (2003c) The distribution of basal motion beneath a High Arctic polythermal glacier. J Glaciol 49:407–414

Copland L, Mueller D, Weir L (2007) Rapid loss of the Ayles Ice Shelf, Ellesmere Island, Canada. Geophys Res Lett 34. https://doi.org/10.1029/2007gl031809

Copland L (2013) Classification of ice masses. In: Shroder JF (ed) Treatise on Geomorphology, Ch. 8.4. Elsevier, Oxford, pp 45–52

Copland L, Mortimer C, White A, McCallum MR, Mueller D (2017) Factors contributing to recent Arctic ice shelf losses. In: Copland L, Mueller D (eds) Arctic Ice Shelves and Ice Islands, Ch. 10. Springer Nature, Dordrecht

Cress P, Wyness R (1961) The Devon Island expedition, observations of glacial movements. Arctic 14:247–259

Dowdeswell JA, Benham TJ, Gorman MR, Burgess D, Sharp M (2004) Form and flow of the Devon Island ice cap, Canadian Arctic. J Geophys Res 109. https://doi.org/10.1029/2003jf000095

Dowdeswell EK, Dowdeswell JA, Cawkwell F (2007) On the glaciers of Bylot Island, Nunavut, Arctic Canada, Arctic, Antarctic and Alpine Research, 39(3):402–411. https://doi.org/10.1657/1523-0430(05-123)%5bDOWDESWELL%5d2.0.CO;2

Dowdeswell JA (2017) Eurasian Arctic Ice Shelves and Tidewater Ice Margins. In: Copland L, Mueller D (eds) Arctic Ice Shelves and Ice Islands, Ch. 3. Springer Nature, Dordrecht

Dowdeswell JA, Jeffries MO (2017) Arctic Ice Shelves: an introduction. In: Copland L, Mueller D (eds) Arctic Ice Shelves and Ice Islands, Ch. 1. Springer Nature, Dordrecht

Dupont F, Royer A, Langlois A, Gressent A, Picard G, Fily M, Cliché P, Chum M (2012) Monitoring the melt season length of the Barnes Ice Cap over the 1979–2010 period using active and passive microwave remote sensing data. Hydrol Processes 26:2643–2652

England JH, Evans DA, Lakeman TR (2017) Holocene history of Arctic Ice Shelves. In: Copland L, Mueller D (eds) Arctic Ice Shelves and Ice Islands, Ch. 7. Springer Nature, Dordrecht

Fisher D, Zheng J, Burgess D, Zdanowicz C, Kinnard C, Sharp M, Bourgeois J (2012) Recent melt rates of Canadian Arctic ice caps are the highest in many millennia. Global Planet Change 84–85. https://doi.org/10.1016/j.gloplacha.2011.06.005

Gardner AS, Moholdt G, Wouters B, Wolken GJ, Burgess DO, Sharp M, Cogley JG, Braun C, Labine C (2011) Sharply increased mass loss from glaciers and ice caps in the Canadian Arctic Archipelago. Nature 473:357–360

Gardner AS, Moholdt G, Cogley JG, Wouters B, Arendt AA, Wahr J, Berthier E, Hock R, Pfeffer WT, Kaser G, Ligtenberg SR, Bolch T, Sharp MJ, Hagen J-O, van den Broeke M, Paul F (2013) A reconciled estimate of glacier contributions to sea level rise 2003–2009. Science 340:852–857

Griggs JA, Bamber JL (2011) Antarctic ice-shelf thickness from satellite radar altimetry. J Glaciol 57:485–498

Hambrey MJ (1994) Glacial environments. UCL Press, London

Hamilton AK, Laval BE, Mueller DR, Vincent WF, Copland L (2017) Dynamic response of an Arctic epishelf lake to seasonal and long-term forcing: implications for ice shelf thickness. The Cryosphere 11:2189–2211

Harig C, Simons FJ (2016) Ice mass loss in Greenland, the Gulf of Alaska, and the Canadian Archipelago: seasonal cycles and decadal trends. Geophys Res Lett 43:3150–3159

Hattersley-Smith G (1969) Recent observations on the surging Otto Glaicer, Ellesmere Island. Can J Earth Sci 6:883–889

Heid T, Kääb A (2012) Repeat optical satellite images reveal widespread and long term decrease in land-terminating glacier speeds. The Cryosphere 6:467–478

Holdsworth G (1973) Evidence of a surge on Barnes Ice Cap, Baffin Island. Can J Earth Sci 10:1565–1574

Holdsworth G (1977) Surge activity on the Barnes Ice Cap. Nature 269:588–590

Holdsworth G (1987) The surface waveforms on the Ellesmere Island ice shelves and ice islands. In: Workshop on extreme ice features, technical memorandum, 141 (NRCC 28003). National Research Council of Canada, pp 385–443

Jacob T, Wahr J, Pfeffer T, Swenson S (2012) Recent contributions of glaciers and ice caps to sea level rise. Nature 482:514–518

Jeffries MO, Krouse HR, Sackinger WM, Serson HV (1989) Stable-isotope (18O/19O) tracing of fresh, brackish and sea ice in multiyear land-fast sea ice, Ellesmere Island, Canada. J Glaciol 35:9–16

Jeffries MO (1991) Massive, ancient sea-ice strata and preserved physical-structural characteristics in the Ward Hunt Ice Shelf. Ann Glaciol 15:125–131

Jeffries MO (1992) Arctic ice shelves and ice islands: origin, growth and disintegration physical characteristics, structural-stratigraphic variability, and dynamics. Rev Geophys 30:245–267

Jeffries MO (2002) Ellesmere Island Ice Shelves and Ice Islands, in Satellite Image Atlas of Glaciers of the world. In: Williams RS, Ferrigno JG (eds) Glaciers of North America. US Geological Survey, Washington DC, pp J147–J164

Jeffries MO (2017) The Ellesmere Ice Shelves, Nunavut, Canada. In: Copland L, Mueller D (eds) Arctic Ice Shelves and Ice Islands, Ch. 2. Springer Nature, Dordrecht

Koenig LS, Greenaway KR, Dunbar M, Hattersley-Smith G (1952) Arctic Ice Islands. Arctic 5:67–103

Koerner RM (1979) Accumulation, ablation, and oxygen isotope variations on the Queen Elizabeth Islands ice caps, Canada. J Glaciol 22:25–41

Koerner RM (2005) Mass balance of glaciers in the Queen Elizabeth Islands, Nunavut, Canada. Ann Glaciol 42:417–423

Lemmen DS, Evans DJ, England J (1988) Ice Shelves of Northern Ellesmere Island, NWT, Canadian landform examples. Can Geogr 32:363–367

Lenaerts JT, van Angelen JH, van den Broeke MR, Gardner AS, Wouters B, van Meijgaard E (2013) Irreversible mass loss of Canadian Arctic Archipelago glaciers. Geophys Res Lett 40:870–874

Leuschen C, Goineni P, Rodriguez-Morales F, Paden J, Allen C (2017) IceBrdige MCords L2 Ice Thickness, Version 1. Boulder, Colorado USA. NASA National Snow and Ice Data Center Distributed Active Archive Center. Acquired 1 April 2014

Løken OH (1969) Evidence of surges on the Barnes Ice Cap, Baffin Island. Can J Earth Sci 6:899–901

Marzeion B, Jarosch AH, Hofer M (2012) Past and future sea-level change from the surface mass balance of glaciers. The Cryosphere 6:1295–1322

Millan R, Mouginot J, Rignot E (2017) Mass budget of the glaciers and ice caps of the Queen Elizabeth Islands, Canada, from 1991 to 2015. Environ Res Lett 12. https://doi.org/10.1088/1748-9326/aa5b04

Mortimer CA, Copland L, Mueller DR (2012) Volume and area changes of the Milne Ice Shelf, Ellesmere Island, Nunavut, Canada, since 1950. J Geophys Res 117. https://doi.org/10.1029/2011jf002074

Mortimer CA, Sharp M, Wouters B (2016) Glacier surface temperatures in the Canadian High Arctic, 2000–2015. J Glaciol 62:963–975

Mueller D, Copland L, Hamilton A, Stern D (2008) Examining Arctic ice shelves prior to the 2008 breakup. EOS, Transactions, American Geophysical Union 89(49)

Mueller D, Copland L, Jeffries MO (2017) Changes in Canadian Arctic ice shelf extent since 1906. In: Copland L, Mueller D (eds) Arctic Ice Shelves and Ice Islands, Ch. 5. Springer Nature, Dordrecht

Müller F (1969) Was the Good Friday Glacier on Axel Heiberg Islad surging? Can J Earth Sci 6:891–894

Müller F, Iken A (1973) Velocity fluctuations and water regime of Arctic valley glaciers. In: Symposium on the hydrology of glaciers, Cambridge, de l'Association Internationale d'Hydrolgie Scientifique

Nutt DC (1966) The drift of Ice Island WH-5. Arctic 19:244–262

Radić V, Bliss A, Beedlow AC, Hock R, Miles E, Cogley JG (2014) Regional and global projections of twenty-first century glacier mass changes in response to climate scenarios from global climate models. Clim Dyn 42:37–58

Reeh N (2017) Greenland Ice Shelves and Ice Tongues. In: Copland L, Mueller D (eds) Arctic Ice Shelves and Ice Islands, Ch. 4. Springer Nature, Dordrecht

Schaffer N, Copland L, Zdanowicz C (2017) Ice velocity changes of Penny Ice Cap, Baffin Island, since the 1950s. J Glaciol 63:716–730

Sharp M, Burgess DO, Cogley JG, Ecclestone M, Labine C, Wolken G (2011) Extreme melt on Canada's Arctic ice caps in the 21st century. Geophys Res Lett 38. https://doi.org/10.1029/2011gl047381

Sharp M, Burgess DO, Cawkwell F, Copland L, Davis JA, Dowdeswell EK, Dowdeswell JA, Gardner A, Mair D, Wang L, Williamson S, Wolken GJ, Wyatt F (2014) Remote sensing of recent glacier changes in the Canadian Arctic. In: Kargel JS, Leonard GJ, Bishop MP, Kääb A, Raup BH (eds) Global land ice measurements from space, Ch. 9. Praxis-Springer, pp 205–228

Shepherd A., Ivins ER, Geruo A, Barletta VR, Bentley MJ, Battadpur S, Briggs KH, Bromwich DH, Forsberg R, Galin N, Horwath M, Jacobs S, Joughin I, King MA, Lenaerts JTM, Li J, Ligtenberg SRM, Luckman A, Luthcke SB, McMillan M, Meiser R, Milne G, Mouginot J, Muir A, Nicolas JP, Paden J, Payne AJ, Pritchard H, Rignot E, Rott H, Sorensen LS, Scambos TA, Scheuchl B, Schrama EJO, Smith B, Sundal AV, van Angelen JH, van de Berg WJ, van den Broeke MR, Vaughen DG, Velicogna I, Wahr J, Whitehouse PL, Wingham DJ, Yi D, Young D, Zwally JH (2012) A reconciled estimate of Ice-Sheet mass balance. Sci 388 (6111):1183–1189. https://www.doi.org/10.1126/science.1228102

Short NH, Gray AL (2005) Glacier dynamics in the Canadian High Arctic from RADARSAT-1 speckle tracking. Can J Remote Sens 31:225–239

Strozzi T, Paul F, Wiesmann A, Schellenberger T, Kääb A (2017) Circum-Arctic changes in the flow of glaciers and ice caps from satellite SAR data between 1990s and 2017. Remote Sens 9. https://doi.org/10.3390/rs909047

Thomson L, Copland L (2017) Multi-decadal reduction in glacier velocities and mechanisms driving deceleration at polythermal White Glacier, Arctic Canada. J Glaciol 63:450–463

Thomson L, Copland L (2018) Changing contribution of peak velocity events to annual velocities following a multi-decadal slowdown at White Glacier. Ann Glaciol 58:145–154

Van Wychen W, Copland L, Gray L, Burgess DO, Danielson B, Sharp M (2012) Spatial and temporal variation of ice motion and ice flux from Devon Ice Cap, Nunavut, Canada. J Glaciol 58:657–664

Van Wychen W, Burgess DO, Gray L, Copland L, Sharp M, Dowdeswell JA, Benham TJ (2014) Glacier velocities and dynamic ice discharge from the Queen Elizabeth Islands, Nunavut, Canada. Geophys Res Lett 41:484–490

Van Wychen W, Copland L, Burgess D, Gray L, Schaffer N (2015) Glacier velocities and dynamic discharge from the ice masses of Baffin Island and Bylot Island, Nunavut, Canada. Can J Earth Sci 52:980–989

Van Wychen W, Davis J, Burgess DO, Copland L, Gray L, Sharp M, Mortimer C (2016) Characterizing interannual variability of glacier dynamics and dynamic discharge (1999–2015) for the ice masses of Ellesmere and Axel Heiberg Islands, Nunavut, Canada. J Geophys Res-Earth Surf 121:39–63

Van Wychen W, Davis J, Copland L, Burgess D, Gray L, Sharp M, Dowdeswell JA, Benham T (2017) Variability in ice motion and dynamic discharge from Devon Ice Cap, Nunavut, Canada. J Glaciol 63:436–449

Van Wychen W, Copland L (2017) Ice island drift mechanisms in the Canadian High Arctic, Ch. 7. In: Copland L, Mueller D (eds) Arctic Ice Shelves and Ice Islands. Springer-Verlag

Vincent WF, Gibson J, Jeffries MO (2001) Ice shelf collapse, climate change, and habitat loss in the Canadian high Arctic. Polar Record 37:133–142

White A, Copland L, Mueller D, Van Wychen W (2015) Assessment of historical changes (1959–2012) and the causes of recent break-ups of the Petersen ice shelf, Nunavut, Canada. Ann Glaciol 56:65–76

Williamson S, Sharp M, Dowdeswell J, Benham T (2008) Iceberg calving rates from northern Ellesmere Island ice caps. Canadian Arctic, 1999–2003. J Glaciol 54:391–4000

Wolken G, Sharp M, Wang L (2009) Snow and ice facies variability and ice layer formation on Canadian Arctic ice caps, 1999–2005. J Geophys Res 114. https://doi.org/10.1029/2008jf001173
Zdanowicz C, Smetny-Sowa A, Fisher D, Schaffer N, Copland L, Eley, Dupont F (2012) Summer melt rates on Penny Ice Cap, Baffin Island: Past and recent trends and implications for regional climate. J Geophys Res 117. https://doi.org/10.1029/2011jf002248

Wesley Van Wychen is the Associate Director of the Canadian Cryospheric Information Network and Polar Data Catalogue at the University of Waterloo where his work focuses on ensuring the long-term integrity of Arctic and Antarctic datasets. His research focuses on the development and use of novel remote sensing techniques for understanding Arctic environmental change at the regional scale. He also has a strong interest in understanding glacier mass balance and the evolution of glacier dynamics in response to climate change.

Luke Copland is a Professor in the Department of Geography, Environment and Geomatics at the University of Ottawa, where he holds the University Research Chair in Glaciology. He is also Director of the Laboratory for Cryospheric Research (https://cryospheric.org), and works with his graduate students on projects across northern Canada to investigate glacier dynamics and the response of ice masses to climate change.

David Burgess is a Research Scientist at the Geological Survey of Canada where he specializes in the use of remote sensing technology and in-situ methods to study the mass balance and dynamics of ice caps in the Canadian Arctic. He leads the Government of Canada's National Glaciology Programme through which he conducts integrated research and assessments of glacier-climate change across Canada's Arctic and Alpine regions. Dr. Burgess is a lead investigator and coordinator for Canada's involvement in calibration and validation of the CryoSat-2 radar altimeter (European Space Agency) over Canada's Arctic ice caps.

Melville, Bathurst, and Cornwallis Islands: Low to Moderate Relief Innuitia

14

Scott F. Lamoureux and Ashley C. A. Rudy

Abstract

Melville, Bathurst, and Cornwallis islands are located within the Innuitian Orogen geological province and find themselves primarily in the moderate relief Parry Plateau and low relief Sverdrup Lowland geomorphological regions. They lie between the high relief Innuitia of Axel Heiberg and Ellesmere islands to the north and the low relief Victoria Lowland of Victoria Island to the south. They experience some of the most extreme cold climates in Canada. Vegetation is limited to low tundra communities and barren ground. The only community is Resolute on Cornwallis Island. The region was influenced by both the Innuitian and Laurentide ice sheets during MIS 2. Glaci-isostatic depression of the regional crust and current submergence of eastern Melville Island are two of the key geomorphic imprints on the landscape. Permafrost and periglacial landforms such as patterned ground are ubiquitous. Ground ice in thermokarst produces active-layer detachments and retrogressive thaw slumps. Limited water availability and a brief melt season, when 90% of the annual discharge occurs, condition fluvial processes. There is little research on the effectiveness of aeolian processes. Widespread sea ice reduction under climate change is responsible for inundation of the coastline and erosion of shoreline bluffs.

Keywords

Innuitian Orogen • Low-moderate relief • Laurentide and Innuitian ice sheets • Permafrost • Periglacial • Thermokarst • Climate change

S. F. Lamoureux (✉)
Department of Geography and Planning, Queen's University, Kingston, ON K7L 3N6, Canada
e-mail: scott.lamoureux@queensu.ca

A. C. A. Rudy
Northwest Territories Geological Survey, Yellowknife, NT, Canada

14.1 Introduction

Extensive lowlands and plateaus form the central and northwestern landscapes of the Arctic Archipelago and contrast with the mountains and uplands of Ellesmere and Axel Heiberg mountains to the north and east, and the more resistant bedrock of the Canadian Shield to the south and east. Within this region, Melville, Bathurst, and Cornwallis islands represent much of the Parry Plateau's low-moderate relief Innuitia (see Figs. 1.1 and 1.10, this volume). These islands are located within the geological province of the Innuitian Orogen (Wheeler et al. 1996). This region is characterized amongst the most extreme cold climate in Canada (Rouse 1993) and is dominated by glacial and deglacial landscapes, overprinted by extensive post-glacial Holocene marine inundation and subsequent cold weather modification by permafrost, rivers and wind. Vegetation is limited to low tundra communities with barrens in many locations, reflective of both the short growing season and extreme aridity (Walker et al. 2005).

Human activity is limited in the region. Inuit have historically occupied the region through highly specialized nomadic hunting and cultural strategies focused on marine mammals and to a lesser extent on land species. The only community is Resolute, a small (c. 250 population) hamlet, formed through government policies in the 1950s. Resource development has occurred sporadically at several mining operations including the Polaris lead-zinc mine on Little Cornwallis Island (now closed), although substantial petroleum exploration and resource delineation occurred in the northwestern and central islands in the 1960–1980s period and base metal exploration continues in some areas. Natural gas reserves located adjacent to Melville, Bathurst, and Lougheed Islands are amongst the largest known in Canada (Fig. 14.1). Additionally, a small oil development at Bent Horn on Cameron Island operated from 1985–96. Hence, the landscape of the low to moderate relief Innuitia remains largely unimpacted by humans and the lack of substantial

© Springer Nature Switzerland AG 2020

O. Slaymaker and N. Catto (eds.), *Landscapes and Landforms of Eastern Canada*, World Geomorphological Landscapes, https://doi.org/10.1007/978-3-030-35137-3_14

vegetation cover results in a remarkable exposure of landforms and surface materials rarely found elsewhere in Canada.

14.2 Setting

14.2.1 Climate and Vegetation

This region has perhaps the most severe climate in Canada. The average annual temperature at Resolute, Nunavut (Fig. 14.1) is −15.7 °C with July mean temperature of 4.5 °C and mean January temperatures routinely below −30 °C. The mean annual precipitation is very low, ranging from 10 to 20 cm, most of which falls as snow. Climate conditions are more severe to the northwest in the islands that border the Arctic Ocean (Edlund and Alt 1989). Strong seasonality of solar insolation results in periods of 3–5 months of continuous daylight in the summer and a corresponding period of winter darkness. Sea ice is usually continuous and lasts 8–12 months, further isolating the land from moderating maritime influences and imparting extensive and persistent summer fog cover. Mean annual summer air temperatures in this region have increased by up to +2.5 °C between 1948 and 2012. This is accompanied by an increase in mean annual winter temperatures of approximately +3.0 °C and a 30% increase in annual precipitation, mostly in the winter months (Vincent et al. 2015).

The combination of severe and persistent cold temperatures greatly reduces the diversity of plant species in this region. Where present, vegetation consists of tundra communities and trees are absent. Vegetation is semi-prostrate to

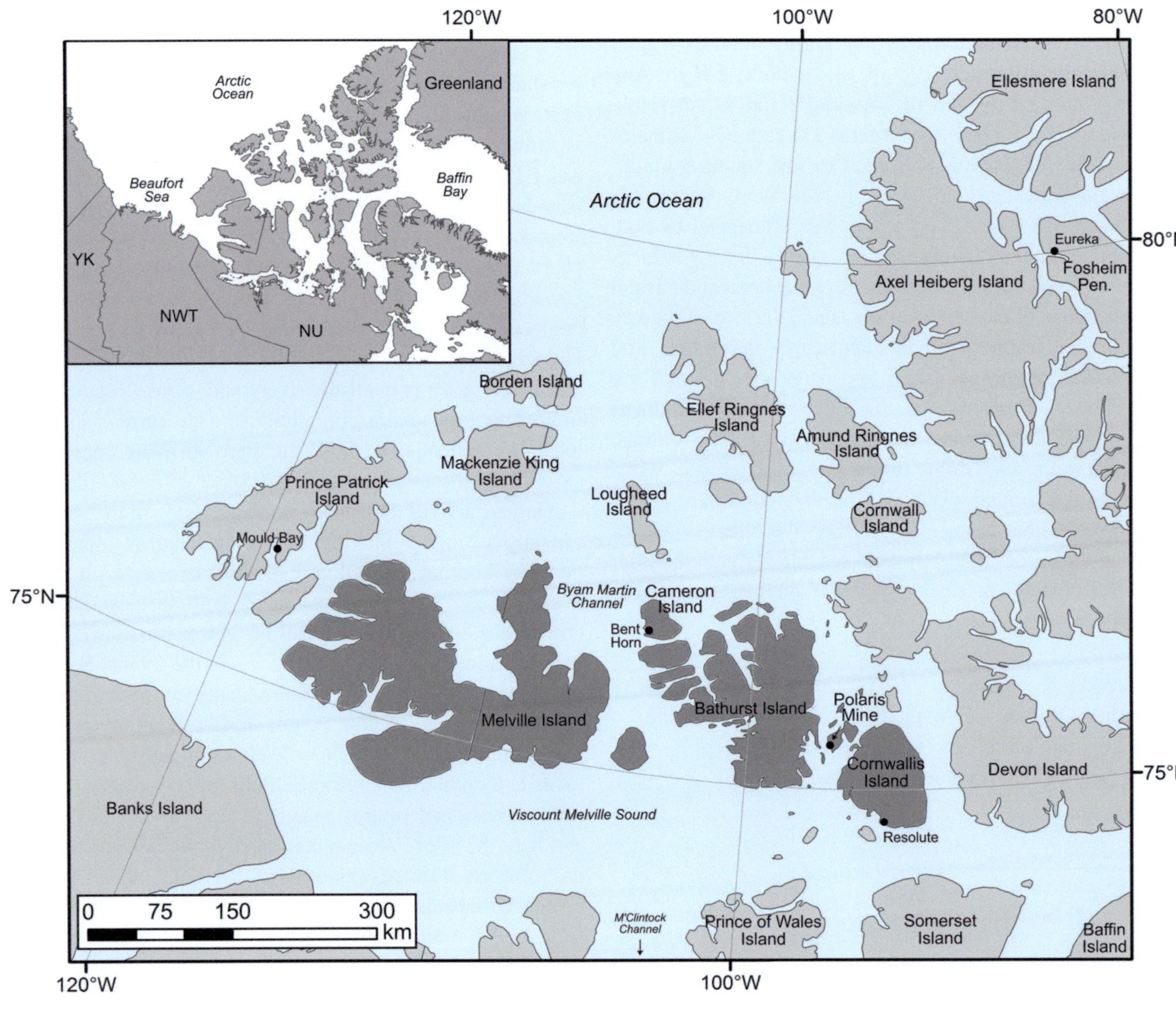

Fig. 14.1 Location of the low to moderate relief Innuitia region. The emphasis of this chapter is on the three major islands: Melville, Bathurst, and Cornwallis. Note that the only community in the region is Resolute, Nunavut

prostrate dwarf shrub tundra, and includes short grasses, lichen, and moss species. Vegetation distribution depends on soil moisture, and vascular plant vegetation is often confined to cracks and depressions in polygonal networks, or in areas supplied by runoff from residual snow patches (Fig. 14.2). Low sedges are limited to areas of increased soil moisture and dominate wet sites.

The presence of semi-prostrate shrubs and well-differentiated plant communities in mires, snowbeds, and stream banks leads to greater species diversity in more southerly and protected sites (Walker et al. 2005). Polar desert communities with patchy plant cover and reduced species diversity occur on upland terrain. In some locations, particularly on Cornwallis and parts of the other islands, carbonate bedrock and coarse surface sediments result in extensive polar desert barrens. More extensive vegetation is limited to moist lower slopes, adjacent to surface water and patchy wetlands.

Permafrost and seasonal active-layer characteristics further influence rooting zone depth and soil moisture conditions, which are important for vegetation succession and growth (Walker et al. 2005). Changes in the surface energy balance that result from the removal of vegetation due to natural processes such as slope disturbance, land

Fig. 14.2 Representative vegetation for the region includes **a** semi-prostrate to prostrate dwarf shrubs, as well as short grasses, lichen and moss species. Photograph is from Cape Bounty located on the southcentral Melville Island, August, 2013. Drier settings are characterized by polar semi-desert **b** communities, with sparse or patch vegetation cover and reduced species diversity. Photograph is from Sabine Peninsula located on northeastern Melville Island, August, 2011. *Photographs* S. Lamoureux

development activities or climate changes can result in increases in ground surface temperature and thaw of permafrost (Smith et al. 2008), resulting in substantial impacts on vegetation growth and distribution.

14.2.2 Physiography

The low to moderate relief Innuitia includes a series of islands that have gently rolling landscapes. The region contains a broad belt of plains and plateaus that include most of the southern and central arctic islands (Dawes and Christie 1991). Lithologies of varying weathering resistance form primary landforms (Fig. 14.3). The Franklinian Basin is a deep basin initiated by the crustal stretching during the breakup of Rodinia at the end of the Precambrian and underlies a belt extending from northern Ellesmere Island southwestward to Devon and Cornwallis islands and then westward to Bathurst and Melville islands (Eyles and Miall 2007). The Mesozoic Sverdrup Basin is composed of nearshore and shallow clastic sedimentary units and overlies the Franklinian Basin with the exception of areas where older rocks have been exposed by tectonism and denudation. The most distinctive feature of the Sverdrup Basin is the presence of thick evaporite formations, the product of precipitation within a hypersaline water body. The evaporite bearing formations consist of cyclically bedded limestone and anhydrite. In the western part of the basin the evaporites have developed diapirs which have risen through upwards of 8 km of sedimentary rock to form distinctive salt dome outcrops at the surface (Fig. 14.3), notably on Melville and Ellef Ringnes islands (Eyles and Miall 2007).

The Innuitian Orogen contains sinuous belts of folded sedimentary rocks with some older metamorphic terrane and includes the moderate relief Parry Plateau and the low relief Sverdrup Lowlands. The Parry Plateau is a rolling surface of moderate elevation and is underlain by mainly clastic rocks on Melville and Bathurst islands and by both clastic and carbonate rocks on Cornwallis, northwestern Devon and southwestern Ellesmere islands (Dawes and Christie 1991). The low relief Sverdrup Lowland that borders the Parry Plateau to the north is a shallow topographic basin that coincides with the less deformed parts of the Sverdrup Basin. The structural style of the rocks has produced a low-lying region of irregular hills and scarps. Dissected gypsum-anhydrite domes and related structures form distinctive landmarks and unusual features in the Sverdrup lowlands (Fig. 14.3). The preservation of these highly soluble minerals at the surface reflects the aridity of the region, and similar features occur only in Saudi Arabia (Dawes and Christie 1991). Localized intrusions of gabbro also outcrop as dykes in some locations including Amund Ringnes and Cornwall Islands. Notably, the region remains seismically active, especially associated with the Gustaf-Lougheed Arch seismic zone immediately northeast of Melville Island that has demonstrated frequent earthquakes (Hasegawa 1977).

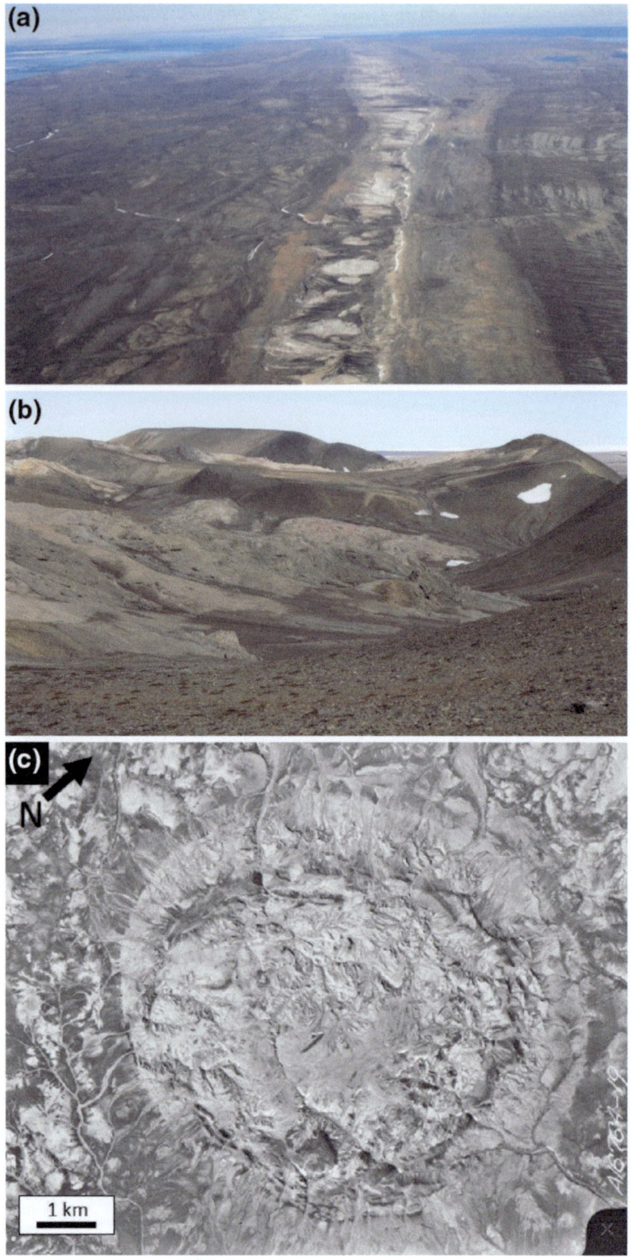

Fig. 14.3 **a** is an oblique aerial photograph looking westward on Melville Island and illustrating the prominent surface expression of the folded geological structures of the Franklinian Basin. **b** illustrates the landscape within Barrow Dome, a dissected anhydrite dome located on the northern Sabine Peninsula, Melville Island, August, 2011. *Photographs* S. Lamoureux. **c** is an aerial photograph of Barrow Dome, indicating prominent surface expression of the surrounding country rock. *Source* The National Air Photo Library—Natural Resources Canada, photograph A16764-19

14.2.3 Glaciation and Sea Level History

MIS 2 glaciation of low to moderate relief Innuitia was characterized by the influence of the larger Laurentide Ice Sheet to the south and the Innuitian Ice Sheet centered over the High Arctic islands (Blake 1970; Dyke et al. 2003; England et al. 2006) (Fig. 14.4). Evidence for earlier MIS glaciations is sparse and usually indicated by the presence of older marine fauna incorporated into MIS 2 glacial deposits. Laurentide ice flow was northward and was increasingly constrained near the ice limit by the broad physiography of the land and inter-island channels.

The MIS 2 Innuitian Ice Sheet developed over the High Arctic Islands and extended as an ice divide along a broad northeast to southwest direction from Greenland to the central islands. Local ice caps on smaller islands developed to the west (Blake 1970; Dyke et al. 2003; England et al. 2006). Innuitian ice coalesced with Laurentide ice in a broad zone along the south coast of the High Arctic islands that make up the low to moderate relief Innuitia (Fig. 14.4). The extent of the interaction of these two ice sheets is poorly defined by stratigraphic evidence, and is often defined by ice marginal landforms and the dispersal of distinctive lithologies such as granite and other igneous rocks from the Canadian Shield on the mainland (England et al. 2009).

Both the Laurentide and Innuitian ice sheets were broadly cold-based, with distinctive landform assemblages associated with polythermal margins (Dyke and Evans 2003). In low to moderate relief Innuitia, cold-based ice generated non-descript landscapes with varying cover of glacial

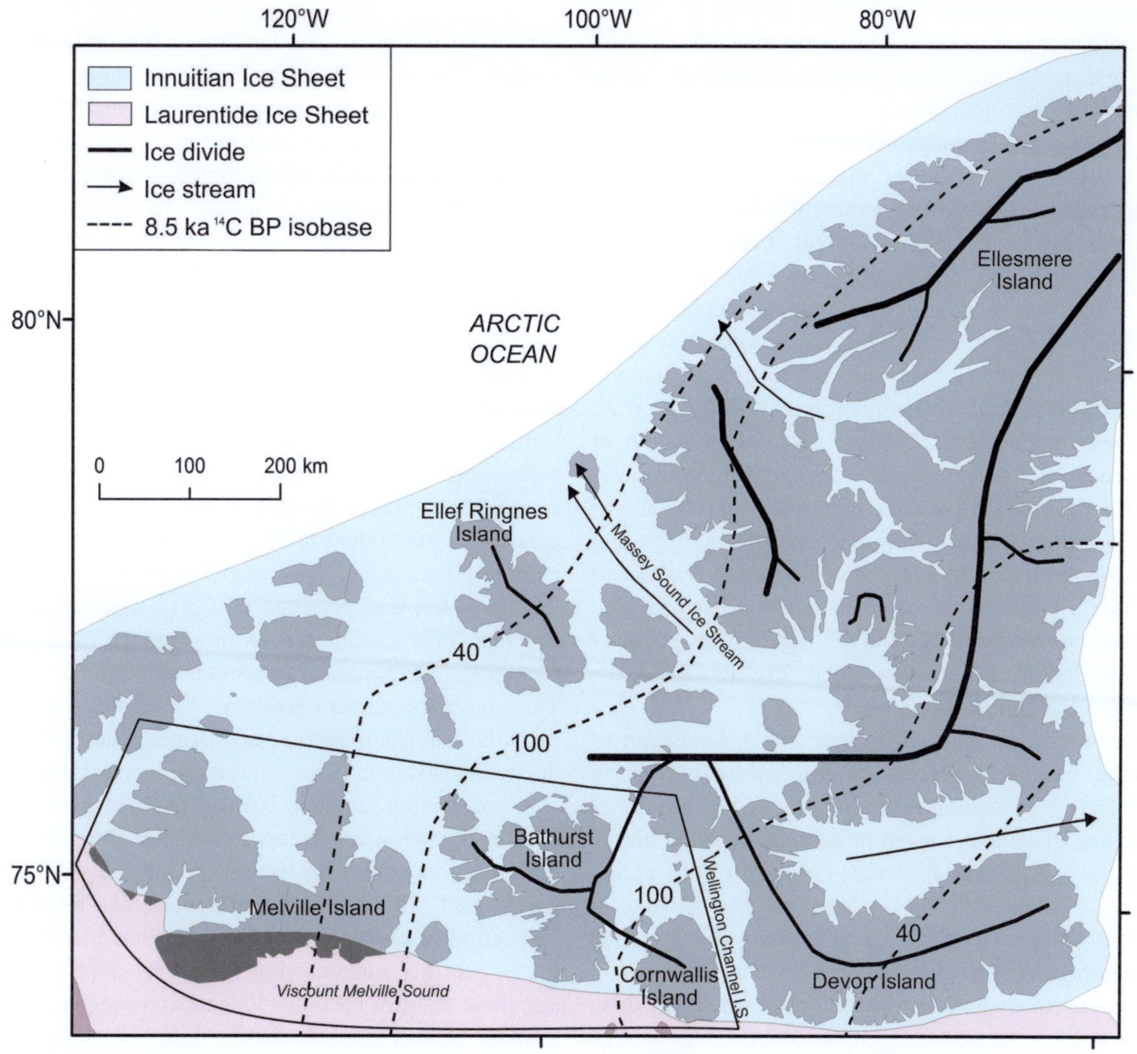

Fig. 14.4 Postglacial isobases drawn on the 8.5 ka ^{14}C BP shoreline are illustrated on a simplified schematic of the extent of the Innuitian and Laurentide ice sheets during the last glaciation. Primary ice divides (thick black lines) and connected secondary ice divides (thinner black lines) identify strong areas of flow convergence recorded by widespread geological observations. Former ice streams are highlighted by arrows. Figure adapted from England et al. (2006)

sediment, often boulder-rich and interspersed by residual features such as tors. Distinctive erratic dispersal plumes and glacial striae provide limited evidence for specific ice flow trajectories (Lamoureux and England 2000). Coastal and ice marginal zones where thinner ice resulted in more focused ice erosion and depositional settings along with increased local relief and incision are found throughout the low to moderate relief Innuitia (Dyke and Evans 2003).

Deglaciation of the region began comparatively late compared to elsewhere in Canada. Ice retreat occurred first in the western islands resulting in increasing confinement to physiographic features, principally the inter-island channels (England et al. 2006). Re-advance phases of the Laurentide ice c. 11 ka resulted in the formation of an extensive ice shelf in Viscount Melville Sound that grounded on the south coast of the newly deglaciated islands, forming broad complexes of glacial and marine sedimentation along the coasts of Melville and adjacent islands (England et al. 2009). Throughout the low to moderate relief Innuitia region, retreat on land generated prominent meltwater channels along successive ice margins, many of which subsequently evolved as Holocene drainage networks.

Glaci-isostatic depression of the crust over the entire region represents one of the key geomorphic imprints of Late Wisconsinan ice in the region. In all areas, ice retreat involved the incursion of marine waters due to regional isostatic depression. Deglacial and early Holocene marine limits range from in excess of 100 m asl around Cornwallis Island where Laurentide ice was thickest, to <30 m asl along the northwest margin of the Innuitian Ice Sheet (Dyke et al. 2003). These marine limits, defined by a combination of relict coastal and deltaic landforms are chronologically constrained by radiocarbon ages of marine fauna, particularly molluscs and less commonly megafauna including bowhead whales found in raised beach deposits (Dyke and Savelle 2000). Mapping of marine inundation has provided critical constraints for ice loads in the region. The extent of marine inundation defines the broad ice divide of the Innuitian Ice Sheet (Fig. 14.4).

Marine inundation resulted in widespread deposition of fine-grained marine sediments in low elevations. Early phases of marine deposition were characterized by proximity to calving ice fronts and contain rafted materials and higher energy deposits with turbid inflows, but most of the marine sediments are clay-rich and drape preexisting glacial deposits (England et al. 2009). Local washing and shoreline processes form clear raised beach deposits that often extend many kilometers inland, while at river mouths, delta sequences indicate the progressive deposition and incision of fluvial processes in response to base level lowering with continued Holocene glaci-isostatic emergence. Where fluvial sediments are particularly abundant such as the central Sverdrup Lowland, some river systems have prograded long linear deltaic features in response to Holocene sea level lowering (Fig. 14.5) (Hodgson 1994).

In addition to Holocene post-glacial emergence throughout the region, western areas of the archipelago have been subject to re-submergence as the distal forebulge has retreated towards the former ice load center. The forebulge represents a fraction of original crustal displacement and is responsible for late Holocene and current submergence along regions including eastern Melville Island (Lajeunesse and Hanson 2008; Forbes and Hansom 2011).

14.3 Permafrost and Periglacial Landforms

Periglacial processes and landforms constitute the primary surface expression and permafrost is an important control of landscape sensitivity to climate change and land development. Continuous permafrost occurs across the region and reaches thicknesses of several hundred metres with shallow permafrost temperatures as low as −15 °C. Recent trends in permafrost temperatures from several monitoring sites in the region indicate that permafrost is warming (Smith 2011). Although long-term permafrost monitoring stations fall outside the extent of this region, in general, changes in shallow permafrost temperatures over the last decade have been greater in the Arctic compared to areas below tree line. Limited vegetation cover and typically thin snow cover means there is a more direct link between air and ground temperatures and variations in mean annual ground temperature in this region reflect local variations in snow cover including timing and thickness (Taylor et al. 2006).

The widespread occurrence of permafrost combined with a severe polar climate results in a number of characteristic landform features that are found throughout the region. There are three primary geometric forms of patterned ground including circles, polygons, and stripes. These features form due to the slow circulation of the soil in a process referred to as cryoturbation (French 2018). Patterned ground features can be separated according to whether they are sorted or non-sorted as controlled by surface materials in which they are formed. Circles, and larger nets that form from the interaction of circles, as well as polygons usually occur on flat to near-flat surfaces (Fig. 14.6). As slope angle increases, these features become elongate and irregular, eventually forming stripes (French 2018). Large-diameter polygonal

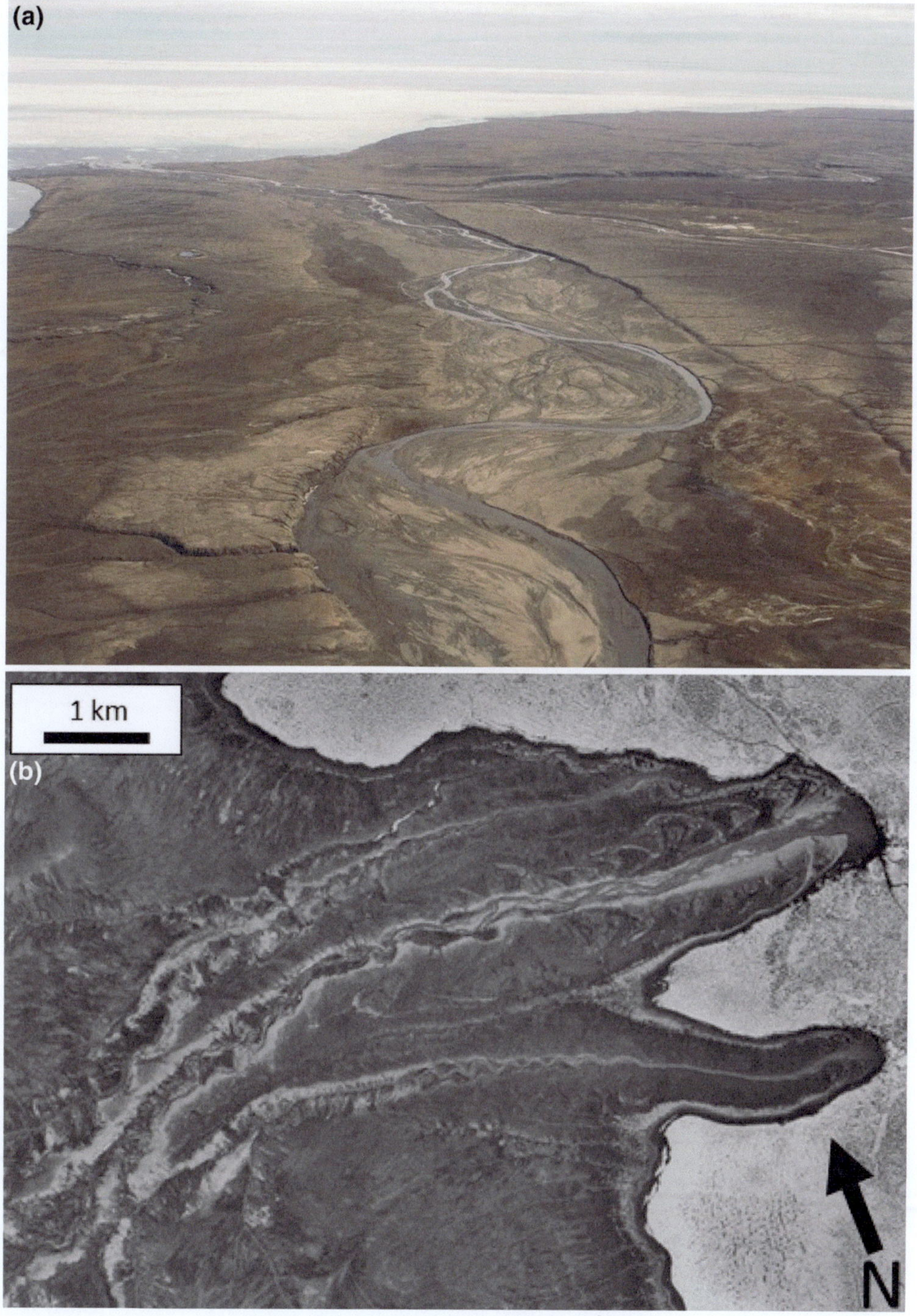

Fig. 14.5 **a** an aerial view of a typical braided large river in the region on the Sabine Peninsula, Melville Island (August, 2011). The river is incised into late Holocene prograded fluvial deposits that formed in the shallow coastal setting. Subsequent glaci-isostatic emergence has resulted in base level lowering and incision. *Photograph* S. Lamoureux. **b** illustrates examples of river deltaic and floodplain deposits prograding into the ocean as prominent lobes, Sabine Peninsula, Melville Island. *Source* The National Air Photo Library—Natural Resources Canada, photograph A16764-5

ground features result from thermal-contraction cracking. The smaller diameter polygons form nets that are delineated by furrows and narrow cracks that represent a surface with hummocky micro-relief generally less than about 0.3 m (French 2018). Parallel lines of vegetation-covered ground associated with seepage and/or surface wash often separate non-sorted stripes occurring on sloping terrain.

Solifluction is a widespread mass- wasting process throughout this region. Solifluction is the broad term used to describe very slow mass movements associated with freeze–thaw action and is an important mass- wasting and denudation process that shapes the periglacial landscape (Ballantyne and Harris 1994). Landforms produced from solifluction are lobes, sheets, terraces, stripes and hummocks. While lobes are more common in alpine slopes with variable surface conditions, sheets dominate on polar slopes where more homogenous surface conditions may prevail (Matsuoka 2001).

Fig. 14.6 **a** shows initial thermokarst pond formation from thaw of ice-rich permafrost and subsequent uneven settlement. **b** is an example of polygonal features formed from ice wedges, including a number of thermokarst ponds and retrogressive thaw slumps (RTSs). Both images from Sabine Peninsula, Melville Island, August, 2011. *Photographs* S. Lamoureux

Ice wedges are wedge-shaped vertical bodies of ice that form when hoar frost develops in an open crack on the surface in the winter and then water from melting snow penetrates the crack in early spring (French 2018). These wedges grow and interconnect into complex polygonal fields that are common throughout this region, especially in poorly drained lowlands (Fig. 14.6).

14.3.1 Thermokarst Processes

Thermokarst refers to a suite of processes and resultant landforms that result from the thaw of ice-rich permafrost (Jorgenson et al. 2008). In contrast to the western Canadian Arctic and southern discontinuous permafrost regions, the cold arid climate of the central and eastern Arctic is thought to be less responsive to thermokarst processes as the active layer is relatively shallow, the period of summer thaw is short and the thermal change required to initiate thermokarst is large (French 2018). Thermokarst features are found throughout this region but are limited to landscapes with increased ground ice content and are often restricted to glacigenic materials and localized areas with ice-rich organic, lacustrine, or marine deposits (French 2018). The primary forms of thermokarst include irregular subsidence features, hillslope processes including active-layer detachments (ALDs) and retrogressive thaw slumps (RTSs) and thermal-erosion gullies that form as ice wedges degrade. Thermokarst ponds are found in this region and originate from poorly drained low-centered ice wedge polygons. The depressed centers promote standing water bodies and with time, the combination of fluvithermal erosion and thermokarst subsidence results in merging of adjacent ponds and possible drainage (French 2018). Thermokarst ponds will vary in shape and size, and in this region usually do not exceed 300 m in diameter (French 2018).

The irregular topography associated with in situ thermokarst processes is due to highly localized variations in ground ice volume and the impact of thermokarst will vary depending on ground ice and drainage conditions. Surface ponding (thermokarst ponds) may occur where settlement of ice-rich terrain impedes drainage and the presence of surface water can accelerate subsequent thaw, resulting in expansion of ponds into larger bodies that frequently coalesce and can abruptly drain (Fig. 14.6).

Ice wedges constitute a large portion of the ground ice throughout the arctic lowlands and commonly occur under polygonal ground and may be concealed under terrain with little to no surface expression (Fig. 14.6). An increase in active-layer thickness can deepen polygonal troughs that lead to changes in the surrounding terrain through alterations of hydrological pathways and initiation of thermo-erosional niching. These features develop intense localized gullying as focused water flow causes thermal incision and rapid headward expansion of the gully (Godin and Fortier 2012). Gully development can be an important driver of local hydrologic change through increased hydrologic connectivity and improved drainage, and gullies are noted for local, often intense, patterns of erosion (Allard and Pollard 2013).

ALDs involve the separation of thawed active-layer overburden and the rapid movement of these materials downslope over the frozen material beneath. ALDs occur on low to moderate-angled slopes and may occur over hours to several days (Fig. 14.7) (Lewkowicz 2007). The morphological characteristics of ALDs will vary depending on the geomorphic and ecological setting but are typified by a head scarp, scar area below the initiation zone, highly disturbed soils along lateral shear zones, and a downslope toe zone where displaced materials accumulate, often forming irregular ridges and troughs (Lewkowicz 2007). Initiation of ALDs occurs following rapid thaw of the ice-rich layer during warm summers, typically in combination with precipitation events (Kokelj and Jorgenson 2013). These conditions lead to high pore water pressures at the base of the active layer, liquefaction of fine-grained sediments, and shear failure (Fig. 14.7). Local and regional observations of ALDs and their distribution shows that they commonly occur in spatial clusters as a result of their association with surficial materials and topographic conditions. Rudy et al. (2017a) noted that land surfaces most susceptible to ALD formation on northern Melville Island occurred primarily on north facing aspects, on slopes less than 15° and below marine limit in frost susceptible materials (Fig. 14.8). Evidence for high pore water pressures caused during warm summers is also evident as ejections of mud slurries to the surface, forming discrete stratiform deposits (Fig. 14.9) (Holloway et al. 2015). These mud ejections appear to occur in association with ALDs, but are most commonly on shallower slope angles and upland settings (Holloway et al. 2017).

RTSs commonly develop in ice-rich glacigenic or glacimarine deposits where massive ice is located close to the surface (Fig. 14.7) (Burn and Lewkowicz 1990). RTSs are initiated through a number of mechanisms that expose the ice including: mechanical erosion from fluvial erosion, thermally driven subsidence along lakeshores or mass wasting triggered by extreme thaw and/or large precipitation events. RTSs are less common in this region than in the ice-rich terrain of the western Canadian Arctic. Their occurrence has been observed on Melville Island and locales with high ice content such as in glacimarine lowlands. These features are often active for many years (Hegginbottom 1984) and can as a result grow to substantial size through continued head scarp retreat. They remain active so long as overburden sedimentary material is removed from the ice face by fluvial erosion (Fig. 14.7). If material accumulates on the ice, the increase in thermal insulation will eventually stabilize the surface through reduced thaw (Burn and Lewkowicz 1990). RTSs are poly-cyclical, and can be re-initiated by further exposure of the massive ice by erosion or deep thaw (Lantuit and Pollard 2008).

Both ALDs and RTSs landforms are associated with substantial impacts on erosion and downstream water quality (Lamoureux and Lafrenière 2017). ALDs generate rapid

Fig. 14.7 **a** presents an example of an active-layer detachment (ALD). View is looking downslope from the headscarp and through the scar zone. Note the shallow depth (<1 m) of the scar zone associated with overburden sliding over the frozen substrate. **b** is an example of a retrogressive thaw slump (RTS). The view, looking upslope towards the headscarp, shows the mud pool and the prominent exposure of massive ice below a shallow sedimentary overburden. Sediment removal through the mud pool and downslope channels maintains a comparatively clean exposed ice face. Note the variation in the exposed overburden thickness in this example. The thicker overburden in the right-side headscarp is likely related to a previous active-layer detachment and subsequent ground ice loss. *Photographs*: S. Lamoureux are from the Sabine Peninsula, Melville Island, August, 2011

increases in sediment erosion and solute flushing, with the magnitude of the impact varying with the morphology of the disturbance, the degree of channelization and downstream connectivity. In some cases, ALDs may form dams on stream and river channels, causing fluvial disruption and altered channel sediment cycling (Lamoureux and Lafrenière 2017). Erosion is particularly enhanced downstream of large ALDs and appears to show evidence of recovery in five years following formation (Lamoureux et al. 2014). By contrast, the ice exposed in the headscarp of RTSs sustains runoff during summer months, and results in high sediment and solute delivery downstream, often lasting for years (or decades) while the headscarp is actively retreating (Rudy et al. 2017b).

14.4 Fluvial Processes and Landforms

Hydrological conditions in the region are conditioned by limited water availability and a short melt season due to the severe polar climate. Through much of the region, winter snow accumulation constitutes the majority of annual

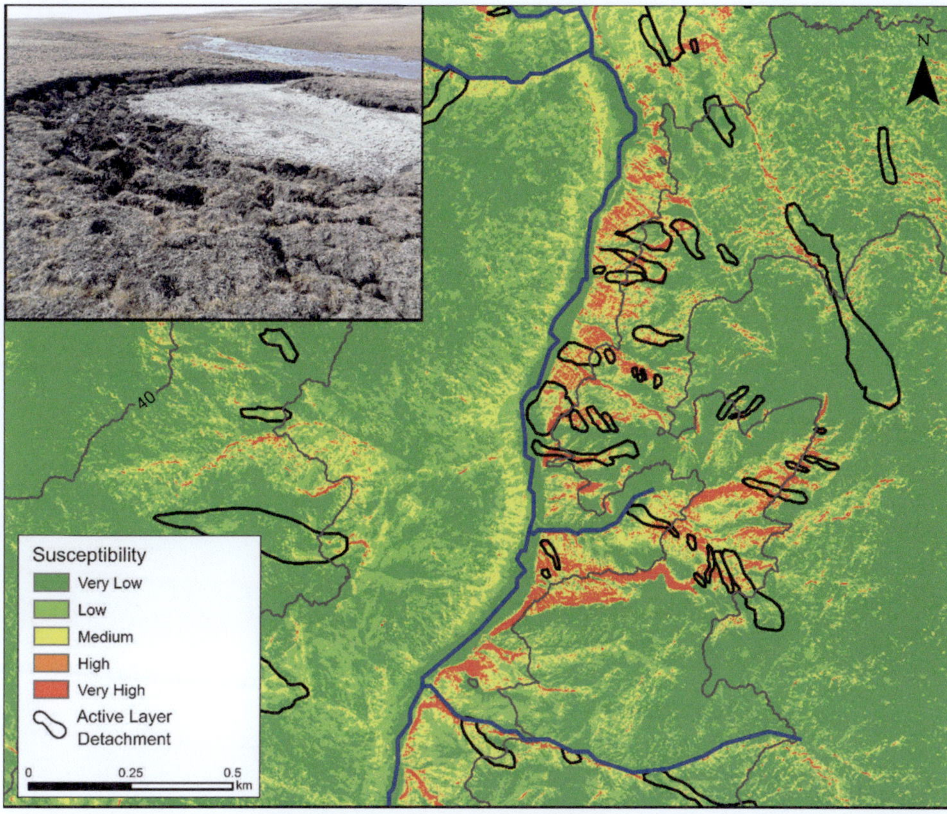

Fig. 14.8 An example of modeling to predict landscape susceptibility to the formation of ALDs from the Sabine Peninsula, Melville Island. This model utilizes mapped ALDs (shown as black outlines, example from map area show as inset photograph), and combines terrain variables (i.e. aspect, slope, elevation, proximity to water). Results predict areas with terrain characteristics with varying susceptibility to disturbance and form the basis of more detailed field investigations (After Rudy et al. 2017a)

Fig. 14.9 An example of a mud ejection formed by high pore water pressures at the base of the active layer during late summer. These mud ejections are often found in clusters in upland and gentle slope settings. Note the lens cap for scale. *Photograph* S. Lamoureux from Cape Bounty, Melville Island, June, 2012

precipitation but is extensively redistributed by wind to accumulate in lee slopes and concavities on the land. This is especially the case in stream and river channels, where accumulations of several metres of snow are common while surrounding slopes have patchy snow cover or none at all.

Snow melt commences in May or June as solar radiation increases and the daylight period is 24 h for all but the southernmost areas. Snow melt runoff generation occurs relatively quickly and water flows from slopes into channels where it is often ponded in deeper snow. Open channel flow

Fig. 14.10 Examples of river channels in the region. **a** shows an example of a small river during waning snowmelt runoff (June, 2012). Note the sediments—including gravel—draped on the channel snowpack from the initial discharge. **b** indicates the same river system in late summer baseflow conditions (August, 2009). Discharge and sediment transport are minimal, and the channel is substantially underfit. Both examples from Cape Bounty, Melville Island. *Photographs* S. Lamoureux

is often accompanied by slush flow activity and initial flow may be located some distance from the channel proper (Woo 2012). Surface flow rapidly incises into the channel snow-pack to reaches the bed and often tunnels under the snow as it reaches the channel bed. In northern areas, channel snow pack can last through much of the initial melt season, and in rare cases may be perennial. This residual channel snow pack, along with snow patches on the slopes, both continues to contribute to runoff, but also isolates flow from portions of the bed (Fig. 14.10).

High discharge from snowmelt often lasts only a few days to several weeks in the largest rivers. As much as 90% of annual discharge can occur during this short interval, resulting in a comparatively intense episode of fluvial activity. Numerous studies have demonstrated that most suspended sediment transport occurs during this period (Lewis et al. 2012) and is associated with high stream power and sediment availability. As discharge wanes, stream power diminishes and sediment fluxes decrease substantially. Rainfall with sufficient intensity to generate increased

Fig. 14.11 Deflation surface formed in sandy glacial sediments, leaving a distinctive armouring of gravel and cobbles. *Photograph* S. Lamoureux

discharge and suspended sediment fluxes is comparatively rare, and may occur as 1–3 events per year, or not at all (Lewis et al. 2012).

The widespread availability of unconsolidated glacial and marine sediments as well as weakly consolidated sedimentary bedrock results in variable, but often high sediment loads and fluxes, particularly compared to the lowlands that form in more resistant rock and irregular drainage patterns of the Canadian Shield. More resistant carbonate bedrock found in some areas of the Arctic Platform appears to have lower suspended sediment loads (Braun et al. 2000). Suspended sediment fluxes are measured at selected locations but are poorly constrained across the region. Strong interannual variability in snowpack and melt season weather results in suspended sediment fluxes that vary by orders of magnitude at a given site (Lewis et al. 2012). Bedload processes and fluxes remain poorly studied in the region but are likely only active during the short intervals of high discharge (Fig. 14.10).

This combination of hydrological regime and sediment availability results in distinctive fluvial features in the region. Channels are often sized for spring snow melt discharge and are underfit for the remainder of the season (Fig. 14.10). The morphology of channels is further altered by limited bank vegetation and irregular erosive action on banks and floodplains by initial flow during snowmelt due to the vagaries of deep channel snowpack. Hence, reworking of channel deposits and irregular deposition of early-season fluvial sediments on channel beds and adjacent surfaces are common (Fig. 14.10). In some cases, channel bank erosion is sustained by exposure of buried ice that supports rapid bank retreat. In general, however, sediment yields appear to depend on solifluction delivery of material to the channel and minimal vegetation to protect bank materials. Interestingly, bank erosion in deeply incised channels may be limited by residual snow that accumulates in these settings, resulting in exposure after discharge has diminished to summer baseflow conditions (Fig. 14.10).

River morphology and landforms broadly reflect these conditions and sediment availability is evident by braided morphologies in many areas of the region, especially where local relief results in well-defined drainage patterns. In coastal lowland areas where beach ridges and ponding affect surface drainage, or where landscape relief is minimal, channel morphology is often limited to single-channel forms and meandering may develop in larger rivers (Fig. 14.10). Widespread thermokarst lakes and ponds that limit fluvial sediment movement characterize these low relief settings.

14.5 Aeolian Processes and Landforms

Strong, persistent winds are a notable feature of the Arctic. Wind-redistribution of snow during the winter is a dominant control over water availability and as a result, vegetation shows affinities to the availability of water, particularly in the most arid and northerly areas of the region. The aridity and strong winds also contribute to aeolian processes in many settings. While extreme wind conditions are capable of transporting gravel-sized materials (Edlund and Woo 1992), depositional features associated with wind are rare. However, deflation appears to be more common and results in

Fig. 14.12 A prominent ventifact formed in poorly lithified sandstone, Melville Island. *Photograph* S. Lamoureux

scouring of exposed soft bedrock and deflation surfaces characterized by gravel-armoured surfaces (Fig. 14.11). The extent to which frost heave and deflation contribute to these surfaces remains speculative and poorly measured. Anecdotal evidence for wind transport of sediment is usually most evident in late winter on snow cover. Plumes of sediment are often observed to extend offshore of exposed fine-grained surfaces and coarse grained materials considered to be of aeolian origin and have been identified in lake sediment records in the region (Lamoureux 1999). Ventifacts also occur in some settings where bedrock lithology is suitable (Fig. 14.12).

14.6 Lake and Coastal Processes and Landforms

The Arctic Archipelago represents several thousand kilometers of coastline and the presence of extensive sea ice through much of the year results in a range of distinct landforms and geomorphic processes. Additionally, lakes are found in coastal and interior settings, and the geomorphic features of many lakes reflect similar processes to the coasts. In general, much of the region is subject to eight or more months of sea ice cover, with some of the region having historically perennial ice cover. Notably, many larger lakes have persistent ice cover, lasting into late July or August, resulting in shoreline processes and landforms similar to coastal settings.

A pronounced decline in sea ice extent in late summer (sea ice minimum usually occurs in late September) measured with satellite imagery since the late 1970s has resulted in many areas of the region having a substantially longer open water periods and increased ice mobility. In addition to recent declines in ice cover, palaeoenvironmental evidence for bowhead whale and walrus along with the incursion of driftwood from the Arctic Ocean indicate that Holocene sea ice conditions varied substantially (Dyke et al. 1997). Glaci-isostatic emergence throughout the region and late Holocene submergence in some northwestern areas has further altered coastal processes, resulting in many areas where the dominant surface landforms are related to coastal activity but are substantially removed from the modern shoreline (Fig. 14.13).

Coastal environments can be classified into transgressive and regressive settings (Forbes 2011). The former settings are associated with inundation of the coastline, flooding of wetlands and lakes, and erosion of shoreline bluffs. In some cases, shoreline erosion may expose ground ice and result in permafrost degradation or disturbances such as retrogressive thaw slumps, but these features are generally rare in the region and are more common in the western Arctic (Lantuit et al. 2012).

Shoreline erosion in lakes forms bluffs. In one detailed study from two lakes at Cape Bounty on southern Melville Island, Normandeau et al. (2016) noted that shoreline erosional bluffs corresponded with subaqueous slump features. Utilizing sidescan swath bathymetry and sub-bottom acoustic survey techniques, they noted similar slump features were widespread and particularly associated with failure along the contact between marine and overlying lacustrine sediments (Fig. 14.14). Slumps and similar mass movements in lakes appear to generate sustained high turbidity in lakes (Fig. 14.14) and are notable for their impacts on aquatic ecosystems in the region (Roberts et al. 2017).

In regressive shoreline settings, wave and ice are the primary geomorphic agents and the presence of pervasive ice reduces wave energy substantially. Tidal flats and low-energy wave washed surfaces are often strewn with boulders that are moved by ice and left isolated (Forbes and Hansom 2011). Sediment deposition is muddy and reflects availability, often through low-energy reworking of tidal flats and longshore drift of fines from river mouths.

Persistent ice cover is mobilized by Currents and wave action move and plow material irregularly on the shoreline in spite of persistent ice cover. In these settings, wave-reworking of shoreline materials is less evident and plough ridges and

Fig. 14.13 Examples of typical coastal landforms in the region. **a** shows regressive raised beaches, formed during glaci-isostatic uplift. **b** shows a raised delta complex where several rivers reach the coast. The delta has prograded and incised with Holocene relative sea level lowering, resulting in a series of prominent abandoned delta lobes. Both aerial photographs of Bathurst Island, August, 2011. *Photographs* S. Lamoureux

other irregular features may persist. Ice-pushed plough ridges are common on lakes as well, and show considerable variation according to lake morphology, fetch, and ice persistence (Fig. 14.15).

Where shoreline erosion and open water conditions are suitable, erosion and reworking of material can winnow fines and leave coarser sorted deposits that form shallow beach ridges (Forbes and Hansom 2011). Beach materials are often ice-rich, and together with limited periods of wave action due to sea ice cover, tend to be resistant to erosion (St-Hilaire-Gravel et al. 2012). Ice plowing enhances these ridges and results in substantial longshore deposits (Fig. 14.15). Where sediment is available and wave activity is sufficient, longshore features including spits may develop.

14.7 Conclusion

Considered one of the most remote regions of Canada, low-moderate relief Innuitia comprises a landscape that reflects the harsh arctic climate and late Quaternary events. In most areas, ice retreat left behind a landscape with subtle

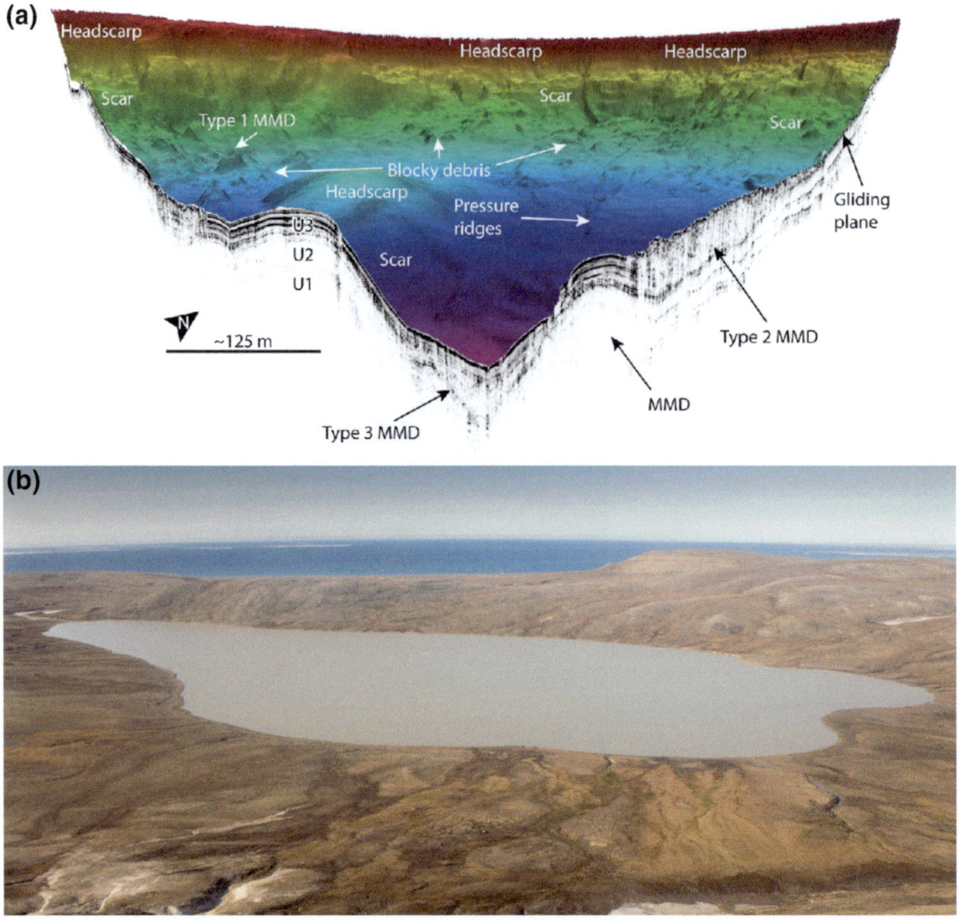

Fig. 14.14 **a** is a reconstructed perspective view of a portion of West Lake, Melville Island indicating primary subaqueous features related to mass movements. Several phases of mass movement deposit (MMD) are evident, as well as headscarp and scar forms. From Normandeau et al. (2016), used with permission. **b** shows an aerial view of West Lake indicating persistent turbidity arising from subaqueous slumps in December 2011 and February 2012 (under ice cover). *Photograph* S. Lamoureux taken August, 2015

Fig. 14.15 Ice push shoreline features along a lake on southern Melville Island (August, 2017). Note the proximal, sharp-crested ridge along with a broader margin of reworked and ploughed cobbles and gravel extending to the left of the photograph. Wave action has largely winnowed these deposits of fines. *Photograph* S. Lamoureuxe

evidence for glacial cover, while subsequent widespread marine submergence further generated conditions conducive for active periglacial, fluvial and coastal processes. This landscape, often only subject to the action of thaw, water and waves for a few months of the year, is especially distinctive amongst the regions of Canada. This distinctiveness is often expressed by a high degree of sensitivity of geomorphic and ecosystem elements to climate change, particularly through processes associated with permafrost degradation, hydrological change, and reduction in sea ice extent and duration. Indeed, this region displays some of the clearest environmental responses to anthropogenic climate change through widespread sea ice reduction and permafrost disturbance, and projected climate changes suggest these responses will continue to expand in the region. Thus, the safety of communities and future resource development in this region will depend on adapting to greater environmental change than that which is occurring further south.

References

Allard M, Pollard W (2013) Permafrost and climate change in Northern Coastal Canada. ArcticNet Annual Research Compendium (2013–14). http://www.arcticnet.ulaval.ca/pdf/compendium2012-13/4.4_permafrost.pdf

Ballantyne CK, Harris C (1994) The Periglaciation of Great Britain. Cambridge University Press, Cambridge, UK

Blake W Jr (1970) Studies of the glacial history in Arctic Canada. 1. Pumice, radiocarbon dates, and different postglacial uplift in the eastern Queen Elizabeth Islands. Can J Earth Sci 7:634–664

Braun C, Hardy DR, Bradley RS, Retelle MJ (2000) Streamflow and suspended sediment transfer to Lake Sophia, Cornwallis Island, Nunavut, Canada. Arct Antarct Alp Res 32:456–465

Burn C, Lewkowicz AG (1990) Canadian landform examples—retrogressive thaw slumps. Can Geogr 34:967–981

Dawes PR, Christie RL (1991) Geomorphic regions. In: Trettin HP (ed) Geology of the Innuitian Orogen and Arctic platform of Canada and Greenland. Geological Survey of Canada, Geology of Canada No 3, pp 29–56

Dyke AS, England J, Reimnitz E, Jetté H (1997) Changes in driftwood delivery to the Canadian Arctic Archipelago: the hypothesis of postglacial oscillations of the transpolar drift. Arctic 50:1–16

Dyke AS, Savelle JM (2000) Major end moraines of Younger Dryas age on Wollaston Peninsula, Victoria Island, Canadian Arctic: implications for paleoclimate and for formation of hummocky moraine. Can J Earth Sci 37:601–619

Dyke AS, Evans DJA (2003) Ice-marginal terrestrial landsystems: northern Laurentide and Innuitian ice sheet margins. In: Evans DJA, Hodder A (eds) Glacial landsystems, pp 143–165

Dyke AS, Moore A, Robertson L (2003) Deglaciation of North America. Geological Survey of Canada, Open File 1574

Edlund SA, Alt BT (1989) Regional congruence of vegetation and summer climate patterns in the Queen Elizabeth Islands, Northwest Territories, Canada. Arctic 42:3–23

Edlund SA, Woo MK (1992) Eolian deposition on western Fosheim Peninsula, Ellesmere Island, Northwest Territories during the winter of 1990–91. Current Research, Part B, Geological Survey of Canada, Paper, pp 91–96

England J, Atkinson N, Bednarski J, Dyke AS, Hodgson DA, O'Cofaigh C (2006) The Innuitian Ice Sheet: configuration, dynamics and chronology. Quat Sci Rev 25:689–703

England JH, Furze MFA, Doupé JP (2009) Revision of the NW Laurentide Ice Sheet: implications for the paleoclimate, the northeast extremity of Beringia, and Arctic Ocean sedimentation. Quat Sci Rev 28:1573–1596

Eyles N, Miall A (2007) Canada rocks: the geologic journey. Fitzhenry and Whiteside Limited, Markham, CA

French HM (2018) The periglacial environment, 4th edn. Wiley, Chichester, UK

Forbes DL (2011) Glaciated coasts. In: Wolanski E, McLusky DS (eds) Treatise on estuarine and coastal science, vol 3. Academic Press, Waltham, pp 223–243

Forbes DL, Hansom JD (2011) Polar coasts. In: Wolanski E, McLusky DS (eds) Treatise on estuarine and coastal science, vol 3, Academic Press, Waltham, pp 245–283

Godin E, Fortier D (2012) Geomorphology and spatio-temporal characterization of a thermo-erosion gully in the valley of glacier C-79, Bylot Island, Nunavut, Canada. Can J Earth Sci 49:979–986

Hasegawa HS (1977) Focal parameters of four Sverdrup Basin, Arctic Canada, earthquakes in November and December of 1972. Can J Earth Sci 14:2481–2494

Hegginbottom JA (1984) Continued head-wall retreat of a retrogressive thaw flow slide, eastern Melville Island, Northwest Territories. Current Research, Part B, Geological Survey of Canada, Paper 84-1B, pp 363–365

Hodgson DA (1994) Episodic ice streams and ice shelves during retreat of the northwesternmost sector of the late Wisconsinan Laurentide Ice Sheet over the central Canadian Arctic Archipelago. Boreas 23:14–28

Holloway JE, Lamoureux SF, Montross SN, Lafrenière MJ (2015) Climatic and landscape controls over mud ejection formation in the Canadian High Arctic. Permafr Periglac Proc 27:204–218

Holloway JE, Rudy ACA, Lamoureux SF, Treitz P (2017) Determining the terrain characteristics related to subsurface water pressurization in permafrost landscapes using susceptibility modelling. The Cryosphere 11:1403–1415

Jorgenson MT, Shur Y, Osterkamp TE (2008) Thermokarst in Alaska. In: Kane DL, Hinkel KM (eds) Proceedings of the 9th international conference on permafrost. Fairbanks, AK, 29 June to 3 July, 2008. Institute of Northern Engineering, University of Alaska Fairbanks, vol 1, pp 869–876

Kokelj SV, Jorgenson MT (2013) Advances in thermokarst research. Permafr Periglac Proc 24:108–119

Lajeunesse P, Hanson MA (2008) Field observations of recent transgression on northern and eastern Melville Island, western Canadian Arctic Archipelago. Geomorph 101:618–630

Lamoureux SF (1999) Catchment and lake controls over the formation of varves in monomictic Nicolay Lake, Cornwall Island, Nunavut. Can J Earth Sci 36:1533–1546

Lamoureux SF, England JH (2000) Late Wisconsinan glaciation of the northwest sector of the Canadian Arctic Archipelago. Quatern Res 54:182–188

Lamoureux SF, Lafrenière MJ, Favaro EA (2014) Erosion dynamics following localized permafrost slope disturbances. Geophys Res Letters 41. https://doi.org/10.1002/2014gl060677

Lamoureux SF, Lafrenière MJ (2017) More than just snowmelt: integrated watershed science for changing climate and permafrost at the Cape Bounty Arctic Watershed Observatory. WIREs Water e1255

Lantuit H, Pollard WH (2008) Fifty years of coastal erosion and retrogressive thaw slump activity on Herschel Island, southern Beaufort Sea, Yukon Territory, Canada. Geomorph 95:84–102

Lantuit H, Overduin PP, Couture N, Wetterich S, Aré F, Atkinson D, Brown J, Cherkashov G, Drozdov D, Forbes DL (2012) The Arctic coastal dynamics database: a new classification scheme and statistics on Arctic permafrost coastlines. Estuaries Coasts 35:383–400

Lewis T, Lafrenière MJ, Lamoureux SF (2012) Hydrochemical and sedimentary responses of paired High Arctic watersheds to unusual climate and permafrost disturbance, Cape Bounty, Melville Island, Canada. Hydrol Proc 26:2003–2018

Lewkowicz AG (2007) Dynamics of active-layer detachment failures: Fosheim Peninsula, Ellesmere Island, Nunavut, Canada. Permafr Periglac Proc 18:89–103

Matsuoka N (2001) Solifluction rates, processes and landforms: a global review. Earth-Sci Rev 55:107–134

Normandeau A, Lamoureux SF, Lajeunesse P, Francus P (2016) Sediment dynamics in paired High Arctic lakes revealed from a high-resolution swath bathymetry and acoustic stratigraphy survey. J Geophys Res Earth Surf Proc 121:1676–1696

Roberts KE, Lamoureux SF, Kyser TK, Muir DCG, Lafrenière MJ, Iqaluk D, Pieńkowski AJ, Normandeau A (2017) Climate and permafrost effects on the chemistry and ecosystems of High Arctic Lakes. Sci Rep 7:13292

Rouse WR (1993) Northern climates. In: French HM, Slaymaker O (eds), Canada's cold environments, pp 65–92

Rudy ACA, Lamoureux SF, Treitz P, Van Ewijk K, Bonnaventure PP, Budkewitsch P (2017a) Terrain controls and landscape-scale modelling of permafrost slope disturbance susceptibility. Permafr Periglac Proc 28:79–91

Rudy ACA, Lamoureux SF, Kokelj SV, Smith IR, England JH (2017b) Accelerating thermokarst transforms ice-cored terrain triggering a downstream cascade to the ocean. Geophys Res Lett 44. https://doi.org/10.1002/2017gl074912

Smith S (2011) Trends in permafrost conditions and ecology in northern Canada. Canadian Biodiversity: Ecosystem Status and Trends 2010. Technical Thematic Report No. 9. Canadian Council of Resource Ministers. Ottawa, ON

Smith SL, Burgess MM, Riseborough D (2008) Ground temperature and thaw settlement in frozen peatlands along the Norman Wells pipeline corridor, NWT Canada: 22 years of monitoring. In: Kane DL, Hinkel KM (eds) Proceedings of the 9th international conference on permafrost. Fairbanks, AK, 29 June to 3 July, 2008, Institute of Northern Engineering, University of Alaska Fairbanks, vol 2, pp 1665–1670

St-Hilaire-Gravel D, Forbes DL, Bell T (2012) Multitemporal analysis of a gravel-dominated coastline in the central Canadian Arctic Archipelago. J Coastal Res 28:421–441

Taylor AE, Wang KL, Smith SL, Burgess MM, Judge AS (2006) Canadian Arctic permafrost observatories: detecting contemporary climate change through inversion of subsurface temperature time series. J Geophys Res-Solid Earth 111:2156–2202

Vincent LA, Zhang X, Brown RD, Feng Y, Mekis E, Milewska EJ, Wan H, Wang XL (2015) Observed trends in Canada's climate and influence of low-frequency variability modes. J Clim 28:4545–4560

Walker DA, Raynolds MK, Daniëls FJA, Einarsson E, Elvebakk A, Gould WA, Katenin AE, Kholod SS, Markon CJ, Melnikov ES, Moskalenko NG, Talbot SS, Yurtsev BA (2005) The circumpolar Arctic vegetation map. J Veg Sci 16:267–282

Woo M (2012) Permafrost hydrology. Springer International Publishing, Berlin

Scott F. Lamoureux is a Professor and the Robert Gilbert Chair in the Department of Geography and Planning at Queen's University. He completed graduate work at the University of Massachusetts Amherst and University of Alberta, and was an NSERC postdoctoral fellow at Queen's University. His primary research interests are focused on climate-water-landscape interactions in permafrost regions. During the past 30 years, he has carried out field research across the North American and European Arctic, including extensive work in the Canadian High Arctic. As part of his research, he initiated the Cape Bounty Arctic Watershed Observatory (CBAWO) in 2003, a long-term experimental watershed site in the High Arctic. His recent work has focused on Arctic watershed processes and the response of these systems to climate variability and permafrost change. He has also carried out extensive work with assessing landscape stability and processes related to permafrost degradation and mass movements.

Ashley C. A. Rudy received her Ph.D. at Queen's University in the Department of Geography and Planning. During this time, she completed research at multiple locations across the Canadian Arctic Archipelago focusing on permafrost disturbance susceptibility, subsidence potential, and thermokarst water quality impacts. She is currently a Geotechnical Data Scientist with the Northwest Territories Geological Survey (Yellowknife). Her research interests continue to focus on permafrost landscape sensitivity, particularly within key northern transportation corridors. Ashley is actively involved in a number of groups/associations, including being a part of an international outreach project (www.frozengroundcartoon.com) that aims to present permafrost research to the public using comic strips, as well serving as the Communications Director for the Canadian Permafrost Association.

The Avalon of Newfoundland: Geomorphology, People and Landscape

15

Norm Catto

Abstract

The geomorphology of the Avalon Peninsula in easternmost Newfoundland is dominated by Ediacaran geology and lithology, local glaciation, and coastal processes developed under a boreal climate. The physiography, particularly the configuration of headlands and embayments, is closely related to the region's tectonic history and structural geology. Repeated glaciation has resulted in landscapes dominated by bedrock erosional features. During MIS 2, numerous local ice caps developed during three phases, and expanded to cover the peninsula. To the west, ice from the main part of Newfoundland flowed into Placentia and Trinity Bays. Deglaciation was triggered by relative sea level rise in St. Mary's Bay c. 14 ka, and was complete by 10 ka. At present, relative sea level is rising c. 3.5 mm/y, primarily as a result of residual glaci-isostatic subsidence. Most shorelines along the Avalon are distinguished by deep indentations open to either the southwest or northeast, aligned parallel to prevailing storm tracks, and steeply sloping nearshore bathymetry. The frequent storm activity, long fetch, and ambient moderate to high energy, in combination with the coastal configurations, produces shorelines dominated by swash activity, shore-normal transport and reflective conditions. A succession of residents has interacted with the geomorphology and landscape since the early Holocene.

Keywords

Avalon • Newfoundland • Ediacaran • MIS 2 • Independent ice caps • Rögens • Boreal • Sea level change • Coastal dynamics

N. Catto (✉)
Department of Geography Memorial, University of Newfoundland, St. John's, Canada
e-mail: ncatto@mun.ca

15.1 Introduction

The main island of Newfoundland, along with 17 larger (≥ 30 km^2) and more than 7,200 smaller islands, has more than 15,000 km of coastline. The easternmost part, the Avalon Peninsula, is connected to the main part of Newfoundland by the Isthmus of Avalon, with a minimum width of 4 km at its northern end (Fig. 15.1). The peninsula and isthmus include approximately 10,360 km^2, and no location on the Avalon is more than 30 km from tidewater.

Knowing the coastline is essential for any understanding of the landscape and culture of eastern Newfoundland. For most of the plant and animal species, and for all the human occupants, the coastal environment has influenced life since deglaciation.

For the succession of indigenous inhabitants, including the Maritime Archaic ($\sim$8–4 ka), Sivullirmiut (formerly Dorset; $\sim$3–1 ka), Beothuk, and Qalipu, the spatial and temporal fluctuations of marine mammals, birds, marine fish, shellfish, salmon and berries were crucial for nutrition (Pastore 1990, 1992). Marine and littoral resources were critical for the indigenous inhabitants, and failure to access or locate either the seal (*Phoca groenlandica, P. vitulina*) or caribou (*Rangifer caribou*) populations would place severe nutritional stress on communities. The Beothuk population, estimated at less than 1000 people c. 1500 (Marshall 1996, 2001–2009), retreated inland from Notre Dame and Bonavista Bays when European fishing communities became established. The Beothuk thus had no access to marine and littoral resources, producing an over-reliance on a small caribou population in a resource-poor area of steadily decreasing size (Pastore 1990, 1992). The combined effects of malnutrition and tuberculosis reduced the Beothuk population to extinction c. 1829 (Marshall 1996, 2001–2009).

Qalipu (pronounced 'hal-la-bu'; the Mi'kmaq word for 'caribou') people have lived along the south coast of Newfoundland for several hundred years, but particularly after 1763 following cession of Acadia to Great Britain. Qalipu coastal communities were reliant on marine and littoral

© Springer Nature Switzerland AG 2020
O. Slaymaker and N. Catto (eds.), *Landscapes and Landforms of Eastern Canada*, World Geomorphological Landscapes, https://doi.org/10.1007/978-3-030-35137-3_15

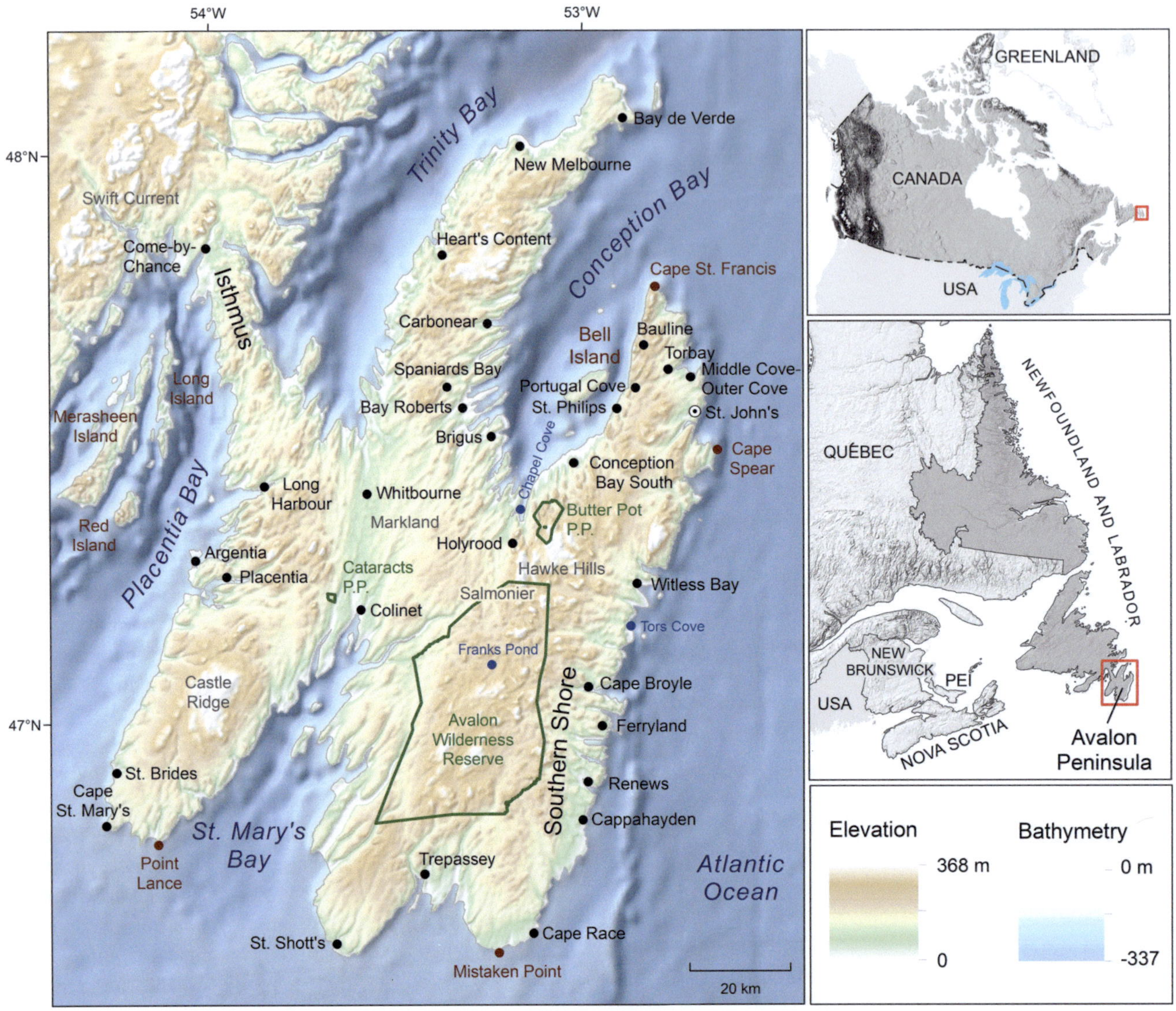

Fig. 15.1 Avalon Peninsula and Isthmus, showing locations mentioned in the text. *Source* backgroup map—Elevation: GMTED2012; Danielson and Gesch (2011). Bathymetry—The GEBCO_2014 Grid, version 20150318, http://www.gebco.net. *Cartography* D. Kovanen

resources in addition to caribou. At present, more than 23,000 people are recognized as Qalipu, and as many as 100,000 people have claimed Qalipu ancestry.

For Newfoundlanders of European descent, harbours have represented economic and cultural centres since the first Basque (prior to 1450), English (1497), and French (1534) voyages, and the first attempts at colonization at coastal localities of Cupids (1610), Renews (1617), and Ferryland (1621), the first permanent settlement (Rowe 1980; Handcock 1989). Most Newfoundlanders trace their ancestry to fishermen who arrived between 1650 and 1780 from the 'west country' of England (Devonshire, Dorsetshire); southeastern Ireland (Waterford, Wexford); or the Channel Islands of Jersey and Guernsey.

Initially, the British government did not support permanent settlement of the coast, wishing to reserve it for a seasonal cod (*Gadus morua*) fishery and the training of naval seamen ('Soe long as there comes no women'; Handcock 1989). In practice, this was an unenforceable regulation, as the numerous embayments and archipelagos afforded many opportunities for those who chose to stay uninvited throughout the year. The 'liveyears' (those who 'live here') gradually began to settle in small 'outports', in sheltered embayments away from the open coastline.

The deeply embayed coasts provided safe harbours for fishing vessels, with suitable onshore areas for drying and processing cod and for construction of the necessary stages, stores, and other buildings. Local fishing grounds supported inshore fisheries for cod, flatfish, herring, and crab, and capelin (*Mallotus villosus*) spawned on gravel beaches.

Although the heritage of the fishery remains central to the culture of Newfoundland, its economic importance has diminished in recent years. The collapse of the northern cod stocks led to the imposition of a moratorium on cod fishing in 1992. The shift to harvesting of previously under-utilized species, such as shrimp and crab, coupled with growing interest in aquaculture, has led to a new evaluation of the significance of the coastal zone and its geomorphology. Newfoundland has increasingly turned to tourism to bolster its economy and stem out-migration. A further factor is the increasing utilization of coastal sites for residential and commercial construction by people not engaged in the fishery, leading to local concerns that the coastline will become inaccessible to other residents and that a valuable link with Newfoundland's heritage will be severed.

Anthropogenic influences on geomorphology are most evident on the northeast Avalon and in the smaller urban centres, but are apparent in all other communities. The potential for petroleum contamination due to accident or offshore discharge is significant (Griffiths 1999; Pink 2004; Etheridge 2005; McNeil 2009; Catto 2011; Duncan 2015). Assessment, utilization and management of the coastal zone have been hampered by a lack of understanding of the geomorphic environment, and its response to the variations in climatic parameters ubiquitous to coastal Newfoundland.

15.2 Bedrock Influences

The configuration of the major embayments flanking the Avalon is a product of structural geology (Fig. 15.2). Lithologic units include Ediacaran clastic and volcaniclastic Harbour Main (oldest), Conception, St. John's, and Signal Hill Groups; granite, granodiorite, diorite and gabbro Ediacaran Holyrood Plutonic Suite intrusions into the Harbour Main Group rocks; Cambrian strata (Random orthoquartzite, clastic and carbonate Adeyton and Harcourt Groups); Ordovician iron-bearing strata (Bell Island and Wabana Groups); and Devonian granitic intrusions on Red Island and Ragged Islands (King 1988, 1990; Colman-Sadd et al. 1990).

The physiography is closely related to the region's tectonic history and structural geology. Ediacaran units were deformed during the Avalonian Orogeny, an early phase influencing the Appalachian orogenic belt, into a series of northeast-southwest trending folds, with faulting. The major faults do not parallel the fold axes, but are aligned NNW–SSE to NNE–SSW. The heights of land follow the axes of the geological structure, with maximum elevations in excess of 330 m asl. Prominent landmarks are underlain by resistant units, such as Signal Hill and Cape Spear (red conglomerate) and Butter Pot and the Hawke Hills (Holyrood granite). Most of the Avalon is below 150 m asl. Drainage systems have short, steep gradient streams with small catchment areas, and storm rainfall rapidly leads to local flooding.

Locally, the alternating ridge and valley pattern is modified due to additional geologic deformation and glacial sedimentation. These areas are marked by rolling uplands, interspersed with small plateaux. Ridges are cut along structural weaknesses in the bedrock, with narrow valleys across the trend of the topography. The Narrows of St. John's Harbour is an example, with adjacent parts of the bisected ridge represented by Signal Hill (elevation 180 m asl) and Fort Amherst (Fig. 15.3).

The axes of the four major embayments follow structural trends, and also parallel the prevailing northeast and southwest storm winds. Structural geology dominates over glacial overdeepening in St. Mary's Bay (>210 m), Conception Bay (>280 m), Placentia Bay (>300 m), and Trinity Bay (>580 m). Submerged moraines are present at the mouths of Trinity and Conception Bays, and numerous glacial landforms have been documented in Placentia Bay (Brushett et al. 2007). St. Mary's Bank, with a maximum depth of 65 m, extends across the mouth of the St. Mary's Bay from St. Shotts to Point Lance. During the Last Glacial Maximum, the lowest point of St. Mary's Bank was above sea level, allowing the development of the terrestrial St. Mary's Bay Ice Centre.

In smaller embayments and coves, the coastal geomorphology of the Avalon Peninsula is largely controlled by differential resistance to erosion. Most prominent cliffs are supported by resistant conglomerates and coarse arkosic sandstones. In contrast, harbours are generally associated with interbedded greywackes and shales, locally accentuated by faulting (Fig. 15.4).

The structural influence is evident in contrasting the shorelines of Conception and Placentia Bays (Fig. 15.5). Along Conception Bay north of Holyrood, embayments trend northeast-southwest, developed along the major faults and fold axes associated with the Avalonian Orogeny. North of Brigus, northeast-southwest trending faults superimposed on SSW-plunging anticlines and synclines result in the development of deep northeast-southwest trending embayments and the excellent harbours of Bay de Grave, Bay Roberts, Spaniards Bay and Harbour Grace (Fig. 15.1). These shorelines are flanked by alternating areas of rock cliffs and low Quaternary coastal bluffs, the result of differential resistance to erosion of bedrock units aligned oblique to the configuration of the embayments.

In contrast, along the Cape Shore of Placentia Bay south from Placentia to Cape St. Mary's, the rolling uplands are aligned northeast-southwest. The alignment of bedrock units

Fig. 15.2 Simplified bedrock geology of the Avalon Peninsula showing principal structures and stratigraphic divisions. Modified from King (1988). Also see Colman-Sadd et al. (1990) and Myrow (1995). *Cartography* D. Kovanen

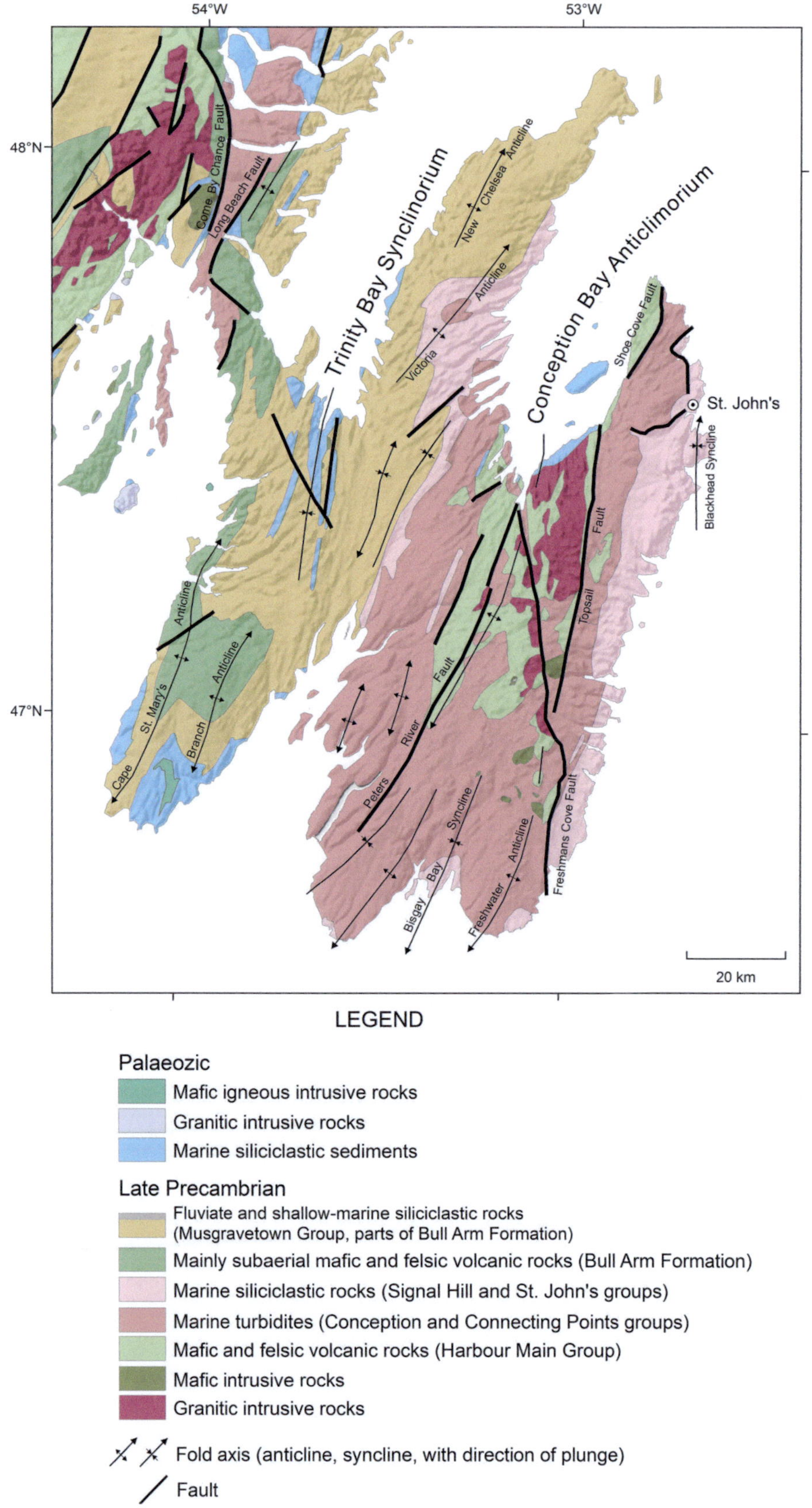

Fig. 15.3 St. John's Harbour and the layout of the downtown are strongly influenced by bedrock geomorphology. **a** Overview of St. John's Harbour from Signal Hill, looking Southwest. **b** Resistant conglomerate of the Signal Hill Group supports Signal Hill and the South Side Hills. The Harbour and downtown are developed over siltstones and argillites of the St. John's Group. Modification by NNE to NE-flowing MIS 2 glaciation from the Witless Bay Barrens was minor. *Photograph* NL Airphotos 01012-108, 109, 110

parallel to the coast has produced a relatively straight shoreline, with cliffs to 65-m high. Embayments do not form deep indentations, and are directly open to strong southwesterly winds, making them unsuitable as harbours. Geomorphology influences human utilization of the Cape Shore, in contrast to the more densely settled coastline of Conception Bay.

15.3 Quaternary Glacial Geomorphology

Newfoundland was glaciated on multiple occasions throughout the Quaternary. On the Avalon, repeated glaciation has resulted in removal of all pre-MIS 2

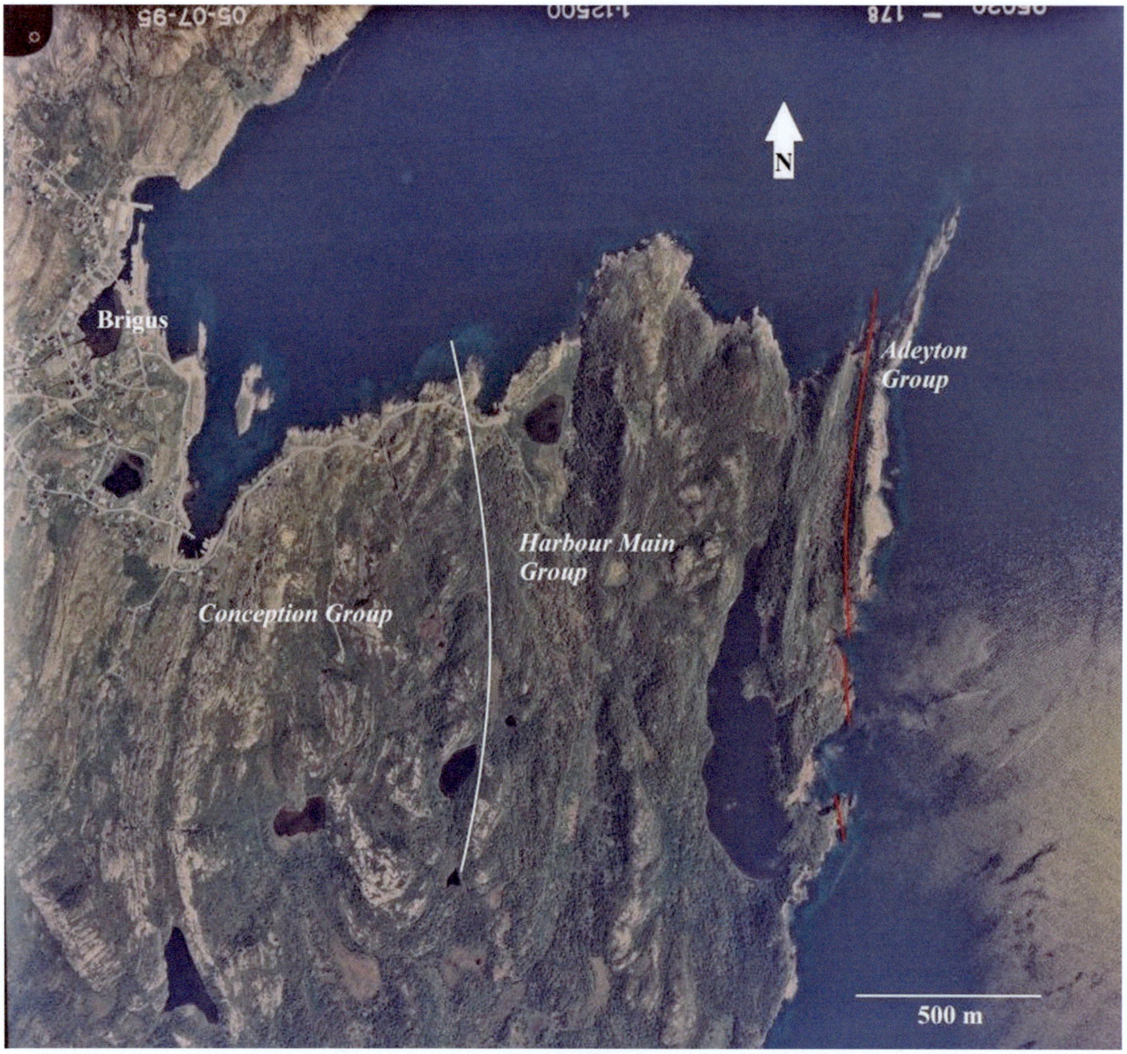

Fig. 15.4 Structural geology is evident as the dominant landscape control at Brigus, Conception Bay. Siltstones, argillites, and sandstones of the Conception Group are flanked along a fault-bounded contact (Brigus Fault, white line) by basic volcaniclastic units of the Harbour Main Group. The resistant Cambrian orthoquartzite and sandstone of the Adeyton Group overlie the Harbour Main Group along an erosional surface (red line). Directions of glacial flow during MIS 2 varied from Northeast to Southeast. *Photograph* NL Airphoto 95020-178

Quaternary sediment, as well as local erosion of the bedrock (Fig. 15.6). Accumulations of glacigenic Quaternary sediment more than 3 m thick are rare, and are confined to low-lying interior regions and sites in the glacial lees of prominent bedrock features.

The independent nature of Avalon glaciation had been recognized by early Quaternary researchers (Chamberlin 1895; Coleman 1926; Summers 1949), based on geomorphic ice flow indicators, striations, and the absence of the distinctive Palaeozoic granite clasts derived from the northwest. During MIS 2, multiple small ice caps developed on the Avalon Peninsula, with sequential shifts in formation and ice flow direction (Henderson 1972; Catto 1998).

In the westernmost Avalon, ice from the main part of Newfoundland flowed southeastward into northern Placentia Bay, carrying Palaeozoic granitic clasts from the Terra Nova and Swift Current areas. Two distinct ice streams developed, with one flowing southward into Placentia Bay and across Merasheen, Long, and Red Islands, producing glacigenic features including DeGeer moraines (Brushett et al. 2007). The other ice stream flowed across the western Isthmus of Avalon into Trinity Bay, carrying Palaeozoic granitic clasts and silty marine sediment from Placentia Bay (Catto 1998). The easternmost extent of silty matrix diamicton indicates the margin of the ice stream across the Isthmus of Avalon. The Trinity Bay embayment funnelled the ice stream to the northeast, terminating in a grounding line moraine (King and Fader 1986).

Three major phases of MIS 2 glaciation, each involving several distinct ice centres, can be recognized. The initial

Fig. 15.5 Bedrock structure has a pronounced influence on harbour configuration, which in turn controlled human settlement. **a** Central Conception Bay's deeply indented coastline, with resistant sandstone and conglomerate ridges flanking friable argillites and siltstones subject to glacial erosion, provided safe harbours for smaller outports and larger communities such as Bay Roberts. *Photograph* NL Airphotos 95017-101, 103. **b** In contrast, the Cape Shore of Placentia Bay south of Placentia is marked by no suitable harbours. Bedrock units parallel the shoreline, and the only indentations develop along small faults and joints at 90° to the bedrock structure. *Photograph* NL Airphoto 95036-100

Phase 1 involved glacial accumulation at several centres along the heights of land, including the Witless Bay Barrens-Hawke Hills, Heart's Content Barrens, Franks Pond, White Hearts Pond (near Markland), and Castle Ridge. Ice flowed seaward, following preexisting structural features, and all parts of the Avalon Peninsula were glaciated.

Phase 2, representing the MIS 2 Last Glacial Maximum, was the dominant geomorphic glacial event on the Avalon. Ice from the St. Mary's Bay centre (Henderson 1972; Catto 1998) flowed across the southern Avalon Peninsula to Placentia Bay, St. Mary's Bay, and the open Atlantic Southern Shore. Additional ice centres persisted over the Heart's Content and Witless Bay Barrens, as well as on the Isthmus.

The concluding Phase 3 of MIS 2 glaciation began c. 15 ka, when sea level reached −65 m, the controlling depth of St. Mary's Bay. The resulting destabilization of the St. Mary's Bay ice centre initiated deglaciation throughout the Avalon. The major ice centres collapsed, punctuated by transitory short-lived centres on heights of land. The effect on the landscape was limited in most instances. Deglaciation of all of the Avalon Peninsula was completed before 10 ka (Macpherson 1995, 1996; Catto 1998).

15.3.1 Glaciated Barrens

In Newfoundland, 'barrens' refers to open landscapes dominated by heathland assemblages, with few or no trees. Many small barrens adjacent to human coastal communities reflect forest clearance between 1650 and 1850, coupled with strong winds which precluded reestablishment of trees (Fig. 15.7). Grazing by sheep and human foot traffic reduced vegetation cover and facilitated frost action, producing thufur and cliff-edge failures through frost wedging. Active and relict periglacial features are only present in areas that show evident disturbance of heathland vegetation, or where European introduced vegetation species are not present.

Glaciated barrens are also present in interior sites within 35 km of the shoreline, subject to prevailing coastal winds. Typical heathland vegetation assemblages include crowberry (*Empetrum nigrum*), partridgeberry (*Vaccinium vitis-idaea*), Saxifrage (*Saxifraga*), bearberry (*Arctostaphylos uva-ursi*), dwarf bilberry (*Vaccinium boreale*), bilberries (*Vaccinium uliginosum*), blueberries (*Vaccinium angustifolium*), partridgeberries (*Vaccinium vitis-idaea*), bakeapple (*Rubus chamemorus*), Labrador Tea (*Ledum groenlandicum*), and other non-arboreal plants.

Landscapes are characterized by exposed bedrock, large erratics (to 3 m^3), discontinuous veneers of fibric peat, patches of porous, highly permeable diamicton in bedrock hollows and lee-side accumulations, and flanking deposits of diamicton successions. Interior barrens areas receive greater snowfall than do coastal regions, and were centres of glacial accumulation during MIS 2. The Witless Bay Barrens, extending from Butter Pot Provincial Park west to the Hawke Hills, is a typical barrens landscape (Fig. 15.8).

Equant granite boulders, with maximum dimensions of 6 m, are scattered across the Witless Bay Barrens (Fig. 15.9). No preferential orientation or pattern is evident in the boulder distribution. The boulders characteristically

Fig. 15.6 **a** Glaciated St. John's Group bedrock exposure at Timber Cove Island, Brigus South, 5 km ENE of Cape Broyle. Glacial flow was eastward, towards the Atlantic Ocean. **b** Basal meltwater grooves on Timber Cove Island have maximum depths in excess of 2 m and show cubic parabolic cross-sections

lack any faceting, striations, or evidence of preferential erosion. A few boulders show orthogonal fracturing or jointing patterns, but most have irregular, non-planar jointing surfaces. Most are perched above thin veneers of diamicton. Where boulders lie directly on bedrock, the bedrock lacks striations or evident faceting. The underlying diamictons contain both granitic and metasedimentary bedrock clasts. The boulders are interpreted as the products of supraglacial transport over short distances from the adjacent Hawke Hills (300 m asl). Deposition resulted from glacial ablation.

In the Witless Bay Barrens area, the distribution of glacial sediments indicates that the summits of the Hawke Hills were deglaciated before the margins (Macpherson 1996). The ice melted from the top down, rather than by frontal retreat. In Conception Bay South, glacial retreat during the lattermost part of the glaciation was followed by a short, minor northward re-advance during the Younger Dryas, approximately 12,000 years ago. Poorly developed periglacial features (ice wedge casts and gravel involutions) are present in coastal glacifluvial gravel exposures in Conception Bay South. Deglaciation was complete c. 10 ka.

After deglaciation, the geomorphology of the Witless Bay Barrens resembled that of today, although there was no vegetation. The climate warmed progressively from 10 to 6 ka, allowing spruce, fir, birch and pine to successively colonize the area (Macpherson 1995). Iron-rich podzolic soils (pH 4–5) developed under conifer forests, grey-green

Fig. 15.7 Coastal barrens east of Bay Roberts, Conception Bay. The landscape reflects the combined effects of Conception Group sandstone, the Avalonian Orogeny, MIS 2 glaciation, 18th century vegetation clearance, frost action, and modern heathland vegetation

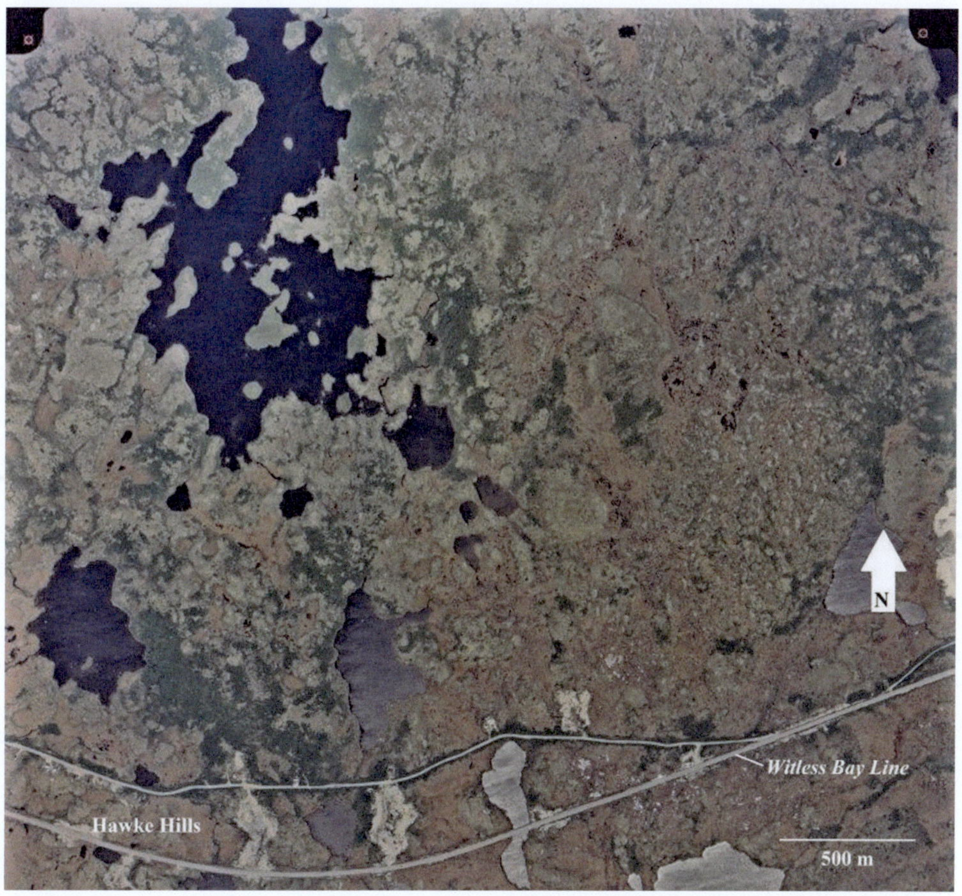

Fig. 15.8 The Witless Bay Barrens extend from the Hawke Hills northward to Butter Pot Provincial Park. Small fens, aapa, and thin veneers of sedge peat and weakly-developed gyttja are present in depressions. Elevated areas are exposed granitic bedrock of the Holyrood Intrusive suite, with scattered glacial erratics. Sediment thicknesses do not exceed 30 cm, and pond depths are typically less than 1 m. The Witless Bay Barrens glacial centre was active during several phases of MIS 2. In winter, the Witless Bay Line (Highway 13) is frequently closed by snowdrifts. *Photograph* NL Airphoto 95024-196

Fig. 15.9 Granite boulder, Witless Bay Barrens. The boulder originated from the Hawke Hills and was transported less than 5 km before deposition. The boulder sits partially on smaller erratics of similar lithology, and partly directly on bedrock. Flanking veneers of organics detritus from heathland vegetation and sedge peat directly overlie bedrock

gleysols developed in groundwater discharge areas, and thin brunisols developed in dry, less favourable sites.

15.3.2 Diamictons, Striations and Landforms

Most diamictons of the Avalon Peninsula are coarse-textured, containing 40–60% cobbles and pebbles (Fig. 15.10). Rounded, subrounded, striated, planar faceted, bullet-shaped and barrel-shaped clasts are all present. Diamictons are dominated by locally (within 2 km) transported clasts. Although many of the metasedimentary Ediacaran bedrock units have similar lithologies, distinctive igneous units of the Holyrood Intrusive Suite, gabbroic mafic intrusions, Cambrian sedimentary strata and the Ediacaran Gaskiers diamictite exposed along the eastern side of St. Mary's Bay provide distinctive clasts for determination of glacial flow directions (Catto 1998). Estimations of glacial flow depend upon analysis of diamictons, striations, and stoss-and-lee features.

Almost 400 striation orientation determinations have been made throughout the peninsula and isthmus. Typically, multiple striae are measured at each site, including separate measurements for each potential population group indicated by cross-cutting or superposition. Orientations determined represent the average values for each population, eliminating small deviations in direction potentially resulting from micro-morphology or clast–clast contact during striation. Nailhead and wedge striations can be used to unequivocally determine flow direction. Coastal erosion and mass movement has resulted in periodic exposure of striated terrain underlying diamicton deposits. However, exposure of striated bedrock can result in rapid destruction of the striations under the Avalon climate.

Stoss-and-lee features range in size from individual bullet-shaped pebbles, cobbles and boulders, to roches moutonnées, to features hundreds of metres to kilometres in length and tens to hundreds of metres high. The geomorphology of roche moutonnées is largely influenced by the lithology and initial configuration of the bedrock. Well-formed roche moutonnées develop from regularly crystalline or textured bedrock, which allows polishing on the stoss side. The most distinctive forms on the Avalon are developed in metamorphosed Ediacaran siltstone, with striated, grooved and polished stoss surfaces and jagged lee surfaces. Bedrock which is more resistant to abrasion develops roche moutonnées which are somewhat less distinctive, and lack striations on the stoss side. Highly fractured bedrock produces less distinct forms, although stoss and lee can still be recognized. Larger features tend to be more irregular in form, reflecting differences in fracture patterns and lithology. On the Avalon Peninsula and Isthmus, more than 200 roche moutonnées indicate ice flow directions.

Numerous diamicton outcrops identified as primary glacial deposits on the basis of other criteria (exclusive of fabric) has allowed almost 900 detailed macrofabric investigations to be conducted. Locally, the density of macrofabric determinations on sediment units identified as primary glacial deposits exceeds 1 analysed outcrop per km^2. The coarse-textured diamictons are generally not suitable for microfabric analysis.

Fig. 15.10 Typical glacial diamicton, Tors Cove, exposed during building construction. The diamicton has a maximum thickness of 5 m. Podzolic and brunisolic soils are typically developed under coniferous boreal forest assemblages

Comparison of macrofabric patterns, striation data and landform orientations at sites separated by 1 km or less reveal general congruence in the inferred flow directions (Fig. 15.11). However, notable variations in mean striation orientation and primary eigenvector orientation in adjacent fabric determinations occur. At approximately 30% of the sites, the difference between striation orientation and fabric trend orientation exceeded 30° azimuth, making correlation between ice-flow directions problematic. The coarse texture of the glacial deposits could allow substantial reorientation and settling during deposition. Macrofabric plunges are generally oriented in the up-ice direction, but multiple examples of down-ice plunging fabrics make use of plunge direction as a definitive indicator of ice-flow unreliable. Frost reorientation is significant to depths of 3 m below the surface in some diamicton outcrops.

15.3.3 Rögens and Hummocks

Alternating belts of rögen moraines in low-lying areas, separated by highs with hummocky topography or veneered bedrock, mark the interior of the Avalon from the western flank of the Hawke Hills to Whitbourne (Fisher and Shaw 1992; Catto 1998; Marich 2006). The rögens are crescentic ridges aligned with their long axes trending northwest-southeast and with horn orientations suggesting northeasterly flow (Fig. 15.12). They have maximum heights of 12 m, and maximum lengths of 100 m.

Texturally, the rögen moraines are composed primarily of coarse diamicton, varying in matrix composition from sandy silt to fine granules. The proportion of pebbles and larger clasts commonly ranges between 35 and 65%. Throughout the region, the pebbles and cobbles are derived from locally exposed Ediacaran rock units, and from Cambrian rock units to the south around and beneath St. Mary's Bay.

The internal structures, clast fabric analysis and lithological assemblages of pebbles within the rögens reveal that the sediments represent several distinct depositional processes and flow events. Sediment in moraines in the eastern (Salmonier) area initially was deposited by ice-flowing northward from the Franks Pond centre in the Avalon Wilderness area, followed by flow north-northeastward from St. Mary's Bay. In the central area, northward flow from the St. Mary's Bay centre followed an earlier eastward flow from the White Hearts Pond ice centre near Markland. Similar complexity is evident in the westernmost extent of the rögen belt, near Whitbourne.

The variations in orientation of the clast fabrics, and the lithological provenances of the clasts, generally correspond to the pattern of glacial flow deduced from striations and erosional geomorphic bedrock features. Flow directions as inferred from the orientation of rögens commonly do not conform with those determined from adjacent striae, and streamlined bedrock features and isolated crag-and-tails on adjacent ridges locally deviate by 10°–40° azimuth from the rögen orientations (Fig. 15.13). The geometry of the sedimentary units within the rögens indicates that they are erosional forms.

The diamicton exposures indicate that initial glacigenic sedimentation was dominated by primary subglacial deposition. Although many tills cannot be precisely classified

Fig. 15.11 Ice-flow determinations rely on combinations of oriented stoss-and-lee features, striations and other microerosional forms, and distinctive bedrock clasts. **a** Roche moutonnée orientation indicates northeasterly flow from the St. Mary's Bay ice centre towards Brigus, Conception Bay. Striations and other microfeatures indicate ice-flows varying from NNE to ENE. **b** Stranded erratic on stoss slope west of Bay de Verde, northernmost Avalon Peninsula. Flow was towards the northeast

genetically, basal melt-out deposits were the most common till type recognized. The primary tills were deposited by glaciers flowing towards the Conception Bay and Trinity Bay shorelines. Most of the primary till deposits can be associated with the MIS 2 glacial maximum (Phase 2). Local variations in primary basal till macrofabric orientation are common. Although many are related to local topographic obstructions, others demonstrate that glacial flow patterns can be variable over short distances.

The majority of the diamicton deposits have been reworked to some degree. Many deposits have been affected by remobilization in supraglacial or subglacial environments,

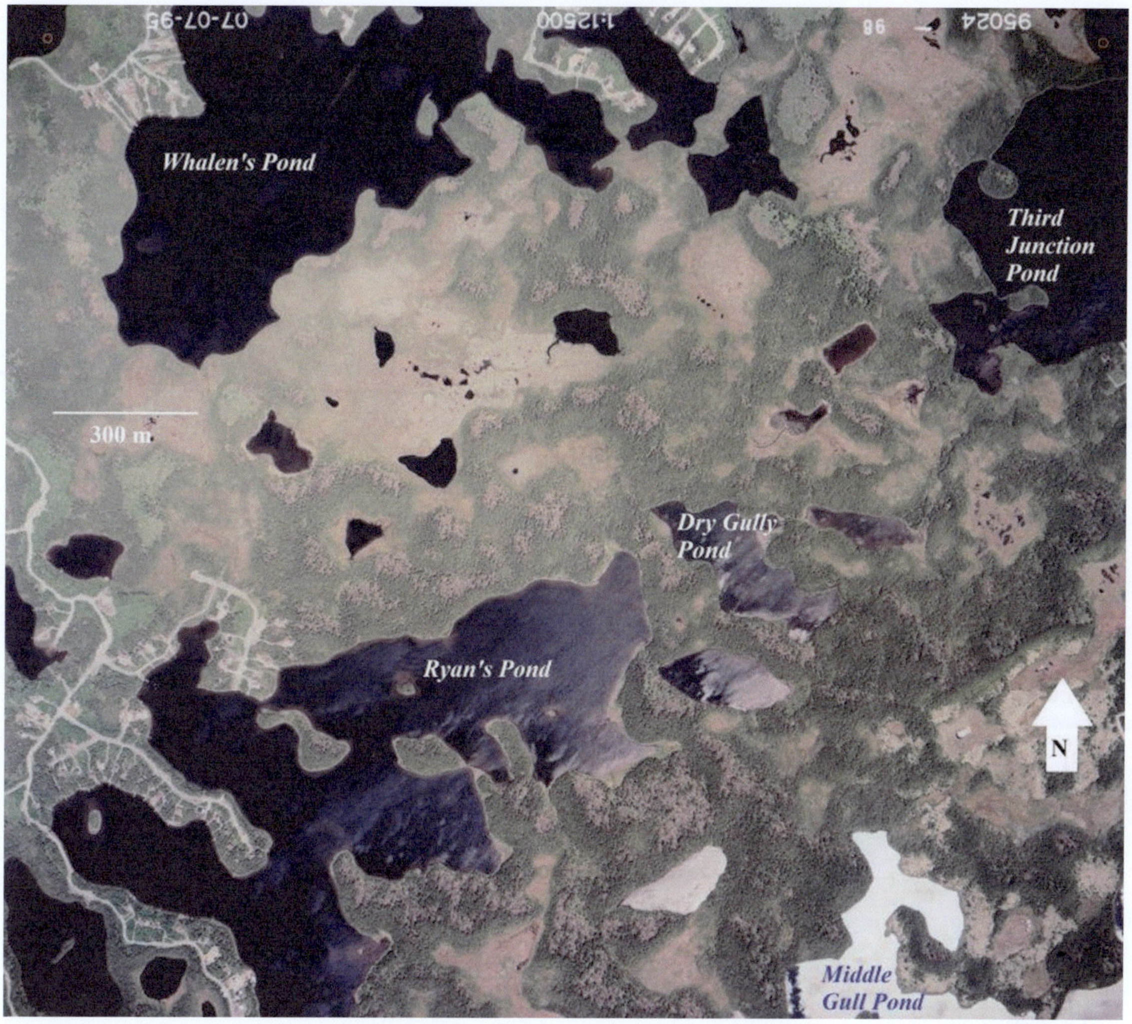

Fig. 15.12 In the central interior (Salmonier area) of the Avalon Peninsula, Rögens occupy lowlands and hummocky terrain marks intervening higher ground. The succession of rögens in the lowlands occupied by Whalen's Pond, Ryan's-Third Junction Pond, and Middle Gull Pond indicate flow towards the Northeast. *Photograph* NL Airphoto 95024-98

or during deglacial and postglacial time. Diamictons influenced by Holocene frost activity contain 10–20% of vertically aligned pebbles.

15.4 Fens and Bogs

Organic soils, primarily fibric sedge peats, cover approximately 15% of the Avalon. The floral emblem of Newfoundland and Labrador is the bog-dwelling carnivorous pitcher plant (*Sarracenia purpurea*). Domed bogs, horizontal to gently undulating fens, and aapa are common organic landforms. Bogs and fens are distinguished by their surface morphology, biological assemblages and hydrology (Fig. 15.14).

Approximately 4.5 ka, the Avalon climate gradually changed from warm, relatively dry conditions to the cooler, wetter conditions prevalent until the 20th century (Macpherson 1995). Precipitation increased by approximately 20%, summer temperatures dropped, the growing season became shorter, and a perhumid moisture regime developed inland. Consequently, wetland development and peat accumulation was favoured (Davis 1985; Macpherson 1995), while decreased forest fire frequency hindered the maintenance of pine forests, favouring spruce and fir.

Horizontal to gently undulating fens are common. Fens are dominated by sedges and woody vegetation, and are underlain by fibric sedge peat and gyttja. The surfaces of the fens are depressed below the surrounding terrain, receive water from both overland and groundwater flow, and are

Fig. 15.13 Cross-section through rögen, Father Duffy's Well, Salmonier area

minerotrophic. Fens are generally mesotrophic or slightly eutrophic, with abundant quantities of nutrients, although their waters are acidic. The thickest fen sediments represent local paludal environments initiated by eutrophication of formerly open basins, and exceed 2 m in thickness.

Aapa fens generally occupy larger areas than slope fens. Aapa are marked by alternating ridges of peat 1–5 m wide, standing 10–40 cm above the general level of the fen, separated by depressed flark areas of standing water. The overall gradients of the aapa are very low, generally less than 1°, but the surface of each successive flark is <10 cm below its upslope neighbour.

Domed bogs are acidic and oligotrophic wetlands without standing water, and are dominated by *Sphagnum* moss vegetation, with less than 25% tree cover. The surface is typically gently convex, with the central part of the bog raised above the edges. Such bogs derive their water supply exclusively from precipitation, and are ombrotrophic. Domed bogs in the Avalon are small, with surface areas generally less than 0.1 km^2.

Fibric peat has accumulated in the domed bogs. Peat development on the Avalon began shortly following deglaciation, and rates of development have accelerated within the past 3,000 years. Most peat accumulations are thin (<1 m) fibric units. The fibric nature of the peat deposits and their limited extent precludes commercial harvesting for use as fuel or for soil fertilization.

15.5 Cascades

Avalon rivers are dominantly cascades, with straight short reaches, steep gradients, waterfalls, high velocity turbulent flow, and generally lacking depositional bar forms (Fig. 15.15). Vortices and whirlpools commonly develop as turbulence and cavitation impact irregular bedrock surfaces.

Ponds and pools within the systems act to store sand and gravel. The pools include both well-defined depressions in the channels, and wider, shallow areas without a defined depression. Infilling of a channel depression will convert a well-defined pool into a shallow; subsequent sweeping of sediment during a flood event can allow a depression to be exhumed and a well-defined pool to be established. Main pool areas are separated by short reaches where gravel moves along reaches. Smaller, temporary pools develop within the straight reaches, as a result of partial obstruction of the channels by accumulations of gravel and sand, brush and litter.

Under natural circumstances, the pools are periodically flushed of sediment by spring flooding and ice jam breakup,

Fig. 15.14 Wetlands occupy 15–20% of the Avalon. **a** Ombrotrophic, oligotrophic bogs support assemblages dominated by bakeapple (*Rubus chamemorus*) and Pitcher Plant (*Sarracenia purpurea*). **b** Extensive fens and bogs are developed throughout the southeastern Avalon between Trepassey and Cape Race

resulting in gradual migration of both fine sediment (silt and medium-fine sand) and coarser sediment (gravel, ranging from granules to fine cobbles) downstream. The largest clasts transported during floods are medium to large cobbles, in addition to litter, brush and debris. Between flood events, these streams carry little or no bedload and almost no suspended load following the spring freshet season (generally mid-April to mid-May).

Floods may occur at any time during the year following heavy rainfalls (Hickman 2006). Rapid runoff from steep slopes and human-modified surfaces results in flooding almost immediately after any storm with more than 40 mm

Fig. 15.15 **a** Cascade reach, Waterford River, Bowring Park, St. John's. **b** Cataracts Provincial Scenic Attraction is a 16 m-high waterfall developed in Ediacaran sandstone, conglomerate, and shale. The site contains 35 bryophyte species

Fig. 15.16 Flooding of the Waterford River directly downstream of Bowing Park during Extratropical Transition Igor. Compare with Fig. 15.15a

of rainfall within 24 h, in the absence of flood-control infrastructure (Fig. 15.16).

'Igor', a hurricane undergoing extratropical transition, impacted much of eastern Newfoundland in September 2010 (Masson 2014). The tropical storm interacted with a low-pressure stationary front previously developed north of the track of the storm, which provided more energy to 'Igor' and allowed it to strengthen as it moved northward. Estimates of damage to human property in eastern Newfoundland exceed $200 million, and one human life was lost. River flooding resulted from rainfall in excess of 100 mm/24 h in numerous locations.

Seasonal transport of river ice affects many Avalon fluvial systems, although the steep gradients preclude significant ice jams. Larger bar surfaces display two forms of transitory depressions attributed to the actions of transported blocks of river ice during spring freshet. Irregular or circular wallows up to 0.5 m deep and 3 m in diameter are formed by tabular blocks of ice that became grounded and were covered with sediment. When the ice melts, depressions are formed. Crescentic depressions, with horns pointing in the downstream direction, are up to 3 m wide, 3 m long and 0.8 m deep. They form where tabular blocks of ice ground on the bar surface, forming obstacles and causing flow divergence, scouring and forming the depressions.

15.6 Coastal Geomorphology

15.6.1 Sea Level History

During the Last Glacial Maximum c. 20 ka, sea level fell in excess of 110 m, exposing the Grand Banks and St. Mary's Bay (Piper et al. 1990; Miller 1999). The largest factor in both previous and projected future sea level change is the interaction between the changing volume of marine water and glaci-isostatic activity.

Following deglaciation, marine inundation affected the South Coast of Newfoundland. Sea level varied from slightly above present elevation near St. Brides to 20 m asl at Swift Current. Around the shoreline of Conception and Trinity Bays, higher sea level formed erosional benches and depositional terraces at elevations between 5 and 20 m above present sea level, with the northwestern shore undergoing the greatest inundation and the southern tips the least (Brückner 1969; Henderson 1972; Catto 1994a, 1998, 2011; Catto et al. 2003).

Along the open Atlantic coastline south of Cape St. Francis, raised marine features have not been recognized. In St. John's Harbour, a freshwater lake existed shortly after deglaciation, c. 12 ka (Lewis et al. 1987), indicating that sea level at this time was at least 14 m below present, the elevation of the controlling sill in The Narrows (prior to blasting to improve access for cruise ships). Marine transgression is recognized by a transition from a brackish (*Centropyxis aculeata*) to a marine foraminiferal assemblage, c. 11 ka. No raised marine deposits have been encountered in excavations in downtown St. John's, although marine deposits at elevations to 8 m above sea level are present along the southern shore of Conception Bay at Portugal Cove-St. Philips and Conception Bay South (Brückner 1969; Catto and St. Croix 1998).

Subsequent glaci-isostatic rebound caused relative sea level to decline. Along the Avalon coast, the fall in relative sea level continued until the coastal areas had rebounded in excess of their original pre-glacial elevation, resulting in glaci-isostatic overcompensation, and relative sea levels lower than present.

In Trinity Bay and Conception Bay, sea level fell to between 10 and 25 m below present during the early Holocene (Grant 1989; Shaw and Forbes 1990, 1995; Liverman 1994; Shaw et al. 2002). Submerged deltas and wave-cut terraces in Placentia Bay indicate lower sea levels offshore of Long Harbour (−18.9 m), and Argentia (−19.6 m). ^{14}C dated terrestrial peat deposits at exposed locations along the Cape Shore and St. Mary's Bay, currently subject to coastal erosion, high winds and salt spray, indicate that sea levels were at or below the present level throughout the mid-Holocene (Catto 1994a).

Approximately 8 ka, subsidence resulted in renewed sea-level rise (Fig. 15.17). Along the Avalon Peninsula, rising relative sea level since the mid-Holocene resulted in the inundation of terrestrial peat deposits and tree stumps at several locations (Catto 1994a, 1999, 2006a, b, 2011). At present, the rate of sea level rise, based on tide gauge data (from St. John's), ^{14}C dates from several Avalon Peninsula locations, and archaeological data, is estimated at 3.0–3.5 mm/y (Catto et al. 2003; Catto 2006a, 2011; Watton 2016).

Although steeply sloping onshore topography limits the potential impact of sea level rise in most Avalon communities, local concerns exist (Shaw et al. 1999; Vasseur and Catto 2008; Catto 2011). Much of Placentia town now lies very near sea level, and the foundations of all structures are below the marine limit. Salt water infiltration is an ongoing problem, as is flooding. Deposition has continued on the landward side of the barrier into the lagoon, following the pattern indicated by the 17 relict beach ridges on the landward side of the gravel flat (Forbes 1985; Shaw and Forbes 1987). Flooding during storm events is usually initiated by waters travelling through the channel under the lift bridge, and then washing over the community from the lagoon. To cope with the problem, coastal infrastructure has been constructed, and contingency and evacuation plans developed.

15.6.2 Coastal Regime

Most shorelines along the Avalon are distinguished by deep indentations open to either the southwest or northeast, aligned parallel to prevailing storm tracks, and steeply sloping nearshore bathymetry. The frequent storm activity in all seasons, long fetch and the ambient moderate to high energy, in combination with the coastal configurations, produces shorelines dominated by swash activity, shore-normal transport and reflective conditions (Fig. 15.18).

Fig. 15.17 During the initial English settlement of Ferryland in 1621, sea level was approximately 1.5 m below the floor level of the latrine, indicating a rise of 3 mm/y since (Watton 2016). *Photograph* G. Catto

Wave energy is controlled by the fetch, the expanse of open water unobstructed by land (or islands) across which winds blow. Prolonged periods of wind activity are necessary to overcome the frictional losses of energy between atmosphere and ocean, and to overcome the inertia of the water, in order to set the waves in motion. For gale-force winds at 20 m/s (40 knots) to produce waves of 25 m height and 30 m/s velocity, they must operate constantly for 70 h over a minimum fetch of 2500 km.

Rogue waves, distinguished both by their height and the relatively unexpected nature of their occurrence, are more prevalent. Heights in excess of 30 m have been recorded offshore of the Avalon. A rogue is a distinctive large wave, not part of a train, and is commonly not associated with storm activity at the point of impact. Beach users may be surprised by the arrival of a single rogue wave, without any weather cues. Rogue waves appear to originate as individual smaller waves or marine swell are focused and combined into larger features.

The geomorphic signature of rogue wave activity can be notable in confined coves backed by Quaternary bluffs. Bluffs are undercut by compressed air impact below the wave surface, resulting in removal of beach face sediment and local bank collapses. As rogue waves are infrequent, undercuts are gradually infilled by debris flows and frost-induced failures from overhanging Quaternary diamicton or gravel.

For wave-dominated shorelines, the angle of attack of the waves, and the resultant direction of sediment movement, can be specified as either shore-parallel, shore-normal (at approximately 90° to the shore), or shore-oblique. The deeply indented Avalon coast is marked by coves with wave action laterally confined by bedrock, focusing energy at approximately 90° to the beach and generating shore-normal transport. Swash-backwash motion dominates, producing features which extend seaward (cuspate spits, tombolos) or indicate that waves move sediment normal to the shore (self-organized cusps and overwash fans).

Directly offshore of most Avalon beaches, steeply sloping bathymetry is common. Most waves reach the vicinity of the shore before breaking, producing relatively high-energy conditions along the beach, strong seaward return flows (including rip currents), and reflective energy regimes. In Atlantic Canada, reflective coastlines are generally associated with deep, indented embayments and ambient moderate to high-energy conditions.

Bedrock-flanked coves commonly contain bayhead bars or pocket beaches at the landward margin, without lagoons. Sediments accumulate on the beach through wave runup and overtopping, unless the beach has been disturbed by human activity. Mid-bay and baymouth bars develop where offshore bathymetry is relatively shallow, allowing accumulation of sediment seaward of a shallow lagoon (Forbes 1984; Boger 1994; Catto 1994a, 2011; Pittman 2004). Locally, these bars are termed barachoix, indicating that the shore is inaccessible to vessels. Barachoix influenced by wave overtopping accumulate sediment to form steep ridges, and are not subject to landward migration. In contrast, where overwashing

Fig. 15.18 Moderate-to-high ambient energy, shore-normal transport, strong swash action, and reflective conditions characterize most beaches around the Avalon. **a** A relatively calm day at Bear Cove, Cappaheden, Southern Shore. Kelp fragments from directly offshore are detached during storm events and stacked to form self-organized cusps. The succession consists of interbedded detritus kelp wrack and coarse pebble-cobble gravel. Kelp is locally utilized as an organic garden supplement. **b** Overwash of this mid-bay barachoix at Clements Cove, north of Carbonear, Conception Bay, was facilitated by lowering and flattening of the crest for road construction. **c** Big Barachoix, Placentia Bay, was developed on a small recessional moraine

dominates, the barachoix crest is flattened and the barrier is subject to landward migration or breaching. Overwashing is associated with combinations of increased storm activity, sea level rise, interruption of sediment flux, or human flattening of the crest for automobile or all-terrain vehicle traffic. All of these processes are actively occurring on the Avalon's barachoix.

Tidal ranges along the Avalon vary from less than 1 to 2.2 m. Tidal geomorphic features are rare, confined to restricted mesotidal areas at the heads of shallow embayments, such as at Renews and Come-by-Chance (Fig. 15.19). Tidal flats are boulder-dominated, with relict clasts transported by glaciation that cannot be moved by tidal or wave activity. Meandering tidal channels, small washover fans and bank collapse features are present. Seaward, the Come-by-Chance tidal flat is confined by a barachoix. As sea level rises, breaching and overwash has led to an increase in storm surge penetration, evident in the uppermost sediments of the tidal flat succession. At Witless Bay, a small tidal flat existed prior to disruption of tidal influx due to road construction.

Newfoundland's prolonged winter also influences coastal dynamics and geomorphology (Forbes and Taylor 1994; Catto 2006a, b, 2011). Snow and ice cover limit erosion and reworking of beach sediment, and collect windblown sediment. In contrast to some other coastal regimes, the period from autumn to spring can see net sediment accumulation or stable beach profiles. Fractional recrystallization of ice and snow can result in the formation of an ice foot, protecting the sediment from erosion by storms, waves and humans (Fig. 15.20). On the Avalon Peninsula, ice foot development

Fig. 15.20 Ice foot at Holyrood, Conception Bay, May 2017. The ice foot had a maximum thickness of 80 cm

Fig. 15.19 Gravelly tidal flat, Come-by-Chance

commonly occurs on beaches north of Carbonear. The presence of offshore ice from February through Spring (locally until mid-June in some years) further hinders wave reworking. When the offshore ice begins to disintegrate, ice shove can modify beaches, moving coarse clasts and forming tracks to channel subsequent wave action.

Differing combinations of geomorphic processes, coastal configurations, weather patterns and human activity operate to produce unique shorelines around the Avalon. Some examples are presented here.

15.6.2.1 Middle Cove and Outer Cove

On 22 May, 2017, residents discovered a humpback whale carcass impaled on Outer Cove Beach. The carcass remained for three weeks, washed gently seaward and returned to shore by each wave event, until it was removed inland three weeks later. The carcass's arrival and movement at Outer Cove, rather than at the popular Middle Cove Beach less than 1 km away, reflect the characteristics, dynamics and geomorphology of the two beach systems.

Middle Cove and Outer Cove are flanked by Ediacaran sedimentary rock of the Conception Group and St. John's Group (Fig. 15.21). The coastal configuration reflects the northeast-southwest trending anticlinal fold system, associated faulting, and the relative resistance of the rock units. All the bedrock units are subject to frost wedging, with north-facing cliffs (southeast sides) more susceptible. Detached blocks line the bases of cliffs.

Middle Cove is a reflective pebble-dominated, generally moderate energy beach (Fig. 15.22). Profiles are planar in the late spring and early summer, becoming slightly concave during late summer and autumn. Most winters are without ice foot development or significant snow cover, inducing concave profiles. Protective snow or ice foot development produces a convex profile near the waterline. Northeasterly storm events increase sediment reworking and erosion.

Fig. 15.21 Middle Cove and Outer Cove. *Source* NL Airphoto 01012-173

Fig. 15.22 Middle Cove. **a** Looking east on an anomalously calm July day. Textural differentiations are visible both laterally and with distance from the shore. The eroding Quaternary diamicton bluff is subject to impact from storm and rogue waves, with winnowed boulders visible on the upper beach. **b** Self-organized cusps forming distinctive tiers, eastern side. Increased summer storm activity, consequent reduction in human beach usage, reduced winter snow and ice cover on the upper part of the beach, and winter storm activity resulted in the production of large, conspicuous, well-formed bowl-shaped cusps during the late autumn and winter, with wavelengths in excess of 12 m, widths to 5 m, and depths to 6 m. **c** Western side, with textural gradations reflecting swash activity and shore-normal movement, and debris input from the Quaternary bluff to the south. Firepit disturbances are also apparent. The western side of the beach is more frequented by visitors, flatter, and primarily exposed to easterly storm waves, which tend to be less energetic than those produced by northerly winds. *Photograph* G. Catto

At the cove head, a Quaternary bluff borders the beach. Sequences of coarse glacial diamictons up to 10 m thick, dominated by pebbles and cobbles and sandy matrices, with less than 15% silt and no clay, flank Middle Cove to the south and west. Sedimentary structures and the geomorphic form of the upper deposits indicate that formation occurred by subglacial and ice-marginal debris and slurry flow activity, terminating at the floating ice margin in Middle Cove. Postglacial disturbances contributed to the complexity of the upper parts of the successions. Coastal and surface winnowing (removing fine sand and silt) and ongoing frost action (reorienting clasts into steep to vertical angles) have contributed to increasing the susceptibility to erosion.

Storm waves impact the base of the bluff flanking the central and western sectors, resulting in erosion and undercutting. After a strong northeast storm surge, the upper part of the beach commonly is stripped of much sediment, exposing the underlying Quaternary sediment and Ediacaran bedrock. Incremental erosion continues between storms, resulting from surface runoff across the bluff face, leakage of groundwater through the sediment, and (to a limited extent) frost activity. Erosion is greatest along the southern edge of the bluff (adjacent to the central sector of the beach). Rogue waves also result in erosion of the Quaternary bluff.

Middle Cove is the most extensively used beach on the Avalon Peninsula, although no systematic visitor counts are available. Typically, 20–30 fire pits of varying freshness are visible on the beach on each day from late May through mid-October (as of July 2019, the largest number recorded was 72). On warm summer days and evenings, and during capelin rolls, more than 150 people can be at Middle Cove Beach. This visitor pressure results in gradual flattening of the upper parts of the beach, which alters profiles, destroys or restricts cusp development, and results in compaction of the sediment.

Samples of the sand on the beach have revealed concentrations of reworked glass reaching 10% of the total volume, although the average concentrations are less than 1%. The glass is predominantly brown (from beer bottles) and clear (not traceable), and other coloured glass is extremely rare. The absence of distal source glass and other beach debris indicates that Middle Cove is not a collector beach.

The changes in cusp morphology and dimensions reflect combinations of tourist pressure, storm activity and sediment input from the eroding Quaternary bluff. Commonly, only one or two tiers of cusps are present, although a stormy, wet summer and autumn may result in the formation of four tiers. A year with increased summer storm activity, consequent reduction in human beach usage, reduced winter snow and ice cover on the upper part of the beach, and winter storm activity results in the production of large, conspicuous, well-formed bowl-shaped cusps during the late autumn and winter, with wavelengths in excess of 12 m, widths to 5 m, and depths to 6 m. Many of these cusps persist over several years, with development and maintenance of large, deep cusps on the eastern side of the beach; less-developed, smaller, shallow cusps on the western-northwestern side; and irregular forms in the central, most heavily tourist-impacted sector.

In the central sector, steady flattening of the upper part of the beach during tourist season constantly redistributes sediment to the waterline. Compaction of the sediment allows wave penetration further landward over the flattened slope, but also leads to more dissipation of wave energy and less backwash.

On the flanks of Middle Cove, the dominant factor in cusp development is the effectiveness of winter storm activity. Winter storms, with waves pushed by northeasterly to northerly winds, impact the eastern-southeastern side of the beach, generating strong swash-backwash activity and forming large cusps. The eastern side of the beach is generally narrower, has coarser sediment and is separated by the stream from the central part of the beach (and parking lot), thus effectively reducing human use. The western side of the beach is more accessible, flatter and primarily is exposed to easterly storm waves, which tend to be less energetic than those produced by northerly winds.

Positive North Atlantic Oscillation conditions result in increased strength and frequency of northeasterly winds along the coastline of the northern Avalon Peninsula, particularly during the autumn and winter. Summer (tropical and extratropical) storms, with dominantly southwesterly (offshore-oriented) winds, are less effective at modifying and reworking deposits at Middle Cove. The only effective erosion by tropical storms is associated with northeasterly (backside) winds. The primary effect of stormy summer conditions is to reduce human use of the beach.

Cusp orientations and clast fabric patterns indicate that shore-normal transport is dominant. A fortuitous concentration of more than 100 golf balls on the western side of the beach on 1 January 2002 exhibited essentially shore-normal movement, with some net transport toward the centre, under a light northerly wind. The sedimentary features, coupled with the wave climate and surf-scaling parameter calculations, indicate that the beach is dominantly reflective and swash-aligned.

Middle Cove is a typical spawning beach for capelin (*Mallotus villosus*). Spawning capelin tend to concentrate on the well-sorted parts of the beach, where fine to medium pebbles are dominant. Areas of the beach with gently concave or planar profiles, and seaward-imbricated pebbles, also are more favoured by the capelin. Steepening of beach fronts as a result of storms or the absence of ice-foot protection, and coarsening of sediment resulting from terrestrial mass movements or enhanced storm-driven transportation, are both less favourable for capelin spawning. Increased storm activity at Middle Cove thus could impact it as a capelin spawning site.

Outer Cove is more exposed to winds from the north than is Middle Cove, but is narrower, precluding any significant shore-parallel transport and forcing a shore-normal regime (Fig. 15.23). It receives far fewer human visits than Middle Cove, but is adjacent to the well-travelled Marine Drive.

As at Middle Cove, Outer Cove receives sediment from erosion and frost splitting of the adjacent Ediacaran bedrock, and from coarse-grained Quaternary diamicton. However, the Quaternary material is only exposed along the western edge of Outer Cove beach. Much of the beach is currently backed by the concrete abutments of a bridge, reconstructed in 2011, and the concrete retaining wall and coarse, angular artificial fill along the western side of Marine Drive. Sediment is also transported to the eastern side of the beach through Outer Cove Brook, although this transport was partially interrupted during bridge construction.

Clasts show a range of roundness, with angular and subangular clasts more common than at Middle Cove, especially in the upper part of the beach. Fragments of riprap, concrete and asphalt are present. Clast shape is also variable, with substantial amounts of equant clasts. Subrounded blades are the most common clast type, but generally represent only 25–35% of the clasts on the upper beach faces and 40–50% on the lower beach face. Coarse pebbles dominate in the lower beach face and the eastern part of the beach, adjacent to the brook mouth. Medium to coarse cobbles dominate in the upper beach face. Cobbles are more common in the western sector of the beach. Coarse sand is present along the waterline and in the subtidal area. The sediment adjacent to the brook mouth and easternmost part of the beach also contains medium and fine sand, in addition to anthropogenic debris transported by the stream.

Beach profiles tend to be gently to strongly concave. Gentle slopes result from periods with limited storm activity, particularly when northerly winds are absent. Steeper profiles, with slope angles up to 36°, are developed by waves driven by northerly storms. The reinforcement of the back of the beach has resulted in steepening of profiles throughout the central and western segments of the beach. The influence of Outer Cove Brook in moving sediment seaward, coupled with the increased velocity of the brook since 2006, causes profiles to remain as gently concave. The sediment fan at the brook mouth is actively prograding, and is only temporarily removed during northeasterly storm events. The strong hydraulic head allows Outer Cove Brook to maintain an open outlet, in contrast to the periodic blockage of the brook discharging at Middle Cove.

Cusps are well-developed at Outer Cove, with as many as 5 tiers present. Cusps are generally deep, symmetrical bowls, indicating strong shore-normal transport and significant backwash. The largest cusps measured had wavelengths of 15 m, widths of 6 m and depths of 4 m. Cusps are generally best-developed in early spring, following autumn and winter storm activity, and tend to degrade throughout the summer.

Erosion of the riprap and undercutting of the retaining wall are evident. The artificially reinforced areas of the system are those most exposed to storm wave action and resulting erosion.

Completion of the bridge restored human access to the beach. Increased use has resulted in modification to the cusps and flattening of beach profiles. However, the generally coarser sediment, narrower beach and lesser human usage significantly limit the impacts in comparison to Middle Cove.

Outer Cove is a reflective, strongly shore-normal, swash-aligned, moderate energy beach system. The cove is less susceptible to wave impact than is Middle Cove, except for waves driven by northeasterly winds. A northeasterly wave event in May 2017 was responsible for the transport of the humpback whale carcass to the beach. Once stranded, the persistent shore-normal pattern precluded any movement of the carcass. In the absence of human intervention, it would have remained on the beach unless subjected to an extremely strong storm event. The steep, high gravel beach, backed by the artificial reinforcement of the bridge abutments and the retaining wall, has accentuated the reflective, shore-normal transport pattern.

15.6.2.2 Chapel Cove

Most beaches along southern Conception Bay are strongly reflective, dominantly shore-normal, swash-aligned and modal moderate energy systems. Influenced by rising sea level, changes in winter storm frequency and intensity, enhanced extratropical transition activity, variations in offshore ice cover, ice foot formation, sediment flux and both accentuation and reduction of human activity, the results are complex amalgams of geomorphic and biological processes and anthropogenic modifications. Adjacent beaches show significant differences in dynamics and response. Chapel Cove provides a good example (Fig. 15.24).

Chapel Cove is a concave coarse sand-coarse pebble beach marked by generally moderate energy conditions. The embayment faces north, and is sheltered from the full force of northeasterly winds. The east side is flanked by a hill of resistant Ediacaran bedrock, and the west side by less resistant sloping bedrock modified by glacial erosion. The beach is a mid-bay bar, with an artificially modified lagoon landward. Sediment flux is limited, supplied by shore-oblique littoral transport from the northwest and from reworking of Quaternary glacial sediments that form the core of the barrier. Removal of sediment from the beach face, either by storm overwash or by human activity, results in enhanced erosion, profile flattening and beach destabilization.

Fig. 15.23 Outer Cove. **a** The western side receives sediment from erosion and frost splitting of the adjacent sandstone and argillite bedrock, and from coarse-grained Quaternary diamicton. Much of the beach is currently backed by a concrete retaining wall and coarse, angular artificial fill along the western side of Marine Drive. Coarse cobbles and boulders dominate. **b** The eastern side is sheltered from all waves except those directly from the north. Fine cobbles and coarse pebbles dominate, and cusp development is less common. Additional sediment is transported to the beach through Outer Cove Brook. The beached whale carcass was transported to this segment of outer Cove by a northerly wave event, and subsequently lodged. *Photograph* G. Catto. **c** Self-organized cusps reflect very local, changing wave dynamics, and also develop in response to interaction of waves and lager clasts. The boulders were displaced from the landward margin of the beach during construction along Marine Drive in 2011

Fig. 15.24 Chapel Cove, Conception Bay. **a** Chapel Cove is exposed to the Northeast, but this collector beach also receives sediment transport from the Northwest along Conception Bay. *Photograph* NL Airphoto 95021-176. **b** Self-organized cusps develop through shore-normal transport. Distinctive gradation between medium-coarse sand (buff) and fine- to coarse pebbles (grey) develops during periods of relative quiescence. Northeasterly and Northerly storm waves rework sediment with shore-oblique transport. Terrestrial sediment influx is minimal. **c** Cribbing, wharves, and retaining walls constructed prior to 1992 continue to influence wave interactions and sediment movement. **d** Artificially-emplaced boulders intended to stabilize the beach for tourism were displaced by a storm in 2013, and now serve as a framework for future beach sedimentation

Between 1990 and 2019, the maximum profile slopes have varied from 11° to 27°. Periods with steeper slopes close to the waterline generally are associated with ice foot development. Prior to 2009, steeper slopes on the higher storm-influenced section generally were associated with late summer and autumn storm activity, resulting from accumulation of sediments by overtopping. Subsequently, overwashing has become more prevalent, especially after 2013. Profiles are generally planar to gently concave in the late spring and early summer, gradually becoming more concave during late summer and autumn.

Cuspate features currently form one or two tiers, with the number reflecting both the amount of time elapsed since the last significant storm event and the absence of significant reworking by seasonal ice. The cusps are asymmetrical in plain view, and are self-organized. Cusp orientations, clast fabric patterns and textural variations indicate that shore-normal transport is dominant, and that the beach is reflective. Between 1992 and 2006, up to five stacked tiers of cusps were developed, with the uppermost formed by the major northeasterly storm of October 1992.

Clasts of beach glass (0.5–2 cm) are notable along the waterline at Chapel Cove. Distinguished by a variety of colours and substantial pitting, frosting and smoothing, the clasts have been reworked by waves over considerable time and have been transported distally. Samples of sand and pebbles from the upper parts of Chapel Cove beach contain only a few angular shards of brown (beer bottle) glass. The glass at the waterline, in addition to other marine debris, indicates that Chapel Cove is a collector beach for southern Conception Bay.

Along Conception Bay, the effects of road construction and introduction of frozen fish processing in the mid-1960s accentuated socio-economic pressures and migration from rural areas. Declining marine resources and particularly the imposition of the Northern Cod Moratorium of July 1992 accelerated the process. However, modifications formerly undertaken by residents, including construction of wooden retaining walls and channels dredged by hand, continue to influence beach dynamics and storm surge effects. The immediate effects of the moratorium were amplified by the northeasterly storm of October 1992 (Catto 1999).

Chapel Cove supported a small-boat nearshore cod fishery prior to July 1992. Open boats were docked in the lagoon, connected to Conception Bay by an artificially maintained channel. Prior to 1992, the channel was regularly dredged to seaward. The channel acted to focus wave energy onto the southeast side of the beach, accentuating shore-normal transport and causing the beach face to steepen. During periods of northeasterly winds, this acts to focus and consequently increase the wave energy affecting the southeastern part of the system, resulting in sediment transport into the artificial channel and the lagoon behind. The channel needed to be re-dredged following each major northeast storm. The lagoon was further protected by a retaining wall, limiting sediment resupply to the system from the land.

With the imposition of the moratorium, there was no reason to maintain the channel, and thus the disruption caused by the October 1992 storm was not repaired. The remnants of the retaining wall remained, however, restricting sediment influx to the beach. In addition, the concentrations of larger clasts used as anchoring material for cribbing and wharf supports remained, influencing sediment transport on the beach. Interruptions in terrestrial sediment flux are evident in combination with significant storm-induced activity.

Throughout Newfoundland, promotion of tourism to enhance economic sustainability of coastal communities such as Chapel Cove has also resulted in changes to beach systems. The effects of relatively low intensity but much increased usage are substantial. Erosion is evident along routes crossing beach crests, and impacts on adjacent cliff-top trails also affect sediment flux. Abundant precipitation, cool spring and autumn temperatures, and wind removal of snow cover, result in numerous freeze-thaw events which promote erosion.

At Chapel Cove, the beach crest was flattened on several occasions beginning in 2009, to provide a flatter surface for tourist usage of the beach, including picnic tables, trash cans and boxes of marigolds for beautification. The result was to promote overwash and beach erosion during autumn storm events, depositing sediment into the lagoon and lowering both the elevation and slope of the beach. Tourist modification of beach systems thus includes both direct pressures of usage and those related to development of tourism as a resource. Despite the lack of large industrial development in most areas, the shorelines of southern Conception Bay have undergone and continue to undergo substantial anthropogenic modification, but not necessarily in uniform, undeviating trends.

15.6.2.3 Point Lance

Bedrock lithology, pervasive frost action, absence of fine-grained Quaternary sedimentary deposits, steep slopes and high-energy conditions (either pervasive or associated with storm activity) characteristic of much of the Avalon shoreline effectively limit the opportunities for developing sandy dune-backed beaches. Only three examples are present: Northern Bay Sands and Salmon Cove (Conception Bay, modified by seasonal tourism); and Point Lance, on the southernmost Avalon Peninsula (Fig. 15.25).

At Point Lance, a sand-dominated flat developed over a rock platform grades laterally to the south into a narrow sandy beach on a narrower rock platform. To the west, the zone grades into an open sand flat. The relative lateral extents of these shore types shift throughout the seasons, and between years.

Based on observations between 1989 and 2016, the sand-dominated flat typically extends for approximately 50 m along the shore, and is 30–40 m wide at low tide. The medium-fine sand is moderately sorted. Derived from aeolian dome dunes backing the flat, sand is transported to the rock platform area by littoral drift, recycling from offshore areas, and less commonly by northwesterly winds that rework the dunes and move sand directly to the platform area. Littoral drift is the dominant process under all but strong hurricane conditions. Sand flux to the beach is high from early spring to mid-winter, and proceeds year-round when seasonal icing or snow cover does not occur. After a period of minimal erosion, the beach is marked by a gentle slope (2–8°) with a planar to very gently concave profile. Slope angles reach a maximum of 12° following storms.

The combination of milder winter conditions and enhanced storm frequency, particularly hurricanes and extratropical transitions, results in more beach reworking and in redistribution of sediment from the dunes to the beach and nearshore. In contrast, years with minimal hurricane frequency and substantial beach icing are marked by lesser erosion. The influence of winter icing is more important in minimizing erosion than are hurricanes in promoting it.

The open sand flat to the west is marked by an extremely gentle slope (<1°–4°). Storm activity results in small areas of undercutting at the dune margins and along the course of the stream that crosses the beach, but these seldom exceed 0.5 m in height and are quickly smoothed by subsequent reworking. Most storm waves travel over the surface of the sand flat without causing significant erosion, and the energy is focused at the dune field margin. Sediment eroded from the dunes is deposited during backwash as a thin, planar, gently dipping stratum. Repetitive swash events destroy the stratification imparted during each backwash period, and the storm deposits are characteristically structureless. During normal wave activity, swash-and-backwash scour patterns

Fig. 15.25 Point Lance. **a** One of very few sand-dominated beaches on the Avalon, Point Lance is exposed to the south. *Photograph* NL Airphoto 95049-159. **b** A broad, gently sloping medium-fine flat develops during periods of relative quiescence. **c** After storm activity, fragmented kelp, eroded backbeach dunes, and the underlying pebble-cobble framework are visible

and low energy ripples are produced, and bioturbation is not in evidence.

The dunes are actively undergoing reworking from coastal onshore winds, fluvial erosion, storm waves, and anthropogenic-induced deflation resulting from overgrazing by sheep and disturbance from all-terrain vehicles (Catto 1994b). Trough blowouts have propagated and deepened across the dunes.

15.6.2.4 Bedrock Shorelines

Bedrock forms the majority of Avalon shorelines. Cliffed shorelines dominate, with narrow rock platforms present in isolated locations (Fig. 15.26).

Ediacaran or Palaeozoic sedimentary bedrock cliff heights vary from less than 5 m to greater than 60 m, and range in slope from 30° to vertical. Numerous examples of overhanging cliffs, caves and offshore arches and stacks are developed, with bedding planes, joint orientation and lithology as the controls. Cliff morphology reflects the lithology, fracture pattern, tectonic history and terrestrial erosional processes (Spellar 2001; Catto et al. 2003; Catto 2011; Thompson 2014). Marine processes play lesser roles. Frost action is a dominant process, facilitated by jointed, fractured and stratified bedrock. Wedging by tree roots and biochemical activity further accentuate erosion.

Although vegetation cover generally inhibits frost action, areas with tuckamore (krummholz) tree vegetation developed on thinly veneered or bare rock surfaces are susceptible to wedging and block toppling. Erosion is particularly accentuated at low cliff sites, where periodic kill or damage of the tuckamore by salt spray or ice storms results in removal of the vegetation, leaving wedged fractures vulnerable to frost activity.

Direct human influence is limited in coastal bedrock sites. In several locations along the Cape Shore of Placentia Bay, northern Conception Bay and northern Trinity Bay, grazing by sheep contributes to erosion. Sheep are responsible for denuding vegetation from cliff edges, causing enhanced runoff and frost wedging and thus generating large quantities of sediment for marine transport and deposition on beaches (Catto 1994a).

Narrow bedrock platforms, largely or totally devoid of sediment, primarily reflect bedrock attitude, fracture pattern and susceptibility to frost action. The Avalon rock platforms are associated with areas characterized by low terrestrial sediment influx.

Rock platforms can form at elevations ranging from mean low tidal position to 12 m asl, so their presence above sea level in an exposed coastal situation cannot be considered as evidence for a former marine high stand.

Along the dominantly microtidal Avalon shore, erosion is effective in the upper intertidal areas, and in exposed areas above the mean high tide line. The distribution of *Ascophyllum*, *Fucus* and other taxa within the rockweed communities (Catto and Hooper 1999) indicates that many platforms are not regularly subjected to sea ice activity or strong wave action, but continue to be eroded in the intertidal and supratidal zones by frost wedging.

15.7 Northeast Avalon—Urban and Suburban

St. John's developed primarily as a trans-shipment port for the cod fishery, with small vessels off-loading catch or salted fish for subsequent trans-Atlantic transport. As the city was not intended to be a naval base, the physical layout and street patterns were established in response to the needs of the fish harvesters and merchants. Longer routes paralleling the topography (Water, Duckworth, Gower Streets) were connected to coves with short, steep streets. The downtown developed over the seventeenth through late nineteenth century, punctuated by several fires, rather than being laid out in a single, engineered episode.

All of the communities of the northeast Avalon have become suburban components of the St. John's metropolitan area. Geomorphological problems influencing development are tied to urban and suburban activities, rather than being related to fishery, aquaculture and tourism issues.

The Avalonian Orogeny's geological structure of northeast-southwest ridges and valleys dominates the geomorphology. The former cascade streams connecting fens, meadows and ponds have been extensively altered (Fig. 15.27), and some reaches have been culvertized. For example, Memorial University of Newfoundland is built on meadows, small wetland areas, and infilled terrain surrounding the artificially maintained Burton's Pond. The eastern part of campus occupies the former Strawberry Marsh, and the western area is built on coarse pebble diamicton overlying the St. John's Group. The University lies over a minor synclinal flexure, with the northeast-southwest oriented domed anticline extending to Kenmount Hill visible to the north. Confinement of O'Leary's Brook directly northwest of campus has resulted in local flooding. The short fluvial systems are susceptible to hurricane-related flooding, typically within three hours of the initiation of severe precipitation events. Numerous modifications to infrastructure have been undertaken since the last major flooding event in 2001.

Block falls and topples resulting from frost action represent local hazards along roadways flanking hillsides (Spooner et al. 2013). The jointed, fractured and bedded nature of much of the bedrock facilitates wedging. Slope failures resulting from frost wedging are common in the Bauline area and would pose a significant hazard to extensive housing development (Fig. 15.28).

Fig. 15.26 Bedrock shorelines form the majority of Avalon's coast. **a** Argillites and siltstones of the St. John's Group form an anticline at Burnt Cove, Northern Conception Bay. **b** Cape St. Mary's Ecological Reserve is home to a large colony of Northern Gannets (*Morus bassanus*). **c** Marine platform, New Melbourne, Trinity Bay, formed where gently dipping silty arkose bedrock has been exposed to erosion, as sea-level rose at ~3 mm/y

Fig. 15.27 Remnant wetland directly behind Confederation Building, the seat of Government of Newfoundland and Labrador, St. John's. Infilling of terrestrial bogs and fens for urban development is common throughout the northeast Avalon. *Photograph* G. Catto

Fig. 15.28 Active block topples and falls at Bauline are indicated by the white spruce trunks

Terrain suitable for agriculture is limited by the combination of topography and coarse, thin Quaternary sediment cover. In the 1800s, agricultural settlement was encouraged along roads radiating from St. John's. After 1960, farmland in the areas closest to existing development was cleared for housing. To prevent urban sprawl and to preserve the rural communities, a freeze on the sale of farmland was imposed, leading to anger among farmers who could no longer farm economically. Eventually, a combination of pressure for development, St. John's' desire for an increased tax base, and removal of government subsidies for some forms of agriculture, resulted in the lifting of the land freeze and rapid suburban development.

Access to the coastline in St. John's is limited by steep cliffs with few indentations. As well, a potential area for suburban development was appropriated in the 1950s for the

Fig. 15.29 **a** Extensive suburban growth has impacted the coastline of Conception Bay South, interfering with sediment flux and maintenance of the barachoix at Long Pond, Manuels, and Chamberlains. *Photograph* NL Airphoto 01012-84). **b** 'The Atlantic Ocean is at your back door': suburban development on glacifluvial gravel at Chamberlains, Conception Bay South

area's waste disposal site at Robin Hood Bay. Consequently, suburban development spread northward and westward to communities offering coastal properties.

The coastline of Conception Bay South (Fig. 15.29) consists of mixed gravel-and-sand flats, steep gravel barachoix and eroding bluffs of Quaternary glacifluvial gravel and sand (Paone 2003; Pittman 2004; Catto 2011). Flats are normally marked by gentle to moderate seaward slopes (2–8°), although following northeast storm events, individual slopes may exceed 20°. The barachoix are steeply sloping, with slopes locally in excess of 30°. Both barachoix and flats display self-organized cusps, indicative of reflective, moderate- to high-energy, shore-oblique, swash-dominated activity.

Sediment is contributed to the shoreline by river flooding, slope failure of the glacifluvial bluffs and marine processes. Sand is moved offshore and towards the northeast by swash-backwash and longshore currents, gradually accumulates in the offshore areas, and is resupplied to the shoreline by wave activity associated with northeast or (less effectively) northwest winds. The sand thus undergoes a net northeast movement along the shoreline from Holyrood to Topsail, as southwesterly winds dominate. Under natural conditions, the shores will be supplied with sand both from offshore and from the adjacent Quaternary shoreline outcrops. Shorelines to the northeast were dominated by finer sediments (greater sand: gravel ratio) and were wider than shorelines to the southwest developed under similar geomorphic conditions.

Anthropogenic activity in Conception Bay South has significantly reduced the supply of sand to the shoreline. The replacement of sheep herding and agriculture with suburban development and the diversion and impoundment of rivers has reduced the erosion of inland Quaternary outcrops. Human stabilization of the Quaternary bluffs and extraction of aggregate have contributed to the reduction of sediment influx from the land. The glacifluvial gravels and sands represent the only source of Quaternary aggregate in the northeast Avalon, and are actively exploited at Seal Cove.

Individual storm events have removed more than 1 m of coastal bluff. Removal of sediment from the base induces undercutting and mass movement of the upper parts of the exposure, resulting in collapse. The modal erosion rate along the Conception Bay South shoreline is 0.3 m/y, with most erosion associated with severe storm activity. Water saturation of sediment flanked by retaining walls on the seaward side has resulted in retaining wall failure at several localities.

15.8 Conclusion

Outside of the urban and suburban northeast, the Avalon appears to be superficially 'untouched' by human activity. Although human modification is evident upon closer examination, the overall appearance of the landscape is dominated by geomorphological elements. Bedrock structure forms the framework, and glacial, glacifluvial, fluvial, wetland and coastal processes have acted upon the terrain over the Quaternary. The result is a complex landscape of geomorphic and biological processes and anthropogenic modifications.

References

Boger R (1994) Morphology, sedimentology, and evolution of two gravel Barachoix systems. Placentia Bay Unpublished MSc thesis, Department of Geography, Memorial University of Newfoundland

Brückner W (1969) Post-glacial geomorphic features in Newfoundland, eastern Canada. Ecol Geol Helv 62:417–441

Brushett D, Bell T, Batterson MJ, Shaw J (2007) Ice-flow history of Placentia Bay, Newfoundland: multibeam seabed mapping. Newfoundland and Labrador Department of Natural Resources, Geological Survey Report 07-1, pp 215–228

Catto NR (1994a) Coastal evolution and sea level variation, Avalon Peninsula, Newfoundland: geomorphic, climatic, and anthropogenic variation. In: Wells PG, Ricketts PJ (eds) Coastal zone Canada 1994, Co-operation in the Coastal Zone, Bedford Institute of Oceanography, vol 4, pp 1785–1803

Catto NR (1994b) Anthropogenic pressures and the dunal coasts of Newfoundland. In: Wells PG, Ricketts PJ (eds) Coastal Zone Canada 1994, Co-operation in the Coastal Zone, Bedford Institute of Oceanography, vol 5, pp 2266–2286

Catto NR (1998) The pattern of glaciation on the Avalon Peninsula of Newfoundland. Géog phys Quat 52:23–45

Catto NR (1999) Embayed gravel coastlines of conception bay, Newfoundland: climate variation, geomorphic response, and management issues. Salzb Geog Arb 34:119–142

Catto NR (2006a) More than 16 years, more than 16 stressors: evolution of a high energy reflective beach, 1989–2005. Géog phys Quat 60:49–62

Catto NR (2006b) Impacts of climate change and variation on Provincial Parks, NL. Newfoundland and Labrador, Ministry of Environment and Conservation, 160 pp

Catto NR (2011) Coastal Erosion in Newfoundland. Newfoundland and Labrador, Ministry of Environment and Conservation, 144 pp

Catto NR, Hooper R (1999) Biological and geomorphological shoreline classification of Placentia Bay, Newfoundland. Can Tech Rep Fish Aqua Sci 2289

Catto NR, Scruton DA, Ollerhead LMN (2003) The coastline of Eastern Newfoundland. Can Tech Rep Fish Aqua Sci 2495

Catto NR, St. Croix L (1998) Urban geology of St. John's, Newfoundland. In: Karrow PF, White OL (eds) Urban geology of

Canadian Cities. Geological Association of Canada, St. John's, pp 445–462
Chamberlin TC (1895) Notes on the glaciation of Newfoundland. Geol Soc Am Bull 6
Coleman AP (1926) Pleistocene of Newfoundland. J Geol 34:193–223
Colman-Sadd SP, Hayes JP, Knight I (1990) Geology of the island of Newfoundland. Newfoundland and Labrador, Department of Mines and Energy, Geological Survey Branch, Map 90-01
Danielson JJ, Gesch DB (2011) Global multi-resolution terrain elevation data 2010 (GMTED2010): U.S. Geological Survey Open-File Report 2011-1073, 26 pp. http://pubs.usgs.gov/of/2011/1073/pdf/of2011-1073.pdf
Davis AM (1985) Causes and character of paludification in Newfoundland. Can Geog 29:361–384
Duncan KN (2015) Geomorphology, sedimentology, and sensitivity to Petroleum contamination in Coastal systems in Newfoundland. Unpublished Masters of Environmental Science thesis, Memorial University of Newfoundland
Etheridge B (2005) The sedimentology, morphology, and sensitivity to petroleum pollution of five Gravel Beaches, Southern Shore, Newfoundland. Unpublished Masters of Environmental Science thesis, Memorial University of Newfoundland
Fisher T, Shaw J (1992) A depositional model for Rogen moraine, with examples from the Avalon Peninsula, Newfoundland. Can J Earth Sci 29:669–686
Forbes DL (1984) Coastal geomorphology and sediments of Newfoundland. Geol Surv Can, Paper 84-1B
Forbes DL (1985) Placentia Road and St. Mary's Bay: field trip guide to coastal sites in the southern Avalon Peninsula, Newfoundland. In: Proceedings, Canadian coastal conference 85, St. John's. National Research Council of Canada, Associate Committee for Research on Shoreline Erosion and Sedimentation, pp 587–605
Forbes DL, Taylor RB (1994) Ice in the shore zone and the geomorphology of cold coasts. Prog Phys Geog 18:59–89
Grant DR (1989) Quaternary geology of the Atlantic Appalachian region of Canada. In: Fulton RJ (ed) Quaternary geology of Canada and Greenland. Geol Surv Canada, Geology of Canada, pp 393–440
Griffiths H (1999) Coastal geomorphology and sedimentology, Whiffen Head-Ship Harbour area, Placentia Bay. Unpublished Masters of Environmental Science thesis, Memorial University of Newfoundland
Handcock G (1989) Soe longe as there comes Noe women: origins of English settlement in Newfoundland. Newfoundland History Series, #6, Breakwater Press, St. John's
Henderson EP (1972) Surficial geology of Avalon Peninsula, Newfoundland. Geol Surv Can Memoir 368, 121 pp
Hickman H (2006) Flood hazard and vulnerability in Newfoundland communities. Unpublished MSc thesis, Environmental Science, Memorial University of Newfoundland
King AF (1988) Geology of the Avalon Peninsula, Newfoundland. Newfoundland and Labrador, Department of Mines and Energy, Geological Survey Branch, Map 88-1
King AF (1990) Geology of the St. John's area, Newfoundland. Newfoundland and Labrador, Department of Mines and Energy, Geological Survey Branch, Map 90-120
King LH, Fader GBJ (1986) Wisconsinan glaciation of the continental shelf, southeast Atlantic Canada. Geol Surv Can Bulletin 363, 72 pp
Lewis CFM, Macpherson JB, Scott DB (1987) Early sea level transgression, eastern Newfoundland. INQUA 1987, Programme with Abstracts, 210
Liverman DGE (1994) Relative sea-level history and isostatic rebound in Newfoundland, Canada. Boreas 23:217–230
Marich AS (2006) A morphological and sedimentological analysis of Rogen moraine on the Avalon Peninsula, Newfoundland. Unpublished MSc thesis, Department of Geography, Memorial University of Newfoundland
Macpherson JB (1995) A 6 ka reconstruction for the island of Newfoundland from a synthesis of Holocene lake-sediment pollen records. Géog phys Quat 49:163–182
Macpherson JB (1996) Delayed deglaciation by downwasting of the northeast Avalon Peninsula, Newfoundland: an application of the early Postglacial pollen record. Géog phys Quat 50:201–220
Marshall I (1996) A history and ethnography of the Beothuk. McGill-Queen's University Press
Marshall I (2001–2009) The Beothuk: a vanished people. Newfoundland and Labrador Historical Society, St. John's
Masson A (2014) The extratropical transition of Hurricane Igor and the impacts on Newfoundland. Nat Haz 72:617–632
McNeil M (2009) Marine debris as an indicator of oil spill vulnerability along selected beaches of Northern Placentia Bay, Newfoundland and Labrador. Unpublished MSc thesis, Department of Geography, Memorial University of Newfoundland
Miller AAL (1999) The quaternary sediments and seismostratigraphy of the Grand Banks of Newfoundland and The Northeast Newfoundland Shelf: foraminiferal refinements and constraints. Unpublished PhD dissertation, George Washington University, Washington
Myrow PM (1995) Neoproterozoic rocks of the Newfoundland Avalon zone. Precambr Res 73:123–136
Paone LC (2003) Hazard sensitivity in Newfoundland coastal communities—impacts and adaptations to climate change: a case study of conception Bay South and Holyrood, Newfoundland. Unpublished MSc thesis, Department of Geography, Memorial University of Newfoundland
Pastore R (1990) Collapse of the Beothuk world. Acadiensis 19:52–71
Pastore R (1992) Shanawdithit's people. Breakwater, St. John's
Pink D (2004) Analysis of Beach Litter volumes, sources and movements on selected coastlines of the Avalon Peninsula, Newfoundland. Unpublished Masters of Environmental Science report, Memorial University of Newfoundland
Piper DJW, Mudie PJ, Fader GB, Josenhans HW, Maclean B, Vilks G (1990) Quaternary geology. In: Keen MJ, Williams GL (eds) Geology of the continental margin of Eastern Canada. Geol Surv Can 2, pp 475–607
Pittman D (2004) Analysis of coastal geomorphological processes on a boreal coarse clastic barrier: long Pond Barachois, Conception Bay, Newfoundland. Unpublished MSc thesis, Department of Geography, Memorial University of Newfoundland
Rowe FW (1980) History of Newfoundland and Labrador. McGraw-Hill Ryerson, Toronto
Shaw J, Forbes DL (1987) Coastal barrier and beach-ridge sedimentation in Newfoundland. In: Proceedings, Canadian Coastal Conference 87, Québec. National Research Council of Canada, pp 437–454
Shaw J Forbes DL (1990) Short- and long-term relative sea-level trends in Atlantic Canada. In: Canadian Coastal Conference Proceedings, National Research Council of Canada, pp 291–305
Shaw J, Forbes DL (1995) The postglacial relative sea-level lowstand in Newfoundland. Can J Earth Sci 32:1308–1330
Shaw J, Gareau P, Courtney RC (2002) Palaeogeography of Atlantic Canada 13–20 kyr. Quat Sci Rev 21:1861–1878
Shaw J, Taylor RB, Forbes DL, Solomon S, Ruz M-H (1999) Sensitivity of the coasts of Canada to Sea-level Rise. Geological Survey of Canada, Bulletin 505

Spellar R (2001) The impact of slope failure and visitation on Cape St. Mary's ecological reserve, Newfoundland. Unpublished Masters of Environmental Science report, Memorial University of Newfoundland

Spooner I, Batterson M, Catto N, Liverman D, Broster BE, Kearns K, Isenor F, McAskill GW (2013) Slope Failure hazard in Canada's Atlantic Provinces: a review. Atl Geol 49:1–14

Summers, W (1949) Physical geography of the Avalon Peninsula of Newfoundland. Unpublished MSc thesis, McGill University, Montréal

Thompson S (2014) Coastal erosion at mistaken point ecological reserve, Avalon Peninsula, Newfoundland. Unpublished MSc Thesis, Department of Geography, Memorial University of Newfoundland

Vasseur L, Catto NR (2008) Atlantic Canada. In: Lemmen DS, Warren FJ, Lacroix J, Bush E (eds) From impacts to adaptation: Canada in a changing climate 2007. Government of Canada, Ottawa, pp 119–170

Watton EC (2016) Coastal geomorphology, processes, and erosion at the Tourist destination of Ferryland, Newfoundland and Labrador. Unpublished MSc thesis, Environmental Science, Memorial University of Newfoundland

Norm Catto joined Memorial University in July 1989 and is now the Head and Professor of Geography at that university. His research and teaching interests include coastal landforms, natural hazards, sea level change and impacts in coastal environments; response of river systems to climate and weather events, and flood risk assessment; aeolian geomorphology and response to climate and human factors in boreal and costal environments; mass movements and slope failures; loess deposition; palaeosol formation; emergency measures policies and effectiveness human adaptations to climate and weather events; the impacts of climate and weather events on agriculture, transportation, fisheries and communities; and the relationships among landscapes and people. His research has included projects and investigations in landscapes and environments in 16 countries, and in all Canadian provinces and territories, particularly in eastern Canada. For 18 years, he served as Editor-in-Chief for *Quaternary International*, a professional international journal dedicated to the study of landscape evolution and climate variation throughout the past 3 million years. He has taught more than 50 different undergraduate and graduate courses focused on numerous aspects of geomorphology, landscape analysis, Quaternary research, natural hazards, ocean sciences, environmental science, climate and climate change, soil science, geoarchaeology and palaeontology. Study of landscapes forms his central focus for understanding of the various areas of Eastern Canada, the combination of physical and cultural constituents that contribute to its identities. For most species, and for all human occupants, life in Eastern Canada has been heavily influenced by geomorphology since deglaciation.

Large-Scale Rock Slope Deformation from the Tablelands and Lookout Hills of Western Newfoundland, Canada

16

Ian Spooner, Dewey Dunnington, John Gosse, Gerald Osborn, Antony R. Berger, and Ian Brookes

Abstract

Five large gravitational slope deformation features (GSDFs) in the Lookout Mountain-Tablelands region of western Newfoundland exemplify bedrock slope instabilities in eastern Canada. The Lookout Hills GSDF (8.3 km^3) on Bonne Bay (glacial trough) may be the largest GSDF in eastern Canada. It appears to be a post-glacial feature, has vertical total displacement up to 100 m, and exhibits a complex arrangement of scarps and fissures. Four other large GSDFs occur along the walls of glacial troughs in the Tablelands of Western Newfoundland with estimated volumes between 1 and 2 km^3. All five features are above steep slopes and have the potential to become catastrophic massive rock slope failures. While it is tempting to link the failure in all 5 GSDFs to the removal of laterally supporting bedrock or glacier ice during the late Pleistocene, the location of these deformations may simply relate to slope steepness. Non-glacial, first order controls on the deformation include changes in the effective shear strength and opening of fractures owing to Holocene permafrost thaw may have induced failure. No discernible movement has occurred in the Tablelands GSDFs in the last 50 years. Nevertheless, the Lookout Hills GSDF requires more detailed study to determine the potential for future movement as rapid collapse into the adjacent fjord may present significant risk to coastal residents, infrastructure and marine ecosystems.

Keywords

Newfoundland • Gravitational slope deformation feature • Sackung • Rock-slope failure • Glaciation • Humber arm allochthon • Long Range Mountains

16.1 Introduction

On June 17, 2017, part of a mountainside, with a volume estimated at between 35 M and 51 M m^3, slid into the sea at Nuugaatsiaq, west Greenland and produced displacement waves that resulted in loss of life (four fatalities) and considerable damage to property (Chao et al. 2018). It was not seismically triggered, and instead was a result of acceleration of mass movement that began months prior to the catastrophic event. Similar conditions to those at Nuugaatsiaq occur in Western Newfoundland where highly jointed and fractured rocks abutting deep fiords are prone to mass movement. Some are in coastal positions where there is potential to generate a displacement wave. A number of factors can promote GSDFs (gravitational slope deformation features), accelerate their motion, and ultimately trigger catastrophic failure. Unloading during deglaciation and post-glacial active isostatic movement may have increased stress and lessened the effective cohesion of these slopes (Vacchi et al. 2018). Seismicity is not uncommon along the passive margin of eastern Canada as well as along the St. Lawrence Trough with recorded earthquakes in the M6.0–7.2 magnitude range (Stein et al. 1979). Stein et al. (1979) contend that large passive-margin earthquakes may occur as

I. Spooner (✉)
Department of Earth and Environmental Science, Acadia University, Wolfville, NS B4P 2R6, Canada
e-mail: ian.spooner@acadiau.ca

D. Dunnington
Department of Civil and Resource Engineering, Dalhousie University, Halifax, NS B3H 4R2, Canada

J. Gosse
Department of Earth Sciences, Dalhousie University, Halifax, NS B3H 4J1, Canada

G. Osborn
Department of Geoscience, University of Calgary, Calgary, AB T2N 1N4, Canada

A. R. Berger
3 Prince Street, Wolfville, NS B4P 1P7, Canada

I. Brookes
York University, Toronto, Canada

© Springer Nature Switzerland AG 2020
O. Slaymaker and N. Catto (eds.), *Landscapes and Landforms of Eastern Canada*, World Geomorphological Landscapes, https://doi.org/10.1007/978-3-030-35137-3_16

far south along the eastern Atlantic coast as ice sheets extended. Besides seismic triggering, climate change induced reduction in effective shear strength, fracture propagation, and freeze–thaw processes may also cause mass wasting.

Collectively these conditions indicate that an examination of the potential for acceleration of mass wasting or catastrophic large rock slope failures in eastern Canada is warranted, particularly along western Newfoundland where subaerial relief exceeds 500 m, including slopes above coastal waters or fjord lakes. We describe and illustrate five rock-slope deformation features in the Lookout Mountains-Tablelands region of western Newfoundland, apparently formed in response to gravitational pull on very large rock masses on plateau margins (Fig. 16.1). These margins were steepened by glaciers and then debuttressed as glaciers receded and disappeared. Our objective is to summarize the geological and geomorphological setting of these sites and indicate where further study is warranted, including hazard genesis and risk analysis. We demonstrate that these gravitational deformation features are common in the region and that three of the five sites have the potential to develop into catastrophic rock slope failures of considerable volume. A failure at the Lookout Hills site will have the potential to be a significant displacement wave hazard owing to its coastal position. While submarine mass transport complexes have been identified from the Scotian Shelf to Pond Inlet and can generate similar tsunami (Jenner et al. 2007; Broom et al. 2017), we are limiting this contribution to subaerial processes.

16.1.1 Terminology

Current terminology relevant to subaerial mass wasting in high-relief glaciated regions is rather unwieldy. The term "rock-slope failure" (RSF) is considered by Evans et al. (2007) and Hermanns and Longva (2012) to encompass rockslides, rock avalanches, catastrophic spreads and rockfalls. They consider very large failures "massive rock slope failure" or MRSF. RSFs, whether or not they are massive, occur when (i) stresses within the rock mass are increased by steepening of the constituent slope, most often by glacial erosion, but in some cases tectonics (Harrison and Falcon 1938; Roberts and Evans 2008), (ii) when stresses applied earlier to the rock-mass are relieved, for example by glacial debuttressing (Holm et al. 2004) or (iii) the effective shear strength of the slope is reduced, for example by permafrost thaw or increased effective porewater pressure. These processes can operate in concert. Although Evans et al. (2007) consider only varieties of rapid, or what we could term catastrophic, mass movement, some failures may begin as very slow, small-scale displacements or sags of mountain slopes which produce gravitational deformation features such as cracks, scarps, and perhaps grabens at the top of the failure (Hermanns and Longva 2012).

"Sackung" (plural "sackungen") is a German term used to describe deep-seated, generally slow-moving slope failures which often exhibit a series of linear scarps facing up-slope (anti-scarps). The term has been assimilated into English, although it may sometimes refer to the geomorphic features at the head of the failure rather than the entire failure. "Deep-seated gravitational slope deformation" (DSGSD) is a more specific term defined by Soldati (2013) as a gravity-induced process affecting large portions of slopes evolving over long periods of time, and gravitational slope-deformation features (GSDF) as the geomorphic manifestations of gravitational slope deformation. These would include the cracks, scarps, and anti-scarps commonly included in the definition of "sackungen".

A simple distillation of current usage is that (a) DSGSD, its shortened version GSD (gravitational slope deformation), and sackungen refer to large rock failures that have begun but have not achieved a state of (rapid) collapse yet; they could be regarded as metastable, (b) GSDF refers to manifestations on the ground of GSD, and (c) RSF, although it technically could include creeping failures that have not collapsed yet, is used for rapid or catastrophic failures. In this paper we will use "GSD" and "GSDF" in the sense that Soldati (2013) defined them, and RSF to mean rapid collapse.

The hazard of GSDFs is that they have the potential to develop into RSFs. According to Soldati (2013), volumes of rock displaced in GSDFs can be up to hundreds of millions of cubic meters, with thicknesses of up to a few hundred meters.

GSD is becoming recognized increasingly as an important contributor to landscape evolution in high-relief terrains (e.g., Hewitt 2009; McColl 2012). As surficial mapping and remote sensing of slope failures have improved, GSDFs have been recognized world-wide as signaling potential hazard to humans in steep mountainous terrains (Crosta and Clague 2006; Evans et al. 2007; Jaboyedoff et al. 2011; Hermanns and Longva 2012).

In eastern Canada, Grant (1974, 1987) and Brookes (1993) identified twenty-four examples of GSD, mostly developed in the ultramafic rocks of western Newfoundland. In western Newfoundland stress on rock masses was likely increased by glacial steepening of valley-walls, and deglacial stress release likely contributed to RSF of plateau-edges and valley-walls (Osborn et al. 2007). GSDFs in this region are characterized by scarps of considerable vertical displacement and the absence of anti-scarps. The lack of anti-scarps may be due in some cases to glacial erosion of the failed mass,

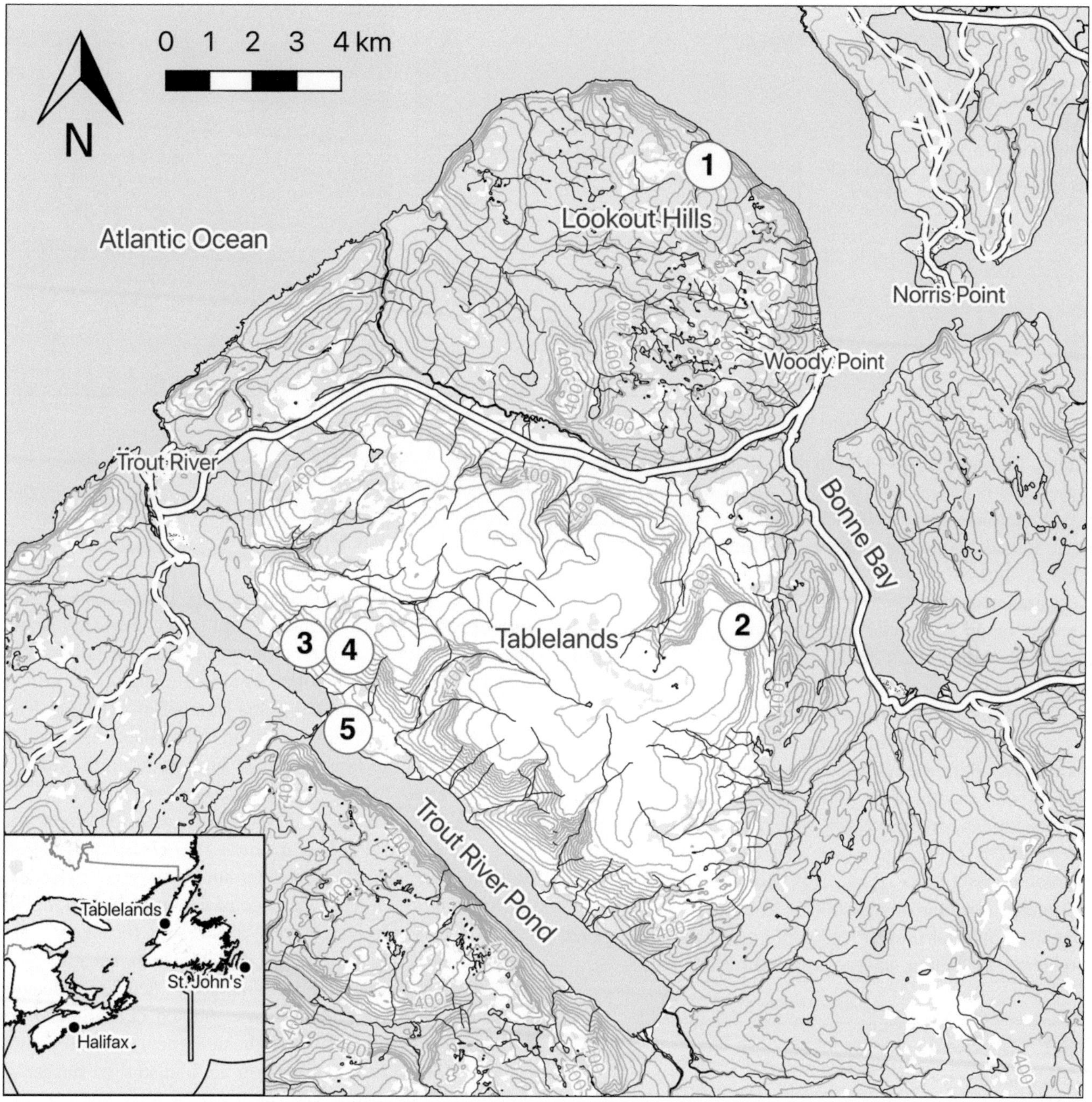

Fig. 16.1 Study area in the Tablelands and Lookout Hills of Western Newfoundland. All GSDF sites mentioned in this paper are situated on glacially steepened valley walls. The region is characterized by highly jointed and fractured ophiolitic rocks; peridotite and gabbro are prominent. Numbers 1–5 refer to GSD sites discussed in text

implying that some failures predated the last glaciation (Grant 1987). At some sites fractures several hundreds of meters long have opened by a few meters, within 100–200 m of, and parallel to, a plateau-edge (Grant 1987). No evidence of the age of these has been gathered and they might have been widening/lengthening at very slow rates since they first formed.

16.2 Geological Setting

The Tablelands and Lookout Hills form the southern part of Gros Morne National Park. The physiography of the study region is dominated by massifs (Fig. 16.1), with steep boundary slopes descending from smoothly undulating or

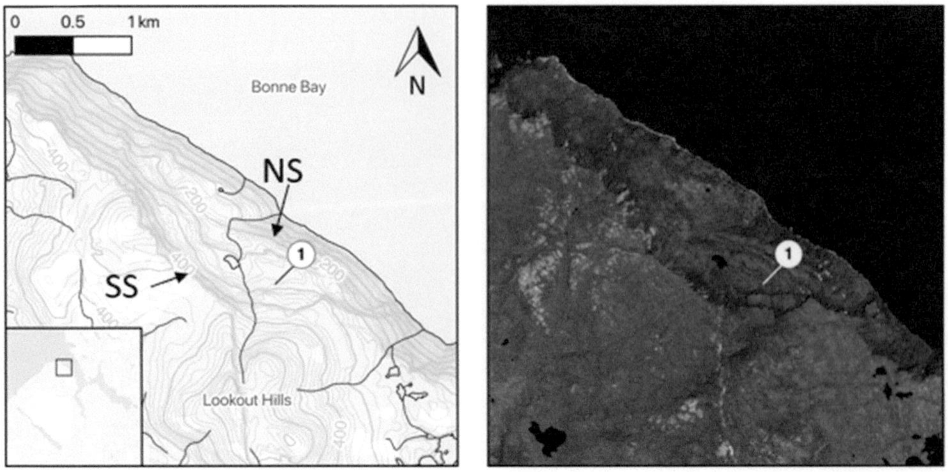

Fig. 16.2 The Lookout Hills GSDF. The Southern Slump which may contain 10 km^3 of material, appears to have been lowered as a coherent block. The southern scarp (SS) is characterized by many narrow scarps which are sub-parallel to the coastline. NS is the northern scarp

rolling plateaus at 600–800 m asl. Between or bordering these massifs are lowland corridors which were occupied by pre-Quaternary rivers and Pleistocene glaciers that drained from inland divides into the proto-Gulf of St. Lawrence. Valleys incised into the margins of these massifs, which were initiated fluvially in the mid-Cenozoic (Eyles 1996), show a range of glacial erosional overprints, particularly U-shaped low-gradient troughs which have resulted from glacial erosion.

The majority of the Long Range Mountains in Gros Morne National Park are underlain by Proterozoic granite and gneiss (Berger et al. 1992). Adjacent to the coast, a 5–10-km-wide piedmont is underlain mainly by tilted Paleozoic continental-margin clastic and carbonate sediments. In the southern part of the park, the subject of this paper, peridotite and gabbro of the Gros Morne ophiolite suite are distinctive for their orange color, barrenness owing to plant toxicities, and flat-topped landscapes controlled by regional structures, glaciation, and post-glacial surface processes.

16.3 Methods

Field work was conducted in 2002, 2003, 2010, and 2016. In 2002 and 2003 geochronology study on the timing of fjord deglaciation was conducted by Osborn et al. (2007). Mapping of glacial and periglacial deposits was completed at a scale of 1:25,000 with the aid of air photos. We reviewed aerial photography and a 50+ year photographic record of landscape change (Berger 2017). In 2016, the sites were again investigated in the field, although the Lookout Hills feature was examined primarily using remote sensing imagery.

16.4 Results

16.4.1 Lookout Hills GSDF

The Lookout Hills GSDF (Site 1, Fig. 16.1 and 16.2) consists of a primary scarp, and several lower secondary scarps that may represent two or more generations of failure. The composite feature is almost 6 km long, contains 8.3 km^3 of material and is sub-parallel to the Bonne Bay coastline. This feature is the largest aerially of the five GSDFs described in this paper and the largest known in eastern North America (Spooner et al. 2013).

The Lookout Hills GSDF has been described by Cumming and Grant (1974) and Grant (1974). Brookes (1993) remarked that the "confused terrain of ridges, cracks, and depressions which, near the edge of the cliff overlooking the entrance to Bonne Bay, is very treacherous for hikers". The Lookout Hills GSDF occurs in an area which, because of its rugged relief, has not yet been mapped in detail. The feature formed in heavily jointed, coarse-grained crystalline rocks composed of gabbro and anorthositic gneiss (Williams and Cawood 1989). Joint patterns on the undisturbed surface to the south of the feature are roughly orthogonal and are clear in both photographs and satellite images. An area of intense crevice development (Fig. 16.3) appears fresh, with some suggestion of alteration since early airphotos have been taken on the site.

The largest headscarp is associated with a feature that Grant (1974) called the Southern Slump. It is 75–130 m in height, and the scarp (Southern Scarp, Fig. 16.2) extends for almost 3 km. The main escarpment is likely of composite origin as it appears to be scalloped and may have formed from two or more discrete movements (Figs. 16.2, 16.3, 16.4 and 16.5). The eastern side of the escarpment is smaller in height and lateral extent and the headscarp is less clearly defined (Fig. 16.2). The surface of the Southern Slump appears to have been lowered as a coherent block. Three small ponds are located on the surface. However, domal features with prominent crevices (tension cracks) are also present and have formed parallel to the main scarp. The more northerly and smaller headscarp (Northern Scarp, Fig. 16.2) delineates an elongated lens-shaped block characterized by many narrow scarps which are sub-parallel to the coastline.

Grant (1973) noted that the southern scarp (Fig. 16.2) appears to truncate a post-glacial marine terrace fringing Bonne Bay at about 60 m asl. He also contended that following retreat of glacial ice from Bonne Bay failure of the over-steepened walls occurred sometime after the deposition of a raised marine terrace. Grant (1973) also suggested that collapse of the underlying soluble strata in the over-deepened floor of Bonne Bay may have contributed to collapse of the Southern Slump (Fig. 16.3).

16.4.2 GSDF in the Tablelands

The "Tablelands" (generally expressed in the plural) refer to a steep-flanked, rolling plateau rising to 711 m asl, south of the Bonne Bay fjord system (Fig. 16.1). The edges of the plateau are steepest and best-defined on the north, where the margin is demarcated by a once glaciated valley and on the southwest, where Trout River Pond fills the bottom of a fjord valley (Fig. 16.1). The plateau is underlain by ophiolitic rocks of the Humber Arm Allochthon (Williams and Cawood 1989). Gabbro is exposed in the northwestern third of the upland, but elsewhere highly fractured peridotite is at the surface. The Tablelands are characterized by barren, frost-shattered rock, brown-weathering where peridotite is exposed and somewhat more vegetated on gabbro surfaces. Soils are thin to non-existent, and evidence of active cryoturbation is common. The summit plains of the Tablelands are generally regarded as remnants of a planation surface(s) uplifted in Eocene time (Berger et al. 1992; Brookes 1993) but unpublished thermochronologic evidence suggests the summit plateaus and some of the relief existed as early as the Jurassic.

16.4.2.1 GSDF Site 2: Shoal Brook

The Shoal Brook GSDF (Site 2, Fig. 16.1) is located on the northeastern escarpment of the Tablelands Plateau where the edge of the plateau is steepest and best-defined (Fig. 16.6). The head scarp of the Shoal Brook GSDF is about 750 m long, slightly concave to the west and exhibits vertical displacement of about 20 m (Fig. 16.7). The failure surface is characterized by many small, stepped failures as well as overlying younger debris cones.

16.4.2.2 GSDF Sites 3, 4, 5: Sandy Top, G4 and Miners Point

On the south side of the Tablelands GSDFs of two possible generations have been recognized—one likely pre-LGM (Last Glacial Maximum) and two post-last glaciation (<12.0 ka; Osborn et al. 2007). Relative ages of these features were first proposed by Brookes (1993). The rock-mass of Miners Point (Site 5, Fig. 16.8) is composed of peridotite on its eastern side, and of foliated peridotite and gabbro at its western end. The same transition occurs at the southern edge and across the width of the Tablelands plateau, some 600–700 m above Miners Point. Previous publications have explained the elevation difference by Early Paleozoic faulting (e.g., Smith 1958). However, while a fault zone must indeed separate the Tablelands massif to the north of Trout River Ponds from mafic volcanic Gregory Plateau to the south, the much more recent origin of the movement of the Miners Point block is indicated by its relationship to glacial features. At best the Paleozoic fault is a fault line scarp.

As shown in Brookes (1993), the Miners Point rock-mass overrode glacially streamlined till. The mass itself is also glacially streamlined across its rocky summit, which indicates that failure occurred pre-LGM (between $\sim$40 and 12 ka). Also noteworthy is the plan shape of the Miners Point mass, which easily fits into an indentation in the cliff edge of Tablelands plateau.

The rock mass NW of Miners Point, known as Sandy Top (Site 3, Figs. 16.8, 16.9 and 16.10), is located vertically half-way between pond level and the edge of Tablelands plateau, at 600 m asl. Its gabbroic rock can be matched to the plateau-edge above. A post-LGM age is suggested by the lack of glacial streamlining of the failed block.

Between Miners Point and Sandy Top, immediately below the plateau-edge, a valley wall failure called G4 (Smith 1958; Brookes 1993) has dropped a third integral mass, this of gabbroic rock at 530 m asl, only $\sim$70 m from the plateau-edge (Site 4, Figs. 16.8 and 16.9). The outward-sloping surface of the block has been roughened by jostling of joint-bound blocks during failure, but there are no glacial features on it. If this surface had been glaciated it would not be as intact as it is. At this elevation glacial over-steepening was likely not a factor in destabilization; dilation of the rock-mass in response to earlier (possibly much earlier) steepening is a possible mechanism.

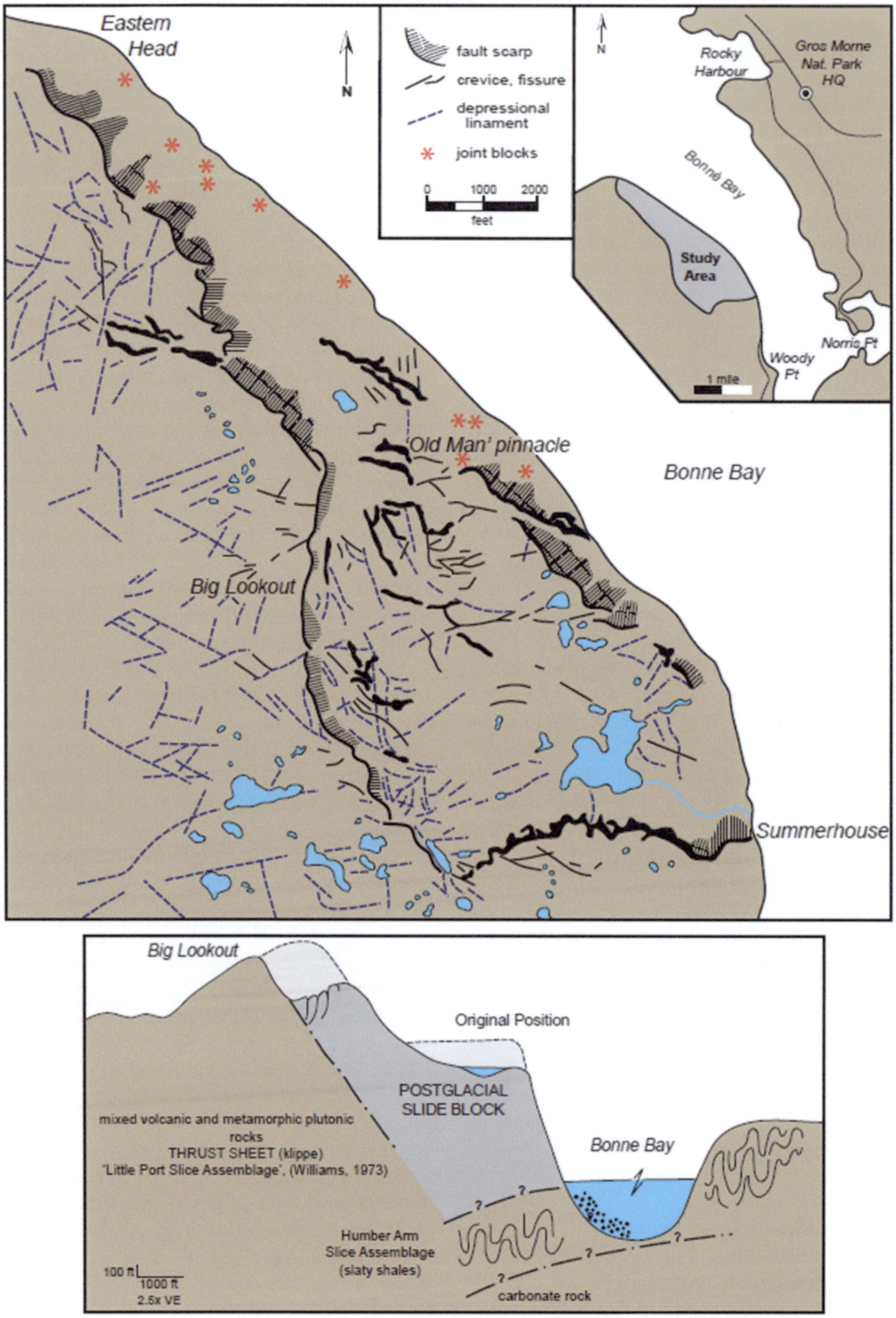

Fig. 16.3 The Lookout Hills GSDF. Grant (1974) hypothesized that the age of the feature was post-LGM and failure was associated with either the removal of the formerly supportive glacier, collapse of soluble underlying strata or by gravity slip of the fractured rock mass. (Figure after Grant 1974)

Fig. 16.4 View across Bonne Bay from Lobster Cove Head, to Lookout Hills showing the main southern failure scarp (black arrow) and the smaller northern failure scarp (white arrow). There is about 80 m of displacement on the southern failure scarp

Fig. 16.5 Significant vertical displacement and crevices in a domal region below the Southern Scarp. These crevices are tension cracks, and most are parallel to the main scarp. This domal area was originally a part of a low hill which rose above the general level of the surrounding surface of the plateau

16.5 Discussion

The causal mechanism for the GSDFs described in this paper is uncertain. As all are located adjacent to glacially over-steepened valleys, the failures may have developed in response to the removal of laterally supporting bedrock during glaciation, triggering lateral expansion when the ice ablated. Similar slope failures, albeit on smaller scales, are associated with valley down-cutting and "valley rebound" in other areas (e.g., Babcock 1977).

Grant (1974) assigned timing of initial failure of the Lookout Hills GSDF to the post-LGM period though apparent glacially-truncated spurs on the lower slopes of the northern headscarp may indicate that some movement either pre-dates the LGM or occurred within this glaciation

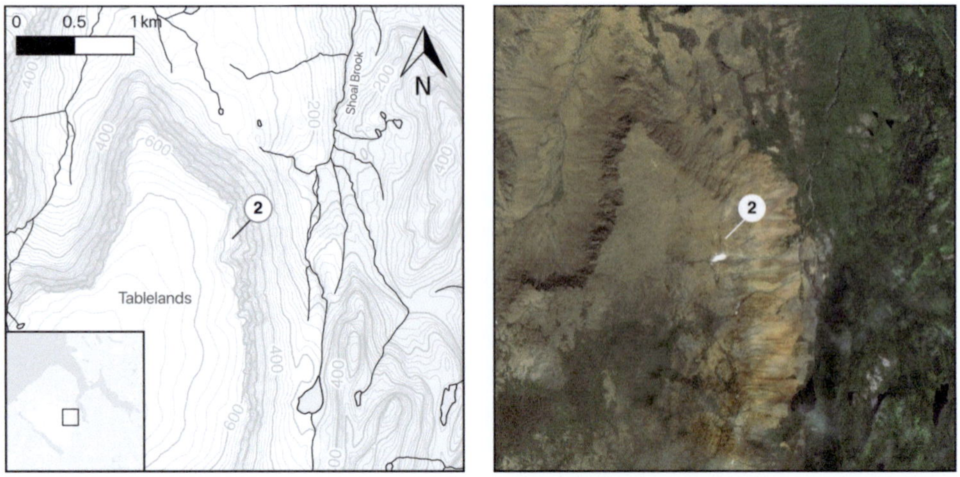

Fig. 16.6 Shoal Brook GSDF (1). The Shoal Brook GSDF is on the easternmost edge of the Tablelands and comprises about 0.7 km^3 of material in highly fractured peridotite. This failure does not present a significant risk or hazard to humans

Fig. 16.7 View southward along the eastern edge of the Tablelands showing the westward-convex scarp of the Shoal Brook GSDF with 20 m vertical displacement. Small, stepped failures are evident as well as younger debris cones (Photograph was taken in July 1988)

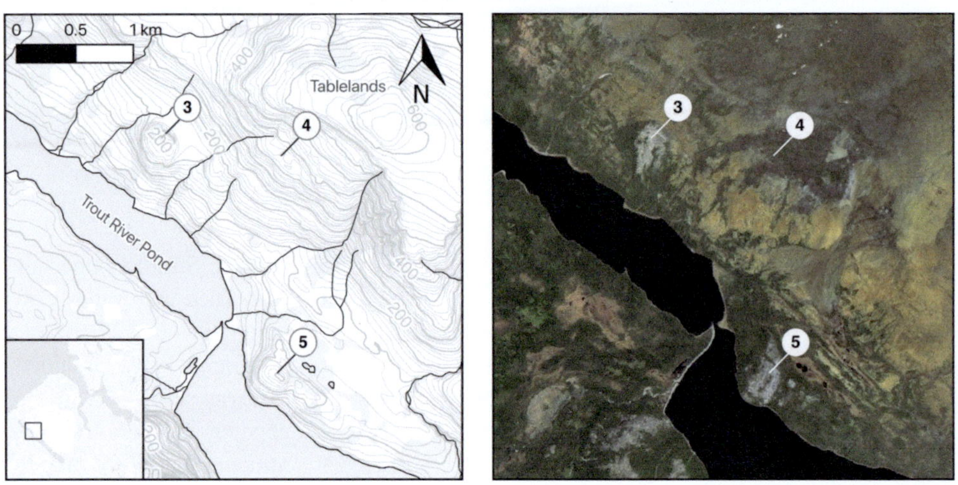

Fig. 16.8 Miners Point failures. The Sandy Top feature (3) is likely pre-LGM whereas the Miners Point Feature (5) is likely post-LGM. Feature 4 is post-LGM and comprises about 1 km^3 of material

Fig. 16.9 Sandy Top GSDF (3) and GSDF G4 (©Google Earth 2019). The Sandy Top feature comprises almost 1 km^3 of material

Fig. 16.10 Sandy Top GSDF

(25–12 ka). Cumming and Grant (1974) also noted that initial glacial rebound followed by more recent regional subsidence may have resulted in "crustal warping" which would be expected to stress the rocks and lessen their effective cohesion.

The role of jointing in the Lookout Hills GSDF is likely very significant. The rocks of the slumped area of the Lookout Hills are highly jointed. Terzaghi (1962) noted that a mechanism in all rock slides is the development of joints after removal of lateral support and that the critical height of slopes on unweathered rock is influenced more so by joints and faults than by the intact strength of the rock. In the Lookout Hills GSDF lateral support was provided by glacial ice during glaciation.

The time and rate of movement of the Lookout Hills GSDF remains uncertain. Though significant post-glacial movement has occurred, all that can be said for certain is that about 80 m of movement has occurred in about 10,000 years (Grant 1987). However, the slope failure at Lookout Hills probably took place within a short interval of time. A future paleolimnological approach in which ponds on the Lookout Hills and the Shoal Brook GSDF are cored and basal dates obtained may provide some insight into the timing of surface deformation associated with movement. Coring in Bonne Bay has the potential to provide insight into the timing of post-glacial movement if it indeed took place.

Periglacial processes have been very active in the study region and may have also played an important role in initiating movement. Osborn et al. (2007) noted that relict rock glaciers are common in the Tablelands and likely contained significant volumes of ice in their cores when they formed. Osborn et al. (2007) suggested that abundant snow-and-debris avalanches during the spring and early summer led to accumulation in talus aprons at a time when snow accumulation was more substantial than it is now. This same mechanism, combined with active freeze–thaw and intense jointing may have been an important contributor to slope instability for all of the GSDFs discussed.

Evaluating the effects of these features may have application to the continuing development of models of deglaciation for eastern North America. An understanding of the timing and causal mechanism of GSDFs in the Tablelands region may provide context to recent research that suggests that some parts of western Newfoundland deglaciated much earlier than others (Osborn et al. 2007). New technologies (drones, high resolution topographical models developed from drone imagers (e.g., Rossi et al. 2015)) offer the opportunity to produce high resolution images of the surfaces of these features, providing greater clarity of geomorphic processes.

Landslide-generated displacement waves are more limited than tsunamis produced by ocean earthquakes. However, the displacement of water can create massive waves (>100 m run-up heights) that can be devastating locally. A landslide-generated wave which occurred in Lituya Bay, Alaska in 1958 was a result of an earthquake triggered rock avalanche (Weiss et al. 2009). The rock avalanche, which had an estimated volume of 30 M m^3, impacted on the waters at the head of the bay and generated a wave inundation distance of 524 m, the highest recorded in history. As noted earlier, a massive rock slope failure in western Greenland produced tsunamic waves that resulted in considerable damage to property and loss of life. The Lookout Hills GSDF contains approximately 8.3 km^3 of material. If even a small fraction of this entered Bonne Bay as a rock avalanche a destructive tsunamic wave could be generated. The proximity of the Lookout Hills GSDF to significant coastal population (approx. 3000) within 10 km and the low to moderate potential for a magnitude 4 or greater earthquake warrants a more quantitative examination of the risk that this feature may pose.

Catastrophic failure into Trout River Pond of either the Miners Point or Sandy Top GSDFs could be equally destructive as the village of Trout River (pop. 600) is constructed upon an ice-marginal marine delta that lies at the mouth of Trout River Pond. However, the potential for the generation of a rapidly moving rock avalanche from either GSDF appears to be minimal due to the low-angled glide plane for both features.

Seismic shaking might well trigger future movements of these GSDFs. Distant earthquakes were felt in Bonne Bay in 1925 and 1929 (Berger 2014). Moreover, the Lower St. Lawrence Seismic Zone is a seismically active region of eastern Canada that is located less than 500 km from the western Newfoundland coastline. On June 23, 1944, an earthquake of magnitude 5.1 on the Richter scale occurred east of Baie-Comeau and on March 16, 1999, an earthquake of magnitude 5.1 occurred in this region, at about 60 km south of Sept-Iles (Adams and Basham 1989).

We recommend follow-up studies of these sites. Monitoring the changes in fracture widths in key locations with annual tensiometer measurements could provide immediate insight into the activity of selected fractures. In addition to chronology of the basal ponded sediments mentioned above, exposure dating of the sliding planes and breakout surfaces would provide direct chronology on the displacements to evaluate whether or not they formed shortly after deglaciation, or if there is any evidence of acceleration of the GSDFs. Exposure dating of boulders in catastrophically failed rock slides and rock avalanches in the region, along with chronology of the slip faces, could also help establish links to climate change or paleoseismicity.

16.6 Conclusion

GSDFs occur along steep bedrock slopes of western Newfoundland but their age, activity, and causes have not been adequately determined. Despite documented seismic triggering of submarine landslides around Newfoundland, there is no evidence that GSDFs in the study region were generated by seismic ground acceleration. Other plausible causes include a re-balancing of stress fields along slopes which were erosionally over-steepened. There is potential that the GSDF in the Lookout Hills and Shoal Brook regions could become catastrophic massive rock slope failures. The Lookout Hills GSDF could pose a considerable hazard considering its potential to generate a displacement wave in Bonne Bay which could affect a number of coastal communities. Opportunities for risk analysis of large scale GSDFs along the Atlantic Canada coastline exist especially in light of new technologies and approaches.

Acknowledgements This paper is based in part on an unpublished review by the late Ian Brookes of rock-slope failures in western Newfoundland. The authors would like to acknowledge the editorial contributions of O. Slaymaker and N. Catto. Helpful comments from D. Liverman were incorporated into the paper. Funding from NSERC to J. Gosse and I. Spooner is gratefully acknowledged.

References

Adams J, Basham P (1989) The seismicity and seismotectonics of Canada East of the Cordillera. Geosci Can 16:3–11

Babcock EA (1977) A comparison of joints in bedrock and fractures in overlying Pleistocene lacustrine deposits, central Alberta. Can Geotech J 14:357–366

Berger AR (2014) The good and beautiful bay. A history of bonne bay to confederation and a little beyond. St. John's NL, Flanker Books

Berger AR (2017) Tracking rapid landscape change with repeated photography, Gros Morne National Park, Newfoundland, Canada. Atl Geol 53:115–126

Berger AR, Bouchard A, Brookes IA, Grant DR, Hay S, Stevens RK (1992) Gros Morne National Park, Newfoundland—geology, topography, vegetation. Geol Surv Can Miscellaneous Report 54, 1:150,000 scale

Brookes IA (1993) Canadian landform examples 26: table mountain, Gros Morne National Park, Newfoundland. Can Geogr 37:69–75

Broom LM, Campbell DC, Gosse JC (2017) Investigation of a Holocene marine sedimentary record from Pond Inlet, Northern Baffin Island, Nunavut. Summary of Activities, pp 93–104

Chao W-A, Wu T-R, Ma K-F, Kuo Y-T, Wu Y-M, Zhao L, Chung M-J, Wu H, Tsai Y-L (2018) The large Greenland landslide of 2017: was a Tsunami warning possible? Seismol Res Lett 89:1335–1344

Crosta GB, Clague JJ (2006) Large landslides: dating, triggering, modelling, and hazard assessment. Eng Geol 83:1–3

Cumming LM, Grant DR (1974) Bedrock landslides of postglacial age in the lookout hills region of Gros Morne National Park. Geol Surv Can, Ottawa, ON

Evans SG, Mugnozza GS, Strom A, Hermanns RL (2007) Landslides from massive rock slope failure. Springer Science & Business Media

Eyles N (1996) Passive margin uplift around the North Atlantic region and its role in Northern Hemisphere late Cenozoic glaciation. Geology 24:103–106

Grant DR (1973) Terrain conditions, Gros Morne National Park, Western Newfoundland. In: Report of activities Part B: November 1972 to March 1973. Geological Survey of Canada, Paper no. 73-1B, pp 121–125

Grant DR (1974) Terrain studies of Cape Breton Island, Nova Scotia and the northern Peninsula, Newfoundland. In: Blackadar RG (ed) Report of activities Part A: April to October 1973. Geological Survey of Canada, Paper no. 74-1A, pp 241–246

Grant DR (1987) Excursion guide book A3/C3: Quaternary geology of Nova Scotia and Newfoundland. In: French HM, Richard P (eds) International union for quaternary research, Ottawa, ON, p 62. https://geoscan.nrcan.gc.ca/starweb/geoscan/servlet.starweb?path=geoscan/shorte.web&search1=R=4758

Harrison JV, Falcon NL (1938) An ancient landslip at Saidmarreh in Southwestern Iran. J Geol 46:296–309

Hermanns RE, Longva O (2012) Rapid rock-slope failures. In: Clague JC, Douglas Stead D (eds) Landslide types, mechanisms and modeling. Cambridge University Press, pp 59–70

Hewitt K (2009) Glacially conditioned rock-slope failures and disturbance-regime landscapes, upper Indus Basin, northern Pakistan. Geol Soc, London, Special Publications 320:235–255

Holm K, Bovis M, Jakob M (2004) The landslide response of alpine basins to post-Little Ice Age glacial thinning and retreat in southwestern British Columbia. Geomorphology 57:201–216

Jaboyedoff M, Crosta GB, Stead D (2011) Slope tectonics: a short introduction. Geol Soc, London, Special Publications 351:1–10

Jenner KA, Piper DJ, Campbell DC, Mosher DC (2007) Lithofacies and origin of late Quaternary mass transport deposits in submarine canyons, central Scotian Slope, Canada. Sedimentology 54:19–38

McColl ST (2012) Paraglacial rock-slope stability. Geomorphology 153–154:1–16

Osborn G, Spooner I, Gosse J, Clark D (2007) Alpine glacial geology of the Tablelands, Gros Morne National Park, Newfoundland. Can J Earth Sci 44:819–834

Roberts NJ, Evans, SG (2008) Seymareh (Saidmarreh) landslide, Zagros Mountains, Iran (EGU2008-A-00764). In: Geophysical research abstracts. European Geosciences Union, Vienna, Austria

Rossi G, Nocentini M, Lombardi L, Vannocci P, Tanteri L, Dotta G, Bicocchi G, Scaduto G, Salvatici T, Tofani V, Moretti S, Casagli N (2015) Integration of multicopter drone measurements and ground-based data for landslide monitoring. In: Proceedings of international symposium of landslides

Smith CH (1958) Bay of Islands Igneous Complex, western Newfoundland. Geological Survey of Canada, Ottawa, ON

Soldati M (2013) Deep-seated gravitational slope deformation. In: Bobrowsky PT (ed) Encyclopedia of natural hazards. Springer, Dordrecht, pp 151–154

Spooner I, Batterson M, Catto N, Liverman D, Broster BE, Kearns K, Isenor F, McAskill GW (2013) Slope failure hazard in Canada's Atlantic Provinces: a review. Atl Geol 49

Stein S, Sleep NH, Geller RJ, Wang S-C, Kroeger GC (1979) Earthquakes along the passive margin of eastern Canada. Geophys Res Lett 6:537–540

Terzaghi K (1962) Stability of steep slopes on hard unweathered rock. Géotechnique 12:251–270

Vacchi M, Engelhart SE, Nikitina D, Ashe EL, Peltier WR, Roy K, Kopp RE, Horton BP (2018) Postglacial relative sea-level histories along the eastern Canadian coastline. Quat Sci Rev 201:124–146

Williams H, Cawood PA (1989) Geology of the humber arm allochthon, Newfoundland. Geological Survey of Canada, Ottawa, ON

Weiss R, Fritz HM, Wünnemann K (2009) Hybrid modeling of the megatsunami runup in Lituya Bay after half a century. Geophys Res Lett 36. https://doi.org/10.1029/2009gl037814

Ian Spooner is Head of the Earth and Environmental Science Department at Acadia University in Nova Scotia, Canada. His primary research interest is investigating environmental change and he has active research programs in northwestern British Columbia, Nova Scotia and Newfoundland. He has consulted in the areas of Environmental and Risk Assessment, groundwater and surface water contamination, coastal erosion and has been involved in environmental risk assessments for resource companies in Nova Scotia, British Columbia and the Northwest Territories. He has served with several professional organizations and was recently awarded the Distinguished Service Award by the Atlantic Geoscience Society.

Dewey Dunnington is an environmental researcher and programmer based in Halifax, Nova Scotia. His research has focused on the accumulation of trace metals in Maritime-Canadian lakes, and the application of emerging data management tools to environmental research. His research has taken him to Nova Scotia, New Brunswick, British Columbia, and Newfoundland, and his technological endeavours have produced mobile applications, data management tools, and open-source mapping solutions to increase the efficiency of environmental research. He is currently pursuing his Ph.D. in Civil and Resources Engineering at Dalhousie University.

John Gosse is a native of western Newfoundland, and is a Killam Professor in Earth Sciences at Dalhousie University, Halifax. His research interests include geohazards and active tectonics in Tibet, Himalaya, Andes, Italy, and USA. In the past decade John has participated in research regarding post-glacial landslide and displacement wave risk in Norwegian fjords with the NGU. He is co-leader of the ULINNIQ Project of MEOPAR, which focusses on terrestrial and marine records of coastal mass wasting, displacement wave and tsunami risk, and seismic hazards for eastern Arctic Canada.

Jerry (Gerald) Osborn is a Professor Emeritus in the Geoscience Department at the University of Calgary. His interests are surficial and Quaternary geology, with forays into geomorphology and engineering geology, and, on the side, interactions between science and society. His main line of research is Holocene climate change using glacial history. Consulting activities have included aggregate searches, mass-movement hazards analysis, flood hazard analysis, and studies of river migration as applied to boundary-law litigation.

Tony (Anthony) Berger graduated with a Ph.D. from the University of Liverpool (UK) has written or co-edited 10 scientific books and monographs on subjects ranging from Irish granites to rapid landscape change, medical geology, the response of past societies to natural disasters, and technical assistance to developing countries. During his varied career he taught geology in Toronto, Sri Lanka, St. John's, and Kuala Lumpur, played a key role in establishing several international scientific organizations, edited EPISODES for International Union of Geological Sciences and headed its Publication Committee, and continues to track rapid geological change in Gros Morne National Park. In 2014 his book on the history of Bonne Bay, Newfoundland, was published by Flanker Books, St. John's.

Ian Brookes (deceased 2015) was Emeritus Professor of Geography at York University, Toronto, where he taught courses in geomorphology and glacial landforms, and in Canadian landscapes and the regional geomorphology of Canada, His research was primarily focused in Newfoundland, but he also carried out field research in New Zealand, Iran, Egypt, and in the Daklah Oasis area in the Sahara Desert. He was founder and editor from 1985 to 1993 of "Canadian Landform Examples," a regular series in the professional quarterly journal, "The Canadian Geographer".

Quaternary Heritage and Landscape in the Témiscouata–Madawaska Valley, Eastern Québec

17

Antoine Morissette, Bernard Hétu, and Patrick Lajeunesse

Abstract

The Quaternary heritage and landscape of the Témiscouata–Madawaska Valley consists of an assemblage of landforms and deposits associated with former stabilizations of glacial margins during the Late Pleistocene development of an ice-free corridor in the Notre-Dame Ice Cap (NDIC), through which Glacial Lake Madawaska developed. This lake occupied an area of more than 500 km^2 in the Témiscouata–Madawaska valley from late-glacial times to the middle of the Holocene, depositing a thick series of varves. The contact between Madawaska Lake and the ice margin of the NDIC contributed to the acceleration of deglaciation and the fragmentation of NDIC into two local ice caps: the Lejeune Ice Cap (LIC) and the Pohénégamook Ice Cap (PIC). In the Bas-Saint-Laurent and particularly in the Témiscouata–Madawaska valley, the Quaternary landscape is the product of the coexistence of an Appalachian ice margin and the glacial lake Madawaska.

Keywords

Quaternary heritage • Laurentide ice sheet complex • Bas-Saint-Laurent • Témiscouata–Madawaska valley • Glacial Lake Madawaska • Notre-Dame Ice Cap

17.1 Introduction

The Bas-Saint-Laurent (BSL) region is located between the towns of La Pocatière and Matane, along the south shore of St. Lawrence estuary (Québec; Fig. 17.1). The landscape consists of stepped plateaux with different altitudes from the sea-level coastal plain to the highlands around 500 m, influenced by past glaciations. A smooth transition occurs from the St. Lawrence Lowlands in the west to the Gaspé high plateau toward the east (Fig. 17.1), with a lower plateau between Lake Témiscouata and the Matane River. There is a clear contrast between the coastal plain (0–250 m) to the north and the Appalachian Highlands (400–900 m) to the south, separated by a steep piedmont (250–400 m) (Fig. 17.1).

This landscape generally follows the regional southwest–northeast Appalachian bedrock structure except for the Témiscouata and Matapédia valleys, which notched the Appalachian Mountains following the transform fault in the Gaspé rift zone (Allen et al. 2009). The BSL was occupied by the Laurentide Ice Sheet Complex (LISC) during the MIS 2 (Late Wisconsinan) glaciation (Gadd 1964; Chauvin et al. 1985; Dyke et al. 2002; Occhietti et al. 2004, 2011), while the eastern Gaspé peninsula was occupied by the regional Gaspé Ice Cap (GIC) (David and Lebuis 1985; Charbonneau and David 1993; Olejczyk and Gray 2007).

During deglaciation, the LISC margin retreated toward the north (Occhietti et al. 2004, 2011) while the St. Lawrence estuary opened (Thomas 1977). Consequently, the Notre-Dame Ice Cap (NDIC) became isolated from the LISC in the BSL, thereby favoring the deposition of moraines in the piedmont of the Appalachians (Martineau and Corbeil 1983; Chauvin et al. 1985). These moraines are the only ones associated with the northern margin of an Appalachian ice in the BSL, with the Saint-Antonin (Martineau and Corbeil 1983; Chauvin et al. 1985; Occhietti et al. 2004, 2011) and St-Jean-Port-Joli (Martineau and Corbeil 1983; Chauvin et al. 1985; Occhietti et al. 2004, 2011) moraines to

A. Morissette (✉) · P. Lajeunesse
Département de géographie & Centre d'études nordiques (CEN), Université Laval, Quebec City, QC G1V 0A6, Canada
e-mail: antoine_morissette@uqar.ca

B. Hétu
Département de Biologie, Chimie et géographie & Centre d'études nordiques (CEN), Université du Québec à Rimouski, Rimouski, QC G5L 3A1, Canada

© Springer Nature Switzerland AG 2020
O. Slaymaker and N. Catto (eds.), *Landscapes and Landforms of Eastern Canada*, World Geomorphological Landscapes, https://doi.org/10.1007/978-3-030-35137-3_17

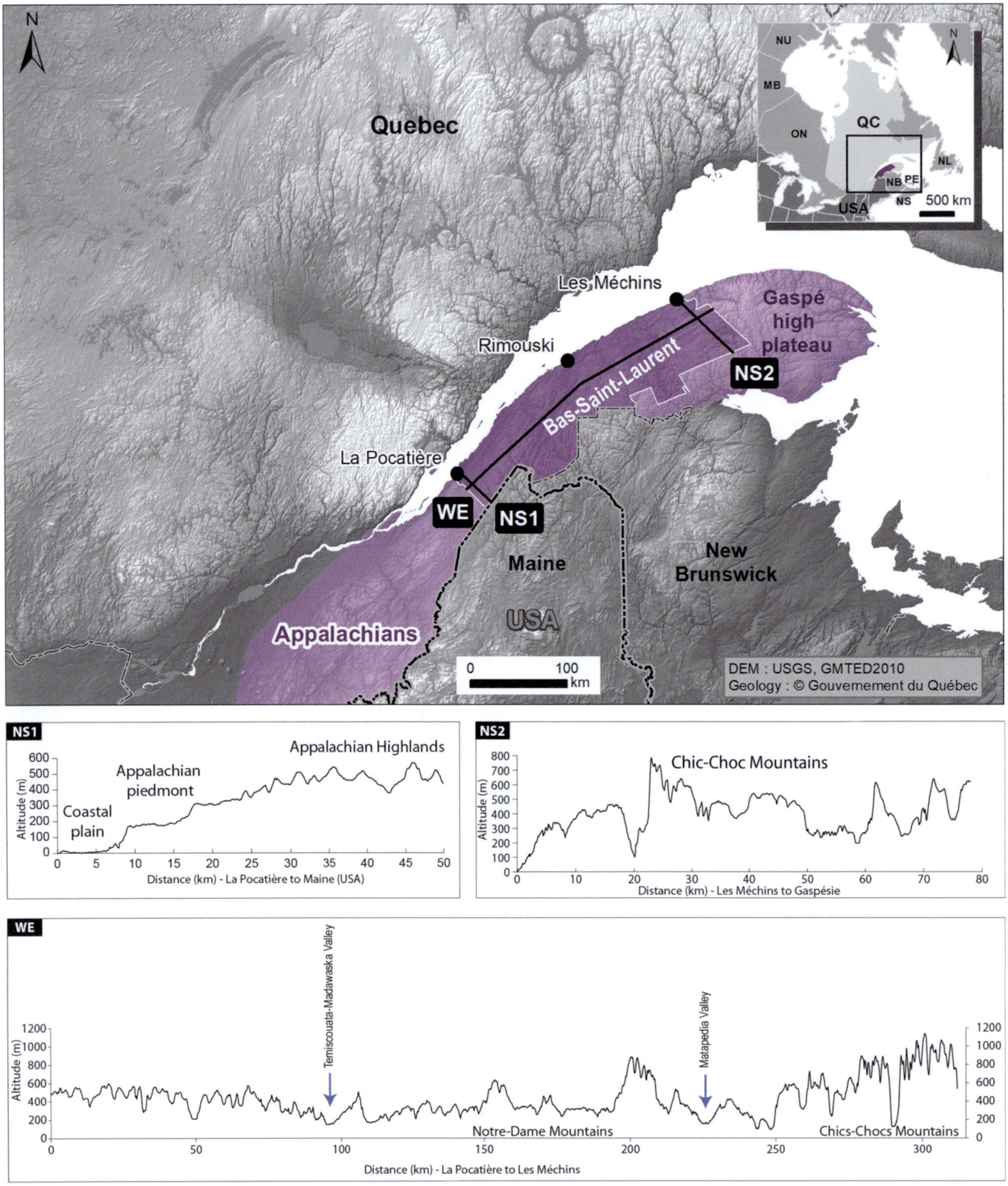

Fig. 17.1 Top: Location and geological and physiographic regions of the Bas-Saint-Laurent, Eastern Québec. NS1: Topographic profile of the Bas-Saint-Laurent along a north-south transect near La Pocatière showing relief contrast between the coastal plain, the Appalachian piedmont and the Appalachian Highlands. NS2: Topographic profile of the Bas-Saint-Laurent along a north-south transect near Les Méchins, where the Chic-Choc Mountains rise. The coastal plain is not represented in this part of the BSL; the structural front of the Chic-Choc replaces the Appalachian piedmont and the Chic-Choc thrust sheet forms the high plateau of the Chics-Chocs Mountains. WE: Topographic profile of the Bas-Saint-Laurent along a west-east transect outlines the transitional landscape between the St. Lawrence Lowlands in the west to the Gaspé high plateau toward the east, with a lower plateau between Lake Témiscouata and the Matane River

the west and the Luceville and Neigette moraines to the east (Hétu 1998). This geomorphological continuum is interrupted by the Trois-Pistoles delta (Lee 1962; Dionne 1968; Rappol 1993) that opens onto the Témiscouata–Madawaska Valley, which was occupied by glacial Lake Madawaska (Kiewiet de Jonge 1951; Lougee 1954; Kite 1983; Rampton et al. 1984; Kite and Stuckenrath 1989). Here, we report the opening of NDIC and describe the glaciated Appalachian landscape of the region by two local ice caps: the Lejeune Ice Cap (LIC) to the east of Témiscouata–Madawaska Valley and the Pohénégamook Ice Cap (PIC) to the west. Glacial Lake Madawaska influenced the processes of deglaciation in the Appalachian Highlands, including local ice cap margins and depositional assemblages.

17.2 Study Area

17.2.1 Topography

The regional physiography played an important role in the growth and decay of local ice caps originating from the NDIC. The flat coastal plain borders the steep piedmont of the northern Appalachian border. South of the piedmont, the Appalachian highlands reach 500 m asl, with only a few important valleys dissecting the landscape. During deglaciation, the ice margin of local ice caps responded to that topography (Gadd 1964). Remnants of these ice margins are revealed by moraine landform assemblages, forming an important heritage on the Quaternary landscape in the BSL.

17.2.2 Geology

The BSL covers an area of 28,401 km^2 with more than 6,000 km^2 located in the St. Lawrence Estuary. This region is delimited to the north by the St. Lawrence Estuary, with over 320 km of coastline between La Pocatière and Les Méchins. To the south, the BSL shares a common boundary with the state of Maine (USA) and the province of New Brunswick (Fig. 17.1).

The BSL is part of the Appalachians geological province. These eroded mountains were formed during three successive orogenic phases (Bourque et al. 2000; Lavoie 2008). The northern part of the Appalachians was formed during two major orogenies: the Taconian orogeny during the late Ordovician (475–450 Ma) and the Acadian orogeny in the middle Devonian (360 Ma) (Bourque et al. 2000). The Taconian orogeny is represented by the Humber and Dunnage zones, with different assemblages of Neoproterozoic Late Ordovician sedimentary rocks (Bourque et al. 2000). The Acadian orogeny is represented by the Gaspé belt, with an assemblage of sedimentary, volcanic and intrusive rocks from the Late Ordovician to the Middle Devonian (Bourque et al. 2000) (Fig. 17.2). At the end of the Devonian, the Acadian orogeny formed large folds and many faults in the bedrock (Bourque et al. 2000). Bedrock geology in the BSL consists of sandstones, mudrocks, conglomerates and limestones with extrusive igneous rocks such as basalt and diorite (Lavoie 2008) (Fig. 17.2). The final stage of the landscape construction is general erosion in cyclic step of the Appalachians that led to paleoplains and peneplains (Grant 1989). Three levels of this relict erosion surfaces are present in the eastern BSL.

17.2.3 Climate and Vegetation

According to the Köppen-Geiger classification, the western part of the BSL has a warm-summer/humid continental climate, while the climate of the eastern part and hinterlands is boreal (Peel et al. 2007). The average annual temperature is 2.5 °C (Deschênes et al. 2010) with strong variability, ranging between 2.5 and 5.0 °C in the west and nearly 0 °C in the east (MDDELCC 2017). The average temperature in January is −14.0 °C (min −19.0 °C/max −7.0 °C); in July it is 16.5 °C (min 11 °C/max 23 °C) (MDDELCC 2017). The mean total annual precipitation is 1,080.30 mm (Deschênes et al. 2010).

The balsam fir (*Abies balsamea*)—yellow birch (*Betula alleghanensis*) bioclimatic domain occupies 82% of the territory. The other bioclimatic domains are balsam fir—white birch (*Betula papyrifera*), sugar maple (*Acer saccharum*)–basswood (*Tilia americana*), sugar maple—yellow birch, and alpine tundra on the highest summits (Saucier et al. 2011).

17.3 Glacial History

17.3.1 Glacial Expansion and Erosion

Current knowledge of the Quaternary paleogeography in the BSL covers only MIS 2 (Late Wisconsinan) (Buffin-Bélanger et al. 2015; Veillette et al. 2017). Nevertheless, the moraines of the Appalachian piedmont provide much information on the process of deglaciation in southern Québec. At the Last Glacial Maximum, the BSL was completely occupied by the LISC (Chauvin et al. 1985; David and Lebuis 1985; Dyke and Prest 1987; Pronk et al. 1989;

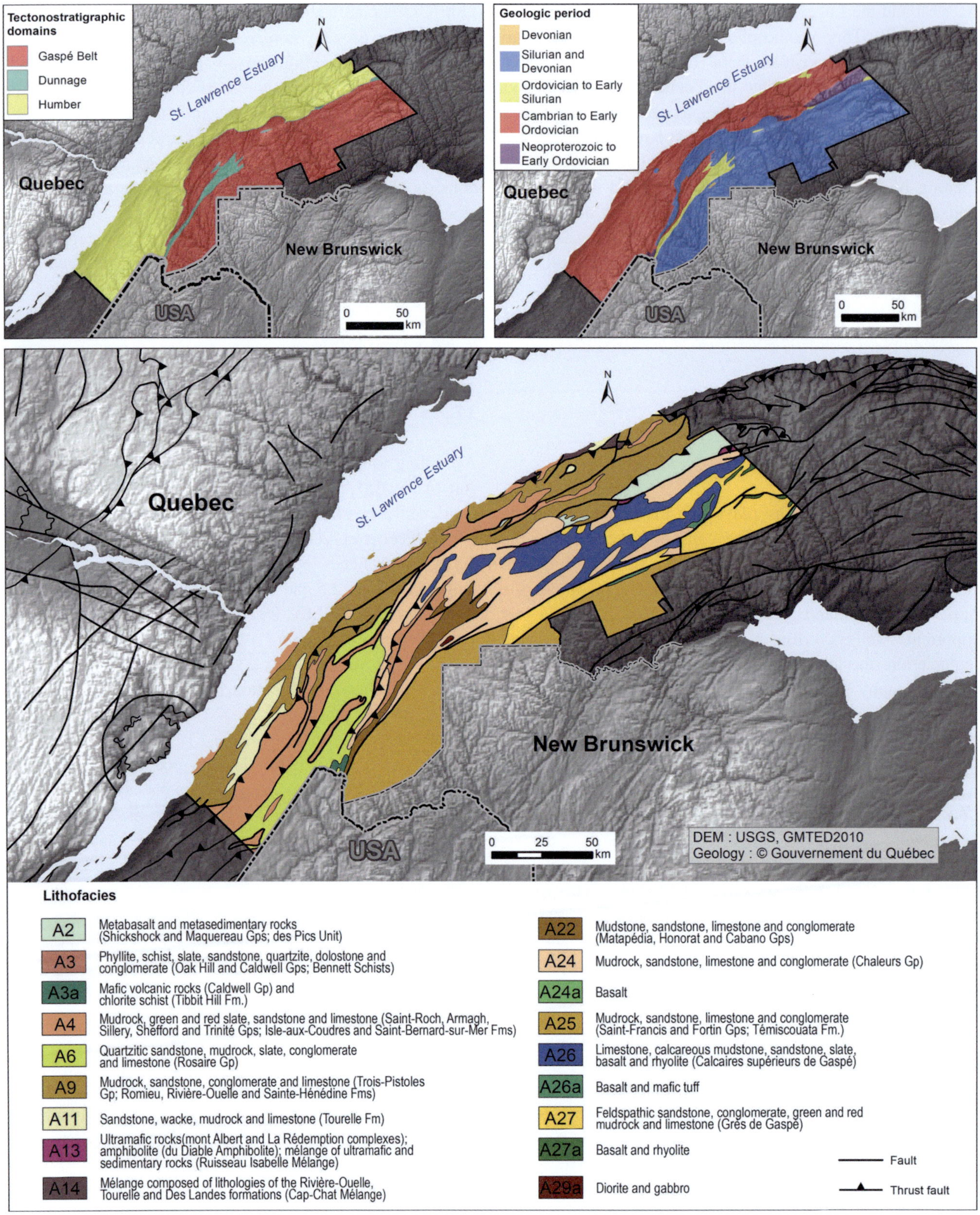

Fig. 17.2 Upper left: Tectonostratigraphic domains of the Appalachians in the Bas-Saint-Laurent. The Humber zone is outlined along the St. Lawrence Estuary; the Gaspé Belt is found toward the south and a narrow part of the Dunnage zone is present near Lake Témiscouata. Upper right: Geological periods of Appalachian orogenies in the Bas-Saint-Laurent. The Taconic orogeny in the north took place in the Cambrian to Early Ordovician period and in the south the Acadian orogeny occurred during Silurian and Devonian periods. A fault network exhumes substratum from the Neoproterozoic and Early Ordovician periods in the Chic-Choc Mountains. Bottom: Rock assemblages and lithofacies present in the Appalachians of the Bas-Saint-Laurent

Rappol 1993; Hétu 1998; Occhietti et al. 2004, 2011; Shaw et al. 2006; Veillette et al. 2017). The GIC occupied the eastern Gaspé peninsula (David and Lebuis 1985; Charbonneau and David 1993; Olejczyk and Gray 2007). In the BSL, ice flows toward the east-southeast are revealed by the dispersal of Precambrian erratics (Rappol 1993; Veillette et al. 2017). The LISC in the BSL affected the landscapes differently than did the local ice cap in the Gaspé highlands (Lebuis and David 1977; Hétu and Gray 1985). The local Gaspé ice cap had a cold-based thermal regime while the LISC was a temperate-based glacier (David and Lebuis 1985).

Lake basins were formed in part by glacial erosion, and thus can be used as a proxy for scour depth and efficiency (Sugden 1978; Hétu 1998; Principato and Johnson 2009). The number of lakes in the territory and their total area indicate the intensity of glacial erosion (Fig. 17.3). The BSL contains 35 lakes/100 km^2, occupying areas of more than 700 ha/100 km^2. On the Gaspé Peninsula, there are fewer than 25 lakes/100 km^2, covering areas of about 100 ha/100 km^2. In the centre of the Gaspé Peninsula, several areas are devoid of water bodies. These indicators allow an evaluation of the intensity of glacial erosion and the thermal regime at the base of the glaciers (Sugden 1978; Principato and Johnson 2009).

The territory that had been covered by the local cold-based thermal regime of the Gaspé ice cap has fewer and smaller lakes than does the territory covered by the temperate-based LISC. The BSL landscapes dotted with lakes are a Quaternary heritage of the higher glacial erosion intensity by LISC. The glacial erosion also accentuated the structural notches (Shilts et al. 1992; Allen et al. 2009) of the trans-Appalachian valleys by forming valleys that cut into BSL landscapes (Fig. 17.1).

17.3.2 The St. Lawrence Ice Stream: Calving and Notre-Dame Ice Cap

The St. Lawrence ice stream (Rappol 1993; Shaw et al. 2006; Margold et al. 2015; Stokes et al. 2016) contributed to the initiation and acceleration of deglaciation in the last glacial period by opening a calving-bay in the axis of St. Lawrence River valley when the ice stream slowed down (Thomas 1977) (Fig. 17.4). This situation led to the splitting of the LISC and development of the NDIC.

The reversal of ice flows from east-southeast to north (Pronk et al. 1989) is one of the most important consequences of the growth of the NDIC. This ice cap occupied the Appalachians between the last phases of MIS 2 and the beginning of the Holocene. At the front of the ice margin a morainic belt is present containing the Luceville-Neigette ice-front deposits to the east, and the St. Antonin-St-Jean-Port-Joli ice-front deposits to the west. This morainic system was produced by the Notre-Dame Ice Cap, formerly incorrectly associated to the LISC Highland Front Moraine (Gadd 1964). Many moraine segments (Lee 1962; Martineau and Corbeil 1983; Hétu 1998) reveal frontal position of the NDIC (Martineau and Corbeil 1983; Rampton et al. 1984).

17.3.3 Notre-Dame Ice Cap: Lejeune Ice Cap and Pohénégamook Ice Cap

According to Rampton et al. (1984), the NDIC split into several small local ice caps (Fig. 17.4) that were anchored on middle and high plateaus. The two main trenches of the Appalachian ice in the BSL correspond to the Témiscouata–Madawaska and the Matapédia valleys. Openings in the NDIC were occupied by glacial lakes (Chalmers 1884; Kiewiet de Jonge 1951; Rampton et al. 1984; Prichonnet 1995), which were responsible for the acceleration of the deglaciation in these areas by increasing iceberg calving at the ice margin.

Morainic systems indicating the presence of local ice caps in the Appalachian Highlands are limited. An important discontinuity in the morainic system occurs near Trois-Pistoles along the Témiscouata–Madawaska axis, where a delta occupies the mouth of the modern river. The assemblage of ice margin landforms and glacifluvial deposits is one of the most important Quaternary heritage features of the BSL.

17.4 Quaternary Heritage and Landscape in Bas-Saint-Laurent

17.4.1 Luceville and Neigette Ice-Front Moraines

In the segments of the Highland Front moraine (Gadd 1964), also named the Appalachian Piedmont moraine (Fig. 17.5), northward ice flows and lithology (Martineau and Corbeil 1983; Chauvin et al. 1985; Hétu 1998) provide clear evidence for the existence the NDIC. However, other than this ice front, knowledge on local Appalachian ice cap dynamics is scarce (Hétu 1998; Buffin-Bélanger et al. 2015), except for ice-flow direction (Rappol 1993; Veillette et al. 2017).

The Luceville moraine segment, deposited around 15 ka cal. BP, and the Neigette moraine segment, deposited around 14 ka cal. BP, represent Appalachian ice margin

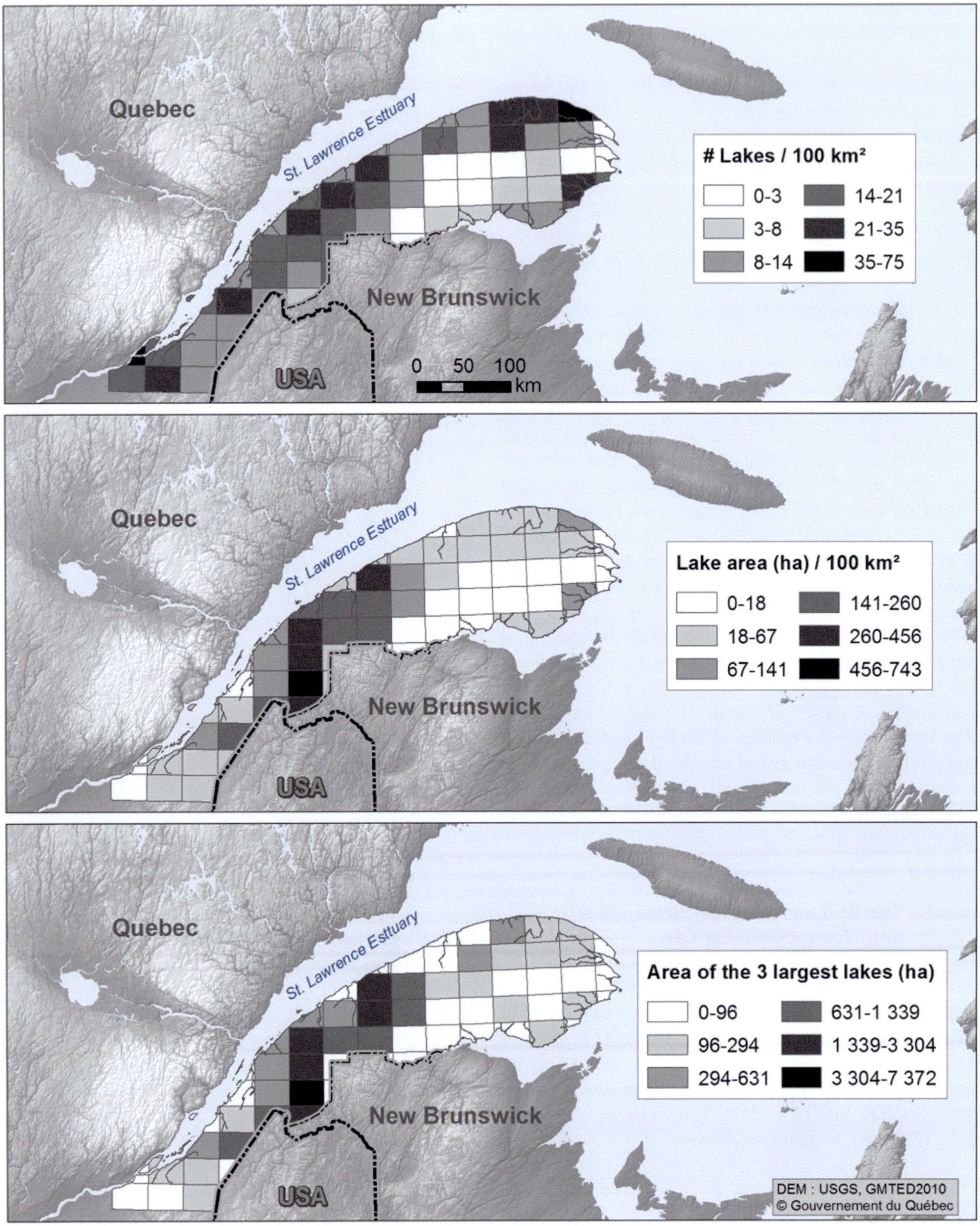

Fig. 17.3 Top: Number of lakes per 100 km^2 determined from 1:50,000 map of eastern Québec. The Bas-Saint-Laurent and the north of the Gaspé Peninsula have the largest number of lakes whereas several areas in the center of the Gaspé Peninsula are devoid of bodies of water. Middle: Lake area in hectares per 100 km^2 determined from 1:50,000 maps of eastern Québec. The largest lakes represent more intense glacial erosion that accentuated structural notches in the Appalachian bedrock structure. Bottom: Area of the three largest lakes in hectares determined from 1:50,000 maps of eastern Québec. The larger lakes also represent more intense glacial erosion

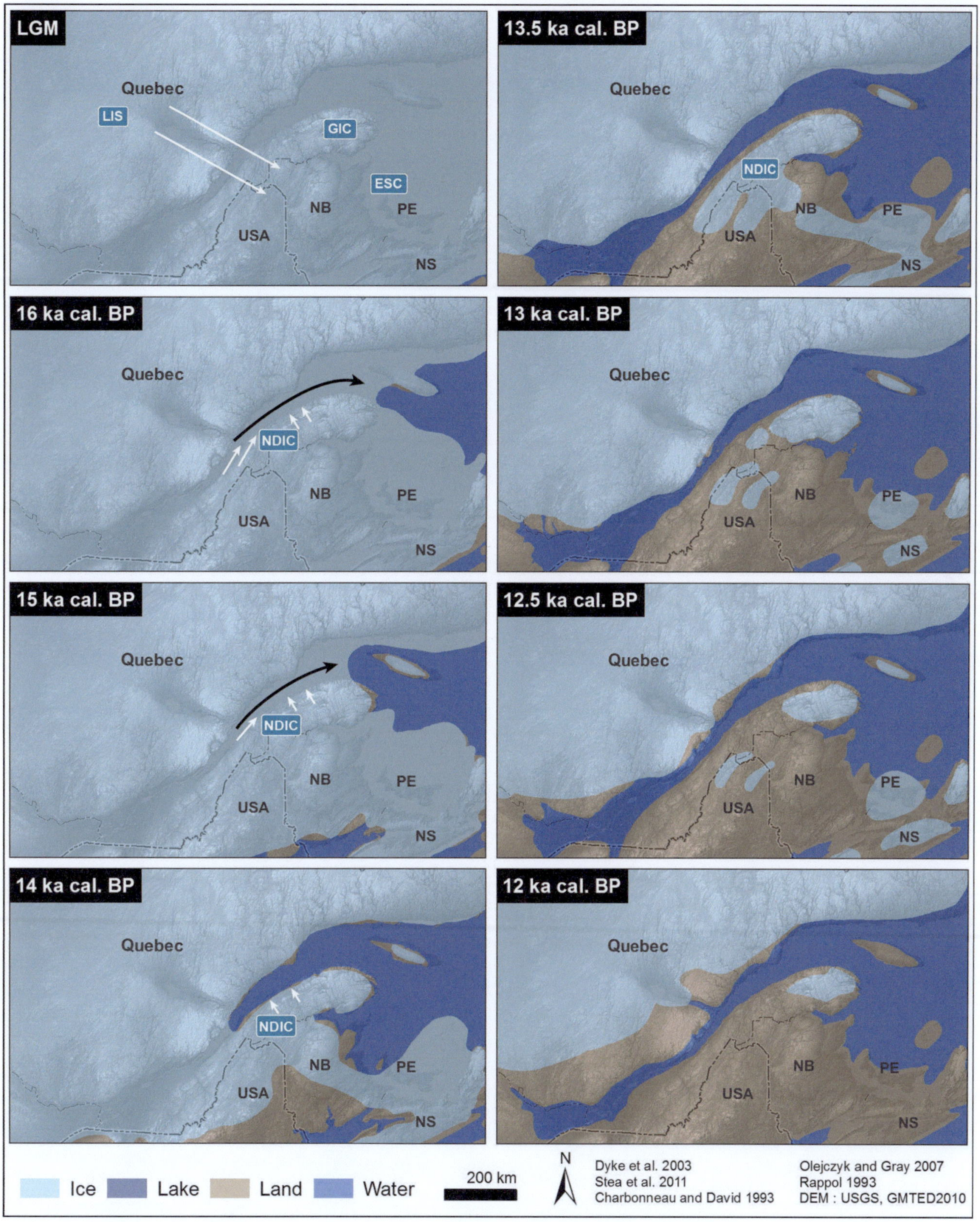

Fig. 17.4 Evolution of deglaciation in eastern Québec from the LGM to the Holocene. In white, ice; dark blue, lakes; light brown, land; light blue, sea. Black arrows outline the St. Lawrence ice stream and white arrows correspond to the direction of ice flow. At the LGM the ice sheets present in eastern Québec were the Laurentide ice sheet (LIS), the Gaspé ice cap (GIC) and the Escuminac ice center (ESC). From LGM to 14 ka cal. BP, the St. Lawrence ice stream was active and then a calving-bay opened the St. Lawrence Estuary, allowing the development of the NDIC. From 13.5 ka cal. BP to 12 ka cal. BP, the NDIC grew and decayed until their final melting. The postglacial seas followed the retreat of ice sheets

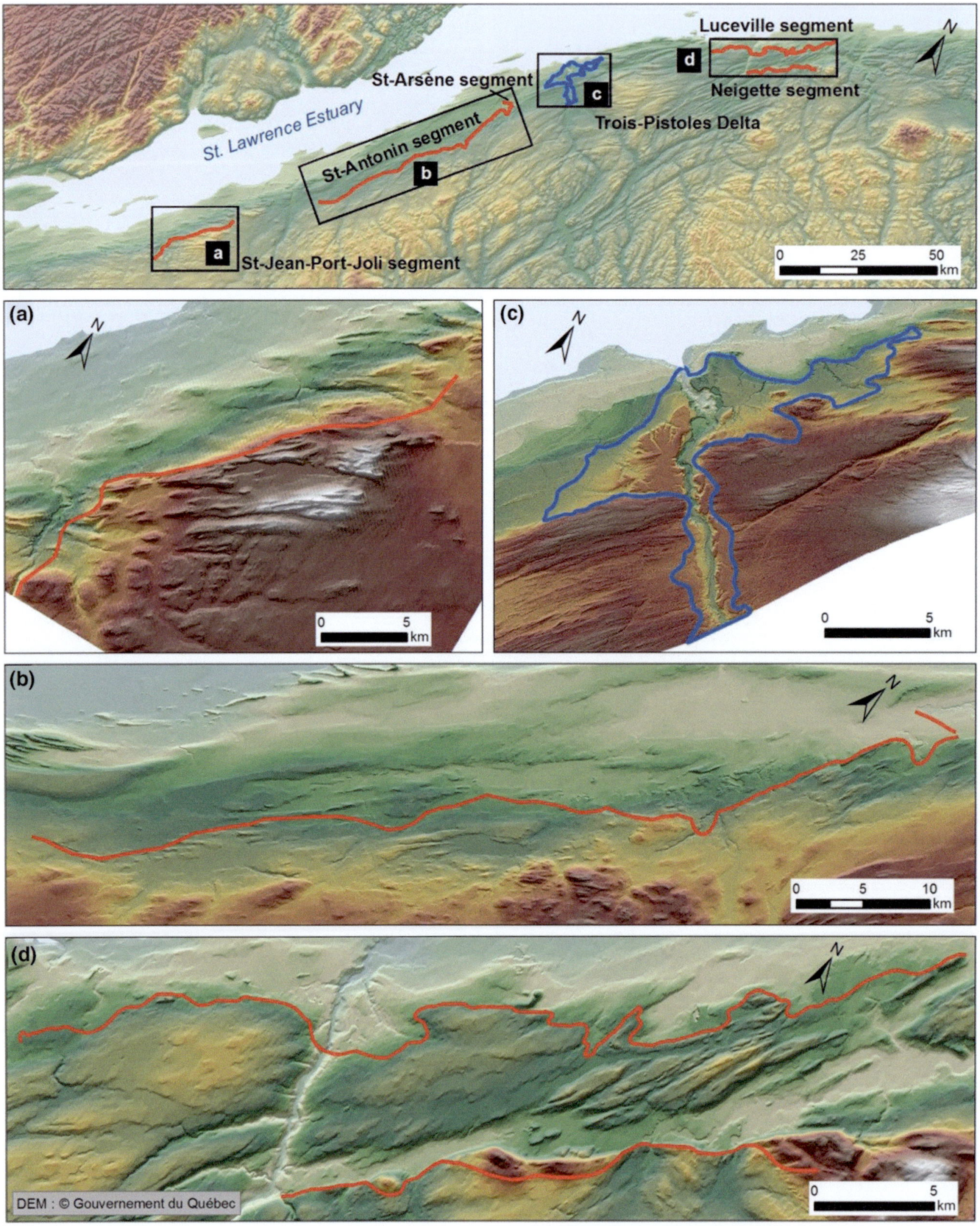

Fig. 17.5 Red line: Moraine segments in the Appalachian piedmont of the Bas-Saint-Laurent determined from Gadd (1964), Martineau and Corbeil (1983), Hétu (1998). Blue line: The Trois-Pistoles delta presents a discontinuity between moraine segments of the western Bas-Saint-Laurent and moraine segments to the east determined from Dionne (1966)

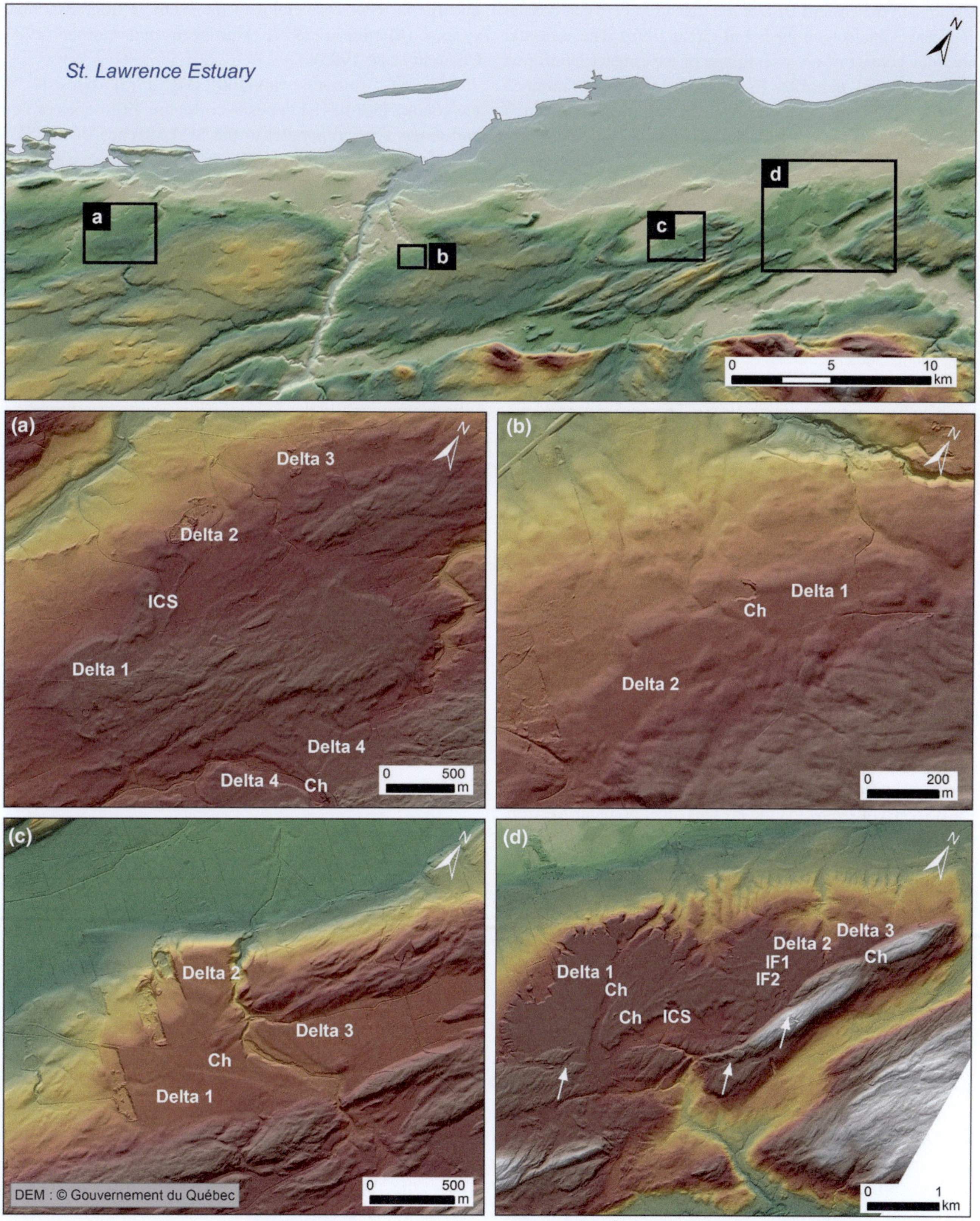

Fig. 17.6 Landform assemblage of the Luceville moraine segment. **a** Bic ice-contact delta. ICS: Ice-contact slope. Ch: Channels. Deltas 1–4 show the time evolution of the deltaic form. **b** St. Odile ice-contact delta. Ch: Channels. Deltas 1 and 2 show the geomorphological parts of the deltaic form. **c** St. Anaclet delta. Ch: Channels. Deltas 1–3 show the time evolution of the deltaic form. **d** Luceville ice-contact delta. ICS: Ice-contact slope; IF: Ice front; Ch: Channels. Glacial fluting in bedrock is shown by white arrows in the bedrock. Deltas 1–3 show the time evolution of the deltaic form (Buffin-Bélanger et al. 2015)

positions in the eastern BSL at the transition between coastal plain and Appalachian piedmont (Hétu 1998). These frontal moraines consist of an assemblage of ice-contact landforms and glacifluvial deposits. In Québec, the term "moraine" includes landforms representing ice-front positions, which may contain a variety of glacial and glacifluvial sediments and features. The Luceville moraine (Fig. 17.6) is composed of the Luceville delta, the St-Odile delta, the Bic delta, and the St-Fabien kames, all of which are ice-contact deposits (Locat 1977, 1978; Rappol 1993; Hétu 1998). The St-Anaclet delta (Fig. 17.6) is also part of this moraine segment, despite no clear evidence of ice-contact (Hétu 1998). The Neigette moraine (Fig. 17.7) is an assemblage of sandur, glacifluvial outwash, kames, esker and an ice-contact delta (Hétu 1998).

The Luceville and Neigette moraines are paleogeographically significant (Hétu 1998). They are the only evidence of the NDIC margin in the eastern BSL. The Luceville moraine, deposited in the Goldthwait Sea around 15 ka cal. BP, is strong evidence of the opening of St. Lawrence River followed by the stabilization of the NDIC on the Appalachian piedmont. This ice generated forms and deposits of mainly glacifluvial origin during glacial decay (Fig. 17.8). Thereafter, the ice retreated from the Neigette Valley and grounded on the escarpment of the Neigette fault around 14 ka cal BP.

The landforms and deposits indicate an episode of glacial recurrence, likely of climatic origin. The short distance between these two morainic segments indicates a slow deglaciation between 15 and 14 ka cal. BP in the eastern part of the BSL. These two morainic segments are the only known positions of the Appalachian ice margin in the eastern BSL (Hétu 1998). At the present time, no other ice front south of the Neigette moraine in the Appalachian Highlands has been reported. Therefore, the regional deglaciation model remains speculative after 14 ka cal. BP.

17.4.2 St. Antonin and St-Jean-Port-Joli Ice-Front Moraines

Although early studies associated segments of the Appalachian piedmont moraine observed in the western BSL with the LISC (Lee 1962; Gadd 1964; Lasalle et al. 1976, 1977a, b), recent studies have associated these moraine segments with Appalachian glaciation (Martineau and Corbeil 1983; Chauvin et al. 1985), specifically with the local ice cap of the BSL, the NDIC. The paleocurrents measured in glacifluvial deposits of the morainic belt show northward sediment transport originating from the NDIC to the south. These glacifluvial deposits belong to the NDIC's frontal moraine system (Martineau 1977; Martineau and Corbeil 1983; Chauvin et al. 1985).

The St. Antonin moraine (Fig. 17.5) consists of ice-contact glacifluvial deposits consisting of a series of fans and deltas aligned parallel to the St. Lawrence over 40 km (Martineau and Corbeil 1983). In some sectors, it is composed of glacial deposits such as till and flowtill (Chauvin et al. 1985). The St. Arsène segment at its northeastern end is a frontal moraine that precedes the deposition of the St. Antonin moraine (Martineau and Corbeil 1983). This segment is built on an Appalachian ridge at marine limit (Fig. 17.9). Part of the sedimentation of this deposit occurred in the Goldthwait Sea (Martineau and Corbeil 1983). It is a glacifluvial outwash with regular sand and gravel stratification. This moraine segment is the same age as the Luceville moraine while the St-Antonin moraine was formed at the same time as the Neigette ice margin deposits.

The St-Jean-Port-Joli moraine extends nearly 50 km between Ste. Louise and Montmagny (Chauvin et al. 1985). It mainly consists of ice-contact stratified drift and flowtill expressed geomorphologically by isolated kames and glacifluvial hills. This moraine is almost devoid of Precambrian clasts. Measurements of paleocurrents towards the north show that this moraine segment is associated with Appalachian ice. This moraine is a dynamic equivalent of the St. Antonin moraine and the Neigette moraine despite its later deposition due to the lag in the opening of St. Lawrence valley (Thomas 1977).

17.4.3 Glacial Lake Madawaska Complex

During the retreat of the NDIC, a network of glacial lakes formed at its margin (Nicholas et al. 1981; Dumais et al. 1998). In the BSL, Glacial Lake Madawaska was a former glacial lake that occupied a larger area during the late-glacial and the early Holocene. This glacial lake was formed by the coalescence of several lakes, such as Lake Témiscouata, Lake Touladi, and Lake des Aigles (Fig. 17.10) (Kiewiet de Jonge 1951; Lougee 1954; Mott 1975; Kite 1983; Rampton et al. 1984; Kite and Stuckenrath 1989). Glacial Lake Madawaska occupied the Témiscouata–Madawaska Valley at the end of MIS 2 for a maximum of 4 ky, starting at marine invasion around 14 ka cal. BP and ended with final draining around 11 ka cal. BP. Radiocarbon dating provides an estimate of the duration of the lake: basal peat over silts along the St. John River gave a maximum age of 11.9 ± 0.5 ka cal BP (I(GSC)-2) and a minimum age of 10.8 ± 0.5 cal ka BP (W-2927) (Kite and Stuckenrath

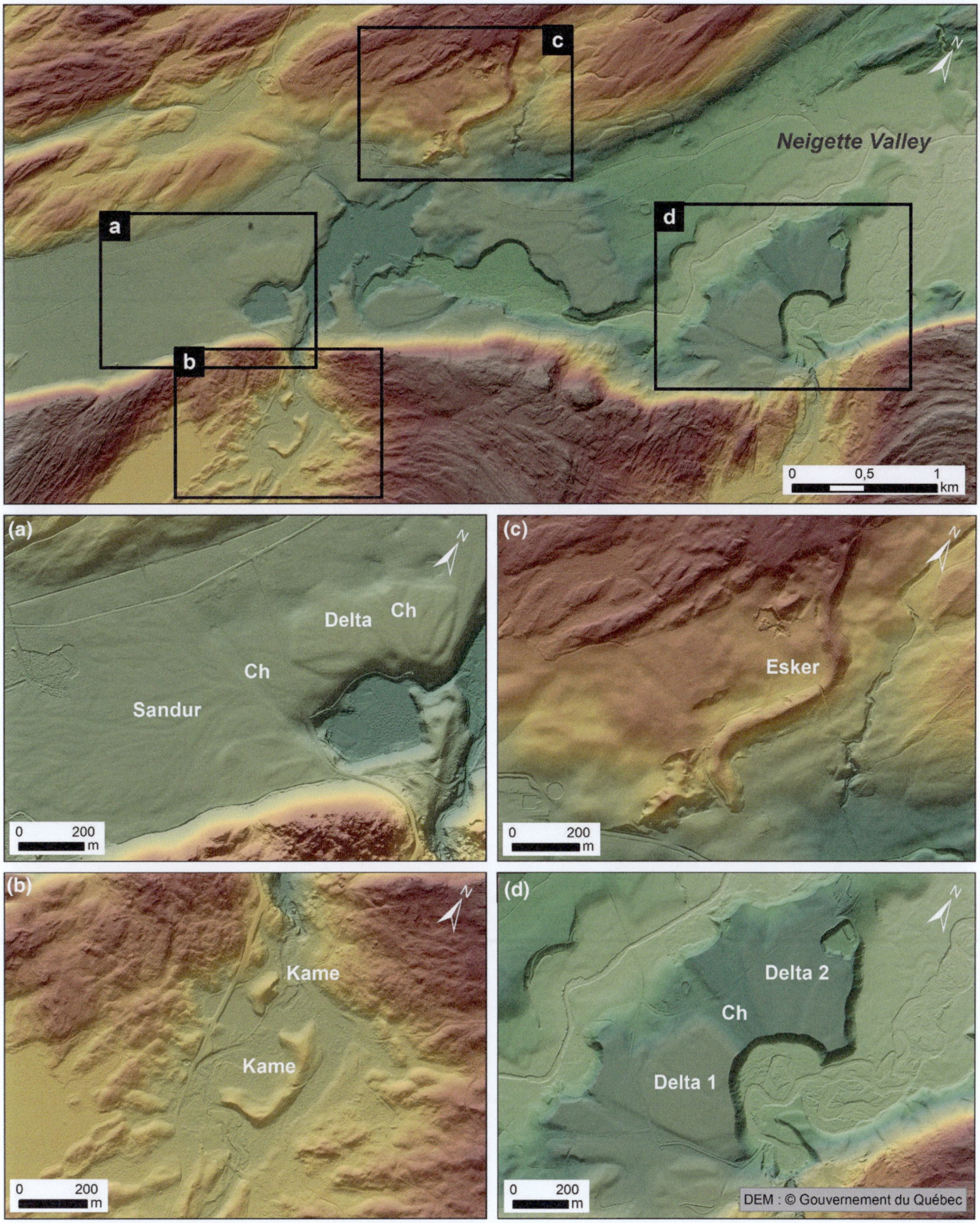

Fig. 17.7 Landform assemblages of the Neigette moraine segment. **a** Neigette ice-contact delta and sandur. Ch: Channels. **b** Ice decayed forms. **c** Neigette esker. **d** Neigette ice-contact delta. Ch: Channels. Deltas 1 and 2 show the time evolution of the deltaic form (Buffin-Bélanger et al. 2015)

Fig. 17.8 Landform and deposits of the moraine segments in the Appalachian piedmont of the Bas-Saint-Laurent. **a** Bic glacifluvial deposits. *Photograph* B. Hétu. **b** Neigette kame. *Photograph* B. Hétu. **c** St. Antonin ice-contact delta. *Photograph* B. Hétu. **d** Trois-Pistoles glacifluvial delta. **e** Luceville ice-contact delta. *Photograph* B. Hétu. **f** Neigette glacifluvial deposit. *Photograph* B. Hétu. **g** St. Antonin ice-contact delta. *Photograph* B. Hétu. **h** Trois-Pistoles glacifluvial delta (Buffin-Bélanger et al. 2015)

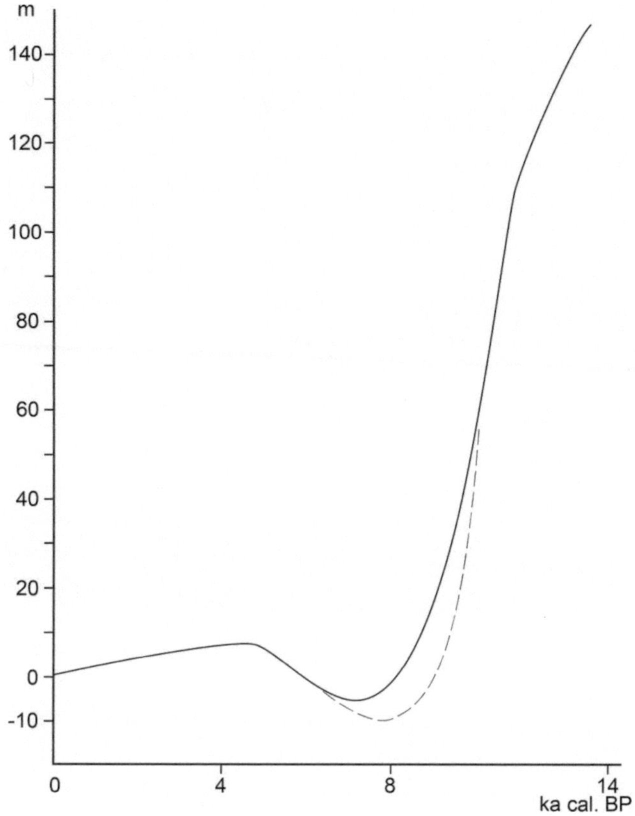

Fig. 17.9 Relative sea-level curve for Rivière-du-Loup area (adapted from Dionne (2002)). See Fig. 17.13 for location

1988). Varves observed in the BSL (Fig. 17.11) confirmed the presence of this lake.

During its maximum extent phase, Glacial Lake Madawaska covered over 500 km^2 in the southern BSL and New Brunswick (Fig. 17.10). During the early stage (proto-Madawaska) at the beginning of deglaciation, the lake drained northward in the St. Lawrence watershed (Nicholas et al. 1981; Rampton et al. 1984). In the first phases, outlets were located toward the north, to the Trois-Pistoles River, and toward the northeast, to the Rimouski River (Fig. 17.10) (Rampton et al. 1984). Formation of the Trois-Pistoles River outlet is consistent with the first stage of formation of the Trois-Pistoles glacifluvial delta. Lake Madawaska, which was in sporadic contact with the NDIC margin (Lougee 1954; Seaman 2004), played a key role in the regional deglaciation (Gadd et al. 1972; Chauvin et al. 1985). The Témiscouata–Madawaska Valley was deglaciated around 13 ka cal BP (Burke and Richard 2010) and was gradually occupied by a lake as the ice margin was retreating. This glacial lake contributed to the complete opening of this valley by increasing ablation through iceberg calving and thus to the opening and separation of the NDIC into local ice caps (PIC and LIC) on the plateaus of the interfluve of the Témiscouata–Madawaska Valley.

17.4.4 The Trois-Pistoles Delta

The Trois-Pistoles delta (Fig. 17.5) is the largest delta on the south shore of St. Lawrence (Buffin-Bélanger et al. 2015). It occupies more than 60 km^2. Despite its size and importance in the BSL landscape, only two authors have documented this large-scale depositional landform (Lee 1962; Dionne 1966, 1968). The construction of this delta overlaps the formation of the Luceville, St. Arsène, Neigette, St. Antonin and St-Jean-Port-Joli moraines, indicating that it was deposited over a long period, overlapping the late-glacial and the Early Holocene.

The upper level of the Trois-Pistoles delta is a glacifluvial delta consisting of coarse beddings of sand and gravel dipped towards the north (Dionne 1966). Deltaic sediments with thicknesses varying between 30 and 50 m (Dionne 1966) were deposited in the Goldthwait Sea from 14 ka cal BP (Lee 1962). The deltaic surface is incised with many marine terraces, which provide evidence for the forced regression of the Goldthwait Sea (Dionne 1966, 1977, 2002). The three main stratigraphic units found in this delta are, from base to top: (1) Goldthwait Sea clays at the base; (2) deltaic beds reworked by coastal processes; and (3) littoral deposits at the top (Dionne 1966). There is no evidence of ice-contact deposits over the entire delta; the deposits are rather associated with ice-distal margin glacifluvial sediments (Fig. 17.8).

17.5 Conclusions

The opening of the St. Lawrence Estuary during the calving phase (Thomas 1977) at the end of MIS 2 led to the formation of an autonomous ice in the Appalachians (Prest 1984), the Notre-Dame Ice Cap. Among the Quaternary heritage landforms and deposits observed in the BSL, the complex of frontal moraines of the Appalachian piedmont (Martineau and Corbeil 1983; Chauvin et al. 1985; Hétu 1998) is undoubtedly the most important. These moraines reveal the position of the northern NDIC margin during the late glacial period, when it had stabilized on the first escarpments south of the Goldthwait Sea.

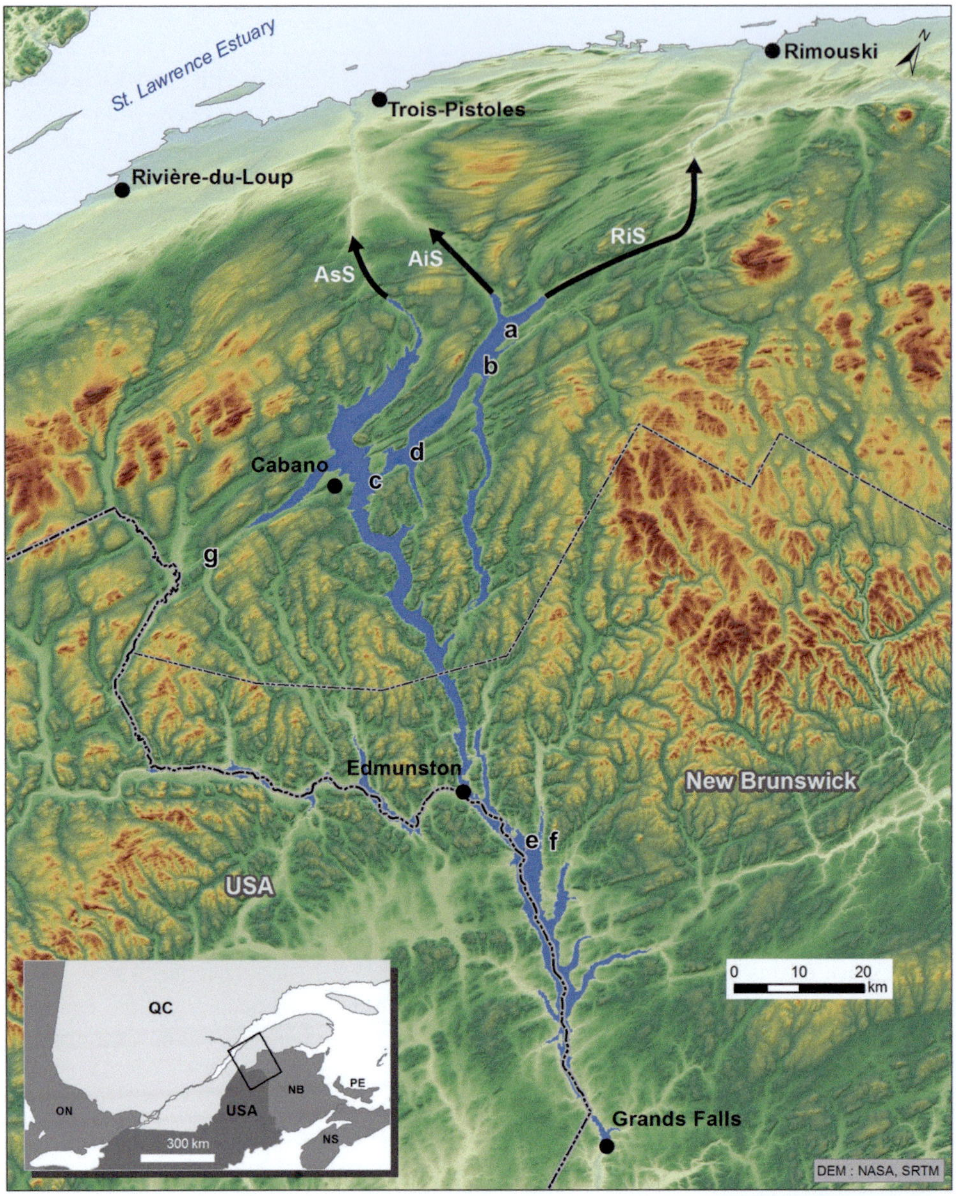

Fig. 17.10 The maximum extent of Lake Madawaska (Kiewiet de Jonge 1951; Rappol 1989). Spillways are shown by black arrows: AsS: Ashberish spillway, AiS: Aigles spillway, RiS: Rimouski spillway. Letters represent the location of photos in Fig. 17.11

There is a geomorphological continuity in landform and deposits assemblages of frontal moraine complex of the Appalachian piedmont between the St. Arsène and Luceville moraine segments and between the St-Jean-Port-Joli, St. Antonin and Neigette moraine segments (Fig. 17.12). These moraines, however, are not geographically connected, because the Trois-Pistoles delta represents a significant discontinuity in the landform assemblage. This delta was supplied by glacial Lake Madawaska in the Témiscouata–Madawaska Valley (Fig. 17.12). Glacial Lake Madawaska contributed to the early opening of this valley and to the rapid separation of the NDIC into two local caps—the Lejeune Ice cap (LIC) and the Pohénégamook Ice cap (PIC)—along the axis of the Témiscouata–Madawaska Valley (Fig. 17.13).

This paper, based on a review of existing research work and field observations, provides new insights on the model of the deglaciation of the Québec Appalachian Highlands.

Fig. 17.11 The varve complex of Lake Madawaska. Location on Fig. 17.10. **a** Lac-des-Aigles. **b** Squatec. **c** Témiscouata. **d** Touladi. **e** Rivière St-Jean. **f** Rivière Verte. **g** Rivière Bleue. *Photographs* A. Morissette

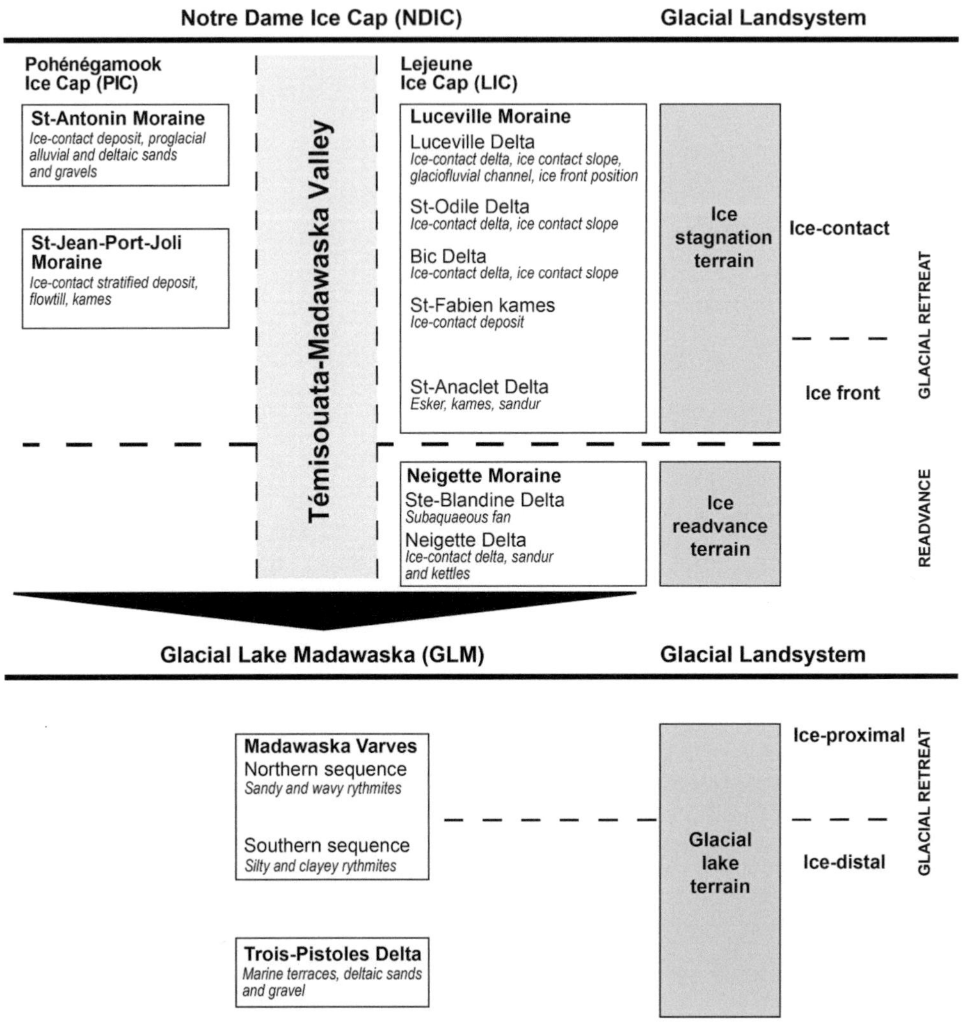

Fig. 17.12 Landform assemblages model accounting for the NDIC and glacial Lake Madawaska. The Témiscouata–Madawaska Valley splits the geomorphologic continuum between moraine segments of the western the Bas-Saint-Laurent belonging to the Pohénégamook Ice Cap (PIC) and moraine segments of the eastern Bas-Saint-Laurent belonging to the Lejeune Ice Cap (LIC). Glacial Lake Madawaska and the Trois-Pistoles delta developed between these two local ice caps

The St. Lawrence Ice Stream and its associated calving-bay contributed to the growth of the NDIC. In the Témiscouata–Madawaska valley, the fault in the Gaspé rift zone, exploited by glacial erosion, allowed rapid flooding by glacifluvial water. The resulting large glacial lake contributed to early deglaciation of this Appalachian valley. These processes activated the disintegration of NDIC into smaller local ice caps, the Lejeune Ice Cap and Pohénégamook Ice Cap, centered on the high plateaus of Notre-Dame Mountains. The Témiscouata–Madawaska valley was the first to be deglaciated due to the development of Glacial Lake Madawaska, while the Lejeune and Pohénégamook plateaus were still occupied by ice centers. The valleys at 90° to the Appalachian folds aided in the initiation and acceleration of deglaciation processes before the top of plateaus.

Acknowledgements This research project was funded by a NSERC Discovery grant to P. Lajeunesse. The accommodation and logistical support of the Parc National du Lac-Témiscouata was greatly appreciated during field campaigns. Several field assistants participated in

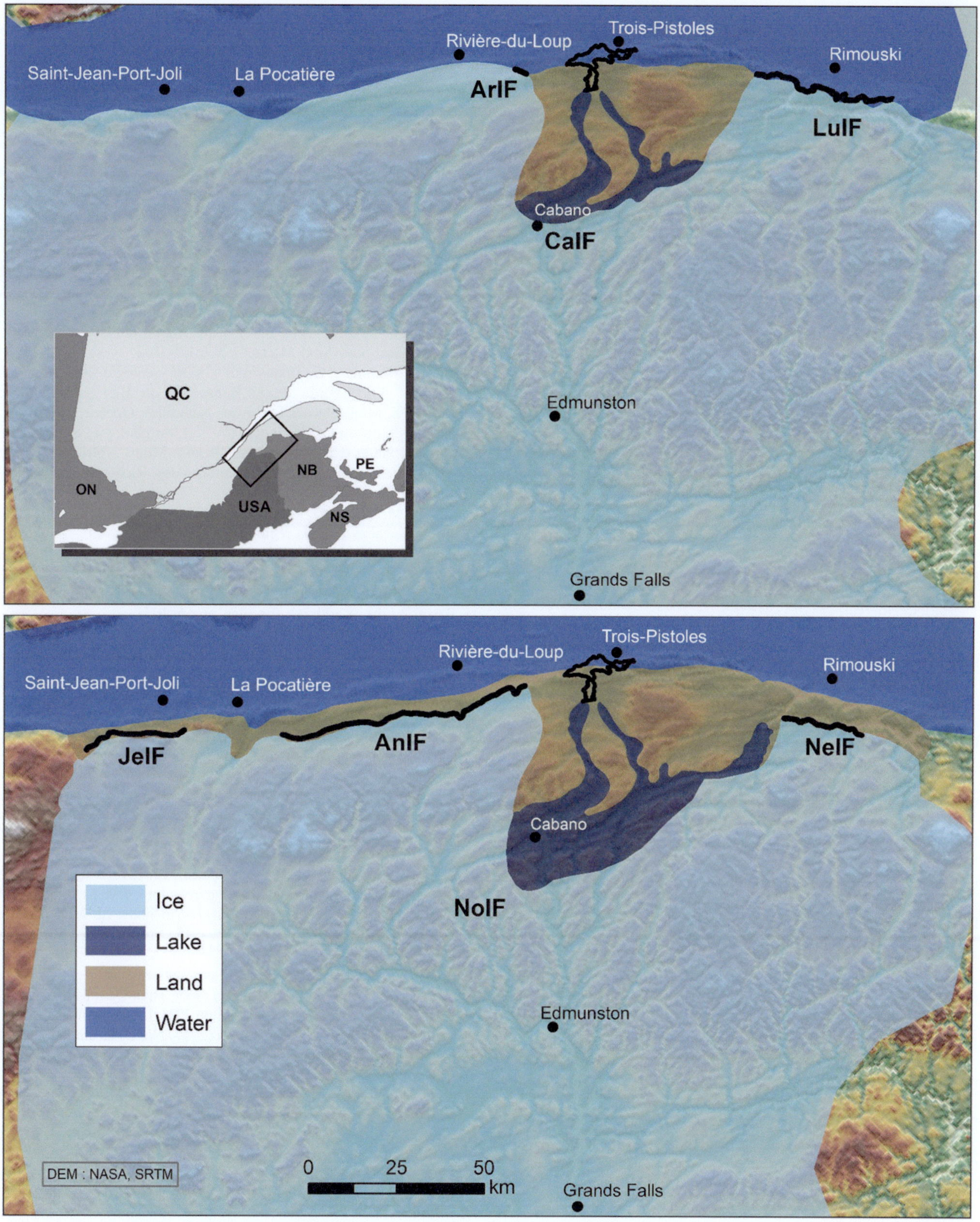

Fig. 17.13 Evolution of deglaciation in the Témiscouata–Madawaska Valley. In white, ice; dark blue, lakes; light brown, land; light blue, sea. ArIF: St. Arsene ice front. CaIF: Cabano ice front. LuIF: Luceville ice front. JeIF: St-Jean-Port-Joli ice front. AnIF: St. Antonin ice front. NoIF: Notre-Dame-du-Lac ice front. NeIF: Neigette ice front. The two phases correspond to the opening of the Notre-Dame Ice Cap by the Témiscouata–Madawaska Valley and the development of the early proglacial Lake Madawaska

field work: M. Boivin, E. Brouard, P.-O. Couette, D. Deschênes, G. Joyal, M. Joyal-Fortier, A.-P. Trottier, and V. Roy

References

Allen JS, Thomas WA, Lavoie D (2009) Stratigraphy and structure of the Laurentian rifted margin in the northern Appalachians: a low-angle detachment rift system. Geology 37:335–338

Bourque PA, Malo M, Kirkwood D (2000) Paleogeography and tectono-sedimentary history at the margin of Laurentia during Silurian to earliest Devonian time: the Gaspé belt, Québec. Bull Geol Soc Am 112:4–20

Buffin-Bélanger T, Chaillou G, Cloutier C-A (2015) Programme d'acquisition de connaissances sur les eaux souterraines du nord-est du Bas-Saint-Laurent (PACES-NEBSL). Rimouski, 199 pp

Burke A, Richard PJH (2010) L'occupation du Témiscouata pendant l'Archaïque: La comparaison du registre archéologique et du registre paléoenvironnemental. In: Loewen B, Chapdelaine C, Burke A (eds) De l'archéologie analytique à l'archéologie sociale. Paléo-Québec 34, pp 103–127

Charbonneau R, David PP (1993) Glacial dispersal of rock debris in central Gaspésie, Quebec, Canada. Can J Earth Sci 30:1697–1707

Chalmers R (1884) Report on the surface geology of western New Brunswick with special reference to the area included in York and Carleton counties. Report of Progress, 1882-83-84, Part GG, Geological and Natural History Survey of Canada, 47 pp

Chauvin L, Martineau G, Lasalle P (1985) Deglaciation of the lower St. Lawrence Région, Quebec. In: Borns H Jr., LaSalle P, Thompson W (eds) Late Pleistocene history of northeastern New England and adjacent Québec. Geol Soc Am Spec Paper 197:111–123

David PP, Lebuis J (1985) Glacial maximum and deglaciation of western Gaspé, Québec, Canada. In: Borns H Jr., LaSalle P, Thompson W (eds) Late Pleistocene history of northeastern New England and adjacent Québec. Geol Soc Am Spec Paper 197:85–110

Deschênes J, Gosselin C, Hardy L (2010) Portrait territorial du Bas-Saint-Laurent. Ressources naturelles et Faune, Gouvernement du Québec, Québec, 117 pp

Dionne J-C (1968) Carte morpho-sédimentologique de la région de Trois-Pistoles. Rev géog Montréal 22:55–64

Dionne J-C (1966) Le delta fini-glaciaire de la rivière Trois-Pistoles. Bureau de l'Aménagement de l'Est du Québec, Rimouski, 2 pp

Dionne J-C (1977) La mer de Goldthwait au Québec. Géogr phys Quat 31:61–80

Dionne J-C (2002) Une nouvelle courbe du niveau marin relatif pour la région de Rivière-du-Loup (Québec). Géogr phys Quat 56:33–44

Dumais P, Poirier J, Rousseau G (1998) La préhsitoire du Témiscouata, trente ans plus tard. Paléo-Québec 27:53–80

Dyke AS, Andrews JT, Clark PU (2002) The Laurentide and Innuitian ice sheets during the last glacial maximum. Quat Sci Rev 21:9–31

Dyke AS, Prest VK (1987) Late Wisconsinan and Holocene history of the Laurentide Ice Sheet. Géogr phys Quat 41:237–263

Gadd NR (1964) Moraines in the Appalachian region of Quebec. Geol Soc Am Bull 75:1249–1254

Gadd NR, McDonald BC, Shilts WW (1972) Deglaciation of Southern Quebec. Geological Survey of Canada, Paper 71-47, 19 pp

Grant DR (1989) Quaternary geology of the Atlantic Appalachian region of Canada. In: Fulton RJ (ed) Quaternary geology of Canada and Greenland. Geological Survey of Canada, pp 391–440

Hétu B (1998) La déglaciation de la région de Rimouski, Bas-Saint-Laurent (Québec): indices d'une récurrence glaciaire dans la Mer de Goldthwait entre 12,400 et 12,000 BP. Géog phys Quat 52: 325–347

Hétu B, Gray JT (1985) Le modelé glaciaire du centre de la Gaspésie septentrionale, Québec. Géog phys Quat 39:47–66

Kiewiet de Jonge EJC (1951) Glacial water levels in the St. John river valley. PhD Thesis, Clark University, 116 pp

Kite JS (1983) Late Quaternary glacial, lacustrine, and alluvial geology of the upper St. John river basin, northwestern Maine and adjacent Canada. PhD Thesis, University of Wisconsin, 339 pp

Kite JS, Stuckenrath R (1988) Postglacial history of the upper St. John drainage basin. Contributions to the Quaternary geology of northern Maine and adjacent Canada. Maine Geological Survey, Department of Conservation, pp 117–132

Kite JS, Stuckenrath R (1989) Postglacial evolution of drainage in the middle and upper St. John River Basin, Maine and New Brunswick. Maine Geol Surv Stud Maine Geol 6:135–142

Lasalle P, Martineau G, Chauvin L (1976) Géologie des sédiments meubles d'une partie de la Beauce et du Bas Saint-Laurent. DPV-438, Ministère des Richesses Naturelles, Québec, 13 pp

Lasalle P, Martineau G, Chauvin L (1977a) Morphologie, stratigraphie et déglaciation dans la région de Beauce-Monts Notre-Dame-Parc des Laurentides. DPV-516, Ministère des Richesses Naturelles, Québec, 74 pp

Lasalle P, Martineau G, Chauvin L (1977b) Dépôts morainiques et stries glaciaires dans la région de Beauce-Monts Notre-Dame-Parc des Laurentides. DPV-515, Ministère des Richesses Naturelles, Québec, 22 pp

Lavoie D (2008) Appalachian Foreland Basin of Canada. Sedimentary Basins of the World 5C, Elsevier, pp 65–103

Lebuis J, David PP (1977) La stratigraphie et les événements du Quaternaire de la partie occidentale de la Gaspésie, Québec. Géogr phys Quat 31:275–296

Lee HA (1962) Géologie de la région de Rivière-du-Loup–Trois-Pistoles, Québec (Dépôts meubles). Geological Survey of Canada, Paper 61-32

Locat J (1977) L'émersion des terres dans la région de Baie-des-Sables/Trois-Pistoles, Québec. Géogr phys Quat 31:297–306

Locat J (1978) Le Quaternaire de la région de Baie-des-Sables-Trois-Pistoles. DPV-605, Ministère des Richesses Naturelles, Québec, 64 pp

Lougee RJ (1954) The role of upwarping in the post-glacial history of Canada. Rev Can Géog 8:3–52

Margold M, Stokes CR, Clark CD, Kleman J (2015) Ice streams in the Laurentide Ice Sheet: a new mapping inventory. J Maps 11:380–395

Martineau G (1977) Géologie des dépôts meubles de la région de Kamouraska-Rivière-du-Loup. DPV-545, Ministère des Richesses Naturelles, Québec, 17 pp

Martineau G, Corbeil P (1983) Réinterprétation d'un segment de la moraine de Saint-Antonin, Québec. Géog phys Quat 37:217–221

Ministère du Développement durable de l'Environnement et de la L contre les changements climatiques (2017) Normales climatiques du Québec 1981–2010. Gouv. du Québec. http://www.mddelcc.gouv.qc.ca/climat/normales/index.asp

Mott RJ (1975) Palynological studies of lake sediment profiles from southwestern New Brunswick. Can J Earth Sci 12:273–288

Nicholas GP, Kite JS, Bonnichsen R (1981) Archaeological survey and testing of late Pleistocene-early Holocene landforms in the Dickey-Lincoln Scholl reservoir area, Northern Maine. Institute for Quaternary studies, University of Maine, Orono, 170 pp

Occhietti S, Govare É, Klassen RA, Parent M, Vincent JS (2004) Late Wisconsinan-Early Holocene deglaciation of Québec-Labrador. Dev Quat Sci 2:243–273

Occhietti S, Parent M, Lajeunesse P, Robert F, Govare É (2011) Late Pleistocene-Early Holocene Decay of the Laurentide Ice sheet in Québec-Labrador. Dev Quat Sci 15:601–630

Olejczyk P, Gray JT (2007) The relative influence of Laurentide and local ice sheets during the last glacial maximum in the eastern Chic-Chocs Range, northern Gaspé Peninsula, Quebec. Can J Earth Sci 44:1603–1625

Peel MC, Finlayson BL, McMahon TA (2007) Updated world map of the Köppen-Geiger climate classification. Hydrol Earth Syst Sci 11:1633–1644

Prest VK (1984) The late Wisconsinan glacier complex. In: Fulton RJ (ed) Quaternary stratigraphy of Canada—a Canadian Contribution to IGCP Project 24, Paper 84-1. Geological Survey of Canada, pp 21–36

Prichonnet G (1995) Géologie glaciaire et géochronologie postglaciaire dans la région limitrophe de la Gaspésie et du Bas-Saint-Laurent, Québec. Commission géologique du Canada, 69 pp

Principato SM, Johnson JS (2009) Using a GIS to quantify patterns of glacial erosion on northwest Iceland: implications for independent ice sheets. Arctic, Antarct Alp Res 41:128–137

Pronk AG, Bobrowsky PT, Parkhill MA (1989) An interpretation of late quaternary glacial flow indicators in the Baie des Chaleurs Region, Northern New Brunswick. Géog phys Quat 43:179–190

Rampton VN, Gauthier RC, Thibault J, Seaman AA (1984) Quaternary geology of New Brunswick. Geological Survey of Canada, Memoir 416, 77 pp

Rappol M (1989) Glacial history and stratigraphy of northwestern new brunswick. Géogr phys Quat 43:191–206

Rappol M (1993) Ice flow and glacial transport in Lower St. Lawrence, Quebec. Geological Survey of Canada, Paper 90-19, 28 pp

Saucier J-P, Robitaille A, Grondin P, et al (2011) Les régions écologiques du Québec méridional. Ressources naturelles et Faune, Gouvernement du Québec, Québec, 1:1,250,000 map

Seaman A (2004) Late Pleistocene history of New Brunswick, Canada. Dev Quat Sci 2, Part B:151–167

Shaw J, Piper DJW, Fader GBJ (2006) A conceptual model of the deglaciation of Atlantic Canada. Quat Sci Rev 25:2059–2081

Shilts WW, Rappol M, Blais A (1992) Evidence of late and postglacial seismic activity in the Temiscouata-Madawaska Valley, Quebec—New Brunswick, Canada. Can J Earth Sci 29:1043–1069

Stokes CR, Margold M, Clark CD, Tarasov L (2016) Ice stream activity scaled to ice sheet volume during Laurentide Ice Sheet deglaciation. Nature 530:322–326

Sugden DE (1978) Glacial erosion by the Laurentide Ice Sheet. J Glacio 20:367–390

Thomas RH (1977) Calving-bay dynamics and ice sheet retreat up the St. Lawrence valley system. Géogr phys Quat 31:347–356

Veillette JJ, Cloutier M, Paradis SJ, Hétu B, Cloutier C-A, Houde-Poirier M, Buffin-Bélanger T (2017) Géologie des formations en surface et histoire glaciaire, Bas Saint-Laurent, Québec. Commission géologique du Canada, Carte géoscientifique du Canada 279 (préliminaire), 1:250,000

Antoine Morissette is a geographer and geomorphologist. He holds a master's degree (M.Sc.) in oceanography from the Institut des sciences de la Mer de Rimouski (ISMER) at the Université du Québec à Rimouski (UQAR) (2007). He is currently completing a Ph.D. at Université Laval on the dynamics and evolution of the proglacial Lake Madawaska in the Appalachian Highlands. He has been teaching geography at UQAR since 2003 as teaching/research assistant and lecturer. His main research interests are in development of geomorphological models of landscape evolution during the last glaciations using the latest geospatial technologies. He also works closely with coastal communities on knowledge transfer projects that promote adaptation to climate change in the coastal zone.

Bernard Hétu was awarded a Ph.D. in geography from the Université de Montréal. Officially retired since September 2013, he was a professor in geography at the Université du Québec à Rimouski (UQAR) for 33 years. During his career at the UQAR, he won the Distinction Pascal-Parent (2006), the Achievement award from the Université du Québec (2007) and the Distinction Alcide-C.-Hort (2012) to highlight his contribution in research and development of teaching programs at the UQAR. He has also been a member-researcher of BORÉAS and the Centre for Northern Studies (CEN) for nearly two decades. Since 2013, he has remained active as Associate Professor at the UQAR and the *Université Laval*, both in teaching and research. He is also part of the scientific council of the Perce Geopark. His past and current research projects are divided between two topics: (1) glacial movements, deglaciation and the reconstitution of long-term changes in relative sea level in Québec's maritime regions (Bas-Saint-Laurent, Gaspésie, Magdalen Islands, and Anticosti); (2) the impact of climate change on the postglacial evolution of the slopes in eastern Québec and the current dynamics of screen slopes in northern Gaspésie.

Patrick Lajeunesse received his Ph.D. in geomorphology from Université Laval (Canada) in 2000. He then undertook a postdoctoral project in glacial geology and geomorphology at the University of Alberta. He has previously worked at the Institut des sciences de la Mer de Rimouski (ISMER) at the University of Québec à Rimouski and the Institut National de la Recherche Scientifique-Eau, Terre, Environnement (INRS-ETE). He now occupies a Professor (Full) position in the Geography Department at Université Laval. His research work focuses on the geomorphological and sedimentary record of glaciation, postglacial environmental changes, and paleoseismicity on formerly glaciated coasts, continental margins and lakes, mainly of Eastern and Arctic Canada.

Atlantic Canada's Tidal Coastlines: Geomorphology and Multiple Resources

18

Norm Catto

Abstract

Coastlines in Atlantic Canada are influenced by microtidal, mesotidal, macrotidal, and hypertidal regimes. Although the hypertidal inner Bay of Fundy has attracted the most attention from geomorphologists and visitors, all atlantic tidal environments are marked by interesting and distinctive coastscapes, influenced by pre-existing topography, coastal bedrock lithology and Quaternary sediment, sea ice activity, and subtidal, intertidal, and supratidal vegetation. The hypertidal-macrotidal Bay of Fundy of New Brunswick and Nova Scotia is marked by extensive tidal flats and channels. Saltmarshes and barrier islands characterize the mesotidal Gulf of St. Lawrence coastlines of New Brunswick and Prince Edward Island. Tides and sea ice influence mesotidal-upper microtidal coastlines of Newfoundland and Labrador. In addition to their geomorphic significance, atlantic Canadian tidal coastlines also are valuable resources from multiple ecological and socio-economic viewpoints. Subtidal and intertidal areas support a variety of organisms and ecological communities. Saltmarshes are marked by sequences of communities conditioned by salinity and terrestrial input. Tide-dominated coastlines have been occupied by humans since shortly after deglaciation, and tidal activity has significant impacts on human coastal utilization. Interest in exploiting tidal energy remains high, but must be balanced against potential ecological impacts. Coastal tourism is a significant economic driver in Atlantic Canada, with tidal coastlines representing major attractions.

Keywords

Tidal geomorphology • Bay of Fundy • New Brunswick • Nova Scotia • Prince Edward Island • Newfoundland and Labrador • Sea ice • Ecology • Human utilization • Tidal energy • Tourism

18.1 Introduction

Atlantic Canada (Fig. 18.1) consists of the four political provinces of New Brunswick (NB), Nova Scotia (NS), Prince Edward Island (PEI), and Newfoundland and Labrador (NL). The Iles-de-la-Madeleine (Magdalen Islands), in the Gulf of St. Lawrence, are part of Québec. Tidal geomorphology dominates much of the coastlines of the three southern Maritime Provinces. To the north, although NL shorelines are primarily wave-dominated, tidal geomorphological features are present in several areas, influenced by pervasive winter sea ice. Thus, although it is included within Atlantic Canada, NL stands somewhat apart from the Maritime Provinces, geomorphologically, culturally and historically.

Most of Atlantic Canada is influenced by semi-diurnal tides (approximately two high and two low tides daily). Diurnal and mixed, mainly diurnal regions are located in Northumberland Strait, separating PEI from NB and NS, and to the southwest of the Iles-de-la-Madeleine.

The Gulf of St. Lawrence has a semi-diurnal amphidromic point located to the west of the Iles-de-la-Madeleine. At the amphidromic point, the semi-diurnal tide is not perceived, as the tidal front pivots on that location. The diurnal tidal range (typically less than 30 cm) is substantially less than the semi-diurnal range, and the semi-diurnal amphidromic point is marked by relatively low tidal influence.

Along the open Atlantic shoreline, the semi-diurnal tidal front moves southwestward, requiring approximately 4 h to progress from Cape St. Francis (NL) to Yarmouth (NS). As the tide enters a restricted embayment, such as the Bay of Fundy (NB/NS), its progression is slowed as the water begins to interact frictionally with the bathymetry and the coastline. In the Bay of Fundy, the tidal range progressively

N. Catto (✉)
Department of Geography, Memorial University of Newfoundland, St. John's, NL A1B 3X9, Canada
e-mail: ncatto@mun.ca

© Springer Nature Switzerland AG 2020

O. Slaymaker and N. Catto (eds.), *Landscapes and Landforms of Eastern Canada*, World Geomorphological Landscapes, https://doi.org/10.1007/978-3-030-35137-3_18

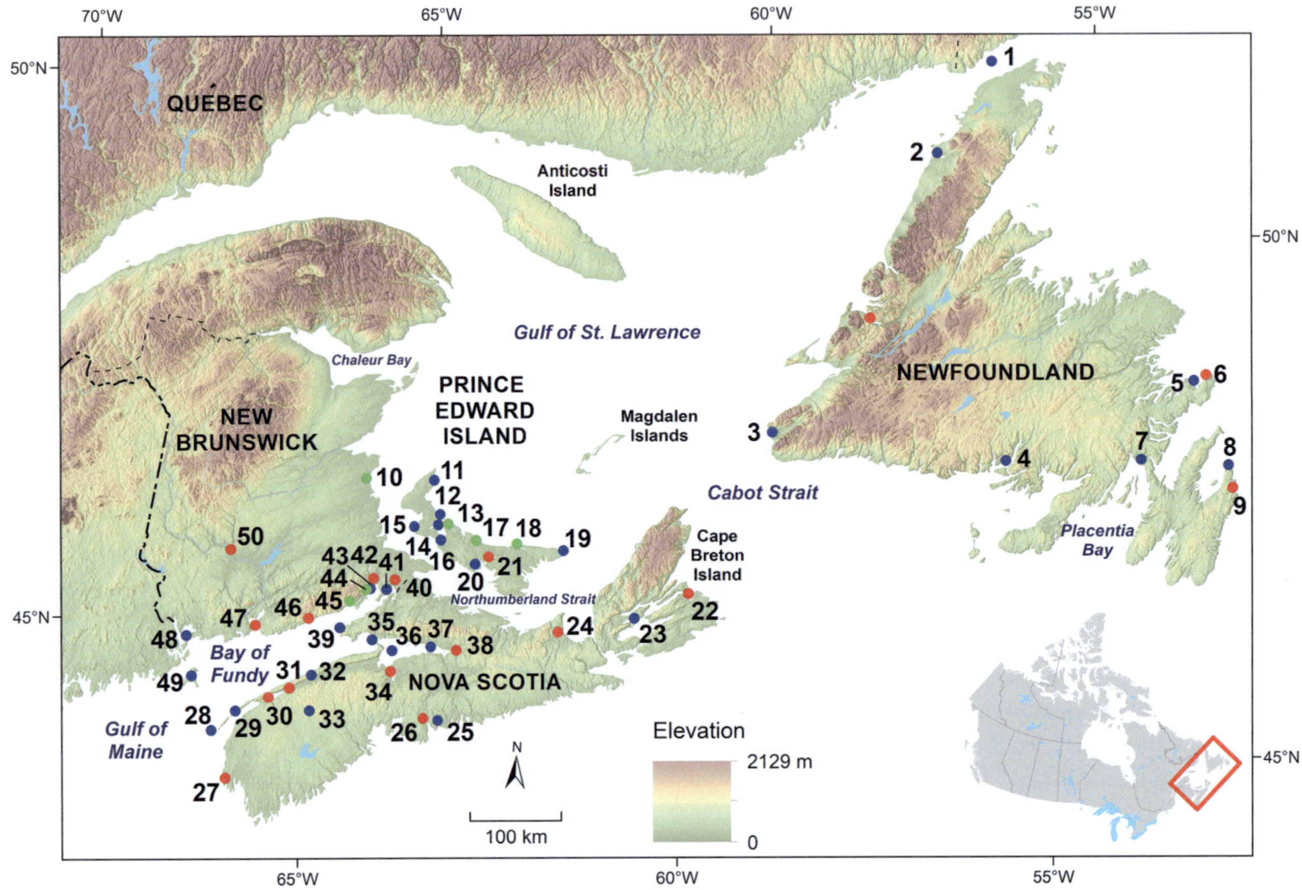

- Landscape features:
  1 - Strait of Belle Isle; 2 - Hawkes Bay; 3 - Grand Codroy; 4 - Bay d'Espoir; 5 - Blackhead Bay; 7 - Come-by-Chance; 8 - Cape St. Francis; 11 - Cape Kildare; 12 - Sandhills -Pītaweikek; 14 - Malpeque Bay; 15 - Cap Egmont; 16 - MacCallums Point; 19 - East Point; 20 - Skmaqn–Port-la-Joye–Fort Amherst National Historic Site; 23 - Bras d'Or; 25 - Chezzetcook Inlet; 28 - Brier Island-Westport; 29 - Digby Neck; 32 - North Mountain; 33 - South Mountain; 35 - Cape Split; 36 - Minas Basin; 37 - Cobequid Bay; 39 - Cape Chignecto; 41 - Cumberland Basin; 43 - Shepody Bay; 48 - Passamaquoddy Bay; 49 - Grand Manan Island
- National and Provincial Parks:
  10 - Kouchibouguac National Park; 13 - Cabot Beach Provincial Park; 17 - Prince Edward Island National Park; 18 - Greenwich unit, Prince Edward Island National Park; 44 - Hopewell Rocks Provincial Park; 45 - Fundy National Park
- Communities:
  6 - Bonavista; 9 - St. John's; 21. Charlottetown; 22 - Sydney (Cape Breton Regional Municipality); 24 - Antigonish; 26 - Halifax; 27 - Yarmouth; 30 - Digby; 31 - Annapolis Royal; 34 - Windsor (Avon River mouth); 38 - Truro; 40 - Sackville; 42 - Moncton (Petitcodiac estuary); 46 - St. Martins; 47 - Saint John; 50 - Fredericton

Fig. 18.1 Atlantic Canada, showing locations discussed in the text. *Cartography* D. Kovanen

increases eastward. Simultaneously, flood-tide velocity decreases. The tidal front requires approximately 1 h under normal conditions to progress from Digby to Truro.

In Atlantic Canada, microtidal (≤ 2 m range) conditions are prevalent along the coastlines of NL, and in western Northumberland Strait. Microtidal shorelines are dominated by wave action. Mesotidal (2–4 m) conditions prevail along the Atlantic coast of NS, and the southern Gulf of St. Lawrence coastlines of NB, PEI, and eastern NS. Isolated mesotidal areas in Newfoundland are present along the Gulf of St. Lawrence coastline, in the heads of Bay d'Espoir and Placentia Bay, and along the Strait of Belle Isle. The Bay of Fundy is macrotidal (range in excess of 4 m) or hypertidal (c.f. Archer 2013, range in excess of 6 m).

Mesotidal, macrotidal, and hypertidal areas are characterized by distinctive coastal landscapes. These environments are marked by characteristic suites of geomorphic features, influenced by pre-existing topography, coastal bedrock lithology and Quaternary sediment, sea ice activity, and subtidal, intertidal, and supratidal vegetation (Savenije 2005). The first part of this discussion focuses on four representative geomorphologies: the hypertidal-macrotidal Bay of Fundy

(NB/NS); the mesotidal Gulf of St. Lawrence coastlines of NB and PEI; the mesotidal-upper microtidal ice-influenced coastline of northeast Newfoundland (island); and the microtidal embayment of Bras d'Or. The discussion then will proceed to consider the geomorphic aspects involved in ecological and socio-economic resources of tidal coastlines.

18.2 Fundy: Macrotidal and Hypertidal Geomorphology

The Bay of Fundy is remarkable for its macrotidal and hypertidal regimes (Fig. 18.2). The name 'Fundy' comes from the French word 'fendu', meaning 'split', derived from Cape Split which separates the Minas Basin from the main, western extent of the bay (Fig. 18.3). The tidal activity in the bay has a major influence not only on the region's geomorphology, but also on human utilization of the coast.

Fundy is connected to the open Atlantic Ocean through the Gulf of Maine. Tidal flow entering the Gulf from the southeast is constricted by the shallow waters (35–100 m) of the superb scallop habitats of Browns Bank and German Bank to the north, and Georges Bank to the southwest. Flow enters the Gulf of Maine through Northeast Channel (width 50 km, controlling depth 200 m) and North Channel (width 20 km, controlling depth 150 m), and then moves northward through intersecting counterclockwise gyres over the Georges Basin (south) and Jordan Basin (north), offshore of German Bank and Yarmouth NS (Townsend et al. 2015).

Northeastward-flowing water enters the Bay of Fundy through Grand Manan Basin (width 40 km, maximum depth 225 m) between Grand Manan Island (NB) and Digby Neck (NS). The dominant current flow proceeds northeastward along the Nova Scotia coastline. The widest expanse of the bay, 80 km, occurs between Digby NS and Saint John NB, with maximum water depth of 150 m midway across the

Fig. 18.2 Blue Beach, Minas Basin (NS) is influenced by a hypertidal regime. **a** Low tide reveals an active sandy tidal flat (right), with the preceding high tide limit indicated by wet red-brown sediment. Blue-grey shale clasts (foreground) are covered during the highest spring tides. Storm surges reach the base of the Mississippian (Lower Carboniferous) succession (left). **b** Sandy tidal flat shows ephemeral meandering rivulets. Boulders are dislodged during storms and frost action. Note exposed bedrock ridge on the left. **c** Overwash fans composed of shale fragments indicate storm surge activity

Fig. 18.3 Cape Split (NS) is composed of columnar jointed tholeiitic North Mountain basalt. The French word 'fendu' ("split') gave Fundy its name

embayment. Fundy shallows northeastward, with maximum water depths of less than 50 m across the bay at Cape Chignecto. The shallow Chignecto Bay, separating NB and NS, extends northeastward into Shepody Bay (west) and Cumberland Basin (east). Minas Channel (maximum depth 125 m) extends eastward from Cape Chignecto, north of Cape Split, and into the very shallow Minas Basin and Cobequid Bay. Return flow follows the northern (NB) shoreline past Grand Manan Island into the Gulf of Maine.

All of Maritime Canada was glaciated (Rampton et al. 1984; Stea and Mott 2005; Shaw et al. 2006; Fig. 18.4). In NS, early phases saw ice flowing across Fundy initially from NB, and subsequently from the Escuminac Ice Centre across western PEI. Later phases involved local ice caps over South Mountain and Antigonish, with flow into the Bay of Fundy. The Annapolis and Avon Valleys and Fundy coastline were deglaciated approximately 12,000 BP (Stea et al. 2011).

Fundy National Park was glaciated several times during the Quaternary by ice originating from the north (Gaspereau Ice Centre) and west (Rampton et al. 1984). Ice flowed along the Petitcodiac River valley to Shepody Bay, following the modern coast to the southwest and filling Chignecto Bay. West of Fundy National Park, ice flowed south along the Saint John River valley and over St. Martins.

Following the initial Gaspereau glaciations, and prior to the MIS 2 glaciation, the area was free of glacial ice and formed a suitable habitat for large Quaternary fauna. A preserved mastodon (*Mammut americanum*) has been recovered from gypsum karst near Hillsborough. At the time of its death, this mastodon was between 15 and 18 years old, and had a mass of 8.3 t.

During MIS 2, parts of Fundy National Park were covered by thin, non-erosive cold-based ice (Rampton et al. 1984). The coastal regions of the park were deglaciated approximately 13,900 BP, and complete deglaciation of the area followed shortly thereafter.

Glaciation thus involved multiple competing flows, alternatively covering the region from several directions. The bathymetry of the shallow Chignecto and Minas embayments does not indicate significant ice stream

Fig. 18.4 Glacial ice covered all parts of the Maritime Provinces. **a** Gently undulating topography underlain by glacial sediments is exposed along the north flank of "The Lookoff", North Mountain (NS). **b** Till sedimentology and mineralogy reflect the underlying bedrock, as in this sediment derived from Permian sandstone at Malpeque Bay (PEI). **c** Drumlinized terrain occurs throughout Nova Scotia. Erosion of this drumlin at Chezzetcook has exceeded 2 m during a single storm event. **d** Striations and grooves are commonly used to determine ice flow directions. Cabot Head Provincial Park (PEI)

development or glacial erosion. Ice streaming, drawdown, and calving bay development occurred at Northeast Channel during deglaciation, and progressed up the bay through the waning phases of MIS 2 (Shaw et al. 2006). However, in contrast to the Laurentian Channel and the major embayments of eastern Newfoundland, glacial activity did not form a deep bathymetric funnel at the mouth of the Bay of Fundy.

Differences in tidal range are controlled predominantly by bathymetry, both nearshore and offshore, with the overall shoreline configuration playing a lesser role. Macrotidal (range 4–6 m) conditions prevail along the NS shoreline at the mouth of the bay (Yarmouth 5 m). The modal range increases towards the northeast, from 6.4 m (Westport, Brier Island, Digby Neck NS) and 7.1 m (North Head, Grand Manan Island, NB) to 14 m (Hantsport, Minas Basin, NS), 14.2 m (Amherst Point, Cumberland Basin, NS) and 16.7 m (Burntcoat Head, Minas Basin, NS). The highest tidal range recorded in the Bay of Fundy during a single event is 31.6 m, at Burntcoat Head.

Constriction of the rising flood tidal flow in some embayments at the head of Fundy causes the development of a foresteepened 'tidal bore'. Along the Petitcodiac estuary (NB), the bore is most pronounced in the narrows at Moncton. The height of the bore varies with the strength of the flood-tide, but generally is between 50 cm and 1 m. Construction of a causeway across the Petitcodiac River at Moncton in 1968 interfered with the water circulation in the upper part of the estuary, causing an overall reduction in bore height. The impounded lagoon suffered a severe loss of biodiversity, with decline of the Atlantic salmon (*Salmo salar*) population and the extirpation of the Canadian dwarf wedge mussel *Alasmidonta heterodon*) population (Locke

Fig. 18.5 **a** Columnar basalt forms distinct coastal landscapes at Westport, Brier Island, Digby Neck (NS). **b** Inland exposures are subject to frost wedging, along with disruptions due to road construction. Dark Harbour, Grand Manan Island (NB)

et al. 2003). In 2010, the causeway was modified to permit tidal influx into the former impoundment, resulting in increased circulation, sediment movement, and bank erosion. No effects were observed in the open estuary (Gerwing et al. 2017).

The combination of tidal range, shoreline bedrock, and Quaternary sediments results in distinctive coastal geomorphologies. In macrotidal areas and hypertidal areas not extensively influenced by seasonal ice shove and grounding, tidal flats are dominantly composed of fine to medium-grained sand. Silt-dominated flats occur under low-energy conditions marked by abundant nearshore-subtidal sediment supply, but with limited fluvial influx and no coarse glacial substrate.

Along the shores of North Mountain and western Grand Manan Island, columnar jointed Triassic (200 Ma) tholeiitic basalts including vertical hexagonal columns (Kontak 2008; Fyffe 2014) form vertical cliffs up to 120 m high (Fig. 18.5). The North Mountain ridge, 6–10 km wide, borders the shoreline from Cape Split south to Digby, continuing to the

Fig. 18.6 Eastern Grand Manan Island (NB). **a** Gravelly tidal flat at low tide, tidal range 6 m, flanked by Neoproterozoic sedimentary rock of the Grand Manan Terrane. **b** High-energy wave-dominated spit at "The Whistle", northeast coastline. **c** "Hole-in-the-Wall" at high tide, developed in the Cambrian Fish Head Gabbro

southwest as Digby Neck to Brier Island. North Mountain isolates the interior areas from the coast, with the exception of two gaps: in the southwest, at Digby; and in the northeast, where the Avon River enters Minas Basin south of Cape Split. Embayments are restricted and dominated by narrow gravel flats and wave-formed beaches.

The coastline of eastern Grand Manan, underlain by Neoproterozoic sedimentary and Ediacaran to Cambrian volcanic-arc and sedimentary rock units of the Grand Manan Terrane (McHone 2011; Fyffe 2014), features sandy and gravelly tidal flats (Fig. 18.6). This coastline developed under a lower hypertidal regime with frequent wave action driven by southwesterly winds, and minimal Quaternary sediment input from inland (Legget 1980). Wave and current activity are effective in shaping the shoreline. Seasonal shifts in currents influence both the development of spits and bars and the distribution of nearshore finfish species. During the winter and spring, the eastern shore is influenced by southerly-flowing currents towards the Gulf Stream, and bars and spits extend southwards. During the summer and autumn, the Gulf Stream-derived current is diverted northward. Spits are re-oriented towards the east along the eastern shore.

To the northeast, Passamaquoddy Bay and the exposed NB coastline, underlain by sedimentary, basic volcanic, and rhyolitic units of Neoproterozoic to Devonian age, with substantial input from exposed Quaternary sediments inland, display sandy to coarse silty tidal flats with channels influenced by fluvial discharge, alternating with exposed rock platforms (Fig. 18.7). Fine sand-dominated tidal flats are also present along Minas Channel, the outer part of Chignecto Bay, and in areas with substantial fluvial input, such as the Kennetcook river estuary. Much sand is derived from erosion of Mississippian-Triassic clastic sedimentary rock, combined with glacial and glacifluvial deposits.

The sandy substrates are suitable habitats for burrowing molluscs (soft-shell clams, quahogs), and the flats provide important resting and feeding areas for seabirds during low tide (Fig. 18.8). Tidal flats with firm silt and sand substrates show the most extensive bioturbation (Dashtgardt et al. 2008). On coarser tidal flats, bioturbation decreases sharply

Fig. 18.7 Passamaquoddy Bay area. **a** Sandy-gravelly tidal flat with interspersed rock platforms. **b** Exposed at low tide, this human-reinforced tombolo is developed on late Devonian volcaniclastic rock. Blacks Harbour ferry terminal (NB)

in the lower intertidal zone. Sediment distribution is characteristically patchy across Fundy tidal flats, leading to variations in water saturation and organic matter contents (Gerwing et al. 2015) linked to differences in bioturbation.

The steep cliffs along much of the shore of Fundy National Park limit the development of tidal flats and salt marshes, which present only along the coastal plain to the east, between Alma and Moncton. Erosion of weak Late Proterozoic rocks is ongoing along the NB coastline, notably at Red Head, near Saint John (Spooner et al. 2013).

Differences in surface water elevation over a tidal cycle can result in reversed flow direction and gradient. At the Reversing Falls in Saint John NB, low tide creates a gradient of 4 m over 3 km, resulting in strong flow into the harbour (Leys 2007). During most of the year, flood-tide flows inland over the Reversing Falls. However, during spring freshet, the elevated water level of the Saint John River is higher than the mean high tide, and seaward water flow continues.

Silt-dominated tidal flats are prevalent in the lowest energy areas of the heads of Shepody Bay (Moncton), Cumberland Basin (Amherst Point), and Cobequid Bay (Truro), where thick Quaternary deposits overlie friable late Palaeozoic clastic successions. These tidal flats are rarely influenced by storm surge, and many have limited fluvial sediment influx. Seasonal ice action is also restricted in these embayments, although ice does form in estuaries and in Minas Channel.

In channels developed in sandy and silt tidal flats, currents result in erosion and transportation of sediments when flow exceeds 10 cm/s (0.36 km/h; Fig. 18.9). Bedforms have heights locally exceeding 1 m and lengths of metres to 10s of metres. Hypertidal and macrotidal channels tend to be

Fig. 18.8 Sandy tidal flats substrates are important for burrowing molluscs and seabirds. **a** Seabirds feeding at low tide, near Windsor, Avon River mouth. **b** Sandy tidal flat surfaces show numerous mollusc burrow and airholes, along with seabird feeding traces

relatively straight, with steep slopes associated with constricted flow (Prandle 2003, 2004; Davies and Woodroffe 2010).

Inbound and outbound tidal flow, influenced by bathymetry, augmentation of ebb flow by rivers, and Coriolis effect, produce distinctive flood-tide and ebb-tide features on opposite sides. Bedforms and deposits can form during either flood-tide or ebb-tide. At high tide, the flats are covered, and powerful currents sweep along the shores and in the main channels. The currents tend to be strongest along the southern bank of the flats. Large ripples created by these currents are frequently visible at low tide.

Inland, tidal channels tend to develop meandering forms, with steep banks and well-defined point bar features. Along the southern shore of the Bay of Fundy, with gently sloping terrain and lack of input from major rivers, tidal channels tend to reflect ebb-tide flow, and commonly develop a dendritic network aligned seaward. Whereas the flood-tide forms a relatively slow-moving sheet, ebb-tide velocities are augmented by gravitational flow, facilitating development of the channel network.

Throughout the Bay of Fundy, tidal coastlines show relationships between storm surge, spring tide height, foreshore and intertidal geomorphology and sedimentology, and resilience (Tibbits and van Proosdij 2013). In the more exposed areas, energetic wave action locally develops bedforms in intertidal areas, resulting in sediment redistribution and local decreases in modal or median texture in intertidal zones (e.g., Hay et al. 2014). Storm surges, generated by winter and spring mid-latitude cyclones, and summer and autumn tropical cyclones, storms, and extratropical transitions, can result in substantial inundation if the storm makes

Fig. 18.9 Tidal channels are generally meandering, but with relatively low sinuosity, and display point (**a**) and (**b**) lateral barforms

landfall at high tide. The 31.6 m record high tide at Burntcoat Head included at least 15 m of storm surge height.

In Atlantic Canada, a northward-moving tropical storm can interact with a mid-latitude storm, merging into a single, larger extratropical transition. The merger provides more energy to the extratropical transition, strengthening it as it tracks. Most hurricanes affecting eastern Canada have undergone partial or complete extratropical transition.

Landfall of the October 4–5, 1869 extratropical transition 'Saxby Gale' corresponded with an anomalously high atmospheric tide (Abraham et al. 1999; Parkes et al. 1999). The resulting storm surge caused water levels to rise more than 15 m above the normal high tide positions, resulting in flooding of areas up to 30 m above mean sea level, notably the Tantamar Marshes area along the NB–NS border near Sackville. Adaptation studies at Sackville NB and Annapolis Royal NS focus on coping with storm surge inundation (Vasseur and Catto 2008).

Bedrock platforms are common along the Fundy coastline (Fig. 18.10). Largely or totally devoid of sediment, they slope seaward at shallow angles (typically <10°) and may exceed 30 m in width (Trenhaile and Mercan 1984; Trenhaile 1987; Trenhaile and Kanyaya 2007). Here, the seaward edge of a shore platform is defined as the seaward limit of surface bedrock erosion (c.f. Kennedy 2015). Platforms are generally associated with gently dipping sedimentary, metamorphosed sedimentary, or extrusive volcanic strata. Platform erosion of clastic rock is facilitated by chemical modification by saline waters, wetting and drying, dissolution of cement, and preferential levering and extraction of

Fig. 18.10 Tidally-eroded Devonian Perry Formation, St. Andrews (NB). **a** Rock platform flanked by hypertidal sandy sediments. **b** Erosion reflects tidal abrasion, frost action, bedding and jointing attitudes, and biological processes

sand and granule clasts (Porter et al. 2010; Trenhaile and Porter 2018). Measured rates approximate 1 mm/y in sites along the Bay of Fundy.

At Hopewell Rocks Provincial Park (NB; Fig. 18.11), constricted incoming tidal flow has eroded red Pennsylvanian conglomerate into 17 stacks (Trenhaile et al. 1998). Frost action and other terrestrial erosion along interconnected fractures aligned both parallel to and at 90° to the coast resulted in the gradual detachment of 25 m high stacks from the cliff. Tidal action eroded the bases, producing the characteristic 'mushroom-shaped' stacks with blocky, rectangular cross-sectional columns supported on slender bases 8–10 m high. Tidal erosion commenced following deglaciation, ~15 ka. Erosion of the stack base is relatively slow, almost undetectable over the past 40 years. Rock falls along the cliff faces, most recently of Elephant Rock in 2016, result from terrestrial frost wedging.

Relative sea level change since MIS 2 deglaciation has impacted tidal regimes and coastal geomorphology. Lower postglacial relative sea level exposed parts of the Bay of Fundy and Gulf of St. Lawrence, allowing development of organic terrestrial and saltmarsh Holocene sediments. Subsequent inundation and microbial activity is producing biogenic methane, marked by linear and circular depressions. More than 10,000 pockmarks were recorded in a single zone in Passamaquoddy Bay (Broster and Legere 2018).

Sea level rise continued through the early and mid-Holocene. The highest points of Georges Bank, Browns Bank, and German Bank were exposed c. 13 ka (Shaw et al. 2002, 2006), but by 10 ka, sea level rise had inundated

Fig. 18.11 Hopewell Rocks Provincial Park. The stacks are produced by tidal erosion of friable Mississippian conglomerate. **a** at low tide, the sand-gravel tidal flat is exposed. The morphology of the pinnacle stacks is controlled by the bedding, vertically aligned joints, and tidal action. High tide-line is located at the top of the narrow supporting columns. **b** Stack exposed at low tide. **c** Stacks at high tide. Spring tide level is indicated by the exposed notches. *Photographs* A. Latocha

Georges Bank. Tidal range has generally increased since deglaciation, particularly during the Late Holocene.

Increased relative sea level throughout the early Holocene gradually converted the Bay of Fundy from a mesotidal to a hypertidal regime. Most locations were hypertidal by 5 ka. In Minas Basin, a sedimentary barrier across Minas Passage may have delayed tidal expansion until c. 4–3.5 ka, when rapid failure occurred (Shaw et al. 2010).

Over the past 200 years, relative sea level and tidal range both have been increasing due to post-glacial isostatic recovery, independent of climate change (Greenberg et al. 2012). In the southern Bay of Fundy, the combined effects of isostatic adjustment and climate-induced sea level change may produce a 0.5–1 m increase in the high water level by 2100. Rising relative sea level would result in increased tidal ranges, increased tidal resonance and amplification throughout Fundy and the Gulf of Maine, and increased potential for storm surge inundation.

18.3 Southern Gulf of St. Lawrence: Mesotidal Geomorphology

Mesotidal and upper microtidal coastlines flank the southern Gulf of St. Lawrence along NB and the northern coastline of PEI. Tidal ranges vary from 1.5 to 3.5 m. These coastlines consist of alternating exposures of friable Pennsylvanian (Late Carboniferous; NB) and Permian (PEI) clastic red bed sequences; Quaternary glacigenic sediments; elongate barrier islands and barred coastlines; and salt marshes.

Rock cliffs and sediment bluffs along the coastlines of NB and PEI are subject to slope failure induced by terrestrial processes (frost action), and wave and tidal current undercutting (Spooner et al. 2013; Fig. 18.12). Relative sea level is currently rising at $\sim$3 mm/y along northern PEI, further accentuating failure. Incremental block failures along joint systems are common, with local rates in excess of 50 cm/y at several locations, including Cape Kildare (Genest and Joseph 1989), Cap-Egmont, Cabot Beach Provincial Park, Skmaqn–Port-la-Joye–Fort Amherst National Historic Site, MacCallums Point, and East Point (Spooner et al. 2013). Joint systems develop from initial structural weaknesses combined with glaci-isostatic deformation, cliff rebound due to loss of lateral support by previous erosion, and frost action. Direct erosion by waves and tides plays a lesser role.

Barrier island complexes developed under mesotidal conditions include Sandhills (Pītaweikek). Flanking the northern side of Malpeque Bay PEI, Hog Island (601 ha), part of the Sandhills, is one of the largest and least disturbed coastal dune complexes in Atlantic Canada (Fig. 18.13). Pītaweikek ('the place where tea is found'), the Mi'kmaq name for the Sandhills, has great traditional importance for the cultural heritage of Mi'kmaq people. The terrain is valued as a source of berries and medicinal plants, as a base for fishing, for camping, and for ceremonial uses. The cultural importance of the Pītaweikek landscape has encouraged the

Fig. 18.12 Block failure of Permian sandstone at Cap-Egmont (PEI). **a** Failure along joint planes results in failure of the overlying Escuminac glacial till. **b** Arch shows the influence of jointing and bedding on block failure. Clasts derived from erosion of bedrock and glacial till are concentrated by spring tides and storm surges

Mi'kmaq Confederacy of PEI to support First Nations-led conservation of the area, particularly as many other areas of the north coast of PEI have been rapidly modified by development. The other three segments of the Sandhills dune-backed coast: Conway Island (461 ha), Cascumpec Island (178 ha), and Kildare shores (104 ha), show greater amounts of disturbance, due both to human activity and rising relative sea level (currently 3.3 mm/y). Erosion between 1974 and 2004 caused the tip of Cascumpec Island to retreat by 115 m, 3.8 m per year (Conroy 2007).

Kouchibouguac National Park (NB) has a barrier island/coastal dune complex, produced by littoral transport from the north (Fig. 18.14). A mesotidal environment, Kouchibouguac can be translated from the Mi'kmaq as 'river of long tides'. Within Kouchibouguac, three barrier islands are present, each of which supports active transverse sand dune complexes. North Kouchibouguac Dune extends 4.3 km in length, has a maximum width of 320 m, and a maximum height of 5 m above sea level. South Kouchibouguac Dune is 6.4 km long, has a maximum width of 600 m, and a maximum height of 4 m. The southernmost of the park's islands is North Richibucto, 7 km long, 6 m high, with a maximum width of 210 m. The dimensions of all three islands change sporadically in response to storm activity and accretion of sediment on both the seaward and lagoon sides.

Under the mesotidal regime, strong tidal currents break the net littoral drift into a cellular pattern. The barrier islands develop relatively high dunes, which serve as sand repositories and also isolate the lagoons from wave and storm activity. Deep, persistent tidal channels develop between the barrier islands, serving as conduits for storm waves and surges. The persistence of the tidal channels facilitates

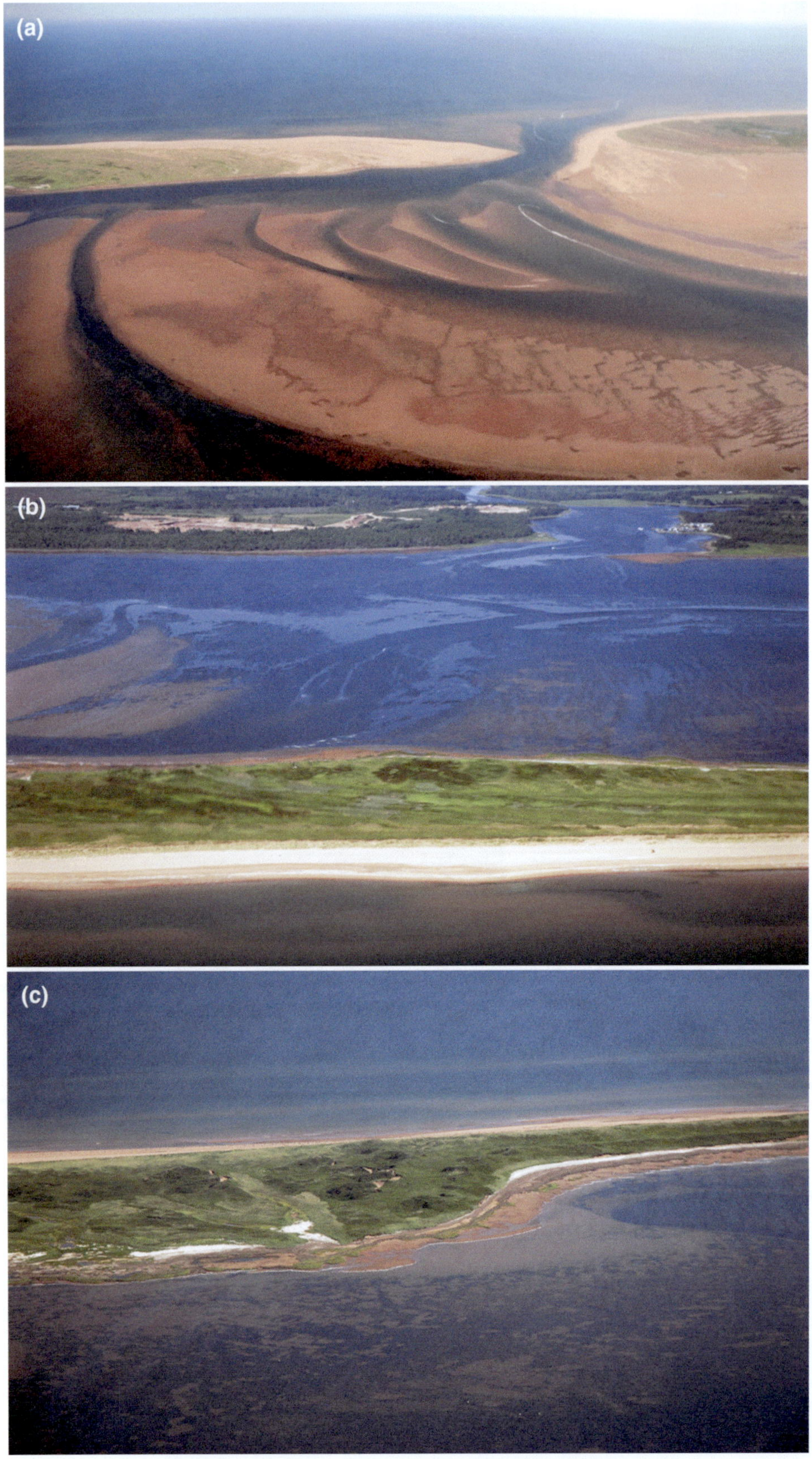

Fig. 18.13 The Sandhills (Pītaweikek) barrier island complex flanks the north shore of Malpeque Bay (PEI). Formed by west-to-east littoral drift, the dissipative system is currently subject to rising relative sea level of 3.3 mm/y. **a** Looking south from the Gulf of St. Lawrence into the mesotidal lagoon, western side of Malpeque Bay. **b** Looking north from Malpeque Bay towards the Gulf of St. Lawrence. **c** Mesotidal inlet into eastern Cascumpec Bay

Fig. 18.14 Kouchibouguac National Park (NB). **a** The dissipative; low-energy, littoral drift-dominated coastline is marked by sand transport from north to south along the Gulf of St. Lawrence. **b** Tourist pressure, combined with decreasing seasonal sea ice and snow cover and increasing storm surge activity, have resulted in local erosion of the coastal dune complex, reducing sand influx for littoral drift and beach replenishment

movement of tidal flux, fish, and human vessel traffic between the lagoons and the Gulf of St. Lawrence.

The Gulf of St. Lawrence's mid-boreal (Köppen-Geiger Dfb) climate influences barrier island development. Extensive winter snow cover effectively prevents dune erosion, minimizing the potential effect of winter storm activity. Winter snow and ice cover along the shoreline prevents erosion and renders littoral drift ineffective as a sediment transport mechanism. Winter ice cover in the gulf prevents tidal and wave energy transfer to the shoreline. Winter snow cover also limits tourist usage, reducing anthropogenic destabilization of dunes: almost all of Kouchibouguac's 120,000 officially recorded visitors come during the summer.

Barrier islands are influenced by their inheritance, developing over pre-existing geomorphology (Storms et al. 2002; Harris et al. 2005). In and along the Gulf of St. Lawrence, relative sea level changes and differing wave conditions have formed multiple sequences of offshore bars and partially submerged barrier islands. Relict features above modern relative sea level are also present.

Kouchibouguac was glaciated by eastward-moving ice from central NB during MIS 2 (Rampton et al. 1984; Shaw et al. 2006). As the ice began to retreat, approximately 14 ka, the isostatically-depressed area was flooded by the Goldthwait Sea, a predecessor of the Gulf of St. Lawrence. Marine waters reached 45 m above modern sea level, and the park area was completely submerged. The presence of the Goldthwait Sea is indicated by abundant shells of marine molluscs at inland locations, and by deposits of marine clay.

Progressive isostatic rebound caused marine regression, and relative sea level fell to approximately −15 m c. 7 ka. Subsequently, transgression resumed, and the barrier island

Fig. 18.15 Grand Codroy salt marsh, southwestern Newfoundland

system developed as sea level rose slowly during the past 2500 years and offshore sediment was moved landward. The Kouchibouguac barrier island system continues to evolve, as relative sea level is rising at $\sim$3 mm/y. Shoreward retreat of the three barrier islands is documented through maps from the 1800s. The overall barrier island morphology is controlled by long-term sea level changes, littoral drift, and mesotidal channels. Similar changes have been documented for the barred littoral drift coastline of Buctouche Spit to the south (Ollerhead and Davidson-Arnott 1995).

Salt marshes are developed in most tidally influenced lagoons and estuaries along the southern Gulf of St. Lawrence coastline, including mesotidal estuaries on the west coast of Newfoundland (Fig. 18.15). The Grand Codroy Ramsar Site, an International Wetlands Migratory Waterfowl Stopover is one of the few salt marshes in Newfoundland, associated with freshwater marshes, deltaic systems, and an estuary with fringing and barrier beaches. Saltmarsh vegetation can reduce incoming wave heights by up to 70%, and tidal energy by more than 90%, resulting in deposition of sediment, gradual build-up of the marsh, and insulation of the shoreline from storm surges.

In Atlantic Canada, saltmarsh development is influenced by sediment influx, sea level change, ice scour, and human activity (Forbes and Syvitski 1995; Shaw and Ceman 1999; Chmura et al. 2001). Typical rates of sediment accumulation are 1–10 mm per year. If sea level rises at a comparable rate and sediment influx is maintained, the saltmarsh will continue to expand inland, as long as there is available and suitable terrain. Saltmarshes can persist in locations with sea level rise in excess of 3 mm/y (e.g. Chezzetcook Inlet, NS, Carter et al. 1989, 1990) if sediment supply is maintained and vegetation cover remains undisturbed. Reduced sediment influx or vegetation removal will trigger erosion. Dykes and embankments isolating the saltmarsh from tidal sediment influx and ebb-tidal flushing will result in erosion and ecological damage.

Typically, the seaward margin of a saltmarsh is dominated by vascular salt-tolerant plants such as *Salicornia maritima* (samphire, sea glasswort), *Spartina alterniflora* (smooth cord-grass), *Festuca* (fescue), and *Suaeda* (sea blite). The higher areas support *Carex* (sedge), *Myrica gale* (sweet gale), *Juncus gerardii* (black-grass rush), *Spartina patens* (saltmeadow cordgrass), *Bolboschoenus (Scirpus)*

Fig. 18.16 The Gravel-dominated tidal flat at Come-By-Chance, head of Placentia Bay (NL), resulted from tidal reworking of glacifluvial and coastal glacimarine sediment. **a** Mesotidal flat at low tide. **b** Large glacially-transported boulders are immobile during the strongest storm events

robustus (sturdy bulrush), *Ranunculus cymbalaria* (seaside buttercup), *Puccinellia maritima* (seaside alkali grass), and *Iris*.

Vegetation cover significantly reduces water flow (typically <0.2 m/s; Davidson-Arnott et al. 2002; Ashall et al. 2016a). Drainage in vegetated areas is oriented normal to channels (Ashall et al. 2016a). Clastic sediment deposition is primarily from suspension, and typically is highest in the lower marsh area, although not at the seaward margin (Davidson-Arnott et al. 2002). Locally, depositional rates per tide can approach 100 g m^{-2} (Poirier et al. 2017), with greater sediment movement in the fall and winter. Increased suspended sediment transport during vigourous wave and spring tidal events increases sediment deposition, as well as transporting sediment into higher marsh areas. Duration of inundation time, transport direction and distance, and relative surface roughness also influence sediment deposition in saltmarshes.

Recently, the use of small, unmanned aerial vehicles and structure-from-motion photogrammetry (e.g. Kalacska et al. 2017) has allowed the production of highly detailed digital surface models of salt marshes, useful for assessing marsh development, surface water flow directions, and pond connectivity. Over longer terms, saltmarsh accretion rates can be estimated through isotopic analysis (^{137}Cs, ^{210}Pb) and palynology (e.g. Chmura et al. 2001). Estimates ranging from 1 to 5 mm/y since 1800 are similar to recorded changes in relative sea level. Higher rates recorded in some marshes, such as those in the southwestern Bay of Fundy, have been attributed to combinations of local neotectonic influences on relative sea level and to increased sediment deposition from ice rafting in the nineteenth century (Chmura et al. 2001).

As water flow and vegetation interact constantly (Ashall et al. 2016a; Silinski et al. 2016), saltmarshes exhibit strong biogeomorphologic effects. Stronger water flow limits establishment of pioneer *Bolboschoenus (Scirpus) maritimus* (Saltmarsh bulrush), *Spartina alterniflora*, and *Festuca.* Organic matter concentrations lead to higher accumulation rates. Saltmarsh accumulation rates decrease with tidal range in mesotidal, macrotidal, and hypertidal environments (Chmura and Hung 2004).

18.4 Newfoundland and Labrador: Areas of Mesotidal Ice-Influenced Geomorphology

Much of the NL coastline is wave-dominated. Mesotidal geomorphological coastscapes are primarily confined to the heads of embayments. Rock platforms develop under mesotidal or upper microtidal regimes with limited sediment cover inland.

Several boulder-dominated tidal flats are present at the heads of mesotidal embayments in NL (Catto 2011; Fig. 18.16). Boulders are scattered across the entire intertidal surface. The boulders were initially transported glacially, and are too large to be moved by tidal action or storm waves. These tidal flats are conditioned by glacial sedimentation, resulting from the reworking of the previously deposited glacial sediments. Coarse sediments dominate the surfaces, with interspersed areas of supratidal and terrestrial vegetation. Slopes are generally very gentle, approximately 1–2°, except where cut by tidal channels.

Along Bay d'Espoir, a small tidal flat area is present at St. Albans (Fig. 18.17). The flat is divided by shallow meandering channels, generally less than 1.5 m deep, with fine to medium-grained sand deposited in the thalwegs. Slopes vary between 1–2°, and the surfaces are mantled with approximately equal proportions of sand and silt, with little clay. The tidal flat is slowly aggrading under a relative sea level rise of ~2.5 mm/y, with abundant sediment input.

Along the Gulf of St. Lawrence and the Bonavista Peninsula of northeastern Newfoundland, thrusted sea ice slabs sweep boulders and cobbles across wide rock platforms (Forbes and Taylor 1994; Forbes et al. 2002, 2011; Figs. 18.18 and 18.19). They extend up to 1 km offshore, and are partially exposed at low tide. Rock platform development is confined to areas with gently to moderately dipping sedimentary or metamorphosed sedimentary bedrock. In some locations, ice shove constructs a boulder rampart, or leaves boulders scattered over the rock substrate. Boulders can be pushed more than 2 m above high tide level. Ice shove elevations are directly related to tidal range.

Typically, during the summer and autumn, the rock platforms act to buffet waves. During the winter, however, the platforms are scoured by ice pushed onshore by tides until freeze-up occurs. Renewed scouring accompanies spring break-up. In areas marked by mesotidal or high microtidal conditions, sediment fluxes from either landward or seaward sources must remain low to keep the platforms exposed. Sea ice must be driven aground, either by waves or tides, in order to be effective as an erosional agent. In years marked by large amounts of ice shove, the lower part of the platform is swept completely clean of sediment, and the boulders form several discontinuous lines across the upper part representing different ice shove events. In years when ice shove activity was less or non-existent, no boulders may be present above mean high tide level.

In the areas above high tide level, the development of rock platforms is primarily the result of frost weathering during intervals when the rock is exposed. The rate of formation is controlled by the number of freeze-thaw cycles. The amount of frost-induced erosion increases in the higher intertidal areas, where sea water only covers the rock for short periods each day at high tide. In contrast, abrasion by tidal action is most effective in the lower parts of the intertidal zone, where the rock is only exposed at low tide.

18.5 Bras d'Or: Brackish Microtidal Embayment

Bras d'Or Lake in Cape Breton (NS) is a UNESCO Biosphere reserve. The 1099 km^2, 32 km^3 embayment is connected to the Atlantic Ocean through the 30 km Great Bras d'Or Channel (controlling depth 16.2 m at the mouth), and the 8 km Little Bras d'Or Channel (controlling depth 5 m) as well as through a small, shallow canal at St. Peters. With no large river discharge, the salinity of Bras d'Or varies between 16 and 27 psu. Bras d'Or Lake is microtidal, with a phase lag of approximately 8 h between the interior and the mouth of Great Bras d'Or Channel (Petrie and Bugden 2002;

Fig. 18.17 Tidal Flat, north of St. Albans, Bay d'Espoir (NL). **a** Overview, looking east. The tidal flat is slowly aggrading under a relative sea level rise of ~2.5 mm/y. **b** Sediment input is largely glacially-derived and fluvially-transported. Large glacial boulders are immobile

Dupont et al. 2003). Tidal currents in the lake's straits and channels average 1–2 m/s, and contribute substantially to water circulation.

18.6 Ecological and Socio-economic Resources: Geomorphic Aspects

In addition to their strictly geomorphological characteristics, the tidal coastlines of atlantic Canada can be considered from a variety of resource perspectives. Port operations are influenced by macrotidal and hypertidal conditions. At low tide, the Nova Scotian port of Windsor (Avon River mouth, Minas Basin) is inaccessible for ships. Historically, tidal ranges were a significant issue for sailing vessels. Limitations in access to Saint John harbour in the 19th century handicapped its growth as a winter railway port for inland Canada. Combining geomorphological and other research highlights the role of geomorphology in ecological and socio-economic resources, including subtidal and intertidal organisms; agricultural dykes; tidal energy potential; and tourism.

18.6.1 Subtidal and Intertidal Organisms

Beds of eelgrass (*Zostera marina*) represent both a geomorphic influence and a key indicator of ecosystem health in atlantic Canada (Hanson 2004; CSAS 2009). Blades grow to 1.5 m height. Eelgrass is a perennial, reproducing both through pollen and vegetatively from rhizome buds. The rhizomes store nutrients and energy required for flowering.

Fig. 18.18 Ice-scoured bedrock platform, Blackhead Bay, west of Bonavista (NL)

Eelgrass grows in shallow, subtidal, sandy to muddy areas with freshwater influence. It requires moderate temperature, moderate salinity, low turbidity, and abundant nutrients. In atlantic Canada, it is commonly found in mesotidal and upper microtidal areas along the Gulf of St. Lawrence, but high turbidity limits its extent along the Bay of Fundy. Eelgrass stabilizes the substrate, reducing erosion and baffling low-energy waves. The beds are suitable areas for algae and diatoms, and provide food and shelter for aquatic invertebrates and fish, including eel (*Anguilla rostrata*) and juvenile cod. Eelgrass is a food source for sea urchins, gastropods, green turtles, black duck, Brant, and Canada goose.

Eelgrass faces several threats, including chemical pollution, petroleum discharges, dredging, increased turbidity, and rising water temperatures. In areas adjacent to harbours, damage from vessel propellers and boat wakes is also significant.

American eel (*Anguilla rostrata*; 'kat' in Mi'kmaq) is an important inhabitant in eelgrass beds from Nova Scotia to southern Newfoundland. Eels occupy the beds in successive juvenile stages as glass eels, elvers, and yellow eels, reach sexual maturity at 20–40 years (in colder Canadian waters), migrate as mature adult silver eels. Eels show marked sexual dimorphism, with adult females typically 80–110 cm long, three times the length of the males. Higher population densities result in increased proportions of males, and the larger females are often preferentially harvested.

Although some arriving glass eels and elvers swim up freshwater streams, most remain and mature in estuarine and coastal waters. In this sense, most *Anguilla rostrata* are not strictly catadromous. Maintenance of the North American eel population thus depends on maintenance of eelgrass beds in their specific environments, including coastal geomorphology. Although *Anguilla rostrata* is not a threatened species, the decline in European eel populations combined with increased demand for glass eels in Europe and Asia is exerting pressure on populations.

Dulse (*Palmaria palmata*) is an edible foliose red-purple algae species (Fig. 18.20). Flat fronds extend 50 cm from a holdfast, attached to gravelly substrates in hypertidal to upper mesotidal intertidal zones. Dulse requires turbulent but clear water to 20 m depth, restricting growth to cobble and boulder substrates open to wave action and winter storms (Garbary et al. 2012). The hypertidal regimes of the Bay of Fundy allow dulse to flourish in areas which undergo winter

Fig. 18.19 Gravel transported by marine ice shove, mesotidal shoreline, Hawkes Bay (NL)

ice cover. In the Bay of Fundy, dulse is present in commercially-harvestable quantities at Dark Harbour, Grand Manan Island, and along the southwest extension of North Mountain, from Digby Neck to Brier Island.

At Dark Harbour, the annual harvest exceeds 400,000 kg. Dulse contains high concentrations of iodine, beneficial for both human health (thyroid) and as fodder for dairy cattle (Gaillard et al. 2018). It is also rich in potassium, and is an antioxidant. Dulse can be used as an additive in Camembert cheeses (Hell et al. 2018).

Irish moss (*Chondrus crispus)* is a clonal rhodophyte, with a fan-like habit extending 15–30 cm from a holdfast (Collén et al. 2014). It inhabits boulder-dominated or exposed bedrock lower intertidal to subtidal zones in mesotidal regions, typically 4–7 m below mean low water, but locally extending below 20 m. Irish moss prefers areas with tidal currents and moderate to low wave activity, and water temperatures not exceeding 28 °C. In atlantic Canada, Irish moss inhabits rocky and boulder-dominated coastlines along the Gulf of St. Lawrence and southwestern Nova Scotia (CSAS 2013). Its preference for horizontally-aligned tabular boulders accounts for its abundance along the western PEI coastline, colonizing displaced Permian clastic redbed slabs. Genetic evidence suggests that both Irish moss and dulse occupied offshore refugia along the Scotian shelf during glaciation (Li et al. 2016). Irish moss is a source of carrageenan, used in foods such as ice cream as an emulsifying, gelling, and thickening agent (Necas and Bartosikova 2013; Pereira et al. 2015).

Ascophyllum nodosum (rockweed) dominates intertidal mesotidal rocky substrates in coastal NS and southern Newfoundland, flourishing under lower wave energy and limited sea ice (Fig. 18.21). It is much less common in areas regularly subjected to storm activity. In southwestern NS, rockweed is harvested commercially for animal fodder and fertilizers.

In Placentia Bay, the inner Ragged Islands area is the best example of a rockweed community in NL, with hundreds of hectares of rock platform shoreline covered by dense bright yellow beds. Formerly, the lower limit of this zone was determined by competition with atlantic kelp (*Laminaria longicruris*), but this latter species has been severely impacted by scallop dragging. The distribution of taxa within the rockweed communities (Catto and Hooper 1999)

Fig. 18.20 Dulse. **a** Harvesting, Dark Harbour, Grand Manan Island (NB). **b** After sun-drying, the dulse will be packaged without further processing

indicates that many platforms are not regularly subjected to sea ice activity or strong wave action, but continue to be eroded in the intertidal and supratidal zones by frost wedging.

Atlantic kelp (*Laminaria (Saccharina) longicruris*) is subtidal to lowest intertidal in distribution (4–18 m), inhabiting waters at 6–14 °C and salinity exceeding 20 psu, with low to moderate turbidity. Bedrock and cobble-boulder substrates are preferred, and kelp does not colonize mud substrates. Kelp includes a holdfast, stipe, and elongated blade, locally exceeding 10 m in length. A mature kelp (2–5 years) may exceed 2 kg. Kelp beds provide refuges for a variety of invertebrate and fish species. Commercial culture and harvesting occurs in the Baie-des-Chaleurs (QC) (Chapman n.d.).

Sugar kelp (*Laminaria saccharina, a.k.a. Saccharina latissima*) has elongate (to 3 m), dark fronds. It occupies shallow subtidal to lower intertidal areas with coarse or rocky substrates and limited storm activity in the southern Maritimes. The distribution of sugar kelp into distinctive geographical and genetic populations has recently been attributed to recolonization from distinctive offshore refugia along the Scotian Shelf (Neiva et al. 2018).

Among the multiple values of saltmarshes is their ability to support populations of mummichog for mosquito control. Brackish waters surrounding Prince Edward Island are inhabited by the Mummichog (*Fundulus heteroclitus*), a small fish 3 cm in length. The mummichog cannot survive in water temperatures below 1 °C, and hence its presence indicates the extent of the warm coastal waters derived from

Fig. 18.21 Marine vegetation, including *Ascophyllum* (rockweed), has a geomorphic impact on rocky and boulder coastlines developed until microtidal to lower mesotidal conditions, in areas where seasonal ice scour is limited. Vegetation reduces the effectiveness of wave action during the summer, although areas above low tide-line are exposed to winter frost action. Biogeochemical breakdown is effective in modifying clastic sedimentary and volcanic rock units

the Gulf Stream. In atlantic Canada, populations are largest in the saltmarshes along the north coast of Prince Edward Island (Fig. 18.22).

Mya arenaria (soft-shell clam) and quahog (*Mercenaria mercenaria*) are important bivalve resources found throughout the Maritime Provinces, occupying flats and tidallyinfluenced coastlines. Soft-shell clams do not exhibit any preference with respect to tidal regime (LeBlanc and Miron 2005), with localized hydrodynamic conditions creating complex patchwork distributions on individual tidal flats.

In silt-dominated intertidal flat zones, bioturbated sediments, burrows, pore water, and organic matter can exceed 90% of the volume. These areas are rich invertebrate habitats. At Mary's Point, population densities of amphipods reach 30,000 m^{-2} in the intertidal zone.

The Bay of Fundy lies along the Atlantic Flyway, used by 2.5 million migratory birds during their southward migration, from early July through late September. The most significant sites are Mary's Point and Upper Shepody Bay (NB), and Amherst Point (Cumberland Basin), Minas Basin (Grand Pre-Wolfville), and the head of Cobequid Bay (NS). More than 360 different bird species have been identified at Bay of Fundy sites, including 228 at the Amherst Point Migratory Bird Sanctuary. The most intensely used sites are geomorphically similar, with fine sand-silt tidal flats flanked by sand-pebble-cobble beaches, developed under hypertidal conditions.

Semipalmated Sandpiper (*Calidris pusilla*) is the most common species, representing more than 75% of migrants. Sandpipers feed on amphipods, especially mudshrimps (*Corophium*). The Fundy sites are the final stopover for sandpipers before they take flight over the open Atlantic, landing in northern South America after four days of non-stop flying.

Extreme high tides submerge mollusc burrows, amphipod habitats, and shorebird roosting sites. Submergence forces the semipalmated sandpiper to flock over water, and to remain on site for longer periods during the migration season. At Cobequid Bay, inundation caused sandpipers to remain for 4 additional days to recoup the food lost and energy expended in coping (Mann et al. 2017). Increased

Fig. 18.22 Fringing salt marshes along Malpeque Bay, north coast of PEI, form suitable habitats for Mummichog (*Fundulus heteroclitus*). Small salt marshes are threatened by relative sea level rise

relative sea level, storm surge, or spring tide activity would have negative impacts on these migratory shorebirds.

At Mary's Point, crescentic depressions in intertidal flat silts are traces produced by atlantic sturgeon (*Acipenser oxyrhynchus*) feeding on amphipods (Pearson et al. 2007). Sediment reworking by sturgeon and other fish feeding during the summer re-suspends silt, reducing sediment density, increasing erosion, turbidity, and subsequent redeposition.

18.6.2 Agricultural Dykes

Beginning in the early 1600s, Acadian settlers established communities around Port-Royal and Grand Pré (NS), Beauséjour (NB, Tantamar Marshes), and on southwestern PEI. Successful settlers adopted farming technology brought from Normandy (mesotidal and macrotidal coastlines) to construct aboiteaux, canals with one-way locks to drain fields at low tide and to exclude water at high tide. In the longer term, the restriction of drainage by the aboiteaux reduced sediment influx and resulted in recession of saltmarshes.

Approximately half of Nova Scotia's coastal wetlands were converted to agricultural dykeland in the 1600s (Sherren et al. 2015). More than 70% of saltmarshes and other coastal wetlands in atlantic Canada have been lost or degraded since 1700 (Austen and Hansen 2007; Fig. 18.23).

The sustainability of agricultural communities throughout atlantic Canada is affected by socio-economic as well as physical environmental factors. Demographic challenges are significant, with more than half the region's agricultural producers over age 55. There is an aging core of agricultural producers, many of whom are planning to leave the industry, contributing to the ongoing decline in the number of atlantic Canadian farms evident since 1981. Ongoing declines and aging populations in farming communities indicate that relatively few young farmers are entering the industry. In

Fig. 18.23 Human pressures on salt marsh terrain. **a** Drainage for agriculture has reduced salt marsh area in atlantic Canada by more than 70% since 1700. **b** Coastal development has encroached on many former marsh areas along the southern Gulf of St. Lawrence. **c** Intensive agriculture in PEI has resulted in enhanced sediment and nitrogen input, altering coastal geomorphology and biological communities. **d** Even when human activity ceases, the legacy continues to influence coastal geomorphology

many instances, farming is seen as an occupation requiring extremes of energy, initiative, and patience for limited economic reward, entailing substantial risk.

Although the dykes protect substantial coastal infrastructure, maintenance is primarily the responsibility of individual farmers. Relative sea level rise, enhanced storm surges, and rising spring tides all represent issues. Demographic changes lessening the ability of farmers and agricultural communities to cope are occurring simultaneously with increased stresses.

At Cumberland Marsh, artificial impoundments on dyked land intended to enhance migratory bird habitat have seen decreased productivity since 1970 (Loder et al. 2018). As the impoundments age, the lack of tidal circulation and isolation from autochthonous nutrient sources result in decreased organic matter accumulation and low productivity. Re-engineering and artificial breaches of the artificial dykes have been attempted to restore tidal circulation and enhance bird and fish habitats. Colonization by vegetation, notably *Spartina alterniflora*, proceeds rapidly.

18.6.3 Tidal Energy

Exploitation of potential tidal power resources of atlantic Canada has been discussed for many years (Clark 1981). Although proposals have been floated for developments at East Point (PEI), the Bras d'Or Lakes (NS), and the Strait of Belle Isle, Rigolet, and Cartwright (NL), most of the interest and controversy has centred on the Bay of Fundy. The experimental 50 GWh/y barrage (dam) facility at Annapolis Royal (NS), operational since 1984, is the only functioning power plant. Numerous proposals have been made to exploit tidal energy in the Western Passage of

Fig. 18.24 Incoming tide at Annapolis Royal, NS, is marked by higher sediment load. The tide is deflected landward by the artificial wall, resulting in enhanced erosion of the saltmarsh (foreground). As this erosion threatened the foundations of Fort Anne National Historic Site, the wall was subsequently modified. Outbound fluvial flow from the Annapolis River, carrying very little sediment, is evident on the far (northern) bank

Passamaquoddy Bay (NB-Maine) and in the Minas Channel (NS). Although initial proposals involved barrage construction, the ecological and financial consequences are considered excessive, and current interest focuses on anchored in-stream turbines. At Annapolis Royal, an existing dam was retrofitted to allow increased tidal influx and generate power (Morris 2013). Modified tidal activity resulted in the harbour (Fig. 18.24) and the ingrown Annapolis River, which flows through weak Triassic clastic bedrock mantled by Quaternary silt.

Sea ice represents a potential issue in some locations. In the Minas Basin, ice cover may exceed 90%, with floes up to 30 cm thick and 100 m long moving at 15 km/h (Sanders and Baddour 2007). Although most ice floes float, the inclusion of salt and sediment may cause differences in density, which under extremely rare circumstances could result in ice being sucked into turbines through turbulence (Sanderson and Redden 2015).

Friction induced by turbines could also influence tidal levels and change phases, although modelling suggests that the effects in the Minas Basin would be minor (Wu et al. 2016). However, tidal current velocities could be reduced by up to 35% along the western head of Minas Basin, while increasing by a similar magnitude in the Southern Bight, east of Cape Split, accompanied by enhanced sediment deposition (Karsten et al. 2008; Ashall et al. 2016b; Wu et al. 2016; Wilson et al. 2017). The combination of rising relative sea level and large in-stream tidal plants in the Bay of Fundy

Fig. 18.25 Accelerated coastal erosion at Skmaqn–Port-la-Joye–Fort Amherst National Historic Site (PEI) threatens the site and coastal trail

could result in substantial changes in tidal amplitude throughout the Gulf of Maine, as the forcing tides approach resonance (Karsten et al. 2008; Pelling and Green 2013).

18.6.4 Tourism

Tourism is a vital component in the economic viability of atlantic Canada, and is seen as a key component in the economic revitalization of formerly fishery-dependent communities (Debaie et al. 2006). Throughout Atlantic Canada, intensive marketing by the provincial governments promotes tourism.

Lengthening of the tourist season and enhanced tourist use of natural areas continue to affect tidal coastal sites throughout atlantic Canada (Catto 2006, 2011). In New Brunswick, tidal tourism sites include Fundy National Park (206 km^2, 22 km of open coastline, approximately 225,000 visitors annually), Hopewell Rocks Provincial Park (0.4 km^2, 1.5 km of open coastline, approximately 190,000 visitors), and Kouchibouguac National Park (238 km^2, 25 km of open coastline, 120,000 visitors). Annual out-of-province visitation to Nova Scotia (1.6 million), PEI (1.5 million), and Newfoundland and Labrador (550,000) includes substantial numbers visiting coastal sites. In all four Atlantic Provinces, visitors outnumber permanent residents, with a ratio of 10:1 in PEI.

Human pressure on bedrock-dominated coastlines has increased, leading to trail erosion and potentially compromising visitor safety (Fig. 18.25). Trail design and maintenance impacted by erosion must be reconsidered in light of the shift in visitor demographics, as seniors represent increasing proportions of visitors.

18.7 Conclusion

Microtidal, mesotidal, macrotidal, and hypertidal regimes, in combination with pre-existing topography, coastal bedrock lithology and Quaternary sediment, sea ice activity, and subtidal, intertidal, and supratidal vegetation, have produced atlantic Canada's geomorphically diverse tidal coastlines.

Geomorphology is not disconnected from other valuable aspects of atlantic Canadian tidal coastlines, from multiple ecological and socio-economic viewpoints. A recent initiative has urged cumulative effects assessment as a desired practice, utilizing the Bay of Fundy as a case study (Sinclair et al. 2017). Geomorphologists have much to contribute to integrative, cumulative, and holistic analysis, while being alert to the geomorphic implications in their fellow researchers' fields of interest.

References

Abraham J, Parkes G, Boye P (1999) The transition of the "Saxby Gale" into an extratropical storm. In: 23rd Conference on hurricanes and tropical meteorology, committee on tropical meteorology and tropical cyclones, American Meteorological Society, 79th annual meeting, January 10–15, Dallas, Texas, Preprints, v. II, paper 11A.3, pp 795–798

Archer AW (2013) World's highest tides: hypertidal coastal systems in North America, South America, and Europe. Sed Geol 284–285:1–25

Ashall LM, Mulligan RP, van Proosdij D, Poirier E (2016a) Application and validation of a three-dimensional hydrodynamic model of a macrotidal salt marsh. Coast Eng 114:35–46

Ashall LM, Mulligan RP, Law BA (2016b) Variability in suspended sediment concentration in the Minas Basin, Bay of Fundy, and implications for changes due to tidal power extraction. Coast Eng 114:102–115

Austen E, Hansen A (2007) An analysis of wetland policy in Atlantic Canada. Can Water Res J 32:163–178

Broster BE, Legere CL (2018) Seafloor pockmarks and gas seepages, northwestern Bay of Fundy, New Brunswick, Canada. Atl Geol 54:1–20

Carter RWG, Forbes DL, Jennings SC, Orford JD, Shaw J, Taylor RB (1989) Barrier and Lagoon coast evolution under differing relative sea-level regimes: examples from Ireland and Nova Scotia. Marine Geol 88:221–242

Carter RWG, Orford JD, Forbes DL, Taylor RB (1990) Morphosedimentary development of Drumlin flank barriers with rapidly rising sea level. Sed Geol 69:117–138

Catto NR (2006) Impacts of climate change and variation on Provincial Parks, NL. NL Ministry of Environment and Conservation, St. John's, NL

Catto NR (2011) Coastal erosion in Newfoundland. NL Department of Environment and Conservation, St. John's, NL

Catto NR, Hooper R (1999) Biological and geomorphological shoreline classification of Placentia Bay, Newfoundland. Can Tech Rep, Fish Aqua Sci 2289

Chapman ARO (n.d.) The wild harvest and culture of *Laminaria longicruris* in Eastern Canada. http://www.fao.org/docrep/X5819E/x5819e08.htm. Accessed 11 July 2018

Chmura G, Hung G (2004) Controls on Salt Marsh accretion: a test in Salt Marshes of Eastern Canada. Estuaries 27:70–81

Chmura G, Helmer LL, Beecher CB, Sunderland EM (2001) Historical rates of salt marsh accretion on the outer Bay of Fundy. Can J Earth Sci 38:1081–1092

Clark RH (1981) Tidal power in Atlantic Canada—the challenge of Fundy. In: Canadian society for civil engineering 5th Canadian hydrotechnical conference, pp 1–32

Collén J, Cornish ML, Craigie J, Fick-Blean E, Hervé C, Krueger-Hadfield SA, Leblanc C, Michel G, Potin P, Tonon T, Boyen C (2014) *Chondrus crispus*—a present and historical model organism for Red Seaweeds. Adv Bot Res 71:53–89

Conroy D (2007) Habitat enhancement in Cascumpec and Tracadie Bays, PEI; M.E.S. thesis, environmental science. Memorial University of Newfoundland, St. John's, Newfoundland and Labrador

CSAS (Canadian Science Advisory Secretariat) (2009) Does Eelgrass (Zostera marina) meet the criteria as an ecologically significant species? Fisheries and Oceans Canada, Science Advisory Report 2009/018

CSAS (Canadian Science Advisory Secretariat) (2013) Assessment of Information on Irish Moss, Rockweed, and Kelp harvests in Nova Scotia. Fisheries and Oceans Canada, Science Advisory Report 2013/004

Dashtgardt SE, Gingras MK, Pemberton SG (2008) Grain-size controls on the origin of bioturbation. Palaeo, palaeo, palaeo 257:224–243

Davidson-Arnott RGD, van Proosdij D, Ollerhead J, Schostak L (2002) Hydrodynamics and sedimentation in salt marshes: examples from a macrotidal marsh, Bay of Fundy. Geomorph 48:209–231

Davies G, Woodroffe CD (2010) Tidal estuary width convergence: theory and form in North Australian estuaries. Earth Surf Proc Land 35:737–749

DeBaie L, Murphy M, Austen A, Fitzgerald N, Daigle R (2006) Bouctouche case study: evaluating integrated effects upon tourism in Bouctouche. In: Daigle R, Forbes DL, Parkes G, Ritchie H, Webster T, Bérubé D, Hanson A, DeBaie L, Nichols S, Vasseur L (eds) Impacts of Sea level rise and climate change on the coastal zone of Southeastern New Brunswick. Environment Canada, Ottawa

Dupont F, Petrie B, Chaffey J (2003) Modelling the tides of the Bras d'Or Lakes. Can Tech Rep Hydrol Ocean Sci 230

Forbes DL, Manson GK, Chagnon R, Solomon S, van der Sanden JJ, Lynds TL (2002) Nearshore ice and climate change in the southern Gulf of St. Lawrence. Ice in the environment. In: Proceedings of the 16th IAHR international symposium on Ice, Dunedin, New Zealand, 2–6 December 2002

Forbes, DL (2011) State of the Arctic Coast 2010—Scientific review and outlook. International Arctic Science Committee, Land-Ocean Interactions in the Coastal Zone, Arctic Monitoring and Assessment Programme, International Permafrost Association. Helmholtz-Zentrum, Geesthacht, Germany, 178 pp. http://arcticcoasts.org

Forbes, DL, Syvitski, JPM (1995) Paraglacial coasts. In: Carter RWG, Woodroffe CD (eds) Coastal evolution: late quaternary shoreline morphodynamics. Cambridge University Press, pp 373–424

Forbes DL, Taylor RB (1994) Ice in the shore zone and the geomorphology of cold coasts. Prog Phys Geog 18:59–89

Fyffe LR (2014) The Grand Manan Terrane of New Brunswick: tectonostratigraphy and relationship to the Gondwanan Margin of the Iapetus Ocean. Geosci Canada 41:483–501

Gaillard C, Bhatti HS, Novoa-Garrido M, Lind V, Roleda M, Weisbjerg MR (2018) Amino acid profiles of nine seaweed species and their in situ degradability in dairy cows. Animal Feed Sci Tech 241:210–222

Garbary DJ, Beveridge L, Flynn A, White KD (2012) Population ecology of *Palmaria palmata* (Palmariales, Rhodophyta) from harvested and non-harvested shores on Digby Neck, Nova Scotia, Canada. Algae 27:33–42

Genest C, Joseph M-C (1989) 88 centimetres of coastal erosion per year: the case of Kildare (Alberton), Prince Edward Island, Canada. Geo J 18(3):297–303

Gerwing TG, Gerwing AMA, Drolet D, Barbeau MA, Hamilton DJ (2015) Spatiotemporal variation in biotic and abiotic features of eight intertidal mudflats in the upper bay of Fundy, Canada. Northeast Natl 22:1–44

Gerwing TG, Hamilton DJ, Barbeau MA, Haralampides K, Yamazuki G (2017) Short-term response of a downstream marine system to the partial opening of a tidal-river Causeway. Estuar Coasts 40:717–725

Greenberg DA, Blanchard W, Smith B, Barrow E (2012) Climate change, mean sea level, and high tides in the Bay of Fundy. Atme-Ocean 50:261–276

Hanson AR (2004) Status and conservation of Eelgrass (*Zostera marina*) in Eastern Canada. Canadian Wildlife Service, Atlantic Region, Technical Report 412

Harris MS, Gayes PT, Kindinger JL, Flocks JG, Krantz DE, Donovan P (2005) Quaternary geomorphology and modern coastal development in response to an inherent geologic framework: an example from Charleston, South Carolina. J Coast Res 21:49–64

Hay AE, Zedel L, Stark N (2014) Sediment dynamics on a steep, megatidal, mixed sand-gravel-cobble beach. Earth Surf Dyn 2:443–453

Hell A, Labrie S, Beaulieu A (2018) Effect of seaweed flakes addition on the development of bioactivities in functional Camembert-type cheese. Int J Food Sci Tech 53:1054–1064

Kalacska M, Chmura GL, Lucanus O, Arroyo-Mora JP (2017) Structure from motion will revolutionize analyses of tidal wetland landscapes. Remote Sens Envir 199:14–24

Karsten RH, McMillan JM, Lickley MJ, Haynes RD (2008) Assessment of tidal current energy in the Minas Passage, Bay of Fundy. Proceedings, IMechE 222A. J Power Energy. https://doi.org/10.1243/09576509jpe555

Kennedy DM (2015) Where is the seaward edge? A review and definition of shore platform morphology. Earth Sci Rev 147:99–108

Kontak DJ (2008) On the edge of CAMP: Geology and volcanology of the Jurassic North Mountain Basalt, Nova Scotia. Lithos 101:74–101

Leblanc S, Miron G (2005) Spatial distribution of adult and juvenile softshell clams (*Mya arenaria*) in two tidally contrasted regimes in New Brunswick. Can Tech Rep Fish Aqua Sci 2602

Legget RG (1980) Glacial geology of Grand Manan Island, New Brunswick. Can J Earth Sci 17:440–452

Leys V (2007) 3D flow and sediment transport modelling at the reversing falls – Saint John Harbour, New Brunswick. Oceans 2007, https://doi.org/10.1109/oceans.2007.4449139

Li J-J, Hu Z-M, Duan D-J (2016) Survival in glacial refugia versus postglacial dispersal in the North Atlantic: the cases of Red Seaweeds. In: Hu Z-M, Fraser C (eds) Seaweed phylogeography. Springer, pp 309–330

Locke A, Hanson JM, Klassen GJ, Richardson SM, Aubé CI (2003) The damming of the Petitcodiac River: species, populations, and habitats lost. Northeast Natl 10:39–54

Loder AL, Mallory ML, Spooner I, McLellan NR, White C, Smol JP (2018) Do rural impoundments in coastal Bay of Fundy, Canada, sustain adequate habitat for wildlife? Wetl Ecol Man 26:213–230

Mann HAR, Hamilton DJ, Paquet JM, Gratto-Trevor CL, Neima SG (2017) Effects of extreme tidal events on semipalmated Sandpiper (*Calidris pusilla*) migratory stopover in the Bay of Fundy, Canada. Waterbirds 40:41–49

McHone JG (2011) Triassic basin stratigraphy at Grand Manan, New Brunswick, Canada. Atl Geol 47:125–137

Morris RKA (2013) Geomorphological analogues for large estuarine engineering projects: a case study of barrages, causeways and tidal energy projects. Ocean Coast Man 79:52–61

Necas J, Bartosikova L (2013) Carrageenan: a review. Veter Medic 58:187–205

Neiva J, Paulino C, Nielsen MM, Krause-Jensen D, Saunders GW, Assis J, Bárbara I, Tamigneaux E, Gouveia L, Aires T, Marbà N, Bruhn A, Pearson GA, Serrão EA (2018) Glacial vicariance drives phylogenetic diversification in the amphi-boreal kelp Saccharina latissimi. Sci Rep 8:112. https://doi.org/10.1038/s41598-018-19620-7

Ollerhead J, Davidson-Arnott RGD (1995) The evolution of Buctouche Spit, New Brunswick, Canada. Mar Geol 124:215–236

Parkes GS, Ketch LA, O'Reilly CT, Shaw J, Ruffman A (1999) The Saxby Gale of 1869 in the Canadian Maritimes: a case study of flooding potential in the Bay of Fundy (extended abstract). In: 23rd conference on hurricanes and tropical meteorology, committee on tropical meteorology and tropical cyclones, American Meteorological Society, 79th annual meeting, January 10–15, Dallas, Texas, Preprints, Vol. II, Paper 11A.2, pp 791–794

Pearson NJ, Gingras MK, Armitage IA, Pemberton SG (2007) Significance of Atlantic sturgeon feeding excavations, Mary's Point, Bay of Fundy, New Brunswick, Canada. Palaios 22:457–464

Pelling HE, Green JAM (2013) Sea level rise and tidal power plants in the Gulf of Maine. J Geophys Res: Oceans 118:2863–2873

Pereira L, Meireles F, Abreu HT, Riberio-Claro PJ (2015) A comparative analysis of Carrageenans produced by underutilized versus industrially utilized macroalgae (Gigartinales, Rhodophyta). In: Marine algae extracts: processes, products, and applications, vol 1–2, pp 277–294

Petrie B, Bugden G (2002) Physical oceanography of the Bras d'Or Lakes. Proc, Nova Scotia Inst Sci 42:9–36

Poirier E, van Proosdij D, Milligan TG (2017) The effect of suspended source sediment concentration on the sediment dynamics of a macrotidal creek and marsh. Cont Shelf Res 148:130–138

Porter NJ, Trenhaile AS, Prestanski KJ, Kanyaya JI (2010) Shore platform downwearing in eastern Canada: the mega-tidal Bay of Fundy. Geomorph 118:1–12

Prandle D (2003) Relationships between tidal dynamics and bathymetry in strongly convergent estuaries. J Phys Ocean 33:2738–2749

Prandle D (2004) How tides and river flows determine estuarine bathymetries. Prog Ocean 61:1–26

Rampton N, Gauthier RC, Thibault J, Seaman AA (1984) Quaternary geology of New Brunswick. Geological Survey of Canada Memoir 416

Sanders R, Baddour E (2007) Tidal power and ocean ice in the Bay of Fundy, 1968–2007. In: Proceedings of the 26th international conference on offshore mechanics and arctic engineering

Sanderson BG, Redden AM (2015) Perspective on the risk that sediment-laden ice poses to in-stream tidal turbines in Minas Passage, Bay of Fundy. Int J Marine Energy 10:52–69

Savenije H (2005) Salinity and tides in alluvial estuaries. Elsevier, Amsterdam

Shaw J, Amos CL, Greenberg DA, O'Reilly CT, Parrott DR, Patton E (2010) Catastrophic tidal expansion in the Bay of Fundy, Canada. Can J Earth Sci 4:1079–1091

Shaw J, Ceman J (1999) Salt-marsh aggradation in response to late-Holocene sea-level rise at Amherst Point, Nova Scotia, Canada. The Holocene 9:439–451

Shaw J, Gareau P, Courtney RC (2002) Paleogeography of Atlantic Canada 13–0 kyr. Quat Sci Rev 21:1861–1878

Shaw J, Piper DJW, Fader GB, King EL, Todd BJ, Bell T, Batterson MJ, Liverman DGE (2006) A conceptual model of the deglaciation of Atlantic Canada. Quat Sci Rev 25:2059–2081

Sherren K, Loik L, Debner JA (2015) Climate adaptation in 'new world' cultural landscapes: the case of Bay of Fundy agricultural dykelands (Nova Scotia, Canada). Land Use Policy 51:267–280

Silinski A, Heuner M, Troch P, Puijalon S, Bouma TJ, Schoelynck L, Schröder U, Fuchs E, Meire P, Temmerman S (2016) Effects of contrasting wave conditions on scour and drag on pioneer tidal marsh plants. Geomorph 255:49–62

Sinclair AJ, Doelle M, Duinker PA (2017) Looking up, down, and sideways: reconceiving cumulative effects assessment as a mindset. Env Impact Assess Rev 62:183–194

Spooner I, Batterson M, Catto N, Liverman D, Broster BE, Kearns K, Isenor F, Gw McAskill (2013) Slope failure hazard in Canada's Atlantic Provinces: a review. Atl Geol 49:1–14

Stea RR, Mott RJ (2005) Younger Dryas glacial advance in the southern Gulf of St. Lawrence, Canada: analogue for ice inception? Boreas 34:345–362

Stea R, Seaman AA, Pronk T, Parkhill MA, Allard S, Utting D (2011) The Appalachian Glacier Complex in Maritime Canada.

In: Ehlers J, Gibbard PL, Hughes PD (eds) Quaternary Glaciations—extent and chronology: a closer look. Developments in quaternary sciences, vol 15, pp 631–659
Storms JEA, Weltje GJ, van Dyke JJ, Geel CR, Kroonenberg SB (2002) Process-response modeling of wave-dominated coastal systems: simulating evolution and stratigraphy on geological timescales. J Sed Res 72:226–239
Tibbits JR, van Proosdij D (2013) Development of a relative coastal vulnerability index in a macro-tidal environment for climate change adaptation. J Coast Cons 17:775–797
Townsend DW, Pettigrew NR, Thomas MA, Neary MG, McGillicuddy DJ, O'Donnell J (2015) Water masses and nutrient sources to the Gulf of Maine. J Marine Res 73:93–122
Trenhaile AS (1987) The geomorphology of rock coasts. Clarendon Press, Oxford
Trenhaile AS, Kanyaya JI (2007) The role of wave erosion on sloping and horizontal shore platforms in macro- and mesotidal environments. J Coast Res 23:298–309
Trenhaile AS, Mercan DW (1984) Frost weathering and the saturation of coastal rocks. Earth Surf Proc Land 9:321–331
Trenhaile AS, Porter NJ (2018) Shore platform downwearing in eastern Canada: A 9–14 year micro-erosion meter record. Geomorph 311:90–102
Trenhaile AS, Pepper DA, Trenhaile RW, Dalimonte M (1998) Stacks and Notches at Hopewell Rocks, New Brunswick, Canada. Earth Surf Proc Land 23:975–988
Vasseur L, Catto NR (2008) Atlantic Canada. In: Lemmen DS, Warren FJ, Lacroix J, Bush E (eds) From impacts to adaptation: Canada in a changing climate 2007. Government of Canada, Ottawa, pp 119–170
Wilson EK, Hill PS, van Proosdij D, Ruhl M (2017) Coastal retreat rates and sediment input to the Minas Basin, Nova Scotia. Can J Earth Sci 54:370–378
Wu Y, Chaffey J, Greenberg DA, Smith PC (2016) Environmental impacts caused by Tidal power extraction in the Upper Bay of Fundy. Atm-Ocean 54:326–336

Norm Catto joined Memorial University in July 1989 and is now the Head and Professor of Geography at that university. His research and teaching interests include coastal landforms, natural hazards, sea level change, and impacts in coastal environments; response of river systems to climate and weather events, and flood risk assessment; aeolian geomorphology and response to climate and human factors in boreal and costal environments; mass movements and slope failures; loess deposition; palaeosol formation; emergency measures policies and effectiveness human adaptations to climate and weather events; the impacts of climate and weather events on agriculture, transportation, fisheries, and communities; and the relationships among landscapes and people. His research has included projects and investigations in landscapes and environments in 16 countries, and in all Canadian provinces and territories, particularly in eastern Canada. For 18 years, he served as Editor-in-Chief for Quaternary International, a professional international journal dedicated to the study of landscape evolution and climate variation throughout the past 3 million years. He has taught more than 50 different undergraduate and graduate courses focused on numerous aspects of geomorphology, landscape analysis, Quaternary research, natural hazards, ocean sciences, environmental science, climate and climate change, soil science, geoarchaeology, and palaeontology. Study of landscapes forms his central focus for understanding of the various areas of Eastern Canada, the combination of physical and cultural constituents that contribute to its identities. For most species, and for all human occupants, life in Eastern Canada has been heavily influenced by geomorphology since deglaciation.

Landforms and Landscapes of the Magdalen Islands: The Role of Geology and Climate

19

Bernard Hétu, Audrey M. Rémillard, Pascal Bernatchez, and Guillaume St-Onge

Abstract

The landscape and landforms of the Magdalen Islands are uniquely varied for such a small archipelago. Geomorphologically they are a part of the Maritime Plain which is located near the center of the Maritime Basin (375–325 Ma). The Magdalen Islands were pushed to the surface by salt-diapirs ($\sim$300 Ma). Formed by volcanic hills (basalt) surrounded by a low sandstone platform, the main islands are interconnected by late Holocene sand spits and tombolos. Located in the southern part of the Gulf of St. Lawrence, the Laurentide Ice Sheet (LIS) did not reach the Magdalen Islands during the last glaciation, but possibly much earlier ($\geq$170 ka). Three different sets of geomorphological processes have impacted the landscape: structural, periglacial, and coastal. Sandstone platforms adjacent to volcanic hills are the structural controls. Superimposed on the sandstone platforms, cryopediments, ice-wedge pseudomorphs and dry valleys are the dominant periglacial landforms. Coastal landforms are dominated by cliffs, tombolos, and littoral spits. Perhaps the single most surprising fact is that eight islands with a cumulative area of 202 km^2 can demonstrate such a remarkable variety of landscapes and landforms. The glacial record remains contentious.

Keywords

Maritime plain • Volcanics, sandstone • Karstic foothills • Cryopediments • Ice-wedge pseudomorphs • Tombolos • Relative sea level

19.1 Introduction

Located in the center of the Gulf of St. Lawrence (Fig. 19.1), the Magdalen Islands offer original landscapes typical of the Maritime Plain (Brisebois 1981; Barr et al. 1985; Dubois 1992; Giles 2008) (Fig. 1.10; this volume). Geomorphology of the islands is the result of a long geological evolution marked by the development of several salt-diapirs, and, since at least 170 ka, by alternating glacial and periglacial periods resulted in inherent large scale relative sea-level (RSL) variations (Rémillard et al. 2016). These RSL changes were produced by the combined effect of global sea-level fluctuations (glaci-eustatic) and vertical adjustments of the Earth's crust (glaci-isostatic). This part of the Gulf of St. Lawrence is currently subsiding owing to the migration and collapse of the peripheral glacial forebulge (Dubois 1992; Koohzare et al. 2008). The resulting exacerbated marine transgression induces rapid coastal submergence and erosion, which represent a major challenge for coastal communities and infrastructures (Bernatchez et al. 2008, 2012).

19.2 Study Area

19.2.1 Geology and Landscapes

The archipelago is 80 km in length measured from the SW of the Havre Aubert Island to the NE of Brion Island and comprises eight rocky islands, six of which are connected to each other by sandy barrier beaches (tombolos) (Figs. 19.1 and 19.2) (e.g., Dubois and Grenier 1993; Bernatchez et al.

B. Hétu (✉) · P. Bernatchez
Département de biologie, chimie et géographie and Centre for Northern Studies (CEN), Université du Québec à Rimouski, 300 Allée des Ursulines, Rimouski, Québec G5L 3A1, Canada
e-mail: Bernard_Hetu@uqar.ca

A. M. Rémillard · G. St-Onge
Canada Research Chair in Marine Geology, Institut des sciences de la mer de Rimouski (ISMER), Université du Québec à Rimouski and GEOTOP, 310 Allée des Ursulines, Rimouski, Québec G5L 3A1, Canada

A. M. Rémillard
Centre de recherche sur les milieux insulaires et maritimes (CERMIM), 37 Chemin Central, Havre aux Maisons, Québec G4T 5G9, Canada

© Springer Nature Switzerland AG 2020
O. Slaymaker and N. Catto (eds.), *Landscapes and Landforms of Eastern Canada*,
World Geomorphological Landscapes, https://doi.org/10.1007/978-3-030-35137-3_19

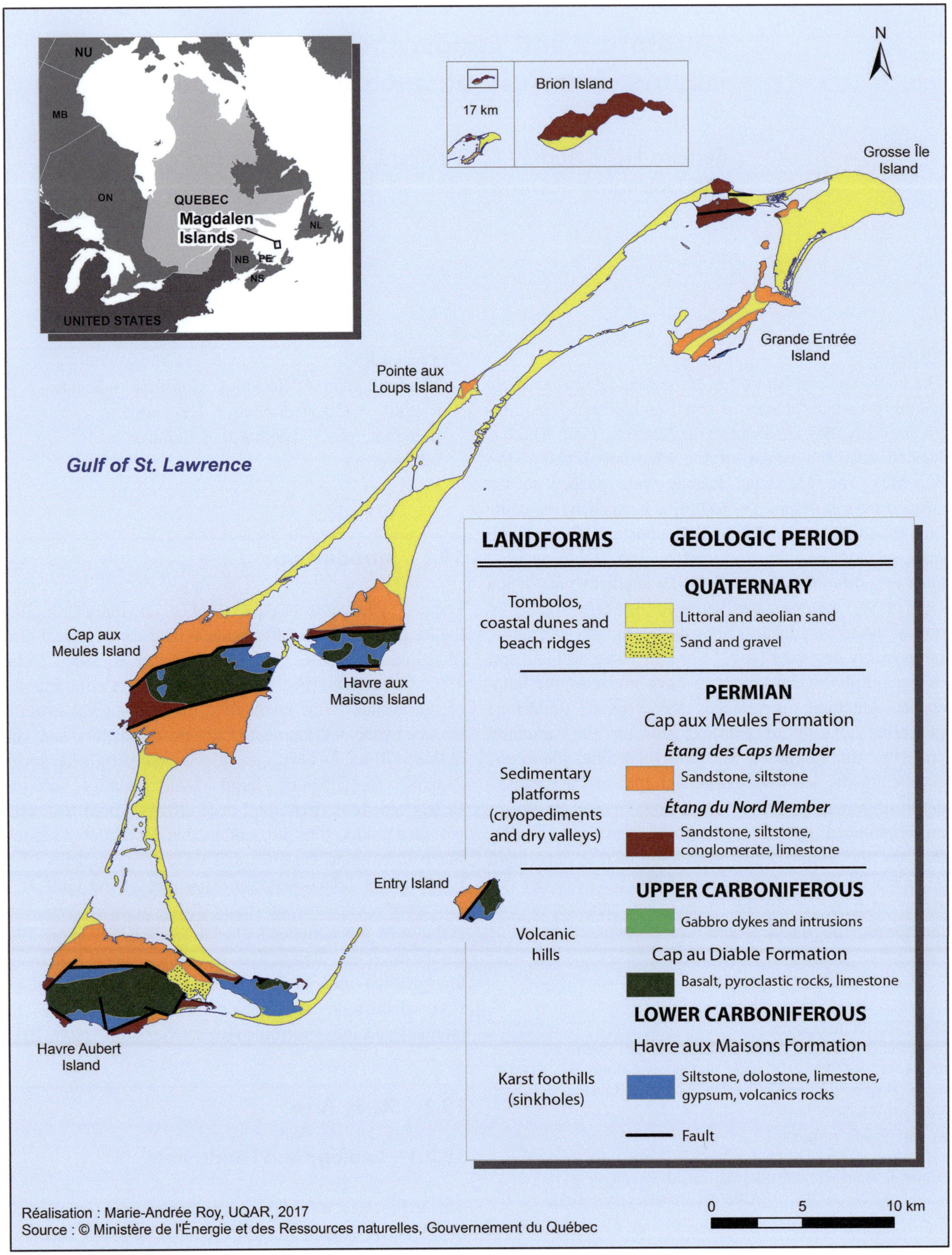

Fig. 19.1 Geology and landforms of Magdalen Islands (modified from Brisebois 1981)

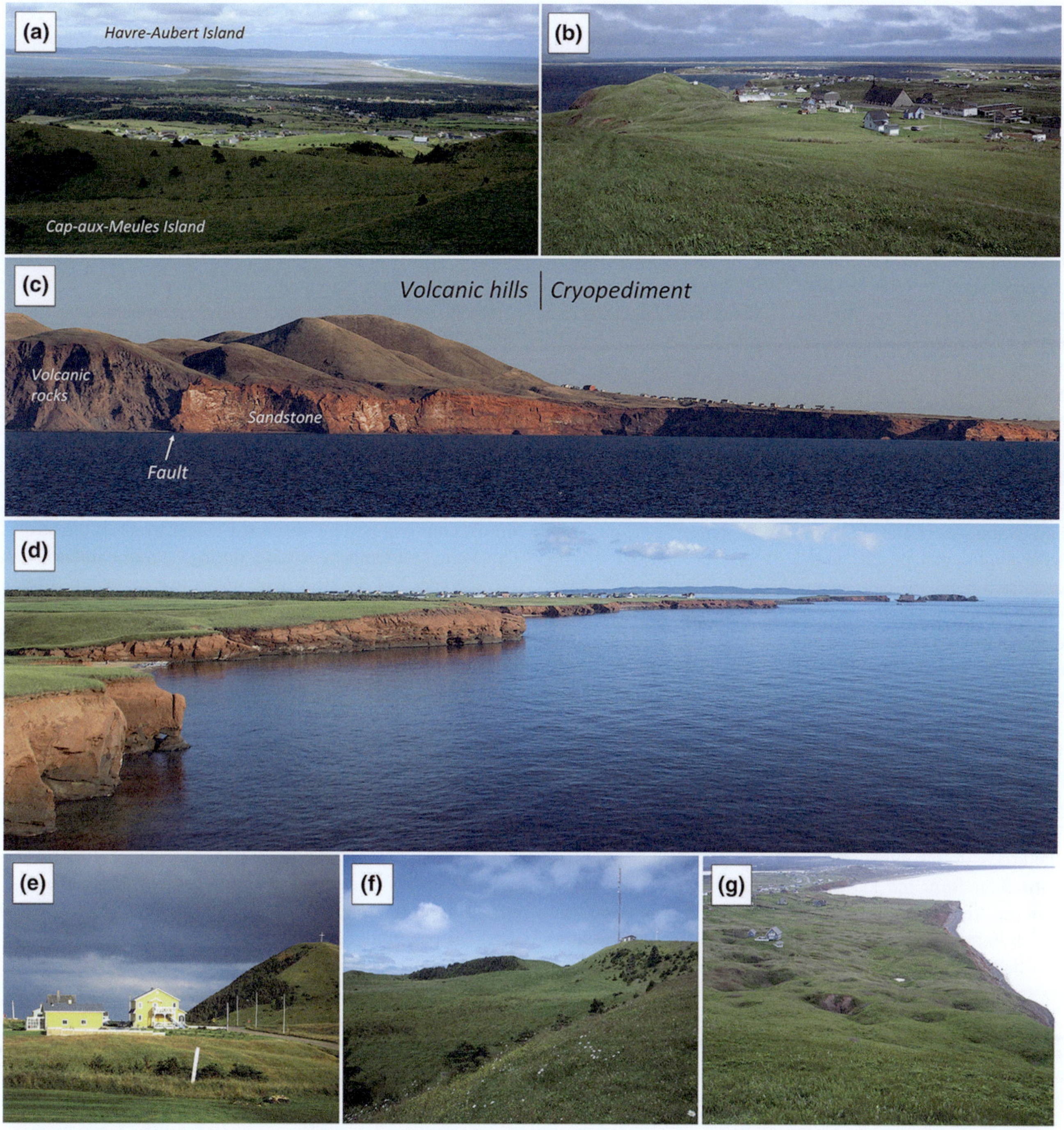

Fig. 19.2 The diverse landscapes of the Magdalen Islands. **a** Double tombolo connecting the Havre Aubert Island to the Cap aux Meules Island. **b** Buttes des Demoiselles (volcanic hills) and the Sandy Hook littoral spit (background) on the Havre Aubert Island. **c** Entry Island: volcanic hills (Big Hill) and the cryopediment that truncates the sandstone platform. **d** Sandstone platform of the Cap aux Meules Island. **e** A volcanic hill of the Havre aux Maisons Island. **f** Relict nivation hollows in the volcanic hills (Butte du Vent, Cap aux Meules Island). **g** Sinkholes group on the Havre Aubert Island. *Photographs* a and g © Antoine Morissette

2012). The terrestrial area is 202 km^2 and reaches 363 km^2 including lagoons and emerged or slightly submerged sandy zones (Quirion 1988). The Magdalen Islands are part of the low relief Maritime Plain (geomorphically) and Maritime Basin (geologically), which corresponds to a wide (approximately 150,000 km^2) system of grabens that appeared just after the Acadian orogeny (375–325 Ma) under the influence of extensional tectonics whose origin is still uncertain (Brisebois 1981; Barr et al. 1985; La Flèche et al. 1998; Giles 2008). Between the end of the Devonian

(~360 Ma) and the beginning of the Permian (~300 Ma), this wide subsiding basin was filled by a thick (~9,000 m) sequence of diverse material including shallow water marine sediments (e.g., salt, gypsum, limestone), volcanic rocks (basalt, andesite, rhyolite), and, finally, terrestrial sediments (sandstone, mudstone) that indicate a semi-arid environment (fluvial sand, playas, petrified dunes) (Brisebois 1981; Barr et al. 1985; Giles 2008).

The present forms of the Magdalen Islands, subdivided into four distinct geomorphological groups (Dubois 1992), are controlled by geology (Figs. 19.3 and 19.4): the volcanic hills, karstic foothills, sandstone platform and Late Holocene barrier beaches. The highest hills on Entry Island (Big Hill: 174 m asl), Havre Aubert Island (140 m asl), Cap aux Meules Island (Butte du Vent: 168 m asl) and Havre aux Maisons Island (105 m asl) are composed of volcanic rocks (basalt, andesite, rhyolite) belonging to the Cap au Diable Formation of Mississippian age (355–325 Ma). More than 4,000 m of volcanic rocks lie above the marine sediments and evaporites of the Windsor Group. These volcanic rocks were diapirically pushed to the surface through the terrigenous sediments of the Cap aux Meules Formation. According to Brisebois (1981), each of the rocky islands composing the archipelago is located at the summit of a salt diapir. The diapirs have also dragged to the surface masses of mixed rocks (mudstone, limestone, gypsum, sandstone), which are intensively deformed, faulted, and brecciated, and some of which are soluble. Brisebois (1981), who interpreted these mixed rocks as collapsed breccia associated with salt dissolution, named the group Havre aux Maisons Formation. After a long period of dissolution that produced numerous sinkholes, the soluble rocks of the Havre aux Maisons Formation form the karstic foothills (Figs. 19.2 and 19.3) between the volcanic hills and the sandstone platforms which extend below (Figs. 19.2, 19.3 and 19.4). The large dunified barrier beaches (double tombolos) (Figs. 19.1 and 19.2), of late Holocene age (Rémillard et al. 2015a), composed the fourth geomorphological group on the Magdalen Islands.

19.2.2 Climate and Vegetation

Located in the center of the Gulf of St. Lawrence and near the Atlantic Ocean, the Magdalen Islands have a boreal climate with a strong maritime influence. Winters are relatively mild (mean temperature in January: −6.4 °C) while summers are relatively cool (mean temperature in July: 17.1 °C). Total precipitation is 1037.3 mm and is well distributed throughout the year; 23% of total precipitations are falling as snow (Environment Canada 2017, Magdalen Islands station, Climate Normals 1981–2010). An analysis undertaken on data collected between 1985 and 2008 reveals an increase of the mean winter temperatures of more than 3 °C (Bernatchez et al. 2008). The archipelago is affected almost continuously by sustained winds from all directions. The wind, particularly strong during winter, blows at an average speed of 31 km/h with peaks that exceed 100 km/h (Agglomération des Îles-de-la-Madeleine 2010). The Magdalen Islands belong to the ecozone of the Acadian Forest of Rowe, but with the exception of the uninhabited Brion Island, there is not much left of the primitive forest. Residual forests occupy only 26% of the total area, concentrated in the hills and a few zones of the tombolos (Roy-Bolduc 2010). These forests are almost exclusively coniferous, including *Abies balsamea*, which accounts for more than 50% of mature trees, 24% *Picea mariana* (black spruce) and 24% *Picea glauca* (white spruce) (Roy-Bolduc 2010). Deforestation started as early as the 16th century and reached its peak around the middle of the twentieth century with wood used for fish smokehouses, shipbuilding, and the manufacture of fishing tackle (Fortin and Larocque 2003).

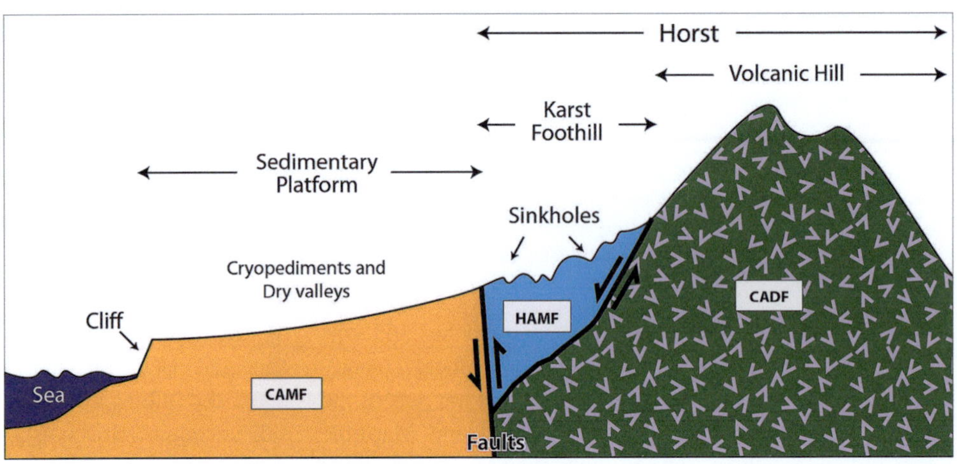

Fig. 19.3 Schematic model showing the relations between the geology and the landforms of the Magdalen Islands. CAMF = Cap aux Meules Formation; HAMF = Havre aux Maisons Formation; CADF = Cap au Diable Formation

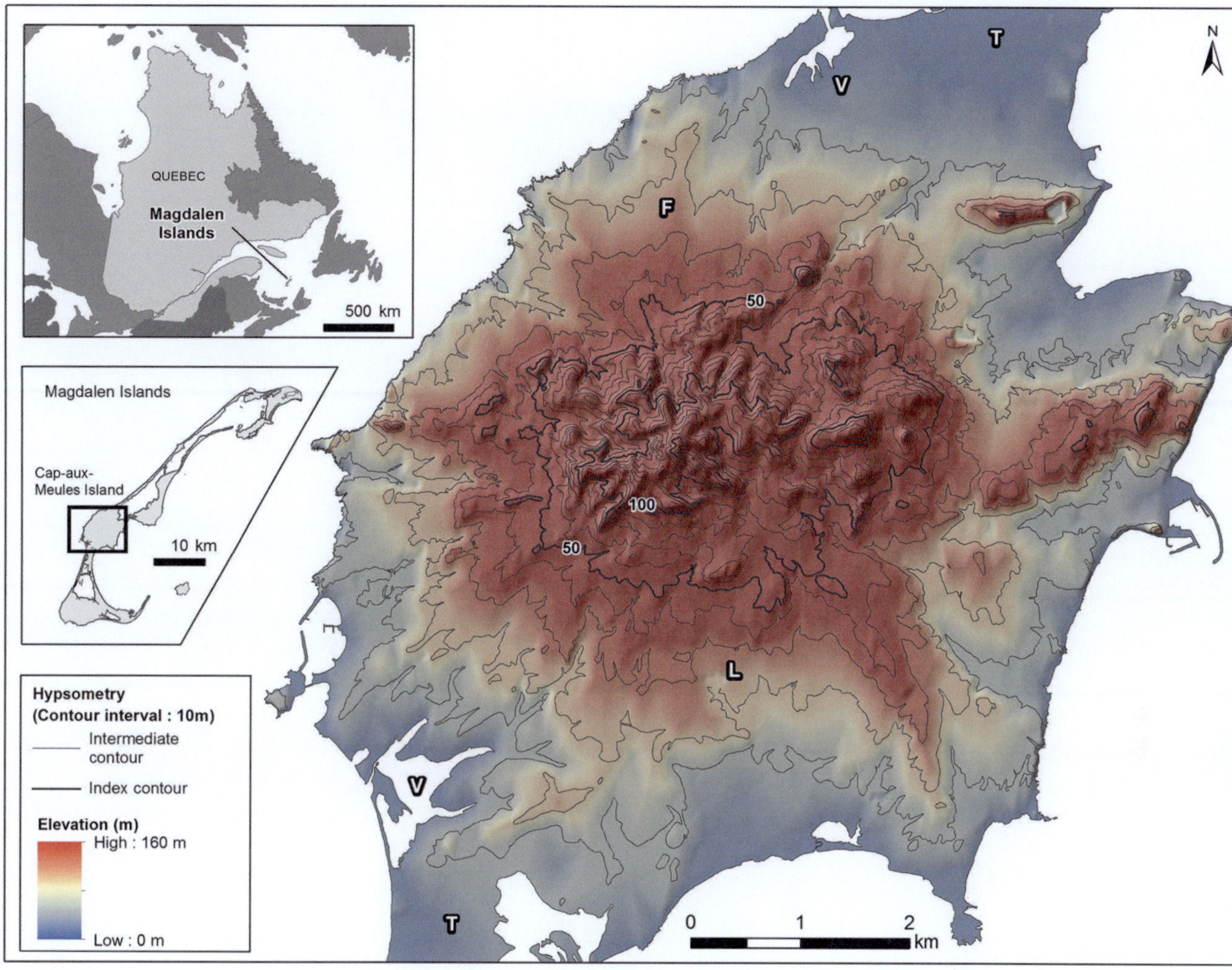

Fig. 19.4 The relief of the Cap aux Meules Island. The volcanic hill correspond to the centre of the island (>50 m). Numerous nivation hollows which were occupied by névé during non-glacial cold periods are observable. Below 50 m, the major part of the island is occupied by cryopediments; the most beautiful are located near Fatima (F) and La Vernière (L). In the lagoons, there are several valleys partially invaded by the sea (rias) (V). To the north and south, we note the beginning of tombolos (T)

19.3 Glaciation and Sea-Level History

19.3.1 Glaciation

Located in the center of the Gulf of St. Lawrence, the Magdalen Islands are a key area to the understanding of the glacial history of the Maritime Provinces (Stea et al. 1998, 2011; Shaw et al. 2006). The glacial history of the archipelago has been the subject of a long debate that started at the end of the 19th century and that persisted until recently. From the first publications of Richardson (1881), Chalmers (1895), and Goldthwait (1915) until the papers of Prest et al. (1976) and Dredge et al. (1992), the subjects of controversy were as numerous as varied (e.g., Stea et al. 2011; Rémillard et al. 2013, 2015b, 2016): origin and age of some deposits (e.g., till or glacimarine), the relative chronology of glacial and periglacial periods, the question of whether the islands were glaciated during the last glacial maximum (LGM) and if so, the origin and nature of the glaciers that reached the archipelago (Laurentide Ice Sheet or regional ice-caps). Using a detailed analysis of the sedimentological units exposed in more than 30 outcrops throughout the islands (Fig. 19.5), added to an extensive chronological dataset including 82 luminescence and 22 radiocarbon ages, Rémillard et al. (2013, 2015b, 2016) reconstructed the principal events of the glacial history since 65 ka. After an ice-free period during the MIS 3 (between approximately 65 and 40 ka), the Magdalen Islands were glaciated during the MIS 2 by two different ice-caps: the southern islands were glaciated by the Escuminac ice-cap located in the western Gulf of St. Lawrence, while the northern archipelago was glaciated at the end of the LGM by an ice flow from Newfoundland. Although it is not possible to make any

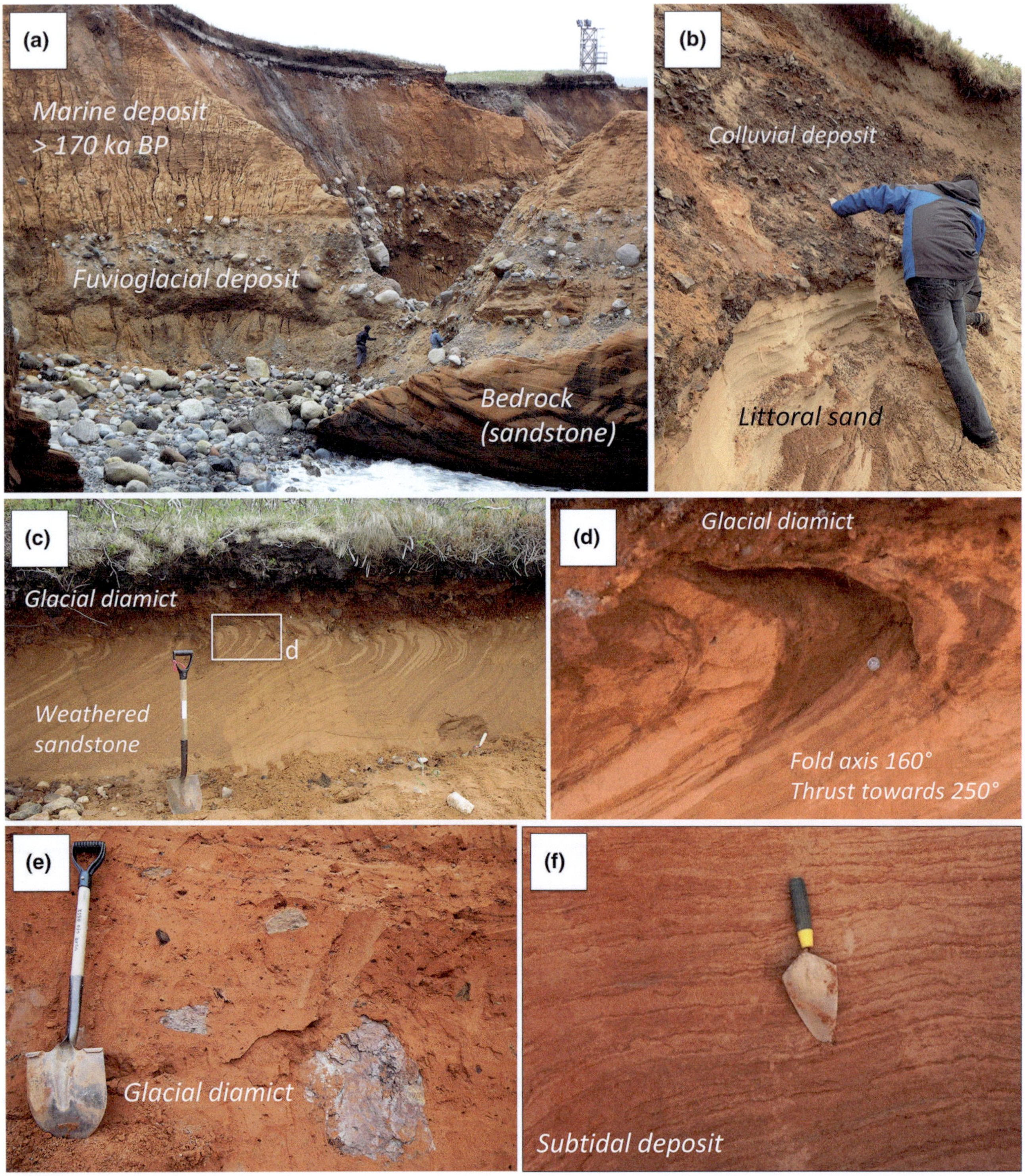

Fig. 19.5 The Quaternary deposits of the Magdalen Islands. **a** The Pointe aux Loups sequence of more than 170 kyr deposited by the LIS. The fluviglacial deposit at the bottom of the sequence contains erratic blocks from the Canadian Shield. **b** Entry Island: the colluvial deposit (cryopediment) overlying the littoral and marine sequence dated between 65 and 51 kyr. **c** and **d** Glacitectonic deformations associated with the ice movement from the Newfoundland ice-cap. **e** Havre Aubert Island: red till of the Escuminac ice-cap that flowed towards south-east. **f** Havre Aubert Island: subtidal deposit located at +24 m and dated to 11.5 kyr

assumptions about whether these ice-caps were synchronous, the observations put forward by Dredge et al. (1992), and more recently Rémillard et al. (2016), suggest that both ice movements from the Escuminac and Newfoundland ice-caps touched the Cap aux Meules Island.

Numerous dates between ~23 and 17 ka from periglacial and coastal deposits located on the southern and central islands indicate that this part of the archipelago was deglaciated early, i.e., immediately after the LGM (~25 ± 2 ka) (Rémillard et al. 2017). Once deglaciated, the southern archipelago was affected by periglacial conditions and a high RSL with different elevations on each island. The northern archipelago, i.e., Pointe aux Loups (PAL) and Grande Entrée islands, was deglaciated later, a little before 15 ka (Rémillard et al. 2016). During the MIS 2, the Magdalen Islands were thus affected by regional ice-caps that were part of the Appalachian Glacier Complex (Stea et al. 1998, 2011; Shaw et al. 2006). The Appalachian glaciers were coalescing with the Laurentian Channel ice stream (Shaw et al. 2006; Rémillard et al. 2016). The Laurentide Ice Sheet (LIS) did not reach the Magdalen Islands since at least the MIS 6. Throughout the archipelago, the only deposit that can be reliably linked to the LIS is located on the west coast of PAL is a proximal fluviglacial unit with varied erratics from the Canadian Shield and Anticosti Island. The fluviglacial unit is overlain by a marine deposit from which luminescence dating produced minimum ages of ≥170 ka (Fig. 19.5; Rémillard et al. 2016). The same deposit was observed elsewhere on the Grande Entrée Island (Sandcove, Bluff east, Old Harry) and all exposures gave minimum ages of 170 ka. These sedimentary bodies of marine origin are the thickest of the entire archipelago and are located only on PAL and Grande Entrée islands (Fig. 19.5).

19.3.2 Sea-Level History

The evolution of the RSL in the Gulf of St. Lawrence during the Quaternary was controlled by the combined sea-level effects of the growth and decay of ice sheets (glaci-eustatic) and the inherent dynamic response of the Earth's crust to the related changes of surface loading (glaci-isostatic). On this basis, Quinlan and Beaumont (1981) subdivided the Gulf of St. Lawrence and the adjacent regions into four zones characterized by distinct types of post-glacial RSL variations (Fig. 19.6). The Magdalen Islands lie within zone B. The latter corresponds to the area affected by an initial high RSL followed by an emergence due to the glaci-isostatic rebound and the migration of the peripheral forebulge, and finally, a submergence that is still ongoing today, caused partly by the collapse of the forebulge. In detail, however, there are important differences between the different areas of the archipelago depending on their respective glacial history. The south of the archipelago, glaciated by the Escuminac ice-cap, was deglaciated much earlier (as early as ~25 ± 2 ka) than the north (~15 ka). The observed marine limit is also much higher on the southern part: >+40 m on Entry Island and >+20 m on Grande Entrée Island (Rémillard et al. 2016). Around 15 ka, the entire archipelago was affected by the post-glacial transgression and the RSL reached minimum elevations between +10 and +20 m asl. Around 10 ka, the RSL passed below the modern level (Fig. 19.7) as indicated by the terrestrial peat dated to 10.7 ± 0.1 cal ka located at modern mean high tide and the organic horizon found by the Canadian Salt Company Seleine Mines at a depth of −17 m asl (Rémillard et al. 2016). Several geomorphological indicators observed between −35 and −200 m asl around the Magdalen Islands in the Gulf of St. Lawrence also attest to a low RSL (e.g., Sanschagrin 1964; Loring and Nota 1973; Josenhans and Lehman 1999; Shaw 2005). The timing of their formation remains unclear and therefore, they do not provide precise information required for RSL reconstructions. However, studies on the western coast of Newfoundland suggest a lowstand that would have reached −25 to −30 m asl (Forbes et al. 1995). After the evidence of a lowstand at ~9.8 ka cal around the Magdalen Islands, there is a lack of data until ~2 ka cal. The late Holocene submergence period is much better documented (Fig. 19.7). Based on multi-proxy analyses (salt-marsh foraminifera, testate amoebae and plant macrofossils; Fig. 19.8), Barnett et al. (2017) demonstrated that the RSL trend rose between 2 mm/yr and 1.3 mm/yr since the two last millennia with an acceleration during the 20th century. According to the tide gauge data, this rate would have doubled between 1964 and 2013 to reach 4.3 mm/yr (Barnett et al. 2016), as the result of the recent acceleration of the global sea-level rise and the still subsiding Earth's crust. This submergence rate, the highest along the Québec coastline, is close to the values measured on adjacent Prince Edward Island and Nova Scotia (Koohzare et al. 2008; Han et al. 2015).

The luminescence and ^{14}C ages obtained on the littoral and subtidal deposits on Entry and Havre Aubert islands report for the first time in the region of the Gulf of St. Lawrence an older transgression-regression cycle, i.e., between 65 and 40 ka (Fig. 19.7). During this period that corresponds to the interval between the end of the MIS 4 and the MIS 3, the RSL changed from >+40 m to ~+15 m asl. At that time, the global sea-level stood between −100 and −50 m below current level (Caputo 2007) which indicates a crustal depression between at least 85 and 135 m (Rémillard et al. 2017). Although there is no evidence of a glaciation on the Magdalen Islands during the MIS 4, such a crustal depression is likely due to the glaci-isostatic adjustment of the Earth's crust. However evidences of a glaciation during the MIS 4 were observed in the adjacent areas (Nova Scotia, New Brunswick) and according to Stea et al. (2011) an

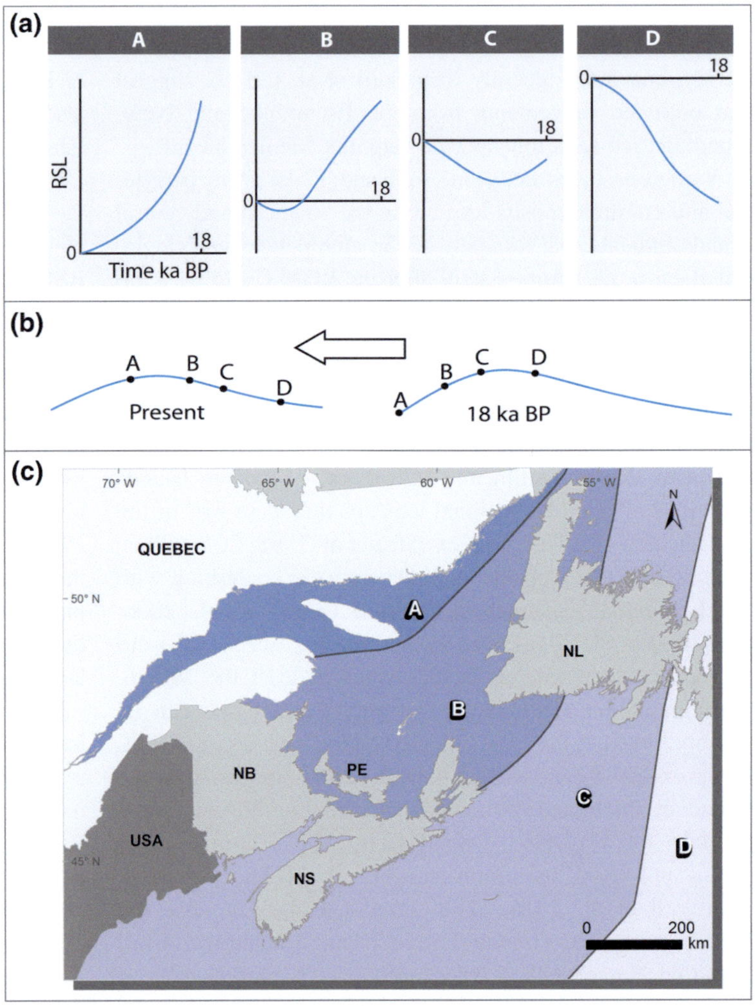

Fig. 19.6 (a) The four scenarios (A, B, C, and D) representing the history of the RSL variations for Atlantic Canada and the northeastern coast of the United States, from 18 kyr BP (~21.7 kyr cal BP) until today. (b) Schematic representation of the position of the peripheral forebulge for 18 18 kyr BP (~21.7 kyr cal BP) and today. The arrow indicates the direction of the forebulge migration and the points correspond to the position of the different scenarios and the influence of the forebulge on the RSL variations. (c) Hypothetical boundaries of the zones affected by each scenario (modified from Quinlan and Beaumont (1981) according to the recent data of Roy and Peltier (2015)

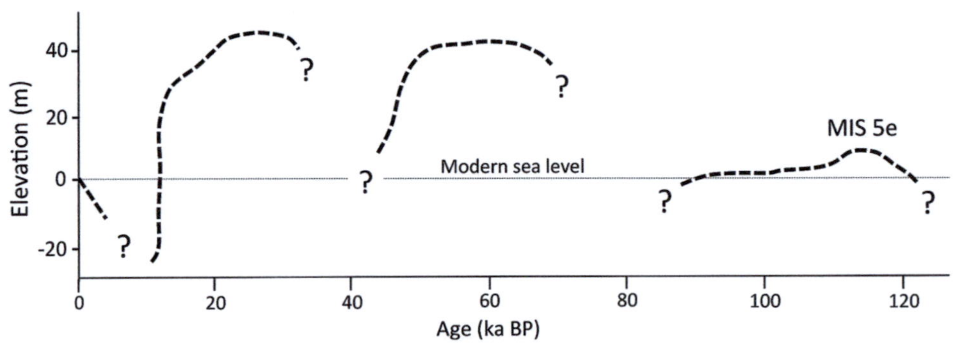

Fig. 19.7 Relative sea level variations on the Magdalen Islands since ~130 kyr based on the luminescence and radiocarbon ages acquired by Rémillard et al. (2017)

extensive Appalachian glacier flowing towards the south-east would have covered the entire Gulf of St. Lawrence up to the edge of the Atlantic continental shelf (Caledonian Phase, 75–50 ka; Stea et al. 2011).

On the Magdalen Islands, evidences of RSL close to the modern level and dated to the MIS 5 were also observed (Fig. 19.7). The stratigraphical context, the luminescence and U/Th ages acquired from sediments and woody remains, and the fossils of thermophilic species (e.g., *Quercus, Corylus, Ostrea virginica*) present in the sediments, indicate that this RSL could correspond to the interglacial climax of the MIS 5.5 (Dredge et al. 1992; Rémillard et al. 2017). Rémillard et al. (2017) also obtained an age of 115 ± 8 ka on a littoral deposit located at +15 m asl on Cap aux Meules Island. This deposit located well above the global sea-level highstand values of +2 up to +9 m asl between 132 and 116 ka (e.g., Caputo 2007; Rohling et al. 2008) suggests a glaci-isostatic depression of ~10 m, presumably associated

Fig. 19.8 Evidence of the recent RSL rise on the Magdalen Islands. In situ tree trunks in submersion or burial were observed at several locations throughout the archipelago (Juneau 2012)

with the MIS 6 glaciation that occurred in eastern Canada (e.g., Stea et al. 2011). However, the hypothesis of a local tectonic uplift as suggested by Paquet (1989) should also be taken into consideration. Other evidence of high RSL during the MIS 5 was observed around the Gulf of St. Lawrence (Grant 1980; Stea et al. 1998).

19.4 Periglacial Landforms

Because of their uniqueness, the periglacial landscape and landforms of the Magdalen Islands, have attracted scientific attention since the mid-20th century (Hamelin 1959). Some researchers even believed that these periglacial features were evidence that the archipelago has not been glaciated throughout MIS 4 to MIS 2 (Hamelin 1959; Ritchot 1968; Laverdière and Guimont 1974). Today, it is known that they developed into two phases, during the deglaciation period of MIS 3 (between 60 and 40 ka), and after the LGM (MIS 2) whereas the deglaciation occurred relatively early (between ~25 and 15 ka) compared to the Maritime Provinces (Shaw et al. 2006; Seaman et al. 2011).

19.4.1 Permafrost Features

Ice-wedge pseudomorphs, composite-wedge casts, and cryoturbations have been observed on the Magdalen Islands since the 1970s (Poirier 1970; Laverdière and Guimont 1974; Dredge et al. 1992). Different ages from MIS 3 to MIS 2 were proposed for these features according to the assumptions made regarding the chronology of glacial events (Laverdière and Guimont 1974; Dredge et al. 1992). Rémillard et al. (2015b) were the first to obtain absolute ages on these features. Generally, the ice-wedge pseudomorphs or composite-wedge casts truncate either a thin subtidal deposit overlying a glacimarine deposit (Havre Aubert Island) or a thin pebbly coastal deposit overlying the highly weathered sandstone (Cap aux Meules Island) (Fig. 19.9). The 10 ages (^{14}C and OSL) acquired from both the infillings and the host material of the ice-wedge pseudomorphs or composite-wedge casts show that they developed between ~12.9 and 11.5 kyr, well after the LGM. This period corresponds to the Younger Dryas, which is a period of climatic cooling well-documented in eastern Canada (e.g., Mott et al. 1986; Mayle and Cwynar 1995; Liverman et al. 2000; Hétu et al. 2003). Since the 1960s (e.g., Lachenbruch 1962), it is generally accepted that ice-wedges form in coarse-grained sediment when the mean annual air temperature (MAAT) is lower than −6 to −8 °C (French 2007). From the temperature thresholds used in the literature, ice-wedges formation would require a ~12 °C cooling relative to present (+5 °C) (See the discussion by Liverman et al. 2000 for an alternative perspective) Although they do not provide an accurate indicator of paleotemperature, the most convincing geomorphic indicator of perennially frozen ground is both ice-wedge pseudomorphs and sand-wedge casts (e.g., French 2007). Rémillard et al. (2015b) also observed involutions interpreted as cryoturbations deforming the highly weathered sandstone and the overlying coastal and colluvial deposits (Fig. 19.9a). The combined depth of the ice-wedge pseudomorphs and composite-wedge casts and cryoturbations suggest a palaeo-active layer thickness ≤ 1 m.

19.4.2 Cryopediments

The term 'cryopediment' is "a gently inclined erosional surface at the foot of valley sides or marginal slopes of geomorphological units developed by cryogenic processes in periglacial conditions" (Czudek and Demek 1970, p. 101).

Fig. 19.9 Permafrost features on the Magdalen Islands. **a** Cryoturbations on the Cap aux Meules Island. **b** and **d** Ice-wedge pseudomorphs (IWP) on the Cap aux Meules Island. **c** and **e** Composite-wedge cast dated to the Younger Dryas on the Havre Aubert Island. Ucf-1 and Ucf-2 = unconformities

While they are prevalent in Europe and Asia (French 1973, 2007; Vandenberghe and Czudek 2008), cryopediments are rather scarce in Canada, the Yukon (Lauriol and Godbout 1988; French and Harry 1992) and the Magdalen Islands (Hamelin 1959; Ritchot 1968) being the only Canadian regions where such landforms have been recognized. On the Magdalen Islands, the cryopediments represent the dominant form on the sandstone platforms (Figs. 19.2 and 19.10). On the Cap aux Meules Island, where they are particularly extensive (Fig. 19.4), they cover approximately 75% of the territory, i.e. $\sim$36 km^2 (Paquet 1989). It is also possible to observe cryopediments on Havre Aubert, Havre aux Maisons and Entry islands where they cover between 10 and 30% of the land.

On the Magdalen Islands, Paquet (1989) produced a detailed geomorphological and statistical study of these original landforms. On Cap aux Meules Island, two levels of cryopediments were determined. The upper (and older) level is observable only in two small areas of the Cap aux Meules Island. The lower level, which generally extends from the foothill to the coast, corresponds to the majority of the cryopediments. The cryopediment gradients of the Magdalen Islands range around 1° while the gradients of the backslope range from 6°–8° (Fig. 19.10). They have the typical profile of the cryopediments described in the literature (Czudek and Demek 1970; Vandenberghe and Czudek 2008). Where the coasts have considerably retreated since their formation (by coastal erosion), the cryopediments are truncated downslope by coastal cliff that reach up to 25 m high (Fig. 19.2). Where the coasts are protected from strong waves (e.g., lagoons), the cryopediments disappear under the Holocene coastal ridges. As they intersect both the inclined strata of the sandstone platform and the faults which delimit the volcanic rocks, cryopediments are clearly erosional landforms.

Fig. 19.10 Cryopediments on the Magdalen Islands. **a** Regular and slightly inclined topography of a cryopediment on the Havre Aubert Island. **b** and **c** Perpendicular outcrop into a cryopediment of the Cap aux Meules Island. Note the channels in the weathered sandstone at the bottom of the colluvial deposit. **d** Colluvial facies: imbricated angular basaltic pebbles from volcanic hills frost-shattering with a reddish sandy matrix that comes from the erosion of the underlying sandstone. The deposit is slightly stratified and comprises lenses of well-sorted sand (**e**)

Moreover, underlying the thin colluvial deposits (±1 m), the outcrops exposing a vertical section of the cryopediments exhibit several channels formed in the weathered sandstone (Fig. 19.10). Colluvial deposits are very similar in different outcrops. They are composed of imbricated angular basaltic pebbles with a reddish sandy matrix that comes from the erosion of the underlying sandstone. They are slightly stratified and comprise lenses composed of well-sorted sand (Fig. 19.10). The colluvial facies implies widespread areal erosion characterized by a dynamics of very mobile divagating channels similar to the pediments of the arid zones (Dylik 1957; Vandenberghe and Czudek 2008). The fact that the colluvial deposits are mainly composed of volcanic pebbles involves intense frost-shattering of the central hills and very efficient transport processes from the slope to the channels. The flows required to the transport of such pebbles over considerable distances (up to 2 km) along the low-angle cryopediments were probably caused by the summer melting of the persistent snow and névé located in the sub-basins of the central hills (Fig. 19.2f), and possibly also by strong summer rains. For such runoffs to occur over very porous sandstone, the presence of perennially frozen ground was needed to block the infiltration of water (Laverdière and Guimont 1974; Vandenberghe and Czudek 2008). The divagating channels also imply a minimal or even absent vegetation cover (Dylik 1957). This entire interpretation scenario is consistent with the absolute ages obtained on the colluvial deposits and the underlying coastal sediment. These ages, added to the stratigraphic context, highlighted two periods of periglacial (colluvial) activity, both contemporaneous of high RSL. A first colluvial period occurred between 65 and 40 ka, while the colluvial deposits are intercalated in coastal deposits located between 15 and 38 m high (Rémillard et al. 2017). A second colluvial period occurred after the deglaciation of the LGM according to the luminescence ages acquired from colluvial deposit at three different sites (19.9, 17.6, and 19.5 ka) on the Cap aux Meules Island (Rémillard et al. 2016). This second period is contemporaneous with a RSL of more than +20 m asl. These ages and the stratigraphic data are consistent with the geomorphological observations of Paquet (1989) regarding the occurrence of two different periods of cryopedimentation on the Magdalen Islands. When the high RSL retreated (owing to the glaci-isostatic rebound), the cryoplanation activity ceased and a dynamic of incision started.

19.4.3 Dry Valleys

The cryopediments of the Magdalen Islands are incised by a system of flat-floored valleys and asymmetrical small valleys extending from the central hills to the coast. They are invaded either by trees or herbaceous plants and are mostly dry; they now drain only thin intermittent water streams fed by snowmelt. Obviously, these dry valleys are relic landforms (Figs. 19.11 and 19.12). Near the coasts, these valleys truncate first glacial and littoral sediments deposited by the last glacial cycle, and secondly, the underlying bedrock (Fig. 19.12). Because of the recent RSL rise, the cliff retreated several tens of meters, transforming the periglacial valleys into perched valleys (Fig. 19.13d). In the lagoons, however, the downstream parts of the valleys, invaded by the RSL rise (rias), extend below modern sea level on the Magdalen Shelf where hydrographic networks have been identified to a depth of −30 to −50 m asl (Shaw et al. 2002; Shaw 2005). These submerged hydrographic networks illustrate that a significant part of the Gulf of St. Lawrence was emerged during early Holocene,

The bottom of the valleys is covered by coarse alluvial deposits which are essentially composed of basaltic pebbles from the central hills (Fig. 19.11). These alluvial deposits are 0.5–1.5 m thick and suggest pervasive frost-shattering activity in the central hills and competent flows sustained by the snowmelt. By blocking water infiltrations, permafrost has certainly supported superficial runoffs (Laverdière and Guimont 1974). These observations indicate that periglacial conditions persisted during the marine regression (between 13 and 10 ka), consistent with the development in littoral deposits of ice-wedge pseudomorphs and composite-wedge casts in littoral deposits dated to the Younger Dryas.

19.5 The Coastal Landscape

The beauty and diversity of the Magdalen Islands coasts, with their succession of vertiginous cliffs, arches, stacks, infinite beaches, and coastal dunes, support a remarkable biological diversity (Hamilton 2002; Attention FragÎles 2009a, b, 2010; Roy-Bolduc 2010), and enchant the 50,000 tourists who visit the archipelago annually.

19.5.1 Wave Regime, Tides, Currents, and Sea Ice

On the Magdalen Islands, the fetch length exceeds more than 300 km in all directions except to the south and south-east where it is approximately 80–90 km. Located 50 km from a amphidromic point, the archipelago has a diurnal mixed tide regime in the north and semi-diurnal in the south, with mean tides of 0.7 m and high tides of 1.1 m. (Service hydrographique du Canada 2017). The waves that form in the Gulf of St. Lawrence generally have greater amplitude than ocean waves. The average height of the waves on the western side of the archipelago is 0.49 m in summer and 1.19 m in winter, whereas it reaches 0.35 and 0.98 m on the east side, respectively (Owens 1977). Between January 1978 and December 2007, Bernatchez et al. (2008) reported 48 storm events with winds with a maximum speed exceeding 60 km/h. Although storm surges originate sometimes from the east or north-east (Bernatchez et al. 2008), storms from the northwest are the most frequent (63% of storms recorded between 1961 and 1990); they also produce the strongest waves, some of which can exceed 4 m (Savard 2008). Historically, sea ice and ice foot form around the archipelago for a period varying generally between 70 and 110 days during winter time.

19.5.2 Cliffs and Beach Environments

On the Magdalen Islands, the shoreline exposed to the offshore waves is 235 km long (Bernatchez et al. 2012). The rocky cliffs, littoral spits, beach ridges, and tombolos account for approximately 99% of the coastline (Fig. 19.2). Luminescence ages acquired by Rémillard et al. (2015a) from different sites of beach ridges suggest that these systems developed recently, i.e. between 2.6 and 0.4 ka, during a period of RSL rise (Fig. 19.14). With a mean erosion rate of −0.3 m/yr between 1963 and 2008 for rocky cliffs, and reaching 1.5 m/yr in some places (Bernatchez et al. 2012), the cliffs of the Magdalen Islands have the highest rocky coastal-cliff erosion rates in the Estuary and Gulf of St. Lawrence (Bernatchez and Dubois 2004). Sandstone cliff erosion due to rising sea level is the main source of sediment to the longshore drift and the building of prograding beaches. Indeed, some low-lying coasts on the archipelago are still prograding, despite a RSL increase of $\sim$4.3 mm/yr over the last few decades (Bernatchez et al. 2012; Barnett et al. 2017).

Fig. 19.11 Periglacial valleys of the Magdalen Islands. **a** Flat-floored valley on Havre Aubert and **b** Cap aux Meules islands. **c** Perched flat-floored valley asl S = sandstone; Ft = fluvial terrace. **d** Cross-section in a fluvial terrace. **e** Zoom on the fluvial deposit of the terrace. **f** Asymmetrical vallon on Cap aux Meules Island

19.5.3 Coastal Erosion and Submergence

Coastal erosion and submergence present a real challenge for the Magdalen Islands community. Shoreline retreat between 1963 and 2008 were measured on vertical aerial photographs using the methodology described by Bernatchez et al. (2012). The erosion rates are presented on Fig. 19.15 for each hydrosedimentary cell. Globally, nearly 70% of the coastline of the archipelago is currently at risk of erosion and this proportion is expected to reach 80% by 2060 (Bernatchez et al. 2012). The average migration rate for the entire archipelago's coastline between 1963 and 2008 was −0.25 m/yr (including prograding and retreating coasts). However, this average hides a great spatial variability especially for the sandy coasts for which accumulation values can reach more than 8 m/yr and erosion values more than 5 m/yr (Fig. 19.15). Erosion is prevalent on the western coast (−0.77 m/yr on average) while on the eastern coast,

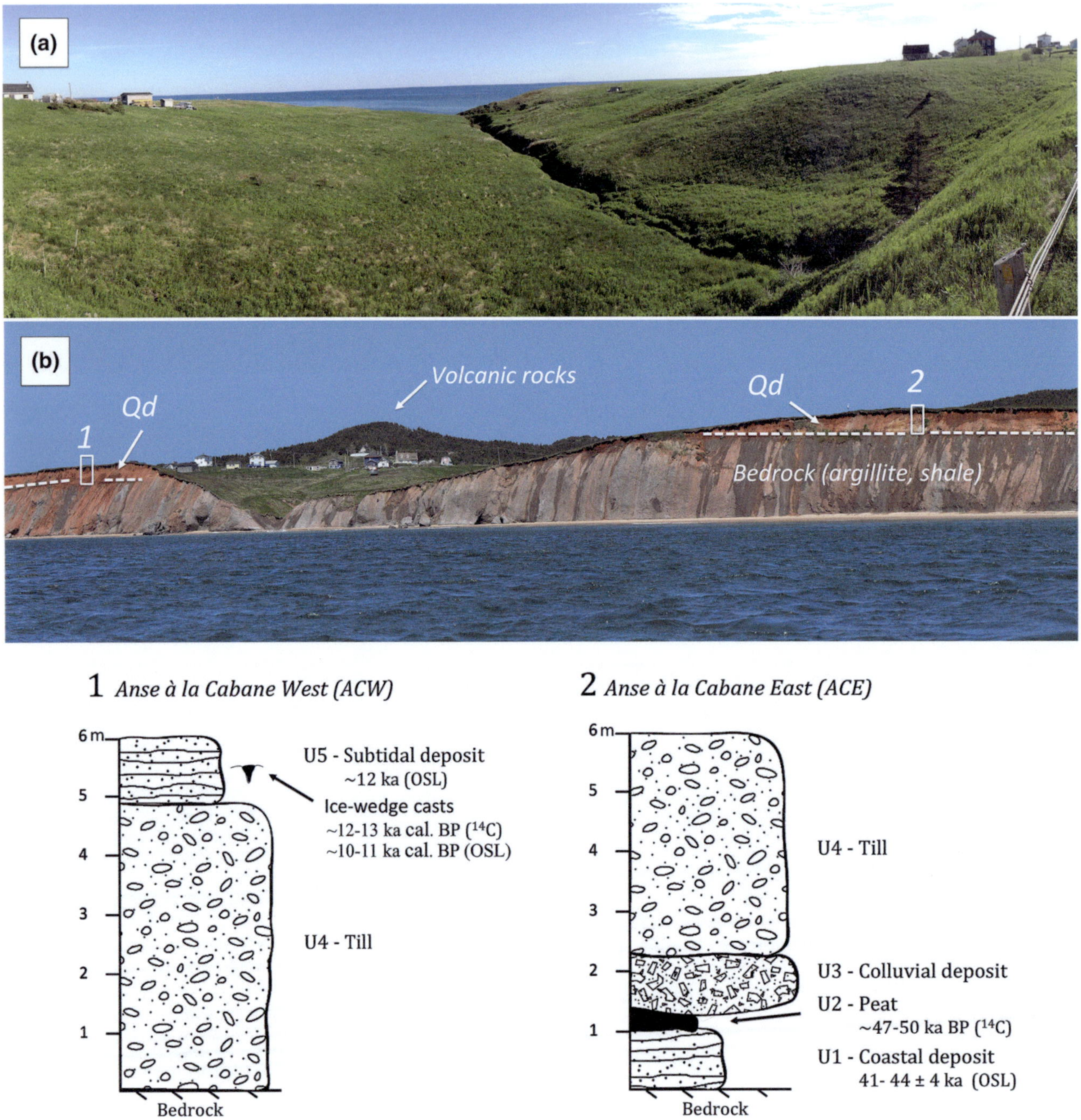

Fig. 19.12 Asymmetrical vallon on the Havre Aubert Island. This vallon truncated the Quaternary deposits exposed in the cliff (see Sect. 1 and 2) during the retreating phase of the RSL at the end of the MIS 2. Outcrops are detailed in Rémillard et al. (2013)

there are some sectors where the balance is positive (in green on Fig. 19.15). This disparity is due to the high exposure of the western coast to the storm surges. More recently, based on the Québec Maritime Coastal Erosion Monitoring Network (Bernatchez and Drejza 2015), that representing 1029 erosion stations on the Magdalen Islands coasts, the average migration rate between 2005 and 2013 was −0.75 m/yr.

Sea ice and ice foot development around the archipelago protect the coasts from the action of waves during the strongest storm surges. However, with climate warming, which especially affects winters, the freezing period is shortened and the ice cover has decreased more and more (Bernatchez et al. 2008). Results from ice foot cover modeling for the 2055 horizon suggest for the Magdalen Islands a reduction of about 40–53 days where the ice foot will no

Fig. 19.13 Rias on the Magdalen Island. **a** Aerial photograph of the northwestern part of the Cap aux Meules Island showing three brackish (1, 2, and 3) representing the downstream part of periglacial valleys that were submerged during the late Holocene RSL rise. These submerged valleys were isolated from the sea by the tombolos and sand spits development during the last two millennia. **b** LiDAR imagery of the same location emphasizing the dry valleys. **c** A valley whose downstream part has been submerged on the Havre Aubert Island. **d** Perched asymmetrical vallon on Havre Aubert Island

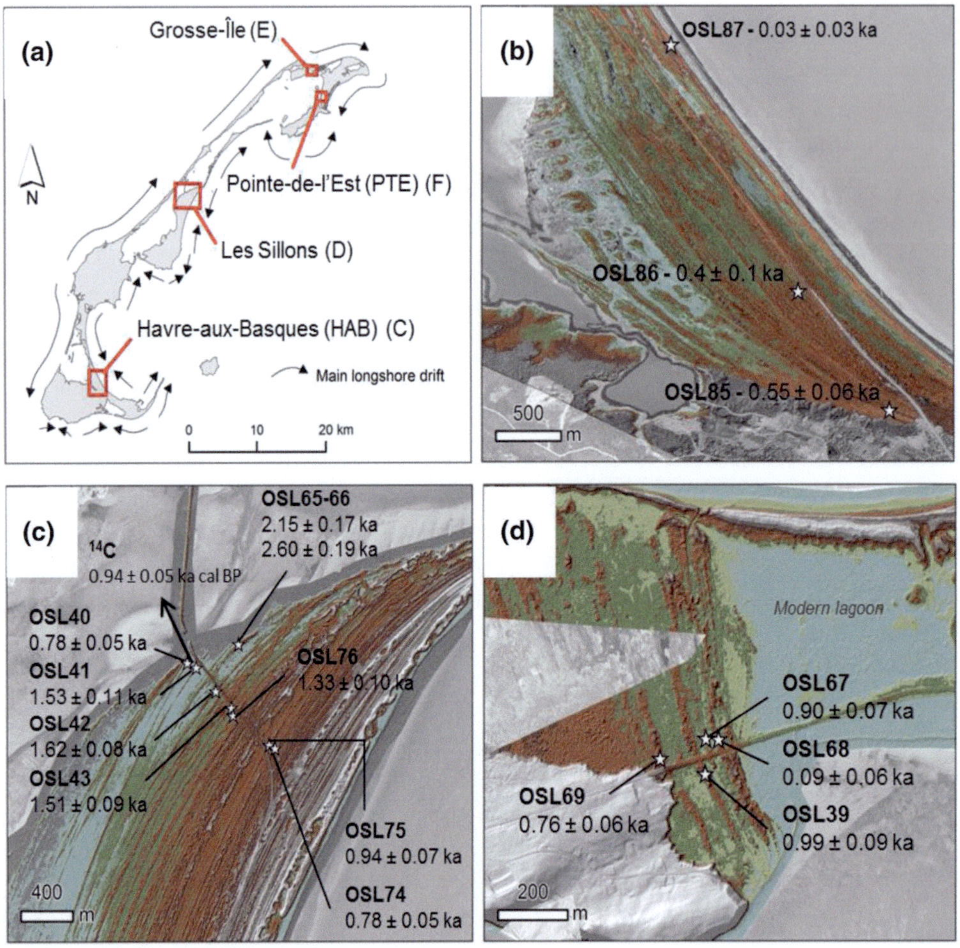

Fig. 19.14 Beach ridges, littoral spits and tombolos of the Magdalen Islands are recent forms that developed during the last two millennia as indicated by the luminescence ages obtained by Rémillard et al. (2015a). **a** Location of the beach ridges sampling sites on the Magdalen Islands. Figures 19.1c–e correspond to LiDAR imagery where white stars represent sampling locations. **b** Les Sillons site. **c** Havre aux Basques (HAB) site. The black star represents sampling location of a modern beach ridge. **d** Grosse-Île site

longer provide coastal protection from storm surges (Senneville et al. 2014). If the warming trend continues, the decrease in ice cover will accelerate coastal erosion. There have been two periods of marked decrease in ice cover, between 1977–1983 and 2001–2007, the latter corresponding to an important reduction of beach width (Bernatchez et al. 2012). The lack of sea ice and ice foot is becoming the norm since the early 2000s. The decrease in sea ice and the global sea-level rise also increase also the risk of submergence. The low coasts, which represent 70% of the total coastline length, are particularly exposed.

By 2065, the number of buildings that may be exposed to coastal erosion if no new adaptation measures are implemented and those already in place are not maintained will be 262 representing value of more than $29 million ($Can) (Bernatchez et al. 2015). There are also 18 km of roads that can be exposed by 2065, representing a value of $ 77.6 million (Bernatchez et al. 2015).

19.6 Conclusions

Figure 19.16 summarizes the major stages of the Magdalen Islands landscape evolution since the MIS 5. Located in the center of the Gulf of St. Lawrence at approximately equal distance from the main ice-caps of the Appalachian Glacier Complex, the Magdalen Islands were deglaciated early ($\sim 25 \pm 2$ ka), at least for the southern part, which has allowed long periods of periglacial processes (MIS 3 and end of MIS 2). Located at the far margins of the Escuminac ice-cap and a lobe of the Newfoundland ice-cap, the archipelago has a complex glaci-isostatic history with marked nuances between the north, the center and the south (Rémillard et al. 2017). However, from roughly 11 ka, the entire archipelago was affected by the migration of the peripheral forebulge that formed by the growth of the regional ice-caps that extended to the continental shelf and

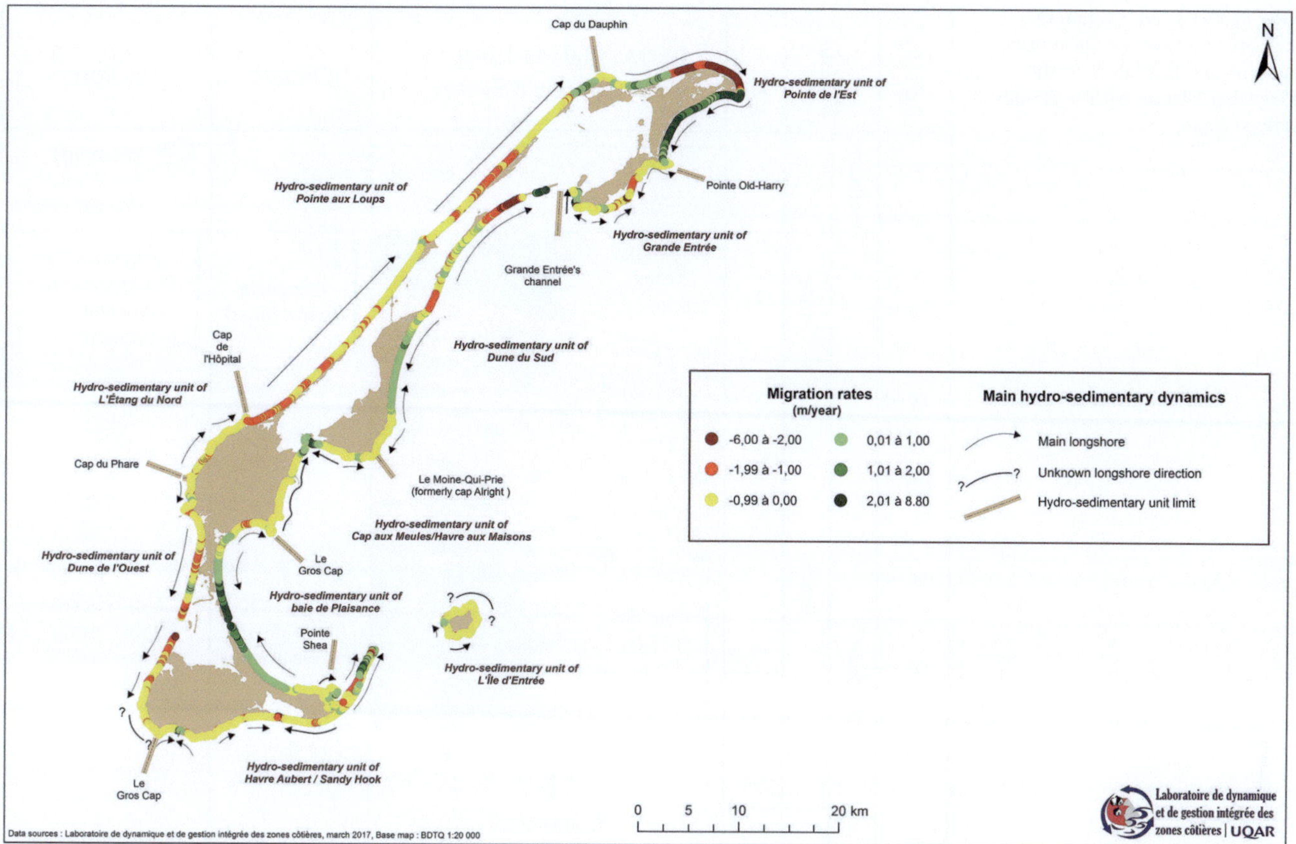

Fig. 19.15 Major hydrosedimentary cells on the Magdalen Islands and coastline migration from 1963 to 2008 measured on vertical aerial photographs using the methodology described by Bernatchez et al. (2012). In green = accumulation rates. In red, orange, and yellow = erosion rates

the LIS which formed an ice stream in the Laurentian Channel (e.g., Stea et al. 1998, 2011; Shaw et al. 2006). Consequently, in eastern Canada, the global sea-level rise due mostly to the thermal expansion and freshwater flux from the melting land-based ice sheets, is exacerbated by the subsidence movement associated with the collapse of the peripheral forebulge that migrated northward after the MIS 2 deglaciation.

Glaciers that covered the Magdalen Islands during the LGM did not leave obvious evidence in the landscape except the thin till exposed in a few cliffs (Figs. 19.5 and 19.12). The landscape of the archipelago originates from three different types of processes: structural, periglacial, and coastal.

For the landforms of structural origin, the volcanic hills adjacent to the sandstone platforms are the most peculiar characteristic (Figs. 19.2 and 19.3). This is the first obvious aspect when approaching the islands by boat. On the sandstone platforms, which are erosion surfaces, the cryopediments and dry valleys are the dominating landforms. Coastal landforms are dominated by cliffs, tombolos, and littoral spits. In spite of their spectacular development, the large double tombolos that connect the rocky islands are very recent and have developed mostly during the last two millennia.

The Quaternary deposits, which are clearly visible at the top of the cliffs, do not have much expression in the surface. A particularly striking aspect when looking at aerial photos is the absence of raised shorelines generally evident in areas that have been submerged by the post-glacial seas. This is easily explained by the intensity of periglacial processes, especially the cryopedimentation that occurred under the influence of the periglacial climate which persisted for more than 10 ka after the LGM on the southern archipelago. It is on this part of the archipelago that the cryopediments, the dry valleys and the ice-wedge pseudomorphs or composite-wedge casts are the best developed. On the northern islands (PAL and Grande Entrée), the evidence of periglacial activity is absent or almost unnoticeable (a few low dry valleys only). The fact that raised shorelines have been obliterated by cryopedimentation does not exclude any influence of the RSL variations on the landscape evolution. Cryopedimentation implies widespread areal erosion characterized by a dynamics of very mobile divagating channels which is incompatible with the valleys incision. The

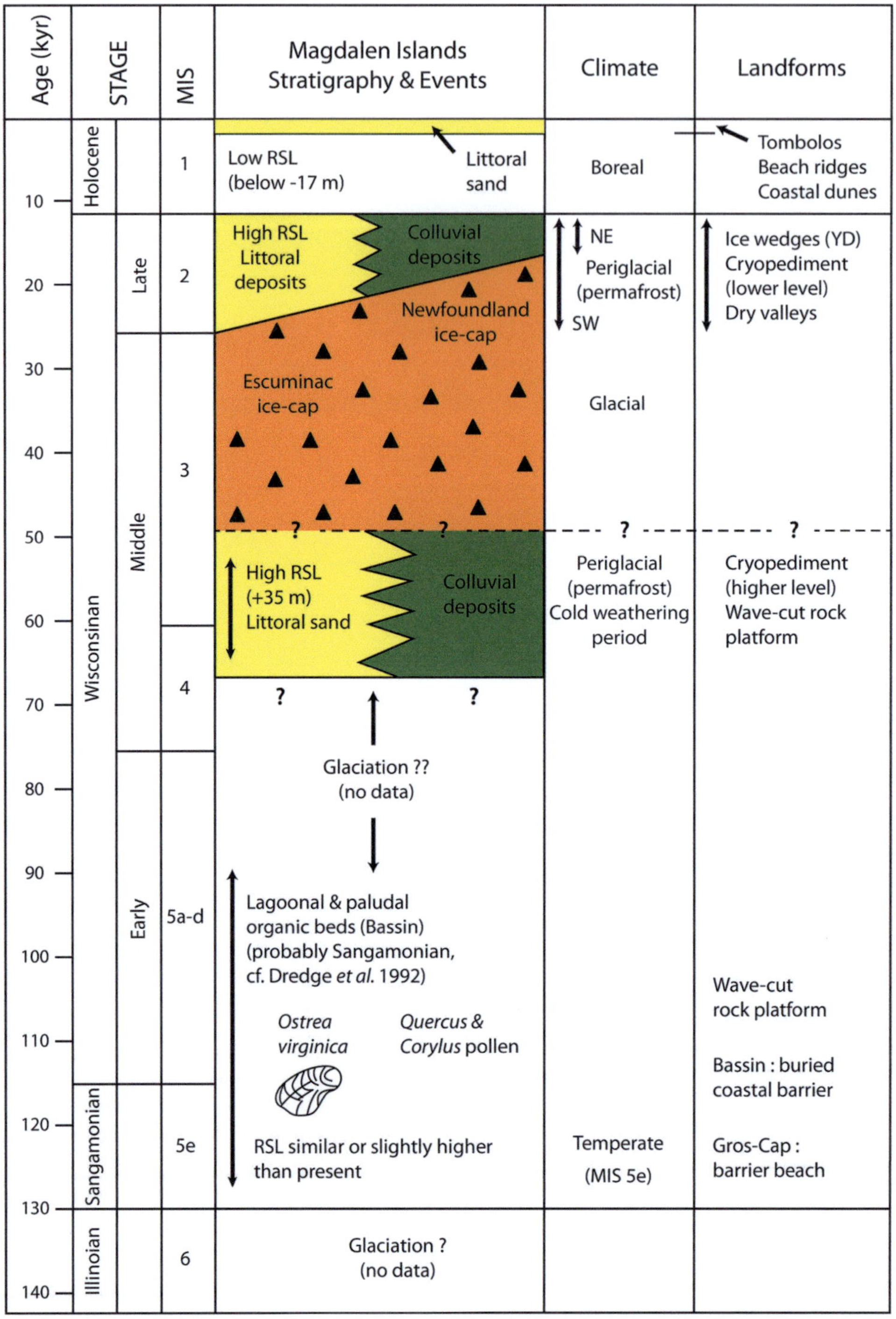

Fig. 19.16 Time diagram of terrestrial and marine Quaternary deposits and landforms on the Magdalen Islands. MIS = Marine isotope stage

formation of cryopediments was done in two phases associated with high RSL: during the MIS 3 and after the MIS 2 deglaciation. When the RSL dropped after 11 kyr owing to the glaci-isostatic rebound and the migration of the fore-bulge, cryopedimentation has been replaced by an incision dynamics (Fig. 19.12). A wide system of flat-floored valleys and asymmetric vallons developed. As the RSL passed below modern level and large parts of the seafloor of the Gulf of St. Lawrence emerged after 10 ka, valleys incised the seafloor to reach the retreating shoreline. With the Holocene RSL rise, the valleys were partly submerged and transformed into rias (Fig. 19.13).

Despite the small size of the archipelago, the landscape of the Magdalen Islands has a uniqueness that can be largely explained by its geographical location in the center of the Gulf of St. Lawrence.

References

Agglomération des Îles-de-la-Madeleine (2010) Schéma d'Aménagement et de développement révisé

Attention FragÎles (2009a) Oiseaux des milieux marins, dunaires, humides et forestiers des Îles de la Madeleine, Îles-de-la-Madeleine

Attention FragÎles (2009b) Plantes des milieux marins, dunaires, humides et forestiers des Îles de la Madeleine, Îles-de-la-Madeleine

Attention FragÎles (2010) Portrait du territoire de la réserve écologique de l'île Brion, Îles-de-la-Madeleine

Barnett RL, Bernatchez P, Garneau M, Juneau M-N (2017) Reconstructing late Holocene relative sea-level changes at the Magdalen Islands (Gulf of St. Lawrence, Canada) using multi-proxy analyses. J Quat Sci 32:380–395

Barnett RL, Garneau M, Bernatchez P (2016) Salt-marsh sea-level indicators and transfer function development for the Magdalen Islands in the Gulf of St. Lawrence, Canada. Marine Micropaleontol 122:13–26

Barr SM, Brisebois D, Macdonald AS (1985) Carboniferous volcanic rocks of the Magdalen Islands, Gulf of St. Laurence. Can J Earth Sci 22:1679–1688

Bernatchez P, Drejza S (2015) Réseau de suivi de l'érosion côtière du Québec maritime—Guide pour les utilisateurs. Laboratoire de dynamique et de gestion intégrée des zones côtières et Chaire de recherche en géoscience côtière. Université du Québec à Rimouski, octobre 2015

Bernatchez P, Dubois J-MM (2004) Bilan des connaissances de la dynamique de l'érosion des côtes du Québec maritime laurentien. Géog Phys Quat 58:45–71

Bernatchez P, Drejza S, Dugas S (2012) Marges de sécurité en érosion côtière : évolution historique et future du littoral des îles de la Madeleine. Chaire de recherche en géoscience côtière, Laboratoire de dynamique et de gestion intégrée des zones côtières, Université du Québec à Rimouski. Rapport remis au ministère de la Sécurité publique du Québec, juillet 2012

Bernatchez P, Dugas S, Fraser C, Da Silva L (2015) Évaluation économique des impacts potentiels de l'érosion des côtes du Québec maritime dans un contexte de changements climatiques. Laboratoire de dynamique et de gestion intégrée des zones côtières, Université du Québec à Rimouski. Rapport remis à Ouranos

Bernatchez P, Fraser C, Friesinger S, Jolivet Y, Dugas S, Drejza S, Morissette A (2008) Sensibilité des côtes et vulnérabilité des communautés du golfe du Saint-laurent aux impacts des changements climatiques. Laboratoire de dynamique et de gestion intégrée des zones côtières, Université du Québec à Rimouski. Rapport de recherche remis au Consiortium OURANOS et au FACC

Brisebois D (1981) Lithostratigraphie des strates permo-carbonnifères, de l'archipel des Îles-de-la-Madeleine. Ministère de l'énergie et des ressources du Québec, DPV-796

Caputo R (2007) Sea-level curves: Perplexities of an end-user in morphotectonic applications. Global Planet Change 57:417–423

Chalmers R (1895) Surface geology of southern New Brunswick, northwestern Nova Scotia and a portion of Prince Edward Island. Geological Survey of Canada, Annual Report

Czudek T, Demek J (1970) Pleistocene cryopedimentation in Czechoslovakia. Acta Geog Lodziensia 24:101–108

Dredge LA, Mott RJ, Grant DR (1992) Quaternary stratigraphy, paleoecology, and glacial geology, Iles de la Madeleine, Québec. Can J Earth Sci 29:1981–1996

Dubois J-MM (1992) Le paysage naturel et son évolution. Infogéographes 1:41–49

Dubois J.-MM, Grenier A (1993) The Magdalen Islands, Gulf of Saint Lawrence. In: Coaslines of Canada. Proceedings of 8th symposium on coastal and ocean management. ASCE, July 19–23, pp 170–182

Dylik J (1957) Tentative comparison of planation surfaces occurring under warm and under cold semi-arid climatic conditions. Biul Peryglac 5:175–186

Environment Canada (2017) Normales climatiques 1981–2010 pour la station des Îles-de-la-Madeleine

Forbes DL, Orford JD, Carter RWG, Shaw J, Jennings SC (1995) Morphodynamic evolution, self-organisation and instability of coarse, clastic barriers on paraglacial coasts. Marine Geol 126:63–85

Fortin JC, Larocque P (2003) Histoire des Îles-de-la-Madeleine. Presses de l'Université Laval

French HM (1973) Cryopediments on the chalk of Southern England. Biul Peryglac 22:149–156

French HM (2007) The Periglacial environment. Wiley, 478 p

French HM, Harry DG (1992) Pediments and cold-climate conditions, Barn Mountains, unglaciated northern Yukon, Canada. Geogr Ann 74A:145–157

Giles PS (2008) Windsor Group (Late Mississipian) stratigraphy, Magdalen Islands, Quebec: a rare eastern Canadian record of late Visean basaltic volcanism. Atl Geol 44:167–185

Goldthwait JW (1915) The Occurrence of Glacial Drift on the Magdalen Islands. Geological Survey of Canada. Museum bulletin no. 14

Grant DR (1980) Quaternary sea-level change in Atlantic Canada as an indication of crustal develelling. In: Morner NA (ed) Earth Rheology, Isostasy and Eustasy. Wiley, London, pp 201–214

Hamelin LE (1959) Sables et Mer aux Îles-de-la-Madeleine. Province de Québec, 1958. Ministère de l'industrie et du commerce, Québec

Hamilton KGA (2002) Îles-de-la-Madeleine (Magdalen Is.): a glacial refugium for short-horned bugs (Homoptera: Auchenorrhyncha)? Le Naturaliste Can 126:25–39

Han G, Ma Z, Chen N, Thomson R, Slangen A (2015) Changes in mean relative sea level around Canada in the twentieth and twenty-first centuries. Atmos Ocean 53:452–463

Hétu B, Gray JT, Gangloff P, Archambault B (2003) Postglacial talus-derived rock glaciers in the Gaspé Peninsula, Québec (Canada). In: Philips SM, Springman SM, Arenson LU (eds) Proceedings of 8th international conference on Permafrost., pp 389–394

Koohzare A, Vanicek P, Santos M (2008) Pattern of recent vertical crustal movements in Canada. J Geodyn 45:133–145

Josenhans HW, Lehman S (1999) Quaternary stratigraphy and glacial history of the Gulf of St. Lawrence, Canada. Can J Earth Sci 36:1327–1345

Juneau M-N (2012) Hausse récente du niveau marin relatif aux Îles-de-la-Madeleine. Mémoire de maîtrise en géographie, Université du Québec à Rimouski

La Flèche MR, Camiré G, Jenner GA (1998) Geochemistry of post-Acadian, Carboniferous continental intraplate basalts from the Maritimes Basin, Magdalen Islands, Québec, Canada. Chem Geol 148:115–136

Lauriol B, Godbout L (1988) Les terrasses de cryoplanation dans le nord du Yukon: distribution, genèse, âge. Géog Phys Quat 42:303–314

Laverdière C, Guimont P (1974) Un froid à sol fendre. GEOS 18–20

Liverman D, Catto N, Batterson M, Mackenzie C, Munro-Stasiuk M, Scott S, Sommerville A (2000) Evidence of late glacial permafrost in Newfoundland. Quat Intern 68:163–174

Loring D, Nota D (1973) Morphology and sediments of the Gulf of St. Lawrence. Service des Pêches et des Sciences de la Mer, Bulletin 182
Mayle FE, Cwynar LC (1995) A review of multi-proxy data for the Younger Dryas in Atlantic Canada. Quat Sci Revs 14:813–821
Mott RJ, Grant DR, Stea RR, Occhietti S (1986) Late-glacial climatic oscillation in Atlantic Canada equivalent to the Allerød/younger Dryas event. Nature 323:247–250
Owens EH (1977) Temporal variations in beach and nearshore dynamics. J Sed Pet 47:168–190
Paquet G (1989) L'évolution de la plate-forme gréseuse de l'île du Cap aux Meules (Îles-de-la-Madeleine). Mémoire de maîtrise, Université de Montréal
Poirier J (1970) Fente de gel fossile aux Îles-de-la-Madeleine. Rev Géog Montréal 24:319–320
Prest VK, Teresme J, Matthews JV Jr, Federovich LS (1976) Late-Quaternary history of Magdalen Islands, Quebec. Marit Sed 12:39–59
Quinlan G, Beaumont C (1981) A comparison of observed and theoretical postglacial relative sea level in Atlantic Canada. Can J Earth Sci 18:1146–1163
Quirion F (1988) Plan quinquennial d'aménagement forestier des Îles-de-la-Madeleine. Ministère de l'Énergie et des Ressources, Québec
Rémillard AM, Hétu B, Bernatchez P (2013) The Drift des Demoiselles on the Magdalen Islands (Québec, Canada): sedimentological and micromorphological evidence of a Late Wisconsinan glacial diamict. Can J Earth Sci 50:1–19
Rémillard AM, Buylaert J-P, Murray AS, St-Onge G, Bernatchez P, Hétu B (2015a) Quartz OSL dating of late Holocene beach ridges from the Magdalen Islands (Quebec, Canada). Quat Geochron 30:264–269
Rémillard AM, Hétu B, Bernatchez P, Buylaert J-P, Murray AS, St-Onge G, Geach M (2015b) Chronology and palaeoenvironmental implications of the ice-wedge pseudomorphs and composite-wedge casts on the Magdalen Islands (eastern Canada). Boreas 44:658–675
Rémillard AM, St-Onge G, Bernatchez P, Hétu B, Buylaert J-P, Murray AS, Vigneault B (2016) Chronology and stratigraphy of the Magdalen Islands archipelago from the last glaciation to the early Holocene: new insights into the glacial and sea-level history of eastern Canada. Boreas 45:604–628
Rémillard AM, St-Onge G, Bernatchez P, Hétu B, Buylaert J-P, Murray AS, Lajeunesse P (2017) Relative sea-level changes and glacio-isostatic adjustment on the Magdalen Islands archipelago (Atlantic Canada) from MIS 5 to the late Holocene. Quat Sci Rev 171:216–233
Richardson J (1881) Report of a geological exploration of the Magdalen Islands. Geological Survey of Canada, Report of Progress (1879–1880), part VIII
Ritchot G (1968) Rapport: cartes géomorphologiques et géotechniques de Rivière-du-Loup, Rimouski-Mont-Joli, Sainte-Anne-des-Monts, Cap-Chat, Gaspé, Iles de la Madeleine. Ministère des affaires municipales, Québec
Rohling EJ, Grant K, Hemleben C, Siddall M, Hoogakker BAA, Bolshaw M, Kucera M (2008) High rates of sea-level rise during the last interglacial period. Nature Geosci 1:38–42
Roy-Bolduc A (2010) Portrait et historique d'utilisation de l'écosystème forestier des Îles-de-la-Madeleine. Centre de recherche sur les milieux insulaires maritimes (CERMIM)
Roy K, Peltier WR (2015) Glacial isostatic adjustment, relative sea level history and mantle viscosity: reconciling relative sea level model predictions for U. S. East coast with geological constraints. Geophys J Intern 201:1156–1181
Sanschagrin R (1964) Magdalen Islands. Ministère de l'Énergie et Ressources du Québec, Geological Report 106
Savard J-P (2008) Étude de la sensibilité des côtes et de la vulnérabilité des communautés du glofe du Saint-Laurent aux impacts des changements climatiques—Synthèse des résultats. Ouranos
Seaman AA, Stea RR, Allard S (2011) The Appalachian Glacier Complex in the Canadian Maritime Provinces during Illinoian and Wisconsinian. Geohydro 2011, CANQUA Meeting, Quebec City
Senneville S, St-Onge-Drouin S, Dumont D, Bihan-Poudec AC, Belemaalem Z, Corriveau M, Bernatchez P, Bélanger S, Tolszczuk-Leclerc S, Villeneuve R (2014) Modélisation des glaces dans l'estuaire et le golfe du Saint-Laurent dans la perspective des changements climatiques, ISMER-UQAR, Rapport final présenté au ministère des Transports du Québec
Service hydrographique du Canada (2017) Marées, courants et niveaux d'eau. http://www.charts.gc.ca/index-fra.asp
Shaw J (2005) Geomorphic evidence of postglacial terrestrial environments on Atlantic Canadian continental shelves. Géog Phys Quat 59:141–154
Shaw J, Gareau P, Courtney RC (2002) Palaeogeography of Atlantic Canada 13–0 kyr. Quat Sci Rev 21:1861–1878
Shaw J, Piper DJW, Fader GBJ, King EL, Todd BJ, Bell T, Batterson MJ, Liverman DJE (2006) A conceptual model of the deglaciation of Atlantic Canada. Quat Sci Rev 25:2059–2081
Stea RR, Piper DJW, Fader GBJ, Boyd R (1998) Wisconsinian glacial and sea-level history of Maritime Canada and the adjacent continental shelf: a correlation of land and sea events. Geol Soc Am Bull 110:821–845
Stea RR, Seaman AA, Pronk T, Parkhill MA, Allard S, Utting D (2011) The Appalachian glacier complex in Maritime Canada. In: Ehlers J, Gibbard PL (eds) Quaternary glaciations—extent and chronology, part II: North America. Elsevier, New York, pp 631–659
Vandenberghe J, Czudek T (2008) Pleistocene cryopediments on variable terrain. Permafr Periglac Proc 19:71–83

Bernard Hétu was awarded a Ph.D. in geography from the Université de Montréal. Officially retired since September 2013, he was a professor in geography at the Université du Québec à Rimouski (UQAR) for 33 years. During his career at the UQAR, he won the Distinction Pascal-Parent (2006), the Achievement award from the Université du Québec (2007) and the Distinction Alcide-C.-Hort (2012) to highlight his contribution in research and development of teaching programs at the UQAR. He has also been a member-researcher of BORÉAS and the Centre for Northern Studies (CEN) for nearly two decades. Since 2013, he has remained active as Associate Professor at the UQAR and the *Université Laval*, both in teaching and research. His past and current research projects are divided between two topics: (1) glacial movements, deglaciation and the reconstitution of long-term changes in relative sea level in Québec's maritime regions (Bas-Saint-Laurent, Gaspésie, Magdalen Islands, and Anticosti); (2) the impact of climate change on the postglacial evolution of the slopes in eastern Québec and the current dynamics of scree slopes in northern Gaspésie.

Dr. Audrey M. Rémillard has received a M.Sc. degree in physical geography from the Université du Québec à Rimouski (UQAR), and a PhD in oceanography at the Institut des Sciences de la mer de Rimouski of the UQAR. Her postgraduate studies focused on the Quaternary history of the Magdalen Islands, located in the center of the Gulf of St. Lawrence, mainly on the last glaciation patterns, the occurrence of periglacial events, and the variations in relative sea-level. She did an internship at Risø National Laboratory (Aarhus University, Denmark) where she developed the first chronological framework for the Magdalen Islands based on optically stimulated luminescence. Owing to the excellence of her postgraduate academic record, she received the Governor General's Gold Medal of Canada in 2017. Since 2016, Audrey works as research associate for the Research Chair in Coastal Geoscience held by Pascal Bernatchez. Her research focuses on past sea-level changes, radiocarbon dating, and Quaternary stratigraphy in maritime Québec and Eastern Canada.

Pascal Bernatchez has been a professor of geography at L'Université du Québec à Rimouski (UQAR) since 2001. He graduated with a PhD in geography from the L'Université Laval with an expertise in coastal geomorphology and remote sensing. He holds the Research Chair in Coastal Geoscience since 2008 and leads the 'Laboratoire de dynamique et de gestion intégrée des zones côtières' since 2003. In 2008, he received the Circle of Excellence Award of the L'Université du Québec. He is also a member-researcher of the Centre for Northern Studies (CEN), Québec-Océan, and BORÉAS. His research focuses on the quantification of sea-level changes, erosion, submersion, and ice cover, and the climate changes effects on the sensitivity and vulnerability of coastal areas. He is also interested in the development of remote sensing and geomatics tools, platforms and methods to quantify and map the coastal conditions for risk and adaptation studies. His team works in collaboration with the Québec government to establish coastal evolution projections that take into account climate change for the zoning of natural hazards.

Guillaume St-Onge is a professor at the Institut des sciences de la mer de Rimouski of the Université du Québec à Rimouski (UQAR). He was awarded a Ph.D. in Natural Resources and he holds the Tier I Canada Research Chair in Marine Geology. He is also the director of the Réseau Québec maritime and the co-director of the Institut France-Québec maritime. He has lead numerous expeditions at sea or on the field and his research focuses on marine geology, sedimentology, stratigraphy, Quaternary geology, natural hazards and paleomagnetism in Eastern and Arctic Canada, the Southern Hemisphere, Europe, and Asia. He is also involved in several national and international programs such as the International Continental Scientific Drilling Program (ICDP) and the International Ocean Discovery Program (IODP). He has setup state-of-the-art marine geology and paleomagnetism laboratories and was awarded several distinctions for the excellence of his research and training.

The Seafloor of Southeastern Canada

20

Alexandre Normandeau, David J. W. Piper, John Shaw, Brian J. Todd, D. Calvin Campbell, and David C. Mosher

Abstract

Quaternary glaciations played a critical role in producing the modern landscape of the seafloor of southeastern Canada. Glacial landscapes such as cross-shelf troughs, fjords, recessional moraines and tunnel valleys were sculpted by the Laurentide Ice Sheet. Following deglaciation, relative sea level rise led to the formation of sandy bedforms and tidal scours on the continental shelf. In contrast, the deep-water margin was not directly modified by the ice sheet, but was significantly influenced by it. Large quantities of glacigenic sediments were distributed along the continental slope during the last glacial maximum through turbidity currents, contour currents and submarine landslides. Turbidity currents eroded submarine canyons and deposited submarine fans, which are among the largest sediment accumulations in southeastern Canada. Although the continental shelf and the deep-water margin have contrasting landscapes, they are genetically linked to tectonics and the pattern of retreat of the Laurentide Ice Sheet. This chapter gives an overview of the landscape and landforms of the seafloor of southeastern Canada and details how its glacial heritage played a major role in sculpting the modern seabed.

Keywords

Marine geology • Sediment transport • Seafloor • Submarine glacial landforms • Continental shelf and slope

A. Normandeau (✉) · D. J. W. Piper · J. Shaw · B. J. Todd · D. Calvin Campbell · D. C. Mosher
Geological Survey of Canada (Atlantic), Natural Resources Canada, Bedford Institute of Oceanography, 1 Challenger Drive, P.O. Box 1006 Dartmouth, NS B2Y 4A2, Canada
e-mail: alexandre.normandeau@canada.ca

20.1 Introduction

Historically, the discipline of geomorphology has focused on subaerial landscapes, as exemplified by this book on landscapes and landforms of eastern Canada. The large number of terrestrial geomorphological studies is explained by the fact that the topography of all of the terrestrial landmass of Canada is mapped at a resolution of <100 m, making it possible to describe its geomorphology in relative detail. Almost half of Canadian territory is submarine, and almost 87% of that territory remains unmapped at sufficient resolution for geomorphological studies similar to those on its terrestrial counterpart. The commercialization of multibeam sonars in the 1990s revolutionized the study of the seafloor by allowing a complete, full-coverage mapping of seabed geomorphology at unprecedented resolution. Although most of the submarine component of the Canadian landmass lacks modern bathymetric mapping, significant efforts were made in the late 1990s and 2000s in southeastern Canada (Fig. 20.1). Use of these new tools led to the detailed characterization of landforms such as recessional moraines, submarine canyons, submarine landslides and bottom-current-related features such as contourite drifts, sandy bedforms and tidal scours, among others. Most of the marine geomorphological vocabulary was borrowed and adapted from the terrestrial community (canyons, dunes, landslides, etc.). For example, a dune, which the Merriam-Webster dictionary defines as 'a hill or ridge of sand piled up by the wind', is now widely used for subaqueous flow-transverse bedforms formed by ocean currents (Ashley 1990).

The gross-scale geomorphology of the submarine portion of southeastern Canada is first and foremost controlled by its tectonic formation and the juxtaposition of continental and oceanic crust. Continental crust sits at a higher elevation and has given rise to the continental shelf whereas oceanic crust lies at a lower elevation and formed the abyssal plain of the deep ocean floor. The transition between the two types of

© Springer Nature Switzerland AG 2020
O. Slaymaker and N. Catto (eds.), *Landscapes and Landforms of Eastern Canada*, World Geomorphological Landscapes, https://doi.org/10.1007/978-3-030-35137-3_20

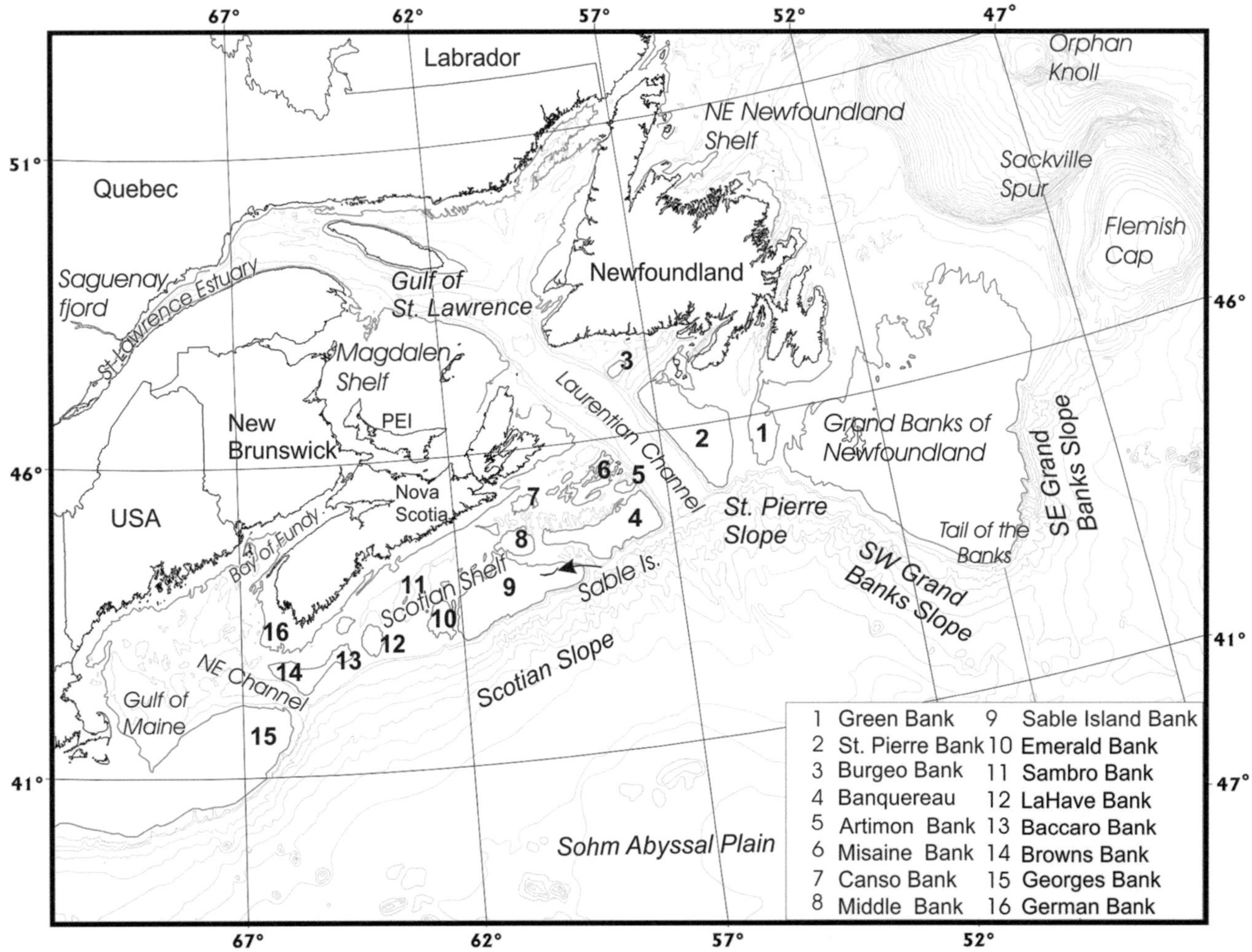

Fig. 20.1 Location of the study area, in southeastern Canada, identifying some of the major geomorphological features on the continental shelf and the deep-water margin

crust forms the continental margin. As a result of this gross morphology, the seafloor off southeastern Canada is divided into two main geomorphological units: (1) the continental shelf and (2) the deep-water margin (Fig. 20.2). The continental shelf is located between the coast and the shelf break and the deep-water margin comprises the upper continental slope ($\sim$100–2500 m deep), the lower continental slope ($\sim$2500–5200 m deep) and the abyssal plain (>5200 m deep). Both major geomorphological units offer a landscape with a wide variety of landforms. Landforms of the continental shelf were mainly sculpted by past glaciations and associated lower sea level and today include bedforms that actively move on the seafloor under the influence of modern tidal, ocean and wave currents. In contrast, the deep-water margin mostly comprises landforms formed by—or related to—gravitational processes such as canyon erosion and landslides and along-slope processes driven by ocean currents depositing contourite drifts.

This chapter highlights the geomorphology of the southeastern Canada seafloor, both on the continental shelf and the deep-water margin, and summarizes the importance of its glacial heritage in the formation and distribution of landforms. Currently, large parts of the Scotian Slope, Flemish Cap, the St. Lawrence Estuary and coastal areas of the Maritime Provinces are well mapped whereas high-resolution bathymetric data are scarce on the Grand Banks and off northeastern Newfoundland. This disparity in availability of high-resolution multibeam-bathymetric datasets introduces a bias in this chapter towards regions that are well mapped and studied.

20.2 The Continental Shelf

The wide, low-gradient (<0.1°) continental shelf of southeastern Canada has a significant gradient change at the shelf edge at an average depth of 210 m, but as shallow as 75 m

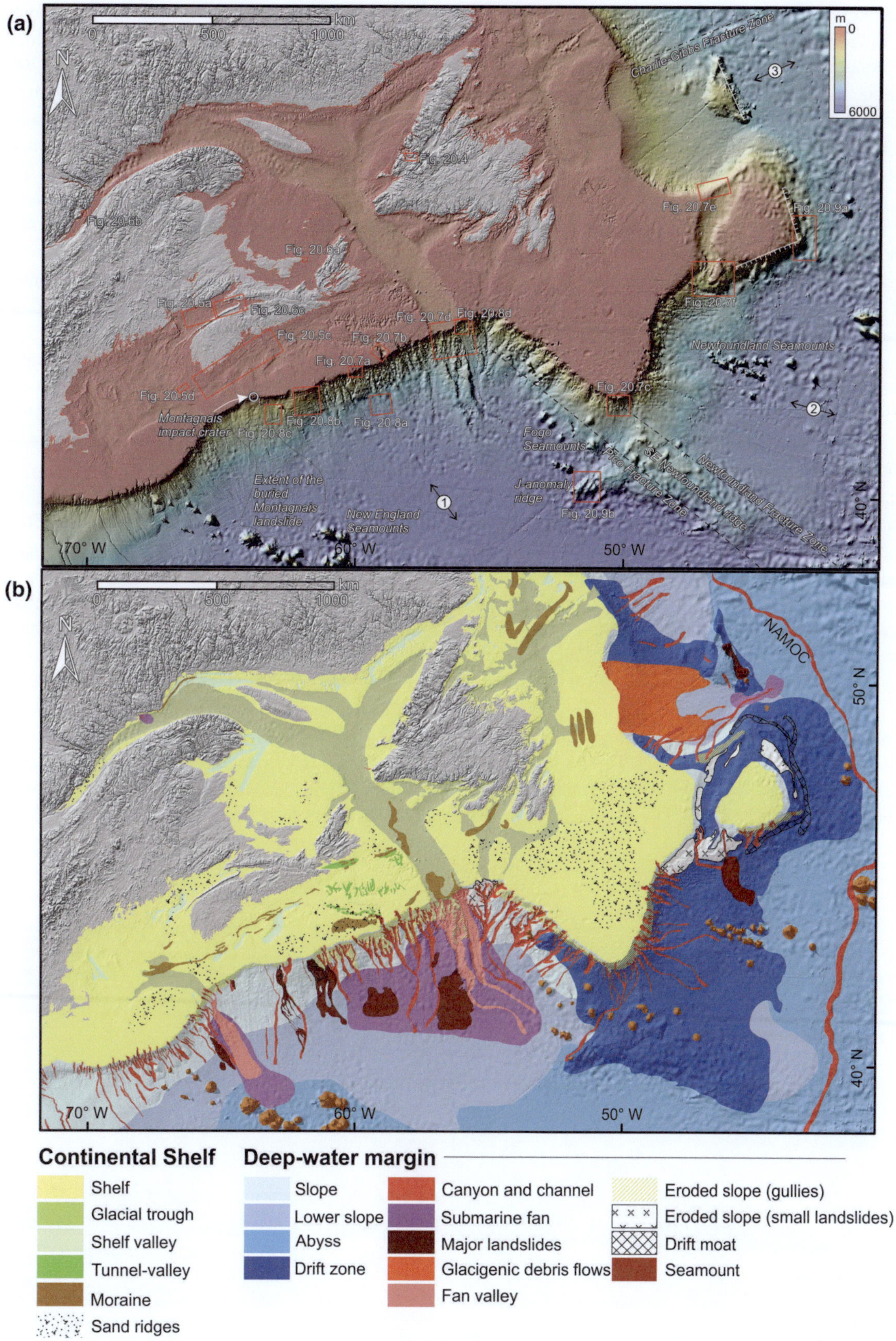

Fig. 20.2 **a** Shaded relief bathymetry of southeastern Canada illustrating the main fracture zones, the Montagnais impact crater and landslide and the locations of Figs. 20.4, 20.5, 20.6, 20.7, 20.8 and 20.9. **b** Interpreted geomorphology of the seafloor southeastern Canada. This map compiles interpretations from Harris et al. (2014) and Geological Survey of Canada (Atlantic) work since the 1970s

on parts of the Scotian Shelf and as deep as 530 m off northeast Newfoundland (Fig. 20.1). The shelf comprises a series of shallow offshore banks separated by both shelf-crossing and shore-parallel troughs. The banks are shallow (<50 m in many areas) and protrude above sea level at Sable Island. In many instances, the shallowest parts are near the shelf break. The shelves have been heavily imprinted by glaciations, in that fluvial valleys have been deepened into fjords and shelf-crossing troughs by ice streams, and erosional basins contain thick sequences of glacial sediments. The Laurentian Channel is the most prominent glacial trough crossing the entire continental shelf for a distance of about 1000 km (Fig. 20.1). The channel originates at the head of the St. Lawrence Estuary and reaches the edge of the continental shelf south of Newfoundland, connecting other small glacial troughs along its way (Fig. 20.3a). The Laurentian Channel drained ice from most of southeastern Canada during the Quaternary (2.58–0 million years ago) (Fig. 20.3a). Northeast Channel drained ice originating from southern Québec and New Brunswick, converging into the Gulf of Maine. Newfoundland ice was mostly drained through Notre Dame and Trinity troughs to the northeast, and Laurentian Channel and Haddock and Halibut troughs to the south (Fig. 20.3a). All of these cross-shelf troughs led to substantial depositional systems seaward of the shelf edge and onto the deep-water margin (Fig. 20.2b).

The eastern Scotian Shelf hosts a large tunnel valley system (Boyd et al. 1988) (Fig. 20.2b). Tunnel valleys are erosional channels generally formed near a former ice margin by subglacial meltwater flows. During glaciation, these tunnel valleys provided pathways for meltwater flows to reach the edge of the continental shelf, and fed submarine canyons on the continental slope (Piper et al. 2007). Relict tunnel valleys are evident on the seafloor of the continental shelf (Fig. 20.2b). As ice streams retreated during deglaciation, glacigenic sediments were reworked by tides and waves to yield thick muds within basins and coarser sediments on banks and in the littoral zone. Near the shelf edge, reworking of sediment resulted in burial of the tunnel valley system.

At the last glacial maximum (LGM), around 24 ka cal BP, the Laurentide Ice Sheet (LIS) comprised a series of catchment basins, with convergent flow into ice streams in over-deepened troughs on the continental shelves (Shaw et al. 2006) (Fig. 20.3a). Deglaciation was initially accomplished by ice calving along deep channels. By c. 16.5 ka cal BP, ice margins lay close to modern coasts and the ice sheet had become a series of isolated ice caps. A retreating calving embayment halted briefly in the Laurentian Channel and rapidly moved northwest into the Gulf of St. Lawrence. At this time, the calving ice margins had reached the south and southwest coasts of Newfoundland. By 11 ka cal. BP, Newfoundland and the Maritimes were ice-free and the LIS continued to retreat to the northwest in Quebec and Labrador (Fig. 20.3b).

Tilted and raised shorelines in southeastern Canada indicate regional variations in sea level history from which the locations of greatest ice-loading are inferred (Grant 1989; Peltier 1998). Reconstructions of the postglacial geography of southeastern Canada shows large islands on the continental shelf during the retreat of the LIS (Fig. 20.3b) (Shaw

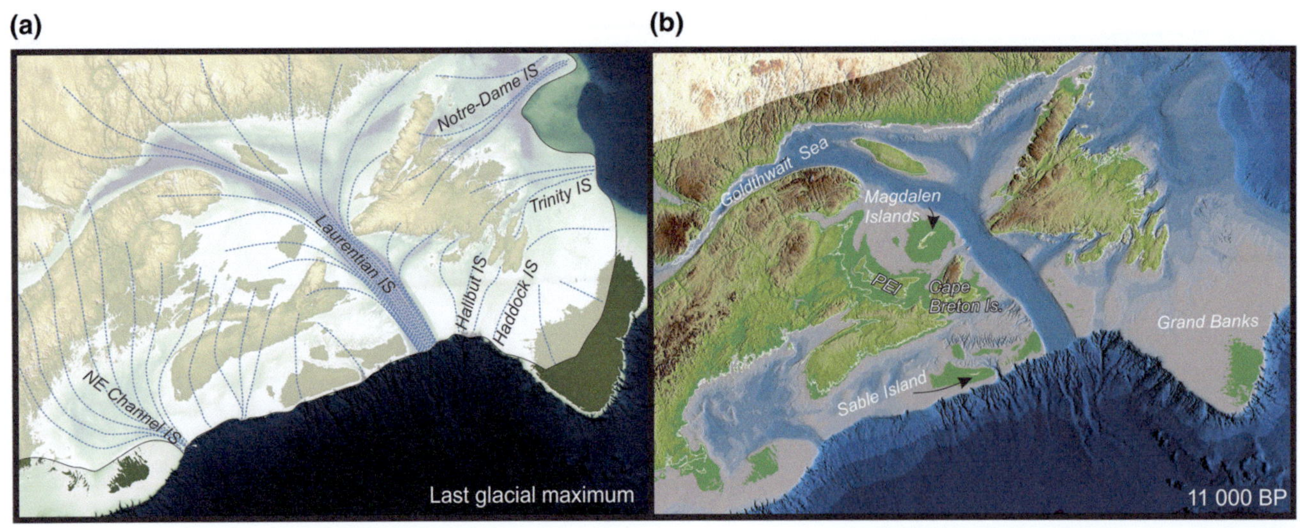

Fig. 20.3 **a** Generalized depiction of ice extent at the Last Glacial Maximum, about 23 ka cal BP. The dashed lines are approximate flow lines. Exposed land shown green. **b** Palaeogeography 11 ka cal BP. Large islands on the continental shelves were shrinking in extent as sea level rose. Concurrently, falling sea levels connected Prince Edward Island with the mainland. In the north, sea levels remained high at this time, and the western Gulf of St. Lawrence coast was submerged

et al. 2002, 2006). For example, at c. 11 ka cal BP Grand Bank and Sable Island Bank were emergent, although reduced in area compared to previous millennia (Fig. 20.3b). In contrast, at the same time emergent areas around Prince Edward Island, the Magdalen Islands, and Cape Breton Island reached their greatest extent.

The rapidity of sea level change, significant wave height and strong currents in exposed areas, the brevity of low-stands, and migration of coastal systems (Shaw et al. 1993) all reduced the preservation of coastal landforms during postglacial sea level rise (Shaw 2005). Along the south coasts of Newfoundland, submerged early Holocene deltas are present in fjords at depths of 10–30 m (Shaw and Forbes 1995). In contrast, glaci-isostatic rebound outpaced glaci-eustatic sea level rise in the western Gulf of St. Lawrence, which led to a relative sea level fall. In the western Gulf of St. Lawrence, sea level was ~150 m higher than today and formed the Goldthwait Sea (Fig. 20.3b). As relative sea level fell, previously inundated marine areas emerged and preserved glacimarine sediments and raised shorelines (Normandeau et al. 2017).

20.2.1 Fjords: The Result of Glacial Erosion

Fjords are long, deep submarine valleys formed primarily as a result of glacial erosion and subsequently inundated during postglacial transgression. In southeastern Canada, the largest example is the 100 km long and 1000 m deep Saguenay Fjord (Fig. 20.1). Other fjords were present along the Quebec coastline during deglaciation but emerged during glaci-isostatically induced relative sea level fall (Dietrich et al. 2017a; Poiré et al. 2018). As a result, most fjords in southern Quebec are now disconnected from the sea and form fjord lakes (Normandeau et al. 2016; Poiré et al. 2018). A few fjords are found in Nova Scotia, but the majority of fjords in southeastern Canada are located in Newfoundland (Fig. 20.4a). Fjords act as sediment traps and typically host thick glacimarine sediments overlain by postglacial mud, as observed in the Bay of Islands Fjord, Newfoundland (Fig. 20.4b–d) (Shaw and Potter 2015). Fjord sidewalls are commonly notched by V-shaped gullies (Fig. 20.4c) formed by late glacial slope failures. They result in stacked lenses of landslide deposits (Fig. 20.4e) that overly stratified glacimarine mud (Shaw and Potter 2015). These events mostly occurred following deglaciation at a time when relative sea level was rapidly falling due to glaci-isostatic rebound and earthquake activity was more prevalent than today. Fjords also contain good records of extreme flood events. In the Saguenay Fjord, a flood in 1996 deposited a 50 cm-thick layer of mud that buried the underlying mercury-contaminated sediment originating from a chlor-alkali plant (Mucci et al. 2015).

In Newfoundland, fjords range in depth to 850 m, and commonly have shallow sills at their seaward margins on the south and west coasts, although not on the north coast. Fjords are variable in morphology, so that the example described here is not necessarily typical. The Bay of Islands fjord in western Newfoundland (Fig. 20.4) is bounded at its seaward end by an arcuate submarine moraine up to 100 m thick (Fig. 20.4b) (Shaw and Potter 2016). The deepest areas of the fjord (295 m) are directly landward of the shallow (30 m) sill. Submarine slides at Corner Brook have a strong expression at the seafloor (Fig. 20.4e). The largest slide complex (Fig. 20.4e) consists of 20 m-deep channels incised on the fjord sidewalls, and several lobes in deeper water that lie about 4 m higher than the surrounding seafloor. A smaller slide nearby (Fig. 20.4e) travelled into deeper water and turned to the left, creating a pair of flanking berms approximately 1 m high.

20.2.2 Submarine Moraines Deposited During Glacial Retreat

Across the continental shelf are suites of curvilinear, sub-parallel sediment ridges interpreted as recessional moraines, deposited during a temporary pause in the retreat of the LIS (Fig. 20.2a). These features were formed near the grounding line of the LIS, and their distribution and seismostratigraphy reveals the dynamics of the regional retreat of the LIS following the LGM (Stea et al. 1998; Shaw et al. 2006, 2014). Large moraines are observed on the Scotian Shelf, the Laurentian Channel and the Grand Banks, whereas smaller moraines are observed in the Gulf of St. Lawrence and nearshore Newfoundland.

The mapping of these ice-retreat features enables the reconstruction of the spatial pattern of ice movement. As an example, a conceptual model of ice-sheet dynamics in southeastern Canada showed that an ice stream occupied the Bay of Fundy (Shaw et al. 2006; Todd and Shaw 2012). The presence of the ice stream is exemplified by multiple, sub-parallel ridges, in places demonstrating pronounced convexity to the southwest (Fig. 20.5a). The ridges are confined to the central section of the Bay of Fundy and do not extend to nearshore regions of outcropping bedrock. Interpretation of seismic reflection profiles indicates that the ridges in the bay are composed of ice-contact sediments overlying glacimarine sediment and bedrock (Todd and Shaw 2012). The ice-contact sediment is over 20 m thick in the small ridges and over 50 m in the large ridges. Based on their geomorphology and seismostratigraphic character, the lobe-shaped ridges with pronounced SW-trending convexity in the Bay of Fundy (Fig. 20.5a, b) are interpreted as recessional moraines marking sequential positions of the

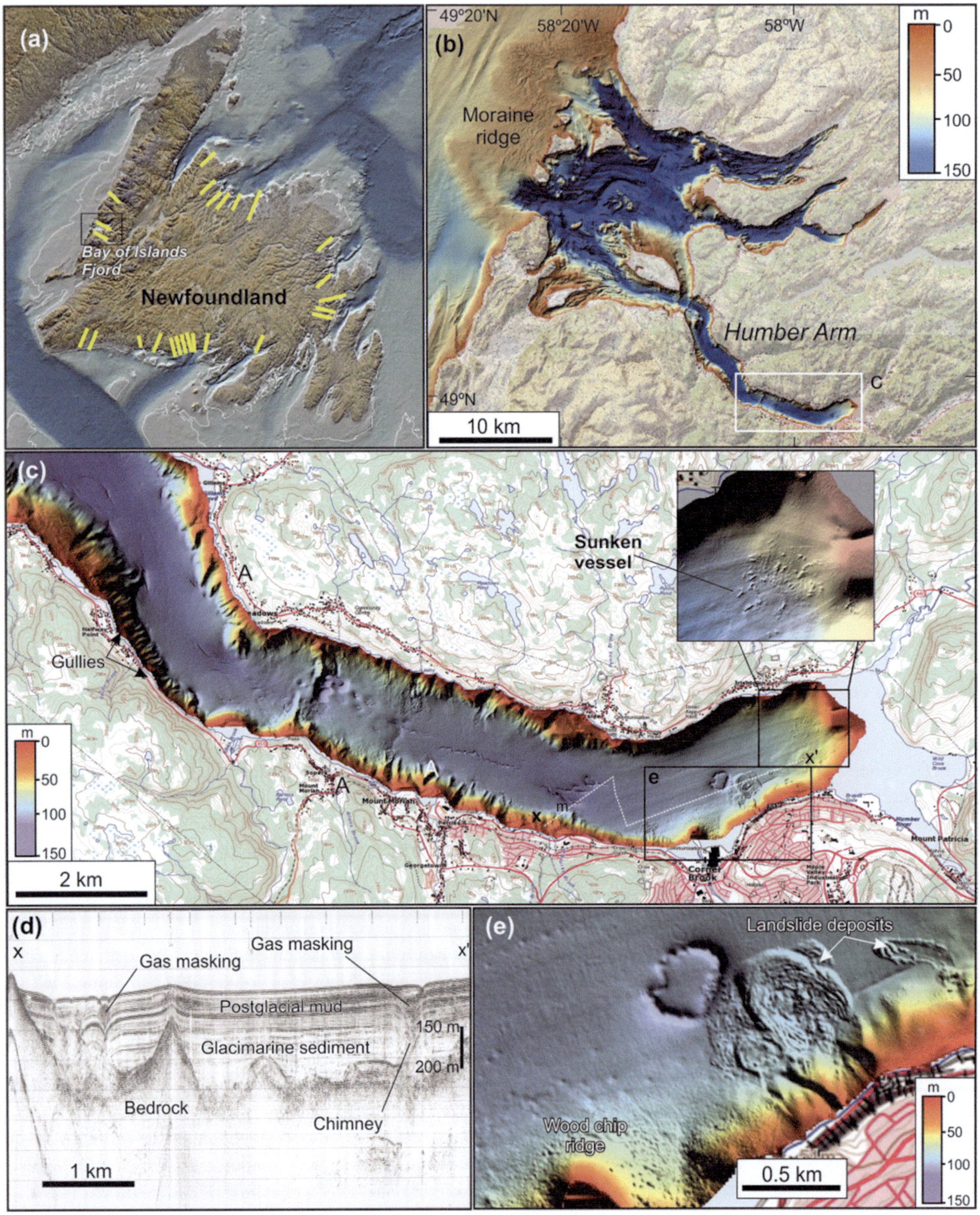

Fig. 20.4 **a** Fjords in Newfoundland (yellow lines). **b** Sun-illuminated bathymetric image of Bay of Islands, Newfoundland. **c** Humber Arm with inset of irregular mounds on the seafloor. **d** Seismic profile of Humber Arm showing thick glacimarine and postglacial sediments, and a gas-escape chimney. Profile is dashed line x–x′ on Fig. 20.4c. **e** Enlarged view of submarine landslides off the city of Corner Brook

complex planform of the LIS terminus as it retreated northeast during the last deglaciation (Todd and Shaw 2012). During regional ice-sheet retreat, continual local advances and retreats resulted in younger moraines overprinted on older moraines, giving a nested appearance to the moraine crests.

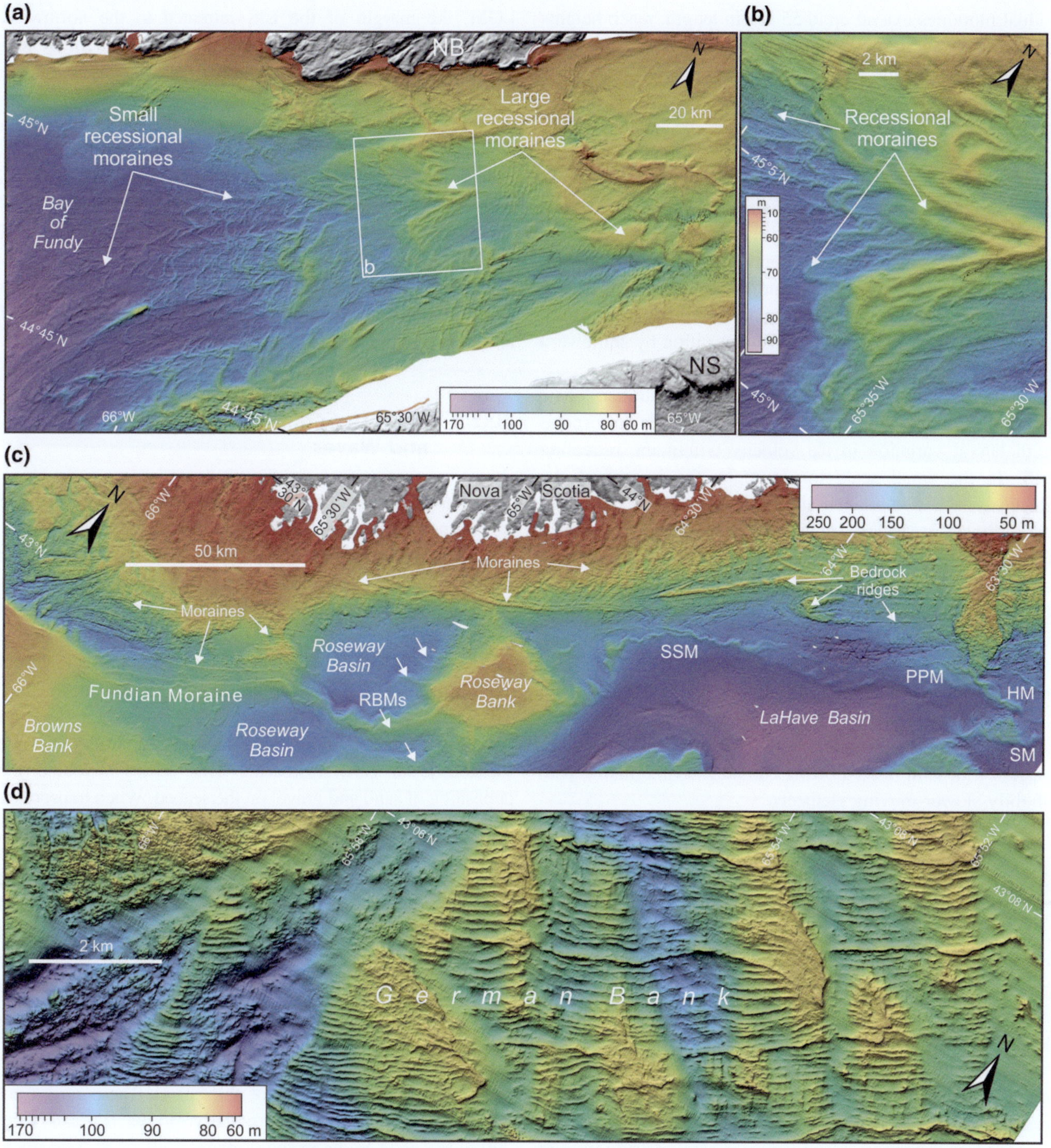

Fig. 20.5 **a** Multibeam sonar image of the central Bay of Fundy showing the curving pattern of recessional moraines. Adapted from Todd et al. (2016). **b** Enlarged image of recessional moraines showing lobate character and overprinting from southwest to northeast. Adapted from Todd et al. (2016). **c** Single-beam Olex bathymetric image showing recessional moraines draping Scotian Shelf banks and basins. RBMs, Roseway Basin Moraines (four white arrows); RB, Roseway Bank; SSM, South Shore Moraine; PPM, Pennant Point Moraine; HM, Halifax Moraine; SM, Sambro Moraine (adapted from Todd (2016a, b)). **d** Sun-illuminated multibeam-bathymetric image showing De Geer moraines on German Bank. Adapted from Todd (2016b)

20.2.3 End Moraine Complexes

Superimposed on the southern Scotian Shelf topography are moraines that are 30–40 km offshore and trend approximately parallel to the present coast of Nova Scotia. Individual moraines extend up to 55 km in length, reach heights of 50–75 m above the seafloor and are arranged in an en échelon pattern, with some moraine crests exhibiting bifurcation. The cross-sectional profile of the moraines exhibits a morphological range from pronounced peaks to broadly rounded humps. In places, the moraines are flat-topped with an exposed boulder pavement (Todd et al. 1999). Within basins on the southern Scotian Shelf, subtle lineations visible in single-beam bathymetric images enable the identification and lateral tracing of buried moraines (Todd 2016a). Collectively this suite of exposed and buried moraines is identified as the Scotian Shelf end moraine complex (King 1969) and represents a time-transgressive deposition by shoreward (i.e. northwestward) retreating ice sheet.

Many of the Scotian Shelf moraines demonstrate a lobate planform in both small and large scales that may correspond with lateral variation in ice velocity within the retreating LIS. In the small scale for example, the largest of the Roseway Basin moraines shows a SW-convex planform approximately 50 km long bisecting Roseway Basin (Fig. 20.5c). In contrast, the South Shore Moraine demonstrates a relatively subtle curvilinear planform in small scale, but in large scale the moraine is composed of a number of SW-convex segments (Fig. 20.5c). The Pennant Point, Halifax and Sambro moraines all demonstrate this segmented style (Fig. 20.5c). These complex moraine planforms revealed by sonar mapping reflect the subtleties of the ice-velocity regime of the retreating ice-front of the southeastern LIS, with faster-flowing components alternating with slightly slower-moving elements.

20.2.4 De Geer Moraines

Numerous regularly spaced, parallel, linear to curvilinear ridges of sediment are recognized to form at or close to the grounding lines of water-terminating glaciers (Lindén and Möller 2005). These subaqueous, ice-flow transverse ridges, known as De Geer moraines, have been interpreted as marking former ice margin positions where sediment is deposited or pushed up during brief stillstands or minor readvances (De Geer 1889).

De Geer moraines extend across a bathymetric range of over 100 m on southern German Bank of the Scotian Shelf (Fig. 20.5d) (Todd et al. 2007). The De Geer moraines display a quasi-linear or curved line in planform and are traced horizontally from short segments of a few hundred metres up to almost 10 km in length. Most moraine crests are simple lines in planform, but in some areas the crests are complex with bifurcating crest lines and (or) short, en échelon crest segments.

The distribution of De Geer moraines provides insight into the subtleties of deglaciation of German Bank. After the LGM, the margin of the LIS retreated to the northwest, punctuated by many small readvances. The pattern of segmented, southeast-convex moraines suggests that the ice sheet margin was gently scalloped. De Geer moraines are absent on exposed bedrock but are numerous on thin deposits of streamlined lineations of till (Fig. 20.5d). In swales between the lineations the moraines are partly to completely buried by sediment. The generally regular horizontal spacing of De Geer moraines suggests that the rate of ice-front retreat was consistent from one cycle of moraine formation to the next. The larger De Geer moraines on German Bank may be the product of several years of ice-front stability or major readvances.

20.2.5 Modern Sediment Transport by Tides and Waves

Today the shallow continental shelf hosts the most active submarine bedforms in southeastern Canada. An understanding of the mobility of these bedforms on the seafloor is important for the sustainable management of seafloor resources and infrastructure. Sandy bedforms are present on the Magdalen Shelf, on Gulf of St. Lawrence deltas, in the Bay of Fundy, on German and Browns banks, around Sable Island and on parts of the Grand Banks, among others (Fig. 20.2b). The eastern part of the continental shelf was exposed during the early deglacial period and transgressed during the late Pleistocene to early Holocene. This setting favoured the deposition of sand and gravel on the seabed. When transported under the influence of waves and currents, these materials formed sandy bedforms. They range in morphology from shore-connected ridges to different types of submarine dunes and megadunes. Large oblique sand ridges and shore-connected ridges have been identified around Sable Island, on the outer Scotian Shelf (Li and King 2007), in shallow nearshore zones around Prince Edward Island (Fig. 20.6a), and on the eroded delta fronts of the Gulf of St. Lawrence (Normandeau et al. 2013). Near Sable Island, ridges 1 km in length and 3.5 m in height are oriented NE–SW along the nearshore environments of the island. These ridges are mobile during major winter easterly storms (Li and King 2007). On delta fronts in the northern Gulf of St. Lawrence, Holocene erosion formed a shallow coastal shelf on which oblique bars form and migrate in the direction of longshore drift during easterly storms (Normandeau et al. 2013).

Tidal currents are responsible for erosional and depositional features in some locations. Tidal forces operate across

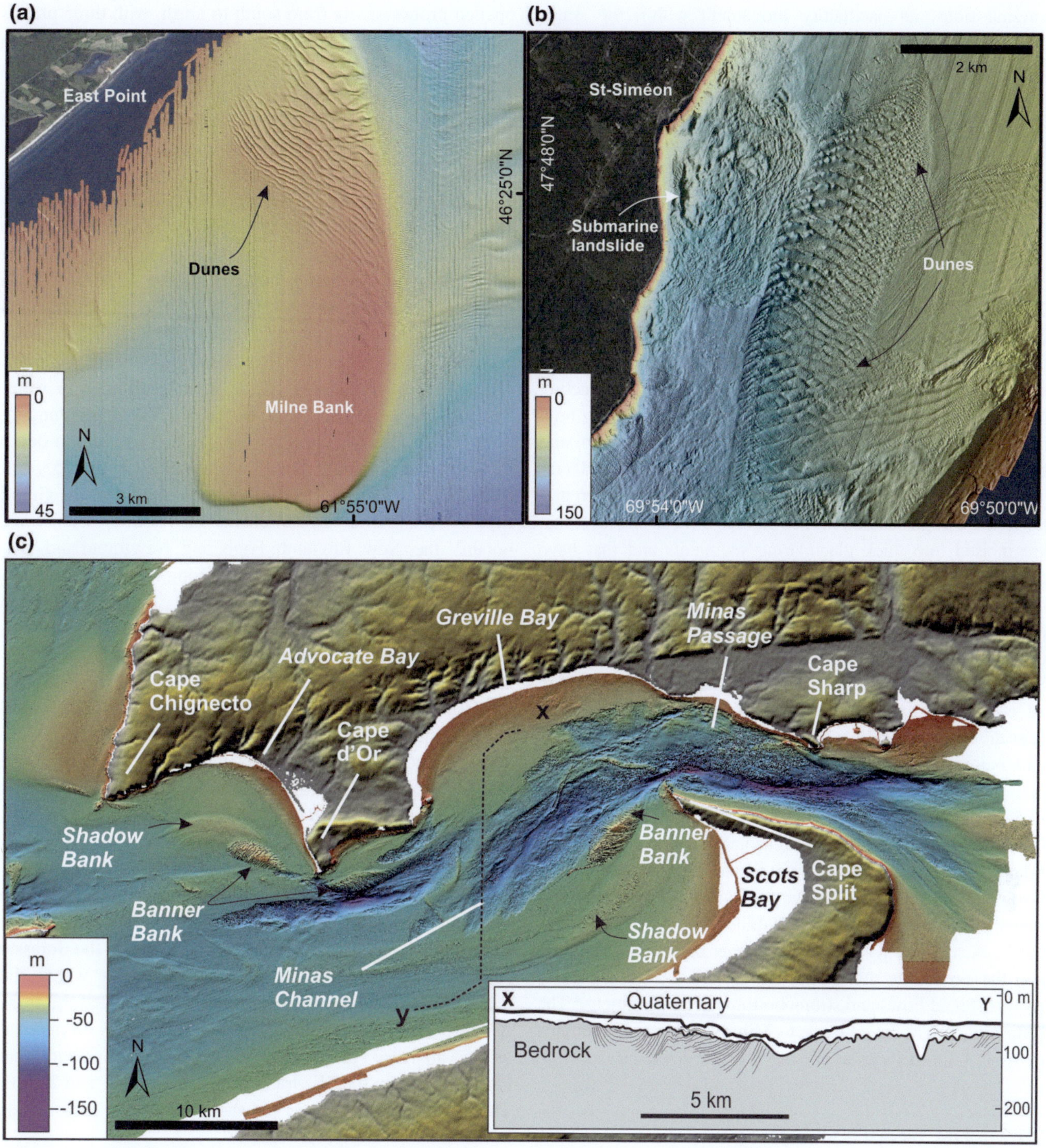

Fig. 20.6 **a** Sun-illuminated bathymetric image of Milne Bank, NE Prince Edward Island, with submarine dunes on top. **b** Sun-illuminated bathymetric image of the St. Lawrence Estuary, near St. Siméon, illustrating submarine dunes migrating with tides and internal tides. Submarine landslides are also present on the steep side of the estuary. **c** Sun-illuminated bathymetric image of Minas Passage, Bay of Fundy, illustrating tidal scours (Minas Channel) along with banner banks, shadow banks and sediment drifts. This landscape reflects the varying tidal current intensity in the Bay of Fundy. See text for details

the continental shelves of southeastern Canada, resulting in oscillatory bottom currents on the shallow, sandy submarine banks off Nova Scotia (Li et al. 1997), and tidal vortices in the lee of headlands in the southern Gulf of St. Lawrence (Fig. 20.6a) (Shaw et al. 2008). The interaction of tides with variable bottom topography can cause the generation of internal tides, an oscillatory movement within the water column at tidal frequencies. The effect of internal tides is probably most evident at the head of the Laurentian Channel in the St. Lawrence Estuary. At that location, large submarine dune fields 10 m high with a wavelength of 200 m form in areas of the seafloor as shallow as 50 m and as deep as 200 m (Fig. 20.6b) (Bolduc and Duchesne 2009; Normandeau et al. 2015). Where currents are stronger, the seabed is eroded with gravel lag deposits.

One of the best examples of the effect of tides on seafloor morphology is located in the Bay of Fundy (Todd et al. 2014). The Bay Fundy has the largest recorded tides in the world (16.3 m) due to near resonance of the Bay of Fundy–Gulf of Maine system, a relatively recent phenomenon (Shaw et al. 2010). Minas Passage is a narrow channel connecting Minas Basin with the rest of the Bay of Fundy. Strong tidal currents (up to 5 m s^{-1}) enter and exit Minas Passage, and have resulted in a system of troughs and banks (Fig. 20.6c). The principal elements in this tidally dominated system are: (1) Scour troughs extending 170 m below mean sea level that are incised into thick glacimarine sediments and that have exhumed bedrock over wide areas; (2) Sets of sand banner banks, which are banks of sand with one end nearly connected to headlands, off Cape d'Or and Cape Chignecto; (3) An atypical set of banner banks (Li et al. 2014) at Cape Split, consisting of the Scots Bay submarine dune field and its counterpart, a large gravel bank trapped in the Minas Passage scour trough; (4) Low-relief banks with sand ribbons and barchan-like dunes (crescentic dunes) alongside some banner banks, and termed 'shadow banks'; (5) A large (0.8 km^3) sediment drift without large bedforms at Cape Chignecto. The location of troughs and banks correlates with tidal-current patterns: troughs are located in regions of strong bi-directional currents. Banner banks have bilateral asymmetry indicative of their location under tidal gyres. 'Shadow' banks exist adjacent to the larger banner banks off Cape d'Or and the Scots Bay dune field and correspond with areas of unidirectional flows and strong bottom stress asymmetry. The sediment drift at the entrance to Minas Channel is in an area of weak bottom stress at all stages of the tides.

20.3 The Deep-Water Margin

The North Atlantic deep-water continental margin comprises a steep upper continental slope (2–4°) and a lower gradient lower continental slope on which deep-sea fans have accumulated, passing seaward onto abyssal plains (<0.1°). In many cases the continental slope has an overall graded profile, resembling the long profile of a river (Mosher et al. 2017) (Fig. 20.2). This morphology is strongly influenced by the Mesozoic tectonic history of the continental margin. The Atlantic Ocean in the region of southeastern Canada opened progressively from south to north, with three phases of rifting, to hyperextension, to seafloor spreading. The process began southwest of the Grand Banks during the Triassic. The second phase began in the late Jurassic between the Grand Banks and Iberia, and the third phase between northern Newfoundland and Ireland in the early Cretaceous (1, 2 and 3 in Fig. 20.2). Thick salt was deposited in the evolving Triassic rift and subsequent salt expulsion and doming has influenced seafloor morphology of the margin. Salt domes form hills on the Scotian Slope, directing sediment pathways that influenced canyon formation. As a result of thermal subsidence, the depth of oceanic crust and hence the water depth of the ocean basins increases from north to south along the southeastern Canadian margin. Many first-order features of the deep-water margin result from this complex tectonic evolution. The high-standing horsts of continental crust at Flemish Cap (minimum depth 123 m) and Orphan Knoll (minimum depth 1533 m) (Fig. 20.1) formed during the Mesozoic rifting of the Grand Banks from Iberia. Linear chains of volcanic seamounts, notably the New England, Fogo and Newfoundland seamounts, formed in the Cretaceous along major tectonic lineaments (Fig. 20.2a). Some of the larger Fogo and Newfoundland seamounts were emergent and coastal erosion and subsidence resulted in guyots, flat-topped seamounts (Pe-Piper et al. 2007). Enhanced melt production at ~130 Ma produced the J-anomaly Ridge (Fig. 20.2a), part of the former mid-ocean ridge that now rises more than 1 km above the ocean floor. The SE Newfoundland Ridge resulted from enhanced volcanism along the Newfoundland–Azores–Gibraltar transform and is now largely blanketed by Cenozoic contourite drift sediments (Fig. 20.2b).

The gross morphology of the continental slope is strongly influenced by the underlying geology and the position of the shelf edge since the Jurassic. The modern shelf edge on the SW Scotian Shelf corresponds to the edge of the Jurassic carbonate bank (Mosher et al. 2017), except where the ~52 Ma Montagnais impact eroded and steepened the slope (Deptuck and Campbell 2012) (Fig. 20.2a). Net-erosional slopes are also present off Banquereau and Laurentian Channel, and on the SE Grand Banks (Figs. 20.1 and 20.2b). The central Scotian Slope and the slope from Flemish Pass northwards are net progradational.

20.3.1 From Continental Shelf to Abyssal Plain: Submarine Canyons and Fans

20.3.1.1 Distribution of Submarine Canyons and Channels

Submarine canyons are steep-sided erosional valleys that deeply (>100 m) incise the continental slope (Shepard 1963). Smaller valleys are referred to as gullies, and large valleys on submarine fans are termed fan valleys. Submarine canyons generally transition downslope to submarine channels, which are part of depositional assemblages that include levees and lobes (Talling 2014). Submarine canyons on the southeastern Canadian margin are most abundant on the eastern part of the Scotian Slope and around the southern Grand Banks (Figs. 20.2 and 20.7). Elsewhere, smaller slope gullies are common.

Submarine canyons are present on steep slopes, commonly of more than 3–4° (Mosher et al. 2004), and are of two main types: (1) slope-confined (or blind) canyons and (2) shelf-indenting canyons (Fig. 20.7). Slope-confined canyons have heads mostly at 500–600 m at the downslope limit of glacial till on the upper slope (Fig. 20.7a). Along the Scotian Slope, only a few canyons are shelf-indenting, among which The Gully (Fig. 20.7b) and Logan (Fig. 20.7a) canyons are the more prominent. Many slope-confined canyons have a well developed dendritic pattern and merge with larger shelf-indenting canyons on the middle slope (Fig. 20.7a; Jenner et al. 2007), before transitioning to submarine channels on the lower slope (Figs. 20.2 and 20.8a). Other slope-confined canyons are strikingly linear (Mosher et al. 2004) (Figs. 20.7c and 20.8b). Canyon walls on the upper parts of canyons commonly have dendritic gullies resulting in a badland-like morphology (Fig. 20.7a).

The types of submarine canyons seaward of transverse troughs on the shelf are variable. Off NE Channel there are three shelf-incising fan valleys (Fig. 20.2) (Piper et al. 2016). Off Laurentian Channel, seven upper slope canyons merge into two main fan valleys (Fig. 20.7d) (Mosher and Piper 2007). In contrast with these net-erosional shelf breaks, off the NE Newfoundland Shelf, seaward of Trinity and Notre Dame troughs, net progradation has occurred and the slope is crossed by numerous gullies.

Variations in submarine canyon and gully morphology around the Grand Banks appears strongly controlled by the glacial and sea level history, together with low sedimentation rates beneath the Labrador Current. When ice sheets reached the shelf break, they favoured the formation of gullies spaced at 0.5–1 km on slopes greater than 1° (Fig. 20.7e). Conversely, in areas where ice sheets did not reach the shelf break, gullies are absent, such as north of Sackville Spur (Fig. 20.7e). On steep slopes off southern and eastern Flemish Cap, submarine canyons have a linear planform and lack dendritic gully tributaries (Fig. 20.7f). Around the Tail of the Banks (Fig. 20.1), old relatively straight canyons have locally been filled by younger proglacial sediment, which in turn is cut by sinuous shelf-indenting canyons that widen downslope and head near the 100 m isobath, interpreted as corresponding to the LGM lowered sea level (Fig. 20.7c). In all these canyons, badland-type dendritic gullies are either absent or blanketed by younger sediment.

The main processes leading to the erosion and deposition of sediment in submarine canyons are turbidity currents. Turbidity currents are sediment-laden flows initiated on steep slopes that result from slope failures, from direct hyperpycnal flow of rivers or glacial meltwater, from fall-out of sediment from freshwater sediment-laden plumes, or from resuspension of seafloor sediments. They were frequent during glacial maxima and deglaciation, when large volumes of sediment were deposited on the shelf or directly transferred to the continental slope. Since the early Holocene, turbidity currents have been infrequent on the slope (Piper 2005). However, most of the shelf-incising canyons have an axial channel, i.e. a small channel within the larger canyon (Fig. 20.7a) resulting from post-glacial turbidity currents. The Gully is a setting in which sediments are currently being transported towards the canyons, as evidenced by large submarine dunes entering the canyon head (Fig. 20.7b). There are also four side gullies that feed the main canyon. However, there is no evidence that turbidity currents are presently active within The Gully. Slumping on the sidewalls of The Gully probably occurs episodically, as small slope failures appear fresh on bathymetric imagery, but it is unlikely that they generate turbidity currents that reach the lower slope.

The highly incised continental slope becomes depositional downslope and the canyons transform into channels with levees. Most canyons merge downslope with one or more other canyons, leading to a reduced number of pathways for turbidity currents (Fig. 20.2). For example, there are more than 100 canyons and smaller gullies incising the uppermost continental slope of the Scotian margin, merging into less than 10 channels on the lower slope. The length of the channels decreases considerably from east to west. Whereas The Gully extends 100s of km offshore, channels to the west terminate near the top of the lower slope (Normandeau et al. 2019a) (Fig. 20.2). Where turbidity currents become unconfined downslope, at a depth of 4000 m, there is also a variety of sediment waves that form on open slopes (Fig. 20.8a). These sediment waves are observed in net-erosional and net-depositional environments, and can be influenced by contour currents, as along-slope currents deflect turbidity currents and create mixed turbidite-contourite systems (Campbell and Mosher 2016; Normandeau et al. 2019a).

20.3.1.2 Submarine Fans, Lower Continental Slope and Abyssal Plains

Deep-sea fans are classically divided into an upper fan, with one or more channels and levees, a mid-fan with complex topography around the channel termination zone and its sandy depositional lobes, and a lower fan with more muted topography. Northeast (NE) Fan, offshore NE Channel (Fig. 20.3a) has abundant landslide deposits on the upper fan, but they are sparse on Laurentian Fan. Smaller fan

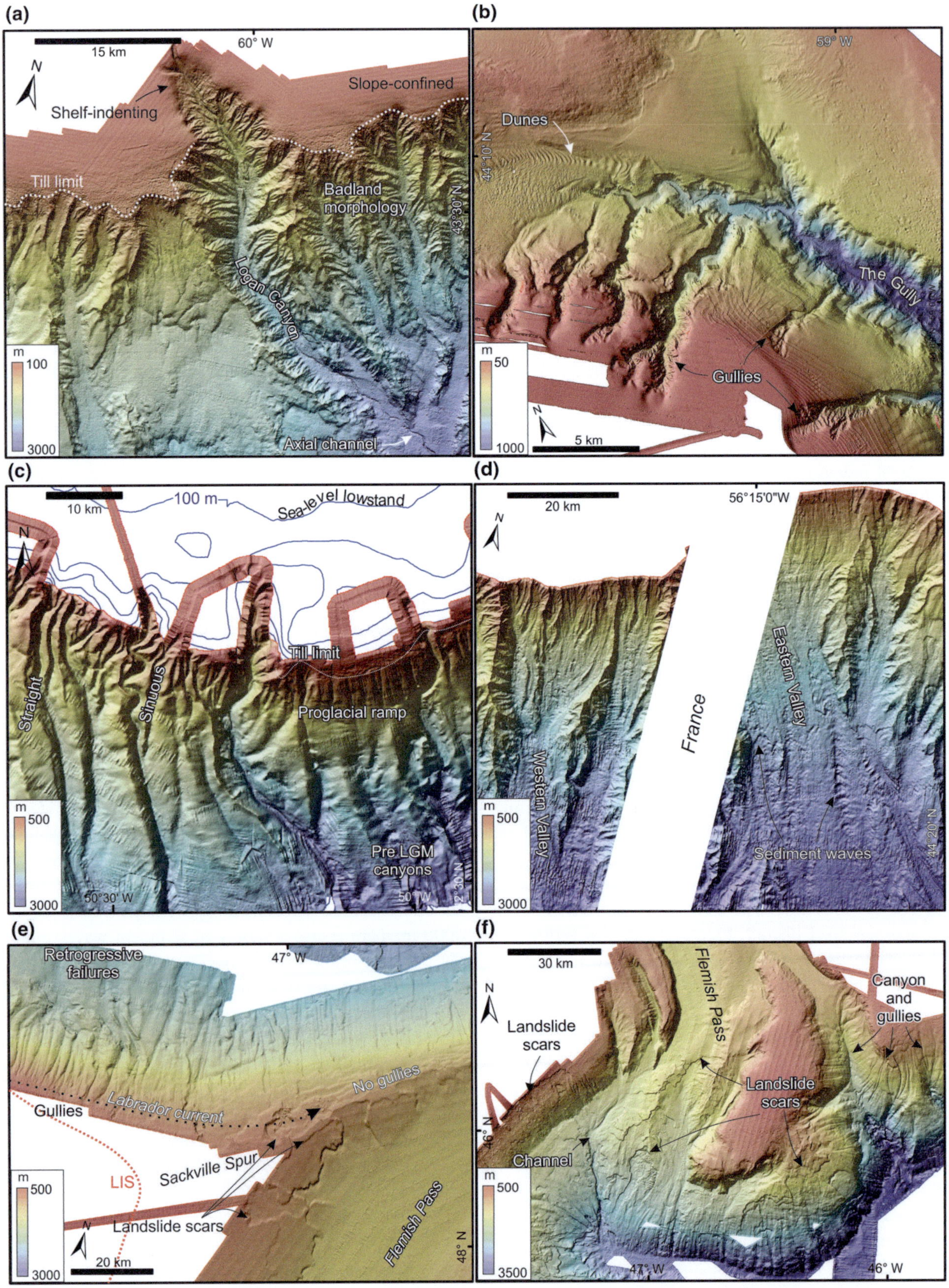

Fig. 20.7 Sun-illuminated bathymetric image of **a** Logan Canyon illustrating shelf-indenting and slope-confined canyons along with a typical badland (gullied) morphology. Note that gullies originate from the limit of the till. **b** Head of The Gully Canyon illustrating side gullies feeding into the main canyon and submarine dunes entering the head. **c** Tail of the Banks Pre LGM canyons being filled by a proglacial ramp along with LGM canyons incising close to the 100 m sea level lowstand. **d** The Laurentian Fan valleys with sediment waves. **e** Sackville Spur gullies and submarine landslides, controlled by the maximum position of the LIS. **f** Southern Flemish Pass illustrating numerous gullies, channels and canyons along with a highly disturbed seafloor with multiple retrogressive landslide scars

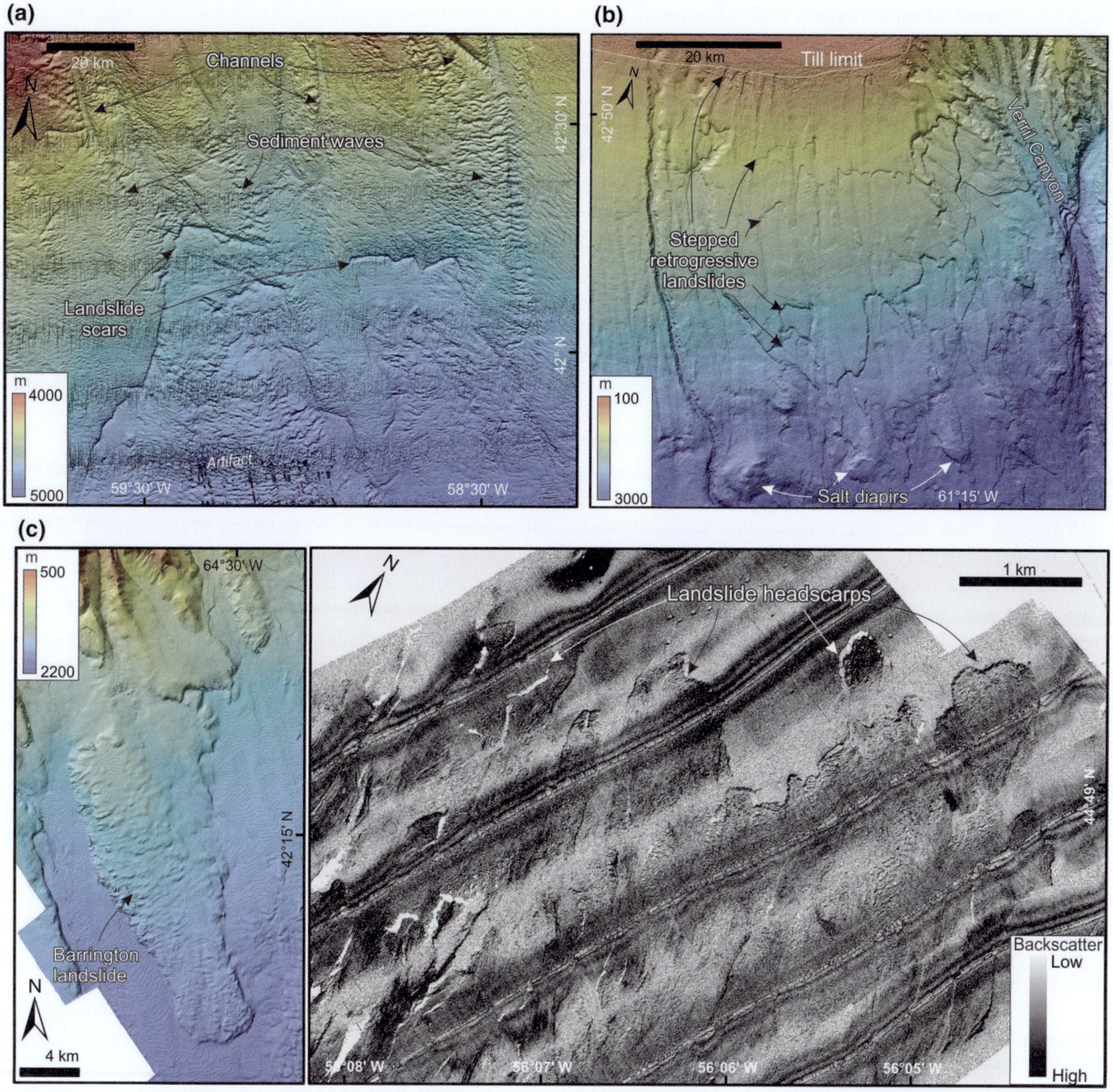

Fig. 20.8 **a** Deep-water submarine landslide scars with overlying sediment waves formed by unchannelized turbidity currents on submarine fans. **b** Sun-illuminated bathymetric image of staircase retrogressive landslides west of Verrill Canyon. **c** Sun-illuminated bathymetric image of Barrington Slide showing frontal thrusting. **d** Sidescan image of the heads of the 1929 submarine landslides that transformed downslope into a turbidity current

deposits are present off some of the larger canyons on the Scotian Slope, although much of the lower slope is constructed of thick landslide units (Fig. 20.8). The Grand Banks was not a major source of sediment during the last glacial cycle: canyons and their continuation in channels have built only small deposits east of Laurentian Fan (Fig. 20.2). Near Orphan Knoll, the Trinity trough-mouth fan is constructed principally by glacigenic debris flows, although some buried erosional channels are present (Fig. 20.2b) (Tripsanas and Piper 2008). Channels crossing Orphan Basin have built a small fan east of Orphan Knoll (Fig. 20.2b).

Lower Laurentian Fan passes imperceptibly southward to the Sohm Abyssal Plain, which is ponded south of 36 °N at about 5400 m water depth (Fig. 20.1). The eastern and western margins of the Sohm Abyssal Plain slope away from the lower Scotian Slope and the Northwest Atlantic Mid-Ocean Channel (NAMOC) respectively (Fig. 20.2).

20.3.1.3 Small Shelf Canyons and Channels: Analogues to the Deep-Water

Although turbidity currents formed large canyons and channels in the deep-water margin, canyons and channels also formed on the continental shelf of southeastern Canada. The steep slopes of the Laurentian Channel favoured the presence of small inner shelf submarine canyons and channels that are the size of gullies on the continental slope, particularly in the St. Lawrence Estuary (Normandeau et al. 2015) and offshore western Newfoundland (Shaw and Potter 2015; Shaw et al. 2019). The St. Lawrence Estuary can be used as a natural laboratory to examine the processes leading to sediment transport within submarine canyons and fans. Normandeau et al. (2015) identified small submarine canyons and channels that are either sediment-starved or that are supplied with sediment by rivers, longshore drift, or former glaciers. This assemblage of small canyons and channels is similar to those found on continental margins, although much smaller. Their behaviour revealed that, in high-latitude regions, submarine canyons connected to the coast, incising the shelf close to a source of sediment from longshore drift, remained active throughout the Holocene. Conversely, channels connected to rivers or glaciers ceased their activity when glaciers retreated inland, as sediment supply ceased (Normandeau et al. 2017; Dietrich et al. 2017b). A notable aspect of the St. Lawrence Estuary is the presence of active sediment-starved canyons (Normandeau et al. 2014, in press). Sediments likely are being suspended within the canyons and are flowing downslope. The triggers of such flows are currently unknown, and it is uncertain whether this type of canyon has a counterpart in deeper water continental margins.

20.3.2 Submarine Slope Instabilities: Landslides

Submarine landslides are among the volumetrically most important form of submarine erosion and transport on Earth (Talling et al. 2014). They are often several orders of magnitude larger than landslides observed on land and are common geomorphological features on the seabed of southeastern Canada. Formation of submarine landslides generally requires steep slopes and a thick sediment accumulation, which makes them far more common on the continental slope than on the continental shelf. More than half of the Scotian margin appears affected by landslides that are sufficiently young to have a seabed expression (Mosher et al. 2004) (Figs. 20.2 and 20.8). The largest landslides in southeastern Canada are located on open slopes, and in deep-water (>500 m) (Fig. 20.2b). For example, a Holocene submarine landslide observed on the Laurentian Fan offshore of Nova Scotia (Fig. 20.2) covers more than 14,000 km^2, and originated at a depth of 4000 m (Normandeau et al. 2018, 2019b). The buried Montagnais landslide, produced by the Montagnais meteorite impact 52 million years ago, is 93,000 km^2 in area, nearly twice the size of mainland Nova Scotia (Fig. 20.2a) (Deptuck and Campbell 2012).

The geomorphological features of submarine landslides are both erosional and depositional. The most prominent erosional features are head and side scarps from where landslides have originated (Fig. 20.8). On parts of the continental slope, such as the Scotian Slope and Orphan Basin, there has been net erosion by landsliding, producing a stepped bathymetry (Figs. 20.7f and 20.8b). The upslope limits of landslides commonly terminate at the downslope limit of glacial till, marked by erosion scarps (Fig. 20.8b). On the middle slope, seaward of the canyon morphology, arcuate headscarps marking the initiation of slumps are common (Fig. 20.8a, b).

Two end-member types of submarine landslide deposit are distinguished based on their geomorphology. Most submarine landslides involve failure of the upper 5–30 m of the seabed, where sediments are weakest prior to compaction. Failure most commonly occurs through rotational slip (slumping) that may be retrogressive. A few submarine landslides involve failure of more deeply buried (>30 m) sediment that has sufficient strength to be transported as large blocks. Such complex blocky landslides may form positive features at the seabed, in some cases with transverse ridges (Fig. 20.8c) (Mosher and Campbell 2011) that resulted from frontal thrusting. Large blocky slides are also known from steep slopes with possible neotectonic fault motion, off Orphan Knoll, southern Flemish Cap, and the SE Grand Banks slope (Fig. 20.2). Some large slides involve failure along discrete weak layers in stratified levees or sediment drifts that may be buried under hundreds of metres of sediment (Fig. 20.9a) (Piper and Ingram 2003; Piper and Campbell 2005).

Small submarine landslides are more frequent on submarine canyon walls because the sediment may have been rapidly deposited and the gradient is steepest in these locations (Fig. 20.7). Additionally, erosion and undercutting of the canyon walls by turbidity currents can lead to wall collapse. The larger open slope landslides are often triggered by earthquakes, sometimes preconditioned by shallow gas that may weaken sediment, which favours failure. The most recent submarine landslide to have affected the seafloor of the continental slope was triggered in 1929 as a result of an M7.2 earthquake on St. Pierre Slope. This large earthquake on the upper continental slope generated multiple landslides (Fig. 20.8d) that transformed downslope into a turbidity current (Piper et al. 1999; Schulten et al. 2019). As a result, 12 deep-sea telegraph cables were broken between the moment the earthquake was triggered and 13 h afterwards. These cable breaks allowed Kuenen (1952) to estimate the speed of the turbidity current at 60–100 km/h, carrying 100–

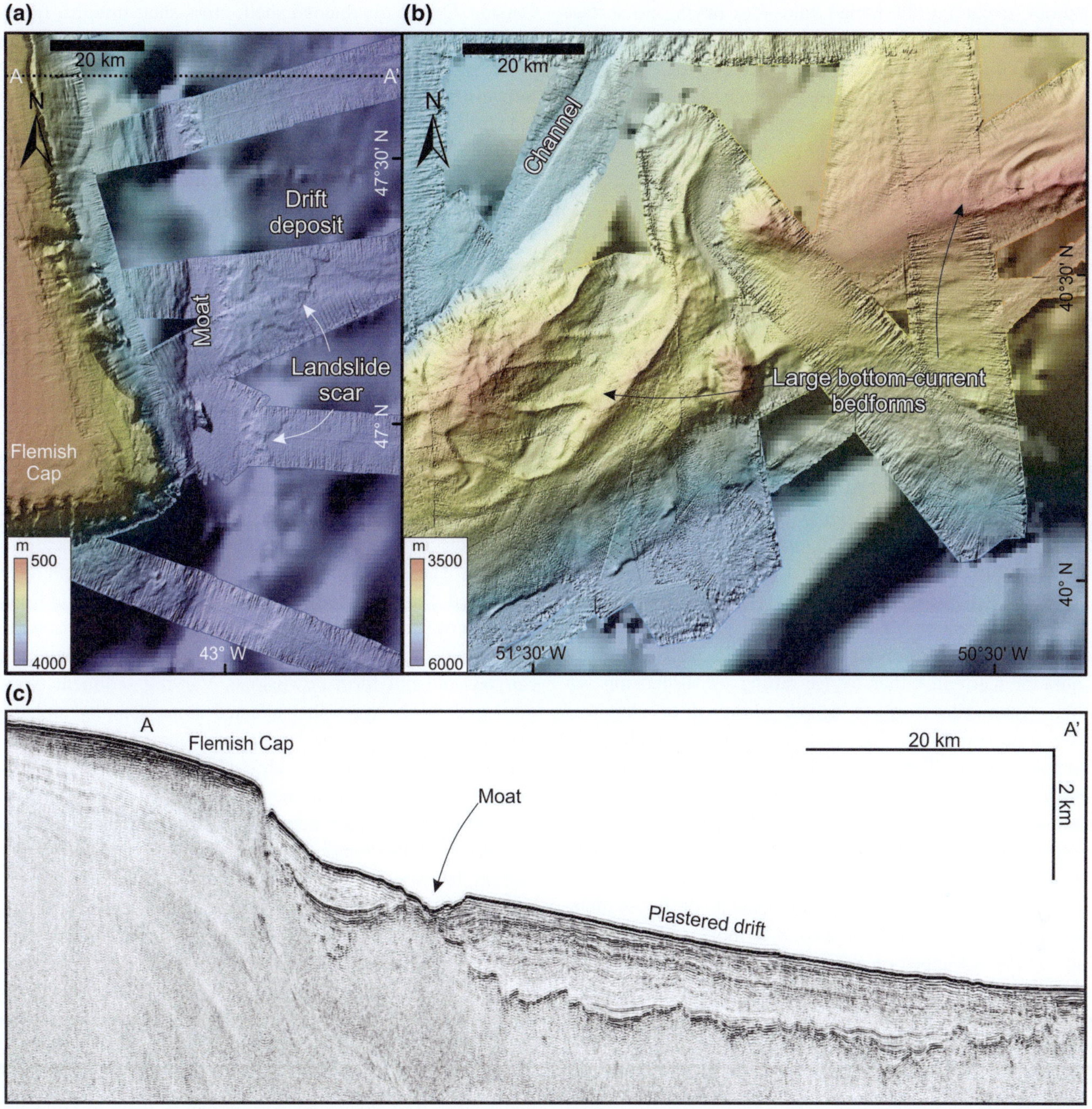

Fig. 20.9 **a** Sun-illuminated bathymetric image of eastern Flemish Cap illustrating an erosive moat at the foot of slope and its associated drift deposit. Landslide scars are present both on the external (low angle slope) and internal (steep angle slope) of the deposit. **b** Sun-illuminated bathymetric image of large bottom-current bedforms at >3500 m water depth on the J-Anomaly Ridge. **c** Seismic profile of the moat and plastered contourite drift illustrated in (**a**)

150 km^3 of sediment. This landslide also generated a 3–7 m high tsunami that impacted the coasts of Newfoundland and Nova Scotia and killed 28 people (Ruffman and Hann 2001).

Many small submarine landslides are observed on the continental shelf, although they are smaller and less common than on the continental slope (Figs. 20.4e and 20.6b). Past Quaternary glaciations have formed large cross-shelf troughs and fjords with steep-sided slopes. During deglaciation, large volumes of proglacial and post-glacial sediments, with distinct lithologies, were deposited on the shelf. The lithology varies between clay and sand, forming weak layers where sliding of material on the steep slopes is facilitated. Landslides are common in the St. Lawrence Estuary. In this region, 96 submarine landslides were identified on the seafloor, excluding the buried landslides having no surface expression (Fig. 20.6b) (Pinet et al. 2015). Landslides are likely to be more present and frequent in this region because it is the most active seismic zone in southeastern Canada and has accumulated over 400 m of sediment during deglaciation in some areas (Duchesne et al. 2010).

20.3.3 Deep-Water Bottom Currents and Contourite Drifts

Bottom currents are important in shaping the lower continental slope of the southeastern Canadian margin (Fig. 20.9a, b). Their action on the seafloor produces contourite drifts, which are sediment deposited or reworked by bottom currents (Rebesco and Stow 2001). Bottom currents are the main process by which sediments are transported and deposited around the Grand Banks, forming different types of drifts (no distinction of types in Fig. 20.2). In order to form, they require sediment supply which, for most of the southeastern Canadian outer margin, was provided by glacial meltwater to the continental margin. However, the southern Grand Banks, Flemish Cap and Orphan Knoll are relatively sediment-starved, receiving sediment principally from the Labrador Current, either by ice-rafting or suspension settling. Contourites form detached, plastered or elongated drifts around Flemish Cap, with a classical example being Sackville Spur (Fig. 20.7e). The low-gradient southeast Grand Banks slope has a complex mix of episodic landslides and drift-controlled deposition, with landslide features dominating the geomorphology (Figs. 20.7f and 20.9a) (Rashid et al. 2017).

Contour currents can also form channels on the seafloor, called moats. A prominent erosional moat (drift moat in Fig. 20.2b) developed off SE Flemish Cap, at the foot of the very steep slope (Fig. 20.9a, c). Where the Western Boundary Undercurrent crosses the SE Newfoundland Ridge, near the wreck of the *Titanic*, Pliocene sediments are exposed at the seabed and barchan-like dunes and gravel ribbons have developed on the erosion surface (Cochonat et al. 1989). Other locations on the J-anomaly Ridge exhibit large, km-long submarine dunes in 3500 m water depth (Fig. 20.9b).

20.4 Conclusions

The tectonic history of the southeastern Canadian margin is the first-order control on the overall geomorphology of the seabed, forming the shelf and deep ocean floor. Subsequent sedimentation modified the original tectonic morphology and Quaternary glaciations played a critical role in producing the modern landscape of the seafloor of southeastern Canada. On the continental shelf, cross-shelf troughs, fjords and tunnel valleys were sculpted below the Laurentide Ice Sheet whereas recessional moraines were deposited by retreating ice margins. The eustatic and relative sea level rise that followed deglaciation remobilized coarse sediment on most of the Scotian and Grand Banks shelves, leading to the formation of sandy bedforms and scours. Closer to Quebec, glaci-isostatic rebound outpaced glaci-eustasy throughout the Holocene, leading to a relative sea level fall, exposing previously underwater environments.

In contrast to the continental shelf, the landscape of the deep-water margin was not directly modified by the ice sheet, but was significantly influenced by it. At the last glacial maximum, large quantities of glacigenic sediments were distributed along the continental slope; failure of these sediments produced landslides and turbidity currents, suspending sediment that was subsequently transported by along-slope currents. The turbidity currents eroded submarine canyons and deposited submarine fans, which are among the largest sediment accumulations in southeastern Canada. The maximum position of the ice margin also controlled the distribution of submarine gullies and the maturity of canyons. Where the ice margin did not reach the shelf break, sediment supply was limited and gullies and canyons did not form downslope. Varying sediment types produced weak layers that affected slope stability and the occurrence of submarine landslides. These deep-water submarine landslides have no equivalent on land in terms of size, being orders of magnitude larger.

A disconnect exists between the timing of sedimentary processes on the continental shelf and on the deep-water margin. Most sediment supply and gravitational processes in deep-water occurred during glaciations, when the southeastern extent of the Laurentide Ice Sheet reached the shelf break. When ice retreated, the deep-water region became much less active, as gravitational processes drastically diminished. Concomitantly, tidal currents and waves on the shallow shelf led to sediment mobility. Although the continental shelf and the deep-water margin have contrasting landscapes, they are genetically linked to tectonics and the pattern of retreat of the Laurentide Ice Sheet.

References

Ashley GM (1990) Classification of large-scale subaqueous bedforms: a new look at an old problem-SEPM bedforms and bedding structures. J Sediment Res 60:160–172

Bolduc A, Duchesne MJ (2009) Découverte de mégadunes dans l'estuaire moyen du fleuve Saint-Laurent, Québec, Canada. Rev des Sci de l'Eau 22:125–134

Boyd R, Scott DB, Douma M (1988) Glacial tunnel valleys and Quaternary history of the outer Scotian shelf. Nature 333:61–64

Campbell DC, Mosher DC (2016) Geophysical evidence for widespread Cenozoic bottom current activity from the continental margin of Nova Scotia, Canada. Mar Geol 378:237–260

Cochonat P, Ollier G, Michel JL (1989) Evidence for slope instability and current-induced sediment transport, the RMS Titanic wreck search area, Newfoundland rise. Geo-marine Lett 9:145–152

De Geer G (1889) Ändmoräner i trakten mellan Spånga och Sundbyberg. Geol Föreningens i Stock Förhandlingar 11:395–397

Deptuck M, Campbell DC (2012) Widespread erosion and mass failure from the 51 Ma Montagnais marine bolide impact off southwestern Nova Scotia, Canada. Can J Earth Sci 49:1567–1594

Dietrich P, Ghienne JF, Schuster M et al (2017a) From outwash to coastal systems in the Portneuf-Forestville deltaic complex (Québec North Shore): anatomy of a forced regressive deglacial sequence. Sedimentology 64:1044–1078

Dietrich P, Ghienne J-F, Normandeau A, Lajeunesse P (2017b) Reconstructing ice-margin retreat using delta morphostratigraphy. Sci Rep 7:16936. https://doi.org/10.1038/s41598-017-16763-x

Duchesne MJ, Pinet N, Bédard K et al (2010) Role of the bedrock topography in the Quaternary filling of a giant estuarine basin: The Lower St. Lawrence Estuary, Eastern Canada. Basin Res 22:933–951

Grant DR (1989) Quaternary geology of the Atlantic Appalachian region of Canada. Quaternary geology of Canada and Greenland 1:391–440

Harris PT, Macmillan-Lawler M, Rupp J, Baker EK (2014) Geomorphology of the oceans. Mar Geol 352:4–24

Jenner KA, Piper DJW, Campbell DC, Mosher DC (2007) Lithofacies and origin of late Quaternary mass transport deposits in submarine canyons, central Scotian Slope, Canada. Sedimentology 54:19–38

King LH (1969) Submarine end moraines and associated deposits on the Scotian Shelf. Geol Soc Am Bull 80:83–96

King LH, Fader GB (1986) Wisconsinan glaciation of the Atlantic continental shelf of southeast Canada. Geological Survey of Canada, Paper 86 1A

Kuenen PH (1952) Estimated size of the Grand Banks turbidity current. Am J Sci 250:874–884

Li MZ, Amos CL, Heffler DE (1997) Boundary layer dynamics and sediment transport under storm and non-storm conditions on the Scotian Shelf. Mar Geol 141:157–181

Li MZ, King EL (2007) Multibeam bathymetric investigations of the morphology of sand ridges and associated bedforms and their relation to storm processes, Sable Island Bank, Scotian Shelf. Mar Geol 243:200–228

Li MZ, Shaw J, Todd BJ et al (2014) Sediment transport and development of banner banks and sandwaves in an extreme tidal system: Upper Bay of Fundy, Canada. Cont Shelf Res 83:86–107

Lindén M, Möller P (2005) Marginal formation of De Geer moraines and their implications to the dynamics of grounding-line recession. J Quat Sci 20:113–133

Mosher DC, Campbell DC (2011) The Barrington submarine mass-transport deposit, western Scotian slope, Canada. Mass-Transport Depos Deep Settings, SEPM Spec Publ No 96:151–159

Mosher DC, Piper DJW (2007) Analysis of multibeam seafloor imagery of the Laurentian Fan and the 1929 Grand Banks landslide area. In: Lykousis V, Sakellariou D, Locat J (eds) Submarine mass movements and their consequences. Advances in natural and technological hazards research, vol 27, Springer, Dordrecht, pp 77–88

Mosher DC, Campbell DC, Gardner JV et al (2017) The role of deep-water sedimentary processes in shaping a continental margin: the northwest Atlantic. Mar Geol 393:245–259

Mosher DC, Piper DJW, Calvin Campbell D, Jenner KA (2004) Near-surface geology and sediment-failure geohazards of the central Scotian Slope. Am Assoc Pet Geol Bull 88:703–723

Mucci A, Bernier G, Guignard C (2015) Mercury remobilization in Saguenay Fjord (Quebec, Canada) sediments: insights following a mass-flow event and its capping efficiency. Appl Geochem 54:13–26

Normandeau A, Lajeunesse P, St. Onge G (2013) Shallow-water longshore drift-fed submarine fan deposition (Moisie River Delta, Eastern Canada). Geo-Marine Lett 33:391–403

Normandeau A, Lajeunesse P, St. Onge G (2015) Submarine canyons and channels in the Lower St. Lawrence Estuary (Eastern Canada): morphology, classification and recent sediment dynamics. Geomorphology 241:1–18

Normandeau A, Lajeunesse P, Poiré AG, Francus P (2016) Morphological expression of bedforms formed by supercritical sediment density flows on four fjord-lake deltas of the south-eastern Canadian Shield (Eastern Canada). Sedimentology 63:2106–2129

Normandeau A, Campbell C, Piper DJW, Jenner KA (2018) New evidence for a major late-Quaternary submarine landslide on the external western levee of Laurentian Fan. Geol Soc London Spec Publ 477

Normandeau A, Lajeunesse P, St. Onge G, Bourgault D, St. Onge-Drouin S, Senneville S, Belanger S (2014) Morphodynamics in sediment-starved inner-shelf submarine canyons (Lower St. Lawrence Estuary, Eastern Canada). Mar Geol 357:243–255

Normandeau A, Dietrich P, Lajeunesse P, St. Onge G, Ghienne J-F, Duchesne MJ, Francus P (2017) Timing and controls on the delivery of coarse sediment to deltas and submarine fans on a formerly glaciated coast and shelf. Geol Soc Am Bull 129:1424–1441

Normandeau A, Calvin Campbell D, Matthieu JBC (2019a) The influence of turbidity currents and contour currents on the distribution of deep-water sediment waves offshore eastern Canada. Sedimentology

Normandeau A et al (2019b) Are submarine landslides an underestimated hazard on the western North Atlantic passive margin?. Geol 47(9):848–852.

Normandeau A, Bourgault D, Neumeier U, Lajeunesse P, St-Onge G, Gostiaux L, Chavanne C (in press) Storm-induced turbidity currents on a sediment-starved shelf: insight from direct monitoring and repeat seabed mapping of upslope migrating bedforms. Sedimentology

Pe-Piper G, Piper DJW, Jansa LF, De Jonge A (2007) Early Cretaceous opening of the North Atlantic Ocean: implications of the petrology and tectonic setting of the Fogo Seamounts off the SW Grand Banks, Newfoundland. Geol Soc Am Bull 119:712–724

Peltier WR (1998) Postglacial variations in the level of the sea: implications for climate dynamics and solid-earth geophysics. Rev Geophys 36:603–689

Pinet N, Brake V, Campbell C, Duchesne MJ (2015) Geomorphological characteristics and variability of Holocene mass-transport complexes, St. Lawrence River Estuary, Canada. Geomorphology 228:286–302

Piper DJW, Cochonat P, Morrison ML (1999) The sequence of events around the epicentre of the 1929 Grand Banks earthquake: initiation of debris flows and turbidity current inferred from sidescan sonar. Sedimentology 46(1):79–97

Piper DJW (2005) Late Cenozoic evolution of the continental margin of eastern Canada. Nor J Geol 85:305–318

Piper DJW, Campbell DC (2005) Quaternary geology of Flemish Pass and its application to geohazard evaluation for hydrocarbon development. Geol Assoc Can Spec Pap 43:29–43

Piper DJW, Ingram S (2003) Major Quaternary sediment failures on the east Scotian Rise, eastern Canada. Curr Res Geol Surv Can 2003-D1:1–7

Piper DJW, Shaw J, Skene KI (2007) Stratigraphic and sedimentological evidence for late Wisconsinan sub-glacial outburst floods to Laurentian Fan. Palaeogeogr Palaeoclimatol Palaeoecol 246:101–119

Piper DJW, Campbell DC, Mosher DC (2016) Mid-latitude complex trough-mouth fans, Laurentian and Northeast fans, eastern Canada. In: Dowdeswell JA, Canals M, Jakobsson M, et al (eds) Atlas of submarine glacial landforms: modern, Quaternary and ancient. Geol Soc London, pp 363–364

Poiré AG, Lajeunesse P, Normandeau A, Francus P, St-Onge G, Nzekwe OP (2018) Late-Quaternary glacial to postglacial sedimentation in three adjacent fjord-lakes of the Québec North Shore (eastern Canadian Shield). Quat Sci Rev 186:91–110

Rashid H, Mackillop K, Sherwin J, Piper DJW, Marche B, Vermooten M (2017) Slope instability on a shallow contourite-dominated continental margin, southeastern Grand Banks, eastern Canada. Mar Geol 393:203–215

Rebesco M, Stow D (2001) Seismic expression of contourites and related deposits: a preface. Mar Geophys Res 22:303–308

Ruffman A, Hann V (2001) The Newfoundland Tsunami of November 18, 1929: an examination of the twenty-eight deaths of the "South Coast Disaster". Newfoundl Labrador Stud 21:1719–1726

Schulten I et al (2019) Surficial sediment failures due to the 1929 Grand Banks Earthquake, St Pierre Slope. Geological Society, London. Spec Publ 477(1):583–596

Shaw J (2005) Geomorphic evidence of postglacial terrestrial environments on Atlantic Canadian continental shelves. Géog Phys Quat 59:141–154

Shaw J, Forbes DL (1995) The postglacial relative sea-level lowstand in Newfoundland. Can J Earth Sci 32:1308–1330

Shaw J, Potter DP (2015) Surficial geology, coastal waters, Island of Newfoundland, Newfoundland and Labrador. Geol Surv Canada Open File Bulletin 6:130

Shaw J, Potter DP (2016) Anthropogenic modification of a fjord: Bay of Islands, Newfoundland. Geol Soc London, Mem 46:121–122

Shaw J, Taylor R, Forbes D (1993) Impact of the Holocene transgression on the Atlantic coastline of Nova Scotia. Géogr Phys Quat 47:221–238

Shaw J, Gareau P, Courtney RC (2002) Paleogeography of Atlantic Canada 13-0 kyr. Quat Sci Rev 21:1861–1878

Shaw J, Todd BJ, Li MZ (2014) Geologic insights from multibeam bathymetry and seascape maps of the Bay of Fundy, Canada. Cont Shelf Res 83:53–63

Shaw J, Piper DJW, Fader GBJ, King EL, Todd BJ, Bell T, Batterson MJ, Liverman DJE (2006) A conceptual model of the deglaciation of Atlantic Canada. Quat Sci Rev 25:2059–2081

Shaw J, Duffy G, Taylor RB, Chasse J, Frobel D (2008) Role of a submarine bank in the long-term evolution of the northeast coast of Prince Edward Island, Canada. J Coast Res 245:1249–1259

Shaw J, Amos CL, Greenberg DA, O'Reilly CT, Parrott DR, Patton E (2010) Catastrophic tidal expansion in the Bay of Fundy, Canada Earth Sciences Sector (ESS) Contribution 20090423. Can J Earth Sci 47:1079–1091

Shaw J, Potter DP, WU Y (2019) Geomorphic diversity and complexity of the inner shelf, Canadian Arctic Archipelago, based on LiDAR and multibeam sonar surveys. Can J Earth Sci

Shepard FP (1963) The Earth Beneath the Sea. History 3:480

Stea RR, Piper DJW, Fader GBJ, Boyd R (1998) Wisconsinan glacial and sea-level history of Maritime Canada and the adjacent continental shelf: a correlation of land and sea events. Geol Soc Am Bull 110:821–845

Talling PJ (2014) On the triggers, resulting flow types and frequencies of subaqueous sediment density flows in different settings. Mar Geol 352:155–182

Talling P, Clare M, Urlaub M, Pope E (2014) Large submarine landslides on continental slopes: geohazards, methane release, and climate change. Oceanography 27:32–45

Todd BJ (2016a) Recessional moraines on the southern Scotian Shelf of Atlantic Canada. Geol Soc London, Mem 46:257–258

Todd BJ (2016b) De Geer moraines on German Bank, southern Scotian Shelf of Atlantic Canada. Geol Soc London, Mem 46:259–260

Todd BJ, Shaw J (2012) Laurentide Ice Sheet dynamics in the Bay of Fundy, Canada, revealed through multibeam sonar mapping of glacial landsystems. Quat Sci Rev 58:83–103

Todd BJ, Shaw J, Valentine PC (2016) Submarine glacial landforms on the Bay of Fundy–northern Gulf of Maine continental shelf. Geol Soc London, Mem 46:429–436

Todd BJ, Fader GBJ, Courtney RC, Pickrill RA (1999) Quaternary geology and surficial sediment processes, Browns Bank, Scotian Shelf, based on multibeam bathymetry. Mar Geol 162:165–214

Todd BJ, Valentine PC, Longva O, Shaw J (2007) Glacial landforms on German Bank, Scotian Shelf: evidence for Late Wisconsinan ice-sheet dynamics and implications for the formation of De Geer moraines. Boreas 36:148–169

Todd BJ, Shaw J, Li MZ, Kostylev VE, Wu Y (2014) Distribution of subtidal sedimentary bedforms in a macrotidal setting: the bay of fundy, Atlantic Canada. Cont Shelf Res 83:64–85

Tripsanas EK, Piper DJW (2008) Glaciogenic debris-flow deposits of Orphan Basin, offshore eastern Canada: sedimentological and rheological properties, origin, and relationship to meltwater discharge. J Sediment Res 78:724–744

Alexandre Normandeau holds a Ph.D. from Université Laval and did a 1-yr post-doc at Queen's University studying High Arctic lakes on Ellesmere and Melville Islands. In 2016, he joined the Marine Environmental Geoscience group at the Geological Survey of Canada (Atlantic). His research focuses on sediment transport processes in marine and lacustrine environments across a broad range of temporal and spatial scales. Specifically, his goal is to understand the triggers and the frequency of extreme events such as submarine landslides and turbidity currents that can severely damage submarine infrastructures. At the GSC, he is involved in the study of marine geohazards along the Eastern Canadian margin, the Western and Eastern Canadian Arctic and the Gulf of St. Lawrence. He uses a wide variety of methods, including high-resolution seafloor mapping techniques (multibeam bathymetry) and geophysics, water column measurements using current profilers and the collection of sediment cores that are analyzed for physical and geochemical properties.

David J. W. Piper is an Emeritus Scientist with the Geological Survey of Canada (Atlantic). He grew up in southern England and was educated at Cambridge University (BA, Ph.D.) before emigrating to Canada to teach at Dalhousie University. His research has been mostly concerned with catastrophic sedimentation processes in the deep sea, particularly turbidity currents and submarine landslides. His understanding of modern oceanic processes has been applied to ancient rock successions seen in outcrop and in seismic profiles in petroleum basins. As Chief Scientist on more than 30 blue water research cruises, principally in the NW Atlantic Ocean, he has developed a broad understanding of the geomorphology and Quaternary history of the continental margin of eastern Canada. He was Editor of the Canadian Journal of Earth Sciences for 5 years, and then became Editor-in-Chief for deep-water papers in Marine Geology for 16 years. His geoscience interests are broad: he was the first to demonstrate the allochthonous nature of the Burgess Shale fauna and he co-authored with G.

Pe-Piper the definitive book on the igneous rocks of Greece. He was recognised with the award of the Shepard Medal for his commitment to mentoring students and young scientists embarking on careers in marine geoscience. He is currently writing up unpublished projects and doing new research on the late Quaternary history of the Labrador Current.

John Shaw holds an M.Sc. and Ph.D. from the University of Ulster, Northern Ireland. He became a post-doctoral fellow at the Atlantic Geoscience Centre in 1986 and has worked for the Geological Survey of Canada (GSC) since 1989 as a research scientist. In the early part of his career he conducted research into sea-level change and coastal response, mostly in coastal Newfoundland. From the mid-1990s onwards, he and his colleagues at GSC Atlantic pioneered the use of multibeam sonar as a geological mapping tool, and developed standards for marine maps. Systematic surveys in Atlantic Canada and British Columbia have been used as the basis of surficial geology, geomorphology, and seascape maps. He is currently conducting research on the continental shelf of British Columbia. He is the author of 60 departmental publications, 75 papers in refereed journals, and 341 miscellaneous un-refereed publications, abstracts, and internal reports.

Brian J. Todd is a marine geoscientist with the Geological Survey of Canada (Natural Resources Canada) at the Bedford Institute of Oceanography in Dartmouth, Nova Scotia. His research has taken him to Canada's three oceans, the Great Lakes, and Siberia. He completed an M.Sc. and Ph. D. in marine geology at Dalhousie University in 1984 and 1988, respectively, studying under the direction of geoscientists at what was then called the Atlantic Geoscience Centre, and is now called the Geological Survey of Canada–Atlantic. After a postdoctoral fellowship at Memorial University of Newfoundland, Dr. Todd joined the Terrain Sciences Division of the Geological Survey of Canada in Ottawa as an engineering geophysicist. While there, his research encompassed the electromagnetic mapping of permafrost distribution and groundwater resources, and high-resolution seismic reflection studies on land and in the Great Lakes. Since his move in 1997 to the Geological Survey of Canada–Atlantic, Dr. Todd has focused on seafloor mapping and was fortunate to be able to participate in the multibeam sonar revolution, driven largely by Canadian innovation. In 2001, Dr. Todd cofounded GeoHab, and international scientific organization dedicated to mapping benthic habitats of the World's oceans. His research has also investigated the glacial and deglacial history of the Scotian Shelf, Labrador shelf, and the Great Lakes basin, and mapping seismic hazard on Canada's Pacific coast.

D. Calvin Campbell joined the Geological Survey of Canada in 1999 as a marine geoscientist after receiving his B.Sc. in Environmental Studies and Geology from Saint Mary's University. In 2012, he graduated with a Ph.D. from the Department of Earth Sciences at Dalhousie University. He is currently a research scientist with the Geological Survey of Canada and an adjunct professor at Dalhousie University. His primary focus has been the study of sedimentary processes and geological history of the eastern Canadian continental margin. He has participated in 30 marine geology expeditions, 17 as chief scientist. His research has applications for marine geological hazard assessment, paleoceanography, offshore resource assessment, and management of the seabed.

David C. Mosher graduated with his Ph.D. from the Department of Oceanography at Dalhousie University in 1993. He is a senior research scientist with the Geological Survey of Canada, a professor with the Dept. Earth Sciences at the University of New Hampshire, an adjunct with the Oceanography Dept. at Dalhousie University and a Commissioner with the Commission on the Limits of the Continental Shelf at the UN in New York. He is a marine geoscientist with 38 years of international research experience in collecting, analyzing and interpreting marine geoscientific data from continental margins around the world. He is the author of 110 refereed publications and editor of five books. He has carried out 50 seagoing expeditions, 28 of these as chief scientist, including co-chief scientist of Ocean Drilling Program expedition 207. His research includes disciplines as diverse as paleoceanography, sedimentology, stratigraphy, structural geology and engineering geology along passive, active and transform margins. His most recent work has involved mapping Canada's margins for determining the limits of its extended continental shelf.

Part III

Anthropocene Landscapes

Human-Modified Landscapes

21

Norm Catto, Olav Slaymaker, and Dori J. Kovanen

Abstract

Although eastern Canada is perceived as dominated by natural landscapes, human modification to geomorphology in many areas is extensive. Agricultural activity has resulted in drainage of wetlands, alteration of stream courses, erosion and sedimentation in coastal areas, and modification of aeolian dunes. Forest clearance, both for initial agricultural expansion and modern forestry, has resulted in erosion and local accentuation of karst topography. Mining, quarrying, dam construction, and canal building have also created geomorphic impacts. Increased tourism has added stresses to several geomorphologically interesting landscapes. Development of urban centres has resulted in significant alterations to drainage systems, with storm-induced flooding of river systems. Construction of cities on glacilacustrine and glacimarine terrains has resulted in relatively few geomorphic hazards. Geotechnical problems include slope stability in ravines, karst dissolution, construction issues on glacilacustrine sand and clay, slope failures in thixotrophic Champlain Sea clays, failures associated with glacifluvial channels, difficulties encountered during subway tunnelling, and glacitectonic-induced fracturing of bedrock. Human impacts on northern and arctic geomorphology, in addition to permafrost ablation resulting from warming climate, focus on modification of coastal landscapes.

Keywords

Eastern Canada • Human impacts • Agriculture • Forestry • Transportation • Coastal • Urban development • Northern communities

21.1 Introduction

Both as researchers and educators, geomorphologists sometimes tend to prefer to focus on 'natural' environments. A geomorphological field trip may invoke images of unpaved roads, limited human populations, and 'unaltered' landscapes. Although eastern Canada affords abundant opportunities for geomorphological appreciation of diverse landscapes, the human presence has resulted in significant changes to geomorphology in many locations.

In 2016, the more than 5.2 million km^2 area considered as 'eastern Canada' in this volume was home to approximately 24.5 million people. Population densities vary substantially by region: Nunavut, Labrador, Nunavik (northern Québec), and northeastern Manitoba have fewer than 1 person/km^2 (Table 21.1). The population of the shield regions of Ontario is concentrated in a few mining- and transportation-related cities along the southern margin (e.g. Thunder Bay, Sault Ste. Marie, Sudbury), with abundant open space between towns and to the north. In contrast, the Maritime Provinces (NB, NS, PEI) and the Appalachian region of Québec have population concentrations in urban and agricultural areas. Human impacts on geomorphology are most evident in the densely populated Great Lakes-St. Lawrence Lowlands (ON, QC), including the metropolitan areas of Toronto, Montréal, and Ottawa-Gatineau.

This chapter takes a sectoral approach, considering both historic and ongoing modifications to geomorphology. Discussions of resource-based and transportation impacts on geomorphology are followed by consideration of effects in urban and northern regions.

21.2 Agriculture

Land currently in agricultural production occupies approximately 95,000 km^2 in eastern Canada, of which about 60% is used as cropland. Based on potential soil productivity and

N. Catto (✉)
Department of Geography, Memorial University of Newfoundland and Labrador, St. John's, NL A1B 3X9, Canada
e-mail: ncatto@mun.ca

O. Slaymaker · D. J. Kovanen
Department of Geography, University of British Columbia, 1984 West Mall, Vancouver, BC V6T 1Z2, Canada

© Springer Nature Switzerland AG 2020
O. Slaymaker and N. Catto (eds.), *Landscapes and Landforms of Eastern Canada*, World Geomorphological Landscapes, https://doi.org/10.1007/978-3-030-35137-3_21

Table 21.1 Regional populations in eastern Canada

Region	Population (2016)	Area (km^2)
Newfoundland & Labrador (NL)	519,716	405,720
Newfoundland (island)	491,799	111,390
Labrador	27,917	294,330
Prince Edward Island (PEI)	142,907	5,686
Nova Scotia (NS)	923,598	52,942
New Brunswick (NB)	747,101	71,389
Québec (QC)	8,164,361	1,356,625
Nunavik and Shield regions	~400,000	~1,290,000
St. Lawrence Lowlands and Appalachia	~7,700,000	~70,000
Ontario (ON)	13,448,494	908,699
Great Lakes- St. Lawrence Lowlands	~12,500,000	~109,000
Shield and Hudson Bay Lowlands regions	~930,000	~800,000
Manitoba (northeast, Hudson Bay Lowlands; MB)	~12,000	~150,000
Nunavut (NU)	35,944	1,877,779

climate, the most suitable terrain (Canada Land Inventory Classes 1, 2, and 3) represents 123,530 km^2. Thus, much land with agricultural potential is currently used for other purposes, particularly urban development, or is not in production. Approximately 50% of Canada's Class 1 farmland is located within a 50 km radius of downtown Toronto, which is also among the most urbanized areas of Canada (Fig. 21.1). Geomorphic impacts of agriculture include drainage of wetlands, forest clearance, changes in erosion rates and sediment flux to river and coastal environments, and utilization of aeolian dunes and adjacent sand plains.

21.2.1 Drainage of Wetlands

In inland areas, historic agricultural clearance and drainage have reduced the proportion of terrain occupied by gleysols, and have also decreased the productivity of all soils. Former wetlands adjacent to developing towns, such as Dow's Great Swamp in Ottawa (geomorphically, marsh and fen terrain) and Strawberry Marsh in St. John's (geomorphically, a fen), were initially drained for agricultural purposes before becoming urbanized.

In southwestern Ontario, alteration of drainage patterns for agricultural purposes has been extensive. The remaining patches of the most isolated wetland areas, often with naturally obstructed drainage, have been converted from the predominantly fens (with fringing marshes) of the early nineteenth century to marshes and swamps, with permanent standing water (Fig. 21.2). In these areas, 'swamp forests' have developed where a combination of elevated water tables and Dutch Elm Disease have killed the groves of *Ulmus americana*, *Ulmus rubra*, and *Ulmus thomasii* that originally were present. These areas also commonly support sycamore, silver maple, pin oak, swamp white oak, and chinquapin oak, which are immune from Dutch Elm Disease and are more tolerant of elevated water tables once they reach maturity. The dead trees of the 'swamp forest' have formed ideal habitats for tree swallows, woodpeckers, great horned owls, and a variety of warblers.

North of Toronto, the Holland Marsh (geomorphically, varying from fens in the south to a freshwater marsh in the north) was an 18 km long, 8500 ha wetland flanking the Holland River in southern Ontario, draining northward into Lake Simcoe. After initial attempts at drainage for agriculture in the 1920s were overwhelmed by the precipitation of Extratropical Transition Hazel (1954), a systematic program of polder construction and drainage was undertaken. The soils are primarily organic Mesisols (Mesic Histosols) and Fibrisols (Fibric Histosols), used initially for growing root crops and lettuce in small market garden farms, and today largely used for bok choy, endive and onions grown on larger commercial farms. As the area is relatively dry (800 mm annual precipitation), particularly in summer, the friable fibrous organic material in drained fields is subject to aeolian erosion.

Along the Bay of Fundy and Gulf of St. Lawrence coasts of Nova Scotia and New Brunswick, and the St. Lawrence estuary of Québec, the relative scarcity of good agricultural land led early settlers to initiate drainage of tidal and brackish marshes (Fig. 21.3). Almost all saltmarsh areas suitable for agriculture were occupied by c. 1850.

In atlantic Canada, saltmarsh development is influenced by sediment influx, relative sea level change, ice scouring and human activity. Typically, organic detritus and sediments accumulate in atlantic Canada's saltmarshes at rates of 1–10 mm per year. At each location, if relative sea level rises at a comparable rate and sediment influx is maintained,

Fig. 21.1 Productive Class 1 farmland, such as this vineyard on the Niagara Peninsula developed on lake sediment, is under pressure from urban development throughout the Toronto-Hamilton-Niagara region

the saltmarsh will continue to develop, as long as there is available and suitable terrain inland. If the saltmarsh is isolated from tidal influx by dyke construction, or if the marsh is drained, the lack of sediment influx and tidal flushing will result in the demise of the marsh.

Over time, hay for fodder became the dominant crop in the drained dykeland areas. Changes in agricultural practice, with the advent of powered machinery, long-distance movement of agricultural products, and consolidation into larger farms, reduced the commercial value of the dykelands. Many of the dykes fell into disuse, leaving the geomorphic and ecological impacts without the agricultural benefit.

In Nova Scotia, legislation states that dykes are intended to protect only agricultural land, rather than commercial or residential developments. Owners of dykelands are encouraged to work with the Nova Scotia Minister of Agriculture and a locally constituted Agricultural Marshlands Conservation Commission. The options available include raising the dyke elevation to compensate for relative sea level rise (currently 2.5–3.5 mm/y) and storm surge activity; intentional decommission; construction of additional dykes inland; and managed realignment, under which a segment of the dyke is allowed to decay, facilitating establishment of a replacement wetland. However, the changed dynamics and offshore slopes usually mean that the replacement wetland does not have the same ecological or geomorphic characteristics as the pre-1600 saltmarsh.

21.2.2 Altered Erosion and Sedimentation

Changes in erosion rates and sediment flux to larger river, estuarine, and coastal environments also result from agricultural activity and related forest clearance. Along the St. Lawrence River in Québec, the seigneurial practice of dividing the land into elongated 'long lots' with short river frontage, subdivided at approximately 90° to the river upon inheritance, resulted in influx of sediment to the St. Lawrence through the numerous ditches. Clearance of all segments of the river bank, to provide each landowner with access to the river through progressively narrowing frontages, increased bank erosion and sediment influx. As the St. Lawrence Valley is dominantly underlain by Champlain Sea glacimarine clay, susceptible to thixotrophic slope failure (Bélanger 1998), small-scale bank collapses of unvegetated areas are common.

Sediment influx to coastlines adjacent to communities also has varied as a result of agriculture, in combination with other human activities (Fig. 21.4). At Topsail Beach (Conception Bay South, NL), major changes in the beach have been affected by sediment mining, interruption of current flow by up-current breakwaters and groynes, changes in land-use practices, damming and regulation of the inflowing streams for hydroelectric power, and recreational usage. Prior to 1940, Topsail Beach was a broad, mixed sand-and-gravel flat, with sand present in excess of gravel.

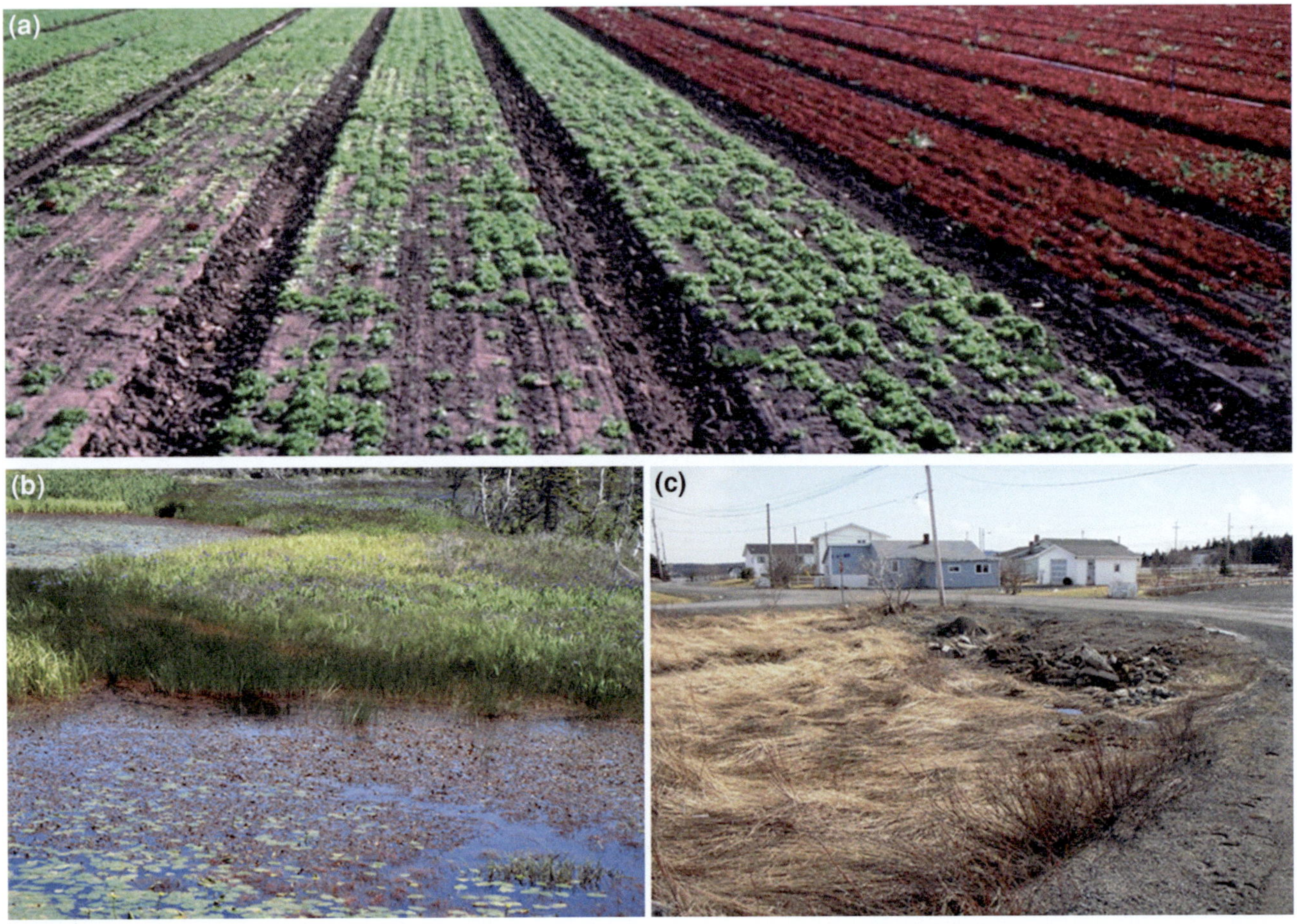

Fig. 21.2 **a** Wetland drainage for agricultural and urban development has been extensive and continues. **b** Alteration and obstruction of drainage has resulted in replacement of fen areas by standing water in marshes and swamps. **c** Infilling of wetlands in suburban areas continues throughout eastern Canada

At the time, it was advertised as 'The Brighton Beach of Newfoundland' and 'The St. John's Riviera', and was a major social gathering place. Photographs taken between 1910 and 1940 indicate that the beach extended more than 25 m seaward of its present (2016) shoreline in the outlet area.

During World War II, large quantities of sand and gravel were removed from Topsail to construct Torbay Airport and other military facilities in St. John's. When extraction ceased in 1943, the beach front was as much as 5 m landward of its present position. Shifting agricultural and land-use practices, in particular the replacement of root crop cultivation, sheep herding, and dairy farming by suburban development, limited the replenishment of sand to the beach system after 1950. Most of the sand had been derived from low bluffs of glacifluvial and glacimarine sediments which line the coast to the southwest. Many cliffs were stabilized by armouring with rip-rap and debris (often illegally dumped). Yachting facilities were constructed up-current, and breakwaters designed to protect these funnelled sediment away from the shore and into the deeper parts of the Bell Island Tickle, thus causing Topsail Beach to be bypassed. The Tickle is underlain by medium and coarse sand throughout much of its length.

These changes have substantially modified the morphology of the beach system. Topsail is now a coarse gravel-dominated beach, characterized by extreme instability of the outlet and high sediment mobility. Seasonal shifts in texture, from pebble-cobble gravel in late autumn to granule-pebble gravel in early summer, are common in the southwestern part of the system, but do not occur to the northeast, in the vicinity of the outlet.

Fig. 21.3 Saltmarsh terrain along the southern Gulf of St. Lawrence, near Shediac NB. **a** This former saltmarsh was drained for agricultural production c. 1740, and abandoned in the 20th century. **b** Coastal development for tourism and summer homes results in ongoing modification of saltmarshes

21.2.3 Aeolian Dunes

Geomorphic consequences of historic agriculture include enhanced erosion and migration of aeolian dunes at sites throughout the Maritimes, southern Québec, and southern Ontario. Migration of dunes at sites such as Sandbanks Provincial Park (Ontario) and Greenwich (Prince Edward Island National Park) was induced by land clearance, and slowed or largely ceased when clearance was discontinued. In some cases, as at Sandbanks, agricultural activity was entirely responsible (Fig. 21.5).

At Sandbanks, Prince Edward County, an esker delta developed through glacilacustrine stages of the Lake Ontario basin supported a transverse complex of linked parabolic and dome dunes, extending from the modern baymouth bar inland. Growth of a white pine-deciduous dominated forest acted to restrict dune migration and height. Prior to 1850, the forested dunes were inactive, with heights of approximately 3 m.

Fuelled by the high price of barley in the United States, the coastline of Lake Ontario at Sandbanks was cleared of trees from 1850 to 1890. Over 15 million bushels of barley were shipped to breweries in New York and New England. Facing southwestward, directly into the prevailing wind from the lake, the dunes at Sandbanks were readily mobilized and began to grow and extend inland. Although the

Fig. 21.4 Human modification to coastal geomorphology has occurred throughout atlantic Canada since 1700. **a** At Point LaHaye, St. Mary's Bay, NL, agricultural clearance, sheep farming, and human use of driftwood beginning 1700 ce is evident in the sharp transition from boreal forest peat accumulated from 7600 BP. **b** Topsail Beach, NL, was a sand-dominated system prior to disruption of sediment supply beginning in World War II. Net littoral drift is towards the northeast (left side of photo). **c** Ongoing silt and sand transportation into Tracadie Bay, PEI, results from agricultural development

barley market collapsed in 1890, leading to the abandonment of farms, dune movement continued. Between 1852 and 1912, the main dune complex migrated 450 m inland, up to 14 m/y. Locally, dune height exceeded 18 m. Several houses were inundated, a hotel was abandoned, demolished, and the site subsequently buried, and the road flanking the complex was moved five times.

Subsequent reforestation efforts between 1911 and the present, combined with infrastructure, have stabilized the dune complex. The coastline is occupied by seasonal

Fig. 21.5 Clearance of forest for barley production in the 19th century triggered expansion of coastal dunes at Sandbanks Provincial Park along Lake Ontario. Several tree-planting programs were required to stabilize the dunes. Today, the landscape is used exclusively for recreation

foredunes, fed by NW–SE transiting littoral drift, subject to strong tourist pressure from May through October. Active aeolian movement is confined to areas along trails and directly behind the tourist beach.

The northern coastline of Prince Edward Island illustrates the geomorphic impact of combined agricultural and forestry practice on coastal dune terrain (Catto et al. 2002: Fig. 21.6). Anthropogenic disturbance became significant in northern PEI following the arrival of European settlers in 1719 (MacArthur 1976). Farming practices included clearing forest for cropland and lumber, and grazing sheep on the dune vegetation. Beginning c. 1800, land clearance by logging and for agricultural purposes by subsequent groups of British settlers transformed the coastal areas of PEI from the dominant white pine stands along the coastal margin to an open landscape of agricultural fields. Although dunes were recognized as agriculturally poor land, the value of white pines for ships' masts led to the rapid clearance of trees.

Parabolic dunes developed rapidly where inland migration was not restricted by wetlands or lagoons. After less than 50 years of intensive agricultural and logging activity, wind-blown sand and migrating dunes were recognized as the consequences of inappropriate development along the PEI shore, and elsewhere in maritime Canada (Dawson 1868).

Throughout the twentieth century, the parabolic dunes at Greenwich migrated landward over forested terrain,

Fig. 21.6 Agricultural clearance resulted in dune migration along the north coast of Prince Edward Island for more than 200 years after initial European settlement. **a** Today, some of this terrain is conserved at the Greenwich unit of Prince Edward Island National Park. **b** Even within the park, tourist traffic continues to process of dune destabilization. **c** Locally, conifers have been planted to stabilize coastal dunes, as here at Cabot Beach Provincial Park

inundating and killing the vegetation. Residents responded to the visible expansion of the dune field by clearing forest from the fringes, resulting in rapid acceleration of inland sand transport and dune migration. MacArthur (1976) estimated that 32 ha of cleared and wooded land and 8 ha of agricultural land had been buried by migrating dunes along the southern margin of the Greenwich dune field, and an additional 5 ha had been affected by sand sheet deposition in the lee of the dunes. In the Greenwich Peninsula, the extensive dune migration during the 1920s led to the abandonment of agriculture. The subsequent re-establishment of vegetation, and the absence of extensive tourist-related infrastructure, resulted in lesser rates of geomorphic modification and dune erosion than are apparent on other areas of the PEI North Shore.

21.2.4 Forest Clearance

Between 1720 and 1920, forest clearance in eastern Canada was intended simultaneously to harvest timber and open the terrain to agriculture. Under this procedure, forestry was not intended to be sustainable: the goal was to remove the trees as rapidly as possible. In many cases, the pioneer farmers also served as lumberjacks.

Throughout eastern Canada, logging of *Pinus strobus* (Eastern white pine) became a significant geomorphic factor in the nineteenth century, with ramifications that continue today (Fig. 21.7). Among the pine species in eastern Canada, *Pinus strobus* requires the greatest degree of soil development and the highest organic content, the best drained conditions, and the least amount of pollution.

Regarded as spiritually significant by the indigenous populations of maritime and central Canada, *Pinus strobus* became the most valuable commercial species following European settlement. As a softwood conifer, it is easy to shape and polish, without splinters or shards, but is also flexible and resilient, and not susceptible to warping. Because it floats well and is water resistant, white pine was the preferred species for ship construction. The tall, straight, knot-free trunks were ideal for ships' masts. Successive exploitation of *Pinus strobus* forests for ship (especially naval) construction began in eastern North America in the 1720s, a process which accelerated after 1795 when the supply of Baltic timber to Britain was interrupted by the Napoleonic wars. As a result, *Pinus strobus*, formerly the dominant pine species, was greatly reduced throughout the region.

Prior to the initiation of widespread agriculture ca. 1800, PEI was covered by the Acadian deciduous forest (Fig. 21.8). Broad-leaved deciduous *Fagus grandifolia* (beech), *Acer saccharum* (sugar maple), and *Quercus rubra* (northern red oak) dominated the assemblage. *Fagus grandifolia* appears to have covered more than 50% of inland areas adjacent to Malpeque Bay in 1800. Other southern boreal species such as *Betula alleghaniensis* (yellow birch), *Pinus strobus* (white pine), *Tsuga canadensis* (hemlock), and *Thuja occidentalis* (eastern white cedar) were also common.

Today, little of this natural assemblage remains. Inland areas have largely been cleared for production of potatoes, corn (maize), and soybeans. Rivers have seen decreased mean discharge, increased silt deposition, local entrenchment, and enhanced meandering. Aeolian erosion and deposition of silty sand veneers and blankets has resulted. Gullying and sheetwash occurs on steeper undulating terrain. Increased silt transport, combined with runoff of nitrogen-rich fertilizers, has changed the geomorphic, sedimentologic and biologic regimes of PEI's numerous embayments.

The moderate- to high-relief shield region of Ontario is marked by glaciated Proterozoic gneiss and granite. In well-drained situations, and in areas of thicker glacial sediment, podzols are developed. Dystric Brunisols (Dystric Cambisols, Dystrochrepts) and Regisols (Entisols) are present in areas with thin or no glacial sediment cover (Fig. 21.9). These soils are highly acidic (typical pH values of 4.0–5.3), organic-poor, and are unsuitable for agriculture. During the period of initial settlement of the Great Lakes-St. Lawrence Lowlands and their bordering regions, many fruitless efforts were made to farm the Canadian Shield, encouraged by governments (for colonization) and logging companies (for harvesting of *Pinus strobus*). Today, most of these forests have been cleared, influencing erosion, river flow, sediment transport, and channel geometry. Local increases in aeolian sedimentation can also result from deforestation (e.g. Steedman and France 2000).

The lack of significant agricultural potential and mineral resources throughout most of the moderate-relief Shield suggests that the most suitable modern uses are recreation and conservation. In most areas, the only lasting legacies of the initial farming efforts are increases in the proportions of *Populus* and *Acer* (as these trees occupied the labouriously cleared land upon abandonment), groups of frustrated would-be farmers, and the geomorphic impacts.

Extensive deforestation generally increases the volume of storm runoff, but the mean annual discharge of a boreal river will drop due to the reduced influx from water retained temporarily as groundwater. Declining water levels following logging are a common occurrence along most small rivers in developed eastern Canada, precluding navigation by canoe. As water systems were defined as 'navigable' (hence with restrictions on development and alteration of fish habitat) based on canoe transit, the decline in water levels has led to changes in legislative and political status of many streams, with geomorphic consequences.

Fig. 21.7 a *Pinus strobus* was the dominant tree in the Great Lakes Lowland and much of maritime Canada prior to European logging. **b** Logging resulted in exposure of glacial sediment, enhancing erosion and downstream sedimentation. Subsequently, mixed deciduous-coniferous assemblages occupied abandoned farmland

Rivers flanked by unstable banks of sand, gravel, or coarse-textured glacial deposits are greatly influenced by the degree of riparian vegetation cover. Multiple channel systems are less common in areas of dense forest. Where they do occur, as a result of enhanced stream power or reduced sediment availability, they are more likely to form anabranching or wandering reaches rather than braided ones.

Deforestation changes the nature and rate of runoff, infiltration, and sediment influx and transport in rivers. Whereas many floods in forested areas tend to be basin-wide events, those in deforested Shield areas can have very localized origins, such as obstruction of channels, ice-damming, alteration of drainage patterns, and expansion of the area of exposed bedrock through erosion of glacial sediment cover. Intense local flooding (e.g. Muskoka, May 2019) can result from deforestation.

During deglaciation c. 10–9 ka, Glacial Lake Barlow-Ojibway expanded across northeastern Ontario and northwestern Québec, as northward drainage was blocked by the receding Laurentide ice (Fig. 21.10). The glacilacustrine Clay Belt was the scene of one of the last efforts to entice faming homesteaders in eastern Canada. Following the discovery of silver at Cobalt to the south, and findings of gold at Kirkland Lake to the east and Timmins to the southwest, governments decided that the new mining communities would benefit from the establishment of agricultural areas on

Fig. 21.8 **a** Prior to European settlement, much of Prince Edward Island was covered by the Acadian Deciduous Forest. **b** Today, the rolling glacial till blanket over Permian sandstone bedrock has been exposed through agricultural activity, with consequences for wetlands, fluvial activity, and coastal sedimentation

the soils of the Clay Belt. Further impetus to agricultural development was provided by railroad construction.

Unfortunately, although additional mineral finds continued (e.g. at Rouyn-Noranda), agriculture was not successful. The Clay Belt offered suitable soil, but the climate was not conducive. Most of the would-be homesteaders rapidly abandoned farming, turning instead to mining, forestry, or emigration to meet their economic needs. Ongoing climate warming may improve agricultural conditions in the future.

Clearance of the forest from the Clay Belt resulted in changes in wetland type and extent, due to changes in sedimentation, hydrology, and forest assemblage. Replacement of *Picea mariana, P. glauca*, and *Abies balsamea* by *Betula papyrifera, Populus balsamifera*, and *P. tremuloides* resulted in increased beaver populations, causing shallow rivers and dry fens to be replaced by lakes and marshes in many areas. Lakes and rivers in the Clay Belt are generally shallow, and are marked by fluctuations in water levels controlled by weather, beaver activity, and hydroelectric manipulation. Rivers are generally wandering or braided, carrying mixed loads of silt, sand, and gravel. Aeolian modification of exposed glacilacustrine silt is locally apparent, with thin veneers of coarse loess-like sediment.

Fig. 21.9 **a** Acidic dystric brunisols (left) and podzols (right) are typical soils developed under Dfb boreal climates and coniferous vegetation over acidic bedrock. These soils are agriculturally limited. **b** Disturbed boreal rivers, such as the Kaministiquia (shown here) show greater discharge variations and increased bedload

21.3 Forestry

Forested terrain in eastern Canada encompasses 200 million ha, approximately 7% of the world total. Mid-boreal and taiga forests represent approximately 75% of the total. Land available for active or potential forestry operations encompasses 110 million ha of boreal forest and 10 million ha of mixed deciduous forests in the Great Lakes-St. Lawrence Lowlands and maritime Canada. Harvesting in these areas included pulp-and-paper manufacture and local consumption, rather than being motivated by agricultural expansion. Softwood (conifer) logging and wood pulp production are the dominant forest industries. Canada has the largest share of the global market for Northern bleached softwood kraft pulp, and is the single largest producer of newsprint (12% of global total).

Extensive logging and deforestation has occurred throughout much of the mid-boreal forests of Newfoundland, Québec, and Ontario. The dominant tree species are *Picea glauca* (white spruce), *Picea mariana* (black spruce), *Abies balsamea* (balsam fir), and *Pinus banksiana* (jack pine, in Ontario and Québec), with associated *Betula papyrifera* (white birch), *Populus balsamifera* (balsam poplar), and *Larix laricina* (tamarack, or larch). Logging debris forms the substrates of many lakes and deeper rivers in eastern Canada.

Fig. 21.10 **a** Thick varved successions deposited in Glacial Lake Barlow-Ojibway formed the sediments of the Clay Belt of northeastern Ontario. **b** The Opasatika River, as others flowing northward across the former lake basin, is a low-gradient, shallow, gently meandering system

Clear-cutting is the dominant technique, representing 85–90% of the logged terrain of the boreal forests of eastern Canada (Fig. 21.11). The rolling topography of the Canadian Shield, combined with the limited cover and thickness of coarse-grained glacial sediment, limits the potential for large slope failures. The dominant geomorphic effects are enhanced frost modification of exposed surfaces, and increased sand and gravel deposition in river systems, resulting in increased channel braiding and shallowing thalwegs. Clear-cut areas are also more susceptible to aeolian disturbance, producing local sandy veneers.

Over the past 75 years, salmon numbers in many atlantic Canadian streams have declined due to disturbances of flow and river bottom sediment related to logging activity, in conjunction with local overfishing and loss of habitat resulting from river regulation by dams. Locally, *Salvelinus fontinalis* (brook trout), which inhabits rivers with sand and fine gravel substrates, requires well-oxygenated, cold (<20° C), clear waters, and does not flourish in streams choked by logging debris, has been replaced by other species.

Development of karst topography can also be influenced by changes in vegetation cover (Fig. 21.12). Removal of vegetation would allow more water to infiltrate the topsoil, resulting in an increase in the rate of limestone or gypsum dissolution. Karst topography is extensively developed on exposed beds of gypsum in western Newfoundland (Codroy Valley and Port-au-Port Peninsula; House and Catto 2008; Catto 2012), New Brunswick (Plaster Rock and Westmorland County; Webb 2002) and in Nova Scotia (Brooklyn, Windsor, and along the Bras d'Or Lakes; Adams 1993; Ford 1997; Falcon-Lang et al. 2007). These areas have also been subject to deforestation, as well as local agricultural development, which may have influenced karst geomorphology.

21.4 Mining and Quarrying

In eastern Canada,, the first commercial attempt at mining was initiated by Martin Frobisher in 1577, near modern Iqaluit. Unfortunately, Frobisher's 'gold' was actually pyrite.

Geomorphic impacts include open pits, tailings accumulations, artificial tailings ponds and impoundments, drainage diversions and artificial islands, and collapse features due to roof failure in underground mines (Table 21.2). Although many of the mine sites are inoperative or abandoned, no large slope failures have occurred in recent years. As elsewhere, problems of dust production, environmental contamination, tailings dam failure, and health impacts for miners and nearby residents have been documented, particularly for older operations.

Active and former quarrying activities impacting geomorphology include extraction of limestone (numerous localities in southern Ontario, Québec, and the Maritimes); and gypsum (e.g. Windsor (beginning in 1770), East Milford, and Cape Breton NS). Dimensional (decorative) stone quarries, including granite, marble, labradorite (anorthosite),

Fig. 21.11 Clear-cut forestry in western Newfoundland. The rolling terrain, with thin glacial sediments mantling bedrock, limits the opportunity for slope failures. *Photograph* E. Edinger

feldspar porphyry, charnockite, diabase, gabbro, gneiss, slate, dolostone and sandstone, are scattered throughout eastern Canada, and together produce ~800,000 t annually.

21.5 Dam Construction and Hydroelectric Development

In eastern Canada, impoundments have been constructed for operation of grist mills, beginning at Port-Royal (NS) in 1607; flood control; navigation; and hydroelectric development. In Canada, the first use of water to produce electricity was for a wheel built by the Ottawa Electric Light Company at Chaudière Falls in 1881. Small dams and impoundments, such as along the Montague River PEI (Knox's Dam, initially built in 1852 to supply a grist mill, and rebuilt in 1917 for hydroelectricity), at Petty Harbour NL (1900), and the St. Margaret's Bay system (NS, 1922), are found along many rivers in agricultural and coastal southern Canada. Québec alone has 1950 registered dams.

Hydroelectric development in eastern Canada is extensive, representing the majority of the country's 993 large dams. Linear reservoirs are developed along the Saint John (NB; Fig. 21.13), Exploits (NL), Nelson (MB), Churchill (NL), Ottawa (ON-QC), Madawaska (ON), Gatineau (QC), St-Francois (QC), and St-Maurice (QC) rivers, among others. Dual purpose (hydroelectric combined with flood control or transportation) reservoirs are present along the St. Lawrence (Long Sault, ON) and Thames (Fanshawe Dam, ON) rivers. Nunavut currently has no hydroelectric dams, although there is a proposal to impound the Armshow River to supply power to Iqaluit.

Sixteen large reservoirs each cover 400 km^2 or more, including raised impoundments of existing lakes. The Smallwood Reservoir, formed by impounding the Churchill River (NL) at Churchill Falls, Labrador, is the world's second

Fig. 21.12 (**a**) Suffusion and collapse (**b**) Dolines developed in gypsum, Woodville, NL. Land clearance for farming and logging (upslope) has contributed to recent expansion of karst activity (House and Catto 2008)

largest in surface area (6527 km^2) and ranks 10th in surface area of all Canadian lakes (Fig. 21.14). Southern Indian Lake, formed by the 85% impoundment of northern Manitoba's Churchill River and the associated diversion of 850 m^3/s into the Nelson River drainage, covers 2247 km^2. Lac Seul, a raised natural lake on the English River (ON) formed by damming at Ear Falls, covers 1657 km^2. Québec has 13 reservoirs exceeding 400 km^2, including Canipiscau (4539 km^2, 16th largest in Canada), Robert-Bourassa (2905 km^2, 20th largest), La Grande 3 (2451 km^2), Manicouagan (1788 km^2), and Gouin (1570 km^2).

All of these large impoundments are located on the Canadian Shield. Lake shorelines reflect the underlying structural geology. Rectangular patterns indicate strong structural control from the underlying bedrock. Contorted shoreline patterns develop in areas where contorted metamorphic or folded sedimentary bedrock is exposed on the surface. The Manicouagan Reservoir shows the impact of the bolide which struck approximately 214 Ma (Spray et al. 2010). The centre of the crater was elevated after impact by resurgence, and the impoundment illustrates the depressed ring of the structure (Fig. 21.15).

The resistant bedrock of the Canadian Shield and the irregular aspects with respect to the prevailing wind limit erosion by wave action or seasonal ice shove. Sediment entrapment and subsequent downstream losses are variable,

Table 21.2 Major mines and geomorphic impacts, eastern Canada

Location	Commodity	Dates of operation[a]	Geomorphic impact
Thetford Mines QC	Asbestos	1877–2011	Open pit mining; tailings piles; asbestos tailings used for railway and roadbeds
Baie Verte NL	Asbestos	1963–1994	Open pit mining; tailings piles; aeolian deposits with asbestos dust
Cape Breton Regional Municipality NS	Coal	1720– (several mines)	Slag piles; contaminated settling ponds; estuarine sedimentation; port infrastructure
Stellarton—Pictou NS	Coal	1807–	Open pit mining; slag piles; port infrastructure
Springhill NS	Coal	1872–1960	Slag piles; water-filled mines used for geothermal energy
Murdochville QC	Copper	1950–1999	Open pit mining; tailings and slag piles; tailings ponds; road infrastructure
Buchans NL	Copper, lead, zinc	1927–1984	Open pit mining; tailings and slag piles; tailings ponds; railway infrastructure
Bathurst NB	Copper, zinc	1964–2013	Tailings piles
Rouyn-Noranda QC	Copper; gold	1917– (several mines)	Open pit mining; tailings and slag piles; tailings ponds
Victor ON	Diamonds	2008–2019	Open pit mining; tailings piles
St. Lawrence NL	Fluorospar	1933–	Open pit mining; tailings piles
Moose River NS	Gold	1860–	Open pit mining; slag piles; tailings ponds
Meadowbank, Kivalliq NU	Gold	2010–	Open pit mining; tailings piles; no road, rail, or water access
Timmins ON	Gold	1910–	Open pit mining; tailings piles; tailings ponds
Kirkland Lake ON	Gold; iron	1914–	Open pit mining; tailings completely infill former Kirkland Lake
Goderich ON	Halite	1866–	Port infrastructure
Windsor ON	Halite	1893–	Collapse (1954)
Labrador City NL	Iron	1962–	Open pit mining; tailings piles; railway infrastructure
Mont Wright QC	Iron	1972–	200 m-deep open pit; railway infrastructure
Temagami ON	Iron	1968–1990	7 open pits; tailings piles; tailings ponds
Steep Rock Lake ON	Iron	1945–1970	Drained lake; open pit mining; tailings piles
Marmora ON	Iron	1952–1978	210 m-deep, 30 ha-area open pit
Raglan, Nunavik QC	Nickel	1997–	Open pit mining; tailings piles; road and port infrastructure
Sudbury ON	Nickel	1886– (several mines)	Slag piles; tailings ponds; widespread acidic pollution
Voisey's Bay NL	Nickel	2005–	Open pit mining; tailings piles; port infrastructure
St-Honoré QC	Niobium	1976–	Open pit mining; tailings piles
Lac des Iles ON	Palladium	1995–	Open pit mining; tailings piles
Long Harbour NL	Phosphate	1968–1989	Open pit mining; tailings piles; port infrastructure; harbour contamination; tailings used for roadbed
Manuels NL	Pyrophyllite	1904–	Open pit mining; tailings piles; port infrastructure
Lac Tio QC	Titanium	1950–	Open pit mining; tailings piles
Elliot Lake ON	Uranium	1955–1997	Tailings piles
Bancroft ON	Uranium	1953–1982	Tailings piles

[a]Initial opening and closing dates; does not necessarily imply continuous operation

Fig. 21.13 **a** Impoundment of the Saint John River at Grand Falls-Grand Sault (NB) allows inspection of the incised meandering channel downstream, carved in Ordovician calcareous slate and limestone. **b** Shallow potholes, maximum dimension 20 cm, excavated in slate

but in many cases the impounded streams carry relatively low sediment loads. The most significant downstream impacts occur where streams are diverted into neighbouring systems to fill reservoirs, or where streams debouch from Shield areas into adjacent sedimentary lowlands.

Dams have also been erected for flood control. In metropolitan Toronto, the three major streams, the Humber, Don, and Rouge, are dammed. During the night of 15–16 October 1954, Extratropical Transition Hazel caused intense flash flooding along the sand-and-gravel-dominated meandering Humber River, killing 81 people and causing more than $850 million (2019 $) in damage. Subsequently, the Humber, Don, and Rouge river valleys and their tributary ravines were designated as conservation (no-build) areas, and the dams constructed. Rouge National Urban Park, established in 2015, encompasses 40 km^2 of river and adjacent ravine terrain in eastern metropolitan Toronto.

21.6 Urban Environments

In 2017, the urban population of eastern Canada was approximately 20 million, representing more than 80% of the total in the region. More than 8 million people live in the

Fig. 21.14 The Churchill River was impounded at Churchill Falls in 1972, creating the Smallwood Reservoir. **a** Landsat image of the impounded Smallwood reservoir. *Source* @ 2018 Google Earth: Landsat/Copernicus. **b** Abandoned channel of the Churchill River. *Source* Wikipedia. **c** Churchill Falls, 1969 and 2017. *Source* Wikimedia commons

urban agglomeration centred around Toronto, extending from Oshawa through Hamilton to St. Catharines-Niagara; 4 million in the Montréal metropolitan area; 1.4 million in Ottawa-Gatineau; and 810,000 in the Québec City metropolitan area (Table 21.3). Other centres with more than 400,000 people include Halifax, Kitchener-Waterloo, and London. Thus, these urban areas, totalling $\sim 20{,}000$ km^2, contain more than 15.7 million people, with 4 million

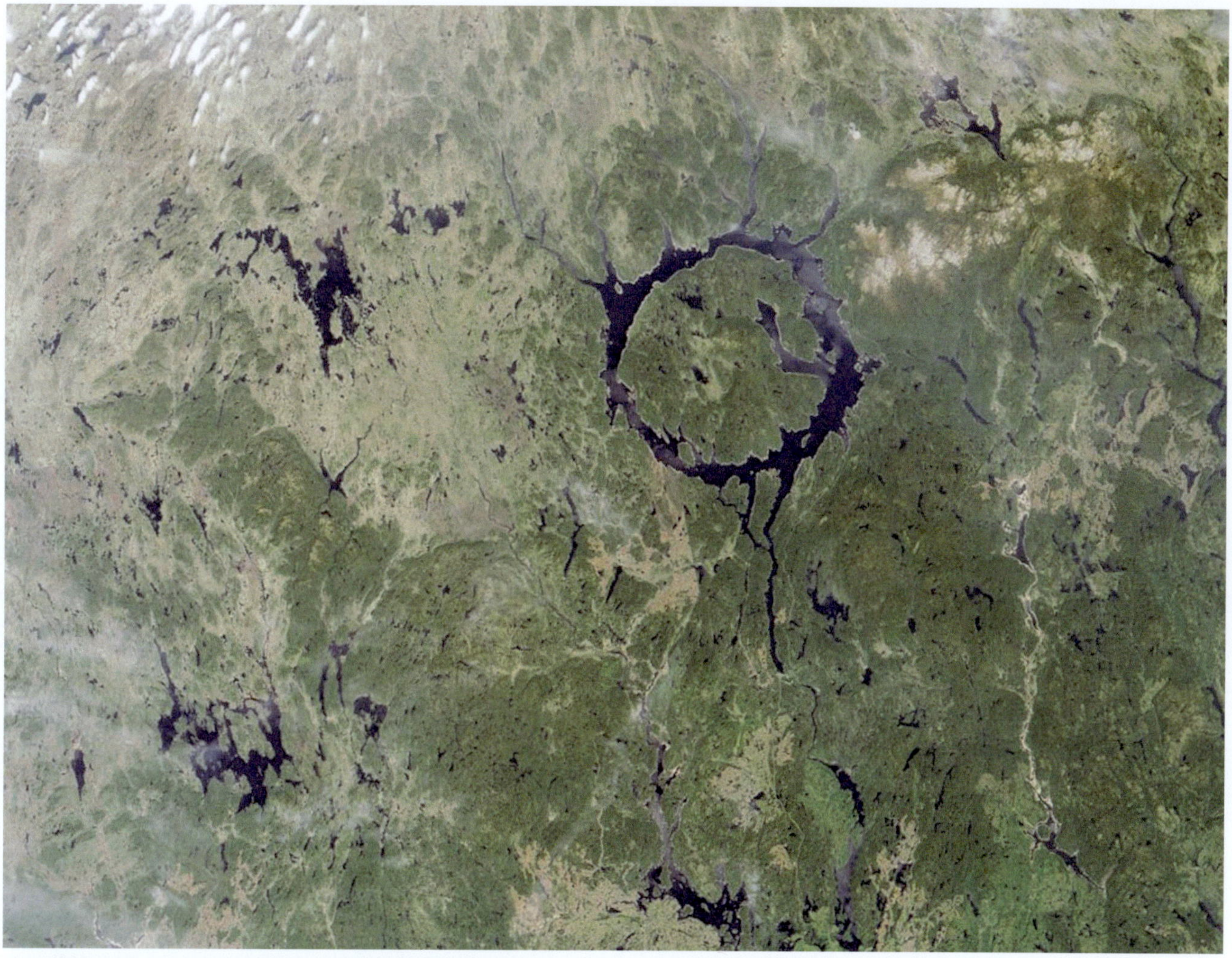

Fig. 21.15 Manicouagan Reservoir. The depression was created by the bolide which struck approximately 214 Ma (Spray et al. 2010). The centre of the crater was elevated after impact by resurgence. (JPL/NASA)

scattered across the remaining 5,250,000 km^2 of eastern and northern Canada. All of the major urban centres in eastern Canada saw population growth in excess of 9% between 2007 and 2017, with growth in the metropolitan Toronto area exceeding 17%. More than 46% of metropolitan Toronto's people were born outside of Canada. Lower or negative population growth rates characterize communities dominated by older heavy industry or mining (e.g. Windsor ON, Sudbury, Cape Breton Regional Municipality).

Cities and towns in eastern Canada developed initially as ports (e.g. Halifax, Saint John, Québec, Toronto), or at waterfalls or rapids where repacking and portaging was required (e.g. Montréal, Ottawa, St. Catharines-Niagara). Falling water also provided one of the few alternatives to muscle power as an energy source. Urban humans interested in eating had to either be near food supplies, or be able to acquire them through trade across and or water. The earliest communities, both indigenous and European, were established where combinations of transport by water and reasonable agricultural conditions (often on fine-grained glacial, glacimarine, glacilacustrine, or glacifluvial deposits) existed.

In atlantic Canada, geomorphology greatly influenced town settings and configurations. Towns which either had a significant British military presence at their founding or were laid out by retired British military engineers are marked by grid systems, including Halifax, Saint John (NB), Charlottetown, and Fredericton. Towns without an initial dominant military presence, such as St. John's, Moncton, and Sydney (Cape Breton Regional Municipality), are marked by street patterns that developed in accordance with combinations of the dominant economic activities (fishing, railroad development, and mining, respectively), and geomorphology.

Halifax was developed primarily as a Royal Navy base in its formative years (c. 1740), a counter to the French Fortress Louisbourg. The grid pattern evident in downtown Halifax,

Table 21.3 Geomorphology of eastern Canadian cities

Census metropolitan area	2007 Pop.	2017 Pop.	% Change 2007–17	Bedrock	Geomorphology	Geomorphic natural hazards
Toronto (ON)[s]	5,418	6,346	17.1	Late Ordovician shale & limestone	Gently undulating glacial terrain (MIS 2); glacilacustrine plain (Lake Iroquois); coastal and fluvial landforms	River flooding
Montréal (QC)[s]	3,715	4,139	11.4	Cambrian sandstone; Ordovician limestone; Early Cretaceous gabbro & syenite (Mont-Royal)	Glacimarine plain (Champlain Sea); Cretaceous intrusion (Mont-Royal)	River flooding; mass movement; earthquakes
Ottawa–Gatineau (ON-QC)	1,188	1,377	15.9	Grenville gneiss; Cambrian sandstone; Ordovician limestone & shale	Glacimarine plain (Champlain Sea); fluvial landforms	River flooding; mass movement; earthquakes
Québec (QC)[m]	737	812	10.2	Grenville gneiss; Ordovician limestone & shale; deformed Lower Palaeozoic sediments	Glacimarine plain (Champlain Sea); fluvial landforms	mass movement; earthquakes
Hamilton (ON)[s]	722	787	9.0	Silurian shale, dolostone, sandstone	Niagara Escarpment; glacilacustrine plain (Lake Iroquois); coastal landforms	Mass movement
Kitchener–Waterloo (ON)	474	528	11.3	Silurian shale & dolostone	Gently undulating to rolling glacial terrain (MIS 2)	River flooding
London (ON)	478	522	9.2	Devonian limestone	Undulating to hummocky glacial terrain with recessional moraines (MIS 2); glacilacustrine plains (Lakes Whittlesey, London); fluvial landforms	River flooding
Halifax (NS)[m]	387	432	11.7	Cambro-Ordovician metasedimentary rocks; Devonian granite	Drumlinized glacial terrain (MIS 2)	Storm surge; coastal erosion
St. Catharines–Niagara (ON)[s]	403	417	3.5	Late Ordovician-Devonian shale, dolostone, sandstone	Niagara Escarpment; rolling glacial terrain (MIS 2); glacilacustrine plain (Lake Iroquois); coastal and fluvial landforms	Mass movement
Oshawa (ON)	348	402	15.5	Ordovician shale & limestone	Gently undulating glacial terrain (MIS 2); glacilacustrine plain (Lake Iroquois)	Mass movement
Windsor (ON)[s]	334	345	3.4	Devonian shale, limestone, halite	Gently undulating glacial terrain (MIS 2); glacilacustrine plain (Lakes Whittelsey-Arkona	Mass movement
St. John's (NL)[m]	186	219	17.8	Ediacaran metasediments and volcaniclastics	Undulating glacial terrain (MIS 2)	River flooding; mass movement; storm surge
Sherbrooke (QC)	195	219	12.1	Grenville (?) metamorphic units	Rolling glacial terrain (MIS 2); proglacial glacilustrine plain; fluvial landforms	River flooding
Sudbury (ON)	165	165	0.3	Proterozoic (Southern) volcanic and metasedimentary units	Undulating glacial terrain (MIS 2); glacilacustrine plain (post-Lake Algonquin)	Mass movement
Guelph (ON)	140	160	14.2	Silurian shale & dolostone	Drumlinized glacial terrain (MIS 2); glacifluvial eskers and plain; fluvial features	River flooding
Saguenay (QC)	157	159	1.3	Grenville igneous and metamorphic units	Undulating glacial terrain (MIS 2); glacimarine plain (Laflamme Gulf); fluvial landforms	River flooding; mass movement; earthquakes
Trois-Rivières (QC)[m]	147	159	8.1	Grenville gneiss; Ordovician limestone & shale	Glacimarine plain (Champlain Sea); fluvial landforms	River flooding; mass movement; earthquakes

(continued)

Table 21.3 (continued)

Census metropolitan area	2007 Pop.	2017 Pop.	% Change 2007–17	Bedrock	Geomorphology	Geomorphic natural hazards
Moncton (NB)	131	152	16.1	Pennsylvanian Red-bed clastics	Gently undulating glacial terrain (MIS 2); glacimarine plain; coastal and fluvial landforms	River flooding; storm surge
Brantford (ON)	136	147	8.2	Devonian limestone	Undulating glacial terrain (MIS 2); glacilacustrine plain (Lakes Whittlesley, Warren); fluvial landforms	River flooding
Saint John (NB)[m]	126	129	2.4	Neoproterozoic igneous units; Cambrian clastic sediments	Gently undulating glacial terrain (MIS 2); glacimarine plain; fluvial landforms	River flooding; earthquakes; storm surge
Thunder Bay (ON)[s]	126	124	−1.5	Archaean- Proterozoic granitic, metamorphic and mafic volcanic units	Gently undulating glacial terrain (MIS 2); glacilacustrine plain (Lakes Minong-Nipissing Great Lakes); deltaic and fluvial landforms	Mass movement; river flooding
Fredericton (NB)	92	105	14.1	Silurian shale; Pennsylvanian shale, sandstone, basalt	Gently undulating glacial terrain (MIS 2); glacifluvial plain; glacimarine plain; fluvial landforms	River flooding; earthquakes
Cape Breton Regional Municipality (NS)[m]	102	94	−9.2	Pennsylvanian sedimentary rocks	Rolling glacial terrain (MIS 2); coastal landforms	River flooding; coastal erosion
Sault Ste. Marie (ON)[s]	80	75	−9.4	Archaean- Proterozoic granitic and metamorphic units	Rolling glacial terrain (MIS 2); glacilacustrine beaches (Lakes Algonquin-Nipissing Great Lakes)	Mass movement
Charlottetown (PEI)[m]	60	70	16.7	Permian Red-bed clastics	Rolling glacial terrain (MIS 2); coastal landforms	Coastal erosion; storm surge
Iqaluit (NU)[m]	6	8	31.1	Archaen metamorphic units	Glacifluvial and glacimarine plain (MIS 2); marine plain; coastal and periglacial landforms	Coastal erosion; storm surge; permafrost failures

[m]year-round marine ports
[s]St. Lawrence Seaway ports

below the military fortification of Citadel Hill, initially was a product of British military surveyors. The contrast is especially evident between Halifax and Boston, USA, as both communities are built in almost identical topographical settings, on the steep flanks of drumlins. As a pre-existing community, Boston's streets do not follow a grid pattern, and their alignment is controlled by the drumlin and harbour.

St. John's downtown layout was also dominated by geomorphology. The modern harbourfront and Harbour Drive are built on infilled ground. At the time of initial settlement, the north side of the harbour consisted of several shallow indentations (coves) separated by spurs of gravel overlying resistant sandstone ledges. The street names, such as Becks Cove and Ayres Cove, preserve this geomorphological memory. Overland transport away from the harbour was limited by the steep hills, which could not be navigated by horses. Consequently, a street pattern developed with short streets at right angles to the coastline (Barter's Hill, Hill O'Chips) linked by long streets aligned northeast-southwest which gradually climbed the faces of the hills towards the northeast (Duckworth Street, New Gower Street-Queen Street).

The shift in transportation to rail in the nineteenth century impacted both existing and new urban form in Ontario (e.g. Guelph, Kitchener, London), Québec (Montréal), and New Brunswick (Moncton), although less so in most of atlantic Canada. The necessity to live directly adjacent to the resource or along the riverbanks no longer existed. Community development tended to move away from riverbanks and towards grid patterns focused on the railway station. In southern Ontario, this trend was accentuated by the existing grid system of concession roads, made feasible by the relatively flat glacial and glacilacustrine (southwestern Ontario) and glacimarine (eastern Ontario) plains. Once a community is established, there are strong obstacles to moving: tradition, property ownership, and inconvenience.

Construction of cities on glacilacustrine and glacimarine terrains, largely plains, in areas with no volcanic and lesser earthquake activity (very limited in southern Ontario, but present in eastern Ontario, Québec, and New Brunswick; Broster 1998), has resulted in relatively few geomorphic hazards (Fig. 21.16). Aside from storm-induced flooding of river systems, geotechnical problems focus on slope stability in ravines, karst dissolution, construction issues and settlement due to interbedded glacilacustrine sand and clay (e.g. Windsor ON; Hudec 1998; Oshawa; Brennand 1998), slope failures in thixotrophic Champlain Sea clays (Ottawa, Bélanger 1998; Montréal, Trois Rivières, Karrow 1998b), failures associated with deep glacifluvial and outlet channels and tunnel valleys (St. Catharines; Menzies and Taylor 1998), difficulties encountered during subway tunnelling through Quaternary successions (Toronto, Baker et al. 1998; Ottawa, Montréal), and glacitectonic-induced fracturing of bedrock (Montréal, Durand and Robillard 1998).

The impact of human intervention is visible in many river systems. Rivers are constrained by artificial ramparts and dykes, and confined to culverts and underground storm drains and sewers in cities. In cities such as London (ON; Dreimanis et al. 1998), Saguenay and Sherbrooke (QC), and Fredericton (NB; Broster 1998), construction in floodplains has restricted sediment availability and input into reaches, causing rivers to become narrower, deeper and straighter. In cities and towns, this has both geomorphic consequences, changing hydrodynamic flow and sediment transport, and human consequences, changing the way that residents see and interact with the river, and their view on what the river 'should' look like and how it 'should behave'.

Many communities in eastern Canada are affected by seasonal river flooding resulting from ice jams (Fig. 21.17). Border ice develops throughout the winter, as freezing extends from the banks towards the channel centre. Slow-moving (<1 km/h) liquid water around the frazil becomes supercooled, congealing to form slush pans, pancake ice and rafts. Gradually, the moving ice becomes attached to river banks, gravel bars, or artificial abutments, forming anchor ice. When ice coverage exceeds 80% of the river surface, continued growth and accumulation at constricted areas can form an ice jam across the river. The increased weight and drag along the keels of the rafts results in adherence to the river bottom, limiting the flow of water and producing flooding behind the dam. Rafts driven into the base of shallow channels form pressure ridges, scours and grooves. Dislodged trees and brush form sites for further ice accumulation. Artificial obstructions, such as bridge abutments and ramparts intended to prevent erosion, can also serve as sites for dynamic ice jams. Continued movement shoves and thrusts the rafts, producing a ridged, hummocky surface.

Thermal ice jams (freeze-up jams) result from rapid changes of water surface temperature in shallow rivers. Thermal ice jam formation is associated with a location where transported rafts can accumulate, such as a bar surface or an artificial feature in the channel (e.g. a bridge abutment).

Dynamic ice jams are triggered by partial break-up of ice cover, either during spring melt (freshet) or as a result of a mid-winter thaw. In large rivers, ice slabs are carried downstream at 10 km/h or more. Drifting river ice lodged or thrust against obstructions or at constrictions in the channel can produce large and destructive floods. The area influenced by a dynamic ice-jam flood is related to the geomorphology of the river system, rather than the amount of winter precipitation. Failure of the ice jam can result in downstream release waves of fast-moving water and ice chunks.

Areas subject to repeated dynamic ice jams can be recognized by disruptions to the vegetation, including broken and ice-scarred trees, and may also show increased diversity of species. Scars preserved in tree rings, and ^{14}C dated accumulations of ice-transported mixtures of woody debris and cobbles, provide longer-term records of ice jams for locations without written records.

Dynamic ice-jam flooding occurs repeatedly along the Saint John River through New Brunswick. Perth-Andover has seen six significant floods between 1987 and 2018, with dynamic ice jamming accentuated by rain-on-snow events and spring freshet. The flood of 23 March 2012 displaced more than 500 people and caused significant property damage.

Perth-Andover is built on a point bar along the west side of the south-flowing river. Water flow drives floating ice towards the point bar across the shallow channel, making it more likely that ice will jam. Erection of protective ramparts could serve as points for border or anchor ice development. Similar effects result from the construction of bridges, with their abutments and supports forming locations for sediment or ice accumulation. Disaggregation of ice cover by blasting has been used in some cases, but unfortunately this often merely exports the problem downstream, and can result in a worse situation if the volume of ice on the move is increased. The best approach could be to relocate the community, but that is probably politically and socially unacceptable. People develop strong attachments to their places, and are usually very reluctant to move. Rezoning or relocation of the most vulnerable structures from the riverfront is a possible compromise, as has been done in Perth-Andover and other riverside communities.

Eastern Canadian communities, including Ottawa-Gatineau, Montréal, and Fredericton, are also affected by freshet flooding, resulting from snowmelt and spring runoff. Freshet-related flooding along the Ottawa and St. Lawrence

Fig. 21.16 Geomorphic setting of Ottawa–Gatineau. The Ottawa River forms the boundary between Ottawa (ON) and Gatineau (QC), and flows eastward. Grenville gneiss and granite of the Canadian Shield is exposed above the 213 m marine limit in Gatineau Park, northwest of Gatineau. Ottawa is underlain by Cambrian sandstone and Ordovician limestone, exposed at the Chaudière (CF), Hog's Back (HB), and Rideau (RF) Falls, and on the flanks of Parliament Hill (P) and Rockcliffe Park (RP). McKay Lake (ML) is a marl lake developed in limestone karst. Most areas below 213 m asl are covered by thick marine clay deposits of the Champlain Sea, resulting in planar topography and permitting development of the grid pattern of streets. Dams at Hog's Back Falls and Chaudière Falls impound the Rideau and Ottawa Rivers respectively. The Rideau Canal was completed in 1832, bypassing Rideau Falls. Dow's Lake (DL), an artificial impoundment along the canal, is the remnant of Dow's Great Swamp (geomorphically, marsh and fen), which extended to the Ottawa River upstream of Chaudière Falls. The Ottawa, Rideau, and lower Gatineau Rivers are flanked by Champlain Sea sediments along low banks. The Rideau River, lower Gatineau, and Ottawa River downstream of the Gatineau confluence are susceptible to ice jamming and spring freshet flooding. Repetitive flooding makes the large longitudinal bars, such as Kettle Island (KI), unsuitable for human habitation

Fig. 21.17 Ice jam flooding is a significant hazard on many eastern Canadian rivers. **a** Developing dynamic ice jam, Ottawa River. **b** Perth-Andover (NB), constructed on a point bar along the Saint John River, is subject to ice-jam flooding. This flood occurred in 2012. **c** Ice-jam flooding along the Exploits River in 2003 inundated Badger (NL), although fortunately without human casualties

Rivers in April–May 2017 caused $225 million damage, and displaced several thousands. Timing, discharge, and duration of spring flooding in several rivers of eastern Canada have shown significant changes resulting from climate warming.

Rainfall on snow-covered terrain can also induce flooding, along with slope failure on steep, urbanized terrain. In Corner Brook NL, 32 mm of rain falling over snow-covered terrain over 48 h is sufficient to trigger slope failures (Hickman 2006). Development of upslope areas, alteration of the four rivers and other natural drainage routes, and inadequate maintenance of drainage infrastructure all contribute to rain-on-snow flooding. Between 1940 and 2005, more than 190 incidences of rain-on-snow flooding were recorded in the community (Hickman 2006).

For 2014–2070, climate change is anticipated to result in an increase in rain-on-snow flood events for most urban areas of eastern Canada. Rain-on-snow events will become a primary flood generating mechanism throughout southern Canada.

21.7 Tourism

Tourism includes intentional and unintentional travel for leisure, recreation, or for a specific non-commercial purpose. Geotourism involves visits specifically intended to have geological and/or geomorphic significance (Rutherford et al. 2014). Although it would be hoped that geomorphologists would consider the consequences of any of their actions to the geomorphology in a geo-ethical manner, most other tourists are not aware of the potential impacts of their visits.

In coastal areas throughout eastern Canada,, including lacustrine coastlines, tourism sites are subject to erosional impacts from coastal processes and anthropogenic activities concurrently. Studies have considered the relationships

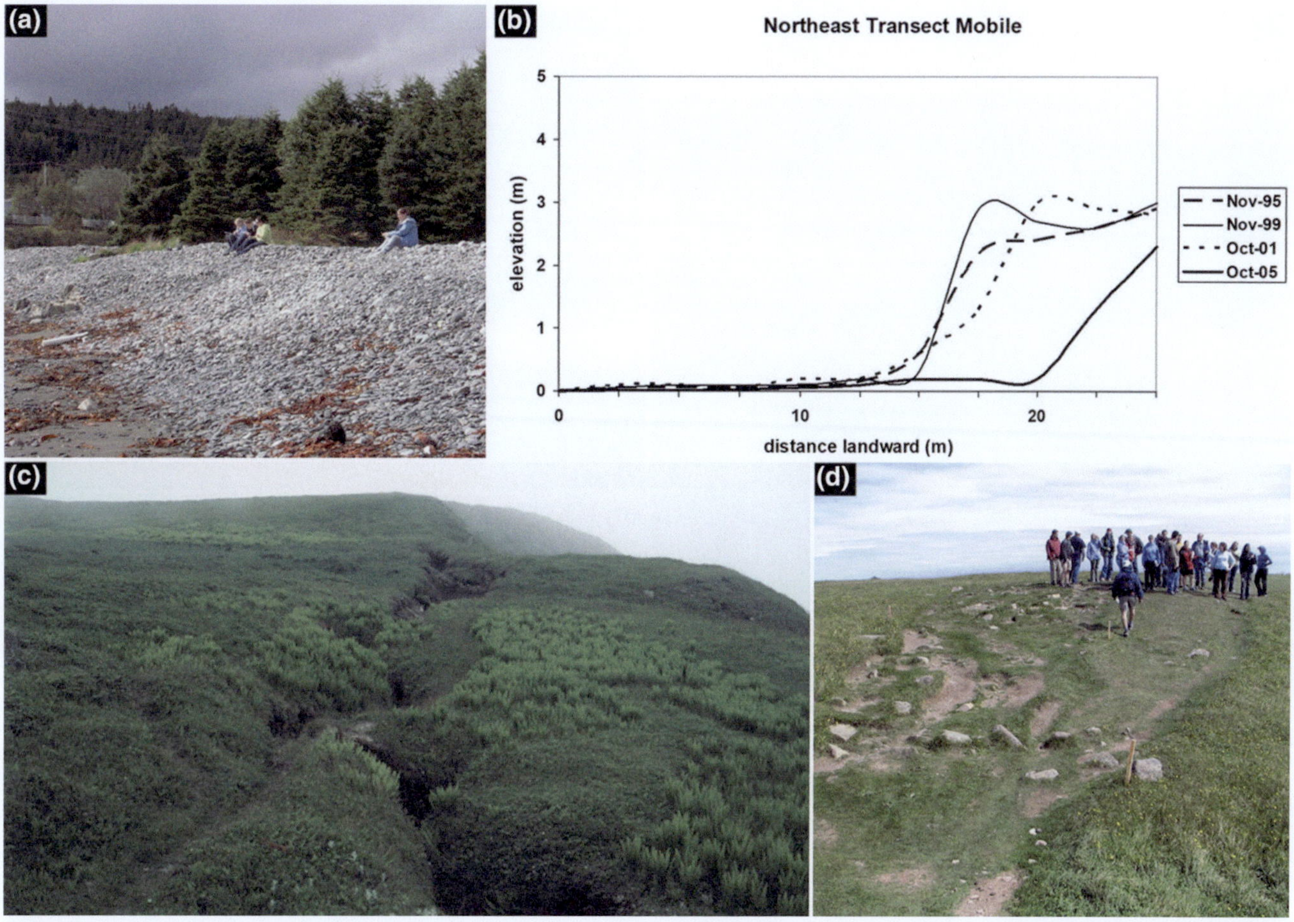

Fig. 21.18 Coastal tourism can have geomorphic impacts. **a**, **b** At Mobile (NL), repeated foot traffic across the barachoix was among the factors contributing to lowering and regression of the crest between November 1995 and October 2005. **c** Foot traffic across this peat blanket bog overlying coarse glacial diamicton resulted in slope failure at Cape St. Mary's Ecological Reserve (NL). **d** Even with the best intentions, geomorphic field trips can also result in vegetation damage, exposing sediment to frost penetration and erosion

between coastal erosion and tourism such as the impacts of weather events on tourism infrastructure, coastal erosion through foot traffic, and coastal landscape response to human impact. Substantial foot traffic over coarse gravel barrier berms can result in barrier coarsening and compaction, facilitating overtopping or overwashing during subsequent storm events. Unmanaged foot traffic can have consequences in sensitive coastal cliff-top settings, causing increased erosion mainly through the denuding of shallow soils and vegetation, exacerbating the impacts of rainfall and freeze-thaw cycles, and locally triggering topples, slumps and slides (Fig. 21.18).

Throughout atlantic Canada, tourism is currently a major economic driver: it is the largest economic industry in Prince Edward Island. It is a vital component in the future economic sustainability of most coastal communities, and in particular is seen as a key component in the economic revitalization of formerly fishery-dependent communities. Intensive marketing by the provincial governments promotes tourism. Lengthening of the tourist season, enhanced tourist use of natural areas, and geomorphic stresses induced by climate change in combination have affected sites throughout the region, and could have an impact on the long-term sustainability of both natural areas and tourism.

Beaches along the North Coast of PEI are among the province's major tourist attractions. Decreased snowfall and limited ice cover over the southern Gulf of St. Lawrence, in combination with sea level rise, has resulted in increased wave erosion and narrowing of beaches. Climate variations also result in increased degradation of coastal dunes. Stronger winter winds sweeping across exposed dunes, some eroded due to human pressure during the previous summers, have resulted in enhanced erosion and lowering and flattening of dune profiles. Increased foot traffic and vehicle access to beaches and backing dunes have accelerated erosion of coastal tourist sites.

The Pītaweikek (Hog Island Sandhills) dune complex along the north shore of Malpeque Bay has no road access, and hence very limited tourist pressure (Fig. 21.19). Comparative geomorphic analysis of this relatively undisturbed

Fig. 21.19 Limited tourist access to Pītaweikek (Hog Island Sandhills) barrier island complex along the Gulf of St. Lawrence PEI coastline has reduced the geomorphic impact. **a** The interior of Pītaweikek is composed of a low-slung parabolic dune complex. **b** The transverse costal dunes are relatively undisturbed by human activity. **c**, **d** In contrast, adjacent areas subject to more human foot traffic are marked by trough blowouts, accentuated overwash channels, and enhanced coastal erosion

complex with other more disturbed dune-backed coastal areas of Prince Edward Island and elsewhere in atlantic Canada indicates that Pītaweikek has retained most of its natural landscape and geomorphology. However, increased tourism without appropriate forethought will result in significant damage from erosion of the dune complex. The potential exists to simultaneously promote tourism and allow beach access without causing significant damage to the dunes, conserving the ecological integrity of the terrain.

In Gaspé, tourist pressure has caused stress on the peninsula's natural areas, particularly in Forillon National Park. The park has a relatively small area of 245 km^2, including preserved offshore marine areas. The majority of visitors come during July and August (at least 80%), and most who come stay for less than 1 day (75%) and come by private vehicle (>95%). Thus, the visitation pressure during the summer months is extreme, and traffic jams on the roads dead-ending at coastal cliffs or dunes are common occurrences during the summer. The balance between providing visitor services and preservation of the natural environment is precarious. This problem is exemplified at the coastal dunes of the Penouille Peninsula, where Parks Canada's desire to preserve the dune environment and the adjacent salt marsh conflicts with the tourist desire to utilize the coast for swimming and sunbathing.

Measures have been put in place to limit visitor impact. All freshwater fishing was banned in park lakes in 1995, and traditional marine fisheries were restricted in the coastal zone. The park management plan identifies forms of recreation to be encouraged, and also identifies forms which will not be promoted. Mountain biking, for example, is described as 'generally detrimental to the natural environment, and is not thought to contribute in any specific way to the authentic experience' of a visit to a national park.

Bruce Peninsula National Park welcomes more than 250,000 visitors annually. This visitation has posed problems, as the terrestrial area of the park is a mere 154 km^2. The park stands on the Niagara Escarpment, and supports the northernmost fringe of the deciduous forest assemblage of Southern Ontario. Its vegetation is further complicated by the calcareous bedrock, thinly veneered with soil and locally exposed directly under vegetation. Coniferous species are diverse, with *Juniperus virginiana* (eastern red cedar) representing the temperate component. *Pinus strobus* (eastern white pine) and *Pinus resinosa* (red pine) represent the southern boreal assemblage, and *Abies balsamea* (balsam fir) is also present. The most distinctive trees in the park are the old-growth *Thuja occidentalis*, (eastern white cedar) found in cliff-top settings over poorly developed highly calcareous soils (and locally, growing directly on limestone). The oldest, stunted cedars are more than 700 years in age, and survived the nineteenth century logging blitz due to their cliff-top positions and their gnarled trunks. Today, the cliff-top cedars are under threat from tourists who come to admire them, and inadvertently destroy the thin soils and trample and compress the near-surface roots, accelerating karst dissolution and toppling.

Tourism and recreational use in interior areas has relatively little geomorphic impact. Interior dune fields and glacifluvial landforms have come under pressure from off-road all-terrain vehicles (ATV). Wetland areas in the southern boreal forest have also been impacted by ATV use (e.g. Trip and Wiersma 2015; Fig. 21.20), although most studies have focused on biological rather than geomorphological impacts. Impacts on sediment budgets in river systems have also been documented (e.g. Fortin et al. 2014).

21.8 Coastal Environments

All port cities in Canada have added space for docks, warehouses, and rail lines by infilling the margins of harbours with sediment and rubble. Infilling the harbour in St John's, NL, has decreased the water area by approximately 30% since the time of European arrival c. 1510. In Halifax, the South End Terminal, Seaport, Naval Dockyard, and Fairview Cove Terminal are all built on infilled 'made ground' (Lewis et al. 1998). In Toronto, Front Street marked the shoreline of Lake Ontario until c. 1840: currently, the shoreline is 700–900 m south, separated by the Union Station rail complex, and Gardiner Expressway. The CN Tower and Rogers Centre are built on infilled terrain. All segments of the 170 km Lake Ontario shoreline from Oshawa to St. Catharines have been modified by human activity, most very extensively. Infill and changes in fluvial influx and lake level have reduced the surface area of Hamilton (ON) Harbour by more than 40% since 1820 (see Karrow and Morin 1998).

In areas where infilling overlies weak organic-rich or water-saturated sediment, slow settlement under the weight of buildings results over time. Geomorphic settlement began shortly after the expansion of human settlement, necessitating infilling of several harbours and drainage of the surrounding salt- or freshwater marshes and fens. In Thunder Bay (ON), construction over up to 50 m of Kaministiquia River deltaic sediment requires the use of deep pilings extending to Precambrian bedrock (Karrow 1998a).

In atlantic Canada, most coastlines away from the major ports appear 'untouched', suitable locations for the investigation of geomorphic phenomena, processes and change. However, combinations of rising sea level, changes in winter storm frequency and intensity, enhanced extratropical transition activity, reductions in offshore ice cover, ice foot formation, sediment flux, and both accentuated and reduced human activity have resulted in complex mosaics of geomorphic and biological processes and anthropogenic modifications.

21.9 Transportation

Diversions and impoundments for canal construction have resulted in geomorphic impacts along the Rideau Canal (Watson 2007), Trent-Severn Waterway (Angus 1988), and at Sault Ste. Marie (Cowan et al. 1998) in Ontario. Smaller canals, including the Shubenacadie and St. Peters (NS), Carillon and Chambly (QC), and Desjardins, Murray, Port Carling, Port Sandfield, and Wolfe Island (ON) have had little or no geomorphic impact.

Construction of the St. Lawrence Seaway includes canals at Sault Ste. Marie and the Welland Canal (including several upgraded versions and route changes since 1826), and impoundments with locks along the St. Lawrence River, including Lakes St. Lawrence, Saint-François (Francis), and Saint-Louis, to accommodate the 68 m drop between Lake Ontario and Montréal. Lake St. Lawrence, impounded by the Long Sault Dam at Cornwall (ON)–Massena (NY), covers 250 km^2 and extends 30 km upstream. Replacement of the original St. Lawrence River by the lake flooded nine villages in Canada and adjacent farmland, displacing 6,500 people. Although farmers and residents of the 'Lost Villages' were compensated, disputes occurred, with some residents being removed forcibly. Most residents were relocated to new, planned communities at South Stormont, Ingleside, and Iroquois. A number of heritage homes and buildings were transported from the inundation area and collected to form the historical tourist attraction of Upper Canada Village.

Fig. 21.20 **a** All-terrain vehicles (ATV) are commonly used throughout atlantic and arctic Canada. Unregulated use has resulted in substantial geomorphic and ecological impacts. **b** Despite being prohibited by legislation, ATV traffic has caused severe damage to this boreal fen

Prior to impoundment, the mean gradient of the 200 km reach of the St. Lawrence River between Brockville and Montréal was 0.00034. The dominant geomorphic effects were local flooding and enhancement of fringing wetlands, notably the Lac Saint-François National Wildlife Area.

Dredging of the St. Lawrence Seaway has impacted local geomorphology. One example is the fluvially dominated St. Clair River delta. The St Clair River flows southward from Lake Huron, discharging into Lake St. Clair, a shallow (modal depth 3 m) basin with limited wave and ice activity. As more than 3000 vessels pass through this segment of the St. Lawrence Seaway annually, dredging is required: a 10 m-deep channel is maintained through the delta and across the lake. Dredging has resulted in the extension of the altered distributaries into the lake, accentuating the digitate configuration of the delta.

21.10 Northern Environments

Although human impacts on geomorphology in Nunavut and the permafrost areas of Labrador, Nunavik, northern Ontario, and northeastern Manitoba are lesser than those felt further south, some effects are registered. In addition to climate change impacts on permafrost and coastal regimes, these include harbour dredging to facilitate seasonal access, local

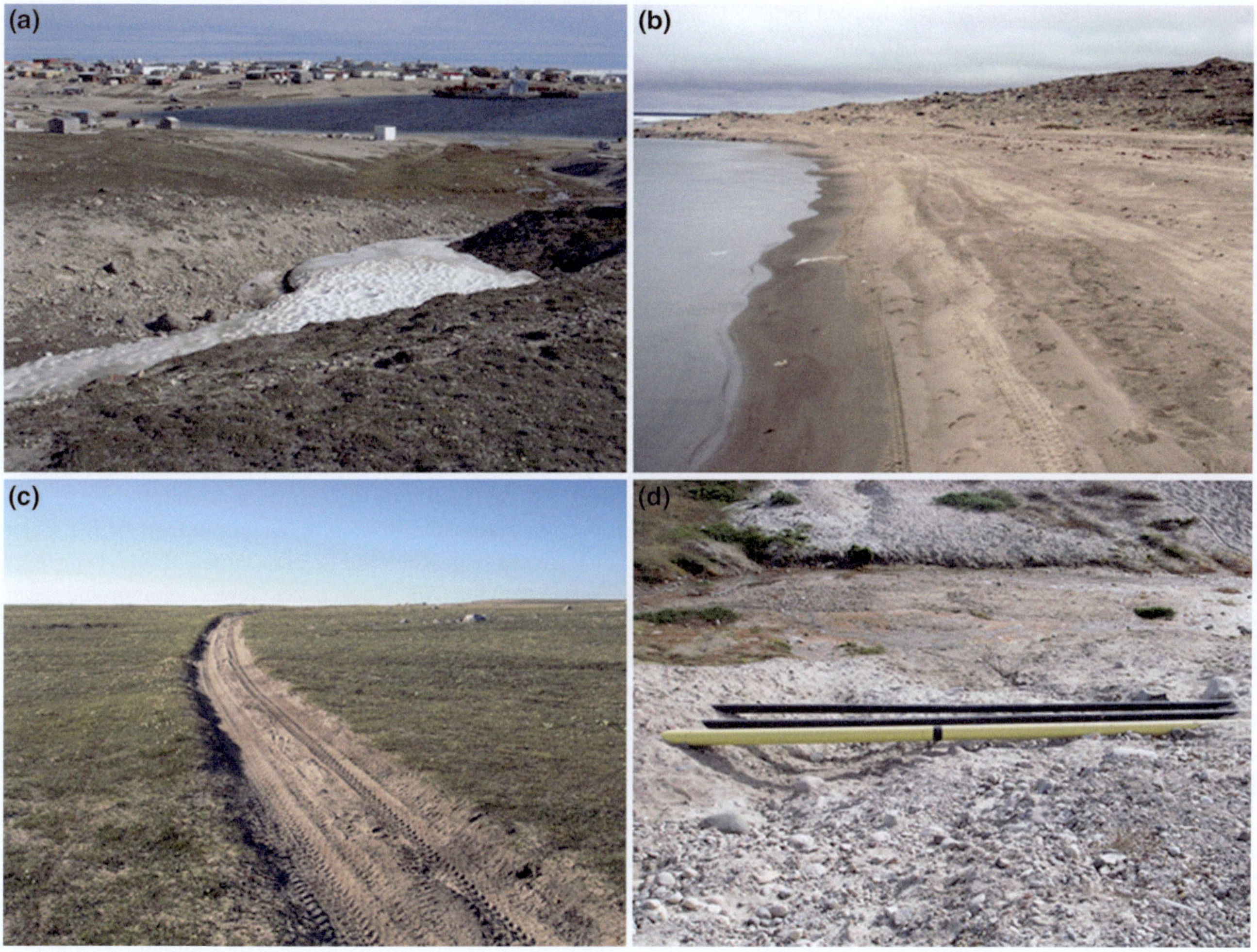

Fig. 21.21 Gjoa Haven, Nunavut, illustrates the influence of geomorphology on an arctic community. **a** The low energy, microtidal harbour is undergoing isostatic uplift and has substantial sediment input resulting from melting permafrost in the sandy substrate. **b** As ship traffic through the Northwest Passage increases, the low energy, sand-dominated, dissipative shoreline is potentially vulnerable to petroleum contamination. **c** All-terrain vehicle traffic destroys vegetation, leading to permafrost ablation in summer and enhanced seasonal frost penetration in winter. **d** Permafrost ablation and sediment erosion has exposed fuel pipelines, creating a hazard

gullying and fluvial erosion, and disturbances induced by construction of military installations (1950s), communities (1960s), and mines (since 1950). The impact of military installations on northern geomorphology was documented by Lackenbauer and Farish (2007).

Gjoa Haven (Uqsuqtuuq) provides an example of human impacts on local geomorphology in Nunavut (Papadimitriou 2007; Fig. 21.21). In common with most Nunavut communities, Gjoa Haven was established as a permanent (non-transitory) community by the Government of Canada beginning in 1960, and Inuit were encouraged (or coerced) to establish residence there. The community was laid out according to town planning standards of the time, with curved streets forming a grid. In 2016, the population was 1324. The recent increase in marine traffic through the Northwest Passage has increased the vulnerability of the coastline to petroleum contamination.

Gjoa Haven, built on glacimarine sand, gravel and silt, is currently undergoing glaci-isostatic emergence (< 1 mm/y). The community is marked by extensive gullying and some thermokarst activity, both of which potentially threaten structures in the community.

Ice in finer-grained materials forms planar layers or lenses of segregated ice, with concentrations locally in excess of 30%. These are 'thaw sensitive' (French, 1996) or 'thaw unstable' (Smith et al. 2001) materials. Thaw flow and slumping has occurred in silt- and fine sand-dominated units. Overland flow due to discharge from poorly consolidated sands and mixed sands and gravels has led to a network of steep-sided channels that poses a risk to infrastructure in the area. In one instance, continued erosion resulting from increased runoff from the community led to the relocation of a house.

Fuel pipelines in areas of the community have become exposed in part due to permafrost thawing as well as to overland gullying flow. Exposed fuel pipelines are crossed by roads and ATV tracks, and thus increase the risk of damage and hazard to the residents. In addition, Gjoa Haven's sewage lagoon is located approximately 1 km away from housing. Due to permafrost ablation, a breach occurred in the containment structure of the lagoon in 2005. Although repaired, ongoing ablation here could pose a risk to the community in the future.

Iqaluit is the largest community in Nunavut, and the only 'urban' centre (2017 population 8,011; Fig. 21.22). A community initiative in 2005 identified several potential hazards to infrastructure, including increased overland flow, failure due to permafrost ablation, and coastal vulnerability (Nielsen 2007; Hatcher and Forbes 2015). Growth of the community and increased ship traffic through the Canadian Arctic has led to expansion of infrastructure along the macrotidal shoreline, along with construction of casual buildings used for storage of fishing equipment and personal gear.

Relative sea level is currently rising, with an estimated upper rate of 7 mm/y between 2010 and 2100 (Hatcher 2014; Hatcher and Forbes 2015). Currently, storm surges result in the inundation of some public and private infrastructure. By 2100, some existing public infrastructure may be only 30 cm above mean spring high tide. Although ice foot development and offshore sea ice dampen wave energy, the ice-free season has lengthened by 30–40 days since 1979, exposing the shoreline to wave action.

Cumberland Sound and Frobisher Bay are marked by macrotidal and hypertidal regimes. At Iqaluit, the mean spring tidal range is 11.3 m (Dale et al. 2002), with tidal currents reaching 3.6 m/s. Extensive silt-sand tidal flats with conspicuous cobbles and boulders mark the bayhead areas, locally extending more than 1 km seaward (Hatcher 2014). The impact of seasonal ice activity is directly related to tidal range, with greater effects in hypertidal regimes (Gilbert 1990; McCann and Dale 1986). Entrained cobbles and small boulders may be moved landward or seaward. Sea ice contributes to the maintenance of tidal flats by hindering offshore sediment transport.

As in many northern coastal settings, winter erosion is limited by the development of a seasonal ice foot, formed by melting and refreezing of snow producing fractional recrystallization. An ice foot acts to protect the underlying sediment from erosion by storm waves, in addition to discouraging human use of the beach. Niveo-coastal features, formed by combinations of refrozen freshwater, snow, congealed saltspray, rime and trapped small fragments of ice (brash), also act to isolate sediments and coastal bedrock outcrops from wave action. At Iqaluit, the ice-free season is currently increasing by ~ 1.5 days per year. Anchorage of large vessels may require periodic dredging of the deltaic and tidal sediments accumulating at the head of the bay.

Major construction projects across discontinuous permafrost terrain in eastern Canada include the Hudson Bay Railway and port facilities at Churchill MB (constructed 1912–1931; see Montison 2015), and the Temiskaming and Northern Ontario Railway (now Ontario Northland Railway) and facilities at Moosonee, Ontario (constructed 1921–1932). Both railways traverse the Hudson Bay Lowlands, although the forested terrain surrounding James Bay provided fewer geomorphological and engineering challenges than did the tundra of northeastern Manitoba. Road construction across the discontinuous permafrost of the Canadian Shield (e.g. Route de la Baie James, Route Transtaïga, Trans-Labrador Highway) poses fewer geomorphic problems, as the terrain is dominated by exposed bedrock and coarse-textured glacial deposits, with few thermokarst features or extensive organic landforms.

Originally, Port Nelson, at the mouth of the Nelson River, was to be the site of Canada's Hudson Bay port. This choice may rank as the worst instance of disregarding geomorphology in Canadian history. The macrotidal estuary is dominated by constantly shifting mudbanks, propelled by sediment influx from the Tyrrell Sea deposits. Influx from the Nelson River, with a mean annual discharge of 70 km^3 (Déry et al. 2005), results in constant migration of the tidal channels, precluding effective dredging. Reworking by seasonal ice shove, enhanced in macrotidal regimes, further compounds the problems of maintain access to the port. Numerous attempts to dredge the channels and construct infrastructure, including an artificial island made of dredged fill connected to the shore by an 800 m-long causeway and 17-span bridge, failed. The difficulties of port construction at Port Nelson led Frederick Palmer, an engineering consultant brought into study matters, to remark, 'You Canadians constantly use two words as if they were one. Port and Harbour. A port is the facilities which you build to work ships. If you've got any damn sense, you put them in a harbour' (Stevens 1960). Consequently, the railway was shifted at Amery MB to extend 240 km due north to Churchill, near the site of the Hudson's Bay Company's Fort Prince of Wales.

The relocated line traverses extensive wetland areas, marked by peat plateaux and local thermokarst. The thickest, most extensive blankets of peat are associated with topographically low terrain, underlain by Tyrrell Sea silt.

Permafrost, peat plateaux, and vegetation assemblages develop together. Undisturbed forest communities over peat plateaux will act to preserve the underlying permafrost.

Fig. 21.22 Iqaluit, Nunavut, is located on a macrotidal embayment, flanked by continuous permafrost. **a** High tide allows ready access to port facilities. Housing construction requires insulation to avoid permafrost degradation. Exposed bedrock shows glacial erosional features, erratics, and frost wedging along joints. **b** Extensive tidal flats with boulders are exposed at low tide. **c** Icefoot development protects the shoreline during winter. *Photographs* Courtesy of Don Forbes

Climate warming will cause the permafrost to melt, but if the process occurs relatively slowly, the differences in relief will continue to support distinctive vegetation assemblages. More rapid increases in temperature result in permafrost melting, causing collapse of the plateau and leading to fenland development. Conversion of peat plateaux to wetland fens has occurred within the last 200 years across north-central Manitoba (e.g. Tarnocai 2009).

The discontinuous permafrost terrain continues to pose difficulties for railway operation, resulting in periodic closures. A combination of flood-induced washouts and permafrost-related failures closed the line from May 2017 through October 2018, interrupting grain shipments from Prairie Canada to Europe. Although ongoing climate warming and reduced sea ice in Hudson Bay may lengthen the shipping season for Churchill, maintenance of the railway remains a problem.

21.11 Conclusion

The study of geomorphology has evolved considerably in recent years. In eastern Canada, as elsewhere, the various roles of past and present human activity are increasingly being recognized. Differentiation of landscapes into purely physical and human components is no longer tenable, requiring us to acknowledge that geomorphology is more than the description, cataloguing, and physical interpretation of landforms. As landscapes are involved in all aspects of terrestrial life in eastern Canada, geomorphologists need to present their input and join the human conversations.

References

Adams GC (1993) Gypsum and anhydrite in Nova Scotia, 3rd edn. Nova Scotia Department of Natural Resources, Mineral Resources Branch, Information Circular ME 16, Halifax

Angus JT (1988) A respectable ditch: a history of the Trent-Severn Waterway 1833-1920. Queen's University Press, Kingston

Baker CL, Lahti LR, Roumbanis DC (1998) Urban geology of Toronto and surrounding area. In: Karrow PF, White OL (eds) Urban geology of Canadian cities. Geological association of Canada special paper 42, pp 323–352

Bélanger JR (1998) Urban geology of Canada's National Capital area. In: Karrow PF, White OL (eds) Urban geology of Canadian cities. Geological association of Canada special paper 42, pp 365–384

Brennand TA (1998) Urban geology note: Oshawa, Ontario. In: Karrow PF, White OL (eds) Urban geology of Canadian cities. Geological association of Canada special paper 42, pp 353–364

Broster BE (1998) Aspects of engineering geology at Fredericton, New Brunswick. In: Karrow PF, White OL (eds) Urban geology of Canadian cities. Geological association of Canada special paper 42, pp 401–408

Catto NR (2012) Natural hazard and vulnerability assessment in Atlantic Canada: review, progress, and challenges. In: Haque CE, Etkin D (eds) Disaster risk and vulnerability. McGill-Queen's University Press, pp 235–274

Catto NR, MacQuarrie K, Hermann M (2002) Geomorphic response to late Holocene climate variation and anthropogenic pressure, northeastern Prince Edward Island, Canada. Quat Int 87:101–118

Cowan WR, McAuley P, Bennett G (1998) Urban geology: City of Sault Ste. Marie, Ontario. In: Karrow PF, White OL (eds) Urban geology of Canadian cities. Geological association of Canada special paper 42, pp 197–206

Dale JE, Leech S, McCann SB, Samuelson G (2002) Sedimentary characteristics, biological zonation and physical processes of the tidal flats of Iqaluit, Nunavut. In: Hewitt K, Byrne ML, English M, Young G (eds) Landscapes in transition: landform assemblages and transformations in cold regions. Kluwer, Dordrecht, pp 205–234

Dawson JW (1868) Acadian geology: the geological structure, organic remains, and mineral resources of Nova Scotia, New Brunswick, and Prince Edward Island. John Lovell, Montreal

Déry SJ, Stieglitz M, McKenna EC, Wood EF (2005) Characteristics and trends of river discharge into Hudson, James, and Ungava Bays, 1964-2000. J Clim 18:2540–2557

Dreimanis A, Winder CG, Aaltonen RA (1998) London, Ontario: geology, geomorphology, geodata. In Karrow PF, White OL (eds) Urban geology of Canadian cities. Geological association of Canada special paper 42, pp 237–260

Durand M, Robillard M, (1998) Geological engineering information management in Montréal. In: Karrow PF, White OL (eds) Urban geology of Canadian cities. Geological association of Canada special paper 42, pp 385–396

Falcon-Lang HJ, Fensome RA, Gibling MR, Malcolm J, Fletcher KR, Holleman M (2007) Karst-related outliers of the cretaceous chaswood formation of Maritime Canada. Can J Earth Sci 44:619–642

Ford DC (1997) Principal features of evaporate karst in Canada. Carb Evapor 12:15–23

Fortin G, LeBlanc M, Schiavone S, Chouinard O, Utzschneider A (2014) Local Perceptions, Ruslefac mapping, and field results: the sediment budget of Cocagne River, New Brunswick, Canada. Env Man 55:113–127

French HM (1996) The periglacial environment. Longman, London

Gilbert R (1990) A distinction between ice-pushed and ice-lifted landforms on lacustrine and marine coasts. Earth Surf Proc Land 15:15–24

Hatcher SV (2014) People at the tidal flats: coastal morphology and hazards in Iqaluit, Nunavut. MSc thesis, Memorial University of Newfoundland, St. John's, NL

Hatcher SV, Forbes DL (2015) Exposure to coastal hazards in a rapidly expanding northern urban centre, Iqaluit, Nunavut. Arctic 68:453–471

Hickman H (2006) Flood hazard and vulnerability in Newfoundland Communities. MSc thesis, Environmental Science, Memorial University of Newfoundland

House K, Catto NR (2008) Gypsum Karst: human initiation and response, Woodville, NL, Canada. European Geosciences Union, Vienna, April 2008; EGU 2008-A-09878

Hudec PP (1998) Geology and geotechnical properties of glacial soils in Windsor. In: Karrow PF, White OL (eds) Urban geology of Canadian cities. Geological association of Canada special paper 42, pp 225–36

Karrow PF (1998a) Urban geology note: Thunder Bay, Ontario. In: Karrow PF, White OL (eds) Urban geology of Canadian cities. Geological association of Canada special paper 42, pp 191–196

Karrow PF (1998b) Urban geology note: Trois-Rivières, Québec. In: Karrow PF, White OL (ed) Urban geology of Canadian cities. Geological association of Canada special paper 42, pp 397–400

Karrow PF, Morin FJ (1998) Urban geology note: Hamilton, Ontario. In: Karrow PF, White OL (eds) Urban geology of Canadian cities. Geological association of Canada special paper 42:279–286

Lackenbauer PW, Farish MW (2007) The cold war on Canadian soil: militarizing a northern environment. Env History 12:920–950

Lewis CFM, Taylor BB, Stea RR, Fader GBJ, Horne RJ, MacNeill SG, Moore JG (1998) Earth science and engineering: urban development in the Metropolitan Halifax region. In: Karrow PF, White OL (eds) Urban geology of Canadian cities. Geological association of Canada special paper 42, pp 409–444

MacArthur MEL (1976) An ecological study of the Greenwich sand Dune system. MSc Thesis, Department of Biology, Acadia University

McCann SB, Dale JE (1986) Sea ice breakup and tidal flat processes, Frobisher Bay, Baffin Island. Phys Geog 7:168–180

Menzies J, Taylor EM (1998) Urban geology of St. Catharines-Niagara Falls, Region Niagara. In: Karrow PF, White OL (eds) Urban geology of Canadian cities. Geological association of Canada special paper 42, pp 287–322

Montison JM (2015) Churchill, Manitoba and the Arctic Gateway: a historical contextualization. Can Geog 59:304–316

Nielsen D (2007) City of Iqaluit's climate change impacts, infrastructure risks and adaptive capacity project. City of Iqaluit, Nunavut

Papadimitriou S (2007) Climate-related hazards and risks to community infrastructure: Gjoa Haven, Nunavut. Master's of Environmental Science thesis, Memorial University of Newfoundland

Rutherford J, Kobryn H, Newsome D (2014) A case study in the evaluation of geotourism potential through geographic information systems: application in a geology-rich island tourism hotspot. Current Iss Tourism 18:267–285

Smith SL, Burgess MM, Heginbottom JA (2001) Permafrost in Canada: a challenge to northern development. In: Brooks GR (ed) A synthesis of geological hazards in Canada. Geol Surv Can Bull 548:241–264

Spray JG, Thompson LM, Biren MB, O'Connell-Cooper C (2010) The Manicouagan impact structure as a terrestrial analogue site for Lunar and Martian planetary science. Planet Space Sci 58:538–551

Steedman RJ, France RL (2000) Origin and transport of Aeolian sediment from new clearcuts into Boreal lakes, northwestern Ontario, Canada. Water Air Soil Poll 122:139–152

Stevens GR (1960) History of Canadian National Railways, vol 2. MacMillan, Toronto

Tarnocai C (2009) The impact of climate change on Canadian peatlands. Can Water Res J 34:453–466

Trip NVV, Wiersma YF (2015) A comparison of all-terrain vehicle (ATV) trail impacts on boreal habitats across scales. Nat Areas J 35:266–278

Watson KW (2007) The Rideau route: exploring the pre-canal waterway. Privately published, Elgin, Ontario

Webb TC (2002) Geology, development history, and exploration alternatives for gypsum resources at Cape Maringouin (parts of NTS 21 H/10, 21 H/15 and 21 H/16), Westmorland County, southeastern New Brunswick. Ministry of Natural Resources, New Brunswick, Fredericton, Open File OF-8

Norm Catto joined Memorial University in July 1989 and is now the Head and Professor of Geography at that university. His research and teaching interests include coastal landforms, natural hazards, sea-level change, and impacts in coastal environments; response of river systems to climate and weather events, and flood risk assessment; aeolian geomorphology and response to climate and human factors in boreal and costal environments; mass movements and slope failures; loess deposition; palaeosol formation; emergency measures policies and effectiveness human adaptations to climate and weather events; the impacts of climate and weather events on agriculture, transportation, fisheries, and communities; and the relationships among landscapes and people. His research has included projects and investigations in landscapes and environments in 16 countries, and in all Canadian provinces and territories, particularly in eastern Canada. For 18 years, he served as Editor-in-Chief for *Quaternary International*, a professional international journal dedicated to the study of landscape evolution and climate variation throughout the past 3 million years. He has taught more than 50 different undergraduate and graduate courses focused on numerous aspects of geomorphology, landscape analysis, Quaternary research, natural hazards, ocean sciences, environmental science, climate and climate change, soil science, geoarchaeology, and palaeontology. Study of landscapes forms his central focus for understanding of the various areas of eastern Canada,, the combination of physical and cultural constituents that contribute to its identities. For most species, and for all human occupants, life in eastern Canada, has been heavily influenced by geomorphology since deglaciation.

Olav Slaymaker is a Member of the Order of Canada, and professor emeritus of geography in the Department of Geography at the University of British Columbia, Vancouver, B.C. His 20+ co-authored and edited monographs and 160+ refereed papers reflect three different styles of work. First, he has pursued original field research for 40 years on sediment systems in the Canadian Cordillera and the European Alps, strongly supported and enhanced by his stellar doctoral and post-doctoral students and international colleagues. Second, he has produced papers and books summarizing trends in land use and environmental change in mountain regions world-wide. Third, he has edited monographs with invited international contributions on major environmental and geomorphological themes. He is former President of the Canadian Association of Geographers and former President of the International Association of Geomorphologists.

Dori J. Kovanen is a professional engineering geologist and Research Associate in the Department of Geography at the University of British Columbia, Vancouver, BC. With an M.Sc. from Western Washington University in Bellingham and a PhD in Quaternary geology from the University of British Columbia she is a specialist in Quaternary geomorphology. She has written extensively on Late Glacial and Holocene history of the Pacific Northwest as well as contemporary environmental changes in the region and was a significant contributor to 'Landscapes and Landforms of Western Canada' (Slaymaker 2017).

22 Glacitectonic Caves in the St. Lawrence Lowlands of Québec

Jacques Schroeder, Michel Beaupré, and Daniel Caron

Abstract

Even though glacitectonic features have been identified in the Montréal area since the 1960s, it is only more recently that their extent and characteristics have been analyzed in greater detail. Exploration and studies have revealed that glacitectonic caves, hundreds of meters long, formed in platform limestones. Two different cases study are briefly discussed: the St-Léonard caves on Montréal Island and the Beauport caves near Québec City. These studies confirm that the weight and displacement of active glaciers can have disturbing effects, including dislocation, folding or stress redistribution, on the underlying bedrock, extending at least a few tens of meters deep.

Keywords

Glacitectonic • Caves • Geomorphology • St. Leonard • Beauport

22.1 A New Geomorphological Object

In Québec province, some 90% of its population lives in its southern part along the St. Lawrence Valley. Over the last half century, the increase in infrastructure has been faster than the increase of population. In some cases, excavation works have encountered voids in the underlying bedrock composed mainly of Palaeozoic bedrock, including limestones. Some 500 caves are known in the province, most of small dimensions with lengths ranging from a few meters to a few kilometers. They are karst caves in that they have been formed by dissolution of limestones by underground waters. However, over the last 50 years, we have found caves that were formed by the dislocation of limestone masses due to deformation by Quaternary glaciers. These mechanical caves, called glacitectonic caves (Schroeder 2004a), and therefore not karst, are paradoxical objects. Earth science specialists responsible for describing bedrock typically relate them to the karst concept because they are found in limestone, and water, running or not, are usually present. This conceptual and deductive approach replaces, for these specialists, the proper study of these caves, which are primarily geomorphological objects with a history distinct from that of the bedrock, and a more complex one than their circulating waters might indicate (Schroeder 1991).

J. Schroeder (✉)
Département de Géographie, Université du Québec à Montréal, CP8888, succ Centre-ville, Montréal, Québec, H3C 3P8 Canada
e-mail: schroeder.jacques@uqam.ca

M. Beaupré
Hatch Engineering, Montréal, Canada

D. Caron
Société québécoise de spéléologie, Montréal, Canada

22.2 Glacitectonic Cave Hypothesis Confirmed

In the 1960s and 1970s, from numerous excavation works, it was noted that the substratum, up to 25 m thick, was perturbed by reverse faults, centimetric openings of vertical joints, both in contact with the surface or at depth, and by compression meso-folds (Schroeder and Beaupré 1985). The assumption from the authors of these field observations was that these dislocations and these deformations of the bedrock resulted from Quaternary glacial flow of the Laurentide Ice Sheet Complex. Glacial movement resulted in the detachment and displacement of slices of the substratum of about 1 km^2. The "glacitectonic" hypothesis was established by Grice (1972), Durand and Ballivy (1974), Ballivy et al. (1977) and Durand (1978) (Fig. 22.1). It was concluded that, without any topographic obstructions, only ice sheet flow could have been able to produce enough energy and the mechanical conditions suitable for dislocation and deformation of the substratum basement. Disruption of the bedrock strata caused problems for progress at infrastructure worksites. Blasting required and planned for excavation work was unsuccessful, as the energy was dissipated in the

© Springer Nature Switzerland AG 2020
O. Slaymaker and N. Catto (eds.), *Landscapes and Landforms of Eastern Canada*, World Geomorphological Landscapes, https://doi.org/10.1007/978-3-030-35137-3_22

Fig. 22.1 Overlapped and displaced limestone blocks mixed with till are present in the upper part of the exposure. Ice-push was from left to right. In the lower part, the limestone bedrock is dislocated and joints are open. Under differential ice push, joints were filled with a breccia injected from the base of the upper part. Photograph taken by Marc Durand in the 1970s

discontinuities of the rock. More recently, some tunnel excavation works have encountered unexpected dislocated saturated limestone masses, stopping excavation work for extensive periods. This entailed additional costs which were not foreseeable in view of the assumed geological characteristics of these rocks. Until recently, knowledge of the "glacitectonic" hypothesis remained largely confined to construction and engineering geologists.

22.3 Favorable Ice Movements

Circa 13.5 ka, the St. Lawrence Valley was entirely covered by the Laurentide Ice Sheet Complex, flowing south along the Lake Champlain Valley and southwest along the St. Lawrence Valley. By 13 ka, the ice front had retreated just north of Montréal Island, while proglacial lake Candona covered parts of the valleys. This lake would be shortly replaced by the Champlain Sea that would completely inundate the Montréal area (Fig. 22.2; see Chap. 2). To the east, the ice front was just south of Québec City by 13 ka. Ice movement was initially to the southeast but turned east when the St. Lawrence valley was deglaciated. The Québec City was then submerged by the Champlain Sea.

22.4 St-Léonard Cave: A Surprising Story

The St-Léonard Cave, in Pie-XII Park, has been known and visited since 1812. It was officially opened to the public during the summer of 1981, after being closed in 1968. Only 35 m of cave was then known. On October 12th 2017, two cavers from the Société québécoise de spéléologie, Daniel Caron and Luc Le Blanc, following excavation work in a tight fissure at the end of the known cave, discovered more than 370 m of new, spectacular galleries. This unexpected discovery would be relayed by a surprising number of media, from local to worldwide. The mapping of this new cave extension during the following months demonstrated that it is, most probably, directly related to the Du Saguenay Street glacitectonic cave (±317 m), discovered and mapped in 1982, some 400 m to the northwest (Fig. 22.3) (Schroeder et al. 1986).

The area around Pie-XII Park is relatively flat with ground elevations around 40 m. A few rocky cuesta-like crests, some 1–3 m high and trending NE–SW, are present over the southern part of the cave. Elsewhere, ground is gently sloping to the northwest and the northeast. In the northern part of the Du Saguenay Street cave, the elevation is around 34 m. The southern part of the cave is in Pie-XII Park, with the remainder under a 50-year-old residential area. The Du Saguenay Street cave in the residential area is accessed through manholes.

Theses caves were formed in limestones belonging to the Rosemont Member of the Montréal Formation (Trenton Group, Middle Ordovician, ±460 Ma). They are located in the geological province and physiographic unit of the St. Lawrence Lowlands. The Rosemont Member is some 75 m thick and is composed of various more or less shaly limestones in irregular and thin (5–15 cm) layers, with minor calcareous shales.

In this part of Montréal Island, limestones are usually sub-horizontal or show very low dip ($\leq 2°$) to the east or northeast. The cave area is located on the southeast side of the Ahuntsic syncline, a SW–NE trending open fold with small northeast plunges. In Pie-XII Park, cave-bearing strata dip to the northwest (6°–9°), toward the Ahuntsic syncline throw. These dip values are unusually high for this area. In the Du Saguenay St. cave area plunge, bedding is almost horizontal.

Joints are sub-vertical to vertical (85°–90°), with metric to pluri-metric spacing and show the following trends: ESE–WNW to SE–NW and N–S (±15°). Some minor joints show E–W and SSE–NNW trends. Joint surfaces usually have millimetric weathering and thin clay fillings. Joint orientation has a major influence on cave configuration, all galleries having formed along these joints.

In this part of the island, dykes and sills, with centimetric to metric thickness, criss-cross limestone formations. They were emplaced during the Lower Cretaceous (±118–138 Ma) and relate to the Monteregian hills intrusions. In the Pie-XII Park cave there is a sill, a few centimeters to 40 cm thick, very badly weathered and locally cross-cutting layers (Fig. 22.4). It is located mainly in the upper parts of the cave galleries. Some millimetric to centimetric thick weathered dykes are also seen in vertical joints.

The Pie-XII Park area shows very dislocated rock outcrops that are covered by a very thin layer of till (≤ 1 m). Till thickness increases gradually to the north and west and can reach 3 m in the Du Saguenay Street cave area. This till is covered, to the north, west and southwest, by silty, clayey, calcareous and sometime fossiliferous Champlain Sea marine deposits. Marine transgression has reached the 170 m level, thus drowning the Pie-XII Park and Du Saguenay Street caves. Some marine clays have been seen in the Du Saguenay St. cave.

Detailed geomorphological mapping of this new extension shows that ceilings and walls are planar, and that walls fit together on small scale features as well as larger ones, suggesting a southwest displacement of the rock mass, in alignment with the glacial southwest movement (Fig. 22.5). These surveys have also shown that cave ceiling bedrock was usually of poor quality and often brecciated, suggesting sliding planes. Rock quality in cave walls is much better, and local dislocations usually produce much larger blocks. The southern part of the new gallery has a floor composed mainly of fallen rock pieces of unknown thickness, while the northern part is partly filled with very slowly flowing water up to 6 m deep. The cave floor in this part is much smoother and contains greyish fine sediments. Ceiling levels going north tend to follow the topography and dip of limestone strata (Fig. 22.3). The Du Saguenay St. cave, with smaller

Fig. 22.2 Simplified DTM of the St. Lawrence Valley with 13.5 and 13.0 ka isochrons. Glacier flow on Montréal Island, slightly before deglaciation, was to the southwest, toward Lake Candona. In the Québec City area, the last ice flow was to the east, toward the St. Lawrence Estuary. Glacial movement induced joint opening in the limestones to some 20 m depth. Displacements of limestone masses, based on morphological study of fitting patterns, correspond to the last recorded ice movement

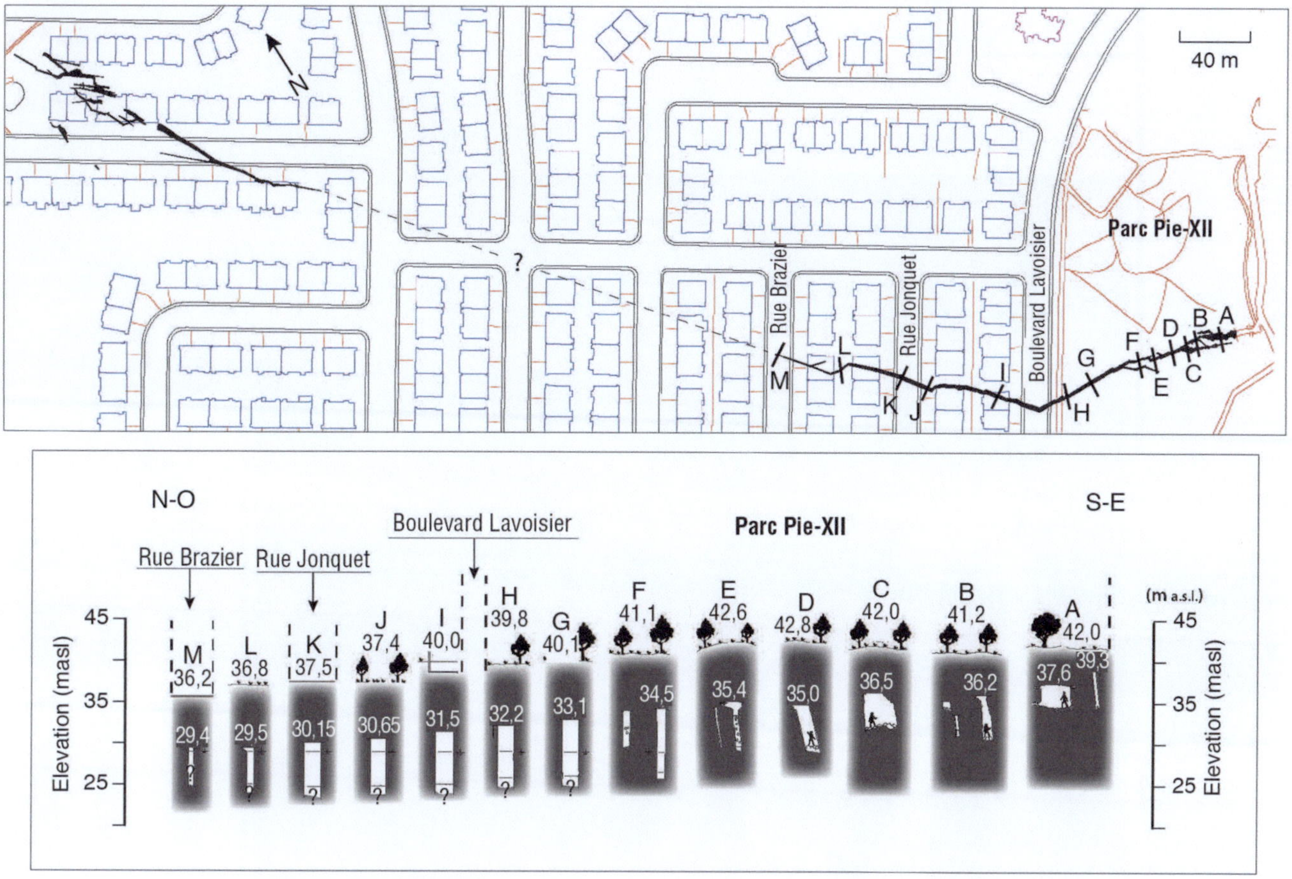

Fig. 22.3 Map of the Montréal area where glacitectonic caves are known. To the left is the cave system discovered and studied in 1982, the Du Saguenay Street Cave. On the right side is the historical Pie-XII Park Cave with its new extension discovered in 2017. The lower figure shows views of its northern part. Cross sections of the Pie-XII cave show that it is some 6 m under the surface. The gallery ceiling corresponds to a massive limestone bed (sections A to M). Those two caves are some 400 m apart. Fitting patterns of joint walls show that rock masses have been displaced toward the southwest

galleries (±317 m), shows the same geomorphological features and slowly flowing underground waters. Its southern gallery is perfectly aligned with the northern gallery of the Pie-XII Park cave, thus strongly suggesting a connection between these two glacitectonic caves.

22.5 Beauport Caves: When a River Gets Involved

The St. Lawrence River downstream of Québec City is overhung by a spectacular escarpment, from which fall the waters of the Montmorency River. The Beauport caves have developed north of this escarpment in a limestone platform whose surface is sloping gently (±3%) to the south, from elevations of ±120 to ±90 m asl. The Montmorency River has entrenched itself in these limestones to reach Precambrian rocks, creating a canyon 15–25 m deep to the east. The limestone platform is also cut, in its southern part, by the Montmorency escarpment, some 90 m high.

In 1981, in the downstream part of that platform, a 190 m long cave was discovered under streets and houses. The underground waters flow there year round and emerge a few tens of meters south, where several springs are located. Dye tracing confirmed that these waters were coming from the left bank of the Montmorency River, just upstream of the Marches Naturelles hydroelectric power plant built in 1908. As the water in this reservoir is some 20 m higher than these springs, it was concluded that this was a classic hydrogeological breakthrough. In 1996, a pit more than 6 m deep was discovered under the lawn of a house located some 275 m WNW from the first discovery (Fig. 22.6). Subsequent explorations through the years have mapped a complex glacitectonic cave system with large galleries, more than 500 m long (Fig. 22.7). Due to subsidence problems experienced in this residential area, Québec City authorities

Fig. 22.4 Photo assembly showing the small fissure that was enlarged by the two cavers in 2017 to reach the new cave extension (see photograph to the left). This spot is located at the limit between brecciated limestone in the ceiling and a zone where the limestone bedrock is open along joints (dislocated). The central part, enlarged on the upper right side, shows that the differential movement between the ceiling and the dislocated underlying limestone was concentrated mainly along a sill, its upper part being deformed and mixed with the overlying breccia. The lower part of the sill remained fixed to the underlying limestones

decided, in 2015, to clog numerous drainage input sinks mapped along the left bank of the Montmorency River. The sinks were feeding those underground galleries and causing suffusion phenomena in specific areas. The distance between sinks and springs is nearly 1 km.

The Beauport glacitectonic caves are formed in the Ordovician limestones of the St. Lawrence Platform, between the igneous and metamorphic Precambrian rocks of the Grenville Province to the north, and the Appalachian sedimentary terrigenous rocks to the south. These limestones rest uncomfortably on Precambrian rocks exposed along the Montmorency River and are truncated in the south by the NE–SW trending Montmorency normal fault. Limestones belong to the Pont-Rouge, Deschambault, and Neuville formations of the Middle Ordovician Trenton Group. Limestones beds are usually sub-horizontal or show small dips

Fig. 22.5 Large gallery found at the beginning of the new section discovered in 2017. Thought quite planar, walls show small protuberances on one wall and hollows in the opposite wall that fit very well together. The cavers indicate the trend of rock mass displacement

mainly to the southeast and the east. Total thickness of this sedimentary sequence is unknown in the cave area but is assumed to be at least 25 m. Total limestone thickness tends to decrease toward the southeast.

Surveys in the area have identified two main sets of sub-vertical ($\geq 70°$) joints: WNW–ESE to NW–SE and ENE–WSW with subsidiary sets of sub-vertical joints trending N–S and E–W. Underground joints of the WNW–ESE set play an important role in underground water transit from upstream inputs to downstream springs. The ENE–WSW set seems to connect those long joints.

Thickness of surface material in the caves area ranges between 0 and 6 m. These deposits are composed of more or less washed till covered by deltaic deposits (sands and gravels) in the northern parts, fining southward (silty and clayey sands) and covering marine deposits of the Champlain Sea in the south. All deposits are of MIS 2 (Late Wisconsinan) age.

The central and downstream parts of this cave complex are clearly of glacitectonic origin, with usually straight and planar walls fitting together and ceilings mostly composed of very dense immature clayey till and brecciated limestones (Fig. 22.8). Some vertical open joints have been tightly filled with till and Precambrian boulders. Most gallery floors are covered with limestone debris of unknown thickness. These initial voids were clearly formed by glacier movements.

Fig. 22.6 In the Beauport area, a 4 × 5 m, 6 m deep pit appeared in a lawn in 1997. The bottom of the collapsed dome was covered with bedrock and till debris. A wood and steel platform was installed at the time, and its remains are visible on the right. By going down between limestone boulders, downstream, one could reach a large cave, some 150 m long. Underground water from the Montmorency River came from the upstream side of the debris cone, slowly eroding the finer part of this collapsed mass. In less than 10 years, underground waters have eroded a significant part of the collapsed debris mass and have given access to the upstream part of the cave. The pit is now 15 m deep

Geomorphological detailed surveys have shown that dislocated limestone blocks were displaced by glacial ice moving east. The numerous limestone breccias and immature very dense till seen in the top of galleries correspond to sliding planes where the upper rock masses were displaced farther east than the underlying levels, thus covering newly made voids (Fig. 22.9). During void formation, blocks would tend to topple from walls or fall from the ceiling (Fig. 22.10).

Following the Champlain Sea retreat, the Montmorency River rapidly entrenched the surface materials, and then, more slowly, the underlying limestones. Some of these surface waters enlarged existing open joints and reached the already existing glacitectonic voids to the south. During this initial phase, some small clearly karst caves were enlarged near the river bank. These were mapped and dated to ±4.0 ka. Underground water circulation probably stopped in large part when the river entrenchment became deeper. All known sinks developed almost at the same level, in the upper 3 m of the Marches Naturelles reservoir (113.5 m). Underground water circulation was restarted when the reservoir was created. At that time, the power company tried to block those

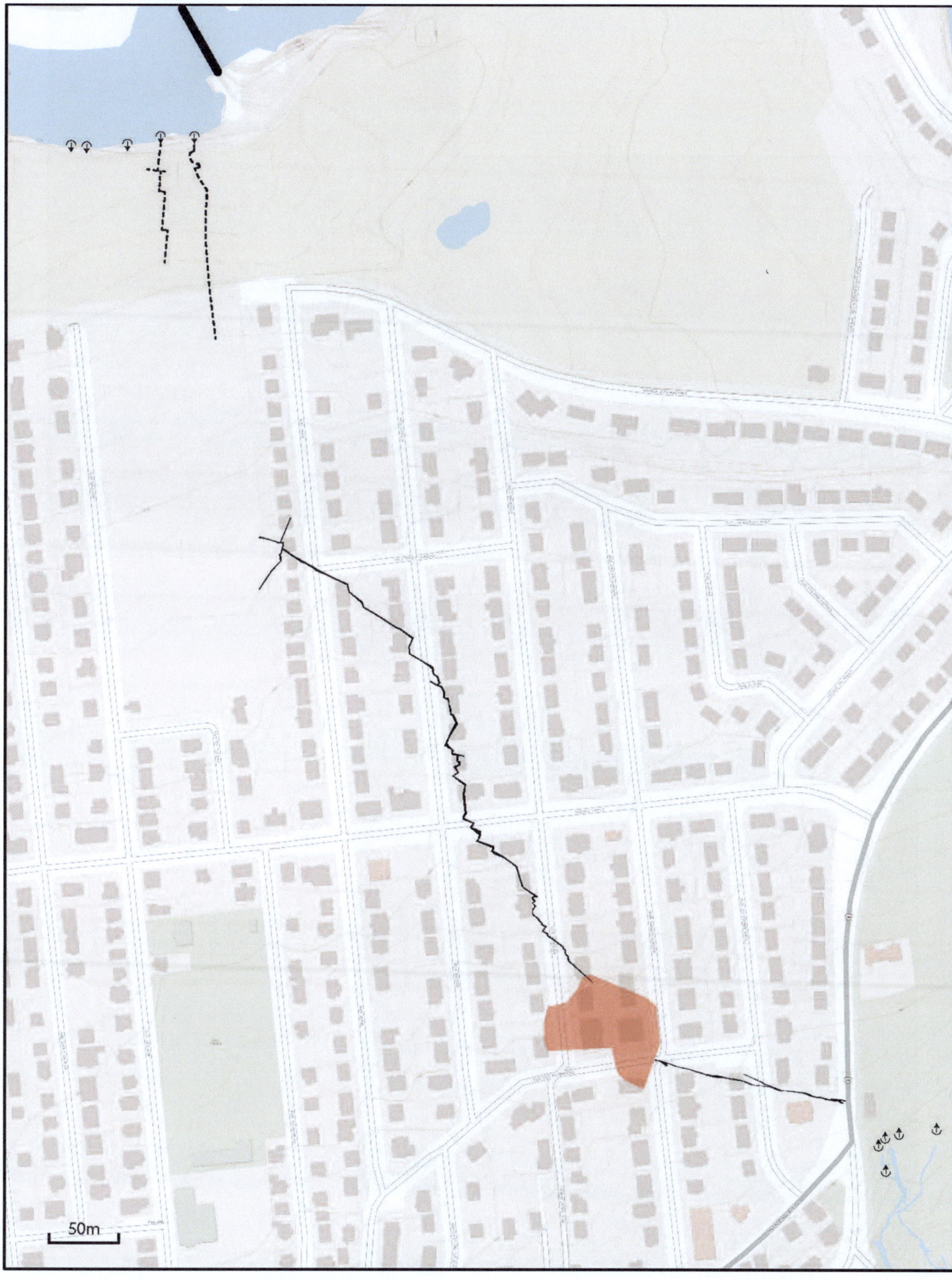

Fig. 22.7 Map of the known Beauport caves, with a filled limestone quarry exploited before residential development. The somewhat saw-tooth shape of the cave and information coming from the study of fit features confirm rock mass displacement toward the east, induced by the last glacier movement. In 2015, Québec City decided to clog the numerous sinks found along the river right bank, upstream of the Marches Naturelles dam, thus reducing underground water flow by nearly 90%. Dotted lines near the river show the location and explorable extent of some small karst caves that brought underground water to the glacitectonic voids to the south

Fig. 22.8 Galleries in the central part of the Beauport cave can reach significant volumes. The wall fitting is apparent. What is more impressive are the ceilings, which in some parts are composed of very dense assemblages of blocks in a finer clayey matrix. Stability of these ceilings is insured if surface conditions are kept stable. Ceiling width is 2.0–2.5 m (top right)

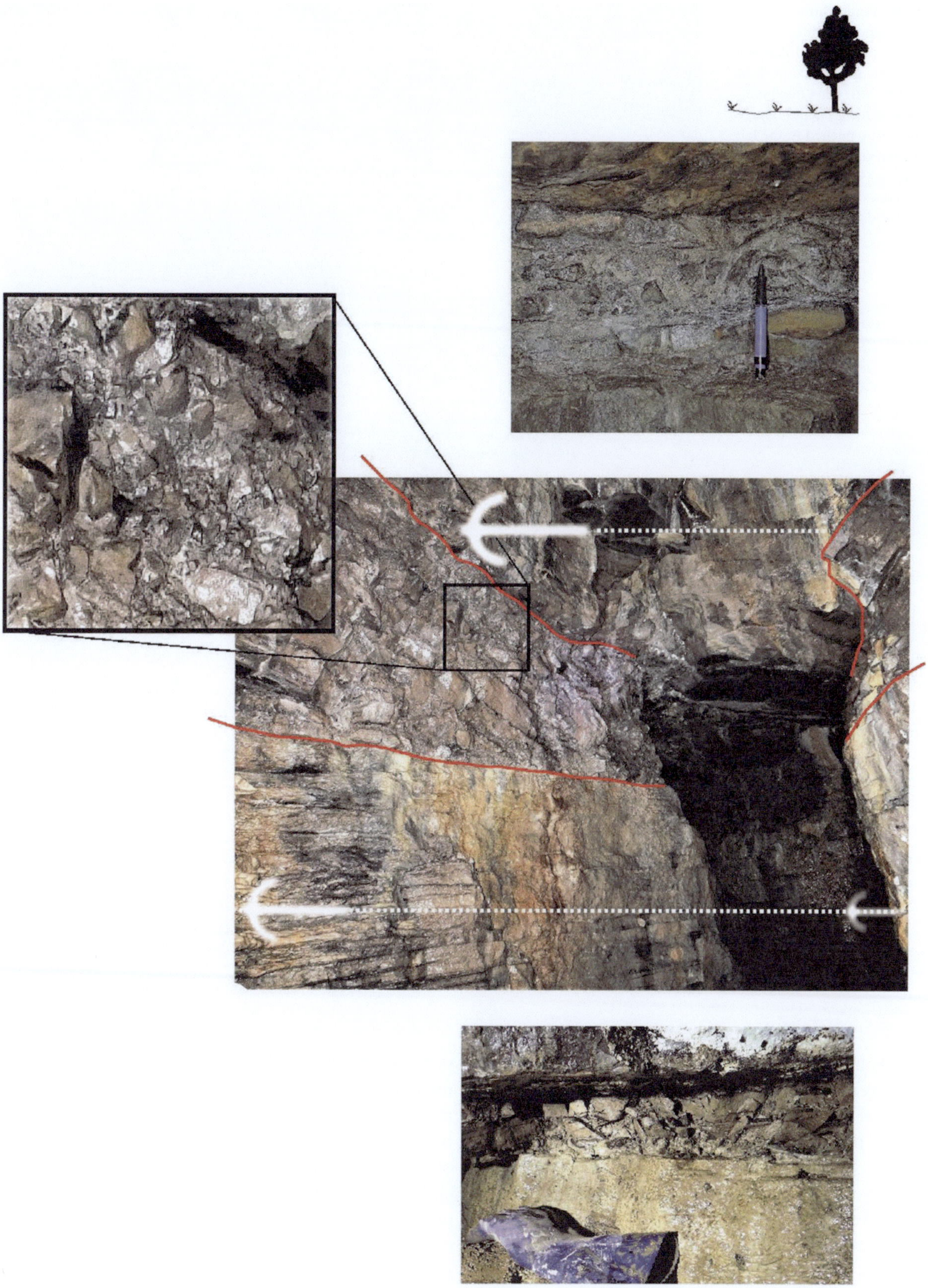

Fig. 22.9 Photo assembly showing different types of slippage horizons associated with joint opening and gallery development. In most cases these horizons are composed of limestone breccia, some 10–15 cm thick. Near the surface, this breccia is mixed with a fine argillaceous matrix. Deeper, the breccia is almost devoid of matrix. All along the large galleries, a breccia horizon some 60–80 cm thick shows pseudo-bedding due to the ice movement (center photo). They are very dense and have very low porosity

Fig. 22.10 Voids show various detailed morphology due to breccia layers found at different levels. On this photo, ice-push is from right to left. Illustrated are flat vertical walls (**A**), and a large limestone block pushed to the left (**B**) by the overlying debris seen in the ceiling (**C**). D is a large limestone mass moved to the left, after the opening of this gallery

sinks, without much success. It was only in 2015 that the clogging operation would prove effective, reducing underground water flow by nearly 90%.

Various phases of fine sediment filling and erosion have been observed along the galleries. Most probably, these relate to the Montmorency River initial entrenchment, followed by the 1908 impoundment of the Marches Naturelles reservoir. The limestone beds on the lower parts of the walls have been more or less dissolved by running underground waters coming from the Montmorency River. The upper part remained flat and unaffected by dissolution, thus confirming its glacitectonic origin. Connection of some vertical open joints with perched underground water or surface waters of anthropic origin also contributed in the enlargement by dissolution of these joints (Fig. 22.11).

Fig. 22.11 In the upstream and downstream parts of the Beauport main glacitectonic cave, more recent underground waters from the Montmorency River have eroded and dissolved the rock mass. This differential erosion highlights less soluble shale beds (top photo). On the lower left photo, underground waters have enlarged the vertical joint enough so that a caver can now enter. On the lower right photo, the left wall shows differential erosion masked by fine silty deposits. The lower part of the right wall has collapsed

22.6 Glacitectonic Caves: A Special Case Worthy of Study

From the 1970s studies made by engineering geologists up to most recent discoveries, it is quite clear that stresses associated with glacial weight and displacement can have significant impact on bedrock strain to at least 20 m depth, especially in layered sedimentary rocks (Schroeder 2004b). Glacial action is also responsible for the relatively higher horizontal stresses observed in glaciated bedrock. Glacitectonic caves constitute a special case of natural hazards, one that is not well known because of difficulties in detection and exploration, and are much more unpredictable than the usual karst caves (Fig. 22.12). They deserve to be studied because they are so special and have had many noticeable impacts on anthropogenic structures.

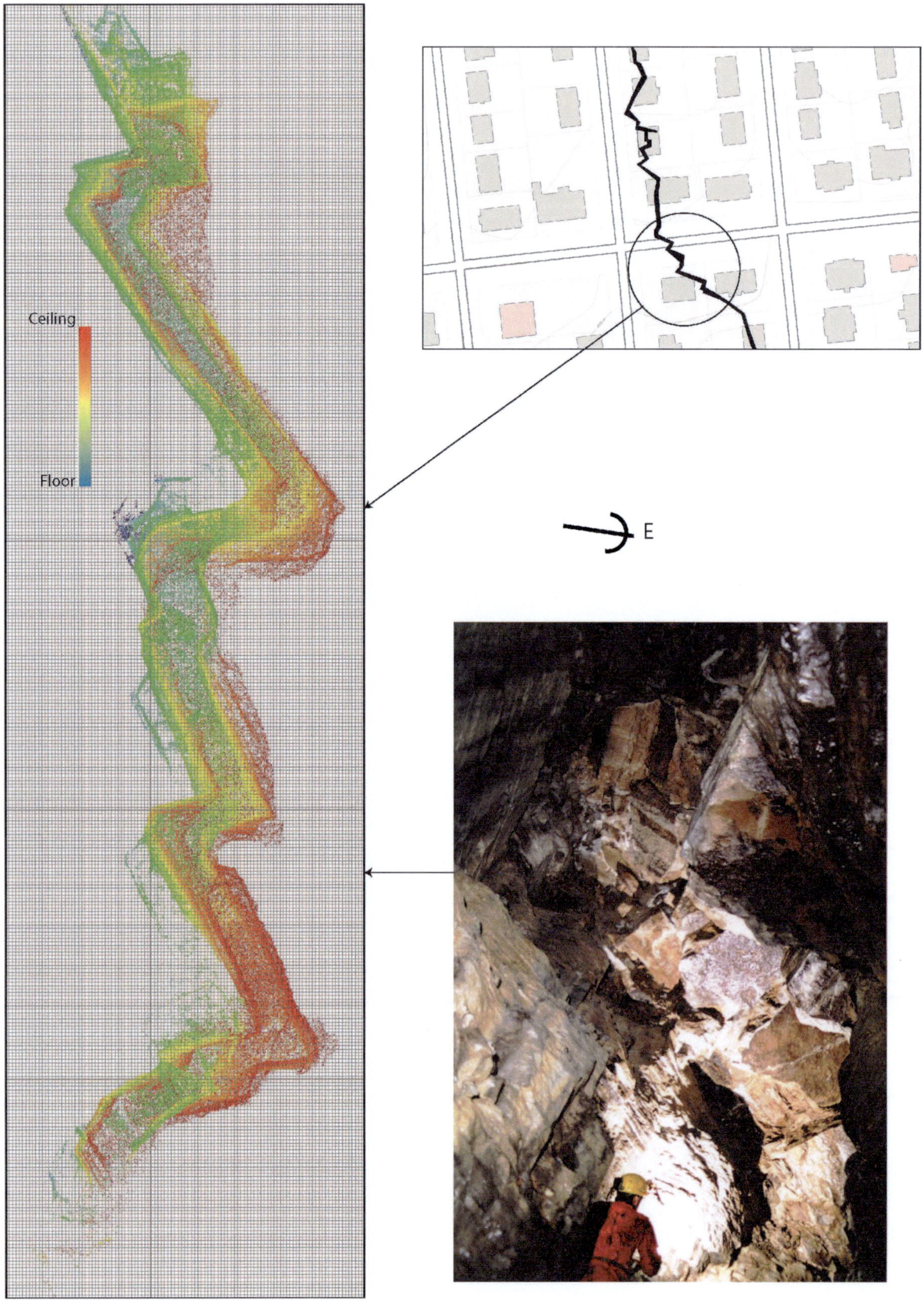

Fig. 22.12 A LIDAR survey using the ZEB-REVO was done along the larger section of the Beauport cave. This plan view shows the close fit of opposite rock walls, especially in their upper parts (red). A photo to the right gives an example of this fit. Rock mass displacement is toward the east

Acknowledgements The authors wish to cheerfully thank Marc Durand for his pioneer work, Luc Le Blanc for his mapping work, and Mourad Djaballah for his dedication and patience in preparing the figures of this paper.

References

Ballivy G, Loiselle A, Durand M, Poirier M (1977) Caractérisations géotechniques du secteur du Parc Olympique, Montréal. Can Geotech J 14:193–205

Durand M (1978) Classification des phénomènes et cartographie des roches rencontrées dans les grands travaux urbains à Montréal, Canada. In: Proceedings of the III international congress, international association for engineering geology, 4–8 Sept 1978, I(1), pp 45–55

Durand M, Ballivy G (1974) Particularités rencontrées dans la région de Montréal résultant de l'arrachement d'écailles de roc par la glaciation. Can Geotech J 11:302–306

Grice RH (1972) Engineering geology of Montréal. In: 24e Congrès international de géologie, Montréal, P. Q. livret-guide, B-18

Schroeder J (1991) Les cavernes à Montréal, du glaciotectonisme à l'aménagement urbain. Can Geogr 35:9–23

Schroeder J (2004a) Glaciotectonic cavity. In: Goudie AS (ed) Encyclopedia of geomorphology. Routledge, London, pp 470–471

Schroeder J (2004b) Les cavernes, un patrimoine gravé par le temps. Actes du premier colloque du Patrimoine géologique du Québec, Ass. Prof. des Géologues et des Géophysiciens du Québec, min. des Ressources naturelles du Québec, MB 2004.05, éd. par G. Prichonnet et M. A. Bouchard, pp 77–84

Schroeder J, Beaupré M (1985) Impact des cavités glaciotectoniques sur l'aménagement urbain de Montréal, Canada. Colloque de karstologie appliquée. Liège 1984. Ann Soc Géol Belg 108:69–75

Schroeder J, Beaupré M, Cloutier M (1986) Ice-push caves in platform limestones of the Montréal area. Can J Earth Sci 23:1842–1851

Jacques Schroeder is a full-time professor at the Department of Geography of Université du Québec à Montréal (UQAM). He has awarded a Ph.D. from Ottawa University and a M.Sc. from Université de Liège, Belgium. His research interests focus on karst in cold region, glaci-tectonism, glacial hydrology and epistemology of geography. He has produced programs related to scientific popularization (www.uqam.tv). Additional biographical information can be found in Canadian Who's Who.

Michel Beaupré is a senior engineering geologist working for Hatch. After an M.Sc. in geology from Montréal University he completed his B. Eng. at L'École Polytechnique de Montréal in engineering geology. He worked for Hydro-Québec for 29 years first as an engineering geologist and head of the geology department then as a project engineer. Since then he has been working for 12 years as a senior engineering geologist for different engineering consulting companies mainly in Canada, Central and South Americas, Africa, Southeast Asia, and China. Main fields of interest are hydroelectric projects and karst and caves studies.

Daniel Caron worked as director of the Québec Speleological Society from 1975 to 1990 and currently serves as vice-president of its board of directors. As national caving instructor and active speleologist for more than 50 years, he has explored hundreds of caves and karst areas in Québec, Canada, from Newfoundland to Vancouver Island including the longest cave in Canada, Castleguard Cave, and around the world, Belgium, Borneo, England, France, Guatemala, Mexico, Spain, and United States. He is the co-discoverer of a major extension in the Saint-Léonard Cave in 2017. He has collaborated in many scientific and technical studies on the caves and karst areas of Québec, including those of the Anticosti Island (North Coast), the Spéos de la Fée (Matapedia), the Courville Cave (Québec), the Saint-Léonard Caves (Montréal) and the Laflèche Cave (Outaouais). He has written dozens of papers on the various aspects of this activity. He is co-author, with Michel Beaupré, of the book 'Découvrez le Québec Souterrain' published in 1986 by Québec Science.

Natural and Human Impacted Landscapes of the Sudbury Mining Region

23

Randy W. Dirszowsky

Abstract

The Sudbury Basin is a unique, regional geologic and geomorphologic feature situated on the southern Canadian Shield at the nexus of three geological provinces and three physiographic zones. Its broad form is attributable to its meteorite impact origin (1,850 Ma) and subsequent northwestward directed stresses and deformation during the Penokean Orogeny. The main basin measures 27 km 60 km with its structural axis aligned northeast-southwest. The basin floor is generally flat and gently sloping toward the southwest due in part to impact/fallback processes and subsequent Precambrian sedimentation, and in part due to more recent Quaternary infilling. The basin rim and surrounding uplands rise to more than 150 m above the basin floor and exhibit a much more rugged topography, numerous lakes and largely deranged drainage patterns. The basin was covered by Laurentide/Labrador ice during the late Wisconsinan and influenced flow rate and direction at least during deglaciation. The basin also acted as a pinning point for the retreating Wisconsinan ice and a low area making up, for a time, an embayment of glacial Lake Algonquin. As a result, a wide range of glacial, glacilacustrine and meltwater-related landforms and deposits are represented in the area despite the relative paucity of surficial materials characteristic of much of the Canadian Shield. The geology and landforms of the area have strongly influenced human settlement based on logging, agriculture and especially mining of the Cu- and Ni-rich ores of the Sudbury Structure. In turn, human activity has left an indelible mark on over 1000 km^2 of the landscape through contamination and acidification of soils, severe vegetation damage or loss, altered hydrology, and enhanced weathering, erosion and sediment redistribution in some areas.

Keywords

Sudbury Structure • Penokean Hills • Abitibi Upland • Wisconsinan glaciation • Glacial Lake Algonquin • Human impact • Mining

23.1 Introduction

The elliptical Sudbury Structure or Sudbury Basin is situated northwest of the Grenville Front along the boundary of the Superior and Southern geological provinces of the southern Canadian Shield and approximately 50 km north of Georgian Bay (Fig. 23.1a). It measures approximately 60 km long and 27 km wide with its long axis oriented southwest-northeast. Although the Sudbury Basin is too small to appear on the 1:5,000,000 scale physiographic map of Bostock (2014), it is important at the regional level due to its direct geological control on topography and effects on the direction of glacier movement and glacial, glacilacustrine and fluvial depositional environments during the Quaternary. The copper-nickel sulphide ore deposits of the Sudbury Structure are among the largest and most productive in the world and their extraction and processing have contributed to significant human impact on the Sudbury landscape and in particular the loss of vegetation and soils from uplands (Fig. 23.1b). The relatively flat basin floor has been exploited historically for agriculture and localized settlement. While most urban development is concentrated to the southeast of the basin itself, the City of Greater Sudbury municipal boundary encompasses most of the feature (Fig. 23.1c). Geological features attributable to the formation of the Sudbury Structure extend to 80 km or more beyond the basin margin (Dressler et al. 1991; Lafrance et al. 2008).

R. W. Dirszowsky (✉)
School of the Environment, Laurentian University, 935 Ramsey Lake Road, Sudbury, ON P3E 2C6, Canada
e-mail: rdirszowsky@laurentian.ca

© Springer Nature Switzerland AG 2020
O. Slaymaker and N. Catto (eds.), *Landscapes and Landforms of Eastern Canada*, World Geomorphological Landscapes, https://doi.org/10.1007/978-3-030-35137-3_23

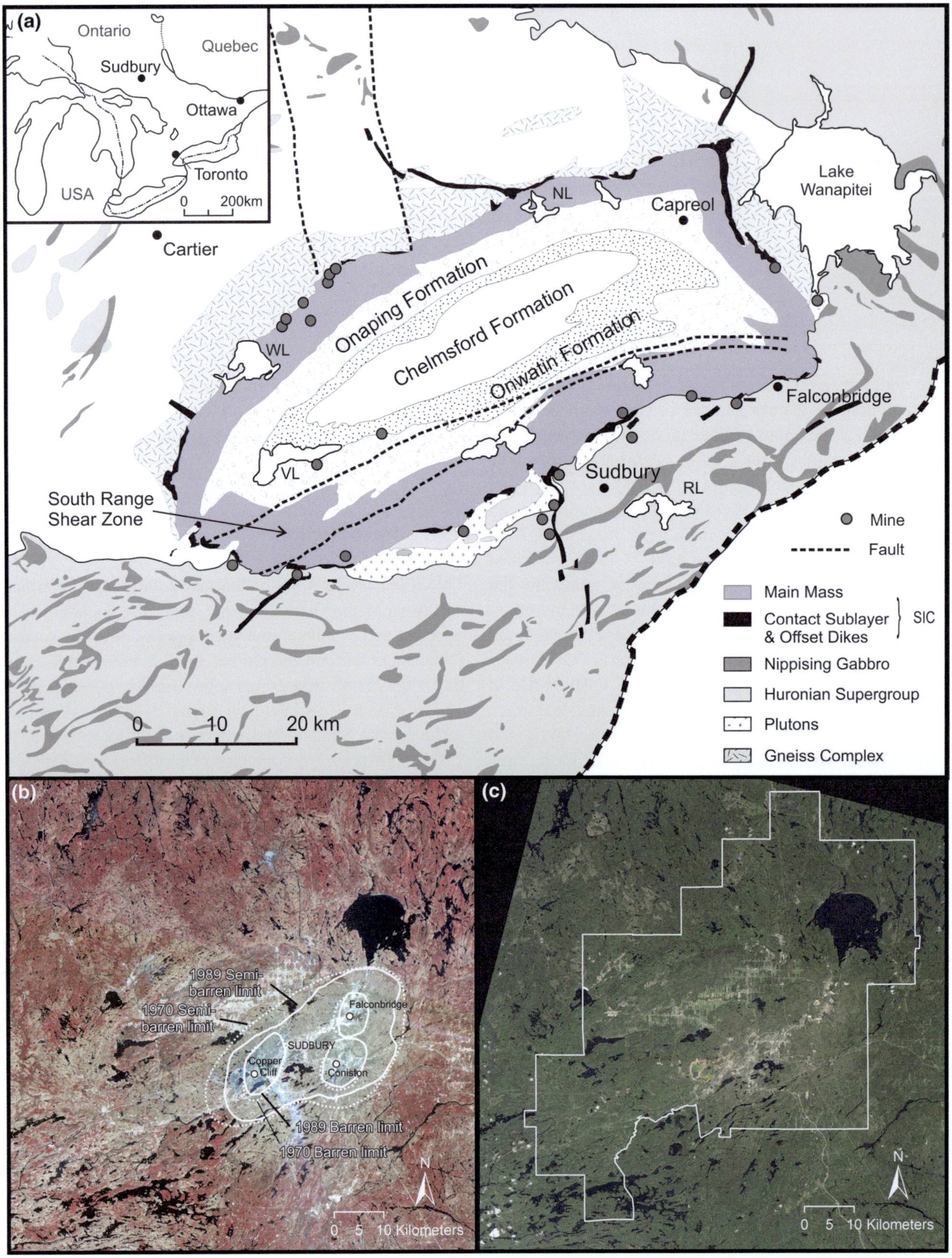

Fig. 23.1 The Sudbury Basin and surrounding area. **a** Location and bedrock geology (after Rousell and Card 2009). SIC = Sudbury Igneous Complex, RL = Ramsey Lake, NL = Nelson Lake, WL = Windy Lake, VL = Vermilion Lake. **b** 1975 Landsat 1 image (bands 2, 3, 4) with 'barren' and 'semi-barren' zones related to historic smelter emissions indicated (after McCall et al. 1995). **c** 2018 Landsat 8 image (panchromatic) showing present land cover and City of Greater Sudbury municipal boundary (satellite data from NASA/U.S. Geological Survey)

23.2 Geology

Archean granite, granite gneiss and metavolcanic rocks of the Superior Province lie to the north, northwest and northeast of the Sudbury Structure, whereas predominantly metasediments of the Huronian Supergroup and the Nipissing Gabbro of the Southern Province lie to the southeast of the structure (Rousell and Card 2009) (Fig. 23.1a). The Sudbury Structure consists of (1) the Sudbury Igneous Complex, in profile a funnel-shaped norite, quartz gabbro, granophyre intrusion and inclusion-rich sublayer; (2) the overlying Whitewater Group which consists of heterolithic breccias, mudstones and wackes of the Onaping, Onwatin and Chelmsford formations respectively; and (3) brecciated footwall rocks which occur along the outer edges of the Sudbury Igneous Complex and within dikes that both radiate from and parallel its outer edges (Dressler et al. 1991). The main part of the structure originated during a meteorite impact and/or endogenic explosion (Sudbury Event $\sim$1,850 Ma) which is believed to have caused the extensive brecciation, deformation and possibly large-scale tectonism of existing country rocks. The basin/structure acquired its present elliptical shape during northwestward thrusting of the Penokean Orogeny which resulted in several (southwest–northeast) strike parallel faults across the Sudbury Igneous Complex and rocks of the Whitewater Group and Huronian Supergroup, as well as some northwestward striking faults cutting across the north-central basin and north range (Fig. 23.1a). Reverse/thrust faulting and additional folding of Chelmsford Formation rocks may account for shortening of 6–13 km which would suggest that the basin began as a more or less circular feature (Shanks 1991).

23.3 Physiography

Physiographic subdivisions of the Sudbury area include part of the Abitibi Upland, Penokean Hills and Cobalt Plain of the physiographic James Region and a small part of the Laurentian Highlands (Laurentian Region) located southeast of the Grenville Front (Fig. 23.2). The Abitibi Upland is underlain by Archean crystalline rocks and characterized by broad, rolling, bedrock-controlled surfaces that rise gently toward the south where elevations exceed 450 m above sea level (Shilts et al. 1987; Dredge and Cowan 1989). Local relief can exceed 90 m where incision has occurred, but is generally much less. The Penokean Hills are composed of folded metasedimentary rocks (Huronian Supergroup) with quartzite ridges reaching elevations of 240–300 m above sea level and local relief as great as 100 m. The Cobalt Plain is underlain by relatively flat-lying clastic metasedimentary rocks with some ridges and hills formed by resistant gabbro sills or older Archean rocks. The Laurentian Highlands consist of low, rounded knobs and ridges with relief on the order of 30 m to 50 m. Elevation is comparable to the Penokean Hills (230–300 m asl).

The Sudbury Basin consists of a central lowland surrounded by high-relief rock ridges that define northern, eastern and southern rims (Barnett and Bajc 2002) (Figs. 23.2 and 23.3a). The lowland is a low-relief plain floored by Quaternary sediments up to 120 m thick (Burwasser 1979) underlain by the Chelmsford and Onwatin formations, and to a much lesser extent the Onaping Formation (in the northeastern end of the basin only). Overall the plain gently slopes to the southwest, dropping in surface elevation from about 295 to 260 m above sea level over a distance of 40 km. The plain is interrupted in places by bedrock ridges (up to 310 m asl) of folded Chelmsford Formation rocks, but overall the Quaternary infill obscures an irregular bedrock topography with relief on the order of 30–50 m.

The ridge zones surrounding the basin are underlain by rocks of the Sudbury Igneous Complex and by the Onaping Formation. The north rim rises to over 105 m immediately above the valley floor about midway across the basin (Fig. 23.3b). Overall it is characterized by bedrock hills and ridges with local relief between 60 and 90 m, especially where cut by streams along faults (Fig. 23.3c), and rises to elevations in excess of 440 m above sea level. The east rim is also dominated by ridges of bedrock, but of lower relief (generally <30–40 m) and at lower elevations (335–385 m asl). The south rim reaches elevations between 305 and 335 m above sea level (relief in the 30–45 m range) and merges more gradually with adjacent landscapes outside the basin. Whereas the Abitibi Upland to the north of the basin has a largely irregular topography cross-cut occasionally by fault lineaments of variable north to west orientation, the Penokean Hills and Laurentian Highlands to the south have a more conspicuous northeast-southwest trending fabric associated with folding/faulting including the Grenville Front (Fig. 23.3d). In the rock-dominated terrain of the Abitibi upland, Penokean hills and the Cobalt plain, Quaternary deposits are very thin except in depressions and along valley bottoms where thicknesses in excess of 60 m have been reported in water well records (Barnett and Bajc 2002). The linear pattern of some thicker Quaternary sediments follows the trend of prominent bedrock structures, and many lake basins and wetlands are underlain by such sediments.

23.4 Quaternary History

The Quaternary deposits and features preserved and exposed within the Sudbury area are of Wisconsinan, m. age (Fig. 23.4). The Laurentide Ice Sheet covered all of Ontario

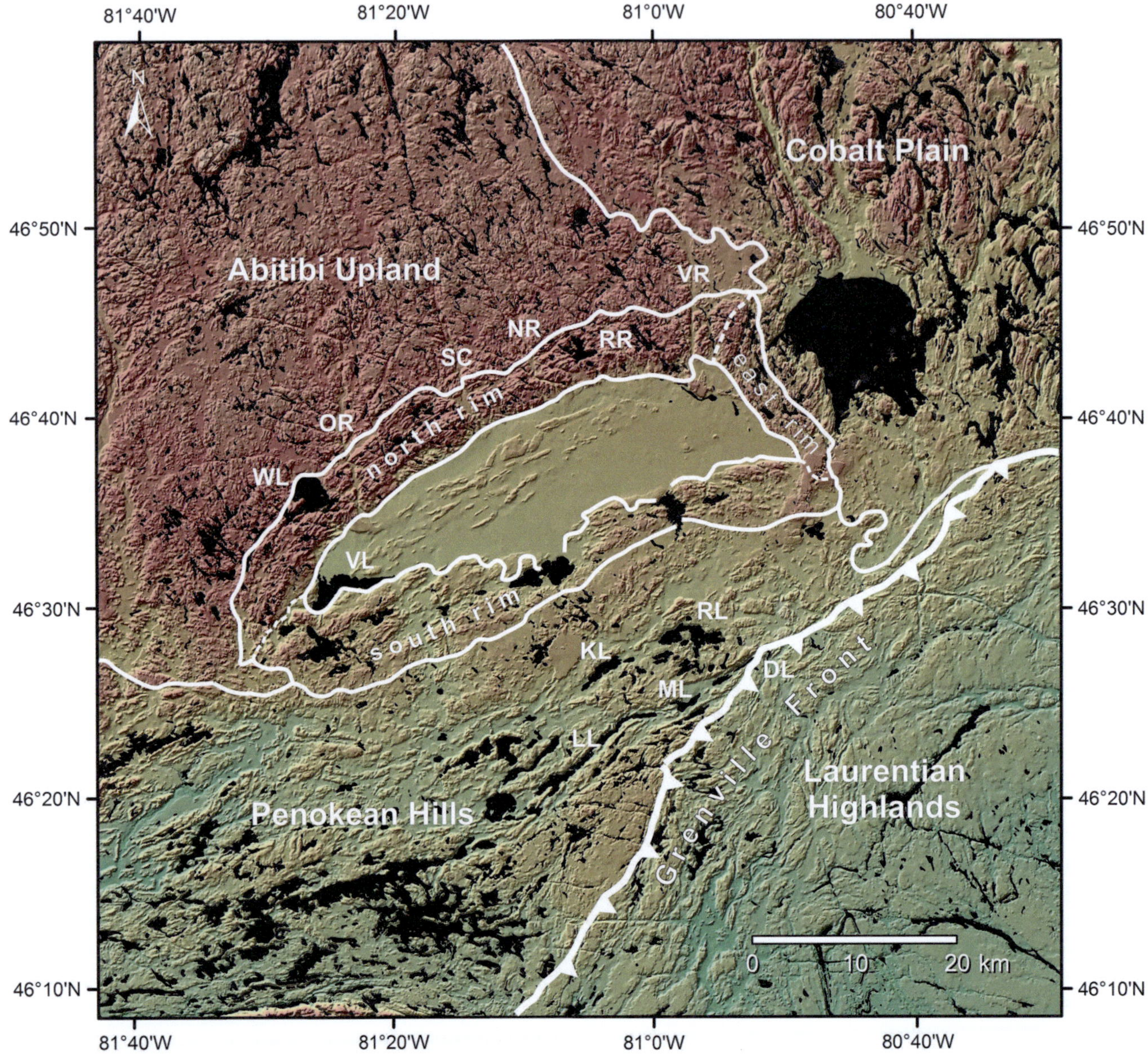

Fig. 23.2 Physiographic subdivisions of the Sudbury Basin area (after Bostock 2014). KL = Kelly Lake, LL = Long Lake, ML = McFarlane Lake, RL = Ramsey Lake, VL = Vermillion Lake, WL = Windy Lake, OR = Onaping River, SC = Sandcherry Creek, NR = Nelson River, RR = Rapid River, VR = Vermilion River (headwater)

and extended into the adjacent northern states of the United States of America about 20–21 ka BP during the Michigan Subepisode/Late Wisconsinan (Dyke et al. 2002). Ice of the Labrador Sector covered the Sudbury region and generally flowed to the south-southwest (Dredge and Cowan 1989). Based on variations within the striation record, Boissonneau (1968) first proposed that two ice lobes met in the vicinity of the Sudbury Basin, an eastern lobe that flowed toward the southwest across the basin and a western lobe that flowed more southward. This may have resulted in a zone of shearing and later ice stagnation in the area west of the basin near Windy Lake (Barnett and Bajc 2002). It appears that the basin itself influenced the rate and direction of flow of the eastern lobe, accelerating and diverting it more toward the southwest.

During deglaciation of the southern parts of Ontario, large proglacial lakes formed against the receding ice margin within the Great Lakes basins (Barnett 1992). Glacial Lake Algonquin was the most extensive of these and occupied the Lake Huron, Lake Michigan and part of the Lake Superior basins and fronted the ice sheet as it retreated across the Sudbury area. Glacial Lake Algonquin had several phases related to ice position and the location and elevation of outlets controlling water level (Lewis et al. 2008). Through

Fig. 23.3 Sudbury Basin landscapes. **a** Basin from north rim looking southeastward toward Sudbury (horizon is south rim). **b** North rim from basin floor north of Chelmsford (view to north). **c** Nelson River valley cut into north rim (view to north). **d** Penokean Hills north of Grenville Front (view to southwest); bedrock (Mississagi Formation) is exposed here due to historic smelting-related erosion

the highest, main Lake Algonquin stage, the Sudbury Basin remained ice covered and drainage occurred through the Kirkfield and Port Huron outlets. Several minor landforms/deposits (i.e. moraine fragments near McFarlane Lake and scattered sub-aqueous fans) mark early ice-marginal positions in the southern part of the region (Barnett and Bajc 1999). A delta at the south end of Fairbank Lake and southernmost end of the basin formed during a high level of Lake Algonquin after drainage had first switched to the Fossmill outlet south of North Bay (Harrison 1972; Chapman and Putnam 1984). The south rim of the basin probably acted as a temporary pinning point for the glacier after which retreat to the north rim and flooding of the basin occurred rapidly and created a large embayment (Heath and Karrow 2007).

The ice margin remained stable on the north rim for a time producing a well-defined segment of the Cartier I moraine while the Fossmill outlet was still in use (Heath and Karrow 2007). Ice-marginal deltas along the moraine were built initially to a level of about 350 m above sea level (Burwasser 1979). Lower surfaces on these deltas (topset beds, residual terraces and other shore features) mark periodic lowering of the lake level (Cedar Point phase, Payette phase, Sheguiandah phase) as successive outlets became ice free in the North Bay area while the Cartier I moraine was still being formed. When ice receded from the Cartier I moraine, meltwater followed a number of southbound valleys in the uplands while lake level continued to drop as lower drainage outlets along the Mattawa River (Korah and post-Korah phases) became available (Heath and Karrow 2007). Flooding of the basin and subsequent phases of lake level drop probably happened quickly, over several centuries between about 10,500 and 10,000 years before present (Lewis et al. 2008). Ultimately nine or more, largely fragmented, east-west-trending moraine systems were produced across north-central Ontario marking ice retreat and standstills/readvances between Lake Superior and the Ottawa River (Boissonneau 1968; Shilts et al. 1987).

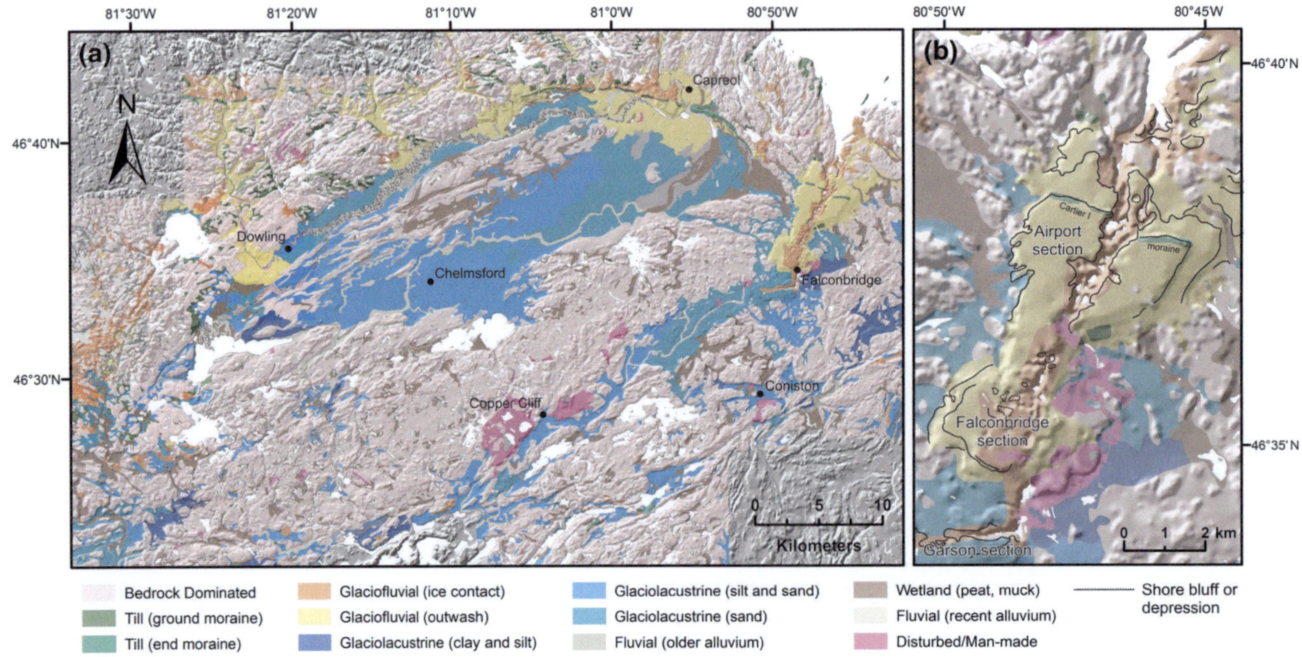

Fig. 23.4 Surficial geology of the Sudbury Basin. **a** Basin-wide deposits. **b** Detail of Sudbury airport glacifluvial system (maps based on Burwasser 1979; Barnett and Bajc 1999) (base data courtesy of the Ontario Geological Survey)

23.5 Glacial Geomorphology

The Sudbury Basin falls within Zone B of the southern Canadian Shield defined by Shilts et al. (1987) as an area lacking well-integrated esker systems (in contrast with their Zone A) due to active ice retreat or only localized ice stagnation during late Wisconsinan deglaciation. Due to the resistant bedrock lithologies of much of the shield, erosion and sediment production were relatively low, resulting in mostly thin, discontinuous drift cover especially in upland areas. It is believed that Quaternary glaciations overall contributed no more than a few metres of surface lowering across the Canadian Shield as a whole.

23.5.1 Glacial Erosional Features

Erosional features on the bedrock surface in the Sudbury area are ubiquitous and commonly exposed, especially where human activities have resulted in soil loss. Erosional forms range from small-scale etchings and polish on rock surfaces to kilometre-scale streamlined bedrock forms that include roches moutonnées and whalebacks (Barnett and Bajc 2002), although smaller features are more prevalent given the strong geologic control. Striae, crescentic marks, grooves, crag and tail features, roches moutonnées, and other streamlined forms (Fig. 23.5a–c) indicate an ice-flow pattern across the Sudbury region that was strongly affected by regional topography. The direction of ice flow across the Abitibi upland and the north rim of the Sudbury Basin was 170°–210°. Within the basin ice-flow direction shifted to between 220° and 245°, parallel to the basin axis. South of the basin along the south rim and in the Penokean hills, ice flow was about 205°–225°. At a few localities along the north rim, evidence for an older, southwestward ice-flow event is preserved on protected bedrock surfaces (Barnett and Bajc 2002).

In addition to erosional forms directly resulting from the movement of glacial ice, bedrock surfaces in the Sudbury region preserve a range of sculptured forms (p-forms or s-forms) that have been attributed to erosion by subglacial meltwater (Fig. 23.5d–f). Kor et al. (1991) mapped similar features across a 70 km wide area along the shore of Georgian Bay (50–120 km to the south of the Sudbury Basin) and argued that they resulted from turbulent subglacial meltwater flows, possibly during a single catastrophic event. Sculpted forms in the Sudbury area are far less abundant and less continuous, indicating more localized, subglacial meltwater flow, though still generally directed toward Georgian Bay. The prevalence of abraded and sculpted forms is clearly indicative of an active warm-based glacier.

23.5.2 Glacial Depositional Features

Till is widespread in the Sudbury area but characteristically discontinuous and very thin (<1 m and therefore unmapped

Fig. 23.5 Glacial erosional features in the Sudbury south end (approximately 2 km east of Kelly Lake). **a** Striated bedrock surface with chatter marks. **b** Roches moutonnées (in school yard). **c** Striae and crescentic gouges. **d** Meltwater furrows and whalebacks (striated). **e** Meltwater furrow (striated/polished). **f** Various s-forms including muschelbruch, sichelwanne, comma forms and partial potholes (shovel is 65 cm in length)

across much of the area; Sado 1977; Burwasser 1979; Gillespie et al. 1982; Barnett and Bajc 1999) (Fig. 23.6a). Thicker accumulations occur sporadically in association with topographic irregularities including bedrock knobs and ridges, lee-side cavities, valleys and depressions (Barnett and Bajc 2002). At very few locations (e.g. along the west shore of Wanapitei Lake northeast of the basin) are deposits thick enough to mask underlying bedrock topography (Fig. 23.6b).

Tills from areas dominated by coarse crystalline rock types are generally sandy and contain moderate to high concentrations of clasts up to large boulder size (Fig. 23.6c, d). Tills associated with Huronian metasedimentary rocks or argillaceous metasedimentary rocks of the Cobalt Plain are usually finer and contain moderate to low concentrations of clasts. The latter make up the thicker deposits found along the western shore of Wanapitei Lake and contain over 60% combined matrix silt and clay (50% silt and 10% clay). Tills coarsen rapidly toward the southwest in response to the addition of material from the Sudbury Igneous Complex. Tills have three main landform associations in the Sudbury area. Widely scattered subglacial tills make up irregular ground moraine (Fig. 23.4a). On the up-ice side of bedrock knobs or ridges this till is commonly massive and compact; however, less compact meltout till, flowtill and stratified sediments occur where greater relief may have caused shearing and detachment of debris-rich ice (Barnett and Bajc 2002). In the lee of bedrock rises subglacial tills tend to be loose and interbedded with variable amounts of sorted, stratified material.

Distinct moraine ridges marking stillstand positions of the retreating Laurentide ice margin are sparsely distributed in north-central Ontario with only two examples in the Sudbury area (Fig. 23.4a). The well-defined Cartier I moraine extends from the Sudbury airport south of Wanapitei Lake, northwest adjacent to the east rim of the basin, then westward along the southern edge of the north rim about 20 km, after which it continues directly west toward the town of Cartier ($\sim$50 km) and beyond. The feature is nearly continuous along the Sudbury Basin, broken locally by younger glacifluvial deposits and scattered bedrock outcrops. It usually occurs as a single ridge 5–10 m in height (Fig. 23.6c, e); however, multiple ridges are observed in places and probably formed in response to slight readjustments of the ice margin. The moraine is composed of loose sand/boulder material with some beds or lenses of stratified sand and gravel, and flowtill, and is flanked on the north side by substantial ice-contact stratified material. The main ridges and outer (south) slopes of the moraine are usually littered with large, sub-angular boulders/erratics (up to 2.5 m) of variable origin (Fig. 23.6f). Similarly composed till/moraine fragments are found somewhat south of the Sudbury Basin in the vicinity of McFarlane and Long lakes.

To the west of the Sudbury Basin and just west of Windy Lake a distinct area of hummocky terrain occurs consisting of ice-contact stratified sediment, stony meltout till and flowtills. The hummocks reach heights of 5–10 m, mostly in low-lying areas within the Archaean upland bedrock. Clasts account for up to 75% of the total volume of this till and are mostly angular in shape despite having a distal origin. The deposit is considered to be ice-stagnation moraine formed within the shear zone between faster flowing ice within the Sudbury Basin and slower-flowing ice on the Abitibi Upland (Barnett and Bajc 2002).

23.5.3 Glacilacustrine and Outwash Features

Massive to laminated deposits of pebbly sand to silt with minor clay associated with glacial Lake Algonquin are found flooring most of the Sudbury Basin and lower lying areas of surrounding uplands to the south (Fig. 23.4a). These glacilacustrine deposits fine laterally away from the main sediment input sources, outwash systems along the north and east rims and to the east of the basin south of Wanapitei Lake. In places the deposits coarsen upward where they are associated with prograded deltas. Primarily sandy materials form broad plains in the eastern part of the basin and thick fill and level ground across the north of the basin along the Vermilion and Onaping rivers and within low-lying areas between Sudbury and Falconbridge. In the eastern part of the basin some of these materials have been reworked by aeolian processes to produce irregular, slightly rolling sheets of fine to very fine sand with longitudinal, transverse and parabolic bedforms in some cases (Barnett and Bajc 2002). Finer textured sediments (silts and clays) dominate the southwestern parts of the basin and isolated depressions and valleys across a wide area south and east of the basin (Fig. 23.7a). Many of these finer deposits are rhythmically bedded with individual couplets thinning and fining upward (Antevs 1925) (Fig. 23.7b).

Abandoned shoreline features, including bars and spits, platforms and shore bluffs occur along the northern edge of the basin where waves and currents eroded newly deposited till and glacifluvial sediments (Fig. 23.8a). Elsewhere in the bedrock-dominated areas, the shoreline features are poorly developed or preserved. Terraced outwash deposits occur within almost all of the structurally controlled river valleys that enter the Sudbury Basin over the north and east rims (i.e. the Vermilion, Rapid, Nelson, and Onaping rivers, Sandcherry Creek, and various tributaries) (Figs. 23.3c and 23.8b) and extend into the basin above lacustrine deposits in some locations. Most of the deposits consist of well-sorted, poorly graded sand and fine gravel, with coarser gravel occurring locally and especially in the upper reaches. Up to six distinct glacifluvial terraces have been identified along

Fig. 23.6 Glacial deposits and landforms. **a** Thin ground moraine in central Sudbury. **b** Thick till deposits on western shore of Lake Wanapitei (shovel is 65 cm in length). **c** Section through crest of Cartier I moraine near Nelson Lake. **d** Boulder surface of Cartier I moraine. **e** Aerial view of Cartier I moraine looking eastward toward Capreol where the Vermilion River enters the basin from the north. **f** Large (~2 m) erratic on south face of Cartier I moraine near Rapid River

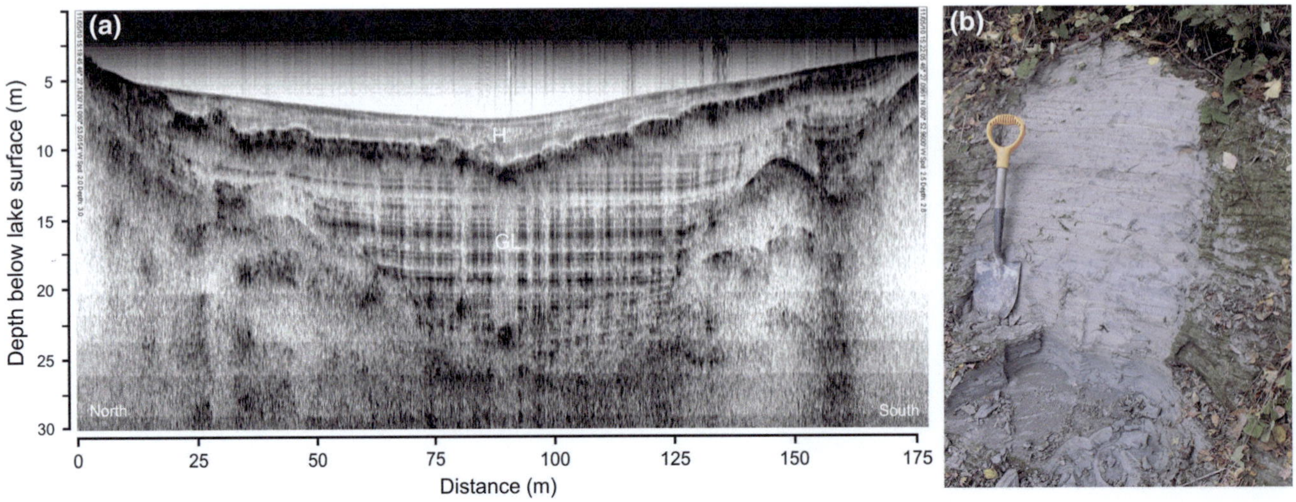

Fig. 23.7 glacilacustrine sediments. **a** Low-frequency, sub-bottom acoustic profile across Daisy Lake just north of the Grenville Front; approximately 2–3 m of Holocene (H) sediment overlie an eroded and channelized surface of acoustically stratified and rhythmically bedded glacilacustrine (GL) infill over bedrock. **b** Rhythmically laminated glacilacustrine sediments exposed in a streambank near Chelmsford (shovel is 65 cm in length)

the larger valleys (Burwasser 1979) and they were apparently formed by downcutting and redeposition of the sediments in response to declining water levels in glacial Lake Algonquin. Large deltas formed where these outwash systems entered the glacial lake and these were in turn incised and terraced, and extended/prograded as water levels continued to drop and the landscape adjusted isostatically (Heath and Karrow 2007) (Fig. 23.8b). The largest delta system occurs where the Vermilion River enters the basin at Capreol and has components both within and beyond the Cartier I moraine (Fig. 23.4a). A large ice-marginal delta also formed at the Sudbury airport site in association with a complex meltwater system that follows a structural depression extending southward from Wanapitei Lake/Massey Bay (see below).

23.5.3.1 The Airport Glacifluvial System

Extensive deposits of glacilacustrine/glacifluvial sand and gravel occur in the area of the Sudbury airport and southward through Falconbridge and Garson into the north end of the main city of Sudbury (Figs. 23.4b and 23.9). These materials appear to have been deposited progressively northward along the receding ice margin. The most northern segments accumulated initially within a subglacial passage and subsequently in an open, ice-walled channel that expanded to about 6 km in width as the ice retreated (Barnett and Bajc 2002). Large ice-block depressions (Fig. 23.10a) probably formed due to the collapse of the tunnel roof and burial of fallen fragments by glacifluvial sediments. The resulting kettle holes and kettle lakes can exceed 45 m in depth (Fig. 23.10b).

The southernmost Falconbridge and Garson segments of the complex (Fig. 23.4b) are thought to have accumulated initially as lobate, sub-aqueous fans within glacial Lake Algonquin once it had flooded the basin. Fan sediments generally fine upward and contain large steep-walled channels and de-watering structures in some locations (Barnett and Bajc 2002). Coarser beds overlying the fan sediments exhibit cross-bedding indicative of deltaic sedimentation and were probably deposited as water level in the basin began to drop. The top of the Falconbridge segment is flat, with a unit of trough cross-bedded sand and gravel that truncates the lower fan and delta sediments (Fig. 23.9a).

The most northern segment, adjacent to the airport, is the youngest part of the system, with elevations slightly lower than the Falconbridge segment. Aside from the central zone of kettled outwash, it is a flat-topped plain graded to a level of glacial Lake Algonquin and probably deltaic (Fig. 23.9b, c). Fragments of the Cartier I moraine mark the northern edge of this segment. North of the moraine there are two additional outwash terraces or delta surfaces that were formed along the receding ice margin as water levels in the basin continued to fall.

23.6 Post-glacial and Recent Drainage

As ice retreated further to the north after 10 ka BP (North Bay Interstade), glacial Lake Algonquin receded permanently from the Sudbury area into Georgian Bay and Lake Huron, with drainage continuing through the North Bay outlet(s) via the French River channel system. Although

(a)

(b)

Fig. 23.8 Ancient shoreline features. **a** Aerial view of glacial Lake Algonquin shore terraces along the northern boundary of the Sudbury Basin south of the Cartier I moraine near Nelson Lake (view to northeast). **b** Aerial view of incised/terraced glacilacustrine delta at Sandcherry Creek; the modern stream exits the valley on the left (northwest) and dissects the delta mid-photo flowing eastward; the lowest visible delta surface supports a conifer plantation

Fig. 23.9 glacifluvial and glacilacustrine deposits of the Sudbury airport area. **a** Delta foreset and topset beds exposed in a pit just east of the airport (view to east). Overlying outwash deposits (brown colour) are visible in the background. **b** Close-up of crudely bedded topset channel fill. **c** Detail of more distal thin-bedded sand and gravel foreset deposits (shovel is 65 cm in length)

regional southward gradients would have been less at the time, initial drainage and continued remote meltwater inputs led to incision of recently deposited lacustrine, delta and outwash deposits, the creation of terraces/ravines and channels now covered by Holocene lake, wetland and floodplain deposits (Fig. 23.7a). Major drainage routes would have existed pre-glaciation, and become re-established during deglaciation with later adjustments due to isostatic rebound (e.g. drainage northward rather than southward from Windy Lake). Modern drainage in the region is largely deranged with major bedrock/structurally controlled channels integrating otherwise smaller isolated basins and creek systems. On the upland areas surrounding the Sudbury Basin, lakes and wetlands are abundant, but associated catchment areas are relatively small and accumulation/infill rates are low in most cases (deep water rates ~0.5–1.5 mm/yr at present; 0.1–0.3 mm/yr over the last few thousand years). Exceptions occur where a lake basin occurs along a larger integrated drainage system, for example Kelly Lake at Sudbury (drainage area 209 km^2) and Vermilion Lake at the western end of the basin (drainage area 2850 km^2); at these sites significant infilling and delta development have taken place in post-glacial and recent times (see below).

Where streams in the area are less constrained by bedrock, they have generally developed a meandering behaviour and planform at a range of scales. The largest and most significant of these, and relatively rare in northern Ontario, is the Vermilion River where it crosses the north side of the basin between the north rim (Onaping Formation) and ridges of the Chelmsford Formation (Fig. 23.11a). Along most of this low gradient reach (~30 km) the river has developed an irregular tortuous sinuosity (~2.3) and complex floodplain architecture of paleochannels and oxbow lakes yet lacking conspicuous meander scrolls (Fig. 23.11b). The channel and floodplain are entrenched up to ~10 m into the Late Glacial substrate especially where they enter the basin at Capreol. Beyond this central reach the river turns southward to cross the Chelmsford Formation ridges in a series of rapids before entering Vermilion Lake at the southwestern end of the basin. Downstream of Vermilion Lake the river continues another 5 km eastward where it exits the basin along a southwest trending fault before turning southwest again and cutting across the structure of the south rim.

Fig. 23.10 Sudbury airport glacifluvial system. **a** Aerial view of airport and Falconbridge sections with kettled outwash materials flanked by flat delta surface and outwash deposits (view to south) (2015 image from Google Earth). **b** Detail of kettle lake adjacent to highway visible in centre of aerial view

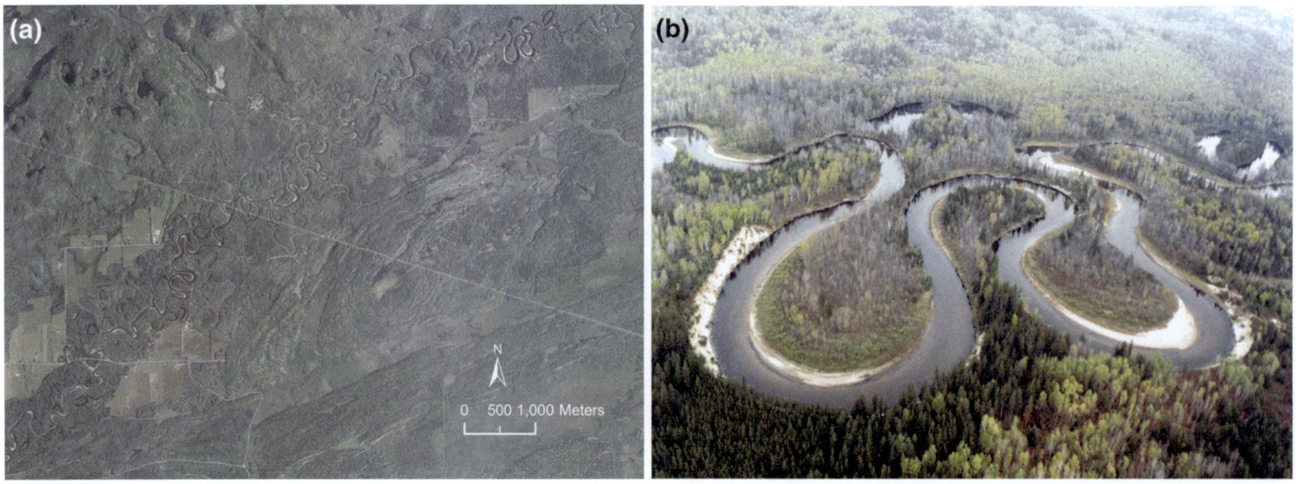

Fig. 23.11 Vermilion River. **a** Vertical aerial view of north-central Sudbury Basin showing the Vermilion River flanked by the north rim Onaping Formation (*upper left*) and ridges of the Chelmsford Formation (lower right); note agricultural land use of the older sections of floodplain. **b** Detailed oblique aerial view of meander loops and floodplain

23.7 Human Impact on Geomorphic Processes

Beginning in the late nineteenth century the combined action of logging, prospecting, fire, and the mining and processing of sulphide ores of copper, nickel and platinum group elements resulted in the localized accumulation of mine waste (Fig. 23.12a) and accelerated erosion rates on larger areas of the landscape adjacent to the Sudbury Basin. Early logging was selective and had minimal impact (Dirszowsky et al. 2013) but as mining became important from the 1880s onward, cutting proceeded less selectively as wood was used for infrastructure and as fuel for the early ore roasting process (Winterhalder 1996). Roasting to reduce ore volume before smelting involved the stacking and ignition of large piles of wood and rock, and open-air combustion over periods of 3–6 months. Through release of sulphur dioxide, land around the roast yards was intensely fumigated and acidified, resulting in barren exposed soils near these sites. In contrast, the smelting process, which completely replaced roasting by 1929, used chimneys to disperse/dilute emissions over a larger area, but also introduced metal particulates to the emissions. At low pH, metals added to the soil were especially toxic to soil biota and plant life. Vegetation removal/loss and acidity led to loss of whole soil, leaching of nutrients, and extreme microclimates that effectively prevented or suppressed natural (re)establishment of vegetation (Courtin 1994).

In the early 1970s, Struik and others (see McCall et al. 1995) defined and mapped areas of 'barrens' (highly acidic, devoid of vegetation and soil) and 'semi-barrens' (impoverished vegetation, eroded ridge-tops, residual soils) totalling more than 1000 km^2 around Sudbury's three major smelting operations (Fig. 23.1b). Soil or residual soil metal concentrations closely correlate with these vegetation/soil zones showing the clear influence of prevailing winds (southwest–northeast) and some indication of topographic effect (Hutchinson and Whitby 1974; Hazlett et al. 1984; Dudka et al. 1995). Barren areas are commonly characterized by relict wood and blackened rock surfaces due to acid weathering and smelter-borne particulate deposition, although these artifacts are becoming rarer as normal weathering and biological decomposition have resumed in recent times (Fig. 23.12b, c). Areas devoid of vegetation but retaining some soil commonly exhibit patterned ground attributable to wetting/drying and/or frost action (Fig. 23.12d, e).

Little direct observation or measurement of erosion has been made in the Sudbury area to date. Pearce (1976) recorded soil erosion rates near smelter sites two orders of magnitude above what would be expected at comparable forested sites, while (exposed) rock weathering rates appeared only slightly elevated. Pearce's work was carried out in the 1970s just after smelter emissions had peaked, pollution control measures initiated, and land restoration efforts had begun (Winterhalder 1996). Paleolimnological and geomorphological studies have confirmed that metal

Fig. 23.12 Human impacted landscapes. **a** Oblique aerial view of the Coniston smelter (closed in 1972) site including slag piles (view to northwest) (photo taken in 1994 courtesy of A. Gallie); note slope failures (rotational slides) along the eastern edge of the landform. **b** Example of eroded and blackened bedrock ridge with residual till soils on lower slopes above road cut in central Sudbury. **c** Striated bedrock with markings enhanced where black coatings related to smelter emissions have been further weathered naturally (Sudbury, south end). **d** Exposed bedrock, residual soils and patterned ground near Daisy Lake, Penokean Hills. **e** Patterned ground in residual soils as in previous photo

deposition and sediment transport rates peaked in the 1960–70s and 1960–80 s respectively (e.g. Dixit et al. 1996; Dirszowsky et al. 2013). Reconnaissance work has revealed numerous sites that have escaped soil erosion and/or where transport/deposition of sediments has occurred instead, especially on lower slopes, valley floor/floodplains and wetland/lake basins and deltas. For example, at Kelly Lake <1 km from the Copper Cliff smelter and just downstream of the Sudbury city core, the Junction Creek delta has increased in area by approximately 30 Ha since the 1880s ($\sim$85% increase) while floodplain vertical accretion has added 20–50 cm or more of contaminated sediment across much of the original area (Fig. 23.13a–d). Lake-basin sedimentation rates have also increased over this time frame and record a detailed history of landscape change in the catchment (Fig. 23.13e).

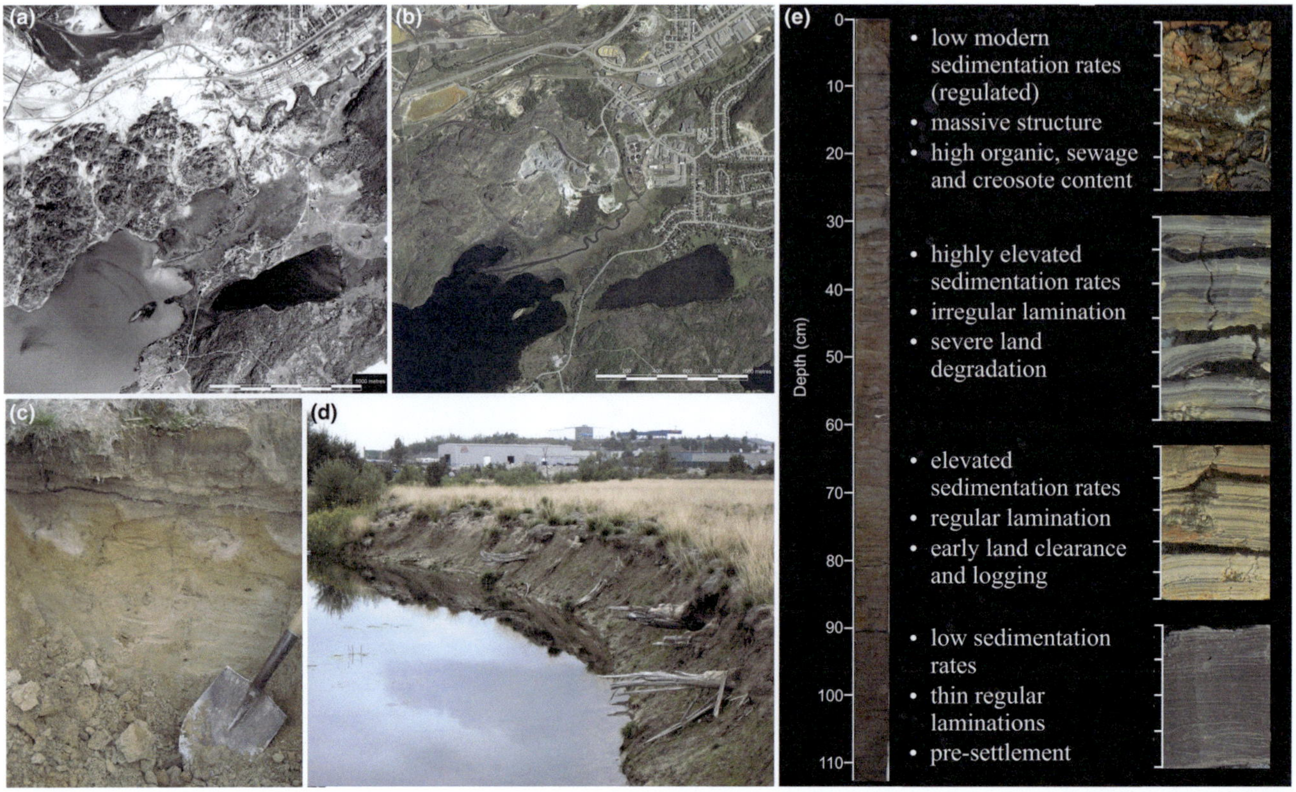

Fig. 23.13 Kelly Lake and Junction Creek delta/floodplain features. Aerial photographs show the extent of delta progradation from 1946 (**a**) to 2003 (**b**). The pre-settlement position of the shoreline occurs where the channel changes from meandering to straight. Ground photos illustrate accelerated vertical accretion resulting in buried tree trunks (**c**) and the original, compressed podzolic soil in glacilacustrine parent materials (**d**). The acidic and metal contaminated floodplain surface supports a simplified vegetation cover dominated by Agrostis scabra (hair grass). Core section photos (**e**) illustrate changing sedimentation in Kelly Lake related to varying human activities in the catchment

23.8 Conclusions

The Sudbury Basin/Structure formed as a result of an ancient meteorite impact and related geologic processes, subject to later deformation, sedimentation and extensive denudation mostly during the Precambrian. Coincidentaly it occurs at the nexus of three geological provinces and three broad physiographic regions. Direct geologic effects of formation are found more than 80 km from the basin rim and direct and indirect effects on landscape development are similarly extensive. During the Quaternary, basin topography influenced ice sheet movement, formation and distribution of streamlined glacial landforms, and dispersal of sediments. The north and south rims of the basin acted as pinning points during Late Wisconsinan deglaciation, resulting in the formation of distinct ice-marginal features (moraines, kame moraine, etc.), outwash, and proglacial lake features around an embayment of glacial Lake Algonquin. Both the geology and geomorphology of the basin have strongly influenced human settlement in the area with agriculture practiced on the low-relief basin floor, and mining around the basin margins. Mineral processing operations have led to extensive vegetation loss, acidification and metal contamination of soils, and severe erosion and redistribution of soils/sediment over a wide area (>1000 km^2) centred on the southeastern margin of the basin. More subtle effects (e.g. acid precipitation) have been felt on a continental scale.

References

Antevs E (1925) Retreat of the last ice-sheet in eastern Canada. Geol Surv Can Memoir 146

Barnett PJ (1992) Quaternary geology of Ontario. In: Thurston PC, Williams HR, Sutcliffe RH, Stott GM (eds) The geology of Ontario. Ontario Geological Survey, Special Volume 4, Part 2, pp 1011–1088

Barnett PJ, Bajc AF (1999) Surficial geology of the Regional Municipality of Sudbury. Ontario Geological Survey Preliminary Map, p 3399, scale 1:100,000

Barnett PJ, Bajc AF (2002) Quaternary geology. In: Rousell DH, Jansons KJ (eds) The physical environment of the City of Greater Sudbury. Ontario Geological Survey Special Volume 6, pp 57–85

Boissonneau AN (1968) Glacial history of northeastern Ontario II. The Temiskaming-Algoma area. Can J Earth Sci 5:97–109

Bostock HS (2014) Geology, physiographic regions of Canada, 2nd edn. Geological Survey of Canada Map 1254A, scale 1:5,000,000

Burwasser GJ (1979) Quaternary geology of the Sudbury Basin area, District of Sudbury. Ontario Geological Survey Report 181
Courtin GM (1994) The last 150 years: a history of environmental degradation in Sudbury. Sci Total Environ 148:99–102
Chapman LJ, Putnam DF (1984) Physiography of Southern Ontario. Ontario Geological Survey Special Volume 2
Dirszowsky RW, Finkelstein SA, Davis AM (2013) Pollen evidence for major shifts in forest composition since the mid-nineteenth century in the vicinity of Sudbury, Ontario. Can Geogr 57:173–185
Dixit AS, Dixit SS, Smol JP (1996) Setting restoration goals for an acid and metal-contaminated lake: a paleolimnological study of Daisy Lake (Sudbury, Canada). J Lake and Reserv Manage 12:323–330
Dredge LA, Cowan WR (1989) Quaternary geology of the southwestern Canadian Shield. In: Fulton RJ (ed) Quaternary geology of Canada and Greenland. Geological Survey of Canada Geology of Canada Publication No. 1, pp 214–249
Dressler BO, Gupta VK, Muir TL (1991) The Sudbury structure. In: Thurston PC, Williams HR, Sutcliffe RH, Stott GM (eds) Geology of Ontario. Ontario Geological Survey Special Volume 2, Part 1, pp 593–625
Dudka S, Ponce-Hernandez R, Hutchinson TC (1995) Current level of total element concentrations in the surface layer of Sudbury's soils. Sci Total Environ 162:161–171
Dyke AS, Andrews JT, Clark PU, England JH, Miller GH, Shaw J, Veillette JJ (2002) The Laurentide and Innuitian ice sheets during the Last Glacial Maximum. Quat Sci Rev 21:9–31
Gillespie JE, Acton CJ, Hoffman DW (1982) Soils of the Sudbury area. Ontario Institute of Pedology Soil Survey Report 49
Harrison JE (1972) Quaternary geology of the North Bay–Mattawa region, 31L/SE (west half) and 31L/SW (east half). Geological Survey of Canada Paper 71–26
Hazlett PW, Rutherford GK, Van Loon GW (1984) Characteristics of soil profiles affected by smelting of nickel and copper at Coniston, Ontario. Geoderma 32:273–285
Heath AJ, Karrow PR (2007) Northernmost (?) Glacial Lake Algonquin series shorelines, Sudbury Basin, Ontario. J Great Lakes Res 33:264–278
Hutchinson TC, Whitby LM (1974) Heavy-metal pollution in the Sudbury mining and smelting region of Canada I. Soil and vegetation contamination by nickel, copper, and other metals. Environ Conserv 1:123–132
Kor PSG, Shaw J, Sharpe DR (1991) Erosion of bedrock by subglacial meltwater, Georgian Bay, Ontario: a regional view. Can J Earth Sci 28:623–642
Lafrance B, Legault D, Ames DE (2008) The formation of the Sudbury breccia in the North Range of the Sudbury impact structure. Precambrian Res 165:107–119
Lewis CFM, Karrow PF, Blasco SM, McCarthy FMG, King JW, Moore TC Jr, Rea DK (2008) Evolution of lakes in the Huron basin: deglaciation to present. Aquat Ecosyst Health 11:127–136
McCall J, Gunn J, Struik H (1995) Photo interpretative study of recovery of damaged lands near the metal smelters of Sudbury, Canada. Water Air Soil Poll 85:847–852
Pearce AJ (1976) Contemporary rates of bedrock weathering, Sudbury, Ontario. Can J Earth Sci 13:188–193
Rousell DH, Card KD (2009) Geological setting. In: Rousell DH, Brown GH (eds) A field guide to the geology of Sudbury, Ontario. Ontario Geological Survey Open File Report 6243, pp 1–6
Sado EV (1977) Quaternary geology of the Coniston (41I/7) and Copper Cliff (41I/6) areas plus portions of the Regional Municipality of Sudbury, District of Sudbury. In: Milne VG, White OL, Barlow RB, Robertson JA (eds) Summary of field work 1977. Ontario Geological Survey Miscellaneous Paper 75, pp 154–155
Shanks WS (1991) Deformation of the central and the southern portions of the Sudbury Structure, Sudbury, Canada. Unpublished PhD thesis, University of Toronto, Toronto, Ontario
Shilts WW, Aylsworth JM, Kaszycki CA, Klassen RA (1987) Canadian shield. In: Graf WL (ed) Geomorphic systems of North America. Geological Society of America Centennial Special Volume 2, pp 119–161
Winterhalder K (1996) Environmental degradation and rehabilitation of the landscape around Sudbury, a major mining and smelting area. Environ Rev 4:18–224

Randy W. Dirszowsky is an associate professor in the School of the Environment at Laurentian University, Sudbury, ON. His research activities and interests include sediment production and movement in watersheds and lacustrine systems, landscape evolution, and natural and anthropogenic environmental change.

Do Landscapes Listen? Wemindji Eeyou Knowledge, Adaptation and Agency in the Context of Coastal Landscape Change

24

Monica E. Mulrennan

Abstract

Drawing on the connections to landscape of members of the Cree Nation of Wemindji, on the east coast of James Bay (Eeyou Istchee), this chapter explores the possibilities for mutually enriching conversations between geomorphologists and Indigenous people. Local placenames, landscape modifications (dyke and tuuhikaan construction), and stewardship of hunting territories are examined to underscore the deep knowledge and attachments Cree maintain to landscape. Wider embrace of ethnogeomorphology as a sub-field that supports intercultural dialogue is advocated.

Keywords

Wemindji Cree • Eeyou • Landscape • Place names • Dykes • Tuuhiikaan • Hunting territory • Hunting law • Stewardship • Spirituality

24.1 Introduction

This chapter builds on recent efforts to reposition geomorphology in relation to the study of human–landscape systems as a more reflexive, engaged discipline open to multiple ways of understanding human–landscape connections (see also Chap. 26). My particular interest is in the potential and prospects for a revised geomorphology that would support more meaningful connections with Indigenous landscapes and the knowledge, rights and interests of the peoples who occupy and maintain responsibilities for those areas. I draw particular inspiration from Wilcock and Brierley (2012) who propose ethnogeomorphology as a framework for intercultural communication and ethical engagement in practice; as they suggest, "Ethnogeomorphology is about finding ways of talking to each other through convergent spaces of dialogue" (ibid 2012). My perspective is informed by doctoral training in coastal geomorphology (Mulrennan 1990) followed by 25 years of partnered research with Indigenous coastal communities on topics related to Indigenous knowledge and institutions of stewardship, local observations and adaptations to environmental changes, and strategies of conservation and environmental protection that align with Indigenous responsibilities and aspirations for their traditional lands and seas. My approach is to give an account of the connections of one particular Indigenous community to their traditional territory in the hope of identifying some common ground for ethnogeomorphology to serve as a frame for mutually beneficial conversations about human–landscape and landform interactions (Wilcock and Brierley 2012). The title of this chapter "Do landscapes listen?" echoes Cruikshank's (2005) exploration in "Do glaciers listen?" of the intersection of nature and culture in the Yukon and Alaska. In one of the few detailed ethnographies addressing a geophysical landscape, glaciers are shown to provide 'scaffolding' for detailed empirical observations about geophysical changes and for the ability of these sentient, highly responsive natural phenomena to make moral judgments and impose serious consequences for disrespectful human behaviour. Landscape is also shown to be deeply implicated in the colonial encounter, such that natural and cultural histories, "are objectively entangled in this place" (Cruikshank 2005, p. 10).

My focus is on the Cree First Nation community of Wemindji, located along the central east coast of James Bay, northern Québec (Figs. 24.1 and 24.2). Through an exploration of Cree connections to landscape, I highlight three points. First, how Indigenous knowledge, perspectives and interests with respect to land have tended to be overlooked by geomorphologists in the past. Second, how profoundly some Indigenous people understand and shape the geomorphic landscape. And third, how landscapes can provide a platform for conversations between geomorphologists and

M. E. Mulrennan (✉)
Department of Geography, Planning and Environment, Concordia University, Montréal, Québec H3G 1M8, Canada
e-mail: Monica.Mulrennan@concordia.ca

© Springer Nature Switzerland AG 2020

O. Slaymaker and N. Catto (eds.), *Landscapes and Landforms of Eastern Canada*,
World Geomorphological Landscapes, https://doi.org/10.1007/978-3-030-35137-3_24

Indigenous people based on shared interests in developing alternative knowledges of landscapes and landform processes. In this respect, the capacity of landscapes to listen is a call to geomorphologists to do likewise.

24.2 Wemindji Cree Landscape Connections

Indigenous cultures have a long history of passing on knowledge from generation to generation through storytelling. Creation stories are an important part of that tradition. They represent symbolic narratives of how the world began. Following 'the two-eyed seeing' principle advocated by Mi'kmaw Elder Albert Marshall—which supports learning to see from one eye with the perspective of Indigenous knowledge and ways of knowing and from the other eye with the perspective of Western knowledge and ways of knowing, and learning to use both eyes together for the benefit of all (Bartlett et al. 2012)—I begin by drawing upon two very different creation stories for James Bay/ Eeyou Istchee.

For earth scientists, the present configuration of eastern James Bay is explained as the outcome of a series of events that began about 10,000 years ago (Dionne 1980; Martini 1986). A rise in global temperatures caused rapid melting of an immense ice sheet that had covered much of what is now North America for several tens of thousands of years. The ice advanced and retreated several times until it finally disappeared about 6000 years ago. Initially the ice sheet thinned and split into smaller ice lobes. As these melted, an enormous glacial lake (Lake Ojibway) formed, held back by a massive ice dam. Sometime around 8500 years ago, the dam broke resulting in the sudden release of lake water into the adjacent sea which rose rapidly to flood all of the ice-free low-lying areas of James Bay up to a ridgeline formed of glacial debris, located 100–250 km from the present shoreline (Martini 1986). Since then, land that was depressed under the enormous weight of the ice-sheets has slowly recovered, resulting in the emergence of this coast at a rate of around one metre per century (Hik et al. 1992).

Eeyou (Cree people) also have stories about the origin of this land.[1] According to one such story, Wiskachachak was given special powers and instructed by the Great Spirit (Chishaaminituu) to watch over all creatures, to keep them safe and ensure they lived harmoniously. But Wiskachachak was a trickster who loved to make trouble. One day he decided to play a trick on Beaver. He built a dam across Beaver's creek so that Beaver would think it was Muskrat who did it and quarrel with him. However, his plan was discovered and when he went to break down the dam, the water began to flow and flow. It kept on flowing until it eventually flooded all of the land. Fearing for his life, Wiskachachak quickly built a raft and tried to save as many animals as he could. After some time the Great Spirit, feeling sorry for Wiskachachak and the animals, reminded them that they had the power to remake the world if they could find sediment beneath the waters from which new land would grow. Several creatures volunteered to dive to the bottom, but all failed to find sediment. Finally, Wolf[2] swam down and succeeded in bringing back a few grains of sediment in his paw. These grew and grew to form smaller, then bigger islands, and eventually a much larger land area until the world was formed again.

The first account does a disservice to the detailed scientific evidence on which such environmental reconstructions are based while the second does an injustice to the rich oral tradition of Creation stories and the careful adjustments for context and audience made by their storytellers. My point in bringing them together here is simply to illustrate the shared interest of both parties in stories about the origins of the land, as well as parallels and differences in their emphases.

Geomorphologists have acknowledged that their knowledge of the morphology and sediments of James Bay is limited to generalized descriptions based on studies in similar settings, aerial photographs and satellite imagery, and a small number of field samples and field observations (Martini 1986). These accounts describe "a complex coast" characterized by numerous embayments, headlands, peninsulas, coves and inlets, fringed by various islands, islets and shoals (Dionne 1980). There is attention to the rapid post-glacial emergence of this coastline and recognition of importance of extensive eelgrass beds, tidal flats, and coastal marshes as habitat for fish and invertebrates and food for passing waterfowl, particularly ducks and geese (Hik et al. 1992). The presence of Crees is often acknowledged, usually in relation to the reliance of subsistence hunters on local wildlife. However, rather than recognizing the dependence of hunting success (and hence Cree survival, until the recent past) on a detailed understanding of this coastal landscape, the rocky shores of James Bay have been characterized as "virtual deserts of knowledge" (Martini 1986, p. 119).

Most maps convey an impression of James Bay as a vast area of unused, unoccupied land, about which little is known. Yet Eeyou people have occupied this land for more than five thousand years (Francis and Morantz 1983). The extent of the area occupied and used is a matter of some controversy for scholars due to the limited material record

[1]There are many versions of this story, which is likely derived from a popular Ojibway legend about the creation of Turtle Island.

[2]In this particular version it was Wolf who brought back sediment. More often, it is the little Muskrat who finally succeeds in reaching the bottom and bringing back sand.

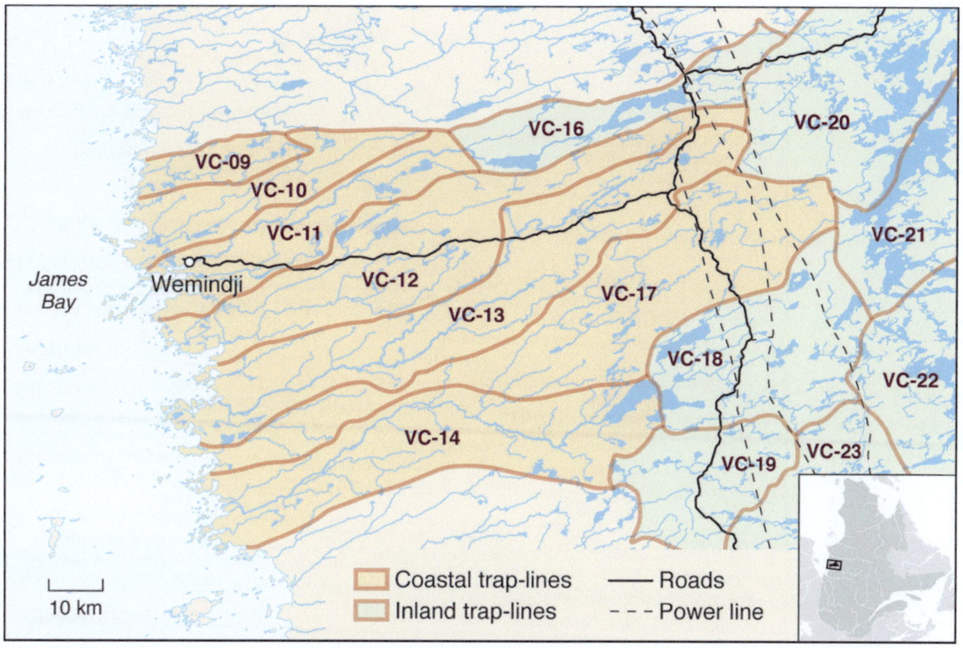

Fig. 24.1 Traditional Wemindji coastal hunting territories. *Cartography* E. Leinberger

Fig. 24.2 Landscape of central east coast of James Bay. *Photograph* **a**—_AAA2287; Government of Québec, MFFP (2017), CC BY 4.0

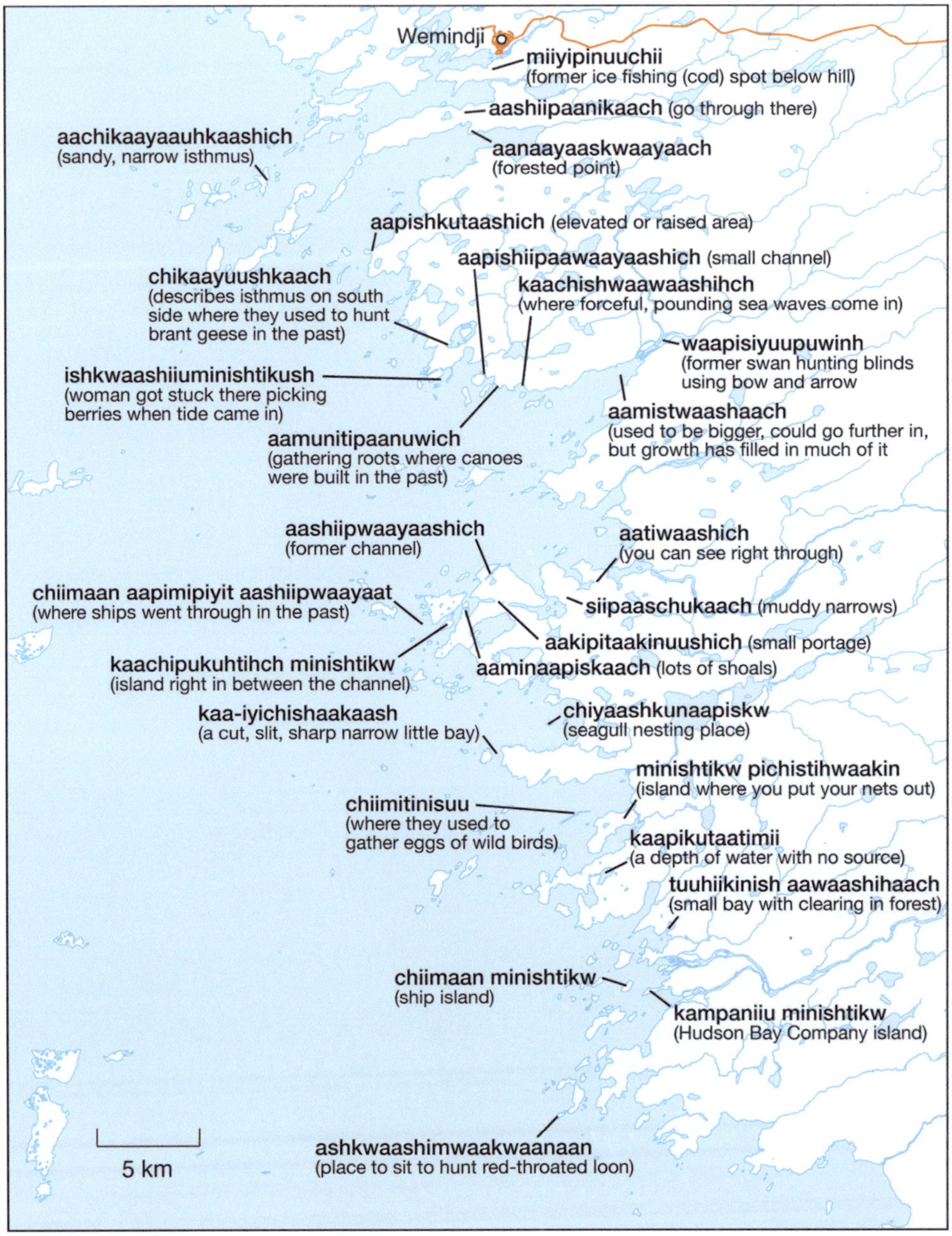

Fig. 24.3 Placenames on the Wemindji coast. *Cartography* E. Leinberger

typically associated with a highly mobile very low-density hunter-gatherer society. However, recent palynological and archaeological evidence from a terrace site on Wemindji territory located 100 km inland from the current shore has revealed "a site of unusual density, size, and organization" (Pendea et al. 2019). Interestingly, this research which is challenging previous assessments of the occupation history of boreal forest peoples (Vaneeckhout et al. 2012), grew out of an invitation from Wemindji community members for a group of archaeologists to visit and explore the site. This fruitful collaboration between archaeologists and Eeyou has been essential to locating the site and unlocking the story of its occupation and use.

24.2.1 Cree Place Names

There are other clues for Cree occupation, for those who take the time and know whom to ask. Placenames, for example, can provide a valuable source of site-specific information confirming not just the presence of Eeyou but the rich observations and experiences of this land that have over generations transformed this vast space into places that are known, named and imbued with memory and meaning (Fig. 24.3, Table 24.1). Most official maps of James Bay maintain the colonial practice of using Europeanized placenames—such as James Bay, Hudson Bay, Old Factory and Cape Hope Islands—thus serving as "intangible wormcasts"

Table 24.1 Selected placenames along Wemindji Coastal territory (see Fig. 24.2 for location)

No.	Cree name	Description
1	miiyipinuuchii	Former ice fishing (cod) spot below hill
2	aashiipaanikaach	Go through there
3	aanaayaaskwaayaach	Forested point
4	aachikaayaauhkaashich	Sandy, narrow isthmus
5	aapishkutaashich	Elevated or raised area
6	chikaayuushkaach	Describes isthmus on south side where they used to hunt brant geese in the past
7	ishkwaashiiuminishtikush	Woman got stuck there picking berries when tide came in
8	aapishiipaawaayaashich	Small channel
9	aamunitipaanuwich	Gathering roots where canoes were built in the past
10	kaachishwaawaashihch	Where forceful, pounding sea waves come in
11	waapisiyuupuwinh	Former swan hunting blinds using bow and arrow
12	aamistwaashaach	Used to be bigger, could go further in but growth has filled in much of it
13	aatiwaashich	You can see right through
14	aashiipwaayaashich	Former channel
15	aakipitaakinuushich	Small portage
16	siipaaschukaach	Muddy narrows
17	aaminaapiskaach	Lots of shoals
18	chiimaan aapimipiyit aashiipwaayaat	Where ships went through in the past
19	kaachipukuhtihch minishtikw	Island right in between the channel
20	chiyaashkunaapiskw	Seagull nesting place
21	kaa-iyichishaakaash	A cut, slit, sharp narrow little bay
22	chiimitinisuu	Where they used to gather eggs of wild birds
23	minishtikw pichistihwaakin	Island where you put your nets out
24	kaapikutaatimii	A depth of water with no source
25	tuuhiikinish aawaashihaach	Small bay with clearing in forest
26	chiimaan minishtikw	Ship island
27	kampaniiu minishtikw	Hudson Bay Company island
28	ashkwaashimwaakwaanaan	Place to sit to hunt red-throated loon

of colonization (Cruikshank 2005, p. 19). It is only in recent years that Eeyou names for the nine Eeyou community villages, including the name Wemindji (although it is often glossed as "Paint Hills"), appear with some consistency.

Some years ago, as part of an effort to document Wemindji Eeyou claims to the offshore, I participated in a project that mapped Eeyou use and occupancy of the coast, including recording Eeyou placenames. Based on interviews with elders, over 300 toponyms were recorded and mapped (Scott and Mulrennan 2000). These placenames are deeply embedded and implicated in how Eeyou know and use the coast, serving as powerful triggers for memory—relating to particular events, circumstances and features—and extending beyond physical connection to take on emotional and political dimensions, such as in the process of renaming places on maps according to Eeyou names (Sieber and Wellen 2011). Table 24.1 provides a small sample of the placenames recorded, hinting at the insights they can provide, particularly with respect to natural features and to processes of landscape change. While the social, cultural and political significance of Indigenous placenaming is widely recognized by social scientists (Basso 1996; Tucker and Rose-Redwood 2015), the rich descriptive information about individual landscape features and how they are used by people, plants and/or animals has received relatively limited attention (for an interesting exception to this, see M'Lot 2001).

Many Eeyou placenames address the nature, extent and/or impact of landscape change along the coast. The relatively low gradient (slopes of 1°–2°) of this emerging

Fig. 24.4 Former settlement at Old Factory. *Photograph* C. Scott

coastal plain (Dionne 1980) renders this coastal landscape among the most dynamic in the world. As a measure of the pace of coastal emergence, the community had to abandon its former settlement site and move 45 km north to the present village site of Wemindji in 1959. The former settlement, in the estuary of the Old Factory River (Fig. 24.4), was situated on an island that had, because of the increasing shallowness of the estuary, become too difficult to reach by supply ship (Francis and Morantz 1983).

Some of my earliest conversations with Eeyou elders and hunters focused on gaining insights into local understandings of isostatic rebound. Once I learned that Eeyou hunters refer to coastal emergence as "the land growing", I initiated an inquiry into how this happens and with what effects. In response, they provided detailed accounts of channels drying up or becoming too shallow to paddle through, of fishing camps (Figs. 24.5 and 24.6), hunting blinds and other coastal sites now stranded some distance inland, and of islands appearing and others joining the mainland along with successive changes to the coastal vegetation and implications for the quality of wildlife habitats and hence hunting success.

24.3 Wemindji Eeyou Response to Landscape Change

Wemindji Eeyou have traditionally been a hunter-gatherer society and these activities remain central today, both economically and culturally (Scott 1996). While a relatively small percentage of Eeyou currently live as full-time subsistence hunters, bush food and time on the land are thoroughly integrated for many community members (Bussières 2005; Sayles and Mulrennan 2019). Goose hunting along the coast during the spring and fall is one of the most important subsistence activities conducted by Wemindji Eeyou and is linked to a set of harvesting practices that respond to environmental variability and complexity (Fig. 24.7). Local adaptations to landscape change include the relocation of hunting and fishing camps, adjustments in travel routes, and revised hunting strategies in response to habitat change and shifts in the abundance and availability of wildlife (Peloquin and Berkes 2009).

A largely undocumented dimension of Eeyou response to landscape change has been the efforts of Eeyou hunters to

Fig. 24.5 Cree fishing camp. *Photograph* C. Scott

resist, retard, delay and/or redirect this change. My initial conversations with Cree hunters about the land growing, expanded into a Master's thesis length research project (Sayles 2008) involving semi-structured interviews supplemented by topographic and vegetation surveys, aerial photography, and satellite imagery measurements and interpretations. We documented several examples of landscape modifications that are no longer practiced today, such as the construction of stone fish weirs on certain rivers, the digging up of marshes to suppress the growth of unpalatable marsh grasses, and some limited use of fire to promote the growth of berries (Sayles and Mulrennan 2010). We focused on two landscape modification practices—the construction of dykes and the cutting of tuuhiikaan or corridors through the coastal forest—that continue to be widely applied today in the context of goose hunting (Figs. 24.8 and 24.9).

24.3.1 Construction of Dykes

Eeyou elders and active hunters identified a total of twenty-two dykes on Wemindji coastal territories (Fig. 24.8). While these vary in form, size and construction material, most are u-shaped features constructed of mud (but sometimes of logs and stones), approximately 15–30 cm in height and ranging from 4 to 250 m in length (Sayles 2008). Most involved a relatively minor investment of time, energy and materials to build but are strategically placed to create or maintain a significant pond (the largest of which is more than 3000 m^2). Some dykes were constructed in recent decades while the oldest date to the late 1700s (Sayles 2008). Dykes function at multiple nested scales, informed by the integrated knowledge of generations of hunters about changes in local ecology and goose habitat in response to the land growing. At the smallest scale, dykes maintain or enhance a wetland area that is attractive to geese by encouraging plant species within an impounded area that sharply contrast those that have colonized adjacent areas. At the larger scale of individual bays, dyke construction takes into account spatial relationships between the dyke and the impoundment it creates or maintains, the dimensions of the pond, and the relative location of hills, ridges and shrub lines. These combine to influence ecological and topographic conditions, including how geese approach the pond and the ability of hunters to strategically position themselves for a successful hunt (Sayles and Mulrennan 2010, 2019).

Finally, at the inter-bay scale the construction of dykes is informed by the relative location of other ponds and the knowledge and experience hunters have accumulated about geese habitat preferences and migration paths in the wider region. As such, these dykes are motivated by a desire to create, maintain and enhance conditions within the coastal landscape that increase the predictability and success of goose hunting activities (Sayles and Mulrennan 2010). By maintaining or at least extending the use of established and familiar hunting sites they also (quite deliberately) retain the relevance of associated place-specific knowledge and hunting strategies and other cultural commitments and attachments to those areas (ibid) (Fig. 24.10).

24.3.2 Cutting of Tuuhiikaan

Eeyou hunters also create tuuhiikaan in response to the observed tendency of geese to fly low to the ground through natural openings in the coastal forest canopy. More than thirty tuuhiikaan have been identified on Wemindji hunting territories (Sayles 2008). These are usually oriented southwest-northeast, aligned with the dominant flight paths of geese during their migration north in the spring. Their location varies but they tend to be cut in areas already favoured by geese, often on headlands between coastal embayments with the southern side more open to channel the approaching geese. They have an average width of 150 m and vary in length from 805 to 1100 m (ibid). As with dykes, cutting a tuuhiikaan is motivated by a desire to create a more predictable location for goose hunting in the context of an emerging coastal landscape (Sayles and Mulrennan 2010). Cutting a tuuhiikaan retards the encroachment of the coastal forest and also improves the ability of hunters to observe and monitor goose flocks. Since substantial investments in time and energy are required to create and maintain them, tuukiikaan are eventually abandoned when the benefits of maintaining them are outweighed by cost (Fig. 24.11).

Fig. 24.6 Drying fish. *Photograph* M. Mulrennan

Fig. 24.7 Canoe on beach. *Photograph* M. Mulrennan

Fig. 24.8 Dykes: artificial mounds in wetland areas intended to impound water and support habitats attractive to migrating geese. *Photograph* J. Sayles

Fig. 24.9 Gathering island. *Photograph* K. Scott

24.4 Eeyou Hunting Territories

Multi-family hunting territories or traplines (Indoh-hoh Istchee) are the primary customary institution of management or stewardship in Eeyou Istchee. Wemindji traditional territory covers a total area of 28,373 km^2 and is divided into twenty hunting territories, seven of which are coastal (Hydro Québec 2012). There is a latitudinal gradient in territory size, with hunting territories in southern parts of Eeyou Istchee smaller than more northern territories reflecting the shorter growing seasons and lower productivity of the latter (Berkes and Freeman 1986; Berkes 2008). Hunting territories are typically defined by watershed boundaries. The long, narrow hunting territories of coastal Eeyou families provide access to coastal and offshore resources while also allowing them to reach inland portions of their territory within three to four days of paddling; this integration of land and sea, supporting a more diversified harvest, is critical to the resilience of subsistence hunting in the region (Mulrennan and Bussières 2018).

The distinctive hunting territory system of Eeyou Istchee has supported Eeyou rights and responsibilities pertaining to land, resources, community, and social relations including

Fig. 24.10 Goose hunting area. *Photograph* C. Scott

Fig. 24.11 Tuuhiikan: an artificial opening in the coastal canopy, intended to influence the flight path of migrating geese *Photograph* J. Sayles

human-animal relationships (Brightman 1993). For every hunting territory, a senior hunting boss or *uuchimaau* is responsible for taking care of the land[3]—this includes: (a) monitoring, overseeing and providing guidance for the Indoh-hoh activities of Eeyou in the Indoh-hoh Istchee; (b) sharing specialized knowledge and educating other Eeyou in the customs, values, principles and practices relevant to Indoh-hoh activities; (c) monitoring access to the Indoh-hoh Istchee; (d) monitoring the use of wildlife and other living resources of the Indoh-hoh Istchee; and (e) managing the habitat, wildlife and other living resources of the Indoh-hoh Istchee to ensure that the land and resources will be available to future generations (Anon 2009). Table 24.2 lists the Eeyou values and principles that guide Indoh-hoh activities; these are interrelated and affirm the importance for Eeyou of promoting and maintaining harmony in their relationships with each other and with the land and living resources of Eeyou Istchee (ibid). Other Eeyou are required to respect and adhere to the expectations of the *uuchimaau*, and consequences, such as social sanctioning, are imposed on those who depart from expected respectful behaviour. For example, shame is bestowed on any who trespass on a hunting territory without permission. Eeyou are also expected to contribute to monitoring as an integral part of being on the land. In addition to information

[3]Baash-chi-cho Ouje-Maaoo: the Kaanoowapmaaakin (or goose hunting boss), is responsible for overseeing the goose hunt activities.

Table 24.2 Eeyou values and principles[a]

Eeyou values	Principles
Souchayimuwiin/Courage:	Eeyou never shy away from doing what is right, as we understand it
Taabwaaoushiiwiin/Honesty:	Eeyou remember that there is an ultimate being that watches what we do
Dibtaiimuwiin/Humility:	Eeyou never brag about our abilities, we earn the responsibility assigned to us as leaders
Souwaayiichiichaawiin/Compassion:	Eeyou show love to our fellow human beings in everything we do
Chishtaiimiiduuwin/Respect:	Eeyou show respect to everyone and everything
Naanahwiikaaduuwiin/Sharing:	Eeyou recognize our limitations and do not shy away from asking for help. Eeyou share their knowledge when requested to do so. Eeyou share the fruits of their harvesting with their families and communities
Kaachaataawaayiitaamuwiin/Wisdom:	Eeyou allow the wisdom gained from our elders and through our experience to guide us

[a]The Eeyou have affirmed in their Hunting Law that their Indoh-hoh (Cree harvesting activities) shall be guided by the following Eeyou values and principles

gathering, monitoring provides a basis for collaborative engagement with other resource users and in so doing promotes opportunities for consensus building, collective sense making and action.

24.4.1 Traditional Eeyou Hunting Law

Eeyou of Eeyou Istchee recently affirmed the Traditional Eeyou Law and adopted the Eeyou Indoh-hoh Weeshou-Wehwun (Traditional Eeyou Hunting Law) as a written document which outlines the "the tools and rules regarding the allocation of hunting territories, the governance of these hunting territories and the conduct of activities within these territories" (Anon 2009, p. 2). Various mechanisms of cultural internalization have traditionally been used to encode knowledge, to remind people of the rules and support compliance, and to reinforce community cohesion (Folke et al. 1998). Certain rituals (e.g. obligations to animals during the hunt) and ceremonies (e.g. the Walking Out ceremony, a rite of passage held for very young children) fulfill these purposes for Wemindji Eeyou (Tanner 1979). Stories play a similar role. For example, the importance of exercising care in the selection of suitable pools and the collection of freshwater is culturally internalized through stories such as the one that tells of a woman who swallowed a tadpole from a puddle and gave birth to a frog-like monster (Colin Scott, personal communication 2016). Stories and superstition are also used to protect or deter access to particular places and resources (see below for example) while Creation stories, as we saw earlier, are used to underscore the importance of maintaining good relations with each other and the land.

24.4.2 Eeyou Stewardship

For Eeyou, stewardship is not simply about taking care of the natural world but involves interdependencies between natural and social worlds. For example, spruce boughs are collected to carpet the floors of *miichiwaahp* (teepee) and the flexible trunks of small willows are used to strengthen the structure of larger *miichiwaahp*; wild berries are collected for preparing fish, and sweetgrass is burned in certain traditional ceremonies. The health of the land is linked to all other aspects of health, the health of soils, the health of plants, the health of animals and the health of Eeyou (Chu et al. 2005). "Health" for Eeyou means "miyupimaatisiiun" or "being alive well" and "is firmly tied to the ideals of living a Cree way of life—more specifically, a way of life imbued with robust connections to the physical and spiritual northern landscape" (Adelson, 2000, p. 62). This way of life is "inseparable from being able to hunt, pursue traditional activities, live well in the bush, eat the right foods, keep warm, and provide for oneself and others" (ibid, p. 97).

24.4.3 Spiritual Connection with the Land

There is a further dimension to Eeyou relations to landscape. This is a spiritual connection that extends beyond connections to burial sites, sacred sites or particular places associated with ancestral ties to one that is fully integrated into Eeyou experience of being on the land and water. Much has been written about human–animal relations in an Eeyou context and how they are informed by principles of respect and reciprocity (Brightman 1993; Tanner 1979; Scott 2006). Animals give themselves to the hunter so that the hunter and his/her family may live. However, with this gift comes with significant responsibilities and obligations. These include not talking or thinking badly about animals, not tormenting or wounding an animal nor killing one if you do not need food, and ensuring proper treatment of animal remains (Nadasdy 2007). Eeyou hunters' ideas about respect for animals grow out of a conviction that animals are other-than-human persons. A hunter's success in hunting is evidence that they have properly maintained their relations (Feit 1986). Eeyou extend this relationality to landscape, for Eeyou landscapes listen, they are sentient, knowing and watchful. For this reason, it is strictly taboo to point at an island and similarly unwise to assume safe passage between certain islands (an offering of tobacco is advised). Hubris, arrogance and carelessness are punishable. Stories of hunters, particularly younger hunters, getting lost or stranded along the coast are interpreted as such or as a warning against inappropriate behaviour. Sometimes this agency is attributed to particular spirits, such as mintuu (sea monsters) or memekweshuu (hairy creatures that are believed to live in the crack of a rock outcrop on a certain island in Moar Bay (K. Scott, interview with Sinclair Mistacheesick, 2017). In these instances people are advised to stay away, take special care and/or be alert for the possibility of their nets being spoiled. Again the maintenance of good relations with all others offers the best guarantee of hunting success, safe travel and general well-being.

Finally, Eeyou responsibilities to care for the land include their obligation to engage with outsiders—their willingness to partner—such as in protected areas creation, forestry and fisheries, and measures to mitigate the impact of dams. Eeyou bring to these conversations values of relationality and responsibility; one example is the insistence of some elders that protected areas not be limited to the protection of intact ecosystems but used to help fulfill responsibilities to restore areas that have been degraded, particularly by hydro-electric development.

24.5 Conclusions

The above account identifies several aspects of Eeyou connections to landscape that may be relevant to an invigorated geomorphology that embraces a broader understanding of human–landscape relations.

Firstly, Eeyou—like many Indigenous peoples—share an interest in landscape and its origins. Creation stories centred on understanding places of origin often articulate the special relationship Indigenous peoples have with the land. While these are easily dismissed as outside the realm of interest of natural scientists, they may hold value, particularly in the context of the Anthropocene, for the development of ethical approaches based on reciprocal moral responsibilities among humans, other living beings, and natural interconnected collectives, such as landscapes (Whyte 2015).

Secondly, through generations of empirical observations, valuable Indigenous knowledge has been developed about landforms and landform processes. In the Eeyou context, success in hunting requires a great deal of experience and knowledge, including a detailed understanding of coastal emergence and its implications for wildlife habitats and hunting conditions. Eeyou hunting knowledge is gathered by experiences as well as communication (dreams, signs, and unanticipated encounters), and is mentally recorded by the *uuchimaau* of every hunting territory (Feit 1986). In western science, knowledge is usually gathered by methodological observations derived from specific research objectives and hypotheses set beforehand. Despite these differences in methodologies, there is much to be gained from a dialogue that involves both (Howitt and Suchet-Pearson 2006; Bartlett et al. 2012).

Thirdly, Indigenous placenames and stories often encode empirical observations about geophysical changes and their consequences. A growing body of research about social memory argues that landscapes are places of remembrance with culturally significant landforms serving as a kind of archive where memories are mentally stored. Attention to the content of this knowledge can tap into valuable information and novel insights about landscape.

Fourthly, Eeyou hunters are not passive observers of landscape change but actively modify the landscape to enhance their success in hunting. Landscape modifications by Eeyou, some of which are at a scale visible on satellite imagery, underscore the presence and agency of Eeyou. Eeyou Istchee is not an empty, unknown landscape but the occupied and known traditional homeland of Eeyou people.

Fifthly, Eeyou responsibilities to the land include responsibilities that integrate natural and social worlds; the health, productivity and renewal of the land is directly and intimately connected with Eeyou health and well-being. Eeyou stewardship is underpinned by values and principles centred on respect and reciprocity. These may yet prove to be a better guide than science in facing the challenges of the Anthropocene. Similarly Eeyou understanding of landscapes as sentient, with the capacity to listen and respond, also supports a more ethical approach to human–nature relations. There is also much to be learned from Eeyou adaptive management approaches applied at multiple time/spatial scales and informed by generations of accumulated knowledge, continued observations and monitoring, and social learning. These approaches have particular relevance to the creation and management of protected areas, biodiversity protection and conservation strategies.

Finally, attention to Indigenous landscapes offers a valuable opportunity to decolonize the discipline of geomorphology. Political reconciliation and ecological restoration are distinct discourses and literatures; yet for Indigenous peoples reconciliation has to be about land and healing includes restoration of the land (Murdock 2018). Conversations between geomorphologists and Indigenous peoples can generate mutually valuable insights. Fulfilling the potential and possibilities of this engagement will require major adjustments, "such as thinking differently about knowledge and scales of relevance, recognizing issues that are important to Indigenous people and the deep ethical challenges associated with this work and its colonial context" (Mould et al. 2018, p. 107). Ethnogeomorphology can serve as a space for us to learn to listen and respond to this exciting opportunity.

Acknowledgements Chiniskumitin (thank you) to members of the Cree Nation of Wemindji who have generously shared their knowledge, observations and teachings about the land with me. Thanks also to Véronique Bussières, Jesse Sayles, Colin Scott, Katherine Scott, and Olav Slaymaker for their contributions to this chapter.

References

Adelson N (2000) Being alive well: health and the politics of cree well-being. University of Toronto Press, Toronto

Anon (2009) Eeyou Indoh-Hoh Weeshou-Wehwun traditional Eeyou hunting Law. http://creetrappers.ca/wp-content/uploads/2014/02/CTAHuntingLaw.pdf

Bartlett C, Marshall M, Marshall A (2012) Two-eyed seeing and other lessons learned within a co-learning journey of bringing together Indigenous and mainstream knowledges and ways of knowing. J Environ Stud Sci 2:331–340

Basso K (1996) Wisdom sits in places: landscape and language among the Western Apache. University of New Mexico Press, Albuquerque

Berkes F (2008) Sacred ecology. Traditional ecological knowledge and resource management, 2nd edn. Taylor and Francis, Philadelphia and London, UK

Berkes F, Freeman MMR (1986) Human ecology and resource use. In: Martini IP (ed) Canadian inland seas. Elsevier, Toronto, Oxford, New York, Tokyo, pp 425–455

Brightman R (1993) Grateful prey: rock Cree and human-animal relationships. University of California Press, Berkeley

Bussières V (2005) Towards a culturally-appropriate locally-managed protected area for the James Bay Cree community of Wemindji, Northern Québec. MA thesis, Concordia University, Montréal

Chu SY, Crystal V, Lefrançois S, Maltais A, Martinson J, Usborne A, Prud'homme G, Maisonneuve J (2005) Aa-Wiichaautuwiihkw: coming together to walk together. Creating a culturally appropriate watershed and marine protected area in Paakumshumwaau (Old Factory) James Bay, Quebec. Final report, Environmental Research ENVR 401, McGill University

Cruikshank J (2005) Do glaciers listen? local knowledge, colonial encounters, and social imagination. UBC Press, Vancouver, BC, p 312

Dionne JC (1980) An outline of the eastern James Bay coastal environments. In: McCann SB (ed) The coastline of Canada. Geological Survey of Canada Paper 80-10, pp 331–338

Feit HA (1986) James Bay Cree Indian management and moral considerations of fur bearers. In: Native people and renewable resource management. 1986 Symposium of the Alberta Society of Professional Biologists (ASPB). ASPB, Edmonton, pp 49–65

Folke C, Berkes F, Colding J (1998) Ecological practices and social mechanisms for building resilience and sustainability. In: Berkes F, Folke C (eds) Linking social and ecological systems: management practices and social mechanisms for building resilience. Cambridge University Press, Cambridge, UK

Francis D, Morantz T (1983) Partners in furs: a history of the fur trade in eastern James Bay, 1600–1870. McGill-Queen's Press, Montréal and Kingston

Hik DS, Jefferies RL, Sinclair ARE (1992) Foraging by geese, isostatic uplift and asymmetry in the development of salt-marsh plant communities. J Ecol 80:395–406

Howitt R, Suchet-Pearson S (2006) Rethinking the building blocks: ontological pluralism and the idea of 'management'. Geogr Ann 88 B:323–335

Hydro-Québec (2012) The community of Wemindji and the project. Eastmain 1-A and Sarcelle powerhouses and Rupert diversion. Part 5 Wemindji, May

M'Lot M (2001) Kâ Isinâkwâk Askîy: using Cree knowledge to perceive and describe the landscape of the Wapusk National Park area. Masters thesis, University of Manitoba

Martini IP (1986) Coastal features of inland seas. In: Martini IP (ed) Canadian inland seas. Elsevier, Toronto, Oxford, New York, Tokyo, pp 117–142

Mould SA, Fryirs K, Howitt R (2018) Practising sociogeomorphology: relationships and dialogue in river research and management. Soc Natur Resour 31:106–120

Mulrennan ME (1990) The geomorphic development of the barrier–beach complexes of the North County Dublin coastline. Unpublished PhD thesis, National University of Ireland

Mulrennan ME, Bussières V (2018) Social-ecological resilience in indigenous coastal edge contexts. Ecol Soc 23:18. https://doi.org/10.5751/ES-10341-230318

Murdock EG (2018) Unsettling reconciliation: decolonial methods for transforming social-ecological systems. Environ Value 27:513–533

Nadasdy P (2007) The gift in the animal: the ontology of hunting and human–animal sociality. Am Ethnol 34:25–43

Pendea F, Costopoulos A, Chmura G, Wren CD, Bracewell J, Vaneeckhout S, Okkonen J, Hulse E, Keeler D (2019) Shoreline displacement and human adaptation in eastern James Bay: a 6,000-year perspective. In: Mulrennan ME, Scott CH, Scott K (eds) Caring for Eeyou Istchee: protected area creation on Wemindji Cree territory. UBC Press, Vancouver

Peloquin C, Berkes F (2009) Local knowledge, subsistence harvests, and social-ecological complexity in James Bay. Hum Ecol 37:533–545

Sayles JS (2008) Tapaiitam: human modifications of the coast as adaptations to environmental change, Wemindji, Eastern James Bay. Master's thesis, Concordia University, Montréal, Québec

Sayles JS, Mulrennan ME (2010) Securing a future: Cree hunters' resistance and flexibility to environmental changes, Wemindji James Bay. Ecol Soc 15:22. http://www.ecologyandsociety.org/vol15/iss4/art22/

Sayles JS, Mulrennan ME (2019) Coastal landscape modifications by Cree hunters. In: Mulrennan C, Scott CH, Scott K (eds) Caring for Eeyou Istchee: protected area creation on Wemindji Cree territory. UBC Press, Vancouver

Scott CH (1996) Science for the west, myth for the rest? the case of James Bay Cree knowledge construction. In: Nader L (ed) Naked science: anthropology inquiry into boundaries, power, and knowledge. Routledge, London, pp 69–86

Scott CH (2006) Spirit and practical knowledge in the person of the bear among Wemindji Cree hunters. Ethnos 71:51–66

Scott CH, Mulrennan ME (2000) Documentation and mapping of Wemindji coastal toponyms. Field Research Notes

Sieber RE, Wellen CC (2011) The role of geospatial technologies for integrating landscape in language: geographic information systems and the Cree of Northern Quebec. In: Mark DM, Turk AG, Burenhult N, Stea D (eds) Landscape in language - transdisciplinary perspectives. Routledge, New York, pp 381–393

Tanner A (1979) Bringing home animals: religious idealogy and mode of production of the Mistassini Cree hunters. ISER Books, St John's, Newfoundland

Tucker B, Rose-Redwood R (2015) Decolonizing the map? toponymic politics and the rescaling of the Salish Sea. Can Geogr 59:194–206

Vaneeckhout S, Okkonen J, Costopoulos A (2012) Paleoshorelines and prehistory on the eastern Bothnian Bay coast (Finland): local environmental variability as a trigger for social change. Polar Geogr 35:51–63

Whyte KP (2015) How similar are indigenous North American and Leopoldian environmental ethics? revisiting Aldo Leopold's land ethic: emerging cultures of sustainability. In: Trusty T, Forbes W, Stephen F (eds) Proceedings from national endowment for the humanities summer institute 'rethinking the land ethic: sustainability and the humanities'. Austin University Press, Austin, Texas

Wilcock DA, Brierley GJ (2012) It's about time: extending time-space discussion in geography through 'ethnogeomorphology' as an education and communication tool. J Sustain Educ 3. http://www.susted.org/

Monica E. Mulrennan is Associate Vice-President of Research and Associate Professor in the Department of Geography, Planning and Environment at Concordia University, Montréal (Canada). She completed a Ph. D. in Coastal Geomorphology at University College Dublin, and spent three years in Australia as Post-Doctoral Fellow (University of Wollongong and Australian National University). During this time she developed a strong interest in the connection of Indigenous peoples to coastal environments. For the past 25 years her research has focused on Indigenous-led strategies of conservation and environmental stewardship that draw upon Indigenous institutions of knowledge and practice. She works closely with James Bay Crees (Eeyou Istchee) on research related to local adaptations to environmental change, protected area creation, and knowledge and use of coastal and marine territories. She has also sustained a long-term research partnership with Indigenous Torres Strait Islanders of northern Queensland and has contributed to the documentation of Islander knowledge, tenure and resource harvesting practices for the successful Torres Strait Islander regional sea claim. Dr. Mulrennan is also involved in a collaborative research project, involving colleagues at the National University of Ireland at Galway (NUIG), which explores the historical and cultural dimensions of seaweed harvesting in Ireland. She is co-editor of the recently published 'Caring for Eeyou Istchee: protected area creation on Wemindji Cree territory Protected Area Creation in Wemindji, Eeyou Istchee' (Vancouver: UBC Press) as well as the author of numerous journal articles and book chapters.

Protecting Geodiversity in Eastern Canada

25

Olav Slaymaker, Norm Catto, and Dori J. Kovanen

Abstract

UNESCO World Cultural and Natural Heritage sites, National Parks and other protected areas in eastern Canada represent a diversity of geomorphic settings. Protected areas of Earth's terrestrial surface are assuming rapidly increasing importance as humans become ever more dominant as geomorphic agents. The initial practice of establishing and designating protected areas as either largely cultural (human) or natural has been superseded by more holistic understandings linking people and land, including geomorphology. Two hopeful signs are identified: the increasing involvement of Indigenous Peoples in the designation and co-management of protected areas; and the more explicit public educational function of parks through mechanisms such as Geoparks. Factors contributing to ongoing changes in eastern Canada's protected areas include environmental, ecological and economic stresses; increased understanding of the role of protected areas in protecting biodiversity; international support for protected areas; and the increased effectiveness and expertise of environmental non-governmental organizations (ENGOs), scientific groups, and communities, particularly Indigenous Peoples.

Keywords

Indigenous Peoples • Geoparks • World Heritage • National Parks • Protected areas • eastern Canada

O. Slaymaker (✉) · D. J. Kovanen
Department of Geography, University of British Columbia, 1984 West Mall, Vancouver, BC V6T 1Z2, Canada
e-mail: olav.slaymaker@ubc.ca

N. Catto
Department of Geography, Memorial University of Newfoundland and Labrador, St. John's, NL A1B 3X9, Canada

25.1 Introduction to Protected Areas

Protected areas of Earth's terrestrial surface are assuming rapidly increasing importance as humans become ever more dominant as geomorphic agents. Over evolutionary time scales, the natural diversity of landscapes, landforms, soils, rocks and minerals ("geodiversity" according to Gray 2004) has made biodiversity possible. Ecosystems have developed, depending almost entirely on interactions among bedrock, landforms, sediments, soils, hydrology and climate. Until the closing decades of the twentieth century the role of society in modifying these landscapes and ecosystems was perceived through the lens of a false nature/culture binary (see Chap. 26, this volume). It was commonly thought that world geoheritage could be conserved by dealing independently with "natural" and "cultural" landscapes. But more recently, the task of geoconservation is seen to involve the protection of geodiversity, biodiversity and cultural diversity through active management that recognizes the interdependence of nature and society (Brocx and Semeniuk 2007; Burek and Prosser 2008).

The recent history of the UNESCO World Heritage site designation process reflects this dramatic change of emphasis. UNESCO World Heritage sites (UNESCO 1972) were originally divided into natural and cultural "properties" at a time when the categories "natural" and "cultural" were intuitively thought to be distinct. Natural landscapes were relatively untouched by human hand and cultural properties were considered the sole result of human activities. The history of the past 50 years has blurred this distinction as the global population has grown from 2 to 7 billion and all landscapes have become increasingly "cultural". The growing realization of multiple relationships between human activities and physical landscapes led to the recognition of cultural landscapes as a significant part of World Heritage in 1992 (Fowler 2003). Landforms form the background of many cultural landscapes and are inseparable components of sites of special cultural, religious, or historical significance

© Springer Nature Switzerland AG 2020

O. Slaymaker and N. Catto (eds.), *Landscapes and Landforms of Eastern Canada*, World Geomorphological Landscapes, https://doi.org/10.1007/978-3-030-35137-3_25

and are thus embedded in the concept of geoheritage (Badman 2010; Kondolf and Piegay 2011).

25.2 World Cultural and Natural Heritage Sites in Eastern Canada

The Convention Concerning the Protection of the World Cultural and Natural Heritage was adopted by the UNESCO General Conference in 1972 (UNESCO 1972) and is generally called the World Heritage Convention. Canada ratified the convention in 1976 and Parks Canada was designated as the lead agency for implementation of the convention in Canada. Places that are of "Outstanding Universal Value" (OUV) are defined by the World Heritage Committee. In order for a site (or property) to achieve OUV, it requires cultural and/or natural significance that is so exceptional as to transcend national boundaries.

25.2.1 Criteria

There is a set of ten specific criteria that in abridged version may be defined as follows:

1. A masterpiece of human creative genius.
2. An important interchange of human values expressed in architecture, technology, town planning or landscape design.
3. Exceptional testimony to a cultural tradition or civilization.
4. Outstanding example of a type of building or landscape that illustrates significant stages in human history.
5. Outstanding example of a traditional human settlement, land use or sea use representative of a culture's interaction with the environment.
6. Directly associated with living traditions or with artistic and literary works of outstanding universal significance.
7. Superlative natural phenomena.
8. Examples representing major stages of Earth history, including the record of life, landform evolution and significant geomorphic features.
9. Significant ecological and biological processes in the evolution of terrestrial, fresh water, coastal and marine systems and communities.
10. The most significant natural habitats for in situ conservation of biological diversity.

A final qualification is added: a property must also meet the conditions of integrity and/or authenticity and must have an adequate protection and management system to ensure its safe guarding. To satisfy the integrity criterion there are three further conditions that must be met:

1. The extent to which it includes all elements necessary to express its OUV
2. Adequate size to ensure the complete representation of the features and processes and
3. No adverse effects from development and/or neglect (UNESCO 2011).

After successfully satisfying these criteria all World Cultural and Natural Heritage Sites in eastern Canada are indicated in Fig. 25.1 and in Table 25.1 (GeoEx 2018).

25.2.2 Geodiversity Sites

Four eastern Canadian World Heritage Sites have been selected on the basis of their superlative representation of major stages of Earth history and/or geomorphological importance.

25.2.2.1 Mistaken Point Natural World Heritage Site, Newfoundland

Mistaken Point Natural World Heritage Site, designated 2016, is located at the southeastern tip of the Avalon Peninsula of Newfoundland (Fig. 25.2; see Chap. 15). It consists of a narrow 17 km strip of rugged coastal cliffs. Of deep marine origin, the rocks in these cliffs date to the middle Ediacaran Period (580–560 million years ago). They are the oldest known assemblages of large fossils from this period in the world. These fossils illustrate a watershed in the history of life on Earth: the appearance of large, biologically complex organisms, including the first ancestral animals after almost three billion years of micro-dominated evolution. Most of the fossils are rangeomorphs, an extinct group of fractal organisms positioned near the base of animal evolution. These soft-bodied creatures lived on the deep sea floor and were buried and preserved in exceptional detail by influxes of volcanic ash each layer creating a virtual "Ediacaran Pompeii". Modern erosion has exhumed more than 100 fossil sea floor surfaces, ranging from small beds with single fossils to larger surfaces with up to 4,500 mega fossils. Under the UNESCO Operational Guidelines, a plan ensuring integrity and appropriate protection of the fossils were established, which required a geomorphic study of prior to site designation (Thompson 2014). Erosion was noted as a result of geomorphic processes including wave impact, storm action, freeze/thaw, mass movement (rock topples and falls), as well as from foot traffic due to increased visitation.

Fig. 25.1 Distribution of world cultural and natural heritage sites in eastern Canada (GeoEX 2018)

Table 25.1 Distribution of world cultural and natural heritage sites in eastern Canada (as of July, 2018)

Location	Year	Criterion	Landscape significance
L'Anse aux Meadows, NL	1978	6	First European settlement in North America
Historic District of Old Québec, QC	1985	4 and 6	Best fortified town north of Mexico
Gros Morne National Park, NL	1987	7 and 8	Glaciated landscape and plate tectonics
Old Town Lunenburg, NS	1995	4 and 5	British settlement displacing Acadians
Miguasha National Park, QC	1999	8	Devonian comprehensive fossil fishes
Rideau Canal, ON/QC	2007	1 and 4	Oldest canal system in North America
Joggins Fossil Cliffs, NS	2008	8	Carboniferous terrestrial fossil record
Landscape of Grand Pré, NS	2012	5 and 6	Acadian 17th century cultural landscape
Red Bay Basque Whaling Sta., NL	2013	3 and 4	Earliest Basque whaling tradition in North America
Mistaken Point, NL	2016	8	Earliest evidence of multi-cellular life
Pimachiowin Aki, MB/ON	2018	5, 7, 9 and 10	Co-managed landscape of four First Nations

Fig. 25.2 Mistaken Point World Heritage Site on Newfoundland's Avalon Peninsula is home to some of the oldest fossils ever discovered. The rocks in these cliffs date to the middle Ediacaran Period (580–560 million years ago) and are exposed to coastal erosion and mass movement of Quaternary diamicton. *Photograph* N. Catto

25.2.2.2 Joggins Fossil Cliffs World Heritage Site, Nova Scotia

Joggins Fossil Cliffs World Heritage Site was designated in 2008. It is located along the hypertidal Chignecto Bay, Bay of Fundy, Maritime Plain (see Chap. 18; Fig. 25.3) and contains the most complete terrestrial fossil record of the Carboniferous Period (354–290 Ma) including tracks of early animals and of the rainforest they lived in. The rocks are the thickest and most comprehensive record of the late Carboniferous (Pennsylvanian) strata. Joggins' 14.7 km of cliffs, low bluffs, rock platforms and beaches include remains of three Carboniferous ecosystems: estuarine bay, floodplain rain forest and fire-prone forested alluvial plain with freshwater pools.

Fossils include some of the first reptiles in Earth history. Upright fossil trees are preserved at a series of levels in the cliffs together with animal, plant and trace fossils that provide environmental context and enable a complete reconstruction to be made of the extensive fossil forests that dominated land at this time, and are now the source of most of the world's coal deposits. Initially discovered through coal mining, Joggins played a vital role in the development of seminal geomorphological and evolutionary principles, including the work of Sir Charles Lyell and Charles Darwin, for which the site has been referred to as the "Coal Age Galápagos". The interior of the Joggins site has been modified by coal mining since the late 1700s. The hypertidal coastal areas are subject to storm surges, particularly during extratropical transitions such as the 1869 Saxby Gale.

25.2.2.3 Gros Morne National Park, Newfoundland

Gros Morne is a Natural World Heritage Site, designated in 1987, and is located on the Great Northern Peninsula of Newfoundland (Fig. 25.4) in Appalachia-Acadia and has some of the world's best examples of the effects of past plate tectonics. The park is also an outstanding demonstration of multiple glaciations in an island setting. Glaciation created fjords, hanging valleys and waterfalls, combining with Holocene geomorphic processes (see Chap. 16) to produce a landscape of high scenic value. A Tertiary erosion surface at c. 700 m asl is a prominent feature of the landscape. Gros Morne portrays the geological events that took place when an ancient ocean and the ancient continental margin of eastern North America were destroyed and uplifted to form a mountain chain through the action of plate tectonics. Tectonic activity during the Late Precambrian resulted in the formation of a mid-ocean rift valley system in the Iapetus Ocean, a predecessor of the modern Atlantic Ocean. Volcanoes erupted along the widening rift system, forming basalts. The widening ocean basin was infilled by limestones, shales, and conglomerates of Cambrian and Ordovician ages, containing brachiopods, graptolites, and trilobites. When the Iapetus Ocean basin closed completely slabs of Iapetus Ocean crust were detached and pushed up during the collision, forming the tablelands of Gros Morne. Pillow basalts, similarly detached and shoved upwards, are today exposed in the Green Gardens area of the tablelands. By the mid-Silurian, approximately 420 million years ago,

Fig. 25.3 Joggins Fossil Cliffs World Heritage Site, Nova Scotia. **a** and **b** View of Joggins Formation (Pennsylvanian) seen from Coal Mine Point. The tilted Joggins beds are overlain by reddish glacial deposits of Pleistocene age. **c** 'In situ' lycopsid from the Joggins Formation. *Photographs* **a**—MC Rygel/Wikimedia Commons, CC BY-SA 3.0. **b**—Alexis/Flickr, CC BY-SA 4.0. **c**—MC Rygel/Wikimedia Commons, CC BY-SA 3.0

all of the bedrock units of Gros Morne had been formed and deformed. Collectively, the sequences of rocks that illustrate this geological evolution are represented by an ancient continental crust composed of intensely metamorphosed granite and gneisses; a continental shelf with tropical carbonate sediments, a continental slope of thick sequences of shales inter-bedded with limestone conglomerates; a complete cross-section of oceanic lithosphere including large exposures of mantle material; and significant sequences of volcanic rocks of oceanic origin.

25.2.2.4 Miguasha Natural World Heritage Site

Miguasha Natural World Heritage Site, designated 1999, is located within the parc national Miguasha, Québec. It is a paleontological site on the southern coast of the Baie des Chaleurs, Gaspé Peninsula, southeastern Québec, Appalachia-Acadia. The rocks of Miguasha are considered to be an outstanding illustration of the Devonian Period known as the "Age of Fishes". Dating from 370 million years ago, the Escuminac Formation contains five of the six fossil fish groups associated with the Upper Devonian. Its significance stems from the discovery of the highest number and best preserved fossil specimens of the lobe-finned fishes that gave rise to the first four-legged, air-breathing terrestrial vertebrates, the tetrapods. With some 13,000 specimens, the park's fossil collection also demonstrates the integrity of this scientific heritage. A large number of these fossils are exceptionally well preserved, with complete 3D specimens and fossilized soft tissues. Miguasha preserves the integrity of the fossil cliff and the Escuminac Formation. Protection of the beach area strengthens this integrity by permitting the recovery of fossils that fall from the cliff and are moved by the tides. Geomorphically, the park is located at the mouth of the Restigouche estuary, and is subject to storm surges and high tides amplified by easterly winds.

25.2.3 Cultural-Historical Sites

Seven locations in eastern Canada have been designated as cultural-historical sites of outstanding value.

25.2.3.1 Pimachiowin Aki World Heritage Site, Manitoba and Ontario

Pimachiowin Aki is a managed landscape (29,000 km^2) on the ancestral lands of four First Nations peoples on the forested shield of westernmost Ontario (north of Kenora) and adjacent Manitoba (northeast of Winnipeg). One of a

Fig. 25.4 Gros Morne National Park, Western Newfoundland. **a** Panorama seen on Route 431, overlooking Bonne Bay. The duff colored plateau, right background, is an erosion surface in ultramafic rock at c. 700 m asl formed in the Tertiary Period. **b** West Brook Pond. In the centre of Gros Morne national park, the erosion surface is intersected by a number of fjords, among them the Western Brook Pond fjord that is 165 m below sea level at its lowest point, a difference in level of more than 850 meters. **c** View of Ten Mile Pond from the top of Gros Morne Mountain (806 m asl). **d** Panoramic view from Lookout Point on the erosion surface. *Photographs* **a**—Tango7174/Wikimedia Commons, CC 4.0. **b**—T. Brent/Flickr, CC BY-SA 4.0. **c**—N. Lucier/Flickr, CC BY 2.0. **d**—T. Brent/Flickr, CC BY-SA 4.0

few UNESCO sites officially designated for both cultural and natural values, it was officially designated a World Heritage site on July 1, 2018. Pimachiowin Aki ("The Land that Gives Life") forms part of the ancestral home of the Anisinaabeg, and Indigenous people living from fishing, hunting, and gathering. The forest landscape includes parts of the lands of four Anishinaabe First Nations: Bloodvein River, Little Grand Rapids, Pauingassi, and Poplar River First Nations. Pimachiowin Aki is an exceptional example of the cultural tradition of Ji-ganawendamang Gidakiiminaan ("keeping the land"), respectful interaction with "aki" (the land and all its life). The Anishinaabe world view of a symbiotic relationship between people and nature attributes animacy to objects in the natural world giving meaning to peoples' existence in this environment over time and through the seasons. The beliefs and values that make up Ji-ganawendamang Gidakiiminaan are sustained by systems of customary governance based on family structures and respect for elders, and through vibrant oral traditions. First Nations and provincial partners have created the Pimachiowin Aki Corporation and developed a consensual, participatory governance structure, financial capacity, and management framework for the property (Lemelin and Bennett 2010; Lemelin and Dawson 2014).

25.2.3.2 L'Anse Aux Meadows World Heritage Site, Newfoundland

L'Anse aux Meadows (Fig. 25.5) contains the excavated remains of a complete 11th-century Viking settlement, the earliest evidence of Europeans in North America. The

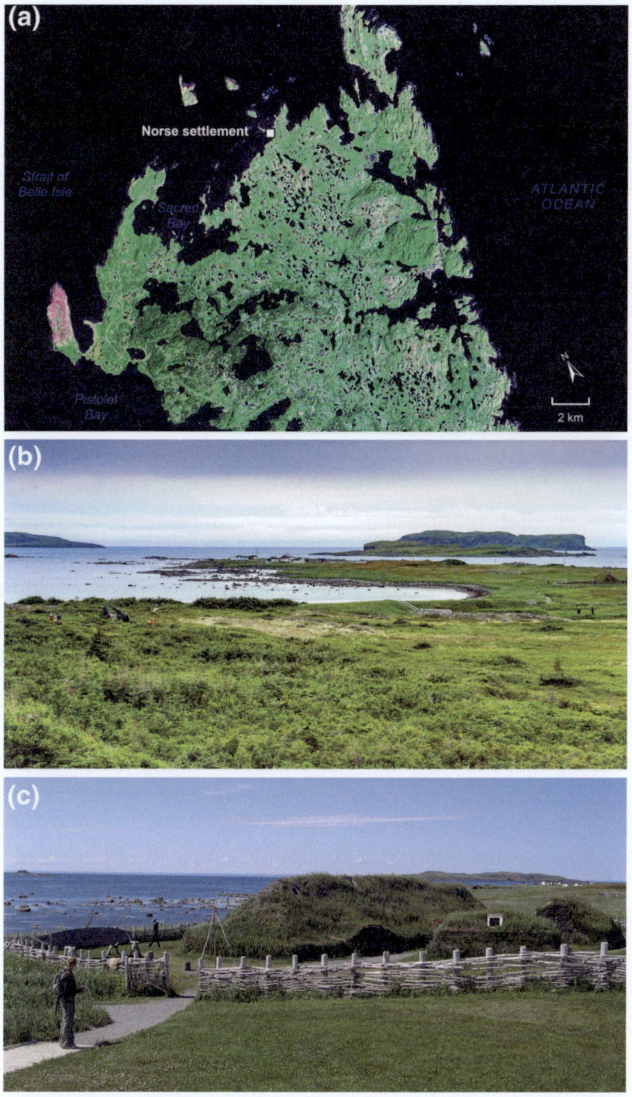

Fig. 25.5 **a** L'Anse aux Meadows National Historic Site, northern tip of Newfoundland. L'Anse aux Meadows; a French-English name meaning the bay with meadows. *Source* MDA Federal (2004). **b** The landscape consists of a mixture of coastal bogs and barrens (plains with low-growing vegetation and stunted trees) and **c** The site harbors the foundations of eight sod-covered buildings and numerous artifacts that were part of a small Norse settlement in the eleventh century. *Photographs* **b**—M. Rathwell/Wikimedia Commons, CC BY 2.0. **c**—D. Kereluk/Flickr, CC BY 2.0

archaeological site consists of eight partially reconstructed timber-framed turf structures built in the same style as those found in Norse Greenland and Iceland from the same period. The buildings include three dwellings, one forge, and four workshops, on a narrow terrace overlooking a wetland and small brook near the shore of Epaves Bay, Strait of Belle Isle. With increased storm activity evident since 2000, L'Anse-aux-Meadows could be vulnerable to storm surge, even under relatively stable relative sea level. The geomorphology of L'Anse-aux-Meadows is discussed in Chap. 7.

25.2.3.3 Historic District of Québec City World Heritage Site, Québec

Founded by the French explorer Champlain in the early seventeenth century, the former capital of Nouvelle-France came under English rule from 1759 until the establishment of Canada. It is the only North American city to have preserved its ramparts, together with the numerous bastions, gates and defensive works which still surround Old Québec (Fig. 25.6). The Upper Town, built on the edge of the St. Lawrence Lowland on the argillaceous Ordovician limestone of Cap Diamant and defended by fortified ramparts, has remained the religious and administrative center, with its churches, convents and other monuments, including the Dauphine Redoubt, the Citadel and Chateau Frontenac. Together with the Lower Town, which grew up around Place Royale, the harbor and its ancient districts, it forms an urban ensemble which is one of the best examples of a fortified colonial city. Québec bears eloquent testimony to important stages in the European settlement of the Americas. Since its inclusion on the World Heritage List in 1985, a number of buildings in Old Québec have been added to the list of properties including the Site historique et archéologique de l'Habitation-Samuel-De Champlain, the Ursuline Convent of Québec and the archaeological reference collection of Place Royale.

25.2.3.4 Red Bay Basque Whaling Station World Heritage Site, Labrador

Red Bay Basque Whaling Station (designated 2013) is located in southeastern Labrador (St. Lawrence Lowland). Between 1550 and the early 17th century, Red Bay was a major Basque whaling area. The site is home to three Basque whaling galleons and four small chalupas used in the capture of whales. The geomorphology of Red Bay is discussed in Chap. 7.

25.2.3.5 Landscape of Grand-Pré World Heritage Site, Nova Scotia

The Landscape of Grand-Pré Cultural World Heritage Site in Appalachia-Acadia is an exceptional example of the adaptation of the first European settlers to the conditions of the North American Atlantic coast. The site is marked by one of the most extreme tidal ranges in the world, averaging more than 11 m. Situated in the southern Minas Basin of Nova Scotia, the Grand Pré marshland and archeological sites are a cultural landscape showing the development of agricultural farmland using dykes and the "aboiteau" wooden sluice system started by the Acadians in the 17th century and further developed and maintained by the Planters and present day inhabitants. The cultural landscape encompasses a large expanse of polder farmland and archeological elements of the towns of Grand Pré and Hortonville, which were built by the British successors of the Acadians. The site is also

Fig. 25.6 Old Québec, Québec. **a** Québec City draws its name from the Algonquin word *kebek*: 'where the river narrows.' *Source* r555_39 satellite image spot5 2.5 m quebec canada 2002, © CNES (2002), Distribution Airbus DS/Spot 5. **b** View from the King's bastion of Old Québec and the St. Lawrence River. **c** The 'Old City,' or *Vieux-Québec*, is rich with 400-year old buildings, narrow cobblestone streets, and blend of old French and English history and culture. It is home to the only fortified city walls in the Americas outside of Mexico. The iconic Chateau Frontenac stands at the top of the city. *Photographs* b—D. Jarvis/Flickr. c—WHC.UNESCO, CC02.0

inscribed as a memorial to the Acadian way of life, and the Grand Dérangement deportation, beginning in 1755. The geomorphology of the Minas Basin, including Grand Pré, is discussed in Chap. 18.

25.2.3.6 Old Town Lunenburg World Heritage Site, Nova Scotia

Lunenburg (designated 1995), NS (Appalachia-Acadia), built on an MIS 2 drumlin, is the best surviving example of a planned British colonial settlement in North America. Established in 1753, it has retained its original layout and overall appearance, based on a rectangular grid pattern drawn up in the home country. The town's identity has been retained by preserving the wooden architecture of the houses, some of which date from the nineteenth century.

25.2.3.7 Rideau Canal World Heritage Site, Ontario/Québec

The Rideau Canal (designated 2007; see Chap. 21) traverses eastern Ontario (Great Lakes–St. Lawrence Lowlands and forested high-relief Shield). The oldest continuously operated canal system in North America, the Rideau Canal system is a monumental early 19th century construction covering 202 km of the Rideau and Cataraqui rivers from Ottawa south to Kingston on Lake Ontario. It was built primarily for strategic military purposes at a time when Great Britain and the United States vied for control of the region. The site played a crucial contributory role in allowing British forces to defend Canada against the USA, leading to the development of two distinct political and cultural entities in the north of the American continent, which can be seen as a significant stage in human history. The Canal cuts through Ordovician limestone along the Cataraqui, Grenville gneiss and granite of the Frontenac Axis in the center, and continues through Ordovican limestone, MIS 2 till, and Champlain Sea clay to the Ottawa River.

25.3 Co-managed Parks and Geoparks

Aldo Leopold's lifelong land ethics journey led from a stewardship of resources management approach to stewardship with an ecological conscience, summarized in his posthumous classic "Sand County Almanac" (Leopold 1949). He searched for ways to preserve the integrity, stability and beauty of the biotic community; that search implied the need to impose certain constraints on how people might act towards the land if they wished to preserve its integrity. In the Anthropocene epoch uncritical acceptance of neoliberal assumptions about the need for continuous and rapid economic growth threatens the survival not only of the biotic community but geodiversity and all living species, including *Homo sapiens*. Two quite different developments can be seen as hopeful signs. They concern the emergence of protected areas managed and co-managed by Indigenous Peoples, and parks that are primarily intended to educate the public.

25.3.1 Participation by Indigenous Peoples

The Indigenous Peoples and Community Conserved Territories and Areas (ICCA) consortium is an international group operating under the aegis of the International Union for the Conservation of Nature (IUCN). They promote indigenous people's conservation areas globally (Borrini-Feyerabend et al. 2004). Such protected areas are a response to the perceived failure of National parks to resist the pressures towards unrestricted commercialization. Vital to the maintenance of Indigenous culture are values such as fishing and hunting, the serenity of First Nations' lands, spiritual practices, and other traditional land uses. In eastern Canada, the most recent parks established in Nunavik, Labrador, and Nunavut are Akami-Uapishk-Kakkasuak-Mealy Mountains (Reserve) NL (2015); Qausuittuq (Bathurst Island), Nunavut (2015). Parc national Kuururjuaq and parc national Ulittaniujalik are Nunavik parks developed and managed by Inuit, in conjunction with the Government of Québec. Most of the park's staff come from nearby Kangiqsualujjuaq and speak three languages.

25.3.1.1 Parc National Kuururjuaq

Parc National Kuururjuaq is a 4,460 km^2 provincial park in northeast Québec, created in 2009 and co-managed by the Kativik Regional Government (Fig. 25.7). It was the second parc national to be established in Nunavik. It stretches all the way from Ungava Bay to Mont D'Iberville, which sits astride the Labrador/Québec border. Mont D'Iberville, at an altitude of 1,646 m, is the highest mountain in eastern continental Canada, dominating over the entire Torngat Mountains. The 166 km Korok River runs through the parc national to Ungava Bay. Kuururjuaq Parc National contains a variety of different ecosystems and comprises three natural regions: the Torngat Mountain Foothills, the George River Plateau and the Ungava Coast. Like other Nunavik parks, Parc National Kuururjuaq is developed and co-managed by the Inuit. The flora and fauna within the park are diverse, owing to its wide variety of habitats. In the Korok River valley, there is also a relatively diverse variety of vegetation including the *Picea mariana* and *larix.*

Fig. 25.7 Parc National Kuururjuaq extends from Ungava Bay to the crest of Mount Caubvick (Mont D'Iberville in Québec)—the highest peak in eastern Canada, rising to 1,646 m asl. **a** and **b** The Korok River, whose waters originate in the Torngat Mountains flows 166 km towards Ungava Bay through a wide glaciated valley. *Photographs* D. Kalant/Wikimedia Commons, CC BY-SA 4.0

25.3.1.2 Parc National Ulittaniujalik

Parc National Ulittaniujalik is a 5,293 km^2 park, created in 2016 in northeast Québec (Fig. 25.8). It was established in partnership with the Inuit communities of Kangiqsualujjuaq and Kuujjuaq, in addition to the Naskapi community of Kawawachikamach. It contains the Pic Pyramide (Pyramid Peak) at an altitude of 457 m asl. The landscapes vary greatly. In the George River valley, there are forests of *picea* and *larix,* and even *betula* and *Populus balsamifera*, which are usually present at lower latitudes. The park protects the

Fig. 25.8 Parc National Ulittaniujalik is a 5,293 km^2 park in northeastern Québec, created in 2016 in partnership with the Inuit and Naskapi Indigenous people. View of the George River Plateau toward Pyramid Peak (457 m asl). *Photograph* Government of Québec; mffp.gouv.qc.ca/CC BY 4.0

George River caribou herd, whose population has dramatically decreased over recent decades. This second largest park in Québec, Ulittaniujalik is the fourth parc national in Nunavik. It protects much of the natural region of the plateau of the George River. After the creation of parcs nationaux Pingualuit (2004), Kuururjuaq (2009) and Tursujuq (2013), Parc National Ulittaniujalik is the latest focus of attention in the region. This protected territory has been shaped by the gradual disappearance of an ancient glacial lake that has formed glacilacustrine terraces. It is also due to these terraces that we owe the name of the park, designated by the Inuit and Naskapis as "the place where there are shorelines". Like the other parks in Nunavik, its almost unaltered natural environment makes it a rare destination in the modern world. But still, the uniqueness of this territory lies in the fact that it is still used by the inhabitants of the region as a site for the practice of traditional activities. Administered by Nunavik Parks, Ulittaniujalik park is the result of collaboration between members of local communities, scientists, officials from northern villages and land corporations involved in Nunavik, the Kativik Regional Government, Makivik Corporation, and the Ministry of Forests, Wildlife and Parks. The Amerindian and Inuit predecessors shaped the identity and history of northern Québec. In Nunavik, it is the Inuit who are most present. However, in the region of Pic Pyramide, there are also Amerindian occupations dating back more than 3,000 years by the Cree, Innu and Naskapi.

As noted above, on July 1, 2018, Pimachiowin Aki was the first Canadian UNESCO World Heritage Site to reflect this greater emphasis on First Nations' participation in developing and co-managing protected areas.

25.3.2 Geoparks: The Role of Education

It can be argued that it was always an important part of the mandate of national parks to educate the public. But concerns about commercialism and pressure from tourism in the major national parks have prompted the establishment of a network of UNESCO Global Geoparks. These Geoparks have an explicit educational function (McKeever and Zouros 2005). Fundamental characteristics are: a focus on geoscience heritage; sustainable management to provide for the social and economic needs of the local populations, protecting the landscape and conserving cultural identity; promotion of geotourism; and membership of a Global Geopark Network for global education. There are two Geoparks in eastern Canada: Percé Rock in Québec and Stonehammer in New Brunswick (Fig. 25.9).

25.3.2.1 Stonehammer Geopark, New Brunswick

Stonehammer Geopark is located on the New Brunswick side of the Bay of Fundy, centered around Saint John and stretching from St. Martins to Lepreau Falls Provincial Park. The location has historical significance as it was the home of the first designated Provincial Geologist in the British Empire in 1838 and the site of the "Steinhammer Club" which was formed in 1857 to explore the geology of the region.

The landscape of the park is the product of the collision of continents, the closing and opening of oceans, volcanoes, earthquakes, ice ages and climate change. The rocks provide evidence of the evolution of life, from the first discovery of Precambrian stromatolite fossils, to the "Cambrian

Fig. 25.9 Stonehammer Geopark lies along the rugged Bay of Fundy on Canada's southeast coast. Centered on Canada's oldest incorporated city, Saint John, New Brunswick, it is the first North American member of the UNESCO Global Geoparks Network. The Reversing Falls are a series of rapids on the Saint John River located in Saint John, New Brunswick, where the river runs through a narrow gorge before emptying into the Bay of Fundy. The semidiurnal tides of the bay force the flow of water to reverse against the prevailing current when the tide is high. *Photograph* J. Binczarowski/Wikimedia Commons, Public Domain

Explosion" of life, to the evolution of vertebrates and the emergence of life on land. Sixty significant geological and fossil locales provide evidence of events from late Precambrian time to the most recent Ice Age (MIS 2).

Stonehammer Geopark covers 2500 km^2 and extends across urban, suburban and rural communities with 150,000 people, with Saint John at its center. The stated goal is "to educate students, community and visitors about the Earth and to provide understanding of energy and mineral resources, changing climates, flood hazards and water shortages". To demonstrate sensitivity to the cultural values of the Mi'kmaq Indigenous people, sites identified as sacred are excluded from public access.

25.3.2.2 Percé Rock Geopark, Québec

Eastern Canada's second UNESCO Global Geopark, is located on the eastern tip of Gaspé Peninsula in Québec. Gaspé, derived from the Mi'kmaq word for "the end", represents the easternmost extension of the Appalachians of Québec. It is a territory particularly rich in fossiliferous rocks and landforms and landscapes. Rocks record 170 million years of the regional geological history within a very small area (40 km^2). These rocks, unlike many of their counterparts elsewhere in the Appalachians, are less metamorphosed, which makes it possible to recognize the textures and structures formed during their deposition. Percé has also witnessed the birth of geology in Canada, since it was this region where Sir William Logan, the first director of the Geological Survey of Canada, started the geological mapping of Canada. The peninsula divides the St. Lawrence estuary to the north from the Baie-de-Chaleurs to the south. Percé Rock itself is a sea arch developed in Devonian limestone. Ongoing erosion has reduced the number of arches from the three reported by Jacques Cartier in 1534 to the single arch and offshore stack existing today. The second arch collapsed on 17 June 1845. At the present rate of erosion, the remaining arch is anticipated to last ~15,000 years, although relative sea level is rising. Ile Bonaventure is composed of Mississippian calcareous conglomerate.

25.4 National Parks in Eastern Canada

National Parks are protected areas under the National Parks Act (2000), owned by the Government of Canada and administered for the benefit, education and enjoyment of the people of Canada and its future generations. The goal of the national parks system is to set aside lands representing the country's 39 natural regions described in the National Parks System Plan, primarily to protect the ecological integrity of the land, and secondarily to allow the public to explore, learn about and enjoy Canada's natural space. In eastern Canada there were as of 2018 some 27 federal National Parks and Park Reserves (Fig. 25.10; Tables 25.2 and 25.3).

Many of the geomorphological landscape regions of eastern Canada do contain protected areas of various kinds. However, it is also noticeable that regions 2a, 2b, 3b, 5a, 5b, 5c, 10, 12b, 13b and 20 (refer to Fig. 1.10 for locations) still have no protected areas, mostly in the Arctic and Subarctic, in spite of the recent increased attention that arctic Canada has received.

The process of establishing a national park in Canada involves initial designation of the terrain as a Reserve, subsequently followed by official proclamation in print by the Government of Canada ("gazetteing"). Significant time may elapse between initial reserve designation and formal gazetteing: Gros Morne was established as a Reserve in 1973, and as a UNESCO World Heritage site in 1987, but was gazetted as an official National Park only in 2005.

In Québec, the designation "*Parc National*" refers to parks established and administered by the provincial Government of Québec, not by the federal Government of Canada. They are not subject to the federal National Parks Act. Miguasha and Percé (Gaspé), Saguenay Fjord (see Chap. 10; Fig. 25.10), and Kuururjuaq, Pingualuit crater (see Chap. 1), and Ulittaniujalik (Fig. 25.8) in Nunavik are examples of "parcs nationaux".

25.4.1 Early Initiatives: Pre-second World War

Prior to the Second World War, there were only five national parks established in eastern Canada: Thousand Islands (1904), Point Pélée (1918), and Georgian Bay Islands (1929) in Ontario; Cape Breton Highlands (1936) in Nova Scotia, and Prince Edward Island (1937).

25.4.1.1 Thousand Islands National Park, Ontario

Thousand Islands National Park in Ontario was the first to be established (1904), some 19 years after the first National Park in Canada, Banff, in Alberta. The park, formerly known as the St. Lawrence Islands National Park, is located along the Thousand Islands Parkway in the Thousand Islands region of the St. Lawrence River. The 6 km^2 park consists of 21 islands, many smaller islets, and 3 km^2 on the mainland. The Thousand Islands are developed in the Frontenac Axis, a narrow belt of Canadian Shield granites and gneisses that extends from the main area of the Grenville Province southeast across the St. Lawrence River to connect with the Adirondack Hills of New York. This Frontenac Axis resembles the remainder of the Grenville Province and

Fig. 25.10 Distribution of National Parks in eastern Canada. National Park boundaries; UNEP-WCMC (June, 2019), available at: www.protectedplanet.net

Laurentian Highlands (forested High-Relief Shield), marked by rolling to rugged topography, isolated hills and depressions, and steep cliffs. Relief is approximately 150 m. Although the terrain is not particularly rugged, it stands in sharp contrast to the lowlands of the St. Lawrence valley to the east and the Great Lakes Lowland to the west. The northeastern edge of the Thousand Islands marks the southwestern marine limit of the Champlain Sea.

25.4.1.2 Point Pélée National Park, Ontario

Point Pélée is the southernmost National Park in Canada, and at 16 km^2 is one of the smallest (Fig. 25.11).

Table 25.2 National parks and national park reserves in eastern Canada

National park	Province/territory	Year established
Thousand Islands	Ontario	1904
Point Pélée	Ontario	1918
Georgian Bay Islands	Ontario	1929
Cape Breton Highlands	Nova Scotia	1936
Prince Edward Island	Prince Edward Island	1937
Fundy	New Brunswick	1948
Terra Nova	Newfoundland and Labrador	1957
Kejimkujik	Nova Scotia	1967
Kouchibouguac	New Brunswick	1969
La Mauricie	Québec	1970
Forillon	Québec	1970
Pukaskwa	Ontario	1971
Gros Morne	Newfoundland and Labrador	1973
Mingan Archipelago Reserve	Québec	1984
Bruce Peninsula	Ontario	1987
Quttinirpaaq	Nunavut	1988
Wapusk	Manitoba	1996
Auyuittuq	Nunavut	2001
Sirmilik	Nunavut	2001
Ukkusiksalik (Kivalliq)	Nunavut	2003
Torngat Mountains (Labrador)	Newfoundland and Labrador	2005
Sable Island Reserve	Nova Scotia	2013
Qausuittuq (Bathurst Island)	Nunavut	2015
Rouge Park	Ontario	2015
Akami-Uaposhk^{u}-KakKasuak-Mealy Mountains Reserve (Labrador)	Newfoundland and Labrador	2015

Established in 1918, primarily as a migratory bird sanctuary, the park attracts 240,000–250,000 paying visitors annually. The small size of the park and the number of visitors places extreme strains upon the area, necessitating restrictions on visitor access. In addition to its importance as the major Canadian staging point on the Eastern North American Flyway, supporting more than 120 regularly visiting bird species, and more than 300 species including casual visitors, Pélée and vicinity also represent the sole Canadian locality for many temperate species of the Carolinian forest. Middle Island, located south of Point Pélée in Lake Erie, is the southernmost point of Canada at 41°41′ N, lying south of the northern boundary of California.

Point Pélée is a low-lying sandspit which extends southerly at almost 90° to the shoreline of Lake Erie. Point Pélée and vicinity were glaciated by the Lake Erie lobe of the Laurentide ice sheet complex during the last glaciation (MIS 2). Point Pélée was initially deglaciated c. 16.5 ka, but a subsequent re-advance buried the area to the east and built the Pélée Moraine system, approximately 16 ka. The ice finally left the Pélée moraine approximately 15.5 ka, allowing the spit to begin to develop. Today, the sandspit extends 15 km into Lake Erie and is up to 70 m thick. Seventy percent of the park consists of marsh, dominated by cattails and ponds, and Carolinian forested areas cover about 21%. This site was named "Pointe-Pelée" (meaning "bald point") by French speaking explorers: the eastern side had no trees, possible as a result of ongoing human usage. The Caldwell First Nation Chippewa people lived on Point Pélée for at least 6000 years before European colonization. In 1790, a treaty with Aboriginal communities "officially" ceded a large tract of land, which included Point Pélée, to the

Table 25.3 Parks in geomorphological landscape regions as defined in Fig. 1.10

Geomorphological landscape region	Geological province	Parks and significant geomorphic areas
Canadian Shield		
1a. Permafrost- dominated Shield (Kivalliq))	Churchill, Bear, Slave	Ukkusiksalik NP
1b. Permafrost- dominated Shield (Nunavik)	Churchill	Pingualuit astrobleme PN
2a. Low to moderate relief Shield (George Plateau)	Churchill	
2b. Low to moderate relief Shield (Larch and Lake Plateaus)	Superior	
2c. Low to moderate relief Shield (Abitibi and Severn Uplands)	Superior, Southern	Pukaskwa NP
3a. Moderate to high-relief Shield (Laurentian Highlands)	Grenville	Thousand Islands NP, La Mauricie NP, Mealy Mts. NP, Algonquin PP
3b. Moderate to high-relief Shield (Hamilton Plateaus and Uplands)	Grenville	
3c. Moderate to high to relief Shield (Mealy Mountains)	Grenville	Mealy Mts. NP
4a. High-relief Shield (Ellesmere and Devon islands)	Churchill	
4b. High-relief Shield (Baffin Island Mountains)	Churchill	Auyuittuq NP, Sirmilik NP
4c. High-relief Shield (Uplands of Baffin Island and Nunavik)	Churchill	
4d. High-relief Shield (Torngat Mountains)	Nain	Torngat Mts. NP
Arctic Lowlands		
5a. Arctic Lowlands (Foxe Plain)	Arctic Platform	The Great Plain of Koukdjuak thermokarst lakes
5b. Arctic Lowlands (Lancaster Plateau and Boothia Plain)	Arctic Platform	
5c. Arctic Lowlands (Victoria Lowland)	Arctic Platform	
Hudson Bay Lowland		
9. Permafrost- dominated Hudson Bay Lowland	Hudson Platform	Wapusk NP, Polar Bear PP
10. Forested Hudson Bay Lowland	Hudson Platform	Attawapiskat River
Great Lakes- St. Lawrence Lowlands		
11a. Gulf of St. Lawrence Lowland	St. Lawrence Platform	Mingan NP
11b. St. Lawrence—Ottawa Valley	St. Lawrence Platform	Monteregian Hills
11c. Great Lakes Lowland	St. Lawrence Platform	Niagara Falls, Point Pélée NP, Bruce Peninsula NP
Innuitia		
12a. Innuitian Mountains	Innuitian Orogen	Quttinirpaaq NP
12b. Eureka Uplands	Innuitian Orogen	
13a. Parry Plateau	Innuitian Orogen	Qausuittuq (Bathurst NP)
13b. Sverdrup Lowland	Innuitian Orogen	

(continued)

Table 25.3 (continued)

Geomorphological landscape region	Geological province	Parks and significant geomorphic areas
Appalachia- Acadia		
19a. Moderate relief Appalachia (Newfoundland)	Appalachian Orogen	Terra Nova NP, Gros Morne NP
19b. Moderate relief Appalachia (Québec and New Brunswick)	Appalachian Orogen	Forillon NP
19c. Low to moderate relief Acadia (New Brunswick and Nova Scotia)	Appalachian Orogen	Fundy NP, Kejimkujik NP, Cape Breton Highlands NP
Maritime Plain		
22. Low relief Maritime Plain (PEI)	Appalachian Orogen	Kouchibouguac NP, Prince Edward Island NP

Fig. 25.11 Point Pélée National Park, Ontario. 70% of the park consists of marsh, dominated by cattails and ponds. *Photograph* Bichengsha/Flickr, CC BY-SA 2.0

Crown. Unfortunately, the Caldwell First Nation was not a signatory of that treaty. Point Pélée therefore remains technically unceded land (Fig. 25.11).

25.4.1.3 Georgian Bay Islands National Park, Ontario

Georgian Bay Islands National Park is among Canada's oldest (1929), smallest, and because it is accessible only by boat, among the least visited (approximately 80,000 visitors each year). However, its small size (total of 25 km^2 scattered over 63 islands and rocks), and its relatively large service infrastructure (particularly docking facilities for pleasure craft) have placed the park under pressure. The park is included in the Georgian Bay Littoral Biosphere Reserve and is a part of the Lake Huron basin.

Most of the park lies included within the eroded terrain of the high-relief Laurentian Highlands, in the southwestern Canadian Shield. The islands are too small to contain lakes or surface streams. Relief on the islands is muted, as the features eroded into the bedrock are generally small in size, and locally they are blanketed with lacustrine clay deposited during Glacial Lake Algonquin. Inland, relief approximates 150 m. North of the park, ragged lake shorelines and isolated, locally rolling erosional hills characterize the landscapes of the tourist and cottage regions of Muskoka and Parry Sound. The southernmost islands in the park lie along the northern fringes of the Great Lakes Lowland. Here, gently rolling plains sloping towards the northeast mark the terrain. Beausoleil Island is a traditional home of the Chippewa people.

25.4.1.4 Cape Breton Highlands National Park, Nova Scotia

Cape Breton Highlands National Park (951 km^2) was established in 1936, partially to preserve the highland environment of northernmost Cape Breton Island and partially to bring an economic boost to the region. Cape Breton lies within the Nova Scotia Uplands physiographic region of

Fig. 25.12 Cape Breton Highlands National Park, Nova Scotia. a In the foreground, the Freshwater Lake. Behind it on the left, the South Bay Ingonish opens up to the Atlantic Ocean beyond the visible Cape Smokey. On the right is partially visible the town of Ingonish Beach. b Cabot Trail. **c** Mackenzie River Valley from the Cabot Trail, Cape Breton. **d** Broad Cove. *Photographs* **a**—Tango7174/Flickr, CC BY-SA 4.0. **b**—and **c**—A. Schaffer/Flickr, CC BY 2.0. **d**—D. Javis/Flickr, CC BY 2.0

Appalachia-Acadia, an area marked by irregular hills, isolated flat-topped to rolling plateaux, and steep escarpments (Fig. 25.12). The core of the park represents one of these plateaux, with a maximum elevation of 532 m above sea level. The plateau is extensively dissected along its margins, and short, steep streams such as the Aspy, Mackenzie, and Cheticamp Rivers plunge abruptly from the plateau's northern, eastern, and western margins to sea level along cataracts. Waterfalls, notably the 40 m high Ben Beulach, are common along these streams. The physiography of the region has contributed to the separate cultural development of the eastern Scottish (Gaelic) and the western Acadian (French) shores of the peninsula.

In contrast to other upland areas of Atlantic Canada, glaciation had very little effect on the northern Cape Breton landscape. Striations and glacially-transported erratics indicate that the Highlands were affected by several glacial events involving ice originating from the west and southwest early in the Quaternary, and all areas were covered by ice at some time. During the MIS 5 interglacial, 125–115 ka, relatively warm conditions prevailed throughout Cape Breton. Fossil plant remains are preserved at Mabou Harbour and Bay St. Lawrence. During MIS 2, ice from the Bras d'Or Lakes area was confined to the region south of Mabou Harbour and Wreck Cove. There are no glacial landforms evident on the surface of the plateau, indicating coverage by cold-based ice.

Following ablation of the glaciers, sea level rose around the Cape Breton coastline, but the extent of rise was minimal (generally less than 5 m). Deposits of marine silt and clay are not present. Throughout the early and mid-Holocene, sea levels remained below present. When the fortress at Louisbourg, south of Sydney (Cape Breton Regional Municipality), was constructed by France beginning in 1721, sea level was approximately 2.5 m below present. Mooring rings used by French vessels now are exposed only at low tide. The rising sea level, currently rising at approximately 3.5 mm/y, has resulted in substantial erosion of the historic remains at Louisbourg. Along the coastline of Cape Breton Highlands, steep slopes have resulted in little change in coastal configuration as sea level has risen, although locally sedimentary bluffs are being undercut by wave action. With its relatively steep topography, Cape Breton Highlands is susceptible to various types of slope failures.

25.4.1.5 Prince Edward Island National Park, PEI

Prince Edward Island National Park (Maritime Plain) was established in 1937, and has subsequently become one of Canada's major tourist attractions, both for its natural attributes and because of its association with "Anne of Green Gables". Visitation is estimated at approximately 500,000 paying visitors annually. The national park encompasses two areas. The older part includes 32 km^2 of beaches and adjacent shoreline areas in the vicinities of Cavendish, Rustico, and Stanhope. With the growth of tourism, additional areas have undergone tourist development, including Cabot Beach Provincial Park, on the northeastern margin of Malpeque Bay. The newer part includes 18 km^2 of terrain to the east on the Greenwich Peninsula, north of St. Peters.

Beaches along the north coast of PEI are among the province's major tourist attractions. Decreased snowfall and limited ice cover over the southern Gulf of St. Lawrence, in combination with sea level rise, has resulted in increased wave erosion and narrowing of beaches. Climate variations also result in increased degradation of coastal dunes. Stronger winter winds sweeping across exposed dunes, some eroded due to human pressure during the previous summers, have resulted in enhanced erosion and lowering and flattening of dune profiles. Increased foot traffic and vehicle access to beaches and backing dunes have accelerated erosion of coastal tourist sites (see Chaps. 18 and 21).

25.4.2 Post Second World War, 1948–1973

Another 8 national parks were designated between 1948 and 1973. Two were created between 1945 and 1968, followed by intense activity during the six year period from 1968–1973. All were in the Atlantic provinces, Ontario and Québec.

25.4.2.1 Fundy National Park, New Brunswick

Fundy National Park (206 km^2) was established in 1948, to preserve part of the Appalachia-Acadia shoreline of Chignecto Bay, to permit the re-establishment of the inland forests which had been removed by logging, and to provide an economic boost to this region of southeast New Brunswick. The limited area of Fundy National Park, combined with the extensive logging activity which continues in the surrounding region and the substitution of spruce monoculture forests for the original mixed vegetation assemblage, has led to concern for the future of the park. Fundy is in danger of having its ecological integrity compromised by developments beyond its borders. In an attempt to deal with this problem, and to understand the resource base and the management issues involved, Fundy National Park personnel initiated a series of consultations with surrounding interest groups, and also carried out a detailed management assessment study. The studies conducted at Fundy National Park are regarded as a model which other Canadian parks continue to emulate. The geomorphology of Fundy National Park and the adjacent Hopewell Rocks Provincial Park are discussed in Chap. 18.

25.4.2.2 Terra Nova National Park, Newfoundland

Terra Nova National Park (399 km^2) was established in 1957, under a federal government plan to establish one national park in each province. Prior to Confederation with Canada in 1949, Newfoundland had established a park in the Serpentine River area southwest of Corner Brook, but the area was allowed to revert to private ownership prior to 1949.

The choice of the Terra Nova site was based primarily on convenience of access from St. John's (no cross-island highway existed at that time), and on the availability of the land. Although Terra Nova National Park does not have any landform features which are unique to it, the region does represent the Appalachia-Acadia coastline of eastern Newfoundland (see Chap. 15 for discussion). The forested areas, particularly those west of the Trans-Canada Highway, have been significantly influenced by logging activity, and no longer represent a natural ecological assemblage. However, no area adjacent to the coastline of eastern Newfoundland has remained untouched by human activity. The small size of the park (399 km^2) means that land use practices in the surrounding area have a direct impact on the park's flora and fauna.

The Terra Nova region was ice-covered throughout the last glaciation (MIS 2). At its maximum extent 20 ka, glacial ice covered all of Bonavista Bay and extended more than

100 km offshore. Roche moutonnées and striations, and deposits of glacial till containing granite boulders from Devonian outcrops west of the park, indicate that the ice flowed northeastward and eastward towards the coast. Terra Nova may be included in the recently proposed Bonavista Peninsula geopark.

25.4.2.3 Kejimkujik National Park, Nova Scotia

Kejimkujik National Park was established in 1974. Located in the center of the Devonian South Mountain granite complex, 48 km southeast of Annapolis Royal, it encompasses 382 km^2. Its establishment was intended to preserve a sample of the Atlantic Coastal Uplands ecological region, much of which has been logged in the past 180 years. A small "adjunct" 25 km west of Liverpool, 2 km^2 in area, is intended to preserve a fragment of the open Atlantic coastline. The word "Kejimkujik" is derived from a Mi'kmaq phrase which can be translated as "tired muscles from paddling to escape enemies".

The park terrain is typical of the Appalachian Uplands, marked by rolling hills developed on granite and metamorphosed sedimentary rocks. All parts of the park are less than 200 m above sea level, and relief throughout is generally less than 30 m. MIS 2 glacial activity centerd over South Mountain has resulted in the deposition of tills and glacifluvial deposits, locally impounding lakes. Numerous drumlins, to 50 m in length and 15 m in height, are scattered across the landscape. The long axes of these features are aligned northwest-southeast. Southeastward glacial movement is responsible for the drumlins exposed at Lunenburg and elsewhere along the Atlantic coast southwest of Halifax. After the glaciers retreated southward from the Annapolis Valley, and northward from Lunenburg and Liverpool, approximately 13.5 ka, ice persisted for approximately 1,000 more years in Kejimkujik, producing an irregular terrain of hummocky moraine. Remnant glaciers may have persisted through the Younger Dryas cold event, but did not readvance, and final ablation of the glaciers occurred prior to 10.5 ka.

The seaside adjunct along Nova Scotia's south shore is marked by a bedrock-dominated coastline, with headlands separating narrow coves. Mixed pebble-cobble-sand beaches are marked by swash-backwash-dominated conditions.

25.4.2.4 Kouchibouguac National Park, New Brunswick

Kouchibouguac National Park (244 km^2) is located along the Gulf of St. Lawrence coast (Maritime Plain) of eastern New Brunswick, north of Moncton. Establishment of the park in 1969 was attended by sharp controversy, as local residents strongly objected to the removal of settlements and the closure of the region to fishing and logging. The issues arising from these problems led Parks Canada to change its style from imposition to community discussion and subsequently co-management, as seen in all subsequently established national parks.

Kouchibouguac National Park was glaciated by eastward-moving ice from central New Brunswick during MIS 2. The park remained ice-covered until approximately 14 ka, but very few glacial landforms were produced. As the ice began to retreat, the area was flooded by the Goldthwait Sea, a predecessor of the Gulf of St. Lawrence. Marine waters reached elevations up to 45 m above modern sea level, and thus the area was completely submerged. Following the withdrawal of the ice, the land began to rebound, and the marine waters started to recede. Sea levels fell to approximately 10–15 m below the present coastline about 6.5 ka. Since then, the marine waters have slowly risen once again to reach their present position. Currently, the sea is rising at $\sim$2.5 mm each year.

As the sea regressed from the land, vegetation began to recolonize the newly exposed terrain. A *juniperus* scrub assemblage had developed by 12 ka. *Picea* and *abies* colonized the region 10.5 ka and 9 ka respectively. Continued climatic warming allowed peatlands to develop, and the Acadian forest began to form as deciduous trees arrived in the area, c. 7 ka. With the arrival of *fagus* approximately 2,900 years ago, the forest had attained the composition which remained intact until European settlement.

25.4.2.5 La Mauricie National Park, Québec

Established in 1970, La Mauricie is located north of Trois-Rivières and Shawinigan, Québec, in the highlands to the west of the St-Maurice River, and has an area of 544 km^2. La Mauricie lies in the Laurentian Highlands of the forested high-relief Canadian Shield. The southern margin of the Canadian Shield is marked by an escarpment located 10–20 km southeast of the park. The park is dominated by a system of southeast–northwest trending valleys developed along faults that are occupied by the major streams and elongated lakes. Relief in the area is approximately 300 m. The terrain between the fault valleys generally consists of rolling to irregularly rugged isolated hills, with interspersed gently rolling plateaux. This physiography is typical of the Laurentian region of the Canadian Shield.

La Mauricie was glaciated by the Laurentide Ice Sheet Complex, which advanced from western Labrador to the northeast (see Chap. 2). Unlike the areas south of the St. Lawrence River, the park remained ice-covered from approximately 100 ka (MIS 5.4) to 11.3 ka (conclusion of MIS 2). During the final phases of this glacial retreat, deglaciation of the outlet of the St. Lawrence River at Québec City allowed marine waters to flood the St. Lawrence Valley. The resulting Champlain Sea embayment of the

Atlantic Ocean reached elevation of 200 m above modern sea level in the Mauricie region, and blanketed all the terrain below this elevation with marine silt and clay.

25.4.2.6 Forillon National Park, Québec

Forillon (245 km^2) was established in 1970, north of Gaspé town in Québec. As in other areas of Appalachian Canada, the physiography and configuration of the coastline is controlled in large measure by bedrock lithology and structure. The rocks in Gaspé are limestones, shales, and sandstones of Ordovician, Silurian, and Devonian age. Subsequently, the rocks were subjected to thrusting from the south-southeast during the Late Palaeozoic (Acadian Orogeny). The deformation characteristic of the strata of Gaspé thus resembles that of the Gros Morne area of Newfoundland, and the Estrie area of Québec to the southwest. The differential resistance of the limestones and shales is evident along the coastline, where coves represent areas of shale outcrops and high cliffs mark limestone exposures. The picturesque sites of Percé Rock, Ile Bonaventure, and the cliffs of Forillon are all associated with limestone exposures.

Following MIS 2 deglaciation, the coastal regions were submerged as a result of isostatic depression. The Goldthwait Sea deposited silt and clay along the south coast to elevations of approximately 25 m asl at Forillon. Marine sediment in these regions is subject to periodic slope failure. Marine regression occurred rapidly, and by 10 ka the coastline essentially had taken on its present configuration.

The balance between providing visitor services and preservation of the natural environment is precarious. This problem is exemplified at the coastal dunes of the Penouille Peninsula, where Parks Canada's desire to preserve the dune environment and the adjacent salt marsh conflicts with the tourist desire to utilize the coast for swimming and sunbathing.

25.4.2.7 Pukaskwa National Park, Ontario

Pukaskwa National Park, established in 1978, encompasses 1,878 km^2 on the shore of Lake Superior, south of Marathon and west of White River. The name comes from the Ojibwe word "pukasu" meaning "cooking the marrow in the bones of animals".

Pukaskwa is situated in the Abitibi Upland, within the James Region of the Canadian Shield. Relief in the area is 457 m, measured to the surface of Lake Superior at 183 m asl. The area is marked by northeast-southwest oriented ridges and valleys, giving a corrugated appearance to the landscape.

The bedrock geology of the Pukaskwa region represents the major influence on its topography. The majority of the park lies within the Superior Geological Province, the oldest area of the Canadian Shield. These rocks consist of a series of belts of alternating resistant acidic gneiss and granite, and weaker volcanic rocks and sedimentary rocks derived from the debris produced by explosive volcanic eruptions The weaker volcanic rocks erode to produce valleys, and thus the characteristic corrugated topography marked by southwest-northeast trending ridges and valleys has developed. Faults have accentuated the pre-existing ridge-and-valley topography.

The southern coastal area of Pukaskwa lies in the younger geological Southern Province, consisting of sedimentary and metamorphosed sedimentary rocks of Palaeohelikian age, which were welded to the Superior Province and deformed during the Elsonian orogeny, 1370 Ma. These rocks are relatively weak gneisses, schists, slates, argillites, and phyllites, and thus do not form the ridge-and-valley topography characteristic of the Superior Province. Consequently, they have been eroded evenly along the coast, which has resulted in the gently curving Superior shoreline, devoid of any harbors. These rocks underlie much of the eastern half of Lake Superior, and their erodability accounts for the location of the lake and its great depth.

25.4.3 Slow Growth, 1984–1996

Four federal National Parks were designated during this time in Québec, Ontario, Nunavut and Manitoba. Bruce Peninsula, Ontario is discussed in Chap. 12 and Wapusk, Manitoba is discussed in Chap. 8 of this volume. Prime Minister Jean Chretien (in office 1993–2003) took a proactive role in supporting growth of national parks in the Arctic.

25.4.3.1 Mingan Archipelago National Park Reserve, Québec

Mingan Archipelago National Park Reserve was established in 1984, and is one of several Canadian parks accessible only by water and air. Located along the Québec North Shore of the Gulf of St. Lawrence offshore of Havre St-Pierre, north of Anticosti Island, the 151 km^2 park represents one of the northernmost outposts of the St. Lawrence Lowlands. The park consists of approximately 30 islands and islets, and numerous skerries and isolated rocks, extending for 175 km along the coast. The region has long been of interest to visitors, due to its unusual coastal "flowerpots" and "pillars" eroded into Ordovician limestone, its unique flora, and the opportunity to view beluga, blue, humpback, and fin whales, dolphins, and other marine organisms.

Laurentide glacial ice advanced from western Labrador and adjacent Québec to cover the Mingan Archipelago and

Anticosti Island during MIS 2, but left few features on the landscape. Deglaciation occurred c. 15 ka. Sea level during the glacimarine Goldthwait Sea phase rose to ±135 m asl, depositing marine clays inland along the embayments and river valleys of the mainland shore. Although all parts of the archipelago were completely inundated, Mingan did not receive substantial quantities of marine sediment. As isostatic rebound progressed, the sea gradually receded from the land, and the archipelago began to develop. The erosional features evident in the limestone cliffs were developed under conditions of modern sea levels, and relict features reflecting former marine shorelines are present at higher elevations.

25.4.3.2 Quttinirpaaq National Park, Nunavut

Quttinirpaaq (Hazen Lake, Ellesmere Island) was established in 1988 (Fig. 25.13). This area of high-relief Innuitia was first settled c. 4500 years ago by palaeo-Inuit of the Independence 1 culture, living in the southern part of the island. These inhabitants remained for c. 400 years, hunting musk oxen, until the supply of game dwindled. Subsequent human occupations occurred c. 3000–2500 years ago (Independence II), 800 AD–1000 AD (Dorset), and 1000 AD–1500 AD (Thule). The Thule people abandoned Northern Ellesmere as the "Little Ice Age" made marine mammal hunting impractical, approximately 500 years ago, and moved south. Thule culture gradually evolved into the Inuit culture of today.

The uniqueness of the landscape suggested preservation was warranted. Establishment was not unanimously accepted, however, as some scientists familiar with the area argued that designation of a National Park would attract more visitors, defeating the purpose of preservation. Eventually, the advocates of designation prevailed, and Quttinirpaaq National Park was established in 1988. Visitation is tightly controlled by Parks Canada, and is limited to approximately 100 visitors/year. The park encompasses 37,775 km^2 of terrain, including Hazen Lake (267 m maximum depth), the largest lake (542 km^2) in the Canadian Arctic.

During the Quaternary, glaciers advanced from the Barbeau and Agassiz Ice Caps, and also from Greenland. The cold, arid conditions limited glacial expansion and nourishment from precipitation, and precluded development of thick successions of glacial sediments. Glacial advances in Ellesmere Island correspond with glacial retreats in some more southerly areas, because "warmer" conditions are necessary to generate enough precipitation to feed the glaciers with snow necessary to sustain their advance. The last glacial ice limit, across the western Hazen Plateau and at the heads of Archer Fjord and Clements Markham Inlet, occurred between 8 ka and 11 ka. At this time, elevated sea levels were developed around Ellesmere's shores, as the island was depressed by the weight of the adjacent Greenland glacier.

Fig. 25.13 Quttinirpaaq National Park ('top of the world'), Ellesmere Island, Nunavut. **a** Hiking alongside the Air Force glacier. **b** *Chamerion latifolium* in foreground, with meltwater stream and Ekblaw Lake in background. **c** Scylla and Charybdis glaciers. *Photographs* P. Gierszewski/Wikimedia Commons, CC BY-SA 3.0

Deglaciation is still ongoing in the area. Lake Hazen's north shore was deglaciated approximately 2,000 years ago, and ongoing climatic changes continue to affect the glaciers of the area. The current landscape of Quttinirpaaq thus resembles that which existed shortly following deglaciation in many areas of Canada.

25.4.3.3 Wapusk National Park, Manitoba

Wapusk National Park is located in the permafrost-dominated Hudson Bay Lowlands, among the least known areas of Canada. The low relief of the terrain, the lack of economic impetus for development (save for the port facilities at Churchill), the harsh winter climate, and the vast squadrons of mosquitoes during the summer have all discouraged visitation and settlement. This region lacks the inspirational terrain of other northern taiga areas, such as the Torngat Mountains. Nonetheless, it contains features of interest to the dedicated professional geomorphologist as demonstrated in Chap. 8 of this volume.

The Hudson Bay Lowland slopes almost imperceptibly towards the bay, and is covered with wetlands (primarily fens) and sluggish meandering and wandering rivers. Total relief throughout the Hudson Bay Lowland approximates 200 m. Within Wapusk, relief is considerably less, and there are no elongate ridges or isolated hills to mar the gently northeastward slope of the plain. The most conspicuous topographic features are the numerous beach ridges, aligned parallel to the modern shore, indicating the successive withdrawal of the Tyrrell Sea marine waters following glaci-isostatic recovery of the land. Rebound continues today at a rate of approximately 3 mm/y, and thus the Hudson Bay Lowland is the fastest growing physiographic region of Canada (in terms of adding new area).

Wapusk encompasses 11,745 km^2, extending from the Nelson River mouth in the south (see Chap. 21) to Churchill in the north, located at the mouth of the Churchill River. The Nelson River serves as the outlet for Lake Winnipeg, and thus carries the entire drainage of the Red, Assiniboine, Saskatchewan, and Bow Rivers. The Broad, Owl, and other rivers drain eastward through Wapusk into Hudson Bay.

25.4.4 Renewal: (2001–2019)

Since 2001, eight more national parks and reserves have been established: four in Nunavut, two in Newfoundland and Labrador, one in Nova Scotia and one in Ontario. They include Sirmilik, Nunavut (see Chap. 2); Torngat Mountains, Newfoundland and Labrador (see Chap. 7); Akami-Uaposhk-Kakkasuak-Mealy Mountains Reserve, Newfoundland and Labrador (see Chap. 7); Qausuittuq (Bathurst Island), Nunavut low-moderate relief Innuitia; (see Chap. 14); and Rouge River, Ontario (see Chap. 21).

25.4.4.1 Auyuittuq National Park, Nunavut

Auyuittuq National Park ("the land that never melts") was established with an area of 21,471 km^2. The park encompasses the western and northern parts of the Cumberland Peninsula, including the head of Pangnirtung (Pannirtuuq) Fjord in the south (Fig. 25.14). The Penny Ice Cap glacial complex covers 5100 km^2, is up to 300 m thick and represents the last remnant of the Laurentide Ice Sheet in Canada.

Fig. 25.14 Auyuittuq National Park, Baffin, Nunavut. **a** Hiking along the Weasel River flood plain toward Mount Thor. Mount Thor is known as Qaiqsualuk in Inuktitut, meaning 'enormous bedrock'. Mount Thor (1675 m asl) is the highest overhanging rock face in the world. **b** Mount Thor from Akshayuk Pass. *Photographs* P. Gierszeski/Flickr, CC BY-SA 3.0

The Cumberland Peninsula of Baffin Island is marked by extremely rugged relief, exhibiting the sharpest contrasts in elevation within the Canadian Arctic. Sharp-crested mountains and deep fjords are aligned SSW–NNE. The original Inuktitut terms for the major landforms are superseding the assigned names from personalities or places taken from Norse mythology. Mount Asgard (Sivanitirutinguak) stands at 2015 m asl Mount Thor (Qaisualuk) has a near vertical cliff more than 1100 m high, and a total relief of 1675 m. Schwartzenbach Falls (Qulitasaniakvik) has a vertical descent of 660 m in several steps. Pangnirtung and other fjords exceed 100 m in depth. Flat-topped or very gently northeastward-sloping plateaus on some Sivanitirutinguak summits are remnants of an erosional plain. The basic landscapes of Cumberland Peninsula were established approximately 10 Ma ago, and glaciation has resulted in only minor widening and deepening of fjords since that time (Ives 2016).

25.4.4.2 Ukkusiksalik, Nunavut

Ukkusiksalik (Wager Bay, Chesterfield Inlet) National Park (20,855 km^2) was established in 2003 in the Kivilliq region of Nunavut, west of Hudson Bay (Fig. 25.15). Establishment of a national park at Ukkusiksalik ("soapstone") reflects the twin desires to preserve the permafrost-dominated shield natural landscape, and the desire to preserve an important heritage site for the Inuit. Management is cooperative through the Ukkusiksalik Park Management Committee, made up of Inuit community members from Naujaat, Coral Harbour, Chesterfield Inlet, Baker Lake and Rankin Inlet.

Ukkusiksalik is located within the Wager Plateau, in the Kazan geologic province of the Canadian Shield. The terrain is dominantly rolling uplands, interspersed with small plateaux. Total relief is approximately 600 m. The major feature is Ukkusiksalik inlet, with a narrow entrance (<10 km wide) at Cape Dobbs, widening westward to 40 km in the deepest part. Tidewater extends 200 km inland from Roes Welcome Sound.

Fig. 25.15 Ukkusiksalik National Park in the Kivallik region of Nunavut. **a** View south along Sila Sila River towards Wager Bay. **b** View of Wager Bay. *Photographs* P. Gierszewski/Wikimedia Commons, CC BY-SA 4.0

At various times during the Quaternary, Ukkusiksalik was glaciated by ice from both the Keewatin Ice Dome area (to the west) and the M'Clintock Ice Dome (to the northwest). Glacial ice from both domes flowed eastward across Ukkusiksalik during MIS 2. The last stage of glaciation involved flow to the northwest from the extension of the Foxe Ice Dome, to the west of Southampton Island, during the early Holocene. Final deglaciation occurred c. 9 ka, as the destabilized ice masses in contact with Tyrrell Sea water in Hudson Bay ablated. The Tyrrell Sea flooded coastal areas to 180 m asl, and isostatic rebound is still in progress.

25.4.4.3 Sable Island, Nova Scotia (Off Shore)

Sable Island Reserve (established 2013) is located 300 km southeast of Halifax in the North Atlantic Ocean. About 12 ka, lower relative sea level due to glaciation exposed the Grand Banks, Georges Bank, and Sable Island Bank. Today, Sable Island is the only one of these regions that remains above relative sea level, although the island is currently undergoing sea-level rise and coastal erosion. Visitation requires a permit from Parks Canada, and all visitors are escorted by Parks Canada staff. Sable Island is noted for its history of shipwrecks (at least 350 since 1583) and its population of feral horses, descended from those confiscated from Acadian settlers during the deportations from Grand-Pré, beginning in 1755.

The approximately 42 km long, 1.5 km wide island, with dimensions subject to change due to erosion and littoral drift, developed on the terminal moraine of the MIS 2 glaciation from mainland Nova Scotia. The island is occupied by a transverse coastal dune complex (Sable = sand), subject to erosion due to storm action and local overgrazing by horses.

25.5 Other Protected Areas

In addition to the federal national parks and the provincially administered *parcs nationals* of Québec, other provincial parks, ecological reserves, wilderness areas, and wilderness river parks have been established within the Atlantic Provinces and Ontario. Many of these are small, intended to conserve a particular local landscape, or to provide an opportunity for recreation. Some of the larger parks are discussed here.

25.5.1 Algonquin Provincial Park, Ontario

The high-relief forested Canadian Shield of Québec and Ontario contains several large parks, including Parc-des-Laurentides, Mont-Tremblant, La Vérendrye, Killarney, and Algonquin. Algonquin Provincial Park is located between Georgian Bay and the Ottawa River in Ontario. Established

Fig. 25.16 Booth's Rock in Algonquin Provincial Park, Ontario. Located between Georgian Bay and the Ottawa River, Algonquin Provincial Park has inspired countless artists over the years, including the Group of Seven. *Photograph* R. Harris/Flickr, CC BY-SA 2.0

in 1893, it is the oldest provincial park in Canada. Additions since its creation have increased the park to its current size of 7,653 km^2. For comparison purposes, this is about a quarter of the size of Belgium. The park is contiguous with several smaller, administratively separate provincial parks that protect important rivers in the area, resulting in a larger total protected area.

Its size, combined with its proximity to the major urban centers of Toronto and Ottawa, makes Algonquin one of the most popular provincial parks in the country. Over 2,400 lakes and 1,200 kms of streams and rivers are located within the park (Fig. 25.16).

The park is considered part of the "border" between northern and southern Ontario, both geomorphically and culturally. The park is in an area of transition between northern coniferous and southern deciduous forest. This mixture of forest types, and the wide variety of environments in the park, allows the park to support an uncommon diversity of plant and animal species.

Algonquin Park was named a National Historic Site in Canada in 1992 in recognition of several heritage values including: it's role in the development of park management; pioneering visitor interpretation programs later adopted by national and provincial parks across the country; its role in inspiring artists, which in turn gave Canadians a greater sense of their country. Algonquin Park is the only designated park within the province of Ontario to allow industrial logging to take place within its borders.

25.5.2 Sir Richard Squires Memorial Provincial Park, Newfoundland

Sir Richard Squires Memorial Provincial Park is located in the upper reach of the Humber River of western Newfoundland. The park landscape is a NNE–SSW-oriented anticline, with Mississippian Windsor Group friable siltstone and sandstone flanked by resistant coarse sandstones and conglomerates. The Little and Big Falls of the Humber River have developed on the contacts between the two geological formations. MIS 2 glaciation flowed southwest, towards Corner Brook and the Gulf of St. Lawrence.

25.5.3 Mount Carleton Provincial Park, New Brunswick

Mount Carleton Provincial Park is located in the interior of northern New Brunswick. The park lies along the International Appalachian Trail. Mount Carleton, composed of Lower Devonian felsic igneous rocks of the Dalhousie Group, is the highest peak in the Maritime Provinces, at 820 m asl. The geomorphology is dominated by late Mesozoic-early Cenozoic erosional features. Although the area was glaciated during the Quaternary, the scarcity of glacial features indicates that cold-based ice prevailed at higher altitudes. MIS 2 glaciation was marked by multiple flow directions from shifting local ice caps.

25.5.4 Parc National Saguenay Fjord, Québec

Parc National Saguenay Fjord lies along the shore of its namesake fjord (Fig. 25.17). The Saguenay–St. Lawrence Marine Park lies to the southeast of the park. Saguenay Fjord's Baie-Éternité visitor center is about 60 km from Saguenay, Québec. Saguenay Fjord has a length of 105 km and a width that varies between 2 and 4 km, making it one of the longest and southernmost fjords in the Northern Hemisphere. The fjord has an average depth of 210 m. The cliffs along the fjord have an average height of 150 m, reaching a maximum height of 350 m asl. Saguenay Fjord flows into the St. Lawrence River; salt water from the river flows under a layer of freshwater originating from Lac St. Jean. The geological origins of Saguenay Fjord park can be traced to the Grenville orogeny during the Precambrian era. Around 200 million years ago, a basal complex between a north fault and a south fault collapsed, creating the Saguenay Graben. The graben was 250 km long and 50 km wide. During MIS2, the last glacial period, the region was covered by ice sheets two to three km thick. The ice further modified the Saguenay graben, gouging the fjord in the process and glaci-isostatic depression caused the region to sink. When the glaciers melted around 10,000 years ago, post-glacial rebound lifted the terrain.

It is patently impossible to describe all of eastern Canada's protected areas. A summary comment would be that Nunavut has the highest proportion of land in protected areas

Fig. 25.17 Parc National Saguenay Fjord. **a** and **b** Majestically encased in the hollow of two mountain massifs, the Saguenay Fjord extends from St-Fulgence to Tadoussac. This valley, initially formed 175 Ma, was deepened and polished by glaciers, and then filled with sea water. One hundred km long and 2 to 3 km wide, this unique geomorphological element is among the longest and southernmost fjords in the world. **c** Baie Éternité on the Saguenay River. *Photographs* **a**—R. Martin/Flickr, CC BY 2.0. **b**—G. Messian/Flickr, CC BY 2.0. **c**—R. Martin/Flickr, CC BY 2.0

(21%), followed by Québec (at 14%). The average for the whole of eastern Canada is 12%. The stated objective of Parks Canada is to cover 17% of Canada's terrestrial surface with national parks by 2025 (National Parks Canada 2019).

25.6 Conclusion

In their discussion of protected areas in Canada, Dearden and Dempsey (2004) identified a number of factors that had contributed to major change in Canada's national parks over the previous fifteen years: (a) the deteriorating planet, livelihoods, and economies; (b) increased understanding of the role of protected areas in protecting biodiversity; (c) international support for protected areas; and (e) the increased effectiveness and expertise of environmental non-governmental organizations (ENGOs), scientific groups, and communities, particularly Indigenous Peoples. Fifteen years later these same factors are still valid and have even intensified to produce continuing growth in numbers of protected areas. They also identified five challenges to the protected area system: (a) large ecosystems that remain unprotected (Table 25.3); (b) stressors that have increased by the exponential increase in numbers of visitors; (c) parks that as isolated ecological islands remain a problem; (d) the National Parks Act (2000) that signals a shift to reinforcement of ecological integrity, an aim which is expensive and difficult to enforce; and (e) parks budgets at provincial and federal level that continue to be subject to pressure.

References

Badman T (2010) World heritage and geomorphology. In: Migon P (ed) Geomorphological landscapes of the world. Springer, Dordrecht, pp 357–368

Borrini-Feyerabend G, Kothari A, Oviedo G (2004) Indigenous and local communities and protected areas: towards equity and enhanced recreation. Best practice protected areas guidelines 11. IUCN, Gland and Cambridge

Brocx M, Semeniuk V (2007) Geoheritage and geoconservation – History, definition, scope and scale. J R Soc West Aust 90:53–87

Burek CV, Prosser CD (2008) The history of geoconservation: an introduction. Geol Soc Lond Spec Publ 300:1–5

Dearden P, Dempsey J (2004) Protected areas in Canada: decade of change. Can Geogr 48:225–239

Fowler PJ (2003) World heritage cultural landscapes: 1992-2002. UNESCO World Heritage Centre, Paris, France

GeoEx (2018) UNESCO's world heritage: what it means and why it matters. www.geoex.com/

Gray M (2004) Geodiversity valuing and conserving abiotic nature. Wiley, Chichester

Ives JD (2016) Baffin Island: field research and high Arctic adventure, 1961-1967. University of Calgary Press, Calgary, Alberta

Kondolf GM, Piegay H (2011) Geomorphology and society. In: Gregory KJ, Goudie AS (eds) Sage handbook of geomorphology. Sage Publications, London, pp 105–117

Lemelin RH, Dawson J (2014) Great expectations: examining the designation effect of marine protected areas in coastal arctic and sub-arctic communities in Canada. Can Geogr 58:217–232

Leopold A (1949) Sand county almanac. Oxford University Press, Oxford

MDA Federal (2004) Landsat GeoCover ETM+ 2000 Edition Mosaics Tile N-21-45 ETM-EarthSat-MrSID, 1.0, USGS, Sioux Falls, South Dakota, 2000. Accessed 2016, at the Global Land Cover Facility

McKeever PJ, Zouros N (2005) Geoparks: celebrating earth heritage, sustaining local communities. Episodes 28:274–278

National Parks Canada (2019) The Canada guide. Accessed at www.thecanadaguide.com/places/national-parks

Thompson S (2014) Coastal erosion at mistaken point ecological reserve, Avalon Peninsula, Newfoundland. Unpublished MSc thesis, Department of Geography, Memorial University of Newfoundland

UNEP-WCMC (2019) Distribution of National Parks of eastern Canada. National Park boundaries; UNEP-WCMC (June, 2019), available at: www.protectedplanet.net

UNESCO (1972) World heritage convention. UNSCO, Paris

UNESCO (2011) Operational guidelines for the implementation of the world heritage convention. UNESCO, Paris

Olav Slaymaker is a Member of the Order of Canada, and professor emeritus of geography in the Department of Geography at the University of British Columbia, Vancouver, B.C. His 20+ co-authored and edited monographs and 160+ refereed papers reflect three different styles of work. First, he has pursued original field research for 40 years on sediment systems in the Canadian Cordillera and the European Alps, strongly supported and enhanced by his stellar doctoral and post-doctoral students and international colleagues. Second, he has produced papers and books summarizing trends in land use and environmental change in mountain regions world-wide. Third, he has edited monographs with invited international contributions on major environmental and geomorphological themes. He is former President of the Canadian Association of Geographers and former President of the International Association of Geomorphologists.

Norm Catto joined Memorial University in July 1989 and is now the Head and Professor of Geography at that university. His research and teaching interests include coastal landforms, natural hazards, sea level change, and impacts in coastal environments; response of river systems to climate and weather events, and flood risk assessment; aeolian geomorphology and response to climate and human factors in boreal and costal environments; mass movements and slope failures; loess deposition; palaeosol formation; emergency measures policies and effectiveness human adaptations to climate and weather events; the impacts of climate and weather events on agriculture, transportation, fisheries, and communities; and the relationships among landscapes and people. His research has included projects and investigations in landscapes and environments in 16 countries, and in all Canadian provinces and territories, particularly in eastern Canada. For 18 years, he served as Editor-in-Chief for *Quaternary International*, a professional international journal dedicated to the study of landscape evolution and climate variation throughout the past 3 million years. He has taught more than 50 different undergraduate and graduate courses focused on numerous aspects of geomorphology, landscape analysis, Quaternary research, natural hazards, ocean sciences, environmental science, climate and climate change, soil science, geoarchaeology, and palaeontology. Study of landscapes forms his central focus for understanding of the various areas of Eastern Canada, the combination of physical and cultural constituents that contribute to its identities. For most species, and for all human occupants, life in Eastern Canada has been heavily influenced by geomorphology since deglaciation.

Dori J. Kovanen is a professional engineering geologist and Research Associate in the Department of Geography at the University of British Columbia, Vancouver, BC. With an M.Sc. from Western Washington University in Bellingham and a Ph.D. in Quaternary geology from the University of British Columbia she is a specialist in Quaternary geomorphology. She has written extensively on Late Glacial and Holocene history of the Pacific Northwest as well as contemporary environmental changes in the region and was a significant contributor to "Landscapes and Landforms of Western Canada" (Slaymaker 2017).

Implications of the Anthropocene Epoch for Geomorphology

26

Olav Slaymaker, Monica E. Mulrennan, and Norm Catto

Abstract

The protection of landscape is a matter of priority in this Anthropocene epoch, as explained in Chap. 25. If we agree that humanity has become the dominant driver of environmental change then figuring out the intellectual and social relevance of geomorphology is a topic that requires equally urgent attention. Classical geomorphology pays scant attention to the social, cultural and political factors that provoke geomorphic change. A modest proposal to recast geomorphology as both a landscape science and a geoscience is a suggested preferred first step. If that step is taken, the contested status of 'landscape' will make broader philosophical and methodological approaches possible in geomorphology. In particular, both the intrinsic and utilitarian value of geomorphology could be assessed more realistically. The net result could be more careful consideration of human well-being in geomorphological research.

Keywords

Geomorphology • Landchange science • Landscape science • Intrinsic and utilitarian value

26.1 The Context Created by Recognition of the Anthropocene Epoch

The Anthropocene epoch has been proposed as a new division of geological time that acknowledges the now dominant role of humans as geomorphic agents (Crutzen and Stoermer 2000; Zalasiewicz et al. 2010; Biermann et al. 2012; Monastersky 2015). This recognition reflects the increasingly precarious state of Earth's terrestrial surface and the civilizations that depend on and interact with the land (Slaymaker 2017). It also has significant implications for the discipline of geomorphology, prompting calls for its repositioning to ensure its role and relevance to efforts to understand the Anthropocene and mitigate the deleterious impacts of human activities on the Earth system (Goudie and Viles 2016).

While Chaps. 24 and 25 (this volume) have alluded to some of these implications, we take up the challenge of addressing them more fully here. In an epoch defined by the dominance of human activities as drivers of environmental change, it is surely the interlinkages between landforms, landscape, land and people that need to be explored in depth and with a sense of urgency (Slaymaker 2009). Hence, in this final chapter of a volume concerned with landscapes and landforms we identify implications of the Anthropocene epoch for geomorphology and identify pathways for its repositioning as a more influential discipline.

26.2 The Root of the Problem

Geomorphology finds its disciplinary origins in a positivist tradition that views "the sense of causality running from the physical environment to its social impacts" (Hewitt 1983, p. 5). This understanding is grounded in a nature/culture binary (Braun and Castree 1998) that sees nature as a domain entirely separate from society. This distinction between the natural and the social world is a product of the

O. Slaymaker (✉)
Department of Geography, University of British Columbia, 1984 West Mall, Vancouver, BC V6T 1Z2, Canada
e-mail: olav.slaymaker@ubc.ca

M. E. Mulrennan
Department of Geography, Planning and Environment, Concordia University, Montréal, QC H3G 1M8, Canada

N. Catto
Department of Geography, Memorial University of Newfoundland and Labrador, St. John's, NL A1B 3X9, Canada

© Springer Nature Switzerland AG 2020
O. Slaymaker and N. Catto (eds.), *Landscapes and Landforms of Eastern Canada*, World Geomorphological Landscapes, https://doi.org/10.1007/978-3-030-35137-3_26

Enlightenment, the practical consequences of which have been four-fold: (a) nature is studied separately from society; (b) fine scale description of the natural world has provided successful positivist models of nature; (c) philosophy and metaphysics have been marginalized as positivism alone is seen as adequate for the needs of such scholars; and (d) phenomena that cannot be seen, measured or sensed have been ignored with impunity. This dualism constrains attempts to appreciate flexible and reflexive interconnections between and within environment and society. For example, Postmodernism with its claims that reality cannot be known nor described objectively, is all too readily dismissed. Indeed some scientists find such claims preposterous, given their access to instrumentation and sophisticated measurement systems. Nevertheless, Postmodernism raises important questions about positivist science's claim to a unique ability to concern itself with "the facts as they are in reality".

It is increasingly clear that a great deal of knowledge depends on the nature of the reality that is being investigated. The sense of causality does not run exclusively from nature to culture. Phenomena that can neither be seen nor measured are as much part of the real world as measurable and visible phenomena (Polanyi 1958). All knowledge claims rely on personal judgment. A great scientist not only identifies patterns but also chooses significant questions likely to lead to successful resolution. Yet only limited attention has been devoted to metaphysical or epistemological issues (Rhoads 1999) and communication between earth scientists and social scientists has been muted.

26.3 Does Geomorphology Need Recasting?

Geomorphology, as the science of the form, evolution and sustainability of Earth's terrestrial surface has generated a body of theory, methodology and applications that has been rigorous and useful to society. As a discipline, it is centrally concerned with landforms and landform processes and has a long history of interest in studying human impacts on these. Unfortunately, too often this has been an ancillary concern focused on human actions as "impacts" upon landforms or "disturbances" to natural systems (Mould et al. 2018). Geomorphology has deep roots as a positivistic empiricist science and we echo the concerns of others in fearing that it could be side-lined as a discipline as social drivers of environmental change become progressively more dominant (Livingstone 2003; Rhoads 2006; Slaymaker 2009). A geomorphology that is narrowly concerned with energy and mass fluxes in and around landforms (Haff 2010) promises to be ineffective in seeking understandings of the rapidly changing landscapes around us if indeed, as we believe, humans have become the dominant driver (Fryirs and Brierley 2013; Slaymaker 2017; Tadaki et al. 2017).

26.4 Geomorphology as Landscape Science

A modest proposal for recasting geomorphology involves stronger ties with landscape science, which in turn would promote greater collaboration and intellectual exchange between earth scientists and social scientists. The word 'landscape' has a variety of meanings: (a) the terrain that is both external and made of material, visible and tangible things, called landforms; (b) a framework through which the world is pictured and represented as, for example, in landscape art; and (c) the product of human labour, as for example in landscaped gardens (Wylie 2011). The combined meanings suggest that the understanding of landscape is both an art and a science, a premise that has already become a central focus in the emergent field of critical physical geography (Lave et al. 2014; Tadaki et al. 2017). If geomorphology was understood to be both a geoscience and a landscape science, the critical interrelationship between humanity and geomorphology would be underscored and more fully embraced (Turner et al. 2007).

Through links to landscape science, engagement with the question of the intrinsic value of our discipline could be promoted. We value landscape for both utilitarian and intrinsic reasons: for what it provides for humans and for the value it has in its own right. Utilitarian values include the many basic needs humans obtain from landscape such as food, fuel and shelter. But landscape evokes deeper and more personal attachments, including cultural values associated with spiritual and religious imperatives. Beyond the contributions geomorphology can make as a scientific discipline, we believe its intrinsic value depends on the extent to which we as geomorphologists connect with these more abstract dimensions of landscape and engage with people who share our interests, concerns and commitment for those landscapes. We affirm that humanity has an important connection with its natural surroundings that is frequently reflected in a deep love of landscape (Goudie and Viles 2010; Slaymaker 2017). We also acknowledge parallel developments by social scientists, who have countered epistemologies that produce a dematerialized and disembodied view of the world, and replaced these with more interconnected views of the world through notions of placemaking, dwelling and engagement with the environment (Ingold 1995, 2000). Connections and attachments to landscape can be translated into art, poetry, spiritual and religious values, and other cultural affirmations of scenery and landforms. They can also inform and support an enhanced sensitivity and commitment to management, enhancement or conservation of the landscape (Gregory 2000).

The ubiquity and speed of contemporary environmental change guarantees that land and landscape are unsustainable in their present form (Blaikie and Brookfield 1987; Fraser et al. 2003; Slaymaker 2009). Inevitably some landscapes that are more threatened than others and critical zones in landscapes will require greater and more urgent measures of protection than others. Making a judgment on the sensitivity of a landscape to change requires deep understanding of the landscape changing processes. But the dominant landscape changer in the Anthropocene has become society itself. As such awareness of the value of geomorphology would likely increase if it became part of the public conscience around that which is happening to landscape (Head 2008). Landscapes may be valued for aesthetic reasons. Areas of outstanding natural beauty, for example, are recognized as part of our geoheritage (UNESCO 1972). Social and cultural views of particular landforms within landscapes also vary widely. Although not necessarily significant on their own, landforms nevertheless are inseparable components of sites of special cultural, religious or historical significance. These linkages are epitomized in the concept of a 'sacred mountain' such as Fuji-san in Japan. The wider cultural significance includes history, architecture, landscape aesthetics, art and religion and is embedded in the concept of geoheritage (Goudie 2002).

26.5 Recasting Geomorphology

We are encouraged by recent calls for an invigorated approach to the geomorphology of human activity and see our proposal for enhanced ties to landscape science as potentially compatible. For example, Ashmore (2015) calls for the adoption of a socio-geomorphic approach that would support a more complete explanation of human-impacted and human created landscapes leading to a broader understanding of landform processes and landscapes in relation to human agency and intention. Meadows and Lin (2016) affirm the need for greater attention to the interdependency of humans and landscapes, while others stress the importance of exploring the heterogeneity, cumulative effects, and feedbacks of these relationships across both time and space (Kondolf and Podolak 2013). Wohl et al. (2013) draw particular attention to the ubiquitous nature of the geomorphic influence of humans which extends to landscapes often erroneously regarded as unoccupied, unaltered and protected (see Mulrennan, in Chap. 24, this volume).

The contribution of place-based knowledge in informing landscape interpretations and management has also gained increased recognition, particularly in the practice of fluvial geomorphology. This shift acknowledges the contingent nature of bio-physical interactions within any given landscape and the inadequacy of transferring insights and interpretations based on limited sample sites. Brierley et al. (2013) propose 'reading the landscape' as an alternative approach to interpreting river landscapes that pays greater attention to place-based understandings of process-form interactions. Others highlight the value of local and Indigenous knowledge (Meadows and Lin 2016). Past tendencies to view Indigenous knowledge as 'cultural heritage' (as well as inaccurate and unreliable) and hence outside the realm of interest of earth scientists are being replaced by a recognition of the rich insights that alternative knowledge can offer. This includes the potential of stories to relay observations of landscape change at spatial and time scales beyond the capacity of geomorphological methods.

Parallel developments in adjacent fields, particularly the emergence of critical physical geography, have prompted calls for the adoption of critical approaches to geomorphology that question the role of science, foster more reflexive approaches and take advantage of methodologies that are consistent with decolonizing the academy. This includes unsettling scientific knowledge claims to objectivity, mentioned earlier as a legacy of geomorphology's positivist tradition, and dismantling imperialist notions of 'the field' as a site of discovery (Lave et al. 2014). Similarly, established commitments to form, process, pattern, connectivity, and evolution are increasingly being brought into productive dialogue with concepts of heterogeneity, emergence, nonlinearity, contingency, and uncertainty (Brierley et al. 2013). This offers new possibilities for theorizing and extending our tool kits as we face the challenge of interpreting landscape change. The notion of 'freedom space for rivers' (Biron et al. 2014) represents an exciting example of a break from traditional management approaches to acknowledge complexity and promote the agency and resilience of the natural system. It also highlights the attention of geomorphologists are devoting to human presence in vulnerable landscapes.

Others have identified a need for geomorphologists to take advantage of opportunities and methodologies that embrace multiple and divergent ways of thinking about landscapes and human–nature connections (Howitt and Suchet-Pearson 2006). This goes beyond the incorporation of Indigenous and local knowledge, to embrace "ontological pluralism" or the possibility "that there are diverse ways of being within and interacting with multiple worlds" (Yates et al. 2017).

26.5.1 Ethnogeomorphology as a Path Forward

There is another conversation in which relatively few geomorphologists have participated, namely the sharing of knowledge between earth scientists and Indigenous people. In order to provide a framework for intercultural communication and ethical engagement (Mulrennan, Chap. 24, this volume) the sub-field of 'ethnogeomorphology' has been proposed (Wilcock et al. 2013). Highlighting the value of local and Indigenous knowledge and rejecting past tendencies to view Indigenous knowledge as 'cultural heritage' (and hence outside the realm of interest of earth scientists) brings First Nations knowledge to the fore. A new recognition of the rich insights alternative knowledge can offer includes the potential of stories to relay observations of landscape change at spatial and time depths previously ignored by traditional geomorphological methods (Muller-Wille 2001; Mulrennan et al. 2012). Such understanding also draws attention to the need for ethical physical geography in research on Indigenous lands (Kershaw et al. 2014). The overall and long-term anticipation would be that human welfare and well-being will be enhanced (Kareiva et al. 2007).

26.6 Conclusion

We believe that there is a need for geomorphologists to embrace multiple and divergent ways of thinking about landscapes and human–nature interactions (Howitt and Suchet-Pearson 2006). This should include greater attention to the interdependence of humans and landscapes, including the heterogeneity, cumulative effects, and feedbacks of these relationships across both time and space (Slaymaker et al. 2009; Kondolf and Podolak 2013; Chin et al. 2013). A range of alternative methodologies, such as storytelling, art and performance, community video, and participatory mapping, are available to foster such a shift and support enhanced engagements between geomorphologists and stakeholders, including community partners and Indigenous peoples. Finally, stronger links are encouraged with fields that share conceptual overlaps—such as resilience, connectivity, and thresholds—facilitating interdisciplinary understanding of complex human-landscape systems by highlighting drivers of change within these systems (Holling 2001). These connections are also essential to more effective management (stewardship) of landscapes and ecosystems (Chin et al. 2013; Wohl et al. 2013). Collaborations with fields such as ethnobotany, ethnoecology and wildlife biology that have a longer tradition of partnered research with Indigenous communities will be particularly valuable (Moller et al. 2004).

References

Ashmore P (2015) Towards a sociogeomorphology of rivers. Geomorphology 251:149–156

Biermann F, Abbott K, Andresen S, Backstrand K, Bernstein S, Betsill MM, Bulkeley H, Cashore B, Clapp J, Folke C, Gupta A, Haas PM, Jordan A, Kanie N, Kluvankova-Oravska Lebel T, Liverman D, Meadowcraft J, Mitchell RB, Newell P (2012) Navigating the anthropocene: improving earth system governance. Science 335:1306–1307

Biron PM, Buffin-Bélanger T, Larocque M, Choné G, Cloutier CA, Ouellet MA, Demers S, Olsen T, Desjarlais C, Eyquem J (2014) Freedom space for rivers: a sustainable management approach to enhance river resilience. Environ Manage 54:1056–1073

Blaikie P, Brookfield H (1987) Land degradation and society. Routledge, New York

Braun B, Castree N (eds) (1998) Remaking reality: nature at the millennium. Routledge, London, UK

Brierley G, Fryirs K, Cullum C, Tadaki M, Huang HQ, Blue B (2013) Reading the landscape: integrating the theory and practice of geomorphology to develop place-based understandings of river systems. Prog Phys Geog 37:601–621

Chin A, O'Dowd AH, Gregory KJ (2013) Urbanization and river channels. In: Shroder JF, Wohl E (eds) Treatise on geomorphology, vol 9. Fluvial geomorphology. Academic Press, San Diego, pp 809–827

Crutzen PJ, Stoermer EF (2000) The 'anthropocene'. IGBP Global Change Newslett 41:17–18

Fraser EDG, Mabee W, Slaymaker O (2003) Mutual vulnerability, mutual dependence. The reflexive relation between human society and the Environment. Global Environ Change 13:137–144

Fryirs K, Brierley GJ (2013) Geomorphic analysis of river systems: an approach to reading the landscape. Blackwell, Oxford

Goudie AS (2002) Aesthetics and relevance in geomorphological outreach. Geomorphology 47:245–249

Goudie AS, Viles HA (2010) Landscapes and geomorphology: a very short introduction. Oxford University Press, Oxford, UK, pp 99–110

Goudie AS, Viles HA (2016) Geomorphology in the anthropocene. Cambridge University Press, Cambridge, UK

Gregory KJ (2000) The changing nature of physical geography. Arnold, London, UK

Haff PK (2010) Hillslopes, rivers, plows and trucks: mass transport on Earth's surface by natural and technological processes. Earth Surf Processes 35:1157–1166

Head L (2008) Is the concept of human impacts past its use-by date? The Holocene 18:373–377

Hewitt K (1983) The idea of calamity in a technocratic age. In: Hewitt K (ed) Interpretations of calamity from the viewpoint of human ecology. Allen and Unwin, Winchester, MA, pp 3–32

Holling CS (2001) Understanding the complexity of economic, ecologic and social systems. Ecosystems 4:390–405

Howitt R, Suchet-Pearson S (2006) Rethinking the building blocks: ontological pluralism and the idea of 'management'. Geogr Ann 88B:323–335

Ingold T (1995) Building, dwelling, living: how animals and people make themselves at home in the world. In: Strathern M (ed) Shifting contexts: transformations in anthropological knowledge. Routledge, London UK, pp 57–80

Ingold T (2000) Epilogue: towards a politics of dwelling. Conservat Soc 3:501–508

Kareiva P, Watts S, McDonald R, Boucher T (2007) Domesticated nature: shaping landscapes and ecosystems for human welfare. Science 316:1866–1869

Kershaw GGL, Castleden H, Laroque CP (2014) An argument for ethical physical geography research on Indigenous landscapes in Canada. Can Geogr 58:393–399
Kondolf GM, Podolak K (2013) Space and time scales in human landscape systems. Environ Manag 53:76–87
Lave R, Wilson MW, Barron ES, Biermann C, Carey MA, Duvall CS, Johnson L, Lane KM, McClintock N, Munroe D, Pain R, Proctor J, Rhoads BL, Robertson MM, Rossi J, Sayre NF, Simon G, Tadaki M, Van Dyke C (2014) Intervention: critical physical geography. Can Geogr 58:1–10
Livingstone DN (2003) Putting science in its place: geographies of scientific knowledge. University of Chicago Press, Chicago
Meadows M, Lin JC (eds) (2016) Geomorphology and society. Advances in geographical and environmental sciences. Springer, Tokyo
Moller H, Berkes F, Lyver PO, Kislalioglu M (2004) Combining science and traditional ecological knowledge: monitoring populations for co-management. Ecol Soc 9:2
Monastersky R (2015) The human age. Nature 519:144–147
Mould SA, Fryirs K, Howitt R (2018) Practising sociogeomorphology: relationships and dialogue in river research and management. Soc Natur Resour 31:106–120
Muller-Wille L (2001) Shaping modern Inuit territorial perception and identity in the Quebec-Labrador Peninsula, Chap. 3. In: Scott C (ed) Aboriginal autonomy and development in Northern Quebec and Labrador. UBC Press, Vancouver
Mulrennan ME, Mark R, Scott CH (2012) Revamping community based conservation through participatory research. Can Geogr 56:243–259
Polanyi M (1958) Personal knowledge. Routledge and Keenan Paul, London, UK
Rhoads BL (2006) The dynamic basis of geomorphology reenvisioned. Ann Assoc Am Geogr 96:14–30
Rhoads BL (1999) Beyond pragmatism. The value of philosophical discourse for physical geography. Ann Assoc Am Geogr 89:760; Ann Assoc Am Geogr 771
Slaymaker O (2009) The future of geomorphology. Geogr Compass 3:329–349
Slaymaker O (2017) Physical geographers' understanding of the real world. Can Geogr 61:64–72
Slaymaker O, Spencer T, Dadson S (2009) Landscape and landscape scale processes as the unfilled niche in the global environmental change debate: an introduction. In: Slaymaker O, Spencer T, Embleton-Hamann C (eds) Geomorphology and global environmental change. Cambridge University Press, Cambridge, UK, pp 1–36
Tadaki M, Slaymaker O, Martin Y (eds) (2017) Critical physical geography. Can Geogr 61:1–148
Turner BL, II, Lambin E, Reenberg A (2007) The emergence of landchange science for global environmental change and sustainability. In: Proceedings of the national academy of sciences of the USA, Washington, DC 104, pp 20666–20671
UNESCO (1972) World Heritage Convention. UNESCO, Paris
Wilcock DA, Brierley GJ, Howitt R (2013) Ethnogeomorphology. Prog Phys Geogr 37:573–600
Wohl E, Gerlak AK, Poff NL, Chin A (2013) Common core themes in geomorphic, ecological and social systems. Environ Manag 53:14–27
Wylie J (2011) Landscape. In: Agnew JA, Livingstone DN (eds) Sage handbook of geographical knowledge. Sage Publications, London, pp 300–315
Yates J, Wilson N, Harris L (2017) Multiple ontologies of water: politics, conflict and implications for governance. Environ Plan D Soc Space
Zalasiewicz J, Williams M, Steffen W, Crutzen PJ (2010) The new world of the Anthropocene. Environ Sci Tech 44:2228–2231

Olav Slaymaker is a Member of the Order of Canada, and professor emeritus of geography in the Department of Geography at the University of British Columbia, Vancouver, B.C. His 20+ co-authored and edited monographs and 160+ refereed papers reflect three different styles of work. First, he has pursued original field research for 40 years on sediment systems in the Canadian Cordillera and the European Alps, strongly supported and enhanced by his stellar doctoral and post-doctoral students and international colleagues. Second, he has produced papers and books summarizing trends in land use and environmental change in mountain regions world-wide. Third, he has edited monographs with invited international contributions on major environmental and geomorphological themes. He is former President of the Canadian Association of Geographers and former President of the International Association of Geomorphologists.

Monica E. Mulrennan is an Associate Professor in the Department of Geography, Planning and Environment at Concordia University, Montréal (Canada). She completed a Ph.D. in Coastal Geomorphology at University College Dublin, and spent 3 years in Australia as Post-Doctoral Fellow (University of Wollongong and Australian National University). During this time she developed a strong interest in the connection of Indigenous peoples to coastal environments. For the past 25 years her research has focused on indigenous-led strategies of conservation and environmental stewardship that draw upon Indigenous institutions of knowledge and practice. She works closely with James Bay Crees (Eeyou Istchee) on research related to local adaptations to environmental change, protected area creation, and knowledge and use of coastal and marine territories. She has also sustained a long-term research partnership with Indigenous Torres Strait Islanders of northern Queensland and has contributed to the documentation of Islander knowledge, tenure and resource harvesting practices for the successful Torres Strait Islander regional sea claim. Her most recent project is focused on Islander women, their fishing activities and attachments to seaspace. Dr. Mulrennan is also involved in a collaborative research project involving colleagues at the National University of Ireland, Galway (NUIG) which explores the historical and cultural dimensions of seaweed harvesting in Ireland. She is senior editor of the forthcoming 'Partnerships, Politics and Perspectives: Protected Area Creation in Wemindji, Eeyou Istchee' (Vancouver: UBC Press) as well as the author of numerous journal articles and book chapters.

Norm Catto joined Memorial University in July 1989 and is now the Head and Professor of Geography at that university. His research and teaching interests include coastal landforms, natural hazards, sea level change, and impacts in coastal environments; response of river systems to climate and weather events, and flood risk assessment; aeolian geomorphology and response to climate and human factors in boreal and costal environments; mass movements and slope failures; loess deposition; palaeosol formation; emergency measures policies and effectiveness human adaptations to climate and weather events; the impacts of climate and weather events on agriculture, transportation, fisheries, and communities; and the relationships among landscapes and people. His research has included projects and investigations in landscapes and environments in 16 countries, and in all Canadian provinces and territories, particularly in eastern Canada. For 18 years, he served as Editor-in-Chief for *Quaternary International*, a professional international journal dedicated to the study of landscape evolution and climate variation throughout the past 3 million years. He has taught more than 50 different undergraduate and graduate courses focused on numerous aspects of geomorphology, landscape analysis, Quaternary research, natural hazards, ocean sciences, environmental science, climate and climate change, soil science, geoarchaeology, and palaeontology. Study of landscapes forms his central focus for understanding of the various areas of Eastern Canada, the combination of physical and cultural constituents that contribute to its identities. For most species, and for all human occupants, life in Eastern Canada has been heavily influenced by geomorphology since deglaciation.

Index

© Springer Nature Switzerland AG 2020
O. Slaymaker and N. Catto (eds.), *Landscapes and Landforms of Eastern Canada*, World Geomorphological Landscapes, https://doi.org/10.1007/978-3-030-35137-3

D

E

MIX
Papier aus verantwortungsvollen Quellen
Paper from responsible sources
FSC® C105338

If you have any concerns about our products,
you can contact us on
ProductSafety@springernature.com

In case Publisher is established outside the EU,
the EU authorized representative is:
Springer Nature Customer Service Center GmbH
Europaplatz 3, 69115 Heidelberg, Germany

Printed by Libri Plureos GmbH
in Hamburg, Germany